AF724971

KENDALL'S

ADVANCED
Theory of
STATISTICS

Volume 1

DISTRIBUTION THEORY

KENDALL'S LIBRARY OF STATISTICS

Advisory Editorial Board

Peter J Green, University of Bristol, UK
Rod J Little, University of Michigan, USA
J Keith Ord, Georgetown University, USA

The development of modern statistical theory is reflected in the history of the late Sir Maurice Kendall's volumes, *The Advanced Theory of Statistics*. This landmark publication began life as a two volume work (Volume 1, 1943; Volume 2, 1946) and grew steadily, as a single authored work, until the late 1950s. At this stage, Alan Stuart became co-author and *The Advanced Theory* was rewritten in three volumes. When Keith Ord joined the authorial team in the early 1980s, plans were laid to expand still further to a four-volume work. However, it became evident that with the ongoing and often rapid pace of developments in statistics there were gaps in the coverage and that it was becoming increasingly difficult to provide timely updates to all volumes, so a new strategy was devised.

The Advanced Theory now forms a core of three volumes in an expanded series of related works, *The Kendall's Library of Statistics*. The other books in this series encompass the areas previously appearing in the old Volume 3 and to some extent in Volume 2 and the series structure allows the inclusion of new titles on other areas which have become established since the publication of the original *Advanced Theory*.

In the preface to the first edition of *The Advanced Theory*, Kendall declared that his aim was 'to develop a systematic treatment of [statistical] theory as it exists at the present time'. This continues to be the guiding principle of the series, which hopes to maintain its role as a standard and comprehensive reference on modern statistics.

KENDALL'S ADVANCED THEORY OF STATISTICS

Kendall's Advanced Theory of Statistics, Volume 1: Distribution Theory, sixth edition (1994), Stuart and Ord

Kendall's Advanced Theory of Statistics, Volume 2A: Classical Inference and the Linear Model (1999), Stuart, Ord and Arnold

Kendall's Advanced Theory of Statistics, Volume 2B: Bayesian Inference (1994), O'Hagan

KENDALL'S LIBRARY OF STATISTICS

Multivariate Analysis, Part 1: Distributions, Ordination and Inference (1994) Krzanowski and Marriott

Multivariate Analysis, Part 2: Classification, Covariance Structures and Repeated Measurements (1995) Krzanowski and Marriott

Multilevel Statistical Models, third edition (2003) Goldstein

The Analysis of Proximity Data (1997) Everitt and Rabe-Hesketh

Robust Nonparametric Statistical Methods (1998) Hettmansperger and McKean

Statistical Regression with Measurement Error (1999) Cheng and Van Ness

Latent Variable Models and Factor Analysis (1999) Bartholomew and Knott

Statistical Inference for Diffusion Type Processes (1999) Prakasa Rao

Statistical Decision Theory (2000) French and Ríos Insua

For further details please see the series website at www.kendallslibrary.com

KENDALL'S

ADVANCED

Theory of

STATISTICS

Volume 1

DISTRIBUTION THEORY

Alan Stuart

Formerly Professor Emeritus of Statistics in the University of London

J. Keith Ord

Professor of Decision Sciences, Georgetown University, USA

John Wiley & Sons, Ltd

Volume 1 of the Sixth Edition
First published in 1994 by Hodder Education
part of Hachette Livre UK,
338 Euston Road, London NW1 3BH

www.kendallslibrary.com

© 1994 Trustees of the late Sir Maurice Kendall, the late Professor Alan Stuart and Professor Keith Ord

Publishing History

Volume 1	Distribution Theory		
		First edition	1943
		New editions and reprints	1945, 1947, 1948, 1952, 1958 1963, 1969, 1977, 1987, 1994, 2002, 2003, 2004
Volume 2	Classical Inference and Relationships		
		First edition	1946
		New editions and reprints	1947, 1951, 1961, 1967, 1973, 1979, 1991
		As Volume 2A	1999, 2004
Volume 2B	Bayesian Inference		
		First edition	1994
		reprint	1999, 2002
		New edition	2004
Volume 3	Design and Analysis and Time Series		
		First edition	1966
		New editions	1968, 1976, 1983
	Now replaced by *Kendall's Library of Statistics*		

John Wiley & Sons Ltd, The Atrium, Southern Gate, Chichester, West Sussex, PO19 8SQ, United Kingdom

For details of our global editorial offices, for customer services and for information about how to apply for permission to reuse the copyright material in this book please see our website at www.wiley.com.

The right of the author to be identified as the author of this work has been asserted in accordance with the Copyright, Design and Patents Act 1988.

All rights reserved. No part of this publication may be reproduced, stored in a retrieval system, or transmitted, in any form or by any means, electronic, mechanical, photocopying, recording or otherwise, except as permitted by the UK Copyright, Designs and Patents Act 1988, without the prior permission of the publisher.

Wiley also publishes its books in a variety of electronic formats. Some content that appears in print may not be available in electronic books.

Designations used by companies to distinguish their products are often claimed as trademarks. All brand names and product names used in this book are trade names, service marks, trademarks or registered trademarks of their respective owners. The publisher is not associated with any product or vendor mentioned in this book. This publication is designed to provide accurate and authoritative information in regard to the subject matter covered. It is sold on the understanding that the publisher is not engaged in rendering professional services. If professional advice or other expert assistance is required, the services of a competent professional should be sought.

British Library Cataloguing in Publication Data
A catalogue record for this book is available from the British Library

Library of Congress Cataloging-in-Publication Data
A catalog record for this book is available from the Library of Congress

ISBN 978 0 470 66530 5

CONTENTS

LIST OF EXAMPLES

INTRODUCTORY NOTE

0.1 The chapter-sections in this work are numbered serially, e.g. **14.13** refers to the thirteenth section of Chapter 14. A similar procedure is followed for tables, Examples (in the text) and Exercises (at the ends of chapters). Similarly, (7.15) refers to the fifteenth equation of Chapter 7. Thus chapter-sections are denoted by bold-face type, equations by ordinary type in parentheses.

All results of importance in the volume may be located from the extensive main index, which covers both subjects and authors. References there are to chapter-sections, which are displayed at the tops of pages. A special index preceding the main Index enables the reader to locate the chapter-section in which any text Example is to be found.

0.2 References are given, by author's name and date of publication, to the alphabetical list of references at the end of the book. References with two authors are listed after those of the first-named author, in alphabetical order of the second author's name and date. These are followed by references to three or more authors (cited in the text as, e.g., Andrews *et al.* (1972)) in a single sequence by date of publication.

Occasionally (e.g. in table-headings), a fuller reference is given in the text; this is always done when no individual author is named, as in extensive sets of statistical tables published by official bodies. Where journal articles are referred to, the number of the volume is given in bold-face type and the number of the first page of the article in ordinary type, e.g. *Ann. Statist.*, **10**, 275, refers to the article beginning on page 275 of volume 10 of the *Annals of Statistics*. Where an exercise is followed by an author's name and a date, the result given in the exercise appears in the article listed in the references under those particulars.

A comprehensive *Bibliography of Statistical Literature* by M. G. Kendall and Alison G. Doig (Oliver & Boyd, Edinburgh), covering about 30,000 items from the sixteenth century up to 1958, is available. Volume 1 (1962) covers the years 1950–58, Volume 2 (1965) the years 1940–49, and Volume 3 (1966) the years up to 1939.

The National Bureau of Standards publication. *An Author and Permutated Title Index to Selected Statistical Journals*, covers seven leading journals throughout the period 1952–69. In 1968, H. O. Lancaster produced the volume *Bibliography of Statistical Bibliographies* (Oliver & Boyd), and he has supplemented this with a series of papers in the *International Statistical Review* (formerly *Review of the International Statistical Institute*) in Volumes 37 (1969), pp. 57–67; 38 (1970), 258–67; 39 (1971), 64–73; 40 (1972); 73–81; 41 (1973), 375–9; 42 (1974), 67–70 and 307–11; 43 (1975), 345–9; 44 (1976), 361–5; 46 (1978), 221–4; 47 (1979), 79–84; 49 (1981), 177–83; 50 (1982), 195–217; 51 (1983), 207–12; 52 (1984), 101–8; 53 (1985), 215–20; 54 (1986), 109–14; 55 (1987), 221–7.

Since 1959, *Statistical Theory and Methods Abstracts* has been published periodically on behalf of the International Statistical Institute by Longman; and from 1975 on, the *Current Index to Statistics* has been published annually by the American Statistical Association and the Institute of Mathematical Statistics. *Abstracts* gives a summary of each paper, whereas the *Index* gives author and keyword listings.

The Royal Statistical Society, in conjunction with the London School of Economics, provides a computerized bibliographical service. For further details, see S. Jones (1974), The London School of Economics Computer Bibliography of Statistical Literature, *J. Roy. Statist. Soc.* **A**, 137, 219.

0.3 The mathematical notation is that in current use, but a few symbols may be explained.

(1) The exclamation mark! written after an integer denotes the factorial of that integer. Some writers give the symbol a more extended use for non-integral numbers x by writing

$$x! = \Gamma(x+1) = \int_0^\infty e^{-t} t^x \, dt.$$

This, of course, accords with the factorial notation, but will not be used in this book.

(2) The use of horizontal lines over expressions to indicate bracketed terms has been avoided; instead, parentheses or brackets are used where necessary.

(3) The combinatorial sign $\binom{n}{r} = \frac{n!}{r!(n-r)!}$ is used.

(4) The summation sign is written as Σ, e.g. $\sum_{j=1}^{j=n} x_j = x_1 + x_2 + \ldots + x_n$. The symbol $\sum_{j=1}^{j=n}$ can as a rule be shortened to $\sum_{j=1}^{n}$ and in many cases to $\sum_{j}$ or merely to Σ, the range of the summation being clear from the context.

Similarly, the product sign Π means the product of what follows it, the range being specified as for Σ.

(5) The ordinary notation for the Γ-function (given above), the B-function, and the hypergeometric function is used, i.e.

$$B(p, q) = \int_0^1 x^{p-1}(1-x)^{q-1} \, dx = \frac{\Gamma(p)\Gamma(q)}{\Gamma(p+q)}$$

and

$$F(\alpha, \beta, \gamma, x) = 1 + \frac{\alpha \cdot \beta}{1 \cdot \gamma} x + \frac{\alpha(\alpha+1) \cdot \beta(\beta+1)}{1 \cdot 2 \cdot \gamma(\gamma+1)} x^2 + \frac{\alpha(\alpha+1)(\alpha+2)\beta(\beta+1)(\beta+2)}{1 \cdot 2 \cdot 3 \cdot \gamma(\gamma+1)(\gamma+2)} x^3 + \ldots$$

(6) Where the exponent is concise, the exponential function is written as a power of e, for example $e^{\frac{1}{2}x^2}$ but where it is lengthy we use the notation exemplified by $\exp\{-\frac{1}{2}(x^2 - 2\rho xy + y^2)\}$ instead of $e^{-\frac{1}{2}(x^2-2\rho xy+y^2)}$.

(7) The notation $O(f(n))$ signifies that an expression is of order not exceeding that of $f(n)$, i.e. that the limit of the ratio of the expression to $f(n)$ is a bounded constant C. If $C=1$, we say that the expression is asymptotically equal to $f(n)$.

If $C=0$, we write $o(f(n))$ instead of $O(f(n))$ to signify that the order is less than that of $f(n)$.

0.4 It is necessary to preserve a distinction between a quantity in a population and that quantity in a sample. Where possible, the former will be denoted by a Greek letter and the latter by a Roman letter, e.g. the product moment correlation coefficient of a population is denoted by ρ and that of a sample by r. It is not, however, always desirable to follow this rule; for the multiple correlation coefficient R, e.g., a Greek capital could be confused with the Roman P, so a different convention is followed. Complete notational consistency can only be achieved at the expense of jettisoning a great deal of accepted statistical usage, and even then would probably result in some cumbrous symbols.

0.5 In order to enable the reader to follow the worked examples and illustrative material, a few tables of the relevant functions are given at the end of each volume. These tables are not substitutes for the comprehensive sets that have been published, which are a necessary adjunct to most practical and a good deal of theoretical work. Frequent reference will be made to the following:-

Biometrika Tables for Statisticians, Volumes I and II, edited by E. S. Pearson and H. O. Hartley, Charles Griffin & Co. Unless otherwise stated, reference is made to Vol. I.

Statistical Tables for use in Biological, Agricultural and Medical Research, by Sir Ronald Fisher and F. Yates. Oliver and Boyd, Edinburgh.

Some useful specialized tables are contained in

Selected Tables in Mathematical Statistics, Vols. I–X, edited for the Institute of Mathematical Statistics by H. L. Harter, D. B. Owen and R. E. Odeh, with J. M. Davenport; published by The American Mathematical Society, Providence, R.I., U.S.A.

There is also a comprehensive

Guide to Tables in Mathematical Statistics, by J. A. Greenwood and H. O. Hartley, Princeton U.P. and Oxford U.P., 1962,

which covers fairly completely all tables published up to 1954 and less completely covers later tables.

Useful tables are described at many points in the text. References to the scope of tables are given in what is now fairly standard form; e.g. $x = 0(0.1)10(1)50(2)100$ means that the function is tabulated for an argument x which proceeds by steps of 0.1 from 0 to 10, thence by steps of 1 from 10 to 50 and thence by steps of 2 from 50 to 100. In describing tables, "decimal places" is sometimes abbreviated to "d.p."

GLOSSARY OF ABBREVIATIONS

The following abbreviations are sometimes used:

ARE	asymptotic relative efficiency
ASN	average sample number
AV	analysis of variance
BAN	best asymptotically normal
BCR	best critical region
c.f.	characteristic function
c.g.f.	cumulant generating function
d.f.	distribution function
d.fr.	degrees of freedom
EM	estimation-maximization (algorithm)
f.f.	frequency function
f.m.g.f.	factorial moment generating function
LF	likelihood function
LR	likelihood ratio
LS	least squares
MCS	minimum chi-square
m.g.f.	moment generating function
ML	maximum likelihood
MS	mean square
m.s.e.	mean square error
MV	minimum variance
MVB	minimum variance bound
$N(a, b)$	(multi-)normal with mean (-vector) a and variance (covariance matrix) b
OC	operating characteristic
p.d.f.	probability density function
p.f.	probability function
p.g.f.	probability generating function
p.m.f.	probability mass function
s.d.	standard deviation
s.e.	standard error
SPR	sequential probability ratio
SS	sum of squares
UMP	uniformly most powerful
UMPU	uniformly most powerful unbiased
UMVU	uniformly minimum variance unbiased

CHAPTER 1

FREQUENCY DISTRIBUTIONS

1.1 The fundamental notion in statistical theory is that of the group or aggregate, a concept for which statisticians use the term *population*. This will be generally employed to denote any collection of objects under consideration, whether animate or inanimate; for example, we shall consider populations of men, of plants, of mistakes in reading a scale, of barometric heights on different days, and even conceptual populations, such as that of the possible ways in which a hand of cards might be dealt. The notion common to all these things is that of aggregation.

The science of Statistics deals with the properties of populations. In considering a population of men we are not usually interested, statistically speaking, in whether some particular individual has brown eyes or is a forger, but rather in how many of the individuals have brown eyes or are forgers, and whether the possession of brown eyes goes with a propensity to forgery in the population. We are, so to speak, concerned with the properties of the population itself. Such a standpoint can occur in physics as well as in demographic sciences. For instance, in discussing the properties of a gas we are, as a rule, not so much interested in the behaviour of particular molecules, as in that of the aggregate of molecules which go to compose the gas. The statistician, like Nature, is mainly concerned with the species and is careless of the individual.

1.2 We may therefore begin an approach to a definition of our subject by the following: Statistics is the branch of scientific method that deals with the properties of populations. This, however, is rather too general. Statistics deals only with the *numerical* properties. A dictionary, for example, sets out a population of words, and among the properties of that population which are a suitable subject for scientific inquiry is that of word-derivation. It is not of statistical concern, however, to know that some words are derived from Latin, some from Anglo-Saxon and some from Hindustani. The subject would only assume a statistical aspect if we were to inquire *how many* words were derived from the different sources.

1.3 As a second approximation to our definition we may then try the following: Statistics is the branch of scientific method that deals with the data obtained by counting or measuring the properties of populations.

This again is a little too general. A set of logarithm tables is a population of numerals, but it is hardly a subject for statistical inquiry, for every numeral is determined according to mathematical laws. The statistician is rather concerned with populations that occur in Nature and are thus subject to the multitudinous influences at work in the world at large. These populations rarely, if ever, conform exactly to simple mathematical rules, and in fact it is in the departure from such rules that we often find topics of the greatest statistical interest. To allow for this factor we may then formulate our definition as follows:—

Statistics is the branch of scientific method that deals with the data obtained by counting or measuring the properties of populations of natural phenomena. In this definition "natural phenomena" includes all the happenings of the external world, whether human or not.

1.4 It may be as well to point out that "Statistics", the name of the scientific method, is a collective noun, has a capital "S", and takes the singular. The same word "statistics" is also applied to the numerical material with which the method operates, and in such a case has no capital letter and takes the plural. Later in this book we shall meet the singular form "statistic", which is defined as a function of the observations in a sample from some population. "Statistic" in this sense takes the plural "statistics".

Frequency distributions

1.5 Consider a population of members each of which bears some numerical value of a variable, e.g. of men with measured height or of flowers classified according to number of petals. This variable we shall call a variate or a random variable. If it can assume only a number of isolated values it will be called discrete or discontinuous, and if it can assume any value in a continuous range, continuous. The population of members will generate a corresponding population of variate values, and it is the properties of this latter population that we have to consider.

If the population consists of only a few members we can without much difficulty consider the population of variate values exhibited by them; but if, as usually happens, the aggregate is large (or, in a sense defined later, infinite), the set of variate values has to be reduced in some way before the mind can grasp their significance. This is done by grouping the individuals into intervals of the variate. So far as possible the intervals should be of equal length, so that the numbers falling into different intervals are comparable. The number of members bearing a variate value falling into a given interval is the frequency in that interval. The manner in which the frequencies are distributed over the intervals is called the frequency distribution (or simply the distribution).

1.6 Tables 1.1 and 1.2 give two frequency distributions of observed populations grouped according to a single variate. Table 1.1 shows the distribution of 1475 local districts in England and Wales for 1953, grouped according to birth-rate (number of births per 1000 of the population). The distribution is shown in this table in a way that would be impossible if each of the 1475 districts were shown separately. More districts fall within the interval 15.5–16.5 than in any other, and the frequencies tail off on either side of this value.

Table 1.2 shows the number of persons liable to surtax in the United Kingdom in 1950 grouped according to annual income. The intervals here are unequal—a common feature of official figures—and the last column of the table makes the frequencies comparable by expressing each as a frequency per £500 range within its interval.

Table 1.1 Local government areas in England and Wales, 1953, grouped according to birth-rate per 1000 of population

(Data from the Registrar-General's Statistical Review of England and Wales for 1953)

Birth-rate	Number of districts	Birth-rate	Number of districts
2.5-	1	15.5-	250
3.5-	0	16.5-	178
4.5-	0	17.5-	129
5.5-	1	18.5-	64
6.5-	0	19.5-	39
7.5-	3	20.5-	20
8.5-	7	21.5-	12
9.5-	19	22.5-	7
10.5-	39	23.5-	1
11.5-	96	24.5-	2
12.5-	151	25.5-	3
13.5-	231	26.5-	0
14.5-	221	27.5-	1
		TOTAL	1475

Note—2.5- means "2.5 and less than 3.5" etc.

Table 1.2 Persons in the United Kingdom liable to surtax in the year beginning 6 April, 1950, grouped according to annual income

(From *Annual Abstract of Statistics*, 1950)

Annual income (£000)	Number of persons	Frequency per £500
2 and less than 2.5	60 336	60 336
2.5 " " 3	41 033	41 033
3 " " 4	45 532	22 766
4 " " 5	23 263	11 632
5 " " 6	13 475	6 737
6 " " 8	13 456	3 364
8 " " 10	6 419	1 605
10 " " 12	3 551	888
12 " " 15	2 926	488
15 " " 20	2 007	201
20 " " 25	820	82
25 " " 30	399	40
30 " " 40	376	19
40 " " 50	134	6
50 " " 75	128	2
75 " " 100	45	1
100 and over	38	?
TOTAL NUMBER OF PERSONS	213 938	–

Looking at this column we see that the maximum frequency per £500 in this case is at the beginning of the frequency distribution.

1.7 The frequency distribution may be represented graphically. Measuring the variate value along the x-axis and frequency per interval along the y-axis, we erect at the abscissa corresponding to the centre of each interval an ordinate equal to the frequency per unit interval in that interval. These ordinates are joined by straight lines, each to the next. The diagram so obtained is called a frequency polygon. Fig. 1.1 shows the frequency polygon for the data of Table 1.1.

As a variant of this procedure we may erect on the abscissa range corresponding to each interval a rectangle whose area is proportional to the frequency in that interval. A diagram constructed in this way is called a histogram. Fig. 1.2 shows such a histogram for part of the data of Table 1.2. It is evident that the histogram is a more suitable form of representation when the intervals are unequal. Making the rectangles' areas, rather than their heights, represent the frequencies is equivalent to making their heights proportional to the frequencies per unit range, which may be used as a vertical scale as in Fig. 1.2.

1.8 A few practical points in the tabulation of observed frequency distributions may be noted.

(1) It has been remarked that wherever possible the intervals should be of equal length. The importance of this will be appreciated more in subsequent chapters, but it is already evident that comparison by inspection is more difficult when the intervals are of unequal length. Scanning the second column of Table 1.2, for example, we note

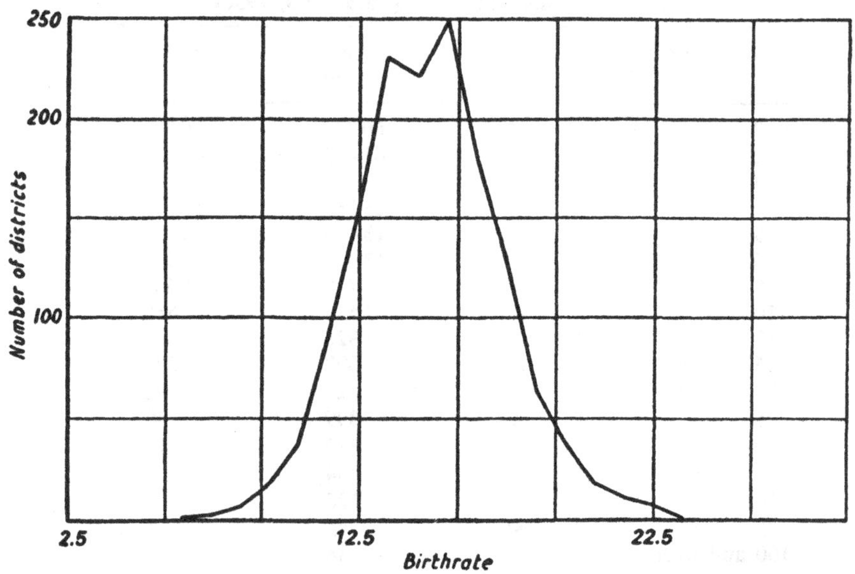

Fig. 1.1 Frequency polygon of the data of Table 1.1

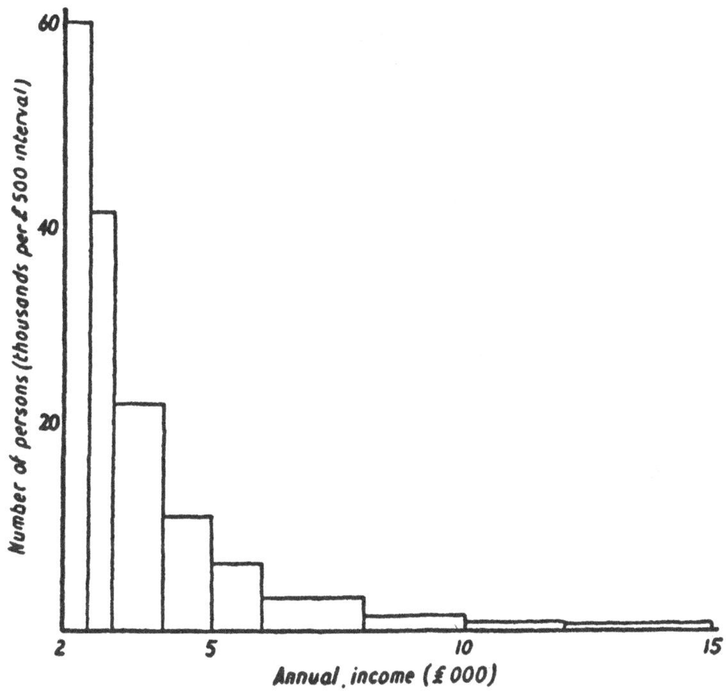

Fig. 1.2 Histogram of the data of Table 1.2

that the frequency in the interval 3–4 is greater than in the immediately preceding interval; but this is merely due to a change in the length of the intervals at that point and, as is seen from the third column, the frequency per unit interval decreases steadily.

If a histogram is to be used, D. W. Scott (1979) gives theoretical reasons for choosing the length of intervals to be approximately $0.6\,N^{-\frac{1}{3}}$ times the range of the variate, where N is the total frequency. Table 1.1 comes close to this, but commonly the intervals are longer than this formula proposes, especially where N is very large—cf. Table 1.8 below.

If a frequency polygon is used instead, the length of interval should be approximately $0.35\,N^{-\frac{1}{5}}$ times the range (cf. D. W. Scott, 1985) which is much smaller when N is very large, but not very different when N is about 100.

For related arguments on optimal interval-length for smoothing, see **1.21**.

(2) It is important to specify the interval with precision. We not infrequently meet with such groupings as "0–10, 10–20, 20–30", etc. To which interval is a member with variate value 10 assigned? Obviously the grouping is ambiguous if such values can in fact arise. We must either take the intervals "greater than or equal to 0 and less than 10, greater than or equal to 10 and less than 20," or make it clear what convention we use to allot a variate value falling on the border between two neighbouring intervals. There are various ways of indicating the interval in practical tables, e.g. "10–, 20–, 30–" means "greater than or equal to 10 and less than 20", and so forth. Sometimes, where

Table 1.3 Deaths from meningitis (except meningo-coccal and tuberculous) in England and Wales in 1953

(From the Registrar-General's Statistical Review for England and Wales for 1953)

Age at death, years	Number of deaths		Age at death, years	Number of deaths
0–	162	208	35–	5
1–	29		40–	11
2–	11		45–	17
3–	6		50–	21
4–	0		55–	32
5–	10		60–	23
10–	8		65–	22
15–	6		70–	15
20–	4		75–	11
25–	3		80–	10
30–	5		85 and over	3
			TOTAL	414

a continuous variate is concerned, there is an element of imprecision in the specification of the fineness to which the measurements are made; for example, if we are measuring lengths in centimetres to the nearest centimetre, an interval shown as "greater than 15 and less than 18" means an interval of "greater than 15.5 and less than 17.5". When the precision of the measurements is known we can specify an interval by its mid-point, in this case, 16.5.

(3) Remark (1) about the importance of equality of length of intervals should not be held to preclude the specification of frequencies in finer intervals where the frequency is changing very rapidly. Table 1.3, for instance, shows the number of deaths from meningitis in England and Wales in 1953 according to the variate "age at death". If the frequencies in the interval "0 and less than 5" were not subdivided and were thus shown as a total 208 for the interval, we may doubt whether the greatest number of deaths occurred in the first year of life. This is so; the individual frequencies in the first five years bring out the relatively heavy mortality in the first year.

(4) Perhaps it is hardly necessary to add that the histogram may be a misleading method of representing data for discrete variates. It shows each frequency uniformly dispersed over its whole interval, whereas if the variate is discrete, frequencies must necessarily be concentrated at certain points.

Discrete variates

1.9 It will be useful at this stage to give some further examples of frequency distributions that occur in practice.

Table 1.4 shows the distribution of digits in numbers taken from a four-figure

Table 1.4 Distribution of digits from a London Telephone Directory
(M. G. Kendall and B. Babington Smith (1938))

Digit	0	1	2	3	4	5	6	7	8	9	TOTAL
Frequency	1026	1107	997	966	1075	933	1107	972	964	853	10 000

telephone directory. The numbers were chosen by opening the directory haphazardly and taking the last two digits of all the numbers on the page except those in heavy type. The distribution is irregular, but from a cursory inspection of the table we are inclined to suppose that the digits occur approximately equally frequently in the larger population from which these 10 000 members were chosen. Using methods to be developed later in this book, we may see that the divergences from the average frequency per digit, 1000, are not accidental sampling effects; but at this stage it is sufficient to note that the data suggest for consideration a population of equally frequent members.

Table 1.5 shows 821 unit samples of foliage on orange trees according to the number of black-scale insects found. The greatest frequency is at zero and the frequencies for higher values of the variate fall off more or less regularly for $x = 1$ to $x = 8$.

Table 1.5 Distribution of black-scale insects on 821 unit samples (ten adjacent leaves and their stem) of orange trees
(W. M. Upholt and R. Craig, 1940, *J. Ecol. Entom.*, 33, 113)

Number of scales	0	1	2	3	4	5	6	7	8	over 8	TOTAL
Frequency	199	124	106	65	42	46	36	30	14	159	821

In Table 1.6, on the other hand, showing the number of suicides among women in eight German states in fourteen years, the distribution reaches its highest frequencies for 1–3 suicides and then falls off rather slowly.

Table 1.6 Showing suicides of women in eight German States in fourteen years
(Von Bortkiewicz, *Das Gesetz der kleinen Zahlen*, 1898)

Number of suicides per year	0	1	2	3	4	5	6	7	8	9	10 and over	TOTAL
Frequency	9	19	17	20	15	11	8	2	3	5	3	112

Continuous variates

1.10 Table 1.7 shows 8585 adult males in the United Kingdom (including, at the time of the collection of the data, the whole of Ireland), distributed according to height. The frequency polygon is shown in Fig. 1.3. It will be seen that the distribution is almost symmetrical, there being a maximum frequency at 67– inches and a steady decrease in frequency on either side of this maximum.

Table 1.7 Frequency distribution of heights for adult males born in the United Kingdom (including the whole of Ireland)

(Final Report of the Anthropometric Committee to the British Association, 1883, p. 256)

As measurements are stated to have been taken to the nearest $\frac{1}{8}$th of an inch, the true intervals are presumably $56\frac{15}{16}$–$57\frac{15}{16}$, $57\frac{15}{16}$–$58\frac{15}{16}$, and so on.

Height without shoes (inches)	Number of men of this height	Height without shoes (inches)	Number of men of this height
57–	2	68–	1230
58–	4	69–	1063
59–	14	70–	646
60–	41	71–	392
61–	83	72–	202
62–	169	73–	79
63–	394	74–	32
64–	669	75–	16
65–	990	76–	5
66–	1223	77–	2
67–	1329		
		TOTAL	8585

This more-or-less uniform tailing off of frequencies is very common in observed distributions, but the symmetry property is comparatively rare. The distribution in Table 1.1 is roughly symmetrical, but Tables 1.8 and 1.9, showing respectively 301 785 Australian marriages distributed according to bridegroom's age, and 679 dairy herds distributed according to costs of production of milk, illustrate that various degrees of

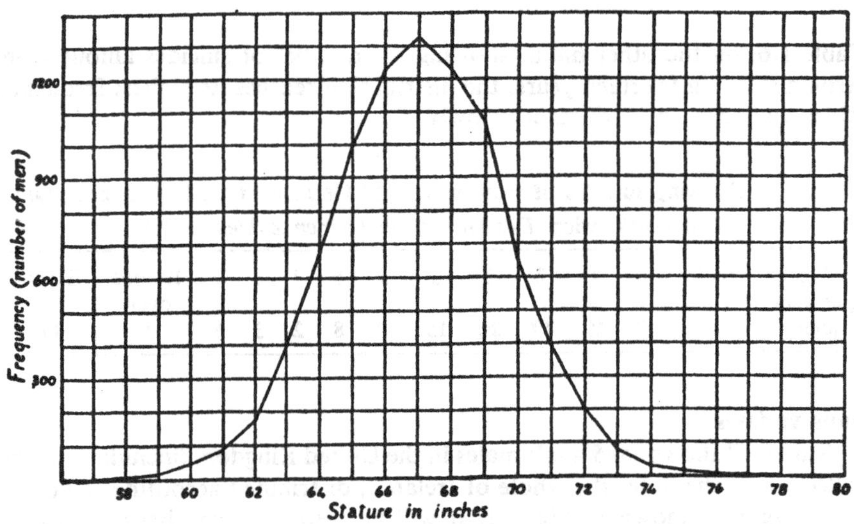

Fig. 1.3 Frequency distribution of the data of Table 1.7. Values of the abscissa correspond to the beginnings of intervals

Table 1.8 Marriages contracted in Australia, 1907–14, grouped according to the age of bridegroom

(From S. J. Pretorius (1930))

Age of bridegroom (central value of 3-year range, in years)	Number of marriages	Age of bridegroom (central value of 3-year range, in years)	Number of marriages
16.5	294	55.5	1 655
19.5	10 995	58.5	1 100
22.5	61 001	61.5	810
25.5	73 054	64.5	649
28.5	56 501	67.5	487
31.5	33 478	70.5	326
34.5	20 569	73.5	211
37.5	14 281	76.5	119
40.5	9 320	79.5	73
43.5	6 236	82.5	27
46.5	4 770	85.5	14
49.5	3 620	88.5	5
52.5	2 190		
		TOTAL	301 785

asymmetry can occur. An extreme form is shown in Table 1.3. In this connection Table 1.10, showing 7749 men distributed according to weight, is of interest for comparison with the height data of Table 1.7. The height distribution is symmetrical but the weight distribution is not.

1.11 When the asymmetry of a distribution such as that of Table 1.3 becomes extreme we may be unable to determine whether, near the maximum ordinate, there is a fall on either side, or whether the maximum occurs right at the start of the

Table 1.9 Dairy herds in a sample of herds from England and Wales, 1948/9 grouped according to cost of production of milk

(From National Investigation into the Economics of Milk Production—Milk Marketing Board, 1950)

Cost in pence per gallon	Number of herds	Cost in pence per gallon	Number of herds
less than 12	3	24–	70
12–	19	26–	49
14–	52	28–	31
16–	96	30–	16
18–	121	32–	6
20–	115	34–	8
22–	86	36 and over	7
		TOTAL	679

Table 1.10 Frequency distribution of weights for adult males born in the United Kingdom

(Source: as in Table 1.7.)

Weights were taken to the nearest pound, so the true intervals are 89.5–99.5, 99.5–109.5, etc

Weight in lb	Frequency	Weight in lb	Frequency
90–	2	190–	263
100–	34	200–	107
110–	152	210–	85
120–	390	220–	41
130–	867	230–	16
140–	1623	240–	11
150–	1559	250–	8
160–	1326	260–	1
170–	787	270–	–
180–	476	280–	1
		TOTAL	7749

distribution. This would have been the case in Table 1.3 if we had not the finer grouping for the first five years of life; and it is the case in Table 1.2, in which the maximum frequency apparently occurs at or very close to the minimum tabulated income of £2000 per annum. Asymmetrical distributions are sometimes called *skew*; and those such as Table 1.2 and Fig. 1.2, in which frequencies decline monotonically as the variate value increases, are called *J-shaped*. Exceptionally, a J-shaped distribution may be reversed, with frequencies increasing monotonically to a terminal maximum.

Table 1.11 Male deaths in England and Wales in 1953, by age at death

(From reference to Table 1.1)

Age at death	Number of deaths	Age at death	Number of deaths
0–	12 244	45–	9 016
5–	1 043	50–	14 507
10–	665	55–	19 204
15–	1 104	60–	26 802
20–	1 640	65–	34 292
25–	1 932	70–	39 644
30–	2 449	75–	40 162
35–	3 068	80–	29 061
40–	5 104	85 and over	17 553
		TOTAL	259 490

Table 1.12 Distribution of trypanosomes from *Glossina morsitans*, by length

(From K. Pearson, 1914–15, *Biometrika*, **10**, 112. Length presumably to nearest micron)

Length (microns)	Frequency	Length (microns)	Frequency
15	7	26	110
16	31	27	127
17	148	28	133
18	230	29	113
19	326	30	96
20	252	31	54
21	237	32	44
22	184	33	11
23	143	34	7
24	115	35	2
25	130		
		TOTAL	2500

1.12 The distributions considered up to this point have one thing in common—they have only one maximum and are therefore called *unimodal*. Distributions also occur showing several maxima, Tables 1.11 and 1.12 being instances in point. The first, showing deaths according to age at death, is typical of death distributions. Near the start of the distribution there is a maximum and a rapid fall in the frequency, which then rises to a pronounced maximum about the age 70–75, the frequencies beyond that point tailing off to zero. It is natural to wonder whether such a distribution can be usefully considered as two or more superposed distributions, e.g. a J-shaped distribution indicative of infantile mortality, and a skew distribution with a maximum at 70–75, the ordinary death curve of senescence.

A similar representation of a complex distribution could be undertaken for the data of Table 1.12, showing a number of trypanosomes from the tsetse fly, *Glossina morsitans*, classified according to length. We are led to suspect that the distribution is composed of several others; such composite distributions are termed *mixtures*; cf **5.20–23**.

1.13 In rare cases the distribution may have maxima at both ends and a minimum between them, as in Table 1.13, showing the distribution of 1715 estimates of degrees of cloudiness. This is known as a U-shaped distribution.

Distributions also occur which in general appearance resemble sections of the types already mentioned. A J-shaped distribution like Fig. 1.2 resembles the tail of the symmetrical distribution of Fig. 1.3. The suicide data of Table 1.6 may be regarded as a roughly symmetrical distribution truncated just below the maximum ordinate by the impossibility of the occurrence of negative values of the variate. This sort of conception is sometimes useful in fitting curves to observed data—a given analytical curve may fit

Table 1.13 Distribution of estimated degrees of cloudiness at Greenwich for the month of July 1890–1904 (excluding 1901)

(Data from Gertrude E. Pearse, 1928, *Biometrika*, 20A, 336)

Degrees of cloudiness	Frequency	Degrees of cloudiness	Frequency
10	676	4	45
9	148	3	68
8	90	2	74
7	65	1	129
6	55	0	320
5	45		
		TOTAL	1715

the data quite well in a certain variate-range, but may also extend into regions where the data cannot, so to speak, follow it.

Stem-and-leaf plots

1.14 Maxima near the ends of the range (1.11), or superposed distributions, (1.13) may be difficult to detect using an ordinary histogram. A simple but effective procedure that allows grouping of data without any loss of information is the stem-and-leaf plot developed by Tukey (1977).

For example, suppose the data of Table 1.3 were reported as age in years ignoring additional months. We might observe a stream of integer ages such as

$$0, 43, 7, 34, 17, 26, 53, 0, 3, 14, 0, 8, \ldots.$$

If these values were grouped into the 22 age-groups of Table 1.3, the frequencies shown there would result. However, we may code the groups as

$$0^- = 0\text{–}4, \quad 0^+ = 5\text{–}9, \quad 1^- = 10\text{–}14, \quad 1^+ = 15\text{–}19, \ldots$$

where the values $0^-, 0^+, 1^-, 1^+, \ldots$, which are the penultimate digits of the ages, are known as *stems*. The final digits of the ages (0, 3, 7, 4, and so on) form the *leaves* to be attached to the stems. Thus the 12 observations above may be summarized by the stem-and-leaf plot

Stem	Leaves	Stem	Leaves
0^-	0003	3^-	4
0^+	78	3^+	
1^-	4	4^-	3
1^+	7	4^+	
2^-		5^-	3
2^+	6	5^+	

Each integer (leaf) is one observation. When all 414 values are entered and ordered on the stems as here, any distinctive features such as the large number of zeros are immediately apparent. The total number of leaves on each stem, when added appropriately into age-groups, will give Table 1.3. Thus, e.g., the age-group "20–" there will consist of the number of leaves on 2^- and 2^+, and so on.

In general, the stems may be selected to give a suitable number of groups, and extreme values outside the main body of the table may be recorded individually rather than grouped into open-ended extreme groups (e.g. "85 and over" in Table 1.3).

Probability functions and distribution functions

1.15 The examples given above illustrate the remarkable fact that the majority of the frequency distributions encountered in practice possess a high degree of regularity. The form of the frequency polygons and histograms above suggests, almost inevitably, that our data may be described, at least approximately, by distributions that can be specified by smooth curves and simple mathematical expressions. This is particularly true when we are dealing with a conceptual population where the set of possible outcomes is non-countably infinite. We now develop the concept of a "theoretical" frequency function by means of same examples. Such "theoretical" frequency functions are better termed *probability functions*, although we defer a formal definition of probability until Chapter 8 after we have introduced distributions.

Consider a discrete distribution such as that of Table 1.4. Let us represent our variate by X. Then we may say that X can take any of the ten values 0, 1, ..., 9 and that the relative frequency of a value of X equal to x, say $f(x)$, is given by the table, that is to say, $f(0) = 1026/10\,000 = 0.1026$, $f(1) = 0.1107$, $f(2) = 0.0997$ and so on. By construction these relative frequencies sum to unity. They define the frequency function. Furthermore, most of the frequencies in the table are approximately 1000, and we may then consider the observed distribution as approximating to that defined by

$$f(x) = 1/10, \qquad x = 0, 1, \ldots, 9, \tag{1.1}$$

where

$$\sum_{x=0}^{9} f(x) = 1. \tag{1.2}$$

This is perhaps the simplest form of discrete distribution, $f(x)$ being a constant for all permissible values of x.

In Table 1.5 we have a discrete variate that can, theoretically, take an infinite number of values, namely, zero and the positive integers. In practice, of course, there must be a limit to the number, but since we do not know that limit we may imagine our variate as infinite in range. The frequency function for the table itself is again simply defined by the relative frequencies therein; but if we wish to proceed to a conceptual generalization of such a table we must define a discrete function $f(x)$ for all positive integral values of x. These values $f(x_j)$ may be defined as the limiting relative frequencies as the number of observations approaches infinity. As we shall see in Chapter 8 this is one way to define probabilities and, indeed, the definition we shall use in Volumes 1 and 2A. By construction, the relative frequencies satisfy $f(x_j) \geqslant 0$ for all j and

$\sum_{j=0}^{\infty} f(x_j) = 1$. Conceptually, we may move to the limiting case, and say that the set of pairs $\{x_j, f(x_j), j = 0,1, \ldots\}$ defines the *probability mass function*, often abbreviated to p.m.f.

1.16 Consider now the case of a continuous variate. In the ordinary data of experience our distributions are invariably discontinuous, because our measurements can only attain a certain degree of accuracy. For instance, we are accustomed to suppose that the height of a man may in reality be any real number of inches in a certain range, say 50 to 80. In fact, we can measure heights only to a certain accuracy, say to the nearest thousandth of an inch. Our measurements thus consist of whole numbers (of thousandths) from 50 000 to 80 000, and such a number as 62 831.85 ($=20\,000\pi$ approximately) cannot appear. All physical measurements are subject to this limitation, but we accept it and nevertheless speak of our variables as continuous, the underlying supposition being that the measurements are approximations to numbers which can fall anywhere in the arithmetic continuum.

1.17 With this understanding we can consider the distribution of grouped frequencies as leading to the concept of a probability function for a continuous variate. If, in one of the distributions above, say that of Table 1.7, we were to subdivide the intervals, we should probably find that *up to a point* the resulting frequencies were smoother and smoother. The reader can verify the appearance of this effect for himself by grouping the data of Table 1.7 in intervals of 8, 4, and 2 inches. We cannot, however, take the process too far, because, with a finite population, continued subdivision of the interval would sooner or later result in irregular frequencies, there being only a few members in each interval and perhaps no members at all in some intervals. But we may suppose that for ranges Δx, not too small, the distribution may be specified by a function $f(x)\ \Delta x$, expressing that in the range $\pm\frac{1}{2}\Delta x$ centred at x the frequency is $f(x)\ \Delta x$, *wherever x may be in the permissible range of the variate.* We may suppose further that as Δx tends to zero the population is perpetually replenished so as to prevent the occurrence of small and irregular frequencies; and in this way we arrive at the concept of the probability function for a continuous variable. We write

$$dF = f(x)dx \tag{1.3}$$

expressing that the element of probability dF between $x-\frac{1}{2}dx$ and $x+\frac{1}{2}dx$ is $f(x)dx$, for all x and for dx however small.

It is evident from (1.3) that the probability function is defined for *finite* intervals of x values, but not for single points. In order to utilize the power of calculus, however, the limit is often taken as $dx \to 0$. Assuming the limit exists for all x, we define $f(x)$ to be the *probability density function*, abbreviated to p.d.f.; the name reflects the fact that the value of the p.d.f. is not a probability, but a convenient mathematical construct from which probabilities may be derived.

1.18 This admittedly somewhat intuitive approach to the concept of continuous distributions appears to be the best for statistical purposes, and is certainly the way in which the concept was originally reached. In formulating the axioms and postulates of a rigorous mathematical theory, however, the mathematician considers a rather more general function.

Let $F(x)$ be a function of x which is defined at every point in a range and is continuous except perhaps at a denumerable number of points. In our convention $F(x)$ is always continuous on the right. We require that F shall be zero at the lower point of the range (which may be $-\infty$) and unity at the upper point (which may be $+\infty$) and that it shall not decrease at any point. Such a function is called a *distribution function*. For observed data, it is the cumulative relative frequency up to and including x. For example, in Table 1.4, $F(x)=0$ for $x<0$; $F(x)=0.1026$ for $0\leqslant x<1$; $F(x)=0.1026+0.1107=0.2133$ for $1\leqslant x<2$; and so on. Here there are ten points of discontinuity for $F(x)$. These points are called "saltuses" (jumps) and $F(x)$ in this case is called a step function. For theoretical distributions, $F(x)$ represents the cumulative probability.

If there is no saltus in the range, $F(x)$ is everywhere continuous and monotonically non-decreasing. If it has a derivative everywhere we have the equation

$$dF/dx = F'(x) = f(x). \tag{1.4}$$

The mathematics of this branch of the subject is then that of the study of functions of the class $F(x)$ and $f(x)$.

1.19 The functions thus defined are more general than those obtained from the statistical approach in two ways: (i) $F(x)$ can be continuous throughout part of the range and elsewhere possess saltuses, e.g. the variate may be continuous up to a point and then discrete; in statistical practice a variate is usually continuous or discrete throughout its

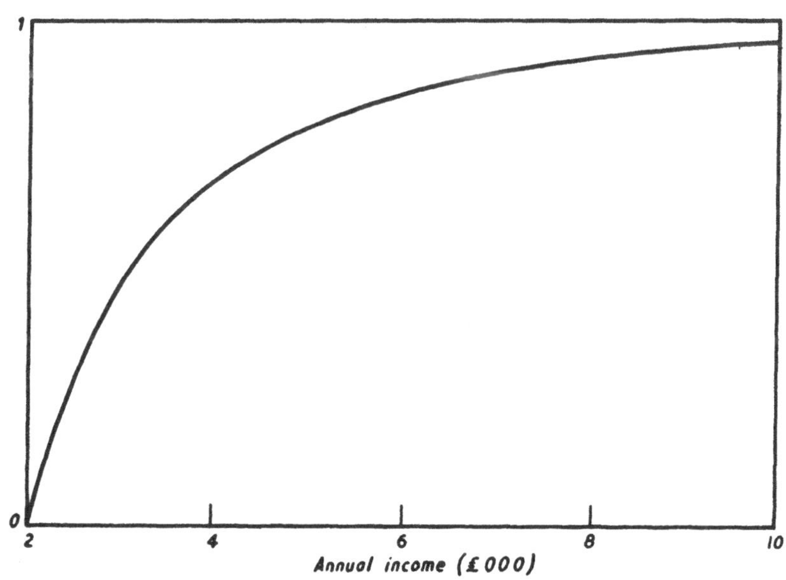

Fig. 1.4 Distribution function of the data of Table 1.2

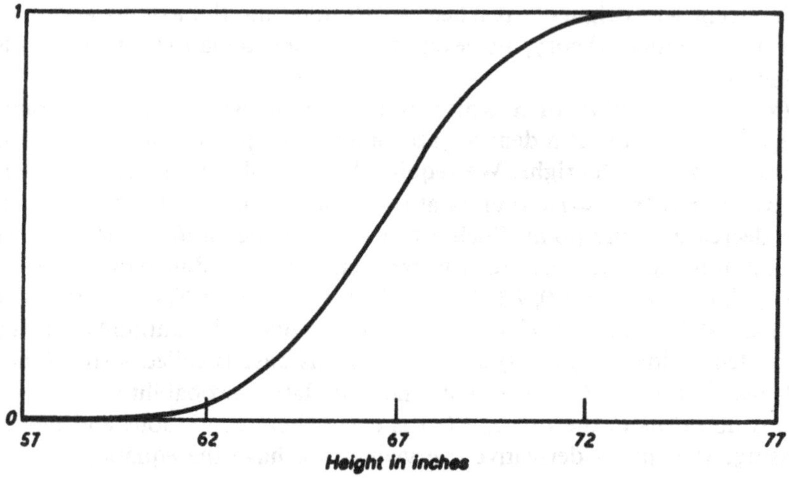

Fig. 1.5 Distribution function of the data of Table 1.7

range (although there are important exceptions—cf. Exercise 5.22 below). (ii) Where no saltus exists $F(x)$ can exist without possessing a density function, since a continuous function need not necessarily possess a derivative. In all the cases we shall consider, the existence of a continuous variate will be accompanied by the existence of a continuous density function.

The function $F(x)$ is sometimes called the cumulative distribution function (c.d.f.), but we shall use the term distribution function only, abbreviated to d.f.

Fig. 1.4 shows the distribution function of part of the data of Table 1.2 in graphical form, $F(x)$ being given as ordinate against x as abscissa. In Fig. 1.5 we show the distribution function of the data of Table 1.7. In both cases we have smoothed the functions to some extent by joining the ends of the ordinates instead of erecting narrow rectangles. For a continuous variate such a process is clearly often justified; it approximates to the continuous distribution that, we imagine, underlies the data.

Whereas in the discrete case we had

$$F(x) = \sum_{a}^{x} f(u), \tag{1.5}$$

in the continuous case it is

$$\begin{aligned} F(x) &= \int_a^x dF(u) \\ &= \int_a^x f(u)du, \end{aligned} \tag{1.6}$$

where x ranges from a to b. We now introduce a convention that simplifies these expressions to some extent. To avoid the constant specification of the limits a and b we may, without loss of generality, suppose that $F(x)$ and $f(x)$ are zero for any x less than a, and that $F(x) = 1$ and $f(x) = 0$ for any x greater than b. With this convention we may write

$$\left.\begin{aligned} F(x) &= \sum_{j=-\infty}^{x} f(u) \\ F(x) &= \int_{-\infty}^{x} f(u)\mathrm{d}u \end{aligned}\right\} \tag{1.7}$$

and

$$\left.\begin{aligned} \sum_{j=-\infty}^{\infty} f(u_j) = F(\infty) - F(-\infty) = 1 \\ \int_{-\infty}^{\infty} f(u)\mathrm{d}u = F(\infty) - F(-\infty) = 1. \end{aligned}\right\} \tag{1.8}$$

Random variables

1.20 In mathematics, a *variable* describes a quantity that is free to take on values in a certain range; no probabilities attach to these values. However, a *random variable*, or *variate*, associates a probability with each possible value in the discrete case, or with each half-open interval in the continuous case. Indeed the random variable is a function that describes a statistical experiment: if a certain procedure is followed, one of the set of possible values will occur, according to the associated set of probabilities. Thus, the random variable is specified by $(\mathscr{X},\mathscr{P})$ where $\mathscr{X}$ denotes the set of possible values, and $\mathscr{P}$ the associated probabilities.

Following common convention, we use upper case letters to denote random variables and lower case letters to denote their variate values. Thus, the statement

$$P(X = x)$$

denotes the probability that the statistical experiment results in an outcome with assigned value x; "$X = x$" is *not* a mathematical statement of equality.

In general, the d.f. is given by

$$P(X \leqslant x) = F(x)$$

where

$$\begin{aligned} P(a < X \leqslant b) &= P(X \leqslant b) - P(X \leqslant a) \\ &= F(b) - F(a), \end{aligned}$$

from which the p.d.f. may be derived as a limit when the variate is continuous.

Notwithstanding the potential nomenclatural dangers, we shall often use lower case letters to denote both variate and value; we revert to the stricter notation whenever necessary. Also for ease of notation, we use the single term $f(x)$ to denote both the p.d.f. for continuous x and the p.m.f. for the discrete case. When no distinction needs to be made, we use the collective term *probability function*.

Since $f(x)$ denotes different concepts in different contexts, the following lexicon serves to relate the particular meaning to a given context:

form of distribution	$f(x)$
numerical (or frequency)	frequency function (f.f.)
theoretical—discrete	probability mass function (p.m.f.)
theoretical—continuous	probability density function (p.d.f.)
theoretical—all	probability function (p.f.)

Density smoothing

1.21 When the variate x has a continuous distribution, it is of interest to determine how one might "smooth" the observed $f(x)$ or $F(x)$ in order to simplify interpretation. For example, such smoothing operations are often applied to actuarial data before life tables are developed.

One approach to such problems is to fit a distribution of known form with adjustable parameters to the data; such distributions are described in Chapters 5 and 6. Another possibility is to use an orthogonal polynomial expansion to approximate $f(x)$; see **6.14–6.24** and Anderson and De Figueiredo (1980). At this stage, however, we shall concentrate upon methods that do not require any distributional assumptions, of which the best-known is the class of kernel methods developed by Rosenblatt (1956) and Parzen (1962).

Intuitively, instead of thinking of a histogram with fixed groups, we might consider a "moving histogram" with interval length $2h$ and set

$$f_s(x) = \frac{F(x+h) - F(x-h)}{2h}, \tag{1.9}$$

where $F(x)$ is the d.f. and the subscript "s" in f_s denotes the smoothed density. Clearly, a large value of h will produce a very flat f_s and, as $h \to 0$, f_s will eventually become very irregular. Rosenblatt (1956) showed that a suitable choice of h, in the sense of minimizing the asymptotic mean-square-error (cf. **17.30**, Volume 2A), is

$$h = h(n) = k(x)n^{-\frac{1}{5}}, \quad \text{where} \quad k(x) = \frac{4.5f(x)}{|f''(x)|^2},$$

recognizing that $k(x)$ will be unknown in practice. Nowadays, it is a straightforward matter to compute f_s for different choices of h and then select a sufficiently smoothed version.

The smoothing function (1.9) may be extended to the more general form

$$f_s(x) = \frac{1}{N}\sum_{i=1}^{N} K_N(x, x_i), \tag{1.10}$$

where K_N is a kernel function. For example, (1.9) corresponds to

$$K_N(x, x_i) = \frac{1}{2h}, \quad |x - x_i| \leqslant h$$

$$= 0, \quad \text{otherwise.}$$

In general, it is assumed that $K_N(u)$ is symmetric about zero and that $\int_{-\infty}^{\infty} K_N(u)\,du = 1$. Although not necessary, it is also usually assumed that $K_N(u)$ is non-negative for all

u and is bounded above. Parzen (1962) demonstrated that, under mild conditions, $f_s(x)$ converges to $f(x)$ "in probability" as $n \to \infty$; see **8.44** for a discussion of this concept. The choice of h is the key factor in the use of such methods; for recent developments in this area, see Woodroofe (1970) and Silverman (1978).

Other approaches include the use of a roughness penalty (Good and Gaskins, 1971; Scott *et al.*, 1980) and the use of splines (e.g. Wahba, 1975). For another approach, see Leonard (1978). A detailed discussion of the problem is given in Silverman (1986).

Excursus on integrals

1.22 We have noted above that the functions which arise in statistical practice are less complicated than the more general distribution function that the mathematical statistician wishes to discuss. This illustrates a point that causes constant difficulty in the systematic development of theoretical statistics. Modern mathematics has devised rigorous theories of great generality, but largely at the expense of added difficulty and abstraction. For most statistical purposes this elaborate apparatus is not required, and statistical functions can be adequately investigated by simpler methods without loss of rigour. In this book we shall, as a rule, prefer these simpler methods, but this course will not overcome all our difficulties. Some results that we require are not obtainable by simple methods; for others, the so-called elementary proofs are so long and tedious as to be worse than the more sophisticated versions; and for others again the simpler treatment may obscure the essence of the statistical argument.

1.23 The modern theory of integration is a case in point. For many purposes it would be enough to consider the integral as defined in introductory treatments, or in the manner of Cauchy. Thus if $f(x)$ is defined in the interval (a, b) we divide the range at points $x_1, x_2, \ldots, x_n$ and consider the sum

$$s_n = f(a)(x_1 - a) + f(x_1)(x_2 - x_1) + \ldots + f(x_n)(b - x_n). \tag{1.11}$$

It may be shown that under certain conditions (such as the continuity of $f(x)$ in the range) this sum tends to a limit as the length of the intervals tends to zero, independently of where the dividing points are drawn or the way in which the tendency to zero proceeds. This limit is the Cauchy integral: $\int_a^b f(x)\,dx$.

A more general type of integration is due to Riemann and Stieltjes. Let $\psi(x)$ be a continuous function of x. Choose ξ_1 in the range a to x_1, ξ_2 in the range x_1 to x_2 and so on. Consider the sum

$$\begin{aligned} s_n' = {} & \psi(\xi_1)\{F(x_1) - F(a)\} + \psi(\xi_2)\{F(x_2) - F(x_1)\} + \ldots \\ & + \psi(\xi_{n+1})\{F(b) - F(x_n)\}. \end{aligned} \tag{1.12}$$

It may be shown that as the length of the intervals tends to zero uniformly s tends to a limit that is independent of the location of the points ξ or the boundary points of the intervals. We then write this limit as

$$\int_a^b \psi(x)\,dF.$$

It is known as the Stieltjes integral of $\psi(x)$ with respect to $F(x)$. Riemann considered the case when $F(x)=x$.

1.24 The advantage of the Stieltjes integral is that it reduces to the Cauchy integral if $f(x)$ is continuous and to an ordinary summation if $f(x)$ is discrete. To save writing all our formulae down twice, we shall often employ an integral of this type, replacing it in special cases by an ordinary integral or a sum as the circumstances require.

Many of the theorems of ordinary integration are true of the Stieltjes integral. We shall frequently require the following:

$$\begin{aligned}\left|\int_a^b \psi\,\mathrm{d}F\right| &\leqslant \int_a^b |\psi|\,\mathrm{d}F \\ &\leqslant M\int_a^b \mathrm{d}F \\ &\leqslant M, \end{aligned} \tag{1.13}$$

where M is the upper bound of $\psi(x)$ in the range (a, b).

$$\int_a^b \psi\,\mathrm{d}F = \psi(\xi)\int_a^b \mathrm{d}F, \tag{1.14}$$

where ξ is a value of x in the range (a, b).

If a and b are finite,

$$\int_a^b \sum_{j=1}^{\infty} f_j(x)\mathrm{d}F = \sum_{j=1}^{\infty} \int_a^b f_j(x)\mathrm{d}F, \tag{1.15}$$

provided that $\sum f_j(x)$ converges uniformly in the range. The theorem is not necessarily true if a or b is infinite.

The ordinary rules of partial integration are also applicable to Stieltjes integrals.

1.25 A more complicated integral is due to Lebesgue. It is equal to the Riemann–Stieltjes integral when the latter exists, but also exists in cases where the Riemann–Stieltjes integral does not. We do not normally require it in the ordinary theory of distributions, but it is essential for a rigorous treatment of Fourier transforms which we need for certain parts of the theory of time-series—for a discussion, see Kingman and Taylor (1966, Chapter 5).

Variate transformations

1.26 Suppose that we have a variable y related functionally to a variable x by the one-to-one relation

$$x = x(y), \qquad y = y(x), \tag{1.16}$$

y being continuous and differentiable in x and vice versa. We then have from (1.7)

$$F\{x(y)\} = \int_{-\infty}^{x(y)} f(u)\mathrm{d}u \tag{1.17}$$

and differentiation with respect to y gives

$$dF(y) = f\{x(y)\} \frac{dx}{dy} dy. \tag{1.18}$$

If X is the variate corresponding to x we may regard this expression as defining a variate $Y(X)$ corresponding to y. Such variate transformations are important in the theory of continuous distributions. By their means many mathematically specified distributions may be reduced to a known form, either exactly or approximately.

For example, a distribution that we shall study in the theory of sampling is

$$dF = \frac{1}{2^{\frac{1}{2}\nu-1}\Gamma(\frac{1}{2}\nu)} e^{-\frac{1}{2}\chi^2}\chi^{\nu-1} d\chi, \qquad 0 \leqslant \chi < \infty. \tag{1.19}$$

It is readily verified that $F(\infty) = 1$.

By the transformation $t = \frac{1}{2}\chi^2$ we reduce this to

$$dF = \frac{1}{\Gamma(\frac{1}{2}\nu)} e^{-t} t^{\frac{1}{2}\nu-1} dt, \qquad 0 \leqslant t < \infty. \tag{1.20}$$

This is a well-known form in analysis, the distribution function being the Incomplete Γ-function:

$$F(t) = \Gamma_t(\tfrac{1}{2}\nu)/\Gamma(\tfrac{1}{2}\nu). \tag{1.21}$$

Again, the distribution

$$dF = \frac{y_0}{(1 + t^2/\nu)^{\frac{1}{2}(\nu+1)}} dt, \qquad -\infty < t < \infty \tag{1.22}$$

(y_0 being chosen so that $F(\infty) = 1$), a symmetrical peaked distribution of infinite range rather like that of Fig. 1.3, may, by the substitution of $t = \sqrt{\nu} \tan \theta$, be transformed into

$$\begin{aligned} dF &= \frac{y_0\sqrt{\nu} \sec^2\theta d\theta}{\sec^{\nu+1}\theta}, \qquad -\tfrac{1}{2}\pi \leqslant \theta \leqslant \tfrac{1}{2}\pi, \\ &= y_0\sqrt{\nu} \cos^{\nu-1}\theta d\theta, \end{aligned} \tag{1.23}$$

a distribution of finite range $-\frac{1}{2}\pi$ to $+\frac{1}{2}\pi$, but still symmetrical. Putting now $\sin \theta = u$, we have

$$dF = y_0\sqrt{\nu}(1 - u^2)^{\frac{1}{2}\nu-1} du, \qquad -1 \leqslant u \leqslant 1, \tag{1.24}$$

and again with $u^2 = x$,

$$dF = y_0\sqrt{\nu} x^{-\frac{1}{2}}(1 - x)^{\frac{1}{2}\nu-1} dx, \qquad 0 \leqslant x \leqslant 1. \tag{1.25}$$

The last substitution is to be noted. u ranges from -1 to $+1$, and as it does so x ranges from $+1$ to 0 and back to $+1$, the relation being two-to-one. (1.18) is applied separately for $-1 \leqslant u < 0$ and $0 \leqslant u \leqslant 1$, with $\left|\frac{du}{dx}\right| = \frac{1}{2}x^{-\frac{1}{2}}$ in each interval. Bringing these two contributions to $f(x)$ together, we obtain (1.25). Whenever substitutions are made which are not one-to-one, points such as this require some watching.

1.27 There is one variate transformation that is worth special attention. In the distribution

$$dF = f(x)dx$$

put

$$y = \int_{-\infty}^{x} f(u)du = F(x).$$

Then

$$\begin{aligned} dF &= f(x)\frac{dx}{dy}dy \\ &= \frac{f(x)}{f(x)}dy \\ &= dy, \qquad 0 \leqslant y \leqslant 1, \end{aligned} \tag{1.26}$$

so that the distribution is transformed into the uniform or rectangular distribution, for which the p.d.f. is constant throughout the interval (0, 1). This is called the *probability integral transformation*. Any continuous distribution can thus be transformed into the uniform distribution; and it follows that there exists at least one transformation that will transform any continuous distribution into any other continuous distribution, namely, the transformation which transforms one into the uniform coupled with the inverse of that which transforms the other into the uniform distribution.

The genesis of distributions

1.28 Up to this point we have not inquired into the origin of the various observed distributions that have been adduced in illustration. Certain of them may be considered apart from any question of origination from a larger population. The death distribution of Table 1.11 is an example; if we are interested only in the distribution of male deaths in England and Wales in 1953 the whole of the population under consideration is before us.

But in the great majority of cases the observations that we are able to examine form only part of a population on which our main interest is centred. The height distribution of Table 1.7 is only a part of the population of men born in the United Kingdom living at the time of the inquiry, and the data are mainly of importance in the light of the information that they provide about that population. Similarly the data on dairy herds of Table 1.9 are mainly of interest in the information they provide about costs of milk production for the whole country.

1.29 In the two cases just mentioned, height and costs of milk production, we have information about a certain sample of individuals chosen from an existing population. Only lack of time and opportunity prevents us from examining the whole population. It sometimes happens, however, that we have data that do not emanate from a finite existent population in this way. Table 1.14 is an example, showing an observed distribution of throws with dice.

Now it is clear that these data do not represent a complete population, for we can add to them by further casting of the dice. But these further throws do not exist in the

Table 1.14 Distribution of the number of successes (throws of 4, 5 or 6) in 4096 throws of 12 dice

(Weldon's data, cited by F. Y. Edgeworth, *Encyclopaedia Britannica*, 11th edn, 22, 39)

Number of successes	Frequency	Number of successes	Frequency
0	0	7	847
1	7	8	536
2	60	9	257
3	198	10	71
4	430	11	11
5	731	12	0
6	948		
		TOTAL	4096

sense that the unexamined men of the United Kingdom or the unexamined dairy herds of England and Wales exist. They have a kind of hypothetical existence conferred on them by our notion of the throwing of the dice.

Even distributions that appear at first sight to be existent may be considered in this light. The trypanosome distribution of Table 1.12, for instance, was obtained from certain tsetse flies. We may consider it as a sample of all the tsetse flies in existence, whether harbouring trypanosomes or not—an existent population; but we may also consider it as a sample of what the distribution would be if all the tsetse flies were infected with trypanosomes—a hypothetical population.

The population conceived of as generating an observed distribution is fundamental to statistical inference. We shall take up this matter again in later chapters when we consider the problem of sampling. The point is mentioned here because it will occasionally arise before we reach those chapters. It must be emphasized that the distinction between existent and hypothetical populations is not merely a matter of ontological speculation—if it were we could safely ignore it—but one of practical importance when inferences are drawn about a population using a sample selected from it.

Multivariate distributions

1.30 So far, we have considered the members of a population according to a single variate, and the frequency distributions may thus be called univariate. This may be readily generalized to include populations of members considered according to two or more variates jointly, yielding bivariate, trivariate, ..., and in general multivariate distributions, which are often referred to as the joint distributions of the variates. Table 1.15, for example, shows the distribution of 9440 beans according to both length and breadth. The border frequencies show the univariate distributions of the beans according to length and breadth separately, and the body of the table shows how the two qualities vary together.

As for the univariate case, the variates may be discontinuous or continuous and we sometimes meet cases in which one variate is of one kind and one of the other.

1.31 As a generalization of the frequency polygon and the histogram we may construct three-dimensional figures to represent the bivariate distribution. Imagine a

Table 1.15 Frequencies of beans with specified lengths and breadths

(Johannsen's data, cited by S. J. Pretorius (1930))

Lengths in millimetres (central values)

Breadth in millimetres (central values)	17	16.5	16	15.5	15	14.5	14	13.5	13	12.5	12	11.5	11	10.5	10	9.5	Totals
9.125	–	2	–	–	3	–	–	–	–	–	–	–	–	–	–	–	5
8.875	4	8	17	19	–	–	–	–	–	–	–	–	–	–	–	–	48
8.625	2	23	101	156	93	23	2	–	–	–	–	–	–	–	–	–	400
8.375	–	18	105	494	574	227	56	9	–	–	–	–	–	–	–	–	1483
8.125	–	4	44	375	956	913	362	73	12	3	–	–	–	–	–	–	2742
7.875	–	–	7	81	385	871	794	330	89	19	3	–	–	–	–	–	2579
7.625	–	–	1	4	65	236	469	361	175	55	27	4	–	–	–	–	1397
7.375	–	–	–	–	6	23	91	137	124	78	37	22	11	–	1	–	530
7.125	–	–	–	–	–	1	13	18	28	35	25	32	11	6	1	–	170
6.875	–	–	–	–	–	–	–	1	9	8	21	12	13	7	1	–	72
6.625	–	–	–	–	–	–	–	–	–	–	2	–	1	4	3	–	10
6.375	–	–	–	–	–	–	–	–	–	1	–	–	–	1	1	1	4
Totals	6	55	275	1129	2082	2294	1787	929	437	199	115	70	36	18	7	1	9440

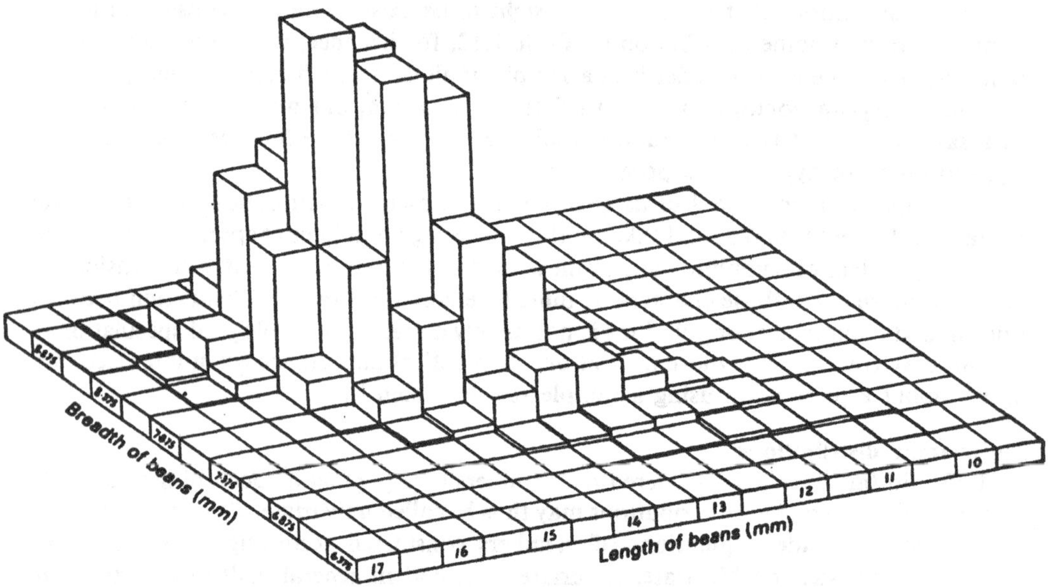

Fig. 1.6 Bivariate histogram of the data of Table 1.15

horizontal plane containing a pair of perpendicular axes and ruled like a chessboard into cells, the ruled lines being drawn at points corresponding to the terminal points of grouping intervals. At the centre of each interval we erect a vertical line proportional in length to the frequency in that interval. The summits of these verticals are joined, each to the four summits of verticals in the neighbouring cells possessing the same values of one or the other variate. The resulting figure is the bivariate frequency polygon.

Similarly we may erect on each cell a pillar proportional in volume to the frequency in that cell and thus obtain a bivariate histogram. Fig. 1.6 shows one for the bean data of Table 1.15. Such representations are now readily constructed using computer graphics packages.

The bivariate case: conditional distributions

1.32 We may write the bivariate probability function for variates x_1, x_2, as

$$dF(x_1, x_2) = f(x_1, x_2)dx_1x_2. \tag{1.27}$$

We then define the bivariate distribution function

$$F(x_1, x_2) = \int_{-\infty}^{x_1}\int_{-\infty}^{x_2} f(u_1, u_2)du_1du_2 \tag{1.28}$$

this integral also being understood in the Stieltjes sense, reducing to ordinary integration if $f(x_1, x_2)$ is continuous and to ordinary summation if it is discrete. Clearly $F(x_1, -\infty) = F(-\infty, x_2) = 0$ and by convention $F(+\infty, +\infty) = 1$ as in **1.20** for the univariate case, while F is a non-decreasing function of each of its arguments since $f(x_1, x_2) \geqslant 0$.

For any fixed value of one variate, say $x_2 = k$, the other variate will, of course, have a univariate distribution. This p.f. is given by

$$p(x_1|x_2 = k) = \frac{f(x_1, k)}{\int_{-\infty}^{\infty} f(x_1, k)dx_1}, \tag{1.29}$$

where the vertical bar on the left is read as "given". (1.29) is called the *conditional probability function* of x_1 for $x_2 = k$. By convention, we take it to be zero when the denominator on the right is zero.

For discrete distributions, (1.29) is well defined at all values of k for which the denominator >0. We shall assume that this also holds for continuous distributions, although this is not a trivial matter—see Kingman and Taylor (1966, pp. 359–64) for details.

Marginal distributions

1.33 The sums in the margins of tables such as Table 1.15 give us the distribution of each variate by itself. For any value of x_2 the total probability is obtained by summing over all values of x_1 to give the probability for that value of x_2. Thus, if the d.f. is $F(x_1, x_2)$ we may express the univariate d.f. of x_1 as $F(x_1, \infty)$; and similarly the univariate d.f. of x_2 is $F(\infty, x_2)$. Then

$$dF_1(x_1) = \left\{\int_{-\infty}^{\infty} f(x_1, x_2)dx_2\right\} dx_1, \tag{1.30}$$

$$dF_2(x_2) = \left\{\int_{-\infty}^{\infty} f(x_1, x_2)dx_1\right\} dx_2. \tag{1.31}$$

These univariate distributions are called *marginal*, for obvious reasons. The term is strictly redundant, but it is convenient and almost universally used.

From (1.31) we see that the denominator on the right of (1.29) is the marginal p.f. of x_2 at $x_2 = k$. Thus the conditional p.f. of x_1 is the ratio of the bivariate p.f. to the marginal p.f. of x_2.

Independence

1.34 If and only if

$$\begin{aligned} F(x_1, x_2) &= F(x_1, \infty)F(\infty, x_2) \\ &= F_1(x_1)F_2(x_2), \quad \text{say,} \end{aligned} \tag{1.32}$$

for all values of x_1 and x_2, the two variates are said to be independent. Where p.d.s exist we have

$$\int_{-\infty}^{x_1}\int_{-\infty}^{x_2} f(u_1, u_2)\mathrm{d}u_1\mathrm{d}u_2 = \int_{-\infty}^{x_1} f_1(u_1)\mathrm{d}u_1 \int_{-\infty}^{x_2} f_2(u_2)\mathrm{d}u_2$$

and hence

$$f(x_1, x_2) = f_1(x_1)f_2(x_2) \tag{1.33}$$

for all x_1, x_2. Conversely, (1.33) leads back to (1.32). Thus the joint d.f. and the joint p.f. factorize into two parts, one for each variate alone. It is readily seen that this definition of statistical independence conforms to the colloquial use of the word and also to its mathematical use. The conditional distribution of x_2 for any fixed x_1 (e.g. the distribution in a row or column of the bivariate table) is the same whatever the fixed value of x_1, that is to say, the distribution of x_2 is independent of x_1. (1.33) must hold for all values of x_1 and x_2 if it is to be sufficient for their independence—cf. Exercise 1.23.

Two variates that are not independent are said to be dependent. Evidently those of Table 1.15 are dependent, for the distributions in rows or in columns are far from similar.

Example 1.1 (The binormal distribution)

Consider the distribution (to be discussed in Chapters 15 and 16) with joint p.d.f.

$$f(x_1, x_2) = z_0 \exp\left\{-\frac{1}{2(1-\rho^2)}\left(\frac{x_1^2}{\sigma_1^2} - \frac{2\rho x_1 x_2}{\sigma_1\sigma_2} + \frac{x_2^2}{\sigma_2^2}\right)\right\}, \qquad -\infty < x_1, x_2 < \infty, \tag{1.34}$$

where z_0 is a constant such that $F(+\infty, +\infty) = 1$. Put

$$y_1 = \frac{1}{(1-\rho^2)^{\frac{1}{2}}}\left(\frac{x_1}{\sigma_1} - \frac{\rho x_2}{\sigma_2}\right)$$

$$y_2 = \frac{x_2}{\sigma_2}.$$

We have for the Jacobian of the transformation (cf. **1.35** below)

$$\frac{\partial(y_1, y_2)}{\partial(x_1, x_2)} = \begin{vmatrix} \dfrac{1}{\sigma_1(1-\rho^2)^{\frac{1}{2}}} & -\dfrac{\rho}{\sigma_2(1-\rho^2)^{\frac{1}{2}}} \\ 0 & \dfrac{1}{\sigma_2} \end{vmatrix}$$

$$= \frac{1}{\sigma_1\sigma_2(1-\rho^2)^{\frac{1}{2}}}.$$

Also

$$y_1^2 + y_2^2 = \frac{1}{1-\rho^2}\left\{\frac{x_1^2}{\sigma_1^2} - \frac{2\rho x_1 x_2}{\sigma_1\sigma_2} + \frac{x_2^2}{\sigma_2^2}\right\}.$$

The p.d.f. for y_1, y_2 then becomes

$$f(y_1, y_2) = z_0\sigma_1\sigma_2(1-\rho^2)^{\frac{1}{2}}\exp\{-\tfrac{1}{2}(y_1^2+y_2^2)\}$$
$$= z_0\sigma_1\sigma_2(1-\rho^2)^{\frac{1}{2}}e^{-\frac{1}{2}y_1^2}e^{-\frac{1}{2}y_2^2}. \tag{1.35}$$

The original variables are clearly dependent, for (1.34) cannot be factorized into a product of a function of x_1 and a function of x_2. From (1.35) we see that the transformed variables are independent.

Incidentally, this gives us a method of evaluating z_0. Using the transformations in Exercise 1.21, we may show that

$$\left(\int_{-\infty}^{\infty} e^{-\frac{1}{2}u^2}du\right)^2 = (2\pi);$$

on integrating the right-hand side of (1.35) with respect to y_1 and y_2 and equating the result to unity we find that

$$z_0 = 1/\{2\pi\sigma_1\sigma_2(1-\rho^2)^{\frac{1}{2}}\}.$$

1.35 In this example we have used a Jacobian to transform the integral from one pair of variables to another. The process is evidently quite general. If

$$x_1 = x_1(y_1, y_2), x_2 = x_2(y_1, y_2),$$

the density function for y_1 and y_2 will be given by

$$g(y_1, y_2) = f\{x_1(y_1, y_2),\quad x_2(y_1, y_2)\}\,J, \tag{1.36}$$

where

$$J = \frac{\partial(x_1, x_2)}{\partial(y_1, y_2)} = \begin{vmatrix} \dfrac{\partial x_1}{\partial y_1} & \dfrac{\partial x_2}{\partial y_1} \\ \dfrac{\partial x_1}{\partial y_2} & \dfrac{\partial x_2}{\partial y_2} \end{vmatrix}$$

and is to be taken with positive sign in (1.36). J is the reciprocal of $J^* = \dfrac{\partial(y_1, y_2)}{\partial(x_1, x_2)}$ for any one-to-one transformation, since transforming from x to y and then from y to x must restore the original distribution, and hence $JJ^* = 1$. Thus we may evaluate J or J^* as convenient.

If J changes sign in the domain of integration the variate transformation requires special attention. We have already met this difficulty in its simplest form in passing from (1.24) to (1.25), and we shall discuss it generally in Chapter 11.

1.36 The extension of the bivariate results to n-variate results is straightforward. For instance, if and only if a d.f. $F(x_1, x_2, \ldots, x_n)$ can be expressed in the form

$$F(x_1, x_2, \ldots, x_n) = F(x_1, \infty, \ldots, \infty)F(\infty, x_2, \ldots, \infty) \ldots F(\infty, \infty, \ldots, x_n) \tag{1.37}$$

the corresponding variates are *completely* independent. It is not sufficient that each variate should be independent of each of the others. That is to say, if

$$F(x_1, x_2, \infty, \ldots, \infty) = F(x_1, \infty, \infty, \ldots, \infty)F(\infty, x_2, \infty, \ldots, \infty) \tag{1.38}$$

and similarly for every pair of xs, equation (1.37) does not necessarily follow, for there are as many marginal distributions of the multivariate distribution as there are subsets of its n variates, i.e. $2^n - 2$ in all. (1.38) states only that each of the $\binom{n}{2}$ *bivariate* marginal distributions has independent constituents, whereas (1.37) ensures this for all the higher-dimensional marginal distributions also. (1.38) will be described as *pairwise* independence. An instance of pairwise independence without complete independence is given as Exercise 1.24; another will occur in Exercise 16.11 below.

Unless otherwise stated, we shall use *independence* to mean *complete independence*.

Similarly, we may consider conditional distributions obtained by holding any subset of the variates constant; the generalization of **1.32–3** is immediate, the conditional distribution being the ratio of the multivariate p.f. to the (possibly also multivariate) marginal p.f. of the variate(s) held constant.

Probability generating functions

1.37 For distributions that are mathematically specified it is often convenient, especially in the discrete case, to summarize the distribution in a generating function. Suppose that a variate x can take values 0, 1, 2, ... and that the probability that $x = r$ is f_r. This probability is then the coefficient of t^r in

$$P(t) \equiv \sum_{r=0}^{\infty} f_r t^r, \tag{1.39}$$

known as the probability generating function.

Example 1.2 (The binomial p.g.f.)

For the binomial distribution, which we shall study in detail in **5.2–7** below, the probability at $x = r (r = 0, 1, \ldots, n)$ is given by

$$f_r = \binom{n}{r} p^r (1-p)^{n-r}, \qquad 0 \leqslant p \leqslant 1. \tag{1.40}$$

Hence the p.g.f. is given by

$$P(t) = \sum_{j=0}^{n} \binom{n}{j} p^j (1-p)^{n-j} t^j$$
$$= \{pt + (1-p)\}^n. \tag{1.41}$$

1.38 In order for $P(t)$ to be a probability generating function we require that $f_r \geqslant 0$, all r and $\sum f_r = 1$. These conditions translate into

$$P(0) \geqslant 0, \qquad P(1) = 1 \quad \text{and} \quad P^{(r)}(0) \geqslant 0, \qquad r \geqslant 1.$$

Example 1.3 (p.g.f. of the geometric distribution)

Consider

$$P(t) = (1-q)/(1-qt), \qquad 0 < q < 1.$$

Clearly,

$$P(0) = 1 - q, \qquad P(1) = 1$$

and

$$P^{(r)}(0)/r! = f_r = (1-q)q^r,$$

the geometric distribution; see Exercise 1.14 with $r = 1$ and **5.16** below.

EXERCISES

1.1 Draw frequency polygons or histograms of the following distributions:—

Table 1.16 Distribution of the number of successes (throws of 6) in 4096 throws of 12 dice

(Weldon's data; source as in Table 1.14)

Number of successes ...	0	1	2	3	4	5	6	7 and over	TOTAL
Number of throws ...	447	1145	1181	796	380	115	24	8	4096

Table 1.17 Distribution of sentence-lengths in passages from Macaulay's essays on Bacon and on Chatham

(From G. Udny Yule, 1939, *Biometrika*, **30**, 363)

Length of sentence in words	Number of sentences	Length of sentence in words	Number of sentences
1-5	46	66-	2
6-	204	71-	4
11-	252	76-	8
16-	200	81-	2
21-	186	86-	2
26-	108	91-	1
31-	61	96-	2
36-	68	101-	1
41-	38	106-	—
46-	24	111-	1
51-	20	116-	—
56-	12	121-	1
61-	8		
		TOTAL	1251

Table 1.18 Distribution of lengths of the left occipital bone in ancient Egyptian skulls

(From T. L. Woo, 1930, *Biometrika*, 22, 324)

Length in mm. (central values)	Frequency	Length in mm. (central values)	Frequency
84.5	12	102.5	74
86.5	12	104.5	68
88.5	32	106.5	36
90.5	48	108.5	18
92.5	79	110.5	7
94.5	116	112.5	4
96.5	104	114.5	4
98.5	126	116.5	–
100.5	123	118.5	1
		TOTAL	864

Table 1.19 Distribution of fecundity (the ratio of the number of yearling foals produced to the number of coverings) for brood-mares (racehorses) covered eight times at least

(Pearson, Lee and Moore, 1899, *Phil. Trans.*, *A*, 192, 303. Where a case fell on the border between two intervals, one-half was assigned to each)

Fecundity	Frequency	Fecundity	Frequency
1/30– 3/30	2.0	17/30–19/30	315.0
3/30– 5/30	7.5	19/30–21/30	337.0
5/30– 7/30	11.5	21/30–23/30	293.5
7/30– 9/30	21.5	23/30–25/30	204.0
9/30–11/30	55.0	25/30–27/30	127.0
11/30–13/30	104.5	27/30–29/30	49.0
13/30–15/30	182.0	29/30–1	19.0
15/30–17/30	271.5		
		TOTAL	2000.0

Table 1.20 Frequency distribution of yield of wheat

(Mercer and Hall, 1911, *J. Agr. Science*, 4, 107)

Yield of grain in pounds per $\frac{1}{100}$th acre (central value of range)	2.8	3.0	3.2	3.4	3.6	3.8	4.0	4.2	4.4	4.6	4.8	5.0	5.2	TOTAL
Number of plots ...	4	15	20	47	63	78	88	69	59	35	10	8	4	500

Table 1.21 Frequency distribution of accidents for 166 London bus-drivers in (*a*) one year, (*b*) five years

(Data of Farmer, E. and Chambers, E. G., 1939, Industrial Health Research Board Report 84. London, H.M. Stationery Office)

Number of accidents	Frequency	
	In one year	In five years
0	45	1
1	36	2
2	40	3
3	19	14
4	12	17
5	8	21
6	3	17
7	2	14
8	1	14
9	—	12
10	—	13
11	—	9
12	—	6
13	—	2
14	—	6
15 or more	—	15
TOTAL	166	166

Table 1.22 Frequency distribution of women aborting at specified term

(From T. V. Pearce, 1930, *Biometrika*, 22, 250)

Term (weeks)	Frequency	Term (weeks)	Frequency
4	3	17	13
5	7	18	14
6	10	19	8
7	13	20	4
8	14	21	2
9	29	22	10
10	22	23	4
11	21	24	4
12	18	25	3
13	28	26	4
14	16	27	6
15	19	28	1
16	10		
		TOTAL	283

1.2 Sketch the following curves and compare their shapes with those of the distributions in the previous exercise:—

$$y = y_0 e^{-\frac{1}{2}x^2}, \qquad -\infty < x < \infty.$$
$$y = y_0 e^{-x}, \qquad 1 \leqslant x < \infty.$$
$$y = y_0 x^{-\gamma}, \qquad \gamma > 1, \quad 0 \leqslant x < \infty.$$
$$y = \frac{y_0}{(1+x^2)^n}, \qquad n > \tfrac{1}{2}, \quad -\infty < x < \infty.$$
$$y = y_0(1-x)^a x^b, \qquad a,b > 1, \quad 0 \leqslant x \leqslant 1.$$
$$y = y_0 e^{-x} x^{\gamma}, \qquad \gamma > 1, \quad 0 \leqslant x < \infty.$$
$$y = y_0(1-x^2)^a, \qquad a > -1, \quad -1 \leqslant x \leqslant 1.$$

1.3 Find the change of variable that transforms each of the following distributions into

$$dF = y_0 x^{p-1}(1-x)^{q-1}dx, \qquad 0 \leqslant x \leqslant 1.$$

(a) $dF = r_0(1-r^2)^{\frac{1}{2}(n-4)}dr, \qquad -1 \leqslant r \leqslant 1.$

(b) $dF = \dfrac{t_0}{(1+t^2/\nu)^{\frac{1}{2}(\nu+1)}}\,dt, \qquad -\infty < t < \infty.$

(c) $dF = \dfrac{z_0 e^{\nu_1 z}}{(\nu_1 e^{2z} + \nu_2)^{\frac{1}{2}(\nu_1+\nu_2)}}\,dz, \qquad -\infty < z < \infty.$

(All these distributions are important in statistical theory. The distribution to which they are reduced is called the Type I distribution or Beta distribution of the first kind, to be studied in Chapter 6.)

1.4 Sketch the bivariate histograms of the following distributions:—

Table 1.23 898 students in the University of London, 1955, grouped by (1) the number of newspapers read and (2) the number of newspapers looked at only, on a particular day

(H. S. Booker, unpublished data)

(1) Number read	(2) Number looked at only: 0	1	2	3	4	5 or more	TOTALS
0	77	75	19	10	1	2	184
1	179	136	65	20	15	2	417
2	86	70	45	18	1	3	223
3	17	21	13	3	4	—	58
4	4	2	2	2	—	—	10
5 or more	2	2	—	2	—	—	6
TOTALS	365	306	144	55	21	7	898

Table 1.24 4912 Ayrshire cows distributed according to (1) age in years and (2) yield of milk per week

(Data from J. F. Tocher, 1928, *Biometrika*, 20B, 106)

(1) Age in years

	3	4	5	6	7	8	9	10	11	12	13	14	15	16	17	18	TOTALS
8	–	–	–	–	–	–	–	–	1	–	–	–	–	–	–	–	1
9	–	2	2	–	1	–	–	–	–	–	–	–	–	–	–	–	5
10	3	5	1	1	3	–	–	–	–	–	–	–	–	–	–	–	13
11	2	10	8	7	1	–	1	–	2	1	–	1	–	–	–	–	33
12	2	25	17	9	5	4	4	2	1	1	–	–	–	1	–	–	71
13	9	76	29	18	9	2	4	1	1	1	–	1	–	–	–	–	151
14	11	76	57	38	23	9	7	6	4	2	3	–	–	–	–	–	236
15	11	115	79	43	34	24	11	8	4	5	1	2	1	–	1	–	339
16	15	149	119	74	59	23	23	16	9	7	4	–	–	–	1	–	499
17	16	148	131	94	58	34	32	15	12	6	5	–	1	–	–	–	552
18	11	146	132	83	73	49	39	22	17	6	5	1	1	–	–	–	585
19	10	117	112	113	87	51	35	33	11	10	2	3	1	–	–	1	586
20	8	97	107	79	69	51	25	30	13	10	3	3	–	–	1	–	496
21	3	63	93	88	70	49	31	29	9	7	4	–	1	–	1	–	448
22	5	42	63	49	45	32	14	18	10	3	1	2	–	–	–	–	284
23	1	19	33	38	38	27	17	17	12	7	1	2	2	–	–	–	214
24	2	20	23	34	27	19	13	9	3	2	1	–	–	–	–	–	153
25	3	10	15	22	17	20	8	10	3	4	–	–	–	–	–	–	112
26	–	7	13	7	4	15	2	4	2	3	1	–	–	–	–	–	58
27	–	2	7	9	5	5	4	2	–	–	–	–	–	1	–	–	35
28	–	–	2	1	4	2	1	1	2	–	–	–	–	–	–	–	13
29	–	–	2	2	4	1	3	–	3	–	–	–	–	–	–	–	15
30	–	–	–	–	–	2	–	–	2	–	–	–	–	–	–	–	4
31	–	–	2	1	–	–	2	–	–	–	–	–	–	–	–	–	5
32	–	–	–	2	–	–	–	–	–	–	–	–	–	–	–	–	2
33	–	–	–	–	–	–	–	–	–	–	1	–	–	–	–	–	1
34	–	–	–	–	–	–	–	–	1	–	–	–	–	–	–	–	1
TOTALS	112	1129	1047	812	636	419	276	223	122	75	32	15	7	2	4	1	4912

Rows show (2) Yield of milk per week (gallons) (central value of intervals)

1.5 Show that the conditions that

$$f(x_1, x_2) = z_0 \exp\{Ax_1^2 + 2Hx_1x_2 + Bx_2^2\} \qquad -\infty < x_1, x_2 < \infty,$$

may represent a density function are

(a) $A \leqslant 0$,

(b) $B \leqslant 0$,

(c) $AB - H^2 \geqslant 0$.

Show further that if these conditions are satisfied and $F(+\infty, +\infty) = 1$, then

$$z_0 = \frac{1}{\pi} \begin{vmatrix} -A & H \\ H & -B \end{vmatrix}^{\frac{1}{2}}.$$

1.6 For the bivariate distribution with p.d.f.

$$f(x, y) = \frac{ky}{(1+x)^4} \exp\left(-\frac{y}{1+x}\right), \qquad 0 \leqslant x,\ y < \infty,$$

show by a transformation $u = y/(1+x)$, $v = 1/(1+x)$ that the p.d.f. of u and v is

$$f(u, v) = kue^{-u}, \qquad 0 \leqslant u < \infty,\ 0 \leqslant v \leqslant 1,$$

and hence that $k = 1$. Verify that u and v are independent.

1.7 For the bivariate distribution with p.d.f.

$$f = \frac{k(1+x+y)}{(1+x)^4(1+y)^4}, \qquad 0 \leqslant x, y < \infty,$$

show that $k = 9/2$. Show also that the marginal p.d.f. of x is

$$f = \frac{3}{4}\frac{(2x+3)}{(1+x)^4}, \qquad 0 \leqslant x < \infty.$$

1.8 A distribution has p.d.f.

$$f = k \operatorname{sech}^n x, \qquad -\infty < x < \infty.$$

Sketch the form of the p.d.f. and show that

$$k = \frac{\Gamma\{\frac{1}{2}(n+1)\}}{\Gamma(\frac{1}{2}n)\surd\pi}.$$

1.9 Three variates are independent and each is distributed in the form

$$f = \frac{1}{\surd(2\pi)} e^{-\frac{1}{2}x^2}, \qquad -\infty < x < \infty.$$

Transforming to new variates by the equations

$$y_1 = (x_1 - x_2)/\surd 2$$
$$y_2 = (x_1 + x_2 - 2x_3)/\surd 6$$
$$y_3 = (x_1 + x_2 + x_3)/\surd 3,$$

show that the ys are also completely independent and are distributed in the same form.

1.10 In the previous exercise, make a transformation of the polar type

$$x_1 = r\cos\theta_1\cos\theta_2$$
$$x_2 = r\cos\theta_1\sin\theta_2$$
$$x_3 = r\sin\theta_1.$$

Hence show that the variates corresponding to r, θ_1 and θ_2 are completely independent and that the distribution of r is given by

$$f(r) = (2/\pi)^{\frac{1}{2}} e^{-\frac{1}{2}r^2} r^2, \qquad 0 \leqslant r < \infty.$$

1.11 For the distribution with p.d.f.

$$f = \frac{1}{2\pi\sigma_1\sigma_2(1-\rho^2)^{\frac{1}{2}}} \exp\left\{\frac{-1}{2(1-\rho^2)}\left(\frac{x^2}{\sigma_1^2} - \frac{2\rho xy}{\sigma_1\sigma_2} + \frac{y^2}{\sigma_2^2}\right)\right\}, \qquad -\infty < x,\ y < \infty,$$

show that the (marginal) distribution of x is

$$f(x) = \frac{1}{\sigma_1\sqrt{(2\pi)}} \exp\left(-\frac{x^2}{2\sigma_1^2}\right), \qquad -\infty < x < \infty.$$

Sketch the bivariate distribution of x and y and consider the limiting case as $\rho \to 1$.

1.12 In the distribution with p.d.f.

$$f = \frac{1}{2\pi\sigma_1\sigma_2(1-\rho^2)^{\frac{1}{2}}} \frac{n-1}{n-2} \frac{1}{\left[1+\frac{1}{2(n-2)(1-\rho^2)}\left(\frac{x^2}{\sigma_1^2} - \frac{2\rho xy}{\sigma_1\sigma_2} + \frac{y^2}{\sigma_2^2}\right)\right]^n} \qquad -\infty < x,\ y < \infty,$$

show that the marginal distribution of x is given by

$$f = \frac{1}{\sigma_1\sqrt{(2\pi)}\sqrt{(n-2)}} \frac{\Gamma(n-\frac{1}{2})}{\Gamma(n-1)} \frac{1}{\left(1+\frac{x^2}{2(n-2)\sigma_1^2}\right)^{n-\frac{1}{2}}}, \qquad -\infty < x < \infty.$$

(K. Pearson, 1923)

1.13 The Poisson distribution is such that the probability at $x = r (r = 0, 1, 2, \ldots)$ is $e^{-\lambda}\lambda^r/r!$. Show that the probability generating function is

$$P(t) = \exp\{\lambda(t-1)\}.$$

1.14 A discrete distribution (the negative binomial) has probabilities

$$f_x = \binom{x+r-1}{r-1} p^r q^x, \qquad x = 0, 1, 2, \ldots; \quad 0 < p < 1, \quad q = 1-p, \quad r \text{ a positive integer.}$$

Show that the probability generating function is

$$P(t) = p^r(1-qt)^{-r}.$$

1.15 A variate is distributed over the interval 0 to ∞ with p.d.f. $f(x)$. Write down the p.d.f. of the variate in the interval $x_0 \leqslant x < \infty\ (x_0 > 0)$. If this is of the same form as the original variate show that $f(x) = ke^{-kx}$. It may be assumed that $f(x)$ is differentiable.

1.16 A n-variate distribution has the joint p.d.f.

$$f(x_1, \ldots, x_n) = k \exp\left\{-\sum_{i,j=1}^{n} a_{ij}x_ix_j\right\}, \qquad -\infty < x_1, \ldots, x_n < \infty,$$

with $a_{ij} = a_{ji}$. By considering x_n show that the exponent can be put in the form

$$-a_{nn}\left(x_n + \frac{1}{a_{nn}}\sum_{i=1}^{n-1} a_{ni}x_i\right)^2 - \sum_{i,j=1}^{n-1} b_{ij}x_ix_j,$$

where $b_{ij} = a_{ij} - \frac{a_{in}a_{jn}}{a_{nn}}$, $\quad i, j = 1, 2, \ldots, n-1$. Assume that a_{nn}, $b_{n-1,n-1}$ etc are strictly positive.

Hence show how to transform the original variates by linear transformation to a completely independent set each of which is distributed in a form of the type

$$f(u)=\frac{1}{\sigma\sqrt{(2\pi)}}\,e^{-u^2/2\sigma^2}, \qquad -\infty<u<\infty.$$

1.17 For a bivariate distribution with a twice differentiable density function $f(x, y)$, show that there exists a linear transformation to new independent variates if and only if

$$\left(A\frac{\partial^2}{\partial x^2}+2H\frac{\partial^2}{\partial x\partial y}+B\frac{\partial^2}{\partial y^2}\right)\log f=0$$

where A, H and B are constants. Hence show that for the distribution of Exercise 1.11 there exists an infinite number of linear transformations to independent variates.

1.18 Three non-negative variates x, y and z have the joint p.d.f.

$$f=kx^{l-1}y^{m-1}z^{n-1}$$

where $x+y+z\leqslant 1$. Show that

$$k=\frac{\Gamma(l+m+n+1)}{\Gamma(l)\Gamma(m)\Gamma(n)}.$$

1.19 n non-negative variates have the joint p.d.f.

$$f=kx_1^{l_1-1}x_2^{l_2-1}\ldots x_n^{l_n-1}g(x_1+x_2+\ldots+x_n),$$

where all $l_i>0$ and $\sum_{i=1}^{n}x_i\leqslant 1$. Show that

$$1/k=\frac{\Gamma(l_1)\Gamma(l_2)\ldots\Gamma(l_n)}{\Gamma(l_1+l_2+\ldots+l_n)}\int_0^1 v^{L-1}g(v)\,dv,$$

where

$$L=\sum_{i=1}^{n}l_i.$$

(The particular case $g(v)=(1-v)^{l_{n+1}-1}$ is called the *Dirichlet distribution,* defined on the n-dimensional simplex $\sum_1^{n+1}x_i=1$. Extensive tables of its d.f. are given by Sobel *et al.* (1976) for x_i all equal and $l_i=l$ $(i=1,2,\ldots,n)$ and $l_{n+1}=m+1-nl$, where m is an integer $\geqslant nl$.)

1.20 Show that if a variate x is distributed symmetrically about a value θ, its d.f. satisfies

$$F(x)=1-F(2\theta-x)$$

and its p.f. correspondingly

$$f(x)=f(2\theta-x).$$

1.21 Two independent variates x_1, x_2 each have the p.d.f.

$$f=\frac{1}{\sqrt{(2\pi)}}\exp(-\tfrac{1}{2}x^2), \qquad -\infty<x<\infty.$$

Show that the variates

$$y_1=\exp\{-\tfrac{1}{2}(x_1^2+x_2^2)\},$$

$$y_2=\frac{1}{2}+\frac{1}{\pi}\,\text{arc}\tan(x_1/x_2)$$

are independently uniformly distributed on the range (0, 1).

(Box and Muller, 1958)

1.22 If $F_1(x_1)$, $F_2(x_2)$ are univariate d.f.s, show that

$$F(x_1, x_2) = F_1(x_1)F_2(x_2)\,[1+\theta\{1-F_1(x_1)\}\,\{1-F_2(x_2)\}]$$

(where θ is a constant, $-1 \leqslant \theta \leqslant 1$) is a bivariate d.f. with marginal d.f.s F_1 and F_2 and p.d.f

$$f(x_1, x_2) = f_1(x_1)f_2(x_2)\,\{1+\theta(1-2F_1(x_1))(1-2F_2(x_2))\}\,.$$

(Cf. Gumbel, 1960; see Exercises 15.24–5 below)

1.23 By considering the bivariate probability function

$$f(x, y) = g(x)h(y) \quad \text{if} \quad \begin{Bmatrix} a \leqslant x \leqslant b \\ a \leqslant y \leqslant b \end{Bmatrix} \quad \text{or} \quad \begin{Bmatrix} b \leqslant x \leqslant c \\ b \leqslant y \leqslant c \end{Bmatrix}$$

$$= 0 \qquad \text{otherwise,}$$

show that factorization of $f(x, y)$ whenever it is non-zero is not sufficient for x to be independent of y.

1.24 If three variates x_1, x_2, x_3 can each only take the values 0 or 1, and their joint distribution is

$$f_{123}(x_1, x_2, x_3) = (1-p)^{3-y}p^y + (-1)^y t,$$

where $y = \sum_{i=1}^{3} x_i$ and $|t| \leqslant \min\{p^3, (1-p)^3\}$, show that the marginal bivariate distribution of x_i and x_j is

$$f_{ij}(x_i, x_j) = (1-p)^{2-w}p^w,$$

where $w = x_i + x_j$, and hence that x_i and x_j are pairwise, although not completely, independent when $t \neq 0$.

CHAPTER 2

MEASURES OF LOCATION AND DISPERSION

2.1 It has been seen in Chapter 1 that the distributions occurring in statistical practice vary considerably in general nature. Some are finite in range and some are not. Some are symmetrical and some markedly skew. Some have only a single maximum and others have several. Amid this variety we may, however, discern four general types: (a) the symmetrical distribution with a single maximum, such as that of Table 1.7; (b) the asymmetrical distribution, or skew distribution, with a single maximum, such as those of Tables 1.8 and 1.9; (c) the extremely skew, or J-shaped, distribution, such as that of Table 1.2; and (d) the U-shaped distribution, such as that of Table 1.13. To make this classification comprehensive we should have to add a fifth class comprising the miscellaneous distributions not falling into the other four.

Location and scale parameters

2.2 It frequently happens in statistical work that we have to compare two distributions. If one is unimodal and the other J-shaped or multimodal a concise comparison is clearly difficult to make, and in such a case it would probably be necessary to specify both distributions completely. But if both are of the same type (and it is in such cases that comparisons most frequently arise) we may be able to make a satisfactory comparison merely by examining their principal characteristics; e.g. if both are unimodal it might be sufficient to compare (a) the whereabouts of some central value, such as the maximum, that locates the distributions; (b) the degree of scatter about this value—the dispersion, which reflects the *scale* of the distribution; and (c) the extent to which the distributions deviate from symmetry—the *skewness*.

The same point emerges when our distributions are specified by some mathematical function. If, for example, we have two distributions of the type

$$f(x) = (2\pi\sigma^2)^{-\frac{1}{2}} \exp\left\{-\frac{(x-\mu)^2}{2\sigma^2}\right\}, \qquad -\infty < x < \infty,$$

symmetrical about $x = \mu$, a complete comparison can be made by comparing the value of the constants μ and σ in the distributions. Such constants are called *parameters* of the distribution.[*]

---

[*] We shall use the term *parameter* to denote quantities, such as μ and σ above, that appear explicitly in the specification of the distribution. However, the term is often used more loosely to describe any population constant, so that, e.g., σ and σ^2 here would both be described as parameters. With our definition, any function of σ may be regarded as a parameter, but there is essentially only one parameter involving σ, and it is better to describe other constants involving σ as functions of that parameter.

For a mathematically specified distribution f, θ is called a *location parameter* if the distribution has the form $f(x-\theta)$, and a *scale parameter* if the form is $\frac{1}{\theta}f\left(\frac{x}{\theta}\right)$, $\theta>0$; while if the form is $\frac{1}{\theta_2}f\left(\frac{x-\theta_1}{\theta_2}\right)$, $\theta_2>0$, we have a location parameter θ_1 and a scale parameter θ_2. It will be seen that in our example, μ is a location parameter and σ a scale parameter.

We now consider various measures of location and of dispersion in turn. In particular cases location and scale parameters will coincide with one or more of the measures that we shall discuss.

Measures of location: the arithmetic mean

2.3 There are three groups of measures of location in common use: the means (arithmetic, geometric and harmonic), the median and the mode. We consider them in turn.

The arithmetic mean is perhaps the most generally used statistical measure, and in fact is far older than the science of Statistics itself. If the probability function of the variate x is $f(x)$, the arithmetic mean μ_1' about the point $x=a$ is defined by

$$\left.\begin{aligned}\mu_1' &= \int_{-\infty}^{\infty}(x-a)f(x)\,dx = \int_{-\infty}^{\infty}(x-a)\,dF\\ &= \int_{-\infty}^{\infty}x\,dF - a\int_{-\infty}^{\infty}dF = \int_{-\infty}^{\infty}x\,dF - a.\end{aligned}\right\} \tag{2.1}$$

The mean is said to be finite (or to exist) if

$$\int_{-\infty}^{\infty}|x|\,dF<\infty. \tag{2.2}$$

(Exercise 2.27 gives a case in which the mean is not finite.) This integral is of the Stieltjes form and thus includes summation in the discrete case (cf. 1.24). If actual frequencies $g(x)$, totalling to N, are used, we have with $a=0$

$$\mu_1' = \sum xg(x)/N. \tag{2.3}$$

The value of the arithmetic mean thus depends on the value of a, the point from which it is measured.

If b is some other arbitrary variate value, (2.1) implies that

$$\mu_1'(\text{about } a) = \mu_1'(\text{about } b) + b - a. \tag{2.4}$$

In other words, we can find the mean about any point very simply when we know the mean about any other. In calculating the arithmetic mean we can then take an arbitrary point as origin and transfer to any other desired point afterwards. Choosing this arbitrary point somewhere near the maximum frequency often serves to improve the accuracy of subsequent numerical computations. Unless otherwise stated, μ_1' will be taken about the origin $(a=0)$.

Exercise 2.24 expresses μ_1' in a less usual form.

Example 2.1 (Calculation of the arithmetic mean)

Consider the distribution of male heights in Table 1.7.

If there were relatively few values in the data set, we should simply sum them and divide by n to obtain μ_1' about zero, but a further point arises in grouped data such as these. We do not know *exactly* the variate values of the individuals within a certain class interval. We therefore assume them to be concentrated at the centre of the interval. Corrections for any distortion thus introduced will be considered in Chapter 3. In fact, no correction is required for the arithmetic mean in the case when the frequency "tails off" at both ends of the distribution.

Table 2.1 Calculation of the arithmetic mean for the distribution of Table 1.7

(1) Height inches	(2) Frequency f	(3) Deviation from arbitrary *value* ξ	(4) Product ξf
57–	2	−10	20
58–	4	−9	36
59–	14	−8	112
60–	41	−7	287
61–	83	−6	498
62–	169	−5	845
63–	394	−4	1576
64–	669	−3	2007
65–	990	−2	1980
66–	1223	−1	1223
67–	1329	0	−8584
68–	1230	+1	1230
69–	1063	+2	2126
70–	646	+3	1938
71–	392	+4	1568
72–	202	+5	1010
73–	79	+6	474
74–	32	+7	224
75–	16	+8	128
76–	5	+9	45
77–	2	+10	20
TOTALS	8585	–	+8763

In this example we take an arbitrary origin at the centre of the interval 67– inches, i.e. at the point $67\frac{7}{16}$ inches, and measure $\xi(=x-a)$ from that point. Column 2 in Table 2.1 shows the frequency, column 3 the value of ξ and column 4 the value of ξf. We find, having due regard to sign,

$$\sum(\xi f) = 8763 - 8584 = 179.$$

Thus the mean about $x=0$ is $67\frac{7}{16} + \dfrac{179}{8585} = 67.46$ inches.

Example 2.2 (Mean of the Beta distribution)

For a distribution specified by a mathematical function, the determination of the mean is a matter of evaluating the integral (2.1), when it exists. For instance, to find the mean of the distribution

$$f(x) = \frac{1}{B(p,q)}x^{p-1}(1-x)^{q-1}, \qquad 0 \leqslant x \leqslant 1; \qquad p,q > 0,$$

we have, about zero,

$$\begin{aligned}\mu_1' &= \frac{1}{B(p,q)}\int_0^1 x^p(1-x)^{q-1}\,\mathrm{d}x \\ &= \frac{B(p+1,q)}{B(p,q)} = \frac{\Gamma(p+1)\Gamma(q)}{\Gamma(p+q+1)}\cdot\frac{\Gamma(p+q)}{\Gamma(p)\Gamma(q)} \\ &= \frac{p}{p+q}.\end{aligned}$$

2.4 If a distribution is specified by a generating function as at (1.39), the mean can be evaluated in the following manner.

Let

$$P(t) = \sum_{j=0}^{\infty} f_j t^j.$$

Then

$$\begin{aligned}\left[\frac{\mathrm{d}P}{\mathrm{d}t}\right]_{t=1} &= \left[\sum f_j j t^{j-1}\right]_{t=1} \\ &= \sum f_j j \\ &= \mu_1' \text{ (about zero).}\end{aligned}$$

Example 2.3 (Mean of the binomial distribution)

From Example 1.2 in **1.37**, the binomial distribution is specified by the p.g.f.

$$P(t) = \{pt + (1-p)\}^n$$

and we have

$$\mu_1' = \left[\frac{\mathrm{d}P}{\mathrm{d}t}\right]_{t=1} = np\{p + (1-p)\}^{n-1}$$
$$= np.$$

The geometric and harmonic means

2.5 Two other types of mean are in use in elementary statistics, though they are of less importance in advanced theory.

The geometric mean of N variate values is the Nth root of their product and is used if the variate values are positive. For discrete $f(x)$ we have probabilities

$$\left.\begin{aligned} G &= \prod_j (x_j^{f_j}) \\ \text{or} \qquad \log G &= \sum_j f_j \log x_j \end{aligned}\right\} \tag{2.5}$$

and for actual frequencies $g(x)$, totalling, N,

$$\left.\begin{aligned} G &= \prod (x_j^{g_j})^{1/N} \\ \log G &= \frac{1}{N}\sum g_j \log x_j \end{aligned}\right\} \tag{2.6}$$

with similar integral expressions for the continuous case. Comparing (2.6) with (2.3) we see that $\log G$ is the arithmetic mean of $\log x$ about zero.

Similarly, the harmonic mean of N positive variate values is the reciprocal of the arithmetic mean of their reciprocals. In the usual notation,

$$\frac{1}{H} = \int_0^\infty \frac{\mathrm{d}F}{x} = \int_0^\infty \frac{f(x)\,\mathrm{d}x}{x} \tag{2.7}$$

or, for actual discrete frequencies,

$$\frac{1}{H} = \frac{1}{N}\sum \frac{g_j(x)}{x_j} \tag{2.8}$$

provided, of course, that this is finite.

Example 2.4 (Geometric and harmonic means of the Beta distribution)

The distribution of Example 2.2 has p.d.f.

$$f(x) = \frac{1}{B(p,q)} x^{p-1}(1-x)^{q-1}, \qquad 0 \leqslant x \leqslant 1; \qquad p, q > 0,$$

so that

$$\log G = \frac{1}{B(p, q)} \int_0^1 x^{p-1}(1-x)^{q-1} \log x \, dx.$$

Now, since by definition

$$\int_0^1 x^{p-1}(1-x)^{q-1} \, dx = B(p, q)$$

we have, differentiating both sides with respect to p, an operation which is legitimate in virtue of the uniform convergence of the integral and the existence of the resulting expressions,

$$\int_0^1 x^{p-1}(1-x)^{q-1} \log x \, dx = \frac{\partial}{\partial p} B(p, q).$$

Thus

$$\begin{aligned} \log G &= \frac{1}{B(p, q)} \frac{\partial}{\partial p} B(p, q) \\ &= \frac{\partial}{\partial p} \log \frac{\Gamma(p)\Gamma(q)}{\Gamma(p+q)} \\ &= \frac{\partial}{\partial p} \{\log \Gamma(p) - \log \Gamma(p+q)\}. \end{aligned}$$

The harmonic mean is given by

$$\frac{1}{H} = \frac{1}{B(p, q)} \int_0^1 x^{p-2}(1-x)^{q-1} \, dx.$$

To keep this expression finite, however, we must require $p > 1$. Then

$$\begin{aligned} \frac{1}{H} &= \frac{B(p-1, q)}{B(p, q)} = \frac{\Gamma(p-1)}{\Gamma(p+q-1)} \cdot \frac{\Gamma(p+q)}{\Gamma(p)} \\ &= \frac{p+q-1}{p-1}, \end{aligned}$$

so that

$$H = \frac{p-1}{p+q-1}.$$

We may note that the arithmetic mean (Example 2.2) is greater than the harmonic mean, for

$$\mu_1' = \frac{p}{p+q} = 1 - \frac{q}{p+q}, \quad \text{and} \quad H = \frac{p-1}{p+q-1} = 1 - \frac{q}{p+q-1}$$

so

$$\mu_1' > H.$$

We shall see in the next section that G lies between H and μ_1'.

2.6 In general it may be shown that for distribution in which the variate values are positive and not all equal

$$H < G < \mu_1'. \tag{2.9}$$

Consider the quantity

$$A(t) = \left[\int_0^\infty x^t \, \mathrm{d}F(x)\right]^{1/t}$$

where the xs are positive real numbers. We shall show that this is an increasing function of t, i.e. $A(t_1) > A(t_2)$ if $t_1 > t_2$. Trivially, this inequality becomes an equality if all the xs are equal, i.e. F has only a single point of increase.

Note that if we put $t = 1$ in $A(t)$ we have the arithmetic mean; when $t = -1$ we have the harmonic mean; and as t tends to zero we have the geometric mean, for

$$\lim_{t \to 0} \log A = \lim t^{-1} \log \int x^t \, \mathrm{d}F$$

and by L'Hôpital's rule, this is

$$= \lim \int x^t \log x \, \mathrm{d}F \Big/ \int x^t \, \mathrm{d}F$$

$$= \int \log x \, \mathrm{d}F = \log G.$$

If the function $h(z)$ is convex and $z \geqslant 0$, Jensen's inequality states that

$$\int_0^\infty h(z) \, \mathrm{d}F(z) \geqslant h\left\{\int_0^\infty z \, \mathrm{d}F(z)\right\}.$$

Putting $h(z) = z^u$ and then $z = x^t$, it follows that when $u > 1$

$$\int_0^\infty x^{tu} \, \mathrm{d}F(x) \geqslant \left\{\int_0^\infty x^t \, \mathrm{d}F(x)\right\}^u.$$

Now put $t = t_2$ and $tu = t_1$; it follows that for $t_1 > t_2$, $A(t_1) \geqslant A(t_2)$, with equality only as stated above.

Hereafter when the "mean" is mentioned without qualification, the arithmetic mean is to be understood.

Exercise 2.2 gives a simple proof that $H < \mu_1'$.

Exercise 2.3 shows that G is approximately the mean of μ_1' and H when dispersion is relatively small; Exercise 2.22 gives an upper bound for μ_1'/H.

Comparison of means: standardization and weighting

2.7 We may wish to compare the arithmetic means of two or more distributions, e.g., of the same variate at different times or in different places. Suppose, for example, that the set of possible values has been subdivided into k mutually exclusive and exhaustive categories $\{I_j = (C_{j-1}, C_j], j = 1, \ldots, k\}$ where $C_0 = -\infty$ and $C_k = +\infty$. Let x_j and y_j denote the values assigned to the jth interval for each variate, with corresponding relative frequencies $f(x_j)$ and $g(y_j)$. Then the means are

$$\mu_x = \sum x_j f(x_j) \quad \text{and} \quad \mu_y = \sum y_j g(y_j)$$

and the difference $\mu_x - \mu_y$ will reflect both differences in the set of (x_j, y_j) values and between the distributions $f(x)$ and $g(y)$. The effect of the differences between f and g can be removed by evaluating the *weighted* arithmetic means:

$${}_w\mu_x = \sum x_j g(y_j) \quad \text{and} \quad {}_w\mu_y = \sum y_j f(x_j), \tag{2.10}$$

recalling that $f(x_j)$ and $g(y_j)$ are the relative frequencies for the same set of values, I_j. From inspection we see that $\mu_y \neq {}_w\mu_x$ in general unless $x_j = y_j$ for all values of j. The weighted means (2.10) are called *standardized* means because each calculates the mean of one variate using the relative frequencies belonging to the other. In each case, the removal of the variation between f and g permits a clearer interpretation of the differences between the sets of values taken by the two variates.

More generally, we may define the weighted mean by

$${}_w\mu_x = \int_{-\infty}^{\infty} x w(x)\, dF(x),$$

where $w(x) \geqslant 0$ is any weighting function with $\int_{-\infty}^{\infty} w(x)\, dF(x) = 1$, $w(x) \equiv 1$ yields the ordinary mean μ_x, *and* $w(x) = g_x(y)/f(x)$ gives (2.10).

Evidently the weighted mean will exceed the ordinary mean (usually called unweighted, though it is strictly equally-weighted) if the values of x and $w(x)$ tend to rise together. In fact,

$$\begin{aligned} {}_w\mu_x - \mu_x &= \int_{-\infty}^{\infty} x\{w(x) - 1)\}\, dF(x) \\ &= \int_{-\infty}^{\infty} (x - \mu_x)\{w(x) - 1\}\, dF(x), \end{aligned}$$

and we shall see later that this is called the covariance of x and $w(x)$. We express this by saying that the weighted mean is the larger if the variate values and the weights are positively correlated.

The practical importance of weighting and standardization may be appreciated from the fact that even if every value of x is greater than the corresponding value of y, the unweighted mean of y may be larger than that of x because of the differences between $f(x)$ and $g(y)$ — the latter will then be the larger for the higher variate values. A striking real-life example is given in Exercise 2.26.

The median

2.8 The median(*) is that value of the variate which divides the total probability into two equal parts, i.e. is the value x_m such that

$$\int_{-\infty}^{x_m} f(x)\,\mathrm{d}x = \int_{x_m}^{\infty} f(x)\,\mathrm{d}x = \tfrac{1}{2}. \tag{2.11}$$

For a continuous distribution, there is a unique solution of (2.11). If the population is finite and has $(2N+1)$ members, where N is a positive integer, the median is the value of the $(N+1)$th ordered member. If there are $2N$ members, any value in the interval between the two middle members satisfies (2.11), and this non-uniqueness is conventionally removed by defining the median to be halfway between the values of the Nth and the $(N+1)$th members.

When the distribution is grouped into intervals there is the usual indeterminacy due to grouping, which may be dealt with in the manner of the following example.

Example 2.5 (Calculation of the median from grouped data)

Consider the distribution of heights examined in Example 2.1.

Half the total frequency of 8585 observation is	4292.5
There are, up to and including the interval ending at $66\frac{15}{16}$ inches,	3589
observations, leaving	703.5
The frequency in the next interval is	1329.

Hence by linear interpolation in that interval, of unit length, we take the median to be

$$66\tfrac{15}{16} + \frac{703.5}{1329} \times 1 = 67.47 \text{ inches}$$

The mean (Example 2.1) is 67.46 inches, practically the same.

If a stem-and-leaf plot is used (cf. **1.14**) the individual variate values are recoverable within the groups, and this linear interpolation is unnecessary.

A graphical method of determining the median is given in **2.15** below.

2.9 From the mathematical viewpoint the non-uniqueness of the median can be removed. For a set of values $x_1, \ldots, x_N$ it may be shown that the sum $\Sigma|\xi - x_i|^p$, considered as a function of ξ, has a unique minimum for some ξ_p if $p > 1$; and further, that as p tends to unity from above ξ_p tends to some unique value, which may be defined as the median. See Jackson (1921) and Exercise 2.1.

(*) The name "median" was first used by Galton in 1883, but the concept was used by him, and by Fechner independently, around 1870.

An extension of the concept of median to the multivariate case is made if we use the result of Exercise 2.1 as the *definition* of the median, and generally define a multivariate median as that point, M, which minimizes the sum of absolute deviations for the distribution. This is computationally more complicated than combining the medians of the univariate marginal distributions, but has theoretical attractions—cf. Haldane (1948) and Gower (1974), who gives a program for computing it in the bivariate case, when it is sometimes called the *spatial median*. Rather surprisingly, the multivariate median is always unique. For suppose that there were two minimizing points, M_1 and M_2, and let M_0 be the point half-way between them on their connecting line L, and the distances from an arbitrary observation (not on L) to these three points be d_1, d_2 and d_0. It is easy to see that d_1, d_2 and $2d_0$ form a triangle, and hence that $d_0 < \frac{1}{2}(d_1 + d_2)$. If an observation is on L, lying between M_1 and M_2, the same inequality holds; if it is on L outside $[M_1, M_2]$, the relation becomes an equality. Summing over all the observations, we see that M_0 has a smaller sum of deviations unless *all* the observations lie on L outside $[M_1, M_2]$. Thus M_1 and M_2 cannot both be minimizing points unless the distribution degenerates onto a line, which is essentially the univariate case.

The mode

2.10 If the probability function has a local maximum at a value x, i.e. if $f(x)$ is greater at x than at neighbouring values below and above x, there is said to be a *mode* of the distribution at x. If there is only one such modal value, the distribution is called unimodal, as we have already seen; if there are several, the distribution is multimodal. If the probability function has a peak at an extreme value of x (e.g. if the distribution is J-shaped, as in Fig. 1.2), it is sometimes called a half-mode.

If a density function is everywhere continuous and twice-differentiable, and there is no terminal peak, a mode must be a solution of

$$f'(x) = \frac{d}{dx}f(x) = 0, \qquad f''(x) = \frac{d^2}{dx^2}f(x) < 0. \tag{2.12}$$

If $f'(x) = 0$ and $f''(x) > 0$ we have a local minimum, and such a point is sometimes called an antimode.

In numerically specified distributions and discrete distributions generally the mode is sometimes difficult to determine exactly. Where the number of observations is large enough to permit grouping, there will usually be an interval containing maximum frequency per unit range, and we may regard the mode as lying in that interval. In the height distribution of Table 1.7, for instance, the mode may be considered as lying somewhere in the interval 67- inches. One way to estimate the mode more accurately is to fit a continuous curve to the observed distribution and determine the mode of that curve—the process of fitting will be considered in Chapter 6. Another way is to smooth the frequencies using a kernel method as in **1.21**, and find the mode of the smoothed density; optimal procedures of this type are considered by Eddy (1980).

2.11 In a symmetrical unimodal distribution the mean, the median and the mode coincide. For skew distributions they differ. There is an interesting approximate relationship between the three quantities which holds for unimodal p.f.s of moderate asymmetry, namely

$$\text{mean} - \text{mode} \doteq 3(\text{mean} - \text{median}). \tag{2.13}$$

A mathematical explanation of this relationship is given in Exercise 6.20 below. Exercise 2.16, **6.30**, Exercise 14.4 and **16.2** below give particular distributions for which (2.13) holds approximately.

It is a useful mnemonic to observe that the mean, median and mode of a unimodal distribution occur in the same order (or the reverse order) as in the dictionary—cf. Exercise 2.18; and that the median is nearer to the mean than to the mode, just as the corresponding words are nearer together in the dictionary.

Mention should also be made of a little-used measure of location, the mid-range. This is the variate value half-way between the terminal points of the distribution if it possesses such points. Its dependence on the terminal points and the use of mathematical distributions that extend to infinity in one or both directions make it unsuitable for many purposes but, like the range (**2.17**, below), it has some uses in sampling theory and is very easily calculated.

Choice of location measure: robustness

2.12 In elementary theory the median and the mode have considerable claims to use as measures of location for unimodal distributions. They are readily interpretable in terms of ordinary ideas—the median is the middle value and the mode is the most frequent value—and the median is usually more easily determined than the mean in numerically specified distributions. What gives the arithmetic mean the greater importance in advanced theory is its superior mathematical tractability and certain sampling properties; but the median has compensating advantages—it is quite insensitive to the configuration of the outlying parts of the probability distribution, unlike the mean.

This problem of sensitivity to outlying observations has received considerable attention, leading to the development of *robust* (i.e. non-sensitive) location measures. Robust methods are discussed in greater detail in a later chapter, but here we outline some of the methods used to obtain protection against excessive influence upon the location measure of a small number of outlying observations.

Robust location measures may be obtained by first sorting the x values into order, and relabelling them so that

$$x_{(1)} \leqslant x_{(2)} \leqslant \dots \leqslant x_{(N)}.$$

Given the actual frequencies $g(x_{(i)})$, a robust location measure is

$$\mu'_R = \frac{1}{N} \sum g(x_{(i)})\theta(x_{(i)}),$$

where θ is a function designed specifically to reduce the influence of certain observations, i.e. form of weighting as in **2.7**. When $f(x)$ is defined mathematically the corresponding definition is

$$\mu'_R = \int_{-\infty}^{\infty} \theta(x)\,\mathrm{d}F. \tag{2.14}$$

The ordinary mean μ'_1 corresponds to $\theta(x) = x$. In general, the function θ may be chosen for the purpose in hand by using the *influence function* of Hampel (1974). Particular versions of (2.14) of interest are the *a-trimmed mean* with

$$\theta(x) = x/(1-2\alpha), \qquad \alpha \leqslant F(x) \leqslant 1-\alpha,$$
$$= 0, \qquad F(x) < \alpha \quad \text{or} \quad F(x) > 1-\alpha,$$

which annihilates a fraction α of the distribution in each tail, and the *Winsorized mean*

$$\theta(x) = x, \qquad \alpha \leqslant F(x) \leqslant 1-\alpha$$
$$= x_L, \qquad F(x) < \alpha, \quad F(x_L) = \alpha$$
$$= x_U, \qquad F(x) > 1-\alpha, \quad F(x_U) = 1-\alpha,$$

which replaces each observation in each tail-fraction α by the value of the nearest unaffected observation. Values of α in the range $0 < \alpha \leqslant 0.25$ are usually considered, depending upon the heaviness of the tails of $f(x)$. The median may thus be viewed as an extreme version of the trimmed mean when only the central, or the two central, observations are retained.

Most interest has focused upon unimodal symmetric distributions where θ is chosen so that $\mu_R' = \mu_1'$ if the whole population is measured, but fluctuates if only a sample from the population is available. The argument in favour of μ_R' is that down-weighting of the extreme observations makes the location measure less sensitive to extreme values, which may be due to recording errors, instrumental failure, etc.

A major study by Andrews *et al.* (1972) contrasted the performance of many such robust location measures and made a persuasive case for using a robust measure if the outlying values are suspect. For further developments, see Huber (1981).

Quantiles

2.13 The concept of median value may be extended to locate a distribution more accurately by the use of several measures. We may, for example, find the three variate values that divide the total probability into four equal parts. The middle one of these will be the median itself; the other two are called the lower and upper *quartiles*(*) respectively. Similarly, we may find the nine variate values that divide the total probability into ten equal parts—the *deciles*—or into a hundred equal parts—the *percentiles*(*) or *percentage points*. Generally we may find the $(n-1)$ variate values that divide the total probability into n equal parts—the *quantiles*. Evidently the knowledge of the quantiles for some fairly high n, such as 10, gives a very good idea of the general form of the distribution. Even the quartiles and the median are valuable general guides. In Chapter 14, we shall study the sampling properties of statistics of this type systematically.

2.14 The determination of the quantiles of a numerically specified distribution proceeds as for the median, indeterminacies due to discreteness being resolved by similar conventions. That of the quantiles of a mathematically specified distribution, say the jth quantile, is a matter of solving for x the equation

$$j/n = \int_{-\infty}^{x} \mathrm{d}F$$

which can be done without difficulty by interpolation when the d.f. has been tabulated.

(*) Percentiles were defined in 1885 by Galton, who also brought quartiles into general use a little earlier.

For a grouped distribution, we proceed as in Example 2.5 for the median.

Example 2.6 (Calculation of the quartiles from grouped data)

Again, we use the height distribution considered in Example 2.1.

One-quarter of the total frequency is 8585/4 =	2146.25
Up to $64\frac{15}{16}$ inches there are	1376
observations leaving	770.25
In the next interval there are	990
Thus the lower quartile is $64\frac{15}{16}+\frac{770.25}{990}\times 1=$	65.71 inches
The upper quartile will similarly be found to be	69.21 inches
We have already found (Example 2.5) that the median is	67.47 inches

Denoting the median by M and the quartiles by Q_1 and Q_3 we see that

$$Q_1 - M = -1.76 \text{ inches}$$

$$Q_3 - M = 1.74 \text{ inches},$$

so that the median is almost half-way between the quartiles, an indication of the symmetry of the distribution—cf. Exercise 3.27 below.

2.15 The quantiles can be easily determined graphically from the graph of the distribution function scaled from 0 to N instead of 0 to 1 as usual. Fig. 2.1 gives the d.f.

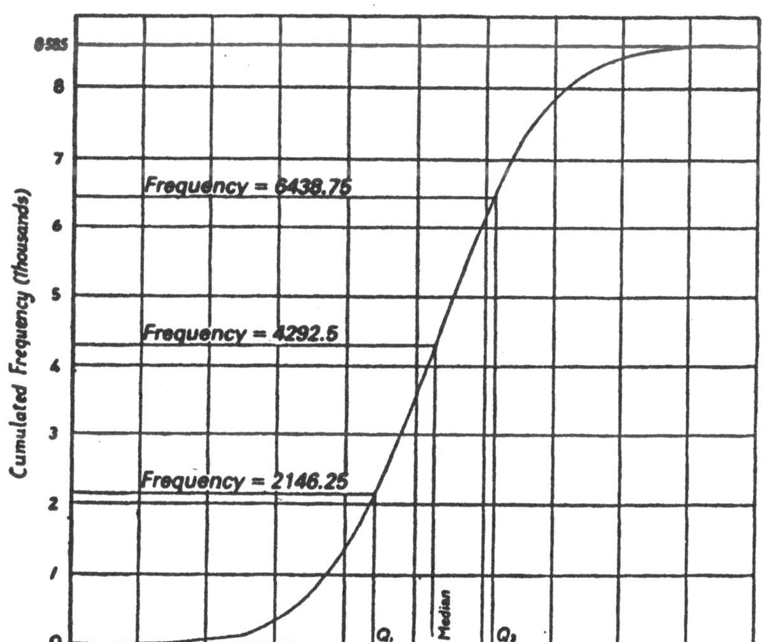

Fig. 2.1 Distribution function of the data of Table 1.7

for the data of Table 1.7. To find the median we determine the ordinate corresponding to the abscissa of $\frac{1}{2}N$ and so on. The positions of the quartiles and the median are shown in Fig. 2.1, and the reader may care to compare the values obtained by reading the graph by eye with those given in Example 2.6.

Measures of dispersion

2.16 We now proceed to the quantities which have been proposed to measure the dispersion of a distribution. They fall into three groups:

(a) Measures of the distance (in terms of the variate) between certain representative values, such as the range, the interdecile range or the interquartile range.

(b) Measures compiled from the deviations of every member of the population from some central value, such as the mean deviation from the mean, the mean deviation from the median, and the standard deviation.

(c) Measures compiled from the deviations of all the members of the population among themselves, such as the mean difference.

In advanced theory the outstandingly important measure is the standard deviation; but they all require some mention.

Range and interquantile differences

2.17 The range of a distribution is the difference between its greatest and least observable values. As a descriptive measure of a population it has very little use. A knowledge of the whereabouts of the extreme values tells us nothing about the way the bulk of the distribution is distributed inside the range; and for distributions of infinite range it is obviously wholly inappropriate.

More useful rough-and-ready measures may be obtained from the quantiles, and there are two such in general use. The interquartile range is the distance between the upper and lower quartiles, and is thus a range which contains one-half of total probability. The interdecile range (or perhaps, more accurately, the 1–9th interdecile range) is the distance between the first and the ninth decile, containing 80% of the total probability. Both these measures evidently give some approximate idea of the "spread" of a distribution, and are easily calculable. For this reason they are fairly widely used in elementary descriptive statistics. In advanced theory they suffer from the disadvantage of being relatively difficult to handle mathematically in the theory of sampling—cf. Chapter 14.

Mean deviation

2.18 The amount of scatter in a population is evidently measured to some extent by the totality of deviations from the mean. It is easily seen from (2.1) that the sum of these deviations taken with appropriate sign is zero. We may however write

$$\delta_1 = \int_{-\infty}^{\infty} |x - \mu_1'| \, dF \tag{2.15}$$

where the deviations are now taken absolutely, and define δ_1 to be a coefficient of dispersion. We shall call it the mean deviation about the mean. It is clearly unaffected by change of origin for x, since μ_1' changes correspondingly by (2.1). We therefore may measure x from zero for convenience.

Because $\int_{-\infty}^{\infty}(x-\mu_1')\,dF = 0$, where μ_1' is taken about zero, (2.15) may be expressed in the forms

$$\delta_1 = 2\int_{\mu_1'}^{\infty}(x-\mu_1')\,dF = -2\int_{-\infty}^{\mu_1'}(x-\mu_1')\,dF.$$

Exercise 2.24 gives another form for δ_1.

Similarly, using the median M we may write

$$\delta_2 = \int_{-\infty}^{\infty}|x-M|\,dF \tag{2.16}$$

and call δ_2 the mean deviation about the median. The reader is asked to prove in Exercise 2.1 that δ_2 is no greater than the mean deviation measured from any other point.

In future the unqualified words *mean deviation* will be taken to refer to the mean deviation about the mean.

Both (2.15) and (2.16) have merits in elementary work, being fairly easily calculable. Once again, however, they are practically excluded from advanced work by their intractability in the theory of sampling.

Variance and standard deviation

2.19 We have seen that the mean about an arbitrary point a is given by

$$\mu_1' = \int_{-\infty}^{\infty}(x-a)\,dF.$$

This is called the first moment(*) of the population. Similarly, the second moment is

$$\mu_2' = \int_{-\infty}^{\infty}(x-a)^2\,dF. \tag{2.17}$$

The second moment about the mean is written without the prime, thus:

$$\mu_2 = \int_{-\infty}^{\infty}(x-\mu_1')^2\,dF \tag{2.18}$$

and is called the variance.(*) The positive square root of the variance is called the standard deviation,(*) and usually denoted by σ, so that we have

$$\sigma = |\sqrt{\mu_2}|. \tag{2.19}$$

(*) The term "moment", used by analogy with its meaning in Statics, appears at least as long ago as the work of A. Quetelet (1796–1874) and has been regularly used since being adopted by K. Pearson, who first used the term "standard deviation" in the 1890s. "Variance" was not used until 1918, when R. A. Fisher defined it in a paper on genetics.

The variance is thus the mean of the squares of the deviations from the mean, and the standard deviation is their root-mean-square. The device of squaring and then taking the square root of the resultant sum in order to obtain the standard deviation may appear a little artificial, but it makes the mathematics of the sampling theory simpler than in the case of the mean deviation, where the absolute values complicate the theory.

The calculation of the variance and the standard deviation proceeds by an easy extension of the methods used for the mean. In particular, if b is some arbitrary value,

$$\begin{aligned}\mu_2'(\text{about } a) &= \int_{-\infty}^{\infty} (x-a)^2\,dF \\ &= \int_{-\infty}^{\infty} \{(x-b)^2 + 2(b-a)(x-b) + (b-a)^2\}\,dF \\ &= \mu_2'(\text{about } b) + 2(b-a)\mu_1'(\text{about } b) + (b-a)^2. \qquad (2.20)\end{aligned}$$

If now b is the mean we have

$$\mu_2' = \mu_2 + (\mu_1' - a)^2 \quad \text{or} \quad \mu_2 = \mu_2' - (\mu_1' - a)^2. \qquad (2.21)$$

Thus the variance can easily be found from the second moment about an arbitrary point, which can be selected to simplify the calculations. Also, we note from (2.21) that μ_2' is minimized when $a = \mu_1'$.

Example 2.7 (Calculation of the mean deviation and standard deviation)

Again consider the distribution of male heights used in Example 2.1 (Table 1.7).

In the case of the mean deviation for a grouped distribution, the sum of deviations should first be calculated from the centre of the interval in which the mean lies and then reduced to the mean as origin. It so happens that in Table 2.1 the mean fell in the interval taken as origin, so that the preliminary arithmetic already exists in the table.

The sum of positive deviations is 8763 and that of negative deviations −8584. Hence the sum of deviations regardless of sign is 17 347, the unit being the class interval and the origin the centre of the interval.

To reduce to the mean as origin, we note that if the number of observations below the mean is N_1 and the number above the mean is N_2, and $d = \mu_1' - a$, we have to add $N_1 d$ to the sum of deviations about the centre of the interval and subtract $N_2 d$. In this case $d = 179/8585 = 0.02$ (Example 2.1), $N_1 = 2 + 4 + \ldots + 1223 + 1329 = 4918$, $N_2 = 8585 - 4918 = 3667$. Hence we add $(4918 - 3667)0.02 = 25$. Hence the mean deviation

$$\delta_1 = \frac{17\,347 + 25}{8585} = 2.02 \text{ inches.}^{(*)}$$

For the standard deviation some further calculation is required, as shown in Table 2.2.

(*) This calculation can be refined slightly—cf. Exercise 2.21.

Table 2.2 Calculation of the standard deviation for the distribution of Table 1.7 *(preliminary calculation already carried out in Table 2.1)*

(1) Height inches	(2) Frequency f	(3) Deviation ξ	(4) $\xi^2 f$
57–	2	−10	200
58–	4	−9	324
59–	14	−8	896
60–	41	−7	2009
61–	83	−6	2988
62–	169	−5	4225
63–	394	−4	6304
64–	669	−3	6021
65–	990	−2	3960
66–	1223	−1	1223
67–	1329	0	0
68–	1230	1	1230
69–	1063	2	4252
70–	646	3	5814
71–	392	4	6272
72–	202	5	5050
73–	79	6	2844
74–	32	7	1568
75–	16	8	1024
76–	5	9	405
77–	2	10	200
TOTALS	8585	–	56 809

Column (4) shows the sum $\sum \xi^2 f$, where f is the actual frequency. We then have, for the second moment about the arbitrary origin,

$$\mu_2' = \frac{56\,809}{8585} = 6.6172.$$

We have already found in Example 2.1 that

$$\mu_1' - a = \frac{179}{8585} = 0.0209.$$

Hence, from (2.21)

$$\begin{aligned}\mu_2 &= 6.6172 - (0.0209)^2 \\ &= 6.6168 \\ \sigma &= \sqrt{\mu_2} = 2.57 \text{ inches.}\end{aligned}$$

It may be noted that in Example 2.7 the mean deviation is about 80 percent of the standard deviation. This relationship often holds approximately for unimodal distributions that are symmetrical or nearly so, including the normal distributions (5.42 below), the logistic distribution (Exercise 4.21 below) and the Laplace distributions (Exercises 4.3, 11.21); even the J-shaped exponential distribution of Exercise 2.6 has $\delta_1/\sigma = 0.74$

Chakrabarti (1948) has shown that, in general, for a set of $n > 2$ values,

$$\left.\begin{aligned}(2/n)^{\frac{1}{2}} \leqslant \delta_1/\sigma &\leqslant (1-n^{-2})^{\frac{1}{2}}, && n \text{ odd} \\ &\leqslant 1, && n \text{ even,}\end{aligned}\right\}$$

the equalities being attainable. Exercise 2.4 is to show that $\delta_1/\sigma \leqslant 1$. Exercise 2.25 gives other relations between δ_1, μ_2 and the d.f. in an important special case.

Example 2.8 (Variance of the Beta distribution)

The distribution of Example 2.2, has p.d.f.

$$f(x) = \frac{1}{B(p,q)} x^{p-1}(1-x)^{q-1}, \qquad 0 \leqslant x \leqslant 1; p, q > 0.$$

We have, about the origin,

$$\begin{aligned}\mu_2' &= \frac{1}{B(p,q)} \int_0^1 x^{p+1}(1-x)^{q-1}\,dx \\ &= \frac{B(p+2,q)}{B(p,q)} = \frac{(p+1)p}{(p+q+1)(p+q)}.\end{aligned}$$

We have already found (Example 2.2) that

$$\mu_1' = \frac{p}{p+q}.$$

Thus

$$\begin{aligned}\mu_2 &= \mu_2' - \mu_1'^2 \\ &= \frac{(p+1)p}{(p+q+1)(p+q)} - \frac{p^2}{(p+q)^2} \\ &= \frac{pq}{(p+q+1)(p+q)^2}.\end{aligned}$$

For discrete distributions we may find the variance, like the mean, by differentiating the probability generating function. Repeating the differentiation in 2.4, we have

$$\frac{d^2P}{dt^2} = \sum \{f_j j(j-1)t^{j-2}\}$$

and on putting $t=1$ we have, for moments about zero,

$$\left[\frac{d^2P}{dt^2}\right]_{t=1} = \sum\{f_j j(j-1)\} = \mu_2' - \mu_1'.$$

Example 2.9 (The variance of the binomial)

In the particular case of the binomial in Example 2.3,

$$\left[\frac{d^2}{dt^2}\{pt+(1-p)\}^n\right]_{t=1} = n(n-1)p^2.$$

Thus

$$\mu_2' = n(n-1)p^2 + np,$$

and

$$\begin{aligned}\mu_2 = \mu_2' - \mu_1'^2 &= n(n-1)p^2 + np - n^2p^2\\ &= np(1-p).\end{aligned}$$

Computational procedures

2.20 One advantage of the mean and variance is the existence of simple updating relations. Suppose that values $x_1, x_2, \ldots, x_N$ are available from which we have computed

$$\mu(N) = \frac{1}{N}\sum x_i, \qquad \sigma^2(N) = \frac{1}{N}\sum\{x_i - \mu(N)\}^2;$$

where the subscript and prime are temporarily dropped from μ_1' for notational ease.

If an additional value, x_{N+1}, is now introduced, it follows that

$$\mu(N+1) = \mu(N) + \frac{1}{N+1}\{x_{N+1} - \mu(N)\} \tag{2.22}$$

and

$$\sigma^2(N+1) = \frac{N}{N+1}\sigma^2(N) + \frac{N}{(N+1)^2}\{x_{N+1} - \mu(N)\}^2. \tag{2.23}$$

Not only are these expressions useful for updating the mean and variance as new values are recorded over time; they also may be used sequentially to compute the mean and variance from a set of values. When high numerical accuracy is important, these formulae are preferable to those discussed earlier.

Sheppard's corrections

2.21 The treatment of the values of a grouped frequency distribution as if they were concentrated at the mid-points of intervals is an approximation, and in certain circumstances it is possible to make corrections for any distortion introduced thereby. These so-called "Sheppard's corrections" will be discussed at length in **3.18–24** below, but at this stage we may indicate without proof the appropriate correction for the second moment.

If the distribution is continuous and has high-order contact with the variate axis at its extremities, i.e. if it tails off smoothly, the crude second moment calculated from grouped frequencies should be corrected by subtracting from it $h^2/12$, where h is the length of the interval. For example, in the height data of Example 2.7, we have $h = 1$, and the corrected second moment is

$$6.6168 - 0.0833 = 6.5335.$$

The corrected value of σ is $\sqrt{6.5335} = 2.56$, as against an uncorrected value of 2.57. Exercise 2.8 provides a general guide to the magnitude of the correction.

Mean difference

2.22 The mean difference (not to be confused with mean deviation) is defined by

$$\Delta = \int_{-\infty}^{\infty} \int_{-\infty}^{\infty} |x-y| \, dF(x) \, dF(y). \tag{2.24}$$

We observe that we may also regard Δ as the *mean deviation* (about its mean of zero) of the variate $(x-y)$, where x and y independently have the d.f. F.

In the discrete case two formulae arise. We have either

$$\Delta = \frac{1}{N(N-1)} \sum_{j=-\infty}^{\infty} \sum_{k=-\infty}^{\infty} |x_j - x_k| f(x_j) f(x_k), \qquad j \neq k, \tag{2.25}$$

the mean difference without repetition, or

$$\Delta = \frac{1}{N^2} \sum_{j=-\infty}^{\infty} \sum_{k=-\infty}^{\infty} |x_j - x_k| f(x_j) f(x_k), \tag{2.26}$$

the mean difference with repetition. The distinction lies only in the divisor and is unimportant if N is large.

The mean difference is the average of the differences of all the possible pairs of variate values, taken regardless of sign. In the coefficient with repetition each value is taken with itself, adding the course nothing to the sum of deviations, but resulting in the total number of pairs being N^2. In the coefficient without repetition only distinct values are taken, so that the number of pairs is $N(N-1)$. Hence the divisors in (2.25) and (2.26).

The mean difference was proposed by Gini (1912), after whom it is usually named, but was discussed by F. R. Helmert and other German writers in the 1870s—cf. H. A. David (1968). It has a certain theoretical attraction, being dependent on the spread of the variate values among themselves and not on the deviations from some central value. Further, its defining integral (2.24) may converge when that of the variance, (2.18), does not converge. Since it is essentially a linear rather than a quadratic measure of dispersion, it converges whenever the mean exists—cf. Exercises 2.9 and 2.19. It is, however, more difficult to compute than the standard deviation, and the appearance of the absolute values in the defining equations indicates, as for the mean deviation, the appearance of difficulties in the theory of sampling. It might be thought that this inconvenience could be overcome by the definition of a coefficient

$$E^2 = \int_{-\infty}^{\infty}\int_{-\infty}^{\infty}(x-y)^2\,dF(x)\,dF(y).$$

This, however, is nothing but twice the variance, for

$$\begin{aligned} E^2 &= \int_{-\infty}^{\infty}\int_{-\infty}^{\infty}\{x^2 - 2xy + y^2\}\,dF(x)\,dF(y) \\ &= 2\mu_2' - 2\mu_1'^2 = 2\mu_2. \end{aligned}$$

This shows that the variance may be defined as half the mean square of all possible variate differences, that is to say, without reference to deviations from a central value, the mean.

Another unusual formula for the variance is given in Exercise 2.24.

Robust dispersion measures

2.23 As might be expected, the variance is more sensitive to extreme values than the mean. Several more robust (less sensitive) alternatives have been proposed, perhaps the simplest of which is to use the mean deviation (2.15):

$$\sigma_{R1} = k\delta_1.$$

When $k = (2/\pi)^{\frac{1}{2}} \simeq 0.8$, $\sigma_R = \sigma$ for the normal distribution; however, k varies somewhat with the underlying distribution—cf. the discussion in **2.19**. Pearson and Tukey (1965) proposed another simple measure

$$\sigma_{R2} = \frac{x(0.95) - x(0.05)}{3.25} \tag{2.27}$$

where $x(\alpha)$ is the solution to the equation $\alpha = F(x)$, using interpolation where necessary. They found that $\sigma_{R2} \simeq \sigma$ for many distributions. A variety of other possibilities exists, including α-trimming of either the variance or the mean deviation (cf. **2.12**). Jaeckel (1972) considered a class of robust dispersion measures which are linear in x.

Care is needed in using these measures. They are valuable when the extreme observations are considered to be unreliable because they contain contaminated or wrongly reported values. On the other hand, when such extreme values are genuine, using one of the σ_R measures may systematically understate the variability in the population.

Coefficients of variation: standardization

2.24 The foregoing measures of dispersion have all been expressed in terms of units of the variate. It is thus difficult to compare dispersions in different populations unless the units happen to be identical; and this has led to a search for measures, independent of the variate scale, that are pure numbers.

Several coefficients of this kind may be constructed, such as the mean deviation divided by the mean or by the median. Only two have been used at all extensively in practice, Karl Pearson's coefficient of variation, defined by

$$V = \sigma/\mu_1', \tag{2.28}$$

and Gini's coefficient of concentration, defined by

$$G = \Delta/(2\mu_1'). \tag{2.29}$$

These coefficients both suffer from the disadvantage of being affected very much by μ_1', the value of the mean measured from some arbitrary origin. They are useful when there is a natural origin of measurement or comparisons are being made between distributions with similar origins.

We shall see at the end of **2.25** that $0 \leqslant G \leqslant 1$. There is no upper bound for V in general, but Exercise 2.23 gives one for a finite set of non-negative observations.

For many purposes, comparability may be attained in a somewhat different way. Let us take σ itself as a new unit and express the p.f. in terms of a new variable y related to x by

$$y = (x - \mu_1')/\sigma. \tag{2.30}$$

Any distribution expressed in this way has zero mean and unit variance. It is then said to be *standardized*. Two standardized distributions can be readily compared in regard to form, skewness, and other qualities, though not of course in regard to mean and variance.

"Standardized" here refers to the relocation and rescaling of a variate, whereas in **2.7** it referred only to the weighted mean of a variate.

Concentration

2.25 Gini's coefficient of concentration (2.29) arises in a natural way from the following approach:

Writing, as usual

let us define

$$\left.\begin{aligned} F(x) &= \int_{-\infty}^{x} f(u)\mathrm{d}u \\ \Phi(x) &= \frac{1}{\mu_1'}\int_{-\infty}^{x} uf(u)\mathrm{d}u. \end{aligned}\right\} \tag{2.31}$$

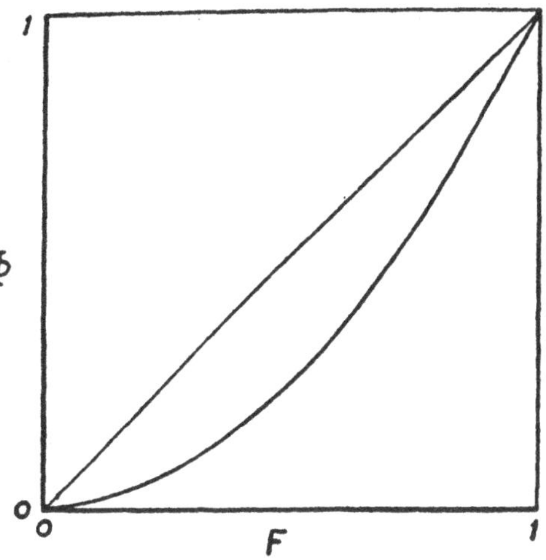

Fig. 2.2 Concentration curve

$\Phi(x)$ exists, of course, only if μ_1' exists. Just as $F(x)$ varies from 0 to 1, $\Phi(x)$ varies from 0 to 1 provided that $x \geqslant a \geqslant 0$. $\Phi(x)$ may be called the *incomplete* first moment and as a function of x is known as the *first moment distribution* of $F(x)$.

Now (2.31) may be regarded as defining a relationship between the variables F and Φ in terms of parametric equations in x. The curve whose ordinate and abscissa are Φ and F is called the concentration curve or Lorenz curve (see Fig. 2.2).

The concentration curve must be convex to the F-axis, for we have

$$\mu_1' \frac{\mathrm{d}\Phi}{\mathrm{d}F} = \mu_1' \frac{\mathrm{d}\Phi}{\mathrm{d}x} \Big/ \frac{\mathrm{d}F}{\mathrm{d}x} = \frac{xf(x)}{f(x)} = x,$$

which is positive since our origin is taken to the left of the start of the distribution. Also

$$\mu_1' \frac{\mathrm{d}^2\Phi}{\mathrm{d}F^2} = \frac{\mathrm{d}x}{\mathrm{d}F} = \frac{1}{f(x)} > 0.$$

Thus the tangent to the curve makes a positive acute angle with the F-axis, and the angle increases as F increases; in other words, the curve is convex to the F-axis.

The distance between the line $\Phi = F$ and the concentration curve is $F - \Phi(F)$. It has a turning point where $\frac{\mathrm{d}}{\mathrm{d}F}\{F - \Phi\} = 0$, i.e. $\mathrm{d}\Phi/\mathrm{d}F = 1$. From above, this is at $x = \mu_1'$; at the mean, the concentration curve is parallel to $\Phi = F$ and most distant from it. This maximum distance is

$$F(\mu_1') - \Phi(\mu_1') = -\frac{1}{\mu_1'} \int_{-\infty}^{\mu_1'} (u - \mu_1')\,\mathrm{d}F(u) = \frac{1}{2}\frac{\delta_1}{\mu_1'}$$

from **2.18**, when δ_1 is the mean deviation from the mean (2.15).

The area between the concentration curve and the line $\Phi = F$ is called the area of concentration. We proceed to show that it is equal to one-half the coefficient of concentration.

In fact, we have from Fig. 2.2

$$2(\text{area of concentration}) = \int_0^1 F\,d\Phi - \int_0^1 \Phi\,dF$$

and thus from above

$$2\mu_1'(\text{area}) = \int_0^\infty F(x)x\,dF(x) - \mu_1'\int_0^\infty \Phi(x)\,dF(x) \tag{2.32}$$

$$= \int_0^\infty x\int_0^x dF(y)\,dF(x) - \int_0^\infty \int_0^x y\,dF(y)\,dF(x)$$

$$= \int_0^\infty \int_0^x (x-y)dF(y)\,dF(x).$$

Now $\int_0^\infty\int_0^\infty (x-y)dF(y)dF(x) = 0$, and hence (as in **2.18** for the mean deviation)

$$2\mu_1'(\text{area}) = \frac{1}{2}\left[\int_0^\infty \int_0^x (x-y)\,dF(y)\,dF(x) + \int_0^\infty \int_x^\infty (y-x)dF(y)dF(x)\right]$$

$$= \frac{1}{2}\int_0^\infty \int_0^\infty |x-y|\,dF(x)dF(y) = \tfrac{1}{2}\Delta.$$

Thus the area of concentration is equal to $\frac{1}{4}\Delta/\mu_1'$, namely one-half of the coefficient of concentration G at (2.29). From Fig. 2.2, the area lies between 0 and $\frac{1}{2}$, so $0 \leqslant G \leqslant 1$.

Reverting to (2.32), we now see that an alternative definition of the mean difference is

$$\Delta = 2\int_0^\infty \{xF(x) - \mu_1'\Phi(x)\}\,dF(x). \tag{2.33}$$

2.26 Various methods have been given for calculating the mean difference. The following is probably the simplest, particularly for distributions specified in equal-length intervals.

Let us, without loss of generality, take an origin at the start of the distribution. We may then write

$$\sum_{j=1}^{N}\sum_{k=1}^{N} |x_j - x_k| = 2\sum{}'(x_j - x_k),$$

the summation $\sum'$ being taken over values such that $j \geqslant k$. We have also

$$x_j - x_k = (x_j - x_{j-1}) + (x_{j-1} - x_{j-2}) + \ldots + (x_{k+1} - x_k).$$

Thus

$$\sum{}'(x_j - x_k) = \sum_{h=1}^{N-1} C_h(x_{h+1} - x_h),$$

where C_h is the number of terms of type $(x_j - x_k)$ in $\sum'$ containing $x_{h+1} - x_h$. Since h is the number of values of j less than or equal to h (the origin being at the start of the distribution) and $N-h$ the number greater than or equal to $h+1$, we have $C_h = h(N-h)$, and thus (2.26) is

$$\Delta_1 = \frac{2}{N^2}\sum{}'(x_j - x_k)$$

$$= \frac{2}{N^2}\sum_{h=1}^{N-1} h(N-h)(x_{h+1} - x_h).$$

This form is particularly useful if all the intervals are equal. F_h being the distribution function of x_h we then have

$$\Delta = \frac{2}{N^2}\sum_{h=1}^{N-1}(NF_h)(N - NF_h) = 2\sum_{h=1}^{N-1} F_h(1 - F_h). \tag{2.34}$$

If the actual cumulated frequency for x_h is G_h we have

$$\Delta = \frac{2}{N^2}\sum_{h=1}^{N-1} G_h(N - G_h),$$

the most convenient form in practice.

Exercise 2.10 gives another form for the calculation.

Δ is often calculated from grouped data, as in the following Example 2.10. Krieger (1979) shows that for unimodal distributions very close upper and lower bounds may be obtained for both Δ and the Lorenz curve, so that grouping effects are rather slight.

Example 2.10 (Calculation of Gini's mean difference)

Returning once more to the height distribution considered in previous examples, we may calculate $\sum G_h(N - G_h)$ as in Table 2.3.

We have, for the mean difference with repetition,

$$\Delta = \frac{2 \times 105\,990\,850}{8585^2}$$

$$= 2.88 \text{ inches}$$

as against a mean deviation of 2.02 inches and a standard deviation of 2.57 inches (Example 2.7).

The mean difference is here about one-eighth larger than the standard deviation. Like the mean deviation's relation to the standard deviation remarked below Example 2.7, this

Table 2.3 Calculation of the mean difference for the height distribution of Table 1.7

Height, inches	Frequency	G_h	$N-G_h$	$G_h(N-G_h)$
57–	2	2	8583	17 166
58–	4	6	8579	51 474
59–	14	20	8565	171 300
60–	41	61	8524	519 964
61–	83	144	8441	1 215 504
62–	169	313	8272	2 589 136
63–	394	707	7878	5 569 746
64–	669	1376	7209	9 919 584
65–	990	2366	6219	14 714 154
66–	1223	3589	4996	17 930 644
67–	1329	4918	3667	18 034 306
68–	1230	6148	2437	14 982 676
69–	1063	7211	1374	9 907 914
70–	646	7857	728	5 719 896
71–	392	8249	336	2 771 664
72–	202	8451	134	1 132 434
73–	79	8530	55	469 150
74–	32	8562	23	196 926
75–	16	8578	7	60 046
76–	5	8583	2	17 166
77–	2	8585	–	–
TOTALS	8585	–	–	105 990 850

relation has its roots in the properties of symmetrical or unimodal distributions like the normal in 5.42 below, the logistic in Exercise 4.21, and the extreme-value distribution in Exercise 14.21—even for the J-shaped exponential distribution of Exercise 2.6, $\Delta/\sigma = 1$. Exercise 2.4 is to show that $\Delta \leqslant \sigma\sqrt{2}$. The sharper inequality $\Delta \leqslant 2\sigma/\sqrt{3}$ is proved in Exercise 2.9

Expected values in general

2.27 The concept of the arithmetic mean of a variate can be easily extended to that of the mean of a function, say $h(x)$. In fact we define

$$E(h) = \int_{-\infty}^{\infty} h(x)\,\mathrm{d}F, \tag{2.35}$$

subject, of course, to the existence of the sum or integral on the right. Mean values such as this play a fundamental role in the theory of Statistics and of probability. The notation E derives from the fact that in probability theory such values are known as *expected values* or *expectations*. It is a remarkable and fruitful fact that in the theory of sampling we are often able to determine certain mean values, especially those of powers of the variate, more easily than the corresponding probability functions.

Generally, for a function $h(x_1, x_2, \ldots, x_n)$ we have

$$E(h) = \int_{-\infty}^{\infty} \cdots \int_{-\infty}^{\infty} h(x_1, \ldots, x_n)\, dF(x_1, \ldots, x_n). \tag{2.36}$$

In particular if h is a function of x_1 only, we have, for the multivariate distribution, on integrating with respect to $x_2, \ldots, x_n$,

$$E(h) = \int_{-\infty}^{\infty} h(x_1)\, dF(x_1) \tag{2.37}$$

where $F(x_1)$ is the (marginal) distribution function of x_1. This is required for consistency with (2.35) and shows us that the mean value of a variate (or some function of it) is the same whether the variate forms part of a multivariate complex or not.

2.28 Two simple but important properties of expected values are to be noted:

(a) The expectation of a sum is the sum of the expectations. If h_1 and h_2 are two functions,

$$E(h_1 + h_2) = E(h_1) + E(h_2). \tag{2.38}$$

This follows at once from the definitions.

(b) The expected value of a product of two functions is the product of their expected values if they are independent. For example, let x_1 and x_2 represent independent variates. Their distribution function then factorizes into two components by **1.34**. Any two functions $h_1(x_1)$, $h_2(x_2)$ will be independent and

$$\begin{aligned} E(h_1 h_2) &= \int_{-\infty}^{\infty} \int_{-\infty}^{\infty} h_1(x_1) h_2(x_2)\, dF(x_1, x_2) \\ &= \int_{-\infty}^{\infty} h_1(x_1)\, dF_1(x_1) \int_{-\infty}^{\infty} h_2(x_2)\, dF_2(x_2) \\ &= E(h_1)E(h_2). \end{aligned} \tag{2.39}$$

Equation (2.38) is true even if the variates are dependent; (2.39) will not be true in general for dependent variates and if it is not, h_1 and h_2 are said to be *correlated*; when it is true, the functions h_1 and h_2 are said to be *uncorrelated*, a weaker property than independence. We shall study correlation in detail in Volume 2A.

2.29 Following the development in 1.32, we may write the conditional p.f. of y given x as

$$f(y|x) = f(x, y)/f_x(x); \tag{2.40}$$

the subscripts on the marginal distributions ensure that no ambiquity arises among the several f functions, provided their complete arguments are stated. The *conditional expectation for y given x* is then defined as

$$E(y|x) = \int y f(y|x)\,dy; \tag{2.41}$$

$E(y|x)$ is also known as the *regression function for y on x*; the theory of regression will be developed in Volume 2A, Chapters 26–28. Since

$$E(y) = \iint y f(x, y)\,dx dy$$

$$= \int y f_y(y)\,dy,$$

we have

$$E(y) = \int f_x(x) \left[\int y \frac{f(x, y)}{f_x(x)} dy \right] dx$$

$$= \int f_x(x) E(y|x) dx \tag{2.24}$$

which may be written as $\underset{x}{E}[E(y|x)]$, the expected value of the conditional expectation, taken over all values of x. The equality (2.42) holds for any distribution for which $E(y)$ exists.

2.30 The conditional variance is given by

$$\text{var}(y|x) = \underset{x}{E}[\{y - E(y|x)\}^2|x]$$
$$= E(y^2|x) - [E(y|x)]^2. \tag{2.43}$$

Taking expectations with respect to x in (2.43) we obtain

$$\underset{x}{E}[\text{var}(y|x)] = \underset{x}{E}[E(y^2|x)] - \underset{x}{E}[g^2(x)],$$

where $g(x) = E(y|x)$. Using (2.42), this expression reduces to

$$E(y^2) - \underset{x}{E}[g^2(x)] = E(y^2) - \mu_y^2 - E[g(x)] + \mu_y^2$$
$$= \text{var}(y) - \text{var}[E(y|x)], \tag{2.44}$$

again using (2.42). Since var $[g(x)] \geqslant 0$, it follows that

$$\underset{x}{E}[\text{var}(y|x)] \leqslant \text{var}(y). \tag{2.45}$$

Example 2.11 (Conditional moments for the binomial distribution)

If x and y follow the binormal distribution given in Example 1.1 with zero means and variances set equal to unity, it follows from (1.35) that

$$z = (y - x)/(1 - \rho^2)^{\frac{1}{2}}$$

has a standard normal distribution, or $z \sim N(0, 1)$, so that the conditional distribution for y given x is normal with mean ρx and variance $(1-\rho^2)$, or

$$y|x \sim N[\rho x, (1-\rho^2)]. \tag{2.46}$$

We see that

$$\underset{x}{E}[E(y|x)] = E(\rho x) = \rho E(x) = 0$$

and that

$$\text{var}(y|x) = 1-\rho^2 < \text{var}(y) = 1, \tag{2.47}$$

for all values of x. Result (2.47) is a feature of the binormal distribution; in general, only the weaker inequality (2.45) holds.

Inequalities

2.31 Jensen's inequality, referred to in **2.6**, may be expressed more succinctly as

$$E[h(z)] \geqslant h[E(z)] \tag{2.48}$$

whenever $z \leqslant 0$ and h is a convex function.

Another inequality that is very useful .in statistical theory is the Cauchy-Schwarz inequality

$$\left(\int xy\,\mathrm{d}F\right)^2 \leqslant \left(\int x^2\,\mathrm{d}F\right)\left(\int y^2\,\mathrm{d}F\right) \tag{2.49}$$

or

$$[E(xy)]^2 \leqslant E(x^2)E(y^2). \tag{2.50}$$

The result is readily demonstrated by noting that

$$E\{(y-cx)^2\} \geqslant 0 \tag{2.51}$$

where c is an arbitrary constant. Putting

$$c = E(xy)/E(x^2)$$

in (2.51) yields (2.50).

The inequality as given by Cauchy in 1821 states that if a_i, b_i $(i = 1, 2, \ldots n)$ are real numbers $\left(\sum_{i=1}^{k} a_i^2\right)\left(\sum_{i=1}^{n} b_i^2\right) > \left(\sum_{i=1}^{n} a_i b_i\right)^2$ unless the *a*s and the *b*s are proportional, in which case the inequality becomes an equality. The corresponding result for integrals was given by Buniakowsky in 1859 and Schwarz in 1885. Many other inequalities can be derived from it, e.g. Hölder's of 1889: if $a_i, b_i, \ldots, l_i$ $(i = 1, \ldots, n)$ are non-negative, $\alpha, \beta, \ldots, \lambda$ are positive with $\alpha+\beta+\ldots+\lambda=1$ then

$$\sum a^{\alpha}b^{\beta}\ldots l^{\lambda}<\sum a^{\alpha}\sum b^{\beta}\ldots\sum l^{\lambda},$$

unless one set of the (a) etc. are all zero or unless one set is proportional to all the others, in which cases the inequality becomes an equality. A very clear and comprehensive account is given in the book *Inequalities* by Hardy, Littlewood and Pólya, 1934, Cambridge University Press; see also Godwin (1964).

EXERCISES

2.1 Show that the mean deviation about an arbitrary point is least when that point is the median.

2.2 By considering the sum $\sum_{i=1}^{n}(x_i-\bar{x})^2x_i^{-1}$, show that for any set of positive values x_i, not all equal, $H<\mu_1'$ as in (2.9).

2.3 Show that, if deviations are small compared with the value of the mean, we have approximately, for the geometric and harmonic means,

$$G \doteq \mu_1'\left(1-\frac{1}{2}\frac{\sigma^2}{\mu_1'^2}\right) \qquad H \doteq \mu_1'\left(1-\frac{\sigma^2}{\mu_1'^2}\right)$$

and hence that

$$\mu_1'-2G+H \doteq 0.$$

2.4 Show that the mean deviation about the mean is not greater than the standard deviation; and that the mean difference cannot exceed $\sqrt{2}$ times the standard deviation.

2.5 Show that for the uniform distribution with p.d.f.

$$f(x)=1, \qquad 0<x\leqslant 1,$$

$$\mu_1'(\text{about the origin})=\tfrac{1}{2}, \mu_2=\tfrac{1}{12}, \delta_1=\tfrac{1}{4} \text{ and } \Delta=\tfrac{1}{3}.$$

2.6 Show that for the distribution (the exponential) with p.d.f.

$$f(x)=ke^{-x/\sigma}, \qquad 0\leqslant x<\infty, \quad \sigma>0,$$

the mean, standard deviation and mean difference are all equal to σ; that the median is $\sigma\log_e 2$ and the mean deviation $\delta_1=2e^{-1}\sigma$; and that the interquartile range is $\sigma\log_e 3$.

2.7 Show that for the distribution (the Gamma)

$$f(x)=\frac{1}{\Gamma(p)}e^{-x}x^{p-1}, \quad p>0; \quad 0\leqslant x<\infty,$$

the mean difference is $\Delta=2/B(p,\tfrac{1}{2})$. (Exercise 2.6 with $\sigma=1$ is the special case $p=1$.)

2.8 Show that if a range of six times the standard deviation contains at least 18 grouping intervals, Sheppard's correction will make a difference of less than 0.5 percent in the uncorrected value of the standard deviation.

2.9 Show (from (2.33)) that for a continuous distribution whose mean exists the mean difference

$$\Delta=2\int_{-\infty}^{\infty}F(x)\{1-F(x)\}\,dx=4\int_{-\infty}^{\infty}x\{F(x)-\tfrac{1}{2}\}\,dF.$$

(Cf. (2.34) in the discrete case.) Hence confirm that $0\leqslant G\equiv\Delta/(2\mu_1')\leqslant 1$. Using the Cauchy-

Schwarz inequality (2.41), show that

$$\Delta = 4\int_{-\infty}^{\infty}(x-\mu)\{F(x)-\tfrac{1}{2}\}\,dF$$
$$\leqslant 4\left[\sigma^2\int_{-\infty}^{\infty}\{F(x)-\tfrac{1}{2}\}^2\,dF\right]^{\frac{1}{2}}$$

whence $\Delta/\sigma \leqslant 2/3^{\frac{1}{2}}$. Exercise 2.5 shows that the uniform distribution attains this bound.

2.10 If the variate values of a distribution are $x_1, \ldots, x_N$ in ascending order of magnitude and

$$s_r = \sum_{j=1}^{r} x_j \qquad U = \sum_{r=1}^{N} s_r$$
$$t_r = \sum_{j=1}^{r} x_{N-j+1} \qquad V = \sum_{r=1}^{N} t_r$$

show that (2.26) is

$$\Delta = \frac{2}{N^2}(V-U) = \frac{2}{N^2}\{2V - N(N+1)\mu_1'\}$$
$$= \frac{2}{N^2}\{N(N+1)\mu_1' - 2U\}.$$

(Cf. 2.25 in the continuous case.)

2.11 Referring to Exercise 1.14, show that the mean of the negative binomial distribution is rq/p and that its variance is rq/p^2.

2.12 Show that the variance of the distribution with p.d.f.

$$f(x) = \frac{k}{1+x^4}, \qquad -\infty < x < \infty,$$

is equal to one.

2.13 With reference to Exercise 1.10 find the mean value of $x_1^2 + x_2^2 + x_3^2$ (a) directly and (b) from the distribution of r^2.

2.14 For the distribution of Exercise 1.11 show that the mean value of x^2 is σ_1^2 and that of y^2 is σ_2^2. By differentiating $\int dF = 1$ with respect to ρ show that $E(xy) = \rho\sigma_1\sigma_2$.

2.15 A variate takes only non-negative values and has mean μ_1'. Show that for any positive t

$$F(t) > 1 - \mu_1'/t.$$

2.16 Show that if the normal distribution

$$f(\xi) = \frac{1}{\sqrt{(2\pi)}}e^{-\frac{1}{2}\xi^2}, \qquad -\infty < \xi < \infty,$$

is transformed by $\xi = \gamma + \delta \log(x-\mu)$, the distribution of x has

$$\frac{\text{mean} - \text{mode}}{\text{mean} - \text{median}} = \frac{\exp\left(\dfrac{1}{2\delta^2}\right) - \exp\left(-\dfrac{1}{\delta^2}\right)}{\exp\left(\dfrac{1}{2\delta^2}\right) - 1},$$

and that this ratio tends to the value 3 as $\delta \to \infty$, in accordance with (2.13).

2.17 A non-increasing density function $f(x)$ on the range 0 to ∞ has mean μ and median m. Show that $\mu \geqslant m$ and hence that $\mu \geqslant 1/\{2f(0)\}$.

2.18 For a unimodal density function $f(x)$ with median m exceeding its mode, show using Exercise 1.20 that if for some value $a > m$

$$\begin{aligned} f(2m-x) &> f(x), \quad x < a, \\ &\leqslant f(x), \quad x \geqslant a, \end{aligned}$$

then $m < \mu$, the mean. (The condition ensures that $f(x)$ falls away faster below than above its mode.)

2.19 Show that for a distribution with finite mean the mean difference Δ also exists. Illustrate by reference to the Pareto distribution

$$f(x) = kx^{-\alpha}, \qquad 0 < a \leqslant x < \infty, \quad \alpha > 1,$$

where the mean may exist but the variance does not exist for $\alpha \leqslant 3$.

2.20 For the distribution of Exercise 1.11 show that if $E(x^2)E(y^2) = E(x^2y^2)$ then the variates are independent.

2.21 In the mean deviation calculation for grouped data in Example 2.7, the interval in which the mean fell made contribution $f|d|$ to the sum of deviations. Show that if the frequency f is split into frequencies above and below μ_1' proportionately to the fractions of the interval length h above and below μ_1', and these frequencies are taken to the concentrated at the centre of their intervals, the sum of deviations is increased by $fh\left(\frac{|d|}{h} - \frac{1}{2}\right)^2$. Confirm that in Example 2.7, this correction increases δ_1 by 0.04.

2.22 Show that for any variate distributed over the range from l (>0) to u,

$$\frac{\mu_1'}{H} \leqslant \frac{(l+u)^2}{4lu},$$

the equality being attainable if and only if half of the probability lies at each extreme variate value.

Show similarly that $\sigma^2 \leqslant \frac{1}{4}(u-l)^2$.

(Jacobson (1969) shows that for unimodal distributions $\sigma^2 \leqslant \frac{1}{9}(u-l)^2$.)

2.23 Show that for a set of n ($\geqslant 2$) non-negative values x_i, not all equal, the coefficient of variation V defined at (2.28) cannot exceed $(n-1)^{\frac{1}{2}}$, attaining this value if and only if all but one of the x_i are zero.

(Katsnelson and Kotz, 1957)

2.24 Show that for any distribution function $F(x)$ with $F(a) = 0$, $F(b) = 1$,

$$\mu_1' - a = \int_a^b (x-a)\,\mathrm{d}F = \int_a^b \{1 - F(x)\}\,\mathrm{d}x$$

and hence that the means of distributions on the same interval may be compared graphically from cumulative values of their distribution functions. Similarly, show from **2.18** that the mean deviation

$$\delta_1 = 2\int_{-\infty}^{\mu_1'} F(x)\,\mathrm{d}x.$$

Show further that the variance of x is

$$\sigma^2 = \int_a^b \left[\int_a^b \{1-F(y)\}\,dy\right]^2 \{1-F(x)\}^{-2}\,dF(x). \qquad \text{(Cf. Pyke (1965))}$$

2.25 If a p.f. may be written in the form

$$f(x,\theta) = a(x)e^{\theta x}/g(\theta)$$

and $F(x, \theta)$ is the corresponding d.f., show that the mean and variance are

$$\mu = \frac{\partial \log g(\theta)}{\partial\theta}, \qquad \mu_2 = \frac{\partial\mu}{\partial\theta}$$

and that

$$\frac{\partial f}{\partial\theta} = (x-\mu)f.$$

Hence show that if x varies continuously, the mean deviation is

$$\delta_1 = -2\int_{-\infty}^{\mu} \frac{\partial f}{\partial\theta}\,dx = 2\mu_2 f(\mu,\theta) - 2\frac{\partial F(\mu,\theta)}{\partial\mu},$$

so that if $F(\mu, \theta)$ does not depend upon θ,

$$\delta_1 = 2\mu_2 f(\mu,\theta). \qquad \text{(A)}$$

In particular show that this holds (a) if f is symmetric about μ, or (b) if θ is a scale parameter not functionally related to μ.

If x takes only non-negative integer values and $m = [\mu]$, show that

$$\delta_1 = -2\sum_{x=0}^{m} \frac{\partial f(x,\theta)}{\delta\theta} = -2\frac{d}{d\theta}F(m,\theta)$$

$$= -2\mu_2 \frac{d}{d\mu}F(m,\theta). \qquad \text{(B)} \qquad \text{(Cf. Kamat (1965))}$$

2.26 In the table below, the value of x is greater than the corresponding value of y for every income group, but y is greater for all groups combined. Resolve the paradox by calculating standardized means.

U.K. PERSONAL INCOME AND TAXATION
(from *Inland Revenue Statistics*, 1974 and 1980)

Income group (£)	1971/72 Income before tax (£m.)	1971/72 Tax paid (£m.)	1971/72 Tax/Income (%) (x)	1977/78 Income before tax (£m.)	1977/78 Tax paid (£m.)	1977/78 Tax/Income (%) (y)
0–999	4 219	382	9.1	523	2	0.4
1000–1999	14 029	2049	14.6	6070	494	8.1
2000–4999	14 145	2682	19.0	40 260	6867	17.1
5000–9999	2124	670	31.5	34 960	6500	18.6
10 000–14 999	547	243	44.4	5070	1475	29.1
15 000 & over	535	330	61.7	4315	2017	46.7
TOTAL	35 599	6356	17.9	91 198	17 355	19.0

2.27 Let

$$f(x) = z_0(1+|x|)^{-2}, \quad -c \leqslant x \leqslant c, \quad c < \infty.$$

Show that the mean of x is zero. However, when $c \to \infty$ show that the integral in (2.2), and hence the mean, does not exist.

2.28 The *lognormal* distribution has density function

$$f(x) = \frac{1}{\sigma x(2\pi)^{\frac{1}{2}}} \exp\left[-\frac{1}{2\sigma^2}(\ln x - \mu)^2\right], \quad 0 < x < \infty.$$

Show that the first moment distribution, $\Phi(x)$ in (2.31), is lognormal with parameters $(\mu + \sigma^2, \sigma^2)$ in place of (μ, σ^2). Hence show that G in (2.29) is $2F(\sigma/\sqrt{2}) - 1$, where F is the d.f. of the standardized normal distribution.

(Aitchison and Brown, 1957)

2.29 For the Pareto distribution in Exercise 2.19, show that G in (2.29) is $(2\alpha - 1)^{-1}$.

2.30 For the lognormal distribution in Exercise 2.28, show that

$$\mu_j' = \exp\{j\mu + \tfrac{1}{2}j^2\sigma^2\},$$

whereas $E(\ln x) = \mu$. Verify (2.48) in this case.

2.31 If

$$f(x|\theta) = x^\theta A(x)B(\theta), \quad x > 0$$

show that

$$E(x^j) = B(\theta)/B(\theta + j).$$

Hence show that the jth moment distribution has p.f. $f(x|\theta + j)$, provided that it exists. Show that the Pareto in Exercise 2.19 is a member of this family with $\theta = -\alpha$, as is the lognormal of Exercise 2.28 with $\theta = \mu/\sigma^2$.

(Patil and Ord, 1976)

CHAPTER 3

MOMENTS AND CUMULANTS

Moments

3.1 In the previous chapter we defined the first moment (arithmetic mean) about an arbitrary point a by the Stieltjes integral

$$\mu_1' = \int_{-\infty}^{\infty} (x-a)\,\mathrm{d}F \tag{3.1}$$

and the second moment about that point by

$$\mu_2' = \int_{-\infty}^{\infty} (x-a)^2\,\mathrm{d}F. \tag{3.2}$$

As a generalization of these equations we may define a series of coefficients μ_r', $r=1$, 2, 3, ..., by the relation

$$\mu_r' = \int_{-\infty}^{\infty} (x-a)^r\,\mathrm{d}F. \tag{3.3}$$

μ_r' is called the moment of order r about the point a. When a is the mean μ_1' (about zero) we write the moment without the prime,

$$\mu_r = \int_{-\infty}^{\infty} (x-\mu_1')^r\,\mathrm{d}F, \tag{3.4}$$

and call it a central moment or a moment about the mean. In particular $\mu_1 = 0$, and we may also define a moment of zero order

$$\mu_0' = \mu_0 = \int_{-\infty}^{\infty} \mathrm{d}F = 1.$$

When reference is made to the rth moment of a particular distribution, it is assumed that the appropriate integral (3.3) converges for that distribution. As will be seen later, some of the theoretical distributions encountered in Statistics do not have finite moments of all orders; some have only a few moments of low order, and some have no finite moments, except of course the moment of order zero. It is easily seen by considering the orders of magnitude of the integrands that if μ_r' exists, so does μ_s' for $s<r$; and that if μ_r' does not exist, nor does μ_s' for $s>r$.

3.2 If a and b are two variate values, let $b-a=c$ and denote the moments about

a and b by $\mu'(a)$ and $\mu'(b)$ respectively. Then we have, by the binomial theorem,

$$(x-a)^r = (x-b+b-a)^r = (x-b+c)^r$$
$$= \sum_{j=0}^{r} \binom{r}{j} (x-b)^{r-j} c^j.$$

Hence

$$\mu'_r(a) = \int_{-\infty}^{\infty} (x-a)^r \, \mathrm{d}F$$
$$= \int_{-\infty}^{\infty} \sum_{j=0}^{r} \binom{r}{j} (x-b)^{r-j} c^j \, \mathrm{d}F$$
$$= \sum_{j=0}^{r} \binom{r}{j} c^j \int_{-\infty}^{\infty} (x-b)^{r-j} \, \mathrm{d}F$$
$$= \sum_{j=0}^{r} \binom{r}{j} \mu'_{r-j}(b) c^j. \tag{3.5}$$

This equation gives the rth moment about a in terms of the rth and lower moments about b. It may be written in a symbolic form which will be found to provide a useful mnemonic, namely

$$\mu'_r(a) = \{\mu'(b) + c\}^r$$

with the convention that the expression on the right is to be expanded binomially and the form $\{\mu'(b)\}^j$ replaced by $\mu'_j(b)$.

The equation (3.5) is of particular importance if one of the values b or a is the mean of the distribution about zero, μ'_1. In this case we have

$$\mu'_r(a) = \sum_{j=0}^{r} \binom{r}{j} \mu_{r-j}(\mu'_1 - a)^j. \tag{3.6}$$

$$\mu_r = \sum_{j=0}^{r} \binom{r}{j} \mu'_{r-j}(b)(b-\mu'_1)^j. \tag{3.7}$$

In particular, (3.6) gives, since $\mu'_1 - a = \mu'_1(a)$ by (2.2),

$$\left.\begin{aligned} \mu'_2 &= \mu_2 + \mu'^2_1 \\ \mu'_3 &= \mu_3 + 3\mu_2\mu'_1 + \mu'^3_1 \\ \mu'_4 &= \mu_4 + 4\mu_3\mu'_1 + 6\mu_2\mu'^2_1 + \mu'^4_1 \end{aligned}\right\} \tag{3.8}$$

and (3.7) similarly gives, since $b - \mu'_1 = -\mu'_1(b)$,

$$\left.\begin{aligned} \mu_2 &= \mu'_2 - \mu'^2_1 \\ \mu_3 &= \mu'_3 - 3\mu'_2\mu'_1 + 2\mu'^3_1 \\ \mu_4 &= \mu'_4 - 4\mu'_3\mu'_1 + 6\mu'_2\mu'^2_1 - 3\mu'^4_1 \end{aligned}\right\} \tag{3.9}$$

where the primed moments in (3.8) and (3.9) may be taken about any fixed origin. The unprimed central moments there are of course taken about the mean, and are unaffected by choice of origin.

Calculation of moments

3.3 For a distribution specified numerically in a frequency table the calculation of moments of third and higher orders is akin to that of the first and second moments. For grouped data the observations are regarded as concentrated at the mid-points of intervals; a convenient arbitrary origin a is chosen, the moments about a calculated, and then if necessary the central moments are ascertained from (3.7). The effect of grouping may be corrected for in certain cases.

In practice numerical moments of order higher than the fourth are rarely required, being so sensitive to sampling fluctuations that values computed from moderate numbers of observations are subject to a large margin of error.

Example 3.1 (Calculation of moments)

To find the first four moments about the mean of the distribution of Australian marriages of Table 1.8.

Until the last stage we work in units of three years, the interval length. A working origin is taken at 28.5 years. To check the arithmetic we use an identity of type

$$(x+1)^3 = x^3 + 3x^2 + 3x + 1,$$
$$(x+1)^4 = x^4 + 4x^3 + 6x^2 + 4x + 1.$$

Thus, for instance, the value of $g(x)(x+1)^r$ is found in addition to that of $g(x)x^r$ and the two checked by identities such as

$$\sum g(x)(x+1)^3 = \sum g(x)x^3 + 3\sum g(x)x^2 + 3\sum g(x)x + \sum g(x),$$

$g(x)$ being the actual frequencies. The arithmetical work is shown in Table 3.1. From this table we find

$$\sum(xg) = 88\,832$$
$$\sum(x^2g) = 2\,155\,838$$
$$\sum(x^3g) = 12\,798\,362$$
$$\sum(x^4g) = 137\,306\,162.$$

The values will be found to check and we have, about the working origin 28.5, on dividing by the total frequency 301 785,

$$\mu_1' = 0.294\,355\,253$$
$$\mu_2' = 7.143\,622\,115$$
$$\mu_3' = 42.408\,873\,867$$
$$\mu_4' = 454.980\,075\,219.$$

Table 3.1 Calculation of the first four moments of the distribution of marriages of Table 1.8

Mid-value of intervals, years		x	xg	$(x+1)g$	x^2g	$(x+1)^2g$	x^3g	$(x+1)^3g$	x^4	$(x+1)^4g$
16.5	294	−4	− 1 176	− 882	4 704	2 646	− 18 816	− 7 938	75 264	23 814
19.5	10 995	−3	− 32 985	−21 990	98 955	43 980	−296 865	− 87 960	890 595	175 920
22.5	61 001	−2	−122 002	−61 001	244 004	61 001	−488 008	− 61 001	976 016	61 001
25.5	73 054	−1	− 73 054	−83 873	73 054	—	− 73 054	−156 899	73 054	—
28.5	56 501	0	−229 217	56 501	—	56 501	−876 743	56 501	—	56 501
31.5	33 478	1	33 478	66 956	33 478	133 912	33 478	267 824	33 478	535 648
34.5	20 569	2	41 138	61 707	82 276	185 121	164 552	555 363	329 104	1 666 089
37.5	14 281	3	42 843	57 124	128 529	228 496	385 587	913 984	1 156 761	3 655 936
40.5	9 320	4	37 280	46 600	149 120	233 000	596 480	1 165 000	2 385 920	5 825 000
43.5	6 236	5	31 180	37 416	155 900	224 496	779 500	1 346 976	3 897 500	8 081 856
46.5	4 770	6	28 620	33 390	171 720	233 730	1 030 320	1 636 110	6 181 920	11 452 770
49.5	3 620	7	25 340	28 960	177 380	231 680	1 241 660	1 853 440	8 691 620	14 827 520
52.5	2 190	8	17 520	19 710	140 160	177 390	1 121 280	1 596 510	8 970 240	14 368 590
55.5	1 655	9	14 895	16 550	134 055	165 500	1 206 495	1 655 000	10 858 455	16 550 000
58.5	1 100	10	11 000	12 100	110 000	133 100	1 100 000	1 464 100	11 000 000	16 105 100
61.5	810	11	8 910	9 720	98 010	116 640	1 078 110	1 399 680	11 859 210	16 796 160
64.5	649	12	7 788	8 437	93 456	109 681	1 121 472	1 425 853	13 457 664	18 536 089
67.5	487	13	6 331	6 818	82 303	95 452	1 069 939	1 336 328	13 909 207	18 708 592
70.5	326	14	4 564	4 890	63 896	73 350	894 544	1 100 250	12 523 616	16 503 750
73.5	211	15	3 165	3 376	47 475	54 016	712 125	864 256	10 681 875	13 828 096
76.5	119	16	1 904	2 023	30 464	34 391	487 424	584 647	7 798 784	9 938 999
79.5	73	17	1 241	1 314	21 097	23 652	358 649	425 736	6 097 033	7 663 248
82.5	27	18	486	513	8 748	9 747	157 464	185 193	2 834 352	3 518 667
85.5	14	19	266	280	5 054	5 600	96 026	112 000	1 824 494	2 240 000
88.5	5	20	100	105	2 000	2 205	40 000	46 305	800 000	972 405
TOTALS OF +VE TERMS	301 785	–	318 049	474 490	2 155 838	2 635 287	13 675 105	19 991 056	137 306 162	202 091 751

For the moments about the mean, substitution in equations (3.9) gives

$$\mu_2 = 7.056\,977$$
$$\mu_3 = 36.151\,595$$
$$\mu_4 = 408.738\,210.$$

These are in units of three years. To express the results in units of one year we multiply the rth moment by 3^r, e.g.

$$\mu_2 = 7.056\,977 \times 9 = 63.512\,79.$$

3.4 If a distribution is specified mathematically the determination of moments is equivalent to the evaluation of certain sums or integrals. Some examples will illustrate the general principles involved.

Example 3.2 (Moments of the binomial distribution)

Consider the binomial distribution of Examples 2.3 and 2.9. All of its moments must be finite, since the range of the variate is finite, from 0 to n. We will write $q = 1 - p$ so that the generating function of the distribution is $P(t) = (pt + q)^n$. In the manner of those examples we have

$$\left[\frac{d^3P}{dt^3}\right]_{t=1} = \sum_{j=0}^{n} j(j-1)(j-2)f_j$$

$$= \sum (j^3 f_j) - 3\sum (j^2 f_j) + 2\sum (jf_j)$$

$$= \mu_3' - 3\mu_2' + 2\mu_1',$$

the moments being measured from zero origin. This gives us

$$\mu_3' = 3\mu_2' - 2\mu_1' + n(n-1)(n-2)p^3.$$

We have already found that

$$\mu_2' = n(n-1)p^2 + np, \qquad \mu_1' = np,$$

and hence we find, after substitution and reduction,

$$\mu_3' = n(n-1)(n-2)p^3 + 3n(n-1)p^2 + np.$$

We are usually more interested in the moments about the mean, and from (3.9) we have

$$\mu_3 = \mu_3' - 3\mu_2'\mu_1' + 2\mu_1'^3$$

which, on substitution on the right, reduces to

$$\mu_3 = np(p-1)(2p-1) \tag{3.10}$$

$$= npq(q-p). \tag{3.11}$$

In a similar manner we find

$$\mu_4 = 3n^2p^2(1-p)^2 + np(1-p)(1-6p+6p^2) \tag{3.12}$$

$$= 3n^2p^2q^2 + npq(1-6pq). \tag{3.13}$$

The process may evidently be extended to give as many moments as are desired. But we shall see later that there are easier ways of finding the moments of the binomial distribution.

Example 3.3 (Moments of Student's t distribution)

Consider the distribution with p.d.f.

$$f(x) = \frac{k}{(1+x^2)^m}, \qquad -\infty < x < \infty; m \geqslant 1,$$

known as Student's t distribution (cf. **16.11**). This is a unimodal distribution symmetrical about $x = 0$. All finite moments of odd order about the origin are therefore zero. The constant k is given by the equation

$$1 = k\int_{-\infty}^{\infty} \frac{dx}{(1+x^2)^m} = k\int_0^{\infty} \frac{dy}{y^{\frac{1}{2}}(1+y)^m}$$

$$= kB(\tfrac{1}{2}, m-\tfrac{1}{2}) = k\frac{\Gamma(\frac{1}{2})\Gamma(m-\frac{1}{2})}{\Gamma(m)}.$$

The central moment of order r, if it exists, is given by

$$\mu_r = k \int_{-\infty}^{\infty} \frac{x^r}{(1+x^2)^m} \, dx,$$

and this integral converges if and only if

$$2m - r > 1.$$

Thus μ_r exists if and only if $r < 2m - 1$.

If $m = 1$, the integral defining the mean, $\lim_{c\to\infty, c'\to\infty} \int_{-c}^{c'} \frac{kx\,dx}{(1+x^2)}$, does not converge, although the principal value

$$I = \lim_{c\to\infty} \int_{-c}^{c} \frac{kx\,dx}{(1+x^2)}$$

does and is equal to zero. Thus even the mean does not exist for $m = 1$.(*) For $m > 1$ the mean exists and is located at the origin.

Making the substitution $z = \frac{1}{1+x^2}$ in the formula for μ_{2r}, we find

$$\mu_{2r} = k \int_0^1 (1-z)^{r-\frac{1}{2}} z^{m-r-\frac{3}{2}} \, dz$$

$$= k \frac{\Gamma(r+\frac{1}{2})\Gamma(m-r-\frac{1}{2})}{\Gamma(\mu)}$$

and on substituting for k,

$$\mu_{2r} = \frac{\Gamma(r+\frac{1}{2})\Gamma(m-r-\frac{1}{2})}{\Gamma(\frac{1}{2})\Gamma(m-\frac{1}{2})}, \qquad 2r < 2m - 1.$$

Example 3.4 (Moments of the normal distribution)

Consider the normal distribution (to be discussed in detail in **5.36–42** below) with p.d.f.

$$f(x) = \frac{1}{\sigma\sqrt{(2\pi)}} e^{-x^2/(2\sigma^2)}, \qquad -\infty < x < \infty. \tag{3.14}$$

This is symmetrical about the origin. All moments exist since $\exp\{-x^2/2\sigma^2)\}$ is the dominant term in the integral defining any moment, those of odd order being zero by symmetry. The even moments

$$\mu_{2r} = \frac{1}{\sigma\sqrt{(2\pi)}} \int_{-\infty}^{\infty} x^{2r} e^{-x^2/(2\sigma^2)} \, dx$$

may be evaluated by partial integration, but a more useful method is as follows.

(*) A disadvantage of the convention that the distribution has a mean for $m = 1$ because I is finite is that it is no longer necessarily true that the mean of a sum is the sum of the means. Cf. Fréchet (1937), page 45, and Exercise 8.29 below.

Consider the integral

$$\begin{aligned} M(t) &= \frac{1}{\sigma\sqrt{(2\pi)}}\int_{-\infty}^{\infty} e^{tx}e^{-x^2/(2\sigma^2)}\,dx \\ &= \frac{e^{\sigma^2t^2/2}}{\sigma\sqrt{(2\pi)}}\int_{-\infty}^{\infty} \exp\left\{-\frac{1}{2}\left(\frac{x}{\sigma}-\sigma t\right)^2\right\}dx \\ &= \exp(\tfrac{1}{2}\sigma^2t^2). \end{aligned} \tag{3.15}$$

We have, for all real values of t,

$$e^{tx}e^{-x^2/(2\sigma^2)} = \sum_{r=0}^{\infty}\left(\frac{t^r}{r!}x^r e^{-x^2/(2\sigma^2)}\right).$$

The series on the right is uniformly convergent in x and may be integrated term by term if the resulting series is uniformly convergent. We then have

$$M(t) = \sum_{r=0}^{\infty}\left(\frac{t^r}{r!}\mu_r\right).$$

In other words, μ_r is the coefficient of $t^r/r!$ in $\exp(\frac{1}{2}\sigma^2t^2)$ and hence

$$\mu_{2r} = \frac{\sigma^{2r}}{2^r}\frac{(2r)!}{r!}. \tag{3.16}$$

Evidently, we may obtain μ_r from $M(t)$ by differentiating r times with respect to t and putting $t=0$.

Moment generating functions and characteristic functions

3.5 The previous example shows that in some cases we can derive from the probability function a function $M(t) = \int_{-\infty}^{\infty} e^{tx}\,dF(x)$ which, when expanded in powers of t, will yield the moments of the distribution as coefficients in the expansion. Such a function is accordingly called a moment generating function. If the probability generating function is $P(t)$, the moment generating function is simply $P(e^t)$. The name is conveniently abbreviated to m.g.f. For continuous distributions, it is usually more convenient to evaluate $M(t)$ rather than $P(t)$, while $M(t)$ is easily obtained from $P(t)$ in discrete cases.

For some distributions the integral $\int_{-\infty}^{\infty} e^{tx}\,dF$ or the sum $\{\sum e^{tx_j}f(x_j)\}$ does not exist for some or all real values of t. This is (cf. Example 3.3) true of the function $f(x) = k(1+x^2)^{-m}$ for finite positive values of m. A more serviceable auxiliary function is

$$\phi(t) = \int_{-\infty}^{\infty} e^{itx}\,dF \quad (t\text{ real}) = M(it) \tag{3.17}$$

where i is the complex operator $(i^2 = -1)$. This is known as the *characteristic function* (often abbreviated to c.f.) of the distribution and is of great theoretical importance. It will be seen in Chapter 4 that under certain general conditions the c.f. determines and is completely determined by the distribution function. It also yields many valuable results in the theory of sampling.

Since by the nature of the distribution function the integral $\int_{-\infty}^{\infty} \mathrm{d}F = 1$, we have

$$|\phi(t)| \leqslant \int_{-\infty}^{\infty} |\mathrm{e}^{itx}| \,\mathrm{d}F = \int_{-\infty}^{\infty} \mathrm{d}F = 1$$

and hence the Stieltjes integral (3.17) converges absolutely and uniformly in t. It may therefore be integrated under the summation sign with respect to t, and may be differentiated provided that the resulting expressions exist and are uniformly convergent. We have, for example, writing D_t for $\mathrm{d}/\mathrm{d}t$,

$$D_t^r \phi(t) = i^r \int_{-\infty}^{\infty} \mathrm{e}^{itx} x^r \,\mathrm{d}F,$$

and hence, putting $t = 0$,

$$\mu_r' = (-i)^r [D_t^r \phi(t)]_{t=0} \tag{3.18}$$

provided that μ_r' exists. If $\phi(t)$ is expanded in powers of t, μ_r' is equal to the coefficient of $(it)^r/r!$ in the expansion. Thus the c.f. is also a moment generating function. For many formal purposes it is sufficient to write, say, θ instead of it and treat it as real.

Our earlier m.g.f., $M(t)$, may similarly be differentiated to obtain the moments if it is finite in an open interval including the origin, but it may not be so even if all moments are finite—an instance is given in Exercise 3.28. If only the first r moments exist, the first r derivatives of $\phi(t)$ will exist at $t = 0$ (cf. Exercise 3.31).

Example 3.5 (M.g.f. of the binomial)

Consider again the binomial distribution of Example 3.2. The probability generating function is $P(t) = (pt + q)^n$ and hence the moment generating function is

$$M(t) = P(\mathrm{e}^t) = (p\mathrm{e}^t + q)^n.$$

Hence

$$\mu_1' = \left[\frac{\mathrm{d}}{\mathrm{d}t} M(t)\right]_{t=0} = np.$$

$$\mu_2' = \left[\frac{\mathrm{d}^2}{\mathrm{d}t^2} M(t)\right]_{t=0} = np + n(n-1)p^2$$

and so on.

Example 3.6 (C.f. of the Gamma distribution)

Consider the distribution with p.d.f.

$$f(x) = \frac{a^\gamma}{\Gamma(\gamma)} \mathrm{e}^{-ax} x^{\gamma-1}, \qquad 0 \leqslant x < \infty; \qquad a > 0, \gamma > 0,$$

known as the Gamma distribution (cf. 6.9 below). The distribution may have a variety of shapes, depending on the value of γ, but moments of all orders exist since the

Γ-function integral $\int_0^\infty e^{-\alpha x}x^r\,dx$ converges when $r>-1$. We have for the c.f.

$$\phi(t)=\frac{a^\gamma}{\Gamma(\gamma)}\int_0^\infty e^{x(-a+it)}x^{\gamma-1}\,dx.$$

By the substitution $z=x(a-it)$ this becomes

$$\begin{aligned}\phi(t)&=\frac{a^\gamma}{\Gamma(\gamma)(a-it)^\gamma}\int_0^\infty e^{-z}z^{\gamma-1}\,dz\\&=\left(1-\frac{it}{a}\right)^{-\gamma}\end{aligned}\tag{3.19}$$

since $\int_0^\infty e^{-z}z^{\gamma-1}\,dz=\Gamma(\gamma)$ whether z is real or complex. Hence

$$\phi(t)=1+\gamma\frac{it}{a}+\frac{\gamma(\gamma+1)}{2!}\left(\frac{it}{a}\right)^2+\cdots$$

and thus

$$\begin{aligned}\mu_1'&=\frac{\gamma}{a},\\\mu_2'&=\frac{\gamma(\gamma+1)}{a^2},\\\mu_3'&=\frac{\gamma(\gamma+1)(\gamma+2)}{a^3},\\\mu_4'&=\frac{\gamma(\gamma+1)(\gamma+2)(\gamma+3)}{a^4}\end{aligned}$$

and so on. We find, using (3.9), the central moments

$$\left.\begin{aligned}\mu_2&=\frac{\gamma}{a^2},\\\mu_3&=\frac{2\gamma}{a^3},\\\mu_4&=\frac{3\gamma(\gamma+2)}{a^4}.\end{aligned}\right\}\tag{3.20}$$

Absolute moments

3.6 The quantity

$$\nu_r'=\int_{-\infty}^{\infty}|x-a|^r\,dF\tag{3.21}$$

is called the absolute moment of order r about a. The absolute moments about the mean are written without primes. Clearly if the moment of order r exists, the absolute moment exists for any order less than or equal to r.

If r is even, the absolute moment is equal to the ordinary moment, and if the range of the distribution is positive the absolute moments about any point to the left of the start of the distribution are equal to the ordinary moments of corresponding order.

There are some interesting inequalities concerning the absolute moments. Referring to the function $A(t)$ of **2.6** and remembering that it is a non-decreasing function of t, we find, on putting $t = 1, 2, \ldots$, that

$$(\nu_1')^{1} \leqslant (\nu_2')^{\frac{1}{2}} \leqslant (\nu_3')^{\frac{1}{3}} \leqslant \cdots \leqslant (\nu_r')^{1/r}.$$

A more general inequality, due to Liapunov (1901), is

$$(\nu_b')^{a-c} \leqslant (\nu_c')^{a-b}(\nu_a')^{b-c}, \qquad a \geqslant b \geqslant c \geqslant 0. \tag{3.22}$$

A proof of this result is given in Exercise 3.15.

Factorial moments

3.7 The factorial expression

$$x(x-h)(x-2h)\ldots\{x-(r-1)h\}$$

may conveniently be written $x^{[r]}$, a notation which brings out an analogy with the power x^r. Taking first differences with respect to x and with unit h, we have

$$\begin{aligned}\Delta x^{[r]} &= (x+h)^{[r]} - x^{[r]}\\ &= (x+h)x(x-h)\ldots\{x-(r-2)h\} - x(x-h)\ldots\{x-(r-1)h\}\\ &= rx^{[r-1]}h,\end{aligned}$$

which may be compared with the derivative

$$\mathrm{d}(x^r)/\mathrm{d}x = rx^{r-1}.$$

Conversely

$$\sum_{u=0}^{x} u^{[r]} = \frac{1}{(r+1)h}(x+h)^{[r+1]}$$

corresponding to

$$\int_0^x u^r\,\mathrm{d}u = \frac{1}{r+1}x^{r+1}.$$

The rth factorial moment about an arbitrary origin may now be defined by

$$\mu'_{[r]} = \sum_{j=-\infty}^{\infty} (x_j - a)^{[r]} f(x_j), \tag{3.23}$$

where we have chosen the summation sign $\sum$ rather than the Stieltjes integral because it is almost entirely for discrete distributions, or continuous distributions grouped in intervals of length h, that the factorial moments are used. In statistical theory they are not very prominent, but they provide very concise formulae for the moments of certain discrete distributions like the binomial, which have probability mass at equally-spaced values. The common interval between adjacent values is taken as unity, and the origin at the lowest value, for convenience. With these conventions, $h = 1$ and the sum is over non-negative integral values.

As usual, when it is necessary to distinguish between factorial moments about the mean and those about an arbitrary point we may write the former without the prime.

3.8 The factorial moments obey laws of transformation similar to those of equation (3.5) governing ordinary moments. In fact we have the expansion(*)

$$(a+b)^{[r]} = \sum_{j=0}^{r} \binom{r}{j} a^{[r-j]} b^{[j]}$$

and hence

$$\begin{aligned}(x-a)^{[r]} &= (x-b+c)^{[r]}, \quad \text{where } c = b-a, \\ &= \sum_{j=0}^{r} \binom{r}{j} (x-b)^{[r-j]} c^{[j]}\end{aligned}$$

so that

$$\mu'_{[r]}(a) = \sum_{j=0}^{r} \binom{r}{j} \mu'_{[r-j]}(b) c^{[j]} \tag{3.24}$$

which may be written symbolically, just as (3.5) could,

$$\mu'_{[r]}(a) = \{\mu'(b) + c\}^{[r]}.$$

3.9 By direct expansion of (3.23) it is seen that

$$\left.\begin{aligned}\mu'_{[1]} &= \mu'_1 \\ \mu'_{[2]} &= \mu'_2 - h\mu'_1 \\ \mu'_{[3]} &= \mu'_3 - 3h\mu'_2 + 2h^2\mu'_1 \\ \mu'_{[4]} &= \mu'_4 - 6h\mu'_3 + 11h^2\mu'_2 - 6h^3\mu'_1\end{aligned}\right\} \tag{3.25}$$

and conversely that

$$\left.\begin{aligned}\mu'_2 &= \mu'_{[2]} + h\mu'_{[1]} \\ \mu'_3 &= \mu'_{[3]} + 3h\mu'_{[2]} + h^2\mu'_{[1]} \\ \mu'_4 &= \mu'_{[4]} + 6h\mu'_{[3]} + 7h^2\mu'_{[2]} + h^3\mu'_{[1]}\end{aligned}\right\} \tag{3.26}$$

Since the first moments are equal the equations remain true when the primes are dropped and terms in first moments omitted.

Frisch (1926) gave general formulae showing the factorial moments about one point in terms of the ordinary moments about another, and vice versa. In fact

$$\mu'_r(a) = \sum_{j=0}^{r} \left\{\binom{r}{j} B_{r-j}^{(-j)}(c) h^{r-j} \mu'_{[j]}(b)\right\} \tag{3.27}$$

$$\mu'_{[r]}(a) = \sum_{j=0}^{r} \left\{\binom{r}{j} B_{r-j}^{(r+1)}(c+h) h^{r-j} \mu'_j(b)\right\} \tag{3.28}$$

where $B_r^{(n)}(x)$ is the Bernoulli polynomial of order n and degree r in x (see **3.24** below).

(*) It is clear that $(a+b)^{[r]}$ will be a polynomial of degree r in a, and may therefore be equated to $\sum_{j=0}^{r} k_j a^{[r-j]}$, where the ks are polynomials in b and h but do not contain a. Putting $a = 0$ we obtain $b^{[r]} = k_r$. Taking first differences with respect to a and putting $a = 0$ we obtain $rb^{[r-1]} = k_{r-1}$. Successive differences give the ks and the above result follows.

Evaluation of factorial moments

3.10 Factorial moments are rarely required as such, but it may be easier to evaluate the factorial moments and then use (3.26) to obtain the moments. Usually this is done for discrete distributions when f_j contains $j!$ in the denominator.

Example 3.7 (Factorial moments of the binomial)

Consider the binomial distribution of Example 3.5 with

$$f_j = \binom{n}{j} p^j q^{n-j}, \quad j = 0, 1, \ldots, n.$$

Then for $r \leqslant n$

$$\begin{aligned}\mu'_{[r]} &= \sum_{j=0}^{n} j^{[r]} \binom{n}{j} p^j q^{n-j}\\ &= \sum_{j=r}^{n} n^{[r]} \binom{n-r}{j-r} p^j q^{n-j}\\ &= n^{[r]} p^r \sum_{s=0}^{n-r} \binom{n-r}{s} p^s q^{n-r-s}\\ &= n^{[r]} p^r.\end{aligned}$$

When $r > n$,

$$\mu'_{[r]} = 0.$$

Factorial moment generating functions

3.11 The probability generating function is given for a discrete variate distributed at equally-spaced values, taken to be $0, 1, 2, \ldots$, by

$$P(t) = \sum_{j=0}^{\infty} f_j t^j.$$

Replacing t by $1 + t$,

$$P(1+t) = \sum f_j (1+t)^j = \sum_{j=0}^{\infty} f_j \sum_{i=0}^{j} \binom{j}{i} t^i$$

and since $j^{[i]} = 0$ for $i > j$, we may write this

$$\begin{aligned}&= \sum_{i=0}^{\infty} \frac{t^i}{i!} \sum_{j=0}^{\infty} f_j j^{[i]}\\ &= \sum_{i=0}^{\infty} t^i \mu'_{[i]}/i!. \qquad (3.29)\end{aligned}$$

Hence $P(1+t)$ is a factorial moment generating function (f.m.g.f.), the factorial moment of order r being the coefficient of $t^r/r!$

The results obtained by differentiating $P(t)$ and putting $t = 1$, in 2.4 and Example 2.9, are also obtained by differentiating $P(1+t)$ and putting $t = 0$. They are, as we now see, the factorial moments.

Example 3.8 (F.m.g.f. of the binomial)

Consider again the binomial distribution of Example 3.5, with $P(t) = \{pt + (1-p)\}^n$. We have

$$P(1+t) = (1+pt)^n = \sum_{j=0}^{n} p^j t^j \binom{n}{j}.$$

Hence, differentiating and putting $t=0$ we obtain the factorial moments given in Example 3.7,

$$\begin{aligned} \mu'_{[r]} &= p^r n^{[r]}, \quad r \leqslant n, \\ &= 0, \quad r > n. \end{aligned}$$

We can now convert to the ordinary moments, if desired, by equations (3.26). For instance

$$\begin{aligned} \mu'_1 &= \mu'_{[1]} = pn, \\ \mu'_2 &= p^2 n(n-1) + pn. \end{aligned}$$

Cumulants

3.12 The moments are a set of descriptive constants of a distribution that are useful for measuring its properties and, in certain circumstances, for specifying it, as we shall see in later chapters. They are not, however, the only set of constants for the purpose, or even the best set. Another set of constants, known as the *cumulants*, have properties that are more useful from the theoretical standpoint.

Formally, the cumulants $\kappa_1, \kappa_2, \ldots, \kappa_r$ are defined by the identity in t

$$\exp\left(\sum_{r=1}^{\infty} \kappa_r t^r/r!\right) = \sum_{r=0}^{\infty} \mu'_r t^r/r! \tag{3.30}$$

It should be observed that there is no κ_0 in (3.30). If we write it for t,

$$\begin{aligned} \exp\left\{\sum_{r=1}^{\infty} \kappa_r (it)^r/r!\right\} &= \sum_{r=0}^{\infty} \mu'_r (it)^r/r! \\ &= \int_{-\infty}^{\infty} e^{itx} \, dF \\ &= \phi(t). \end{aligned} \tag{3.31}$$

Thus, whereas μ'_r is the coefficient of $(it)^r/r!$ in $\phi(t)$, the c.f., κ_r is the coefficient of $(it)^r/r!$ in $\log\phi(t)$, if an expansion in power series exists. The function $\psi(t) = \log\phi(t)$ is known as the cumulant generating function, denoted by c.g.f.

3.13 If in equation (3.31) the origin is changed from a to b, where as usual $b - a = c$, the effect on $\phi(t)$ is to multiply it by e^{-itc}, for $\int e^{itx} dF$ becomes $\int e^{it(x-c)} dF$. Hence the effect on $\log\phi(t)$ is merely to add the term $-itc$, and consequently the coefficients in $\log\phi(t)$ are unchanged, except the first, which is decreased by c.

Thus the cumulants except the first are invariant under change of origin. So are all the central moments, but the property is not shared by the moments about an arbitrary point.

Both cumulants and moments have another property of an invariantive kind, namely, that if the variate values are multiplied by a constant a, μ'_r and κ_r are multiplied by a^r. This is at once evident from their definitions. Thus any linear transformation of the kind

$$\xi = lx + m \tag{3.32}$$

leaves the cumulants unchanged so far as the constant m is concerned and multiplies κ_r by l^r. The sole exception is the first cumulant, which is equal to the mean. In particular, if we standardize a distribution, the only effect is to multiply κ_r by σ^{-r}, σ being the standard deviation and, as we shall see in a moment, being equal to $\kappa_2^{\frac{1}{2}}$.

The invariantive properties of the cumulants were the origin of their original name of seminvariants (Thiele, 1903). In accordance with the theory of algebraic invariants, however, it seems best to reserve the word "seminvariant" for any constant λ_r which, under the transformation (3.32), is multiplied by l^r. The cumulants and the central moments are thus particular cases of seminvariants.

Relations between moments and cumulants

3.14 Provided that the expansions are permissible, we have, from (3.30),

$$\sum_{r=0}^{\infty} \mu'_r t^r/r! = \exp\left(\sum_{r=1}^{\infty} \kappa_r t^r/r!\right) = \prod_{r=1}^{\infty} \exp(\kappa_r t^r/r!)$$

$$= \prod_{r=1}^{\infty} \sum_{s=0}^{\infty} \left(\frac{\kappa_r t^r}{r!}\right)^s / s!$$

Picking out the terms in the exponential expansions which, when multiplied together, give a power of t^r, we have

$$\mu'_r = \sum_{m=1}^{r} \sum \left(\frac{\kappa_{p1}}{p_1!}\right)^{\pi_1} \left(\frac{\kappa_{p2}}{p_2!}\right)^{\pi_2} \cdots \left(\frac{\kappa_{p_m}}{p_m!}\right)^{\pi_m} \frac{r!}{\pi_1!\pi_2!\ldots\pi_m!}, \tag{3.33}$$

where the second summation extends over all non-negative values of the πs such that

$$p_1\pi_1 + p_2\pi_2 + \ldots p_m\pi_m = r. \tag{3.34}$$

It is worth noting that the rather tedious process of writing down the explicit relations for particular values of r may be shortened considerably. In fact, differentiating (3.30) with respect to κ_j we have

$$\left(\sum_{r=0}^{\infty} \mu'_r t^r/r!\right) t^j/j! = \sum_{r=1}^{\infty} \frac{\partial \mu'_r}{\partial \kappa_j} t^r/r!, \quad j \geqslant 1,$$

and hence, identifying powers of t,

$$\frac{\partial \mu'_r}{\partial \kappa_j} = \binom{r}{j} \mu'_{r-j}. \tag{3.35}$$

In particular

$$\frac{\partial \mu'_r}{\partial \kappa_1} = r\mu'_{r-1} \tag{3.36}$$

and thus, given any μ'_r in terms of the κs, we can successively write down those of lower orders by differentiation.[(*)]

The first ten of these expressions are, for moments about an arbitrary point:

$$\left.\begin{aligned}
\mu'_1 &= \kappa_1, \\
\mu'_2 &= \kappa_2 + \kappa_1^2, \\
\mu'_3 &= \kappa_3 + 3\kappa_2\kappa_1 + \kappa_1^3, \\
\mu'_4 &= \kappa_4 + 4\kappa_3\kappa_1 + 3\kappa_2^2 + 6\kappa_2\kappa_1^2 + \kappa_1^4, \\
\mu'_5 &= \kappa_5 + 5\kappa_4\kappa_1 + 10\kappa_3\kappa_2 + 10\kappa_3\kappa_1^2 + 15\kappa_2^2\kappa_1 + 10\kappa_2\kappa_1^3 + \kappa_1^5, \\
\mu'_6 &= \kappa_6 + 6\kappa_5\kappa_1 + 15\kappa_4\kappa_2 + 15\kappa_4\kappa_1^2 + 10\kappa_3^2 + 60\kappa_3\kappa_2\kappa_1 + 20\kappa_3\kappa_1^3 \\
&\quad + 15\kappa_2^3 + 45\kappa_2^2\kappa_1^2 + 15\kappa_2\kappa_1^4 + \kappa_1^6, \\
\mu'_7 &= \kappa_7 + 7\kappa_6\kappa_1 + 21\kappa_5\kappa_2 + 21\kappa_5\kappa_1^2 + 35\kappa_4\kappa_3 + 105\kappa_4\kappa_2\kappa_1 \\
&\quad + 35\kappa_4\kappa_1^3 + 70\kappa_3^2\kappa_1 + 105\kappa_3\kappa_2^2 + 210\kappa_3\kappa_2\kappa_1^2 + 35\kappa_3\kappa_1^4 \\
&\quad + 105\kappa_2^3\kappa_1 + 105\kappa_2^2\kappa_1^3 + 21\kappa_2\kappa_1^5 + \kappa_1^7, \\
\mu'_8 &= \kappa_8 + 8\kappa_7\kappa_1 + 28\kappa_6\kappa_2 + 28\kappa_6\kappa_1^2 + 56\kappa_5\kappa_3 + 168\kappa_5\kappa_2\kappa_1 + 56\kappa_5\kappa_1^3 \\
&\quad + 35\kappa_4^2 + 280\kappa_4\kappa_3\kappa_1 + 210\kappa_4\kappa_2^2 + 420\kappa_4\kappa_2\kappa_1^2 + 70\kappa_4\kappa_1^4 \\
&\quad + 280\kappa_3^2\kappa_2 + 280\kappa_3^2\kappa_1^2 + 840\kappa_3\kappa_2^2\kappa_1 + 560\kappa_3\kappa_2\kappa_1^3 + 56\kappa_3\kappa_1^5 \\
&\quad + 105\kappa_2^4 + 420\kappa_2^3\kappa_1^2 + 210\kappa_2^2\kappa_1^4 + 28\kappa_2\kappa_1^6 + \kappa_1^8, \\
\mu'_9 &= \kappa_9 + 9\kappa_8\kappa_1 + 36\kappa_7\kappa_2 + 36\kappa_7\kappa_1^2 + 84\kappa_6\kappa_3 + 252\kappa_6\kappa_2\kappa_1 \\
&\quad + 84\kappa_6\kappa_1^3 + 126\kappa_5\kappa_4 + 504\kappa_5\kappa_3\kappa_1 + 378\kappa_5\kappa_2^2 + 756\kappa_5\kappa_2\kappa_1^2 \\
&\quad + 126\kappa_5\kappa_1^4 + 315\kappa_4^2\kappa_1 + 1260\kappa_4\kappa_3\kappa_2 + 1260\kappa_4\kappa_3\kappa_1^2 + 1890\kappa_4\kappa_2^2\kappa_1 \\
&\quad + 1260\kappa_4\kappa_2\kappa_1^3 + 126\kappa_4\kappa_1^5 + 280\kappa_3^3 + 2520\kappa_3^2\kappa_2\kappa_1 + 840\kappa_3^2\kappa_1^3 \\
&\quad + 1260\kappa_3\kappa_2^3 + 3780\kappa_3\kappa_2^2\kappa_1^2 + 1260\kappa_3\kappa_2\kappa_1^4 + 84\kappa_3\kappa_1^6 + 945\kappa_2^4\kappa_1 \\
&\quad + 1260\kappa_2^3\kappa_1^3 + 378\kappa_2^2\kappa_1^5 + 36\kappa_2\kappa_1^7 + \kappa_1^9, \\
\mu'_{10} &= \kappa_{10} + 10\kappa_9\kappa_1 + 45\kappa_8\kappa_2 + 45\kappa_8\kappa_1^2 + 120\kappa_7\kappa_3 + 360\kappa_7\kappa_2\kappa_1 \\
&\quad + 120\kappa_7\kappa_1^3 + 210\kappa_6\kappa_4 + 840\kappa_6\kappa_3\kappa_1 + 630\kappa_6\kappa_2^2 + 1260\kappa_6\kappa_2\kappa_1^2 \\
&\quad + 210\kappa_6\kappa_1^4 + 126\kappa_5^2 + 1260\kappa_5\kappa_4\kappa_1 + 2520\kappa_5\kappa_3\kappa_2 + 2520\kappa_5\kappa_3\kappa_1^2 \\
&\quad + 3780\kappa_5\kappa_2^2\kappa_1 + 2520\kappa_5\kappa_2\kappa_1^3 + 252\kappa_5\kappa_1^5 + 1575\kappa_4^2\kappa_2 + 1575\kappa_4^2\kappa_1^2 \\
&\quad + 2100\kappa_4\kappa_3^2 + 12\,600\kappa_4\kappa_3\kappa_2\kappa_1 + 4200\kappa_4\kappa_3\kappa_1^3 + 3150\kappa_4\kappa_2^3
\end{aligned}\right\} \tag{3.37}$$

[(*)] The coefficients of μs in terms of κs are those of the unitary symmetric functions (1^r) in terms of the augmented symmetric functions and can be read from the tables of David and Kendall (1949) as far as order 12. The inverse relations may also be obtained from the tables on multiplication by appropriate factorials. See 12.5 below.

$$\left.\begin{aligned}&+9450\kappa_4\kappa_2^2\kappa_1^2+3150\kappa_4\kappa_2\kappa_1^4+210\kappa_4\kappa_1^6+2800\kappa_3^3\kappa_1\\&+6300\kappa_3^2\kappa_2^2+12\,600\kappa_3^2\kappa_2\kappa_1^2+2100\kappa_3^2\kappa_1^4+12\,600\kappa_3\kappa_2^3\kappa_1\\&+12\,600\kappa_3\kappa_2^2\kappa_1^3+2520\kappa_3\kappa_2\kappa_1^5+120\kappa_3\kappa_1^7+945\kappa_2^5+4725\kappa_2^4\kappa_1^2\\&+3150\kappa_2^3\kappa_1^4+630\kappa_2^2\kappa_1^6+45\kappa_2\kappa_1^8+\kappa_1^{10};\end{aligned}\right\} \tag{3.37}$$

and, for central moments ($\kappa_1 = 0$ in (3.37)),

$$\left.\begin{aligned}\mu_2&=\kappa_2,\\ \mu_3&=\kappa_3,\\ \mu_4&=\kappa_4+3\kappa_2^2,\\ \mu_5&=\kappa_5+10\kappa_3\kappa_2,\\ \mu_6&=\kappa_6+15\kappa_4\kappa_2+10\kappa_3^2+15\kappa_2^3,\\ \mu_7&=\kappa_7+21\kappa_5\kappa_2+35\kappa_4\kappa_3+105\kappa_3\kappa_2^2,\\ \mu_8&=\kappa_8+28\kappa_6\kappa_2+56\kappa_5\kappa_3+35\kappa_4^2+210\kappa_4\kappa_2^2+280\kappa_3^2\kappa_2+105\kappa_2^4,\\ \mu_9&=\kappa_9+36\kappa_7\kappa_2+84\kappa_6\kappa_3+126\kappa_5\kappa_4+378\kappa_5\kappa_2^2+1260\kappa_4\kappa_3\kappa_2+280\kappa_3^3\\&\quad+1260\kappa_3\kappa_2^3,\\ \mu_{10}&=\kappa_{10}+45\kappa_8\kappa_2+120\kappa_7\kappa_3+210\kappa_6\kappa_4+630\kappa_6\kappa_2^2+126\kappa_5^2\\&\quad+2520\kappa_5\kappa_3\kappa_2+1575\kappa_4^2\kappa_2+2100\kappa_4\kappa_3^2+3150\kappa_4\kappa_2^3\\&\quad+6300\kappa_3^2\kappa_2^2+945\kappa_2^5.\end{aligned}\right\} \tag{3.38}$$

Conversely we have from (3.30)

$$\sum_{r=1}^{\infty}\kappa_r t^r/r! = \log\left(\sum_{r=0}^{\infty}\mu_r' t^r/r!\right). \tag{3.39}$$

Expanding the logarithm and picking out powers of t^r as before, we have

$$\kappa_r = r!\sum_{m=1}^{r}\sum\left(\frac{\mu_{p1}'}{p_1!}\right)^{\pi_1}\cdots\left(\frac{\mu_{pm}'}{p_m!}\right)^{\pi_m}\frac{(-1)^{\rho-1}(\rho-1)!}{\pi_1!\ldots\pi_m!}, \tag{3.40}$$

the second summation extending over all non-negative πs and ρs, subject to (3.34) and the further condition

$$\pi_1+\pi_2+\ldots+\pi_m=\rho. \tag{3.41}$$

The first ten formulae are, in terms of moments about an arbitrary point:

$$\left.\begin{aligned}\kappa_1&=\mu_1',\\ \kappa_2&=\mu_2'-\mu_1'^2,\\ \kappa_3&=\mu_3'-3\mu_2'\mu_1'+2\mu_1'^3,\\ \kappa_4&=\mu_4'-4\mu_3'\mu_1'-3\mu_2'^2+12\mu_2'\mu_1'^2-6\mu_1'^4,\\ \kappa_5&=\mu_5'-5\mu_4'\mu_1'-10\mu_3'\mu_2'+20\mu_3'\mu_1'^2+30\mu_2'^2\mu_1'-60\mu_2'\mu_1'^3+24\mu_1'^5,\end{aligned}\right\} \tag{3.42}$$

$$
\left.
\begin{aligned}
\kappa_6 = \mu_6' &- 6\mu_5'\mu_1' - 15\mu_4'\mu_2' + 30\mu_4'\mu_1'^2 - 10\mu_3'^2 + 120\mu_3'\mu_2'\mu_1' - 120\mu_3'\mu_1'^3 \\
&+ 30\mu_2'^3 - 270\mu_2'^2\mu_1'^2 + 360\mu_2'\mu_1'^4 - 120\mu_1'^6, \\
\kappa_7 = \mu_7' &- 7\mu_6'\mu_1' - 21\mu_5'\mu_2' + 42\mu_5'\mu_1'^2 - 35\mu_4'\mu_3' + 210\mu_4'\mu_2'\mu_1' \\
&- 210\mu_4'\mu_1'^3 + 140\mu_3'^2\mu_1' + 210\mu_3'\mu_2'^2 - 1260\mu_3'\mu_2'\mu_1'^2 + 840\mu_3'\mu_1'^4 \\
&- 630\mu_2'^3\mu_1' + 2520\mu_2'^2\mu_1'^3 - 2520\mu_2'\mu_1'^5 + 720\mu_1'^7, \\
\kappa_8 = \mu_8' &- 8\mu_7'\mu_1' - 28\mu_6'\mu_2' + 56\mu_6'\mu_1'^2 - 56\mu_5'\mu_3' + 336\mu_5'\mu_2'\mu_1' \\
&- 336\mu_5'\mu_1'^3 - 35\mu_4'^2 + 560\mu_4'\mu_3'\mu_1' + 420\mu_4'\mu_2'^2 - 2520\mu_4'\mu_2'\mu_1'^2 \\
&+ 1680\mu_4'\mu_1'^4 + 560\mu_3'^2\mu_2' - 1680\mu_3'^2\mu_1'^2 - 5040\mu_3'\mu_2'^2\mu_1' \\
&+ 13\,440\mu_3'\mu_2'\mu_1'^3 - 6720\mu_3'\mu_1'^5 - 630\mu_2'^4 + 10\,080\mu_2'^3\mu_1'^2 \\
&- 25\,200\mu_2'^2\mu_1'^4 + 20\,160\mu_2'\mu_1'^6 - 5040\mu_1'^8, \\
\kappa_9 = \mu_9' &- 9\mu_8'\mu_1' - 36\mu_7'\mu_2' + 72\mu_7'\mu_1'^2 - 84\mu_6'\mu_3' + 504\mu_6'\mu_2'\mu_1' \\
&- 504\mu_6'\mu_1'^3 - 126\mu_5'\mu_4' + 1008\mu_5'\mu_3'\mu_1' + 756\mu_5'\mu_2'^2 - 4536\mu_5'\mu_2'\mu_1'^2 \\
&+ 3024\mu_5'\mu_1'^4 + 630\mu_4'^2\mu_1' + 2520\mu_4'\mu_3'\mu_2' - 7560\mu_4'\mu_3'\mu_1'^2 \\
&- 11\,340\mu_4'\mu_2'^2\mu_1' + 30\,240\mu_4'\mu_2'\mu_1'^3 - 15\,120\mu_4'\mu_1'^5 + 560\mu_3'^3 \\
&- 15\,120\mu_3'^2\mu_2'\mu_1' + 20\,160\mu_3'^2\mu_1'^3 - 7560\mu_3'\mu_2'^3 + 90\,720\mu_3'\mu_2'^2\mu_1'^2 \\
&- 151\,200\mu_3'\mu_2'\mu_1'^4 + 60\,480\mu_3'\mu_1'^6 + 22\,680\mu_2'^4\mu_1' - 151\,200\mu_2'^3\mu_1'^3 \\
&+ 272\,160\mu_2'^2\mu_1'^5 - 181\,440\mu_2'\mu_1'^7 + 40\,320\mu_1'^9, \\
\kappa_{10} = \mu_{10}' &- 10\mu_9'\mu_1' - 45\mu_8'\mu_2' + 90\mu_8'\mu_1'^2 - 120\mu_7'\mu_3' + 720\mu_7'\mu_2'\mu_1' \\
&- 720\mu_7'\mu_1'^3 - 210\mu_6'\mu_4' + 1680\mu_6'\mu_3'\mu_1' + 1260\mu_6'\mu_2'^2 \\
&- 7560\mu_6'\mu_2'\mu_1'^2 + 5040\mu_6'\mu_1'^4 - 126\mu_5'^2 + 2520\mu_5'\mu_4'\mu_1' \\
&+ 5040\mu_5'\mu_3'\mu_2' - 15\,120\mu_5'\mu_3'\mu_1'^2 - 22\,680\mu_5'\mu_2'^2\mu_1' + 60\,480\mu_5'\mu_2'\mu_1'^3 \\
&- 30\,240\mu_5'\mu_1'^5 + 3150\mu_4'^2\mu_2' - 9450\mu_4'^2\mu_1'^2 + 4200\mu_4'\mu_3'^2 \\
&- 75\,600\mu_4'\mu_3'\mu_2'\mu_1' + 100\,800\mu_4'\mu_3'\mu_1'^3 - 18\,900\mu_4'\mu_2'^3 \\
&+ 226\,800\mu_4'\mu_2'^2\mu_1'^2 - 378\,000\mu_4'\mu_2'\mu_1'^4 + 151\,200\mu_4'\mu_1'^6 - 16\,800\mu_3'^3\mu_1' \\
&- 37\,800\mu_3'^2\mu_2'^2 + 302\,400\mu_3'^2\mu_2'\mu_1'^2 - 252\,000\mu_3'^2\mu_1'^4 + 302\,400\mu_3'\mu_2'^3\mu_1' \\
&- 1\,512\,000\mu_3'\mu_2'^2\mu_1'^3 + 1\,814\,400\mu_3'\mu_2'\mu_1'^5 - 604\,800\mu_3'\mu_1'^7 \\
&+ 22\,680\mu_2'^5 - 567\,000\mu_2'^4\mu_1'^2 + 2\,268\,000\mu_2'^3\mu_1'^4 - 3\,175\,200\mu_2'^2\mu_1'^6 \\
&+ 1\,814\,400\mu_2'\mu_1'^8 - 362\,880\mu_1'^{10};
\end{aligned}
\right\} \quad (3.42)
$$

and, for moments about the mean ($\mu_1' = 0$ in (3.42)),

$$
\left.
\begin{aligned}
\kappa_2 &= \mu_2, \\
\kappa_3 &= \mu_3, \\
\kappa_4 &= \mu_4 - 3\mu_2^2, \\
\kappa_5 &= \mu_5 - 10\mu_3\mu_2,
\end{aligned}
\right\} \quad (3.43)
$$

$$\left.\begin{aligned}
\kappa_6 &= \mu_6 - 15\mu_4\mu_2 - 10\mu_3^2 + 30\mu_2^3, \\
\kappa_7 &= \mu_7 - 21\mu_5\mu_2 - 35\mu_4\mu_3 + 210\mu_3\mu_2^2, \\
\kappa_8 &= \mu_8 - 28\mu_6\mu_2 - 56\mu_5\mu_3 - 35\mu_4^2 + 420\mu_4\mu_2^2 + 560\mu_3^2\mu_2 - 630\mu_2^4, \\
\kappa_9 &= \mu_9 - 36\mu_7\mu_2 - 84\mu_6\mu_3 - 126\mu_5\mu_4 + 756\mu_5\mu_2^2 + 2520\mu_4\mu_3\mu_2 \\
&\quad + 560\mu_3^3 - 7560\mu_3\mu_2^3, \\
\kappa_{10} &= \mu_{10} - 45\mu_8\mu_2 - 120\mu_7\mu_3 - 210\mu_6\mu_4 + 1260\mu_6\mu_2^2 - 126\mu_5^2 \\
&\quad + 5040\mu_5\mu_3\mu_2 + 3150\mu_4^2\mu_2 + 4200\mu_4\mu_3^2 - 18\,900\mu_4\mu_2^3 \\
&\quad - 37\,800\mu_3^2\mu_2^2 + 22\,680\mu_2^5.
\end{aligned}\right\} \tag{3.43}$$

Exercise 3.9 gives a more direct way of obtaining cumulants in terms of moments, and moments in terms of cumulants and moments, of the same or lower order.

3.15 The formal expression (3.30) defines the cumulants in terms of the moments, and it is thus evident that the cumulant of order r exists if the moments of orders r and lower exist. Analytically, we may expand the c.f. $\phi(t)$ in (3.31) and obtain

$$\begin{aligned}
\phi(t) &= \int_{-\infty}^{\infty} \left\{ \sum_0^r \frac{(itx)^j}{j!} + O\{|tx|^{r+1}/(r+1)!\} \right\} \mathrm{d}F \\
&= \sum_{j=0}^{r} \mu_j' (it)^j/j! + R_r
\end{aligned} \tag{3.44}$$

where R_r is a remainder term not greater than $\rho \nu_{r+1} |t|^{r+1}/(r+1)!$ with $|\rho| \leqslant 1$.

We may write this as

$$\phi(t) = \sum_{j=0}^{r} \mu_j' (it)^j/j! + o(t^r).$$

For some small t we may then take logarithms and expand to obtain

$$\log \phi(t) = \sum_{j=1}^{r} \kappa_j (it)^j/j! + o(t^r), \tag{3.45}$$

the coefficients κ being the cumulants by definition. Thus if the absolute moment ν_{r+1} exists, κ_r and lower cumulants exist.

Evaluation of cumulants

3.16 The cumulants are not, like the moments, directly ascertainable by summatory or integrative processes, and to find them it is necessary either to find the moments and employ equations (3.42–3) or to derive them from the c.f. via the c.g.f., its logarithm. The following examples will illustrate the processes involved.

Example 3.9 (Cumulants of the binomial)

Since the c.g.f. is

$$\psi(t) = \log \phi(t);$$

its derivatives with respect to *it* are

$$\psi'(t) = \phi'(t)/\phi(t),$$
$$\psi''(t) = \phi''(t)/\phi(t) - \{\phi'(t)/\phi(t)\}^2,$$

and so on. At $t = 0$, where $\phi(0) \equiv 1$, we obtain

$$\kappa_1 = \phi'(0) = \mu_1',$$
$$\kappa_2 = \phi''(0) - \{\phi'(0)\}^2 = \mu_2' - \mu_1'^2 = \mu_2.$$

Thus, for the binomial distribution of Example 3.2,

$$\phi(t) = M(it) = (pe^{it} + q)^n,$$
$$\psi(t) = n \log (pe^{it} + q),$$

and for the mean and variance

$$\kappa_1 = np$$
$$\kappa_2 = np + n(n-1)p^2 - (np)^2 = np(1-p)$$

as in Example 2.9.

Example 3.10 (Cumulants of the Poisson)

Consider the discrete (Poisson) distribution, to be discussed in **5.8–9** below, whose probabilities at $x = j$ ($j = 0, 1, 2, \ldots$) are $e^{-\lambda}\lambda^j/j!$. The c.f. is given by

$$\begin{aligned}\phi(t) &= e^{-\lambda}\sum_{j=0}^{\infty} \frac{\lambda^j}{j!} e^{itj} \\ &= e^{-\lambda}\exp(\lambda e^{it}) \\ &= \exp\{\lambda(e^{it}-1)\}.\end{aligned}$$

Since the variate is non-negative, for any r the absolute moment is the same as the ordinary moment, and we have

$$\mu_r' = e^{-\lambda}\sum_{j=0}^{\infty} \frac{\lambda^j j^r}{j!};$$

and since this converges(*) cumulants of all orders exist. They are therefore given by the expansion of $\log \phi(t)$ as a power series in t. Thus the c.g.f. is

$$\begin{aligned}\psi(t) = \log\phi(t) &= \lambda(e^{it}-1) \\ &= \lambda\sum_{j=1}^{\infty} (it)^j/j!\end{aligned}$$

(*) For the ratio of the $(n+1)$th term of the series to the nth is

$$\frac{\lambda^{n+1}(n+1)^r}{(n+1)!} \bigg/ \frac{\lambda^n n^r}{n!} = \frac{\lambda}{n+1}\left(1+\frac{1}{n}\right)^r = O\left(\frac{\lambda}{n}\right),$$

and thus the series converges for all finite values of λ.

and hence

$$\kappa_r = \lambda$$

for all r. Thus all cumulants of the distribution are equal to λ.

Example 3.11 (Cumulants of the normal)

In Example 3.4 we found the c.f. of the normal distribution,

$$f(x) = \frac{1}{\sigma\sqrt{(2\pi)}} \exp\left(-\frac{x^2}{2\sigma^2}\right), \qquad -\infty < x < \infty,$$

to be

$$\phi(t) = M(it) = \exp\left(-\tfrac{1}{2}\sigma^2 t^2\right).$$

Thus

$$\log \phi(t) = -\tfrac{1}{2}\sigma^2 t^2$$

from which the cumulants are

$$\kappa_1 = 0, \quad \kappa_2 = \sigma^2 \qquad \text{and} \quad \kappa_r = 0, \quad r > 2.$$

Example 3.12 (Cumulants of the Gamma)

In Example 3.6 it was found that for the distribution with p.d.f.

$$f(x) = \frac{a^\gamma}{\Gamma(\gamma)} e^{-ax} x^{\gamma-1}, \qquad 0 \leqslant x < \infty; \quad a, \gamma > 0,$$

the c.f. is given by

$$\phi(t) = \left(1 - \frac{it}{a}\right)^{-\gamma}.$$

It is readily verified that cumulants of all orders exist and hence

$$\begin{aligned}\kappa_r &= \text{coeff. of } \frac{(it)^r}{r!} \text{ in } -\gamma \log\left(1 - \frac{it}{a}\right) \\ &= \gamma(r-1)!\,a^{-r}.\end{aligned}$$

Example 3.13 (C.f. of Student's t)

Consider again the distribution of Example 3.3,

$$f(x) = \frac{k}{(1+x^2)^m}, \qquad -\infty < x < \infty; \quad m \geqslant 1.$$

The c.f. is given by

$$\phi(t) = k\int_{-\infty}^{\infty} \frac{e^{ixt}}{(1+x^2)^m}\, dx,$$

which, since sin xt is an odd function, reduces to

$$k\int_{-\infty}^{\infty}\frac{\cos xt}{(1+x^2)^m}\,dx.$$

This integral may be evaluated for integral m by complex integration round a contour consisting of the x-axis, the infinite semicircle above the x-axis and the infinitely small circle round the point $x = i$. It is found that

$$\phi(t)=\frac{k\pi}{2^{2m-2}(m-1)!}\,e^{-|t|}\sum_{j=0}^{m-1}(2|t|)^{m-1-j}(m-1+j)^{[2j]}/j!.$$

For $m = 1$, $k = \pi^{-1}$ by Example 3.3, and $\phi(t) = \exp\{-|t|\}$.

If $r < 2m-1$ the moment of order r exists by Example 3.3 and hence so does the cumulant of order r. But in this case we cannot expand $\log\phi(t)$ to an *infinite* series of powers of t, though this might perhaps be thought possible from the form of $\phi(t)$. In fact, we can only expand $\log\phi(t)$ in powers of t up to the point at which the derivatives of $\phi(t)$ exist at $t = 0$.

To simplify the discussion, consider the case $m = 2$. We have then, since $k = 2/\pi$ by Example 3.3,

$$\phi(t) = e^{-|t|}\{|t|+1\}$$
$$\log\phi(t) = -|t| + \log\{1+|t|\}.$$

If t is positive this equals

$$-\tfrac{1}{2}t^2+\tfrac{1}{3}t^3-\ldots$$

but if t is negative it equals

$$-\tfrac{1}{2}t^2-\tfrac{1}{3}t^3-\ldots$$

the two expressions differing in the sign of the term in t^3 and every second term thereafter. There is thus no unique expansion of $\log\phi(t)$ in powers of t about the point $t = 0$. There are two forms of the function expressing $\log\phi(t)$ according as t is positive or negative.

However, these expressions coincide as far as their terms in t and t^2, and the first and second derivatives of $\log\phi(t)$ are uniquely defined when $t = 0$. Thus the first and second cumulants exist and are given by

$$\kappa_1 = 0, \quad \kappa_2 = 1.$$

Cumulants of higher orders do not exist.

Factorial cumulants

3.17 Analogously to the relation (3.39) between cumulants and moments we may define a factorial cumulant $\kappa_{[r]}$ as the coefficient of $t^r/r!$ in the expansion of the logarithm of the factorial moment generating function. Thus if $P(t)$ is the probability generating function the factorial cumulant generating function is, from 3.11,

$$\omega(t) = \log P(1+t). \tag{3.46}$$

Like the factorial moment generating function, $\omega(t)$ is mainly of use for certain classes of discrete distribution.

The relations between cumulants and factorial cumulants are formally similar to those between moments and factorial moments. Taking a unit interval between values we have, analogously to (3.25) and (3.26),

$$\left.\begin{aligned}\kappa_{[1]} &= \kappa_1\\ \kappa_{[2]} &= \kappa_2 - \kappa_1\\ \kappa_{[3]} &= \kappa_3 - 3\kappa_2 + 2\kappa_1\\ \kappa_{[4]} &= \kappa_4 - 6\kappa_3 + 11\kappa_2 - 6\kappa_1\end{aligned}\right\} \tag{3.47}$$

Conversely

$$\left.\begin{aligned}\kappa_2 &= \kappa_{[2]} + \kappa_{[1]}\\ \kappa_3 &= \kappa_{[3]} + 3\kappa_{[2]} + \kappa_{[1]}\\ \kappa_4 &= \kappa_{[4]} + 6\kappa_{[3]} + 7\kappa_{[2]} + \kappa_{[1]}\end{aligned}\right\} \tag{3.48}$$

Example 3.14 (Factorial cumulants of the binomial)

Reverting once more to the binomial distribution of Example 3.8, we have

$$\omega(t) = n \log(1 + pt)$$

and hence

$$\kappa_{[r]} = (-1)^{r+1} n(r-1)!p^r.$$

Corrections for grouping (Sheppard's corrections)

3.18 When moments are calculated from a numerically specified distribution that is grouped, the results are approximate because the frequencies are assumed to be concentrated at the mid-points of intervals. It is possible to correct for this effect under certain conditions.

Let $f(x)$ be continuous and of finite range a to b; and let the range be divided into n intervals of length h. The probability in the jth interval, centred at $x_j = a + (j - \frac{1}{2})h$, is then given by

$$f_j = \int_{-\frac{1}{2}h}^{\frac{1}{2}h} f(x_j + \xi)\,d\xi. \tag{3.49}$$

The moments calculated from the grouped values, which we will denote by a bar, are then given by

$$\bar{\mu}'_r = \sum_{j=1}^{n} x_j^r \int_{-\frac{1}{2}h}^{\frac{1}{2}h} f(x_j + \xi)\,d\xi. \tag{3.50}$$

Now from one version of the Euler–Maclaurin sum formula(*) we have, for a function $y(x)$ with $2m$ derivatives,

$$\frac{1}{h}\int_a^b y(x)\,dx = [y(a + \tfrac{1}{2}h) + y(a + \tfrac{3}{2}h) + \ldots + y\{a + (n - \tfrac{1}{2})h\}]$$
$$- \sum_{j=1}^{m} \frac{h^{2j-1}}{(2j)!} B_{2j}(\tfrac{1}{2})[y^{(2j-1)}(x)]_a^b - S_{2m} \tag{3.51}$$

(*) See, for example, Milne-Thomson, *The Calculus of Finite Differences*, section 7.5.

where $y^{(r)}$ is the rth derivative of y, $B_{2m}(\frac{1}{2})$ is the value of the $2m$th Bernoulli polynomial (defined in 3.24 below) for argument $\frac{1}{2}$ and S_{2m} is a remainder term which can be expressed as

$$S_{2m} = \frac{nh^{2m}}{(2m)!} B_{2m}(\tfrac{1}{2}) y^{(2m)}(a+nh\theta), \qquad 0<\theta<1. \tag{3.52}$$

Putting

$$y(x) = x^r \int_{-\frac{1}{2}h}^{\frac{1}{2}h} f(x+\xi)\,d\xi$$

in (3.51) and neglecting S_{2m} we have, if the first $2m-3$ derivatives of y vanish at the terminals a and b,

$$\frac{1}{h}\int_a^b x^r \int_{-\frac{1}{2}h}^{\frac{1}{2}h} f(x+\xi)\,d\xi\,dx = \bar{\mu}_r'.$$

Hence

$$\begin{aligned}\bar{\mu}_r' &= \frac{1}{h}\int_a^b \int_{-\frac{1}{2}h}^{\frac{1}{2}h} (u-\xi)^r f(u)\,du\,d\xi \\ &= \frac{1}{h}\int_a^b \frac{(u+\frac{1}{2}h)^{r+1}-(u-\frac{1}{2}h)^{r+1}}{r+1} f(u)\,du \\ &= \sum_{j=0}^{[\frac{1}{2}r]} \binom{r}{2j} (\tfrac{1}{2}h)^{2j} \frac{1}{2j+1} \mu_{r-2j}'\end{aligned} \tag{3.53}$$

where $[\frac{1}{2}r]$ is the integral part of $\frac{1}{2}r$. This gives the raw (i.e., uncorrected) moments in terms of the actual moments. In practice we require the latter in terms of the former, and it is easy to find from (3.53) the following expressions:

$$\left.\begin{aligned}\mu_1' &= \bar{\mu}_1' \\ \mu_2' &= \bar{\mu}_2' - \tfrac{1}{12}h^2 \\ \mu_3' &= \bar{\mu}_3' - \tfrac{1}{4}\bar{\mu}_1' h^2 \\ \mu_4' &= \bar{\mu}_4' - \tfrac{1}{2}\bar{\mu}_2' h^2 + \tfrac{7}{240}h^4 \\ \mu_5' &= \bar{\mu}_5' - \tfrac{5}{6}\bar{\mu}_3' h^2 + \tfrac{7}{48}\bar{\mu}_1' h^4 \\ \mu_6' &= \bar{\mu}_6' - \tfrac{5}{4}\bar{\mu}_4' h^2 + \tfrac{7}{16}\bar{\mu}_2' h^4 - \tfrac{31}{1344}h^6\end{aligned}\right\} \tag{3.54}$$

It will be seen by putting $\bar{\mu}_1' = 0$ that (3.54) gives for the central moments

$$\left.\begin{aligned}\mu_2 &= \bar{\mu}_2 - h^2/12, \\ \mu_3 &= \bar{\mu}_3, \\ \mu_4 &= \bar{\mu}_4 - \bar{\mu}_2 h^2/2 + 7h^4/240.\end{aligned}\right\} \tag{3.55}$$

The general expression for these formulae is

$$\mu_r' = \sum_{j=0}^{r} \left\{ \binom{r}{j} (2^{1-j}-1) B_j h^j \bar{\mu}_{r-j}' \right\} \tag{3.56}$$

where B_j is the Bernoulli number[*] of order j (Wold, 1934a). (See also Exercise 3.17.)

[*] See 3.24 below.

3.19 We have made various assumptions in arriving at these formulae: that there is high-order contact at the terminals of the range, that the range itself is finite and that the remainder term may be neglected. In absolute magnitude $B_{2m}(\frac{1}{2})/(2m)!$ is less than $4/(2\pi)^{2m}$ and thus $|S_{2m}|$ in (3.52) is less than $4nh^{2m}/(2\pi)^{2m}$ multiplied by some value of $y^{(2m)}(x)$ in the range a to b. The extent to which the remainder term is negligible then depends on the behaviour of $f^{(2m-1)}(x)$ in the range a to b.

In practice the corrections result in improvements if there is reasonably high-order contact, but may break down if there is not; see Exercise 3.10. For distributions of infinite range the frequency tails off near the ends and we may still apply the corrections with some confidence to those distributions (in practice, of course, necessarily of finite range) where the major part of the distribution is concentrated and the frequencies at the extremes are small.

Example 3.15 (Use of Sheppard's corrections)

Consider the distribution

$$f(x) = \frac{1}{B(12, 6)} x^{11}(1-x)^5, \qquad 0 \leqslant x \leqslant 1,$$

known as the Beta or Type I distribution, to be discussed in **6.5–6** below. It is unimodal and touches the x-axis at its terminals. The exact probabilities for intervals of $h = 0.1$ may be obtained from the Tables of the Incomplete B-Function, and are as follows:-

Centre of interval	Probability
0.05	0.000 000 0
0.15	0.000 009 2
0.25	0.000 646 8
0.35	0.009 938 2
0.45	0.061 137 4
0.55	0.192 199 6
0.65	0.332 887 7
0.75	0.297 479 9
0.85	0.101 033 7
0.95	0.004 667 5
Total	1.000 000 0

The raw moments about $x = 0$ are shown in the second column of the following table:

Moment	Raw	Exact	Corrected
μ_1'	0.666 662 8	0.666 666 7	0.666 662 8
μ_2'	0.456 965 5	0.456 140 4	0.456 132 2
μ_3'	0.320 952 3	0.319 298 2	0.319 285 7
μ_4'	0.230 335 1	0.228 070 2	0.228 053 2
μ_5'	0.168 512 9	0.165 869 2	0.165 848 0
μ_6'	0.125 433 2	0.122 599 0	0.122 574 0

The exact values in column 3 are calculated by direct integration. It is to be noted that the corrected second and third moments in column 4 are more accurate than the order of the terms, $h^2/12$ and $\bar{\mu}_1' h^2/4$, used in making the corrections. The corrected fourth moment is in error by a term of order 2×10^{-5}, which is of the same magnitude as one of the correcting terms, $7h^4/240$, used in arriving at it; similarly for the errors in the corrected fifth and sixth moments.

Example 3.16 (Invalid use of Sheppard's corrections)

As an illustration of the way in which Sheppard's corrections break down when the condition for high-order contact is violated, an example is taken from a paper by Pairman and Pearson (1919). The following table shows the probabilities for the normal distribution truncated to the range $1.25 \leqslant x \leqslant 5.25$ and grouped into 8 intervals of length 0.5 centred at the values shown:

$$f(x) = 3.77607\, e^{-\frac{1}{2}x^2}$$

Interval centred at	Probability
1.5	0.620 83
2.0	0.263 45
2.5	0.087 51
3.0	0.022 74
3.5	0.004 63
4.0	0.000 74
4.5	0.000 09
5.0	0.000 01

These probabilities may be obtained from tables of the integral like Appendix Table 2. The distribution has high-order contact at the upper terminal but not at the lower, being in fact J-shaped and very abrupt at that point.

The following table shows the raw moments about the mean up to the fourth, the exact moments calculated from the continuous distribution, and the corrected moments using (3.55):

Moment	Raw	Exact	Corrected
μ_2	0.158 524	0.172 222	0.137 691
μ_3	0.104 226	0.098 612	0.104 226
μ_4	0.149 090	0.156 405	0.131 097

It will be noted that the two non-zero corrections operate in the wrong direction. For the fourth moment they increase the difference between calculated and true values from about 4 percent to about 16 percent. It is clear that, at least for fairly coarse grouping as in this example, Sheppard's corrections may fail completely.

Average corrections

3.20 There is a distinct type of problem that also leads to the Sheppard corrections. Consider a distribution with probabilities falling into intervals of known equal length but unknown location. One may ask what are the corrections to be applied to the raw moments so as to bring them *on the average* into closer relation with the true moments. In other words, supposing that the interval mesh is located at random on the distribution, what are the average values of the raw moments?

Let $x_j \pm h/2$ be the end-points of the intervals and $f(x)$ be the density function. By definition,

$$\bar{\mu}_r' = \sum_{j=-\infty}^{\infty} \left\{ x_j^r \int_{-\frac{1}{2}h}^{\frac{1}{2}h} f(x_j + \xi)\, d\xi \right\}.$$

We need only consider the variation of the interval mesh over a length h, since any variation can be reduced (mod h) without affecting the problem. Denoting by $E(\bar{\mu}_r')$ the average as x_j varies from $X_j - \frac{1}{2}h$ to $X_j + \frac{1}{2}h$, we have

$$\begin{aligned} E(\bar{\mu}_r') &= \frac{1}{h} \int_{X_j-\frac{1}{2}h}^{X_j+\frac{1}{2}h} \sum \left\{ x_j^r \int_{-\frac{1}{2}h}^{\frac{1}{2}h} f(x_j + \xi)\, d\xi \right\} dx_j \\ &= \frac{1}{h} \sum_{j=-\infty}^{\infty} \int_{X_j-\frac{1}{2}h}^{X_j+\frac{1}{2}h} x_j^r \int_{-\frac{1}{2}h}^{\frac{1}{2}h} f(x_j + \xi)\, d\xi\, dx_j \\ &= \frac{1}{h} \int_{-\infty}^{\infty} x^r \int_{-\frac{1}{2}h}^{\frac{1}{2}h} f(x + \xi)\, d\xi\, dx, \end{aligned} \qquad (3.57)$$

which is the same as above (3.53) with the substitution of $E(\bar{\mu}_r')$ for $\bar{\mu}_r'$. Thus the Sheppard corrections apply for the average grouped moments, whatever the nature of the terminal contact.

3.21 They cannot, however, be applied indiscriminately on that ground. In place of the conditions about terminal contact and bounded derivatives, which ensure the applicability of Sheppard's corrections to any particular distribution, there is the condition that the grouping intervals are located at random on the range, which implies that although the corrections may be wrong in any given instance, the average effect in a large number of cases will be correct. The condition about the random location of grouping does not operate very frequently for J- and U-shaped distributions, where the Sheppard corrections would not ordinarily apply; for instance, in a distribution of incomes or deaths at given ages it is almost inevitable to begin the grouping at zero.

3.22 It is also illegitimate to drop the primes in (3.57) in order to obtain average corrections for moments about the observed mean. If the mean of the grouped distribution is denoted by m, the average value of the rth moment about the mean is given by

$$E(\bar{\mu}_r) = \frac{1}{h} \int_{-\infty}^{\infty} (x - m)^r \int_{-\frac{1}{2}h}^{\frac{1}{2}h} f(x + \xi)\, d\xi\, dx,$$

where m is a function of x and we may not proceed as before. Explicit expressions for average corrections to moments about the mean have not been developed. From a consideration of some particular distributions, however, Kendall (1938) concluded that for all ordinary purposes it is sufficient to use equations (3.54) as if the mean were a fixed point.

Average corrections may also be applied to discrete data that have been grouped in wider intervals, but are different from those of the continuous case. Cf. Exercise 3.13 and C. C. Craig (1936b). For corrections when the Sheppard conditions are violated see Pairman and Pearson (1919), Sandon (1924), Martin (1934), Elderton (1938) and Exercise 3.10.

Sheppard's corrections for factorial moments

3.23 It has been shown by Wold (1934a) that for factorial moments the Sheppard corrections are as follows:-

$$\left.\begin{aligned}
\mu'_{[1]} &= \bar{\mu}'_{[1]} \\
\mu'_{[2]} &= \bar{\mu}'_{[2]} - \frac{h^2}{12} \\
\mu'_{[3]} &= \bar{\mu}'_{[3]} - \frac{h^2}{4}\bar{\mu}'_{[1]} + \frac{h^3}{4} \\
\mu'_{[4]} &= \bar{\mu}'_{[4]} - \frac{h^2}{2}\bar{\mu}'_{[2]} + h^3\bar{\mu}'_{[1]} - \frac{71}{80}h^4 \\
\mu'_{[5]} &= \bar{\mu}'_{[5]} - \frac{5}{6}\bar{\mu}'_{[3]}h^2 + \frac{5}{2}\bar{\mu}'_{[2]}h^3 - \frac{71}{16}\bar{\mu}'_{[1]}h^4 + \frac{31}{8}h^5 \\
\mu'_{[6]} &= \bar{\mu}'_{[6]} - \frac{5}{4}\bar{\mu}'_{[4]}h^2 + 5\bar{\mu}'_{[3]}h^3 - \frac{213}{16}\bar{\mu}'_{[2]}h^4 \\
&\quad + \frac{93}{4}\bar{\mu}'_{[1]}h^5 - \frac{9129}{448}h^6
\end{aligned}\right\} \tag{3.58}$$

and in general are given by

$$\mu'_{[r]} = \sum_{j=0}^{r} \binom{r}{j} B_j^{(j+2)}(\tfrac{3}{2})h^j\bar{\mu}'_{[r-j]}, \tag{3.59}$$

where the Bernoulli polynomial $B_j^{(j+2)}(\frac{3}{2})$ defined at (3.65) below is equal to

$$(-1)^{j+1}\frac{(2j)!}{2^{2j}(j+1)!}\left(\tfrac{1}{3}+\tfrac{1}{5}+\ldots+\frac{1}{2j-1}\right), \quad j>1,$$

and

$$B_0^{(2)}(\tfrac{3}{2}) = 1, \qquad B_1^{(3)}(\tfrac{3}{2}) = 0.$$

Benoulli numbers and polynomials and Gamma function formulae

3.24 The Bernoulli numbers and (to a smaller extent) the Bernoulli polynomials have several important applications in statistics. The number of order j, B_j, is defined as the coefficient of $t^j/j!$ in $t/(e^t-1)$. Thus

$$\frac{t}{e^t-1} = \sum_{j=0}^{\infty}\frac{B_j t^j}{j!} = e^{\mathbf{B}t}, \tag{3.60}$$

where $\mathbf{B}$ is an operator such that $\mathbf{B}^j = B_j$. From (3.60),

$$t + e^{\mathbf{B}t} = e^{(\mathbf{B}+1)t}$$

so that on equating coefficients of $t^j/!$ we find the recurrence relation

$$B_j = (\mathbf{B}+1)^j, \qquad j>1.$$

Clearly, $B_0 = 1$, and the recurrence relation gives $B_2 = B_2 + 2B_1 + 1$, so $B_1 = -\frac{1}{2}$. Now, writing (3.60) as

$$\frac{t}{e^t - 1} + \tfrac{1}{2}t = 1 + \sum_{j=2}^{\infty} \frac{B_j t^j}{j!},$$

the left side is seen to be an even function of t, so all odd $B_{2j+1} = 0$ for $j > 0$. The recurrence relation gives the first few even B_{2j} as

$$B_2 = 1/6; \quad B_4 = -1/30; \quad B_6 = 1/42; \quad B_8 = -1/30; \quad B_{10} = 5/66;$$
$$B_{12} = -691/2730; \quad B_{14} = 7/6.$$

Now putting (3.60) in the form

$$\frac{1}{e^t - 1} - \frac{1}{t} + \frac{1}{2} = \sum_{j=2}^{\infty} \frac{B_j t^{j-1}}{j!}$$

and integrating from 0 to t, we find

$$\log\{(\sinh \tfrac{1}{2}t)/\tfrac{1}{2}t\} = \sum_{j=2}^{\infty} B_j t^j/(j!j) = \sum_{j=1}^{\infty} B_{2j} t^{2j}/\{(2j)!2j\}. \tag{3.61}$$

An expansion of frequent use is the so-called Stirling's series:

$$\log(x-1)! = \log\Gamma(x) = (x - \tfrac{1}{2})\log x - x + \tfrac{1}{2}\log(2\pi) + \sum_{j=1}^{\infty} \frac{B_{2j}}{2j(2j-1)x^{2j-1}}. \tag{3.62}$$

This is an asymptotic expansion in the sense that the true value of $\log\Gamma(x)$, for x real and positive, always lies between the sum of n and $n+1$ terms, whatever n may be. In particular

$$\log\Gamma(x) = (x - \tfrac{1}{2})\log x - x + \tfrac{1}{2}\log(2\pi) + \frac{1}{12x} - \frac{1}{360x^3} + \frac{1}{1260x^5} - \frac{1}{1680x^7} + \frac{1}{1188x^9} - \ldots \tag{3.63}$$

and

$$(x-1)! = \Gamma(x) = e^{-x}x^{x-\frac{1}{2}}\sqrt{(2\pi)}\left\{1 + \frac{1}{12x} + \frac{1}{288x^2} - \frac{139}{51\,840x^3} - \frac{571}{2\,488\,320x^4}\cdots\right\}. \tag{3.64}$$

The first term is often called Stirling's approximation to the factorial.

The Bernoulli polynomial of order n and degree j in x, $B_j^{(n)}(x)$, is defined by

$$e^{xt}t^n/(e^t - 1)^n = \sum_{j=0}^{\infty} t^j B_j^{(n)}(x)/j! \tag{3.65}$$

If the order is unstated, it is $n = 1$.

Starting from $\{\Gamma(p)\}^2/\Gamma(2p) = B(p,p) = 2\int_0^{1/2}\{u(1-u)\}^{p-1}\,du$, the substitution $v = 4u(1-u)$ yields $2^{1-2p}B(p, \frac{1}{2})$, so we find Legendre's duplication formula for Gamma functions

$$\Gamma(p)\Gamma(p + \tfrac{1}{2}) = \Gamma(2p)\Gamma(\tfrac{1}{2})/2^{2p-1}. \tag{3.66}$$

Finally, the integral

$$I = \Gamma(z)\Gamma(1-z) = B(z, 1-z) = \int_0^{\infty} \frac{x^{z-1}\,dx}{1+x}$$

may be evaluated by a complex integration over a contour consisting of two circles c, C

with radii r, R centred at the origin, and the segment of the positive real axis, $r<x<R$, which is traversed once in each direction, so that the contour consists of rR, C counter-clockwise, Rr and c clockwise. As $R\to\infty$, the integral on $C\to 0$ if $z<1$, and as $r\to 0$ the integral on $c\to 0$ if $z>0$. The integral on rR then $\to I$ and that on $Rr\to -e^{(z-1)2\pi i}I$ owing to the previous circuit of the origin. The only singularity is at $x=-1$, with residue $(-1)^{z-1}=e^{(z-1)\pi i}$. Thus

$$\Gamma(z)\Gamma(1-z)=\frac{2\pi i\cdot e^{(z-1)\pi i}}{1-e^{(z-1)2\pi i}}=\frac{\pi}{\sin \pi z}. \tag{3.67}$$

This proof holds only for $0<z<1$, but (3.67) is valid much more generally, for complex $z\neq 0$ or an integer. A different method of proof is necessary for the general result.

Sheppard's corrections for cumulants

3.25 As in **3.18** we have, writing θ for it, the c.f. of the grouped distribution

$$\begin{aligned}\sum\left\{e^{\theta x_j}\int_{-\frac{1}{2}h}^{\frac{1}{2}h}f(x_j+\xi)\,d\xi\right\}&=\frac{1}{h}\int_{-\infty}^{\infty}e^{\theta x}\,dx\int_{-\frac{1}{2}h}^{\frac{1}{2}h}f(x+\xi)\,d\xi\\&=\frac{1}{h}\int_{-\frac{1}{2}h}^{\frac{1}{2}h}e^{-\theta\xi}\,d\xi\int_{-\infty}^{\infty}e^{\theta x}f(x)\,dx\\&=\frac{\sinh\frac{1}{2}\theta h}{\frac{1}{2}\theta h}\int_{-\infty}^{\infty}e^{\theta x}f(x)\,dx\end{aligned}$$

the integral being the true c.f. Taking logarithms and recalling (3.61) we have, from the coefficients of $\theta^r/r!$,

$$\kappa_r=\bar{\kappa}_r-B_r h^r/r,\qquad r>1,$$

an attractively simple result for the Sheppard corrections to cumulants. Since $B_{2j+1}=0$ for $j>0$ and κ_1 is equal to the mean, no cumulant of odd order needs any correction: μ_1' and μ_3 in (3.54–5) are now seen to be special cases of this general rule. For the even cumulants we have

$$\left.\begin{aligned}\kappa_2&=\bar{\kappa}_2-\frac{h^2}{12}\\\kappa_4&=\bar{\kappa}_4+\frac{h^4}{120}\\\kappa_6&=\bar{\kappa}_6-\frac{h^6}{252}\end{aligned}\right\}$$

which of course agree with the results derived from (3.54–5).

Negative and fractional moments

3.26 We saw in **2.27** that expected values may be defined quite generally, provided the expectations exist. In particular, we may define $\mu_c'=E(x^c)$ to be a *negative* (*integer*) moment when c is a negative integer and a *fractional* moment when c is non-integral. Thus $c=-1$ corresponds to the harmonic mean whereas $\lim_{c\to 0}\left\{\frac{1}{c}(\mu_c'-1)\right\}$ gives the geometric mean (cf. 2.6 above).

Although these moments are much less used than the ordinary moments, they may be valuable when the ordinary moments do not exist.

When $x>0$, the negative and fractional moments may be obtained from the moment-generating function for x, provided that it exists for $-\infty<t\leqslant 0$. Let $y=x^{-1}$ and let the m.g.f. of x be $M_x(t)$. Since

$$y=x^{-1}=\int_{-\infty}^{0} e^{tx}\,dt,$$

it follows that

$$\begin{aligned} E(y) &= \int_0^\infty x^{-1}\,dF=\int_0^\infty\int_{-\infty}^{0} e^{tx}\,dt\,dF \\ &= \int_{-\infty}^{0}\int_0^\infty e^{tx}\,dF\,dt=\int_{-\infty}^{0} M_x(t)\,dt \\ &= \int_0^\infty M_x(-t)\,dt. \end{aligned} \tag{3.68}$$

Similarly

$$E(y^c)=\int_0^\infty t^{c-1}M_x(-t)\,dt/\Gamma(c),$$

provided the expectation exists. This result is due to Cressie *et al.* (1981).

J. F. Hannan (1985) points out that $\lim_{x\downarrow 0}|\log x|^n f(x)<\infty$ for some $n>1$ is a sufficient condition for the existence of $E(x^{-1})$.

Example 3.17 (Negative moments of the Gamma)

The Gamma distribution considered in Example 3.12 has m.g.f.

$$M_x(t)=\phi\left(\frac{t}{i}\right)=a^\gamma(a-t)^{-\gamma}$$

so that

$$E(y^c)=\int_0^\infty \frac{t^{c-1}a^\gamma}{(a+t)^\gamma}\,dt/\Gamma(c).$$

If we make the transformation $u=t/(a+t)$, this becomes

$$\begin{aligned} E(y^c) &= a^c\int_0^1 u^{c-1}(1-u)^{\gamma-c-1}\,du/\Gamma(c) \\ &= a^c\Gamma(\gamma-c)/\Gamma(\gamma), \end{aligned}$$

provided $\gamma>c$. In particular, $E(x^{-1})=a(\gamma-1)^{-1}$.

When $M_x(t)$ does not exist, the fractional moments may still be derived directly.

Example 3.18 (Negative moments of the Cauchy)

The Cauchy distribution (the special case $m = 1$ in Examples 3.3 and 3.13) has p.d.f.

$$f(x) = \frac{1}{\pi(1+x^2)}, \quad -\infty < x < \infty.$$

For $|c| < 1$,

$$\nu_c' = E\{|x|^c\} = \frac{2}{\pi}\int_0^\infty \frac{x^c}{(1+x^2)}\,dx.$$

Putting $u = (1+x^2)^{-1}$, we obtain

$$\begin{aligned}\nu_c' &= \pi^{-1}\int_0^1 u^{-(1+c)/2}(1-u)^{(c-1)/2}\,du \\ &= \pi^{-1}\Gamma\{\tfrac{1}{2}(1+c)\}\Gamma\{\tfrac{1}{2}(1-c)\} \\ &= \left[\sin\left\{\frac{\pi}{2}(1+c)\right\}\right]^{-1},\end{aligned}$$

using (3.67). Laue (1980) considers relationships between fractional moments and fractional derivatives of the c.f. Wooff (1985) gives some bounds on reciprocal moments.

Multivariate moments and cumulants

3.27 The foregoing results in this chapter may be readily generalized to the multivariate case. To avoid complicating the algebraic expressions, we shall deal mainly with two variates x_1 and x_2; generalizations to more variates are straightforward.

The bivariate moment μ_{rs}' about an origin a_1 for x_1 and a_2 for x_2 is defined by

$$\mu_{rs}' = \int_{-\infty}^{\infty}\int_{-\infty}^{\infty}(x_1-a_1)^r(x_2-a_2)^s\,dF. \tag{3.69}$$

Evidently, μ_{r0}' and μ_{0s}' denote respectively the rth moment of (the marginal distribution of) x_1 and the sth moment of x_2. If $r \neq 0 \neq s$, μ_{rs}' is called a *product moment*. By (2.39), $\mu_{rs}' = \mu_{r0}'\mu_{0s}'$ if x_1 and x_2 are independent.

As in the univariate case, bivariate moments about certain points can be expressed in terms of those about other points. If the x_1 origin is transferred from a_1 to b_1, where $c_1 = b_1 - a_1$, and the x_2 origin from a_2 to b_2, where $c_2 = b_2 - a_2$, we have

$$\mu_{rs}'(a_1, a_2) = (\mu' + c_1)^r(\mu' + c_2)^s \tag{3.70}$$

where the product $\mu'^j\mu'^k$ on the right is to be replaced by $\mu_{jk}'(b_1, b_2)$. This corresponds to (3.5) in the univariate case. If the variate means are taken as origins, μ_{rs} is written without the prime as before. μ_{11} is called the *covariance* of x_1 and x_2 and is important in the theory of correlation.

Methods of calculating the product moments for numerically specified distributions will be considered in Volume 2A when we discuss the theory of correlation. The determination of bivariate moments for a mathematically specified population is a matter of evaluating double sums or double integrals, and no new statistical points call for comment.

Example 3.19 (M.g.f. of the binormal)

Consider again the bivariate distribution of Example 1.1,

$$f(x_1, x_2) = \frac{1}{2\pi\sigma_1\sigma_2(1-\rho^2)^{\frac{1}{2}}} \exp\left\{-\frac{1}{2(1-\rho^2)}\left(\frac{x_1^2}{\sigma_1^2} - \frac{2\rho x_1 x_2}{\sigma_1\sigma_2} + \frac{x_2^2}{\sigma_2^2}\right)\right\},$$

$$\rho^2 < 1; \quad \sigma_1, \sigma_2 > 0; \quad -\infty < x_1, x_2 < \infty.$$

Let us evaluate the bivariate m.g.f.

$$M(t_1, t_2) = \int_{-\infty}^{\infty}\int_{-\infty}^{\infty} e^{x_1 t_1 + x_2 t_2}\,dF.$$

Making the substitution

$$\xi = x_1 - \sigma_1^2 t_1 - \rho\sigma_1\sigma_2 t_2$$
$$\eta = x_2 - \rho\sigma_1\sigma_2 t_1 - \sigma_2^2 t_2$$

we find

$$M(t_1, t_2) = \exp\{\tfrac{1}{2}(t_1^2\sigma_1^2 + 2t_1t_2\sigma_1\sigma_2\rho + t_2^2\sigma_2^2)\}$$
$$\times \frac{1}{2\pi\sigma_1\sigma_2(1-\rho^2)^{\frac{1}{2}}}\int_{-\infty}^{\infty}\int_{-\infty}^{\infty} \exp\left\{-\frac{1}{2(1-\rho^2)}\left(\frac{\xi^2}{\sigma_1^2} - \frac{2\rho\xi\eta}{\sigma_1\sigma_2} + \frac{\eta^2}{\sigma_2^2}\right)\right\} d\xi\, d\eta$$
$$= \exp\{\tfrac{1}{2}(t_1^2\sigma_1^2 + 2\rho\sigma_1\sigma_2 t_1 t_2 + t_2^2\sigma_2^2)\}.$$

Now μ_{rs} is the coefficient of $\dfrac{t_1^r}{r!}\dfrac{t_2^s}{s!}$ in $M(t_1, t_2)$ and thus we find, for instance,

$$\left.\begin{aligned}
&\mu_{20} = \sigma_1^2, \quad \mu_{11} = \rho\sigma_1\sigma_2, \quad \mu_{02} = \sigma_2^2;\\
&\mu_{30} = \mu_{21} = \mu_{12} = \mu_{03} = 0;\\
&\mu_{40} = 3\sigma_1^4, \quad \mu_{31} = 3\rho\sigma_1^3\sigma_2, \quad \mu_{22} = (1+2\rho^2)\sigma_1^2\sigma_2^2,\\
&\mu_{13} = 3\rho\sigma_1\sigma_2^3, \quad \mu_{04} = 3\sigma_2^4.
\end{aligned}\right\} \tag{3.71}$$

3.28 The bivariate analogue of equation (3.30) is

$$\exp\left\{\sum_{r=0}^{\infty}\sum_{s=0}^{\infty} \kappa_{rs} t_1^r t_2^s/(r!\,s!)\right\} = \sum_{r=0}^{\infty}\sum_{s=0}^{\infty} \mu'_{rs} t_1^r t_2^s/(r!\,s!) \tag{3.72}$$

where we define $\kappa_{00} \equiv 0$. The joint c.f. of the two variates is

$$\phi(t_1, t_2) = \int_{-\infty}^{\infty}\int_{-\infty}^{\infty} e^{ix_1t_1 + ix_2t_2}\,dF \tag{3.73}$$

and, as before,

$$\phi(t_1, t_2) = \sum_{r,s=0}^{\infty} \mu'_{rs} \frac{(it_1)^r}{r!} \frac{(it_2)^s}{r!}$$
$$= \exp\left\{\sum_{r,s=0}^{\infty} \kappa_{rs} \frac{(it_1)^r}{r!} \frac{(it_2)^s}{s!}\right\}, \quad (3.74)$$

provided that the expansions are permissible. If $r \neq 0$, $s \neq 0$, κ_{rs} is called a *product cumulant*.

From these equations the bivariate moments can be expressed in terms of bivariate cumulants and vice versa. It is also possible to derive bivariate equations from the univariate equations by the following simple symbolic process.

3.29 Consider the equation

$$\mu'_4 = \kappa_4 + 4\kappa_3\kappa_1 + 3\kappa_2^2 + 6\kappa_2\kappa_1^2 + \kappa_1^4 \quad (3.75)$$

in (3.37). Write this formally as

$$\mu'(r^4) = \kappa(r^4) + 4\kappa(r^3)\kappa(r) + 3\{\kappa(r^2)\}^2 + 6\kappa(r^2)\{\kappa(r)\}^2 + \{\kappa(r)\}^4.$$

Treat this as a function of r and operate by $s\dfrac{\partial}{\partial r}$, giving

$$4\mu'(r^3s) = 4\kappa(r^3s) + 12\kappa(r^2s)\kappa(r) + 4\kappa(r^3)\kappa(s) + 12\kappa(rs)\kappa(r^2)$$
$$+ 12\kappa(rs)\{\kappa(r)\}^2 + 12\kappa(r^2)\kappa(r)\kappa(s) + 4\{\kappa(r)\}^3\kappa(s).$$

Divide through by a factor 4 and replace r and s by suffixes referring to the first and second variate. We obtain

$$\mu'_{31} = \kappa_{31} + 3\kappa_{21}\kappa_{10} + \kappa_{30}\kappa_{01} + 3\kappa_{11}\kappa_{20} + 3\kappa_{11}\kappa_{10}^2 + 3\kappa_{20}\kappa_{10}\kappa_{01} + \kappa_{10}^3\kappa_{01}.$$

This is the expression of μ'_{31} in terms of cumulants. The process is quite general and can be justified—cf. Kendall (1940c). It also applies to the expression of cumulants in terms of moments and to formulae involving central moments. For example, from

$$\mu_4 = \kappa_4 + 3\kappa_2^2 \quad (3.76)$$

in (3.38) we derive, in the same manner,

$$\mu_{31} = \kappa_{31} + 3\kappa_{11}\kappa_{20}. \quad (3.77)$$

A further operation of the same kind will yield

$$\mu_{22} = \kappa_{22} + \kappa_{20}\kappa_{02} + 2\kappa_{11}^2. \quad (3.78)$$

Similarly, we may use the symbolic process to generate, e.g., the trivariate result

$$\mu_{211} = \kappa_{211} + \kappa_{200}\kappa_{011} + 2\kappa_{110}\kappa_{101} \quad (3.79)$$

from either (3.77) or (3.78) by using the operator $t\dfrac{\partial}{\partial r}$ on $\mu(r^3s)$ or $\mu(r^2s^2)$.

The reverse process is much simpler, for suffixes to μ and κ need only be added correspondingly to produce the univariate from the bivariate formulae, or the bivariate from the trivariate. Thus, adding suffixes in (3.77) or (3.78) gives us (3.76), while addition of pairs of suffixes in (3.79) gives (3.77) and (3.78). For the reason, see **13.11–14** below.

The following formulae are sometimes required:

$$\left.\begin{aligned}
&\mu_{11}=\kappa_{11};\quad \mu_{21}=\kappa_{21};\quad \mu_{31}=\kappa_{31}+3\kappa_{20}\kappa_{11};\\
&\mu_{22}=\kappa_{22}+\kappa_{20}\kappa_{02}+2\kappa_{11}^2;\quad \mu_{41}=\kappa_{41}+4\kappa_{30}\kappa_{11}+6\kappa_{21}\kappa_{20};\\
&\mu_{32}=\kappa_{32}+\kappa_{30}\kappa_{02}+6\kappa_{21}\kappa_{11}+3\kappa_{20}\kappa_{12};\\
&\mu_{51}=\kappa_{51}+5\kappa_{40}\kappa_{11}+10\kappa_{31}\kappa_{20}+10\kappa_{30}\kappa_{21}+15\kappa_{20}^2\kappa_{11};\\
&\mu_{42}=\kappa_{42}+\kappa_{40}\kappa_{02}+8\kappa_{31}\kappa_{11}+4\kappa_{30}\kappa_{12}+6\kappa_{22}\kappa_{20}+6\kappa_{21}^2+3\kappa_{20}^2\kappa_{02}+12\kappa_{20}\kappa_{11}^2;\\
&\mu_{33}=\kappa_{33}+3\kappa_{31}\kappa_{02}+\kappa_{30}\kappa_{03}+9\kappa_{22}\kappa_{11}+9\kappa_{21}\kappa_{12}+3\kappa_{20}\kappa_{13}+9\kappa_{20}\kappa_{11}\kappa_{02}+6\kappa_{11}^3.
\end{aligned}\right\}\quad (3.80)$$

$$\left.\begin{aligned}
&\kappa_{31}=\mu_{31}-3\mu_{20}\mu_{11};\quad \kappa_{22}=\mu_{22}-\mu_{20}\mu_{02}-2\mu_{11}^2;\\
&\kappa_{41}=\mu_{41}-4\mu_{30}\mu_{11}-6\mu_{21}\mu_{20};\\
&\kappa_{32}=\mu_{32}-\mu_{30}\mu_{02}-6\mu_{21}\mu_{11}-3\mu_{20}\mu_{12};\\
&\kappa_{51}=\mu_{51}-5\mu_{40}\mu_{11}-10\mu_{31}\mu_{20}-10\mu_{30}\mu_{21}+30\mu_{20}^2\mu_{11};\\
&\kappa_{42}=\mu_{42}-\mu_{40}\mu_{02}-8\mu_{31}\mu_{11}-4\mu_{30}\mu_{12}-6\mu_{22}\mu_{20}-6\mu_{21}^2+6\mu_{20}^2\mu_{02}+24\mu_{20}\mu_{11}^2;\\
&\kappa_{33}=\mu_{33}-3\mu_{31}\mu_{02}-\mu_{30}\mu_{03}-9\mu_{22}\mu_{11}-9\mu_{21}\mu_{12}-3\mu_{20}\mu_{13}+18\mu_{20}\mu_{11}\mu_{02}+12\mu_{11}^3
\end{aligned}\right\}\quad (3.81)$$

The reader should develop familiarity with the symbolic process by verifying some of these formulae, e.g. by deriving those for κ_{42} and κ_{33} in (3.81) from that for κ_6 in (3.42).

The formulae have been given for μ'_{rs} in terms of κ_{rs} and vice versa for orders $r+s\leqslant 6$ by Cook (1951). The symbolic process has been computerized and used by Kratky *et al.* (1972) to give the moments and the central moments in terms of cumulants for all bivariate cases up to order 9 and 10 respectively, and for many other multivariate cases up to order 10.

3.30 Wold (1934b) has given the following expressions for Sheppard corrections to bivariate moments and cumulants, the variates being grouped in intervals h_1, h_2.

$$\mu'_{rs}=\sum_{j=0}^{r}\sum_{k=0}^{s}h_1^j h_2^k\binom{r}{j}\binom{s}{j}(2^{1-j}-1)(2^{1-k}-1)B_jB_k\bar{\mu}'_{r-j,s-k},\quad (3.82)$$

the generalization of (3.56). In particular

$$\left.\begin{aligned}
&\mu'_{20}=\bar{\mu}'_{20}-\tfrac{1}{12}h_1^2,\quad \mu'_{11}=\bar{\mu}'_{11},\quad \mu'_{02}=\bar{\mu}'_{02}-\tfrac{1}{12}h_2^2;\\
&\mu'_{30}=\bar{\mu}'_{30}-\bar{\mu}'_{10}h_1^2/4,\quad \mu'_{21}=\bar{\mu}'_{21}-\bar{\mu}'_{01}h_1^2/12;\\
&\mu'_{40}=\bar{\mu}'_{40}-\bar{\mu}'_{20}h_1^2/2+7h_1^4/240,\quad \mu'_{31}=\bar{\mu}'_{31}-\bar{\mu}'_{11}h_1^2/4;\\
&\mu'_{22}=\bar{\mu}'_{22}-\bar{\mu}'_{20}h_2^2/12-\bar{\mu}'_{02}h_1^2/12+h_1^2h_2^2/144.
\end{aligned}\right\}\quad (3.83)$$

For cumulants we have

$$\left.\begin{aligned}\kappa_{rs} &= \bar{\kappa}_{rs} && r,s>0\\ \kappa_{r0} &= \bar{\kappa}_{r0}-B_r h_1^r/r, && r\geqslant 2\\ \kappa_{0s} &= \bar{\kappa}_{0s}-B_s h_2^s/s && s\geqslant 2\end{aligned}\right\} \tag{3.84}$$

so that the product cumulants require no corrections.

Measures of skewness

3.31 We have considered measures of location and dispersion in Chapter 2. With the aid of the moments we can now proceed to consider measures of other qualities of the population, and in particular its departure from symmetry.

In a symmetrical population, mean, median and mode coincide. It is thus natural to take distance from mean to mode or mean to median as measuring the skewness of the distribution. K. Pearson proposed the measure

$$\frac{\text{mean}-\text{mode}}{\sigma}$$

which is subject to the inconvenience of determining the mode. For a wide class of distributions known as Pearson's (cf. Chapter 6), this measure may, however, be expressed exactly in terms of the first four moments. We define

$$\beta_1=\frac{\mu_3^2}{\mu_2^3} \tag{3.85}$$

$$\beta_2=\frac{\mu_4}{\mu_2^2}. \tag{3.86}$$

It will be seen in 6.2 that for Pearson distributions

$$\frac{\text{mean}-\text{mode}}{\sigma}=\frac{\sqrt{\beta_1}(\beta_2+3)}{2(5\beta_2-6\beta_1-9)} \tag{3.87}$$

and this equation may be taken as defining a measure of skewness applicable to all distributions with μ_4 finite.

N. L. Johnson and Rogers (1951) show that for unimodal distributions, (mean − mode)/$\sigma\leqslant\sqrt{3}$ and give a condition for unimodality in terms of β_1 and β_2. Exercise 3.22 concerns the measure of skewness (mean − median)/σ which lies between −1 and +1. Exercise 3.27 gives a measure based on median and quartiles which has the same bounds.

Generally we may define

$$\beta_{2n+1}=\frac{\mu_3\mu_{2n+3}}{\mu_2^{n+3}},\qquad \beta_{2n}=\frac{\mu_{2n+2}}{\mu_2^{n+1}}, \tag{3.88}$$

quantities that are not in general use but will be found to occur occasionally in the statistical literature.

More convenient quantities than β_1 and β_2 for certain purposes are

$$\gamma_1 = \sqrt{\beta_1} = \frac{\mu_3}{\mu_2^{3/2}} = \frac{\kappa_3}{\kappa_2^{3/2}}, \tag{3.89}$$

$$\gamma_2 = \beta_2 - 3 = \frac{\mu_4}{\mu_2^2} - 3 = \frac{\kappa_4}{\kappa_2^2}. \tag{3.90}$$

If the distribution is standardized, γ_1 and γ_2 are its third and fourth cumulants.

The coefficient β_1 itself is also a measure of skewness. Clearly, if the distribution is symmetrical, β_1 vanishes with μ_3. In general the ratio of μ_3 to $\mu_2^{3/2}$ (i.e. γ_1) will give some indication of the extent of departure from symmetry. γ_1 also takes the sign of μ_3, and therefore gives sign to the skewness: positive γ_1 usually means that the upper tail of the distribution is the heavier, and that mean > median > mode; negative γ_1 that the lower tail is heavier and mode > median > mean. However, Ord (1968b) gives some asymmetrical distributions with as many zero odd-order moments as desired, and Exercise 3.26 gives one with odd-order moments all zero, so the value of γ_1 must be interpreted with some caution. MacGillivray (1981) gives sufficient conditions for mean > median > mode to hold.

Kurtosis

3.32 In the normal distribution of Example 3.4,

$$f(x) = \frac{1}{\sigma\sqrt{(2\pi)}} e^{-\frac{1}{2}x^2/\sigma^2}, \qquad -\infty < x < \infty$$

$\beta_2 = 3$ and γ_2 is zero. Clearly, β_2 will be larger for distributions with greater relative probabilities in one or both tails, which contribute disproportionately to μ_4. Distributions for which $\gamma_2 = 0$ are called mesokurtic. Those for which $\gamma_2 > 0$ are called leptokurtic; those for which $\gamma_2 < 0$ are called platykurtic.

These names were originally conferred because among certain regular symmetric distributions, those with larger tail areas are also more sharply peaked, and those with smaller tail areas are also flatter-topped. This is not necessarily so for other symmetrical or for asymmetrical distributions, and although the terms are useful they are better regarded as describing the sign of γ_2 rather than the shape of the distribution. Cf. Exercises 3.20–21 and 3.25.

By putting $a_i = (x_i - \mu_1')^2$, $b_i \equiv 1$ in the Cauchy–Schwarz inequality in **2.29**, we see that $\mu_4 \geqslant \mu_2^2$, with equality only when the squared deviations a_i do not vary at all, i.e. when the distribution is wholly concentrated at two values of x. Thus $\beta_2 \geqslant 1\,(\gamma_2 \geqslant -2)$ always.

Exercise 3.18 shows that for regular unimodal symmetric distributions, $\beta_2 \geqslant 1.8$, while Exercise 3.19 shows that $\beta_2 > 1 + \beta_1$ always.

Exercise 3.16 shows that, like the variance in **2.22**, skewness and kurtosis coefficients can be defined without reference to the mean of a distribution.

Example 3.20 (Calculation of skewness and kurtosis measures)

For the distribution of Australian marriages considered in Example 3.1 we found, for the raw moments about the mean in units of three years,

$$\bar{\mu}_2 = 7.056\,977, \quad \bar{\mu}_3 = 36.151\,595, \quad \bar{\mu}_4 = 408.738\,210.$$

With Sheppard's corrections these become

$$\mu_2 = 6.973\,644, \quad \mu_3 = 36.151\,595, \quad \mu_4 = 405.238\,888.$$

From these values we find

$$\beta_1 = 3.854, \quad \gamma_1 = 1.963$$

$$\beta_2 = 8.333, \quad \gamma_2 = 5.333$$

indicating considerable skewness and leptokurtosis.

Example 3.21 (Skewness and kurtosis of the binomial)

From the formulae for the moments of the binomial distribution considered in Example 3.2 we find

$$\gamma_1 = \frac{q-p}{\sqrt{(npq)}}$$

$$\gamma_2 = \frac{1-6pq}{npq}$$

so that, as $n \to \infty$, γ_1 and $\gamma_2 \to 0$. This is in accordance with a result we shall prove later, that the standardized binomial distribution tends to the normal form as n tends to infinity.

Example 3.22 (Skewness and kurtosis of the Gamma)

Similarly, from Example 3.6 we find for the Gamma distribution there

$$\gamma_1 = 2/\gamma^{\frac{1}{2}}, \quad \gamma_2 = 6/\gamma,$$

which again $\to 0$ as the parameter $\gamma \to \infty$. Here too we shall see that this is because the distribution tends to the normal.

Measures of shape

3.33 Thus far, we have defined skewness and kurtosis intuitively and this does not take us very far. Van Zwet (1964) first provided more general definitions as follows. Given two distribution functions, F and G:

(1) G is more skewed to the right than F if $G^{-1}(F)$ is a convex function for all x such that $0 < F(x) < 1$, written as $G >_s F$

(2) when F and G are both symmetric, G has greater kurtosis than F if $G^{-1}(F)$ is a convex function for all $x > M$, the median, written as $G >_k F$.

Balanda and MacGillivray (1990) have extended (2) to cover all distributions by first defining the spread function

$$S_F(u) = F^{-1}(0.5+u) - F^{-1}(0.5-u), \qquad 0 \leqslant u < \tfrac{1}{2}$$

and then specifying that G has greater kurtosis than F if $S_G(S_F^{-1}(u))$ is a convex function for all $u \geqslant 0$.

Example 3.23 (Shape of the Weibull distribution)

The Weibull distribution has $F(x) = 1 - \exp(-\theta x^{c_1})$; let G have the same distribution but with index c_2. Then

$$F^{-1}(u) = [-\theta^{-1}\log(1-u)]^{1/c_1}$$

thus

$$G^{-1}(F) = [-\theta^{-1}\log\{1-F(x)\}]^{1/c_2}$$
$$= x^{c_1/c_2}$$

which is convex if and only if $c_1 > c_2$. Thus the family of Weibull distributions is well-ordered according to this definition of skewness.

It is natural to inquire how these more general definitions relate to the measures defined earlier. Groeneveld and Meeden (1984) show that the ordering $>_s$ implies the same ordering of distributions as using any member of the class of measures

$$b_1(u) = \frac{F^{-1}(1-u) + F^{-1}(u) - 2M}{F^{-1}(1-u) - F^{-1}(u)}, \qquad 0 < u < \tfrac{1}{2},$$

but that this does not hold for (mean − median)/(mean absolute deviation). MacGillivray (1986) explores such orderings in detail and shows that the orderings implied by β_1 and, more generally, β_{2n+1} are consistent with $>_s$. The measures $b_1(u)$ are consistent with $>_k$, but β_2 is not (MacGillivray and Balanda, 1988).

Other orderings are considered by Oja (1981) and Lawrence (1975); see also Exercise 3.33.

Moments as characteristics of a distribution

3.34 The use of moments and cumulants in determining the nature of a distribution will be abundantly illustrated in later chapters, but some general remarks may be made at this stage.

It has been noted that the c.f. determines the moments when they exist, and it will be proved in Chapter 4 that the c.f. also determines the d.f. It does not, however, follow that the moments completely determine the distribution, even when moments of all orders exist. Only under certain conditions will a set of moments determine a distribution uniquely, but, fortunately for statisticians, those conditions are satisfied by most of the distributions commonly arising in statistical practice; for counterexamples, see Exercises 3.12 and 6.21. A full discussion of the conditions is given in **4.20–4**. For most ordinary purposes, therefore, knowledge of the moments is equivalent to a knowledge of the d.f., in the sense that it should be possible *theoretically* to exhibit all the properties of the distribution in terms of the moments.

In particular we expect that if two distributions have a certain number of moments in common they will bear some resemblance to each other. If moments up to those of order n are identical we expect that as n tends to infinity the distributions approach each other, and consequently we expect that by equating the lower moments of two distributions we bring them to approximate equality. Some mathematical support for this may be derived from the following approach.

It is known that a function that is continuous in a finite range a to b can be represented in that range by a uniformly convergent series of polynomials in x, say $\sum_{n=0}^{\infty} P_n(x)$ where $P_n(x)$ is of degree n. Suppose we wish to represent such a function approximately by the *finite* series of powers $\sum_{n=0}^{s} a_n x^n$. The coefficients a_n may be determined by the principle of least squares, i.e. so as to make

$$\int_a^b \left(f - \sum a_n x^n\right)^2 dx \qquad (3.91)$$

a minimum. Differentiating by a_n we have

$$2\int_a^b \left(f - \sum a_n x^n\right) x^j dx = 0$$

or

$$\int_a^b f x^j dx = \mu_j' = \int_a^b \sum a_n x^{n+j} dx. \qquad (3.92)$$

If now two distributions have moments up to order s equal they must have the same least-squares approximation, for the coefficients a_n are determined by the moments in (3.92). Furthermore, if the distribution f_1 differs from $\sum a_n x^n$ by ε_1 and f_2 by ε_2, then f_1 differs from f_2 by not more than $\varepsilon_1 + \varepsilon_2$.

A similar approach may be adopted when the range is infinite, the distributions in such cases being, under certain conditions, capable of representation by a series of terms such as $e^{-x^2}P_n(x)$.(*) The same conclusion is reached.

Thus distributions that have a finite number of the lower moments in common will, in a sense, be approximations to one another. We shall encounter many cases where, although we cannot determine a distribution function explicitly, we may ascertain at least some of its moments; and hence we shall be able to approximate to the distribution by finding another distribution of known form that has the same lower moments. In practice, approximations of this kind often turn out to be remarkably good, even when only the first two, three or four moments are equated.

Inequalities of the Chebyshev type

3.35 It is also possible, given certain moments or expected values of a distribution, to make precise statements about the distribution function in terms of inequalities. Let F represent a distribution with a finite second moment. For any real t we have

(*) A theorem of Vera Myller-Lebedeff (1907, *Math. Ann.*, 64, 388) states that a function can be expanded in a series of derivatives of $\exp(-x^2)$. This is not to be confused with the Gram–Charlier series of Chapter 6, where a function is expanded in terms of derivatives of $\exp(-\frac{1}{2}x^2)$.

$$\int_{-\infty}^{\infty}\left(1-\frac{x^2}{t^2}\right)dF=\left\{\int_{-\infty}^{-t}+\int_{t}^{\infty}\right\}\left(1-\frac{x^2}{t^2}\right)dF+\int_{-t}^{t}\left(1-\frac{x^2}{t^2}\right)dF.$$

In the first two integrals on the right $1-x^2/t^2$ is $\leqslant 0$ and hence

$$1-\frac{\mu_2'}{t^2}=\int_{-\infty}^{\infty}\left(1-\frac{x^2}{t^2}\right)dF\leqslant\int_{-t}^{t}\left(1-\frac{x^2}{t^2}\right)dF\leqslant\int_{-t}^{t}dF.$$

Thus the probability from $-t$ to t is, no less than $1-\mu_2'/t^2$. In particular, if we take an origin at the mean of the distribution,

$$F(t)-F(-t)\geqslant 1-\mu_2/t^2. \tag{3.93}$$

This is more usually written in the form, taking $\mu_2=\sigma^2$ and $t=\lambda\sigma$,

$$P\{|x-E(x)|\leqslant\lambda\sigma\}\geqslant 1-1/\lambda^2 \tag{3.94}$$

or

$$P\{|x-E(x)|\geqslant\lambda\sigma\}\leqslant 1/\lambda^2. \tag{3.95}$$

This is the inequality given by I. J. Bienaymé in 1853. It is usually named after P. L. Chebyshev, who published it in 1867, but it is better called the Bienaymé–Chebyshev inequality. It gives us a lower limit to the probability that x lies within a range of $\lambda\sigma$ on either side of the mean. Because of its great generality, this limit is usually too low to be more than a rough guide in determining probabilities in particular cases, but such inequalities are invaluable in proving limiting properties.

3.36 There are by now a large number of inequalities of the Chebyshev type and later we shall meet several of them. Here we will record a few of the commoner ones for reference.

If ν_r is the absolute moment of order r it is easy to show by a direct extension of the foregoing method that

$$P\{|x-E(x)|/\nu_r^{1/r}\leqslant\lambda\}\geqslant 1-1/\lambda^r. \tag{3.96}$$

This inequality, given by K. Pearson (1919), gives (3.94) when $r=2$. More generally, if $g(x)$ is a non-negative function of x and R the region for which $g(x)\geqslant k$, we have

$$E\{g(x)\}=\int_{-\infty}^{\infty}g(x)dF\geqslant\int_R g(x)dF\geqslant k\int_R dF=kP\{g(x)\geqslant k\}. \tag{3.97}$$

If the variate x itself is non-negative, this yields for its d.f.

$$F(t)>1-\mu_1'/t \tag{3.98}$$

which is sometimes called Markov's inequality. If we put $g(x)=\{x-E(x)\}^2$ in (3.97) and $k=\lambda^2\sigma^2$, we obtain (3.95).

For general distributions these inequalities cannot be improved upon, but it is rather remarkable that by imposing quite broad restrictions on the form of the distribution we can improve them considerably. For example, if a continuous density function has a single mode, say m_0, and the origin is taken there,

$$P\{|x-m_0|\leqslant\lambda\tau\}\geqslant 1-4/(9\lambda^2) \tag{3.99}$$

where τ is the square root of the second moment about the mode, as compared with (3.94). This expression, in a different form, is traceable to Gauss.

3.37 When a random sample of size $n \geqslant 2$ is available from a population for which μ and σ are unknown, Saw *et al.* (1984) show that for a further independent observation x_{n+1},

$$P[|x_{n+1}-\bar{x}|>\lambda Q] \leqslant \frac{g_{n+1}\{n\lambda^2/(n-1+\lambda^2)\}}{(n+1)} \tag{3.100}$$

where

$$n\bar{x}=\sum_{i=1}^{n} x_i, \quad \lambda \geqslant 1,$$

$$Q=(n+1)\sum_{i=1}^{n}(x_i-\bar{x})^2/n(n-1)$$

and $g_{n+1}(u)$ is the largest integer less than $(n+1)/u$. The right-hand side of (3.100) is less than $\left(\frac{1}{n}+\frac{1}{\lambda^2}\right)$, a simpler bound. Slightly stronger versions of (3.100) are possible in some cases. As $n\to\infty$ (3.100) tends to (3.95).

3.38 By making further assumptions (e.g., that several moments are known), we may obtain more refined inequalities. For an account of these inequalities see a comprehensive review of the subject by Godwin (1955) and the monograph by the same author (1964). Beesack (1984) provides an historical account of inequalities on absolute moments.

Multivariate inequalities

3.39 The simplest multivariate extension to (3.94) may be derived as follows. Let $B_j=\{x_j: |x_j-\mu_j| \geqslant \lambda_j\sigma_j\}$ for $j=1,2,\ldots,k$, where (μ_j, σ_j^2) denote the mean and variance of the jth variate. Boole's inequality (see 8.6)) states that

$$\sum_{j=1}^{k} P(B_j) \geqslant P[\text{at least one } x_j \in B_j]. \tag{3.101}$$

Since

$$P[\text{no } x_j\in B_j]=1-P[\text{at least one } x_j\in B_j] \tag{3.102}$$

it follows from (3.94), (3.95), (3.101) and (3.102) that

$$\begin{aligned} P[\text{no } x_j\in B_j] &= P[|x_j-\mu_j| \leqslant \lambda_j\sigma_j \text{ for all } j] \\ &\geqslant 1-\sum_{j=1}^{k} P(B_j) \\ &= 1-\sum_{j=1}^{k} \lambda_j^{-2}. \end{aligned} \tag{3.103}$$

when $\lambda_j=\lambda$ for all j, (3.103) reduces to $1-(k/\lambda^2)$, so it is evident that the lower bound may be poor. However, it has been derived without making any assumptions about the form of the multivariate distribution other than that the means and variances exist; as such, it is of considerable value in the development of limiting results.

Tong (1981, 1984) and Eaton (1982) provide detailed coverage of other multivariate inequalities.

Information measures

3.40 The information contained in a probability function may be defined in the discrete case as

$$I(f) = -\sum [f(x)]^{a+1}/a + a^{-1} \tag{3.104}$$

and in the continuous case as

$$I(f) = -\int [f(x)]^{a+1}\,dx/a + a^{-1}. \tag{3.105}$$

The distinction between the continuous and discrete forms is important, since the discrete version is invariant under transformations of the variate, whereas the continuous form is not.

Information measures arise naturally in communications theory, as formulated in the seminal work of Shannon and Weaver (1949). In particular, Shannon's measure takes $a \to 0$, so that

$$I(f) = -E_f\{\log f(x)\} \tag{3.106}$$

and is also known as the *(negative) entropy* measure.

Example 3.24 (Information measures for uniform distributions)

Let $f(x) = N^{-1}$, $x = 1, 2, \ldots, N$ and take $M = \log_2 N$. If, as is common practice, we use logarithms to base 2 in (3.106), it follows that

$$I(f) = M;$$

further, this value is unchanged for any choice of N distinct x values. By contrast, if we set

$$f(x) = B^{-1}, \qquad 0 < x < B$$

and let $A = \log B$, we have

$$I(f) = A;$$

putting $y = cx$ yields $I_y(f) = A + \log c$.

It may be shown that $I(f)$ is maximized, for all distributions with N points of support, when the distribution is discrete uniform and minimized when $f(x)$ is degenerate. In signalling terms, the minimum corresponds to perfect signal transmission, whereas the maximum corresponds to the worst case where all possible signals are equally likely so that no discrimination is possible.

The measures (3.104) and (3.105) correspond to the frequency moments described in **6.39** when a is an integer.

3.41 If we wish to compare the p.f. $g(x)$ with $f(x)$, where $f(x)$ represents a 'standard' (eg. an original signal), an appropriate measure is

$$I(f, g) = -E_f\{\log f(x)/g(x)\}. \tag{3.107}$$

Clearly $I(f, g) \equiv 0$ when $f \equiv g$.

Example 3.25 (Information measures for comparison of two normals)

Let f and g denote normal densities with common means μ and variances σ^2, ω^2 respectively. Since

$$\log(f/g) = -\tfrac{1}{2}(x-\mu)^2[\sigma^{-2} - \omega^{-2}]$$

it follows directly that

$$I(f, g) = +\tfrac{1}{2}[1 - (\sigma^2/\omega^2)]$$

If g represents a deterioration from f, such as the addition of a random error term so that $\omega^2 > \sigma^2$, clearly $I(f, g) > 0$.

If there is no particular ordering of g with respect to f, but a direct comparison between the two is required, the symmetric measure

$$J(f, g) = I(f, g) + I(g, f)$$

is often used. For the two normals defined above

$$J(f, g) = -(\sigma^2 - \omega^2)^2/(2\sigma^2\omega^2);$$

in general $J(f, g) < 0$ unless $f \equiv g$.

Distance between distributions

3.42 The distance between two distributions may be defined as

$$D(f, g) = \sum |f(x) - g(x)| \tag{3.108}$$

$$\text{or} \quad \int |f(x) - g(x)|\, dx \tag{3.109}$$

according to whether the variate is discrete or continuous, provided that $f(x)$ is absolutely continuous with respect to $g(x)$. Under this condition, Kemperman (1969) shows that

$$D(f, g) \leqslant [2I(f, g)]^{\frac{1}{2}}. \tag{3.110}$$

When g is an approximation to f, (3.110) provides an indication of the closeness of that approximation.

The method of moments

3.43 There are many occasions when we may identify a particular process as leading to a certain distribution; numerous examples are given in Chapter 5. However, the particular parameter values for the distribution are usually unknown. One general approach to estimating the parameters from a data set is to equate the sample moments to their population counterparts. This approach is known as the *method of moments* and was introduced by Karl Pearson (1895) as a way of fitting data to the family of curves that bears his name; see 6.2 below. The method is of general applicability provided that the theoretical moments exist, as the following examples illustrate.

Example 3.26 (The method of moments for the binomial and the normal)
The method of moments applies to the distributions introduced in Examples 3.2 and 3.4 as follows:

(a) *binomial*: observe x successes in n trials, and $E(x) = np$ so the estimator is $\hat{p} = x/n$; the 'hat' notation is in general use to denote the estimator for a parameter.
(b) *normal*: observe $x_1, x_2, \ldots, x_n$ where

$$E(x) = \mu \quad \text{and} \quad E(x^2) = \sigma^2 + \mu^2;$$

it follows that the moments estimators are

$$\hat{\mu} = \sum x/n \quad \text{and} \quad \hat{\sigma}^2 = \left(\sum x^2 - \hat{\mu}^2\right)\Big/ n$$

although the n is $\hat{\sigma}^2$ is often replaced by $(n-1)$ to make $E(\hat{\sigma}^2) = \sigma^2$. Generally, when $E(\hat{\theta}) = \hat{\theta}$ for some parameter, the estimator is said to be *unbiased*.

We shall see in Volume 2A that the moments estimators may not be very efficient, but they are widely applicable and often easy to compute.

3.44 There are occasions where the constant of integration, k, may be very difficult to determine, particularly for multivariate distributions. In those cases, we can finesse the problem by using ratios of moments; however, the reader should be warned that the resulting estimators may be of very low efficiency.

Example 3.27 (Moments estimator for the zero-truncated Poisson)
Consider the distribution

$$f(x) = ke^{-\lambda}\lambda^x/x!, \qquad x = 1, 2, \ldots, \tag{3.111}$$

known as the truncated Poisson distribution (c.f. Exercise 3.29). Suppose, for the sake of argument, that k cannot be determined. The first two factorial moments are

$$\mu'_{[1]} = E(x) = k\lambda \quad \text{and} \quad \mu'_{[2]} = E(x^{[2]}) = k\lambda^2$$

so that a possible estimator, given $(x_1, x_2, \ldots, x_n)$, is

$$\hat{\lambda} = \sum x(x-1)\Big/\sum x. \tag{3.112}$$

In this example, since $k^{-1} = 1 - e^{-\lambda}$, a better estimator is given by the solution to the equation:

$$\sum x/n = \lambda/(1 - e^{-\lambda}); \tag{3.113}$$

however, the ratio of moments approach is useful for otherwise intractable problems.

EXERCISES

3.1 Show that the discrete (Poisson) distribution with probabilities at $x = j\,(j = 0, 1, 2, \ldots)$ given by

$$e^{-\lambda}\lambda^j/j!$$

has mean λ and central moments

$$\mu_2 = \lambda, \quad \mu_3 = \lambda, \quad \mu_4 = \lambda(1 + 3\lambda).$$

3.2 For the distribution with p.d.f.

$$f(x) = kx^{-p}e^{-\gamma/x}, \qquad 0 \leqslant x < \infty; \quad \gamma > 0, \quad p > 1,$$

show that, about the origin,

$$\mu_r' = \frac{\gamma^r\Gamma(p - r - 1)}{\Gamma(p - 1)}$$

if $r < p - 1$, and show that $y = x^{-1}$ has a Gamma distribution.

3.3 In the distribution with p.d.f.

$$f(x) = k\left(1 + \frac{x^2}{a^2}\right)^{-m} \exp\{-\nu \arctan(x/a)\}, \qquad -\infty < x < \infty,$$

show that, about the origin,

$$\mu_r' = ka^{r+1} \int_{-\frac{1}{2}\pi}^{\frac{1}{2}\pi} \cos^{2m-r-2}\theta \sin^r\theta e^{-\nu\theta} d\theta$$

and hence that

$$\mu_r' = \frac{a}{2m - r - 1}\{(r - 1)a\mu_{r-2}' - \nu\mu_{r-1}'\}.$$

3.4 $f(x)$ is a p.d.f., symmetrical about its unique mode at $x = 0$ and ranging from $-a$ to $+a$. Show that

$$\mu_{2r} < a^{2r}/(2r + 1), \qquad r \geqslant 1.$$

3.5 Show that for any symmetrical distribution the finite cumulants of odd order (except κ_1) are zero.

3.6 Show that for the exponential distribution

$$f(x) = e^{-x/\sigma}/\sigma, \qquad 0 \leqslant x < \infty; \quad \sigma > 0,$$
$$\kappa_r = \sigma^r(r - 1)!$$

so that the skewness and kurtosis coefficients (3.89–90) are $\gamma_1 = 2$, $\gamma_2 = 6$.

3.7 Show that for a discrete distribution defined at 0, 1, 2, ..., the factorial moments may be obtained from the c.f. $\phi(t)$ as

$$\mu_{[r]}' = [D^r\phi(t)]_{t=0},$$

where

$$D = \frac{d}{d(e^{it})}.$$

Hence verify, for the binomial distribution in Example 3.8,

$$\mu_{[r]}' = n^{[r]}p^r.$$

3.8 Show that for the distribution with p.d.f.

$$f(x) = \frac{k}{1+x^{2r}}, \qquad -\infty < x < \infty, \ r \text{ a positive integer,}$$

the moments μ_p exist for $p \leqslant 2(r-1)$, with

$$\mu_{2s-1} = 0, \qquad s = 1, 2, \ldots$$

$$\mu_{2s} = \sin\left(\frac{\pi}{2r}\right) \Big/ \sin\left\{\frac{(2s+1)\pi}{2r}\right\}.$$

3.9 Show from (3.39) that

$$\mu'_r = \sum_{j=1}^{r} \binom{r-1}{j-1} \mu'_{r-j} \kappa_j$$

and hence that

$$\kappa_r = (-1)^{r-1} \begin{vmatrix} \mu'_1 & 1 & 0 & 0 & \cdot & \cdot & 0 \\ \mu'_2 & \binom{1}{0}\mu'_1 & 1 & 0 & \cdot & \cdot & 0 \\ \mu'_3 & \binom{2}{0}\mu'_2 & \binom{2}{1}\mu'_1 & 1 & \cdot & \cdot & 0 \\ \cdot & \cdot & \cdot & \cdot & \cdot & \cdot & 1 \\ \mu'_r & \binom{r-1}{0}\mu'_{r-1} & \binom{r-1}{1}\mu'_{r-2} & \cdot & \cdot & \cdot & \binom{r-1}{r-2}\mu'_1 \end{vmatrix}.$$

3.10 Show that for the uniform distribution with $f(x) = 1, 0 \leqslant x \leqslant 1$, grouped into an integral number of intervals of equal length h, the corrections to the second and fourth moments about the mean are

$$\mu_2 = \bar{\mu}_2 + \frac{h^2}{12}, \qquad \mu_4 = \bar{\mu}_4 + \bar{\mu}_2\frac{h^2}{2} + \frac{h^4}{80}.$$

(Cf. Elderton, 1938. Note that the first is exactly, and the second approximately, the Sheppard correction *with sign reversed.*)

3.11 If ∂_p is an operator such that

$$\partial_p \mu'_r = r^{[p]} \mu'_{r-p} \qquad r \geqslant p$$
$$= 0, \qquad r < p,$$

and

$$\partial_p(AB) = B(\partial_p A) + A(\partial_p B),$$

show that ∂_p annihilates every cumulant (considered as a function of the moments) except κ_p, and that $\partial_p \kappa_p = p!$

3.12 If $f(x)$ is an odd function of x of period $\frac{1}{2}$, show that

$$\int_0^\infty x^r x^{-\log x} f(\log x) dx = 0$$

for all integral values of r. Hence show that the distributions

$$f(x) = x^{-\log x}\{1 - \lambda \sin(4\pi \log x)\}, \qquad 0 \leqslant x < \infty; \quad 0 \leqslant \lambda \leqslant 1,$$

have the same moments whatever the value of λ. (Stieltjes, 1918)

3.13 Show that if the frequencies of a discrete distribution are distributed with m observations equally spaced at intervals h/m within each grouping interval h, the average grouping corrections to the cumulants are given by

$$\kappa_r = \bar{\kappa}_r - \frac{B_r h^r}{r}\left(1 - \frac{1}{m^r}\right).$$

(C. C. Craig, 1936b)

3.14 Show that for the bivariate distribution of Example 3.19 with central moments μ_{rs},

$$f(x_1, x_2) = \frac{1}{2\pi\sigma_1\sigma_2(1-\rho^2)^{\frac{1}{2}}}\exp\left[-\frac{1}{2(1-\rho^2)}\left\{\frac{x_1^2}{\sigma_1^2} - \frac{2\rho x_1 x_2}{\sigma_1\sigma_2} + \frac{x_2^2}{\sigma_2^2}\right\}\right], \qquad -\infty < x_1, x_2 < \infty,$$

the cumulants $\kappa_{rs} = 0$ if $r+s > 2$; and further, if $M(t_1, t_2)$ is the m.g.f. in Example 3.19 with $\sigma_1 = \sigma_2 = 1$ and the standardized moments are

$$\lambda_{rs} = \frac{\mu_{rs}}{\sigma_1^2 \sigma_2^s},$$

show that

$$\frac{\partial^2 M}{\partial t_1 \partial t_2} = \left\{\rho\left(t_1\frac{\partial M}{\partial t_1} + t_2\frac{\partial M}{\partial t_2} + M\right) + (1-\rho^2)t_1 t_2 M\right\}$$

and hence that

$$\lambda_{rs} = (r+s-1)\rho\lambda_{r-1,s-1} + (r-1)(s-1)(1-\rho^2)\lambda_{r-2,s-2},$$

$$\lambda_{2r,2s} = \frac{(2r)!(2s)!}{2^{r+s}}\sum_{j=0}^{t}\frac{(2\rho)^{2j}}{(r-j)!(s-j)!(2j)!},$$

$$\lambda_{2r+1,2s+1} = \frac{(2r+1)!(2s+1)!}{2^{r+s}}\rho\sum_{j=0}^{t}\frac{(2\rho)^{2j}}{(r-j)!(s-j)!(2j+1)!},$$

$$\lambda_{2r,2s+1} = \lambda_{2r+1,2s} = 0,$$

where t is the smaller of r and s. In particular

$$\lambda_{11} = \rho, \quad \lambda_{31} = 3\rho, \quad \lambda_{51} = 15\rho, \quad \lambda_{71} = 105\rho, \quad \lambda_{91} = 945\rho$$
$$\lambda_{22} = (1+2\rho^2), \quad \lambda_{24} = 3(1+4\rho^2), \quad \lambda_{26} = 15(1+6\rho^2),$$
$$\lambda_{28} = 105(1+8\rho^2), \quad \lambda_{2,10} = 945(1+10\rho^2);$$
$$\lambda_{33} = 3\rho(3+2\rho^2), \quad \lambda_{35} = 15\rho(3+4\rho^2),$$
$$\lambda_{44} = 3(3+24\rho^2+8\rho^4).$$

3.15 Regarding the absolute moment ν_r' at (3.21) as a function of r, show that $\dfrac{\partial^2 \log\nu_r'}{\partial r^2} > 0$ in non-degenerate cases, and hence that for any $a \geqslant b \geqslant c \geqslant 0$

$$\log\nu_b' \leqslant \{(a-b)\log\nu_c' + (b-c)\log\nu_a'\}/(a-c),$$

(Belz, 1947)

whence Liapunov's inequality (3.22) follows.

3.16 Show that if x, y, z are independent with the same d.f. F and central molvents μ_r,

$$E\{(x-y)^r\} = \sum_{s=0}^{r}(-1)^s\binom{r}{s}\mu_s\mu_{r-s},$$

and that this is zero if r is odd, while

$$E\{(x-y)^{r-1}(x-z)\}=\sum_{s=0}^{r-1}(-1)^s\binom{r-1}{s}\mu_s\mu_{r-s}.$$

Hence shows that as in **2.22**

$$\tfrac{1}{2}E\{(x-y)^2\}=E\{(x-y)(x-z)\}=\mu_2,$$

while

$$E\{(x-y)^2(x-z)\}=\mu_3$$

and

$$E\{(x-y)^4\}=2(\mu_4+3\mu_2^2),$$

so that the skewness and kurtosis coefficients γ_1 and γ_2, like the variance, may be defined without reference to the mean.

3.17 A distribution is grouped in intervals of length h. Show that as the grouping grid moves along the variate axis the raw moments have period h. Hence, writing

$$\bar{\mu}_r'=\sum_{j=-\infty}^{\infty}\zeta^r\int_{\zeta-\frac{1}{2}h}^{\zeta+\frac{1}{2}h}f(x)\,\mathrm{d}x$$

$$=A_0+\sum_{j=1}^{\infty}A_j\sin j\theta+\sum_{j=1}^{\infty}B_j\cos j\theta$$

where

$$\zeta=(j+\theta/2\pi)h,$$

show that

$$A_0=\frac{1}{h}\int_{-\infty}^{\infty}x^r\int_{-\frac{1}{2}h}^{\frac{1}{2}h}f(x+\zeta)\,\mathrm{d}\zeta\,\mathrm{d}x$$

and hence that this leads to Sheppard's corrections.

Show also that

$$A_s=\frac{2}{h}\int_{-\infty}^{\infty}f(x)\mathrm{d}x\int_{x-\frac{1}{2}h}^{x+\frac{1}{2}h}\zeta^r\sin(2\pi s\zeta/h)\mathrm{d}\zeta$$

with a similar expression for B_s.

For the p.d.f. $f(x)=\dfrac{1}{\sqrt{(2\pi)}}\mathrm{e}^{-\frac{1}{2}x^2}$, $-\infty<x<\infty$, show that, for corrections to the mean, the Bs vanish and $A_s=(-1)^{s+1}\dfrac{h}{\pi s}\exp\left(-\dfrac{2s^2\pi^2}{h^2}\right)$ and hence that, even for a coarse grouping $h=1$, the correction to the mean is not greater than $\mathrm{e}^{-2\pi^2}/\pi$ approximately.

(Fisher, 1921a)

3.18 *The Gauss-Winckler inequality*. By considering $1-f(x)/f(\hat{x})$ as a distribution function and using Liapunov's inequality, show that for a density function that is differentiable and has a single mode $\hat{x}$, the absolute moments about $\hat{x}$ obey the relation

$$\{(r+1)\nu_r'\}^{1/r}\leqslant\{(n+1)\nu_n'\}^{1/n},\quad r<n.$$

In particular show that, about the mode, $\mu_4'/\mu_2'^2\geqslant 1.8$.

3.19 Show that

$$\begin{vmatrix}\mu_0' & \mu_1' & \mu_2'\\ \mu_1' & \mu_2' & \mu_3'\\ \mu_2' & \mu_3' & \mu_4'\end{vmatrix}\geqslant 0$$

and hence that $\beta_2 \geqslant 1+\beta_1$. [Extension: given any set of n values, show that $\beta_2 \leqslant n$ and hence that $\beta_1 \leqslant n-1$.

(M. E. Johnson and Lowe, 1979)]

3.20 Show that if two distributions $f(x)$, $g(x)$ are symmetrical with zero means and unit variances and

$$f(x) < g(x) \quad \text{for} \quad a < |x| < b,$$
$$f(x) > g(x) \quad \text{otherwise},$$

then μ_4 is greater for $f(x)$ than for $g(x)$. (Finucan, 1964)

3.21 The following four unimodal symmetrical p.d.f.s have zero mean and unit variance. Show that they have the values of μ_4 and the maximum ordinates indicated.

	Value of μ_4	Maximum ordinate
(1) $\frac{1}{3\sqrt{\pi}}(\frac{9}{4}+x^4)e^{-x^2}$	2.75	0.423
(2) $\frac{3}{2\sqrt{(2\pi)}}e^{-\frac{1}{2}x^2}-\frac{1}{6\sqrt{\pi}}(\frac{9}{4}-x^4)e^{-x^2}$	3.125	0.387
(3) $\frac{1}{6\sqrt{\pi}}(e^{-\frac{1}{4}x^2}+4e^{-x^2})$	4.5	0.470
(4) $\frac{3\sqrt{3}}{16\sqrt{\pi}}(2+x^2)e^{-\frac{3}{4}x^2}$	2.667	0.366

Comparing these with the normal values $\mu_4 = 3$, max. ordinate $= 0.399$, show that any combination of higher or lower ordinate at the centre and lepto- or platykurtosis is possible.

(Kaplansky, 1945)

3.22 x is a variate with mean μ, median M, variance σ^2, $\int_{\mu+}^{\infty} dF = p$, $\int_{-\infty}^{\mu-} dF = q$, $p+q \leqslant 1$. Putting $a = \int_{\mu+}^{\infty}(x-\mu)dF = -\int_{-\infty}^{\mu-}(x-\mu)dF$, show that $a^2\left(\frac{1}{p}+\frac{1}{q}\right) \leqslant \sigma^2$, that $a \geqslant \frac{1}{2}(M-\mu)$ if $M > \mu$, $a \geqslant -\frac{1}{2}(M-\mu)$ if $M < \mu$, and hence that

$$\left|\frac{\mu - M}{\sigma}\right| \leqslant 2\left(\frac{1}{p}+\frac{1}{q}\right)^{-\frac{1}{2}} \leqslant 1.$$

(Majindar (1962); he also shows that the left-hand equality is not attainable. Hotelling and Solomons (1932) gave $\left|\frac{\mu - M}{\sigma}\right| \leqslant 1$.)

3.23 Using (3.61) show that the c.f. of the uniform distribution $f(x) = h^{-1}$, $-\frac{1}{2}h < x \leqslant \frac{1}{2}h$ is given by

$$\phi_x(t) = \sinh(\tfrac{1}{2}ith)/(\tfrac{1}{2}ith)$$

with cumulants

$$\kappa_{2r-1} = 0, \quad \kappa_{2r} = B_{2r}h^{2r}/(2r), \qquad r = 1, 2, \ldots$$

Similarly for the discrete distribution

$$P(y = r) = 1/n, \qquad r = 1, 2, \ldots, n,$$

show that the c.f. about the mean is

$$\phi_y(t) = \sinh(\tfrac{1}{2}itn)/\{n\sinh(\tfrac{1}{2}it)\}$$

with cumulants about zero

$$\begin{aligned}\kappa_1 &= \tfrac{1}{2}(n+1)\\ \kappa_{2r+1} &= 0,\\ \kappa_{2r} &= B_{2r}(n^{2r}-1)/(2r), \qquad r = 1, 2, \ldots\end{aligned}$$

Deduce $\phi_y(t)$ from $\phi_x(t)$ by considering the sum of independent variates x (with $h = 1$) and y.

3.24 Show that if the variate takes integer values $0, 1, 2, \ldots, n$ only, (3.29) may be inverted to give

$$f_j = \frac{1}{j!}\sum_{r=0}^{n-j}(-1)^r \mu'_{[j+r]}/r!$$

Invert the factorial moments obtained in Example 3.8 by this means.

3.25 A sequence of variates x_K is defined by the d.f.s

$$F_K(x) = \left(1 - \frac{1}{K^2-1}\right)\Phi(x) + \frac{1}{K^2-1}\Phi\left(\frac{x}{K}\right), \qquad K = 2, 3, \ldots,$$

where $\Phi(x) = \int_{-\infty}^{x}(2\pi)^{-\frac{1}{2}}e^{-\frac{1}{2}u^2}du$ is the d.f. of the standardized normal distribution. Show that each x_K is symmetric about the origin, with variance 2, and that its kurtosis coefficient $\gamma_2 = \frac{3}{4}(K^2-2)$, increasing monotonically to infinity as $K \to \infty$, when $F_K(x) \to \Phi(x)$.

(Ali, 1974)

3.26 Using the result for $p > 0$

$$\int_0^\infty x^{p-1}e^{-x}\sin x\,dx = 2^{-\frac{1}{2}p}\Gamma(p)\sin(p\pi/4),$$

show that the distribution

$$dF = \tfrac{1}{48}\{1 - \operatorname{sgn} x \sin(|x|^{1/4})\}\exp\{-|x|^{1/4}\}, \qquad -\infty < x < \infty,$$

has all odd-order moments zero, but that it is not symmetrical.

(Cf. Churchill (1946)—the result is due to Stieltjes.)

3.27 If Q_1, Q_3 and M are the quartiles and median of a distribution, show that the ratio $W = (Q_1 + Q_3 - 2M)/Q_3 - Q_1)$ is a measure of skewness lying between the attainable limits ± 1. Show that for the half-normal distribution $f(x) = \left(\frac{2}{\pi}\right)^{\frac{1}{2}}e^{-\frac{1}{2}x^2}$, $0 \leqslant x < \infty$, Appendix Table 2 yields the values $Q_1 = 0.3186$, $Q_3 = 1.1503$ and $M = 0.6745$, so that $W = 0.14$; show further that this distribution has mean $(2/\pi)^{\frac{1}{2}}$ and variance $\left(1 - \frac{2}{\pi}\right)$, so that Pearson's skewness measure (mean − mode)/σ equals 1.32, while the measure of Exercise 3.22 equals 0.20.

3.28 If $f(x) = k\exp(-x^2)$ for all integer values x, show that the moments of $y = \exp(2px)$, where p is a positive integer, are $\mu'_r = \exp(p^2r^2)$, whereas the m.g.f.

$$M_y(t) = k\sum_{-\infty}^{\infty}\exp(te^{2px} - x^2)$$

does not converge for $t > 0$, so that the moments cannot be obtained from it.

(Cf. Fu, 1984)

3.29 In the zero-truncated Poisson distribution, with p.m.f.

$$f_j = \frac{e^{-\lambda}\lambda^j}{j!(1-e^{-\lambda})}, \quad j = 1, 2, \ldots$$

show that

$$E\left(\frac{1}{j+1}\right) = \frac{1}{\lambda} - \frac{1}{(e^{\lambda}-1)}.$$

3.30 If $f(x) = \exp\{x\theta + C(x) + D(\theta)\}$, $-\infty < x < \infty$, with x discrete or continuous, show that the m.g.f. is

$$M(t) = \exp\{D(\theta) - D(\theta + t)\}.$$

Hence find the mean and variance of x when they exist.

(This p.f. defines the exponential family of distributions which plays an important role in statistical inference; see 5.47–8 below.)

3.31 Show that the stable distributions (c.f. 4.36) with c.f.

$$\phi(t) = \exp[ait - c|t|^{\alpha}], \quad 0 < \alpha \leqslant 2$$

have a finite mean for $\alpha > 1$, but finite variance only if $\alpha = 2$ (the normal case).

3.32 If x is a variate whose first four moments are known, show that

$$P(x \geqslant \mu + t\sigma) \leqslant \frac{(\beta_2 - \beta_1 - 1)}{(\beta_2 - \beta_1 - 1)(1 + t^2) + (t^2 - t\gamma_1 - 1)^2}$$

where $\gamma_1^2 = \beta_1$. Further show that strict equality is attained when

$$P(x = \pm\beta_2^{\frac{1}{2}}) = (2\beta_2)^{-1}, \qquad P(x = 0) = 1 - \beta_2^{-1}.$$

(Bhattacharyya, 1987)

3.33 Suppose that $f(x)$ and $g(x)$ are continuous p.d.f.s, standardized to have zero mean and unit variance, and that there exist constants $a_1 < a_2 < a_3 < a_4$ such that the p.d.f.s *cross* at these four values:

$$f(x) \leqslant g(x), \quad \text{when} \quad x < a_1,\ a_2 < x < a_3 \quad \text{or} \quad x > a_4$$

$$f(x) \geqslant g(x), \quad \text{when} \quad a_1 < x < a_2 \quad \text{or} \quad a_3 < x < a_4.$$

If, in addition, $(a_1 + a_2 + a_3 + a_4)[\mu_3(f) - \mu_3(g)] \leqslant 0$, show that the kurtosis coefficient γ_2 given by (3.90) satisfy $\gamma_2(f) \leqslant \gamma_2(g)$.

(Dyson, 1943; for further discussion see Balanda and MacGillivray, 1988).

CHAPTER 4

CHARACTERISTIC FUNCTIONS

4.1 In 3.5, we introduced the characteristic function

$$\phi(t) = \int_{-\infty}^{\infty} e^{itx} dF \tag{4.1}$$

as a moment generating function. It has many other useful and important properties which give it a central role in statistical theory and in this chapter we shall give an account of them. To establish our general theorems we shall, however, have to use somewhat advanced mathematics; and the reader who is interested principally in the statistical applications may prefer to omit the proofs, especially at first reading, and merely note the results. Broadly speaking, this part of our theory may be regarded as a special case of the general theory of Fourier transforms—special because our functions are probability and distribution functions. Several of the proofs in this chapter are based on Lévy (1925). Lukacs (1970) has published an elegant monograph on the subject. See also Lukacs and Laha (1964) and Lukacs (1983).

4.2 We recall, in the first instance, that $\phi(t)$ always exists, since

$$|\phi(t)| = \left| \int_{-\infty}^{\infty} e^{itx} dF \right| \leqslant \int_{-\infty}^{\infty} |e^{itx}| dF = \int_{-\infty}^{\infty} dF = 1, \tag{4.2}$$

so that the defining integral converges absolutely. Further, $\phi(t)$ is uniformly continuous in t and differentiable j times under the integral sign if the resulting expressions exist and are uniformly convergent, for which it is sufficient that ν_j exists. For then

$$\begin{aligned} |\phi^{(j)}(t)| &= \left| \int_{-\infty}^{\infty} x^j e^{itx} dF \right| \\ &\leqslant \int_{-\infty}^{\infty} |x^j| dF = \nu_j. \end{aligned} \tag{4.3}$$

$\phi(t)$ is real if and only if the distribution is symmetrical—cf. Exercise 4.1.

The Inversion Theorem

4.3 We now prove the fundamental theorem of the theory of characteristic functions, which will be called the Inversion Theorem, namely that the characteristic function (c.f.) uniquely determines the distribution function; more precisely, if $\phi(t)$ is given by (4.1) then

$$F(x) - F(0) = \frac{1}{2\pi} \int_{-\infty}^{\infty} \frac{1 - e^{-itx}}{it} \phi(t) dt, \tag{4.4}$$

the integral being understood as a principal value, i.e. as

$$\lim_{c\to\infty}\frac{1}{2\pi}\int_{-c}^{c}\frac{1-e^{-itx}}{it}\phi(t)dt.$$

Further, if $F(x)$ is continuous everywhere and $dF=f(x)\,dx$

$$f(x)=\frac{1}{2\pi}\int_{-\infty}^{\infty}e^{-itx}\phi(t)\,dt, \tag{4.5}$$

the integral, as before, being a principal value if there is not separate convergence at the limits. Equation (4.5) may be compared with the form

$$\phi(t)=\int_{-\infty}^{\infty}e^{itx}f(x)\,dx, \tag{4.6}$$

the comparison exhibiting the kind of reciprocal relationship that exists between $f(x)$ and $\phi(t)$.

Equation (4.4) contains $F(0)$ and we may avoid introducing this quantity by an alternative form

$$F(x)=\frac{1}{2}+\frac{1}{2\pi}\int_{0}^{\infty}\frac{e^{itx}\phi(-t)-e^{-itx}\phi(t)}{it}\,dt. \tag{4.7}$$

We require as a preliminary result the theorem that for a real number ξ,

$$\frac{1}{\pi}\int_{-\infty}^{\infty}\frac{e^{it\xi}}{it}\,dt=\frac{1}{\pi}\int_{-\infty}^{\infty}\frac{\sin t\xi}{t}\,dt=\left.\begin{matrix}-1, & \xi<0\\ 0, & \xi=0\\ 1, & \xi>0\end{matrix}\right\}=\mathrm{sgn}(\xi), \tag{4.8}$$

which we read as "signum ξ". Consider for $\xi>0$ complex integral $\int\frac{e^{iz\xi}}{iz}\,dz$ around an anticlockwise contour above the real axis consisting of (a) a large semicircle of radius R; (b) a small semicircle of radius r; and (c) the segments of the real axis connecting them. The integral is equal to zero, for there is no pole inside the contour, and of its three constituents, the integral over (a)$\to 0$ as $R\to\infty$; that over (b), a clockwise semicircle, $\to -\pi$ as $r\to 0$; and that over (c) then tends to the first integral in (4.8), which therefore equals π. If $\xi<0$, this has the effect of multiplying the second integral in (4.8) by -1, so that the same argument gives that integral the value $-\pi$; while if $\xi=0$, the first integral in (4.8), a principal value, is zero since the integrand is an odd function.

We note that for a variate u and some fixed x

$$\begin{aligned}\int_{-\infty}^{\infty}\mathrm{sgn}(u-x)dF(u)&=-\int_{-\infty}^{x}dF(u)+\int_{x}^{\infty}dF(u)\\&=1-2F(x).\end{aligned} \tag{4.9}$$

Now we have, for a positive number c, the uniformly convergent integral

$$I_c = \int_{-c}^{c} \frac{1-e^{-ixt}}{it}\phi(t)\mathrm{d}t = \int_{-c}^{c} \frac{1-e^{ixt}}{it}\int_{-\infty}^{\infty} e^{itu}\mathrm{d}F(u)\,\mathrm{d}t$$

$$= \int_{-c}^{c}\int_{-\infty}^{\infty} \frac{e^{itu}-e^{it(u-x)}}{it}\,\mathrm{d}F(u)\,\mathrm{d}t$$

$$= \int_{-c}^{c}\int_{-\infty}^{\infty} \frac{\sin tu - \sin\{t(u-x)\}}{t}\,\mathrm{d}F(u)\,\mathrm{d}t.$$

Because the integral is uniformly convergent, we may change the order of integration to obtain

$$I_c = \int_{-\infty}^{\infty}\int_{-c}^{c}\left\{\frac{\sin tu}{t} - \frac{\sin\{t(u-x)\}}{t}\right\}\mathrm{d}t\,\mathrm{d}F(u). \tag{4.10}$$

The integral with respect to t is continuous in c and bounded. We may therefore let c tend to infinity to obtain, from (4.8) and (4.9),

$$\lim_{c\to\infty} I_c = \pi\int_{-\infty}^{\infty}\{\operatorname{sgn} u - \operatorname{sgn}(u-x)\}\,\mathrm{d}F(u)$$

$$= \pi[1-2F(0) - \{1-2F(x)\}] = 2\pi\{F(x)-F(0)\}.$$

Equation (4.4) follows. Equation (4.7) may be obtained in a similar way by use of the result

$$\frac{2}{\pi}\int_{0}^{\infty}\frac{\sin t\xi}{t}\,\mathrm{d}t = \operatorname{sgn}\xi. \tag{4.11}$$

Further, if $f(x)$ exists we may differentiate (4.4) under the integral sign, since the integral converges uniformly, to obtain (4.5). We also have, on differentiating (4.7),

$$f(x) = \frac{1}{2\pi}\int_{0}^{\infty}\{e^{itx}\phi(-t) + e^{-itx}\phi(t)\}\,\mathrm{d}t. \tag{4.12}$$

4.4 From the definition of the c.f. we note that $\phi(t)$ and $\phi(-t)$ are conjugate quantities and hence, if $\mathscr{R}(t)$ and $\mathscr{I}(t)$ are the real and imaginary parts of $\phi(t)$ we have, from (4.4),

$$F(x) - F(0) = \frac{1}{2\pi}\int_{-\infty}^{\infty}\frac{\mathscr{R}(t)\sin xt + \mathscr{I}(t)(1-\cos xt)}{t}\,\mathrm{d}t, \tag{4.13}$$

and, from (4.7),

$$F(x) = \frac{1}{2} + \frac{1}{\pi}\int_{0}^{\infty}\frac{\mathscr{R}(t)\sin xt - \mathscr{I}(t)\cos xt}{t}\,\mathrm{d}t. \tag{4.14}$$

Similarly we have

$$f(x) = \frac{1}{2\pi}\int_{-\infty}^{\infty}\{\mathscr{R}(t)\cos xt + \mathscr{I}(t)\sin xt\}\,\mathrm{d}t. \tag{4.15}$$

4.5 Consider now the expression J_c defined by

$$J_c = \frac{1}{2c}\int_{-c}^{c} e^{-itx}\phi(t)\,dt. \tag{4.16}$$

If the distribution function $F(x)$ has a derivative $f(x)$ we have, from (4.5),

$$\lim_{c\to\infty} J_c = 2\pi \lim f(x)/2c = 0$$

and hence J_c tends to zero at all points where $F(x)$ is continuous and differentiable.

If the distribution function is discrete, consider one point of discontinuity, say where the probability is f_j. The contribution of this part of the probability to $\phi(t)$ is $f_j \exp(itx_j)$ and thus the contribution to J_c is

$$\frac{1}{2c}\int_{-c}^{c} f_j \exp(itx_j - itx)\,dt = \frac{1}{2c} f_j \left[\frac{\exp\{it(x_j - x)\}}{i(x_j - x)}\right]_{-c}^{c}.$$

If $x_j \neq x$ this clearly tends to zero as $c \to \infty$. But if $x_j = x$ the integral is simply

$$\frac{1}{2c}\int_{-c}^{c} f_j\,dt = f_j.$$

Thus J_c tends to f_j at $x = x_j$. Hence, if J_c tends to zero at a point, there is no discontinuity in the distribution function at that point; but if it tends to a positive number f_j the d.f. is discontinuous at that point and the probability is f_j. This gives us a criterion whether a given c.f. represents a continuous distribution or not.

4.6 Looking similarly to (4.6) we see that if the distribution has discrete probabilities at x_j ($j = 1, 2, \ldots$) the c.f. is given by

$$\phi(t) = \sum f_j \exp(itx_j)$$

and

$$\begin{aligned} F(x) &= \frac{1}{2} + \frac{1}{2\pi}\int_0^\infty \frac{\sum f_j \exp\{it(x - x_j)\} - \sum f_j \exp\{-it(x - x_j)\}}{it}\,dt \\ &= \frac{1}{2} + \frac{1}{\pi}\sum f_j \int_0^\infty \frac{\sin t(x - x_j)}{t}\,dt \\ &= \tfrac{1}{2} + \tfrac{1}{2}\sum f_j \operatorname{sgn}(x - x_j). \end{aligned} \tag{4.17}$$

This has a saltus at $x = x_j$ equal to f_j. We note, however, that for $x = x_k$

$$\begin{aligned} F(x_k) &= \tfrac{1}{2} + \tfrac{1}{2}\sum_{j=1}^{k-1} f_j - \tfrac{1}{2}\sum_{j=k+1}^{\infty} f_j \\ &= \tfrac{1}{2}\left(1 - \sum_{k+1}^{\infty} f_j\right) + \tfrac{1}{2}\sum_{1}^{k-1} f_j \\ &= \sum_{1}^{k-1} f_j + \tfrac{1}{2} f_k. \end{aligned} \tag{4.18}$$

The value assigned to $F(x)$ at the saltus is thus the probability up to that point plus one-half of the increase in probability at that point. On our ordinary definition, with $F(x)$ continuous on the right, the value of $F(x_k)$ would be $\sum_1^k f_j$. This is a point to remember when the distribution function of a discrete distribution is obtained from the characteristic function.

Example 4.1 (C.f. of the normal distribution)

In Example 3.4, replacing t by it shows that the c.f. of the normal distribution

$$f(x) = \frac{1}{\sigma\sqrt{(2\pi)}}\, e^{-x^2/(2\sigma^2)}, \qquad -\infty < x < \infty,$$

is

$$\phi(t) = \exp(-\tfrac{1}{2}t^2\sigma^2).$$

Suppose we are given this function and require to find the distribution, if any, of which it is the c.f.

In the first place we note that the distribution, if any, is continuous. For, from (4.16),

$$J_c = \frac{1}{2c}\int_{-c}^{c} \exp\{-\tfrac{1}{2}t^2\sigma^2 - itx\}\, dt$$

and the integral is less in modulus than $\int_{-c}^{c}\exp(-\tfrac{1}{2}t^2\sigma^2)dt$, which is less than $\int_{-\infty}^{\infty}\exp(-\tfrac{1}{2}t^2\sigma^2)dt = \sqrt{(2\pi)}/\sigma$. Thus $J_c \to 0$ everywhere. We have for the density function, if any,

$$\begin{aligned} f(x) &= \frac{1}{2\pi}\int_{-\infty}^{\infty} \exp\{-\tfrac{1}{2}t^2\sigma^2 - itx\}\, dt \\ &= \frac{1}{2\pi}\, e^{-x^2/(2\sigma^2)} \int_{-\infty}^{\infty} \exp\{-\tfrac{1}{2}(t\sigma + ix/\sigma)^2\}\, dt. \end{aligned}$$

This may be regarded as an integral in the complex plane along a line parallel to the real axis. Taking $t\sigma + ix/\sigma = u$ as a new variable we find that the integral is $\sqrt{(2\pi)}/\sigma)$. Thus

$$f(x) = \frac{1}{\sigma\sqrt{(2\pi)}}\, e^{-\frac{1}{2}x^2/\sigma^2}.$$

This is everywhere positive and $\int_{-\infty}^{\infty} dF$ converges to unity. Hence it is, in fact, a density function with the given expression for its c.f.

It will be seen that, with $\sigma = 1$, $f(x)$ and $\phi(t)$ are here of the same functional form apart from the constant $(2\pi)^{-\frac{1}{2}}$. This property is not unique to the normal distribution—Exercise 4.9 contains another instance.

Example 4.2 (C.f. of the Cauchy distribution)

To find the distribution, if any, for which

$$\phi(t) = e^{-|t|}.$$

We note that J_c at (4.16) tends to zero and that the distribution, if any, is continuous. We then have for $f(x)$, if it exists,

$$f(x) = \frac{1}{2\pi}\int_{-\infty}^{\infty} e^{-|t|}e^{-itx}dt$$

and using symmetries, we therefore have

$$\begin{aligned}\pi f(x) &= \int_0^{\infty} e^{-t}\cos tx\,dt \\ &= [-e^{-t}\cos tx]_0^{\infty} - x\int_0^{\infty} e^{-t}\sin tx\,dt \\ &= 1 - x\left\{[-e^{-t}\sin tx]_0^{\infty} + x\int_0^{\infty} e^{-t}\cos tx\,dt\right\} \\ &= 1 - x^2\pi f(x).\end{aligned}$$

Thus

$$f(x) = \frac{1}{\pi(1+x^2)}, \qquad -\infty < x < \infty.$$

As before, this function can represent a p.d.f., and Example 3.13 showed that $f(x)$ has, in fact, the required c.f. This p.d.f. is known as the Cauchy distribution, although it was discussed by Poisson in 1824, nearly 30 years earlier than by Cauchy.

Example 4.3 (Degenerate characteristic function)

Does there exist a probability function for which $\phi(t) = e^{it}$?

We have

$$J_c = \frac{1}{2c}\int_{-c}^{c} e^{it-itx}dt = \frac{1}{2c}\int_{-c}^{c} e^{it(1-x)}dt.$$

If $1-x$ is not zero the integral is

$$\int_{-c}^{c}[\cos\{(1-x)t\} + i\sin\{(1-x)t\}]dt.$$

Since sin t is an odd function this is equal to

$$\int_{-c}^{c} \cos\{(1-x)t\}\mathrm{d}t = \left[-\frac{\sin(1-x)t}{1-x}\right]_{-c}^{c}.$$

This does not converge, but it is bounded and hence $J_c \to 0$.

If, however, $x = 1$, the integral is simply

$$\int_{-c}^{c} \mathrm{d}t$$

and thus $J_c = 1$.

Thus there is unit probability at $x = 1$ and it is seen at once that this accounts for the whole distribution. The distribution thus consists of a unit at $x = 1$. This is otherwise evident from the consideration that $\log \phi(t) = it$, so that the mean is 1, the second cumulant is zero and there is no dispersion.

Example 4.4 (Inversion of c.f. for the gamma distribution)

For what distribution, if any, are the cumulants given by $\kappa_r = (r-1)!$?

The series

$$\sum_{j=1}^{\infty} \kappa_j \frac{(it)^j}{j!} = \sum \frac{(it)^j}{j}$$

converges absolutely for $|t| < 1$ and is thus equal to the c.g.f., say $\psi(t)$, if such a function exists. We have

$$\psi(t) = \sum \frac{(it)^j}{j} = -\log(1-it),$$

and thus

$$\phi(t) = \frac{1}{1-it}.$$

If the density function exists we have

$$f(x) = \frac{1}{2\pi}\int_{-\infty}^{\infty} \frac{\mathrm{e}^{-itx}}{1-it}\,\mathrm{d}t.$$

This integral may be evaluated by integrating the complex function $\dfrac{\mathrm{e}^{-ixz}}{1-iz}$ round a contour consisting of the real axis and the infinite semicircle below that axis. The first part reduces to the integral we are seeking with its sign reversed. On the semicircle of radius R we have $z = R(\cos\theta + i\sin\theta)$ and the integrand becomes

$$\frac{\exp(-ixR\cos\theta + xR\sin\theta)}{1 - iR\cos\theta + R\sin\theta}.$$

θ here lies between π and 2π and hence $\sin\theta$ is negative. Hence *if x is positive* the expression is less in modulus than

$$\frac{e^{-xR|\sin\theta|}}{R},$$

i.e. tends to zero as $R \to \infty$.

Now the function $\dfrac{e^{-ixz}}{1-iz}$ has a pole within the domain of integration at $z=-i$ and the residue there is ie^{-x}. Hence

$$f(x) = -\frac{1}{2\pi} \cdot -2\pi e^{-x}$$

$$= e^{-x}, \qquad 0 \leqslant x < \infty.$$

More generally, if $\kappa_r = p(r-1)!$, $p>0$, will be found that the residue of $\dfrac{e^{-ixz}}{(1-iz)^p}$ is $\dfrac{ix^{p-1}e^{-x}}{\Gamma(p)}$, so that the distribution is

$$f(x) = \frac{e^{-x}x^{p-1}}{\Gamma(p)}, \qquad 0 \leqslant x < \infty; \quad p>0.$$

Example 4.5 (Inversion of a c.f. of given form)

For what distribution, if any, are all cumulants of odd order zero and those of even order a constant, say $2a$?

We have for the c.g.f.

$$\psi(t) = 2a\left\{\frac{(it)^2}{2!} + \frac{(it)^4}{4!} + \dots\right\}.$$

This series converges and

$$\psi(t) = 2a(\cos t - 1).$$

Hence

$$\phi(t) = e^{2a(\cos t - 1)}$$

Using (4.16), we investigate the continuity of the d.f.

We have

$$J_c = \frac{e^{-2a}}{2c}\int_{-c}^{c} e^{2a\cos t}e^{-itx}dt$$

$$= \frac{e^{-2a}}{2c}\int_{-c}^{c}\sum_{j=0}^{\infty}\frac{(2a)^j}{j!}\cos^j t\, e^{-itx}dt.$$

The series is uniformly convergent and hence

$$J_c = \frac{e^{-2a}}{2c}\sum_{j=0}^{\infty}\frac{(2a)^j}{j!}\int_{-c}^{c}\cos^j t\, e^{-itx}dt$$

$$= \frac{e^{-2a}}{2c}\sum_{j=0}^{\infty}\frac{a^j}{j!}\int_{-c}^{c} 2^j\cos^j t\cos xt\, dt,$$

since sin xt is an odd function. Now,

$$2^j \cos^j t \cos xt = \tfrac{1}{2}(e^{it}+e^{-it})^j(e^{ixt}+e^{-ixt})$$
$$= \tfrac{1}{2}(e^{ixt}+e^{-ixt})\sum_{r=0}^{j}\binom{j}{r}e^{it(j-2r)}.$$

Since this integrand is an even function of t, the important term upon integration will be the constant term, which is zero unless x is an integer, and in that case is

$$\frac{1}{2}\left\{\binom{j}{\frac{1}{2}(j-x)}+\binom{j}{\frac{1}{2}(j+x)}\right\}=\binom{j}{\frac{1}{2}(j-x)},$$

so J_c tends to zero unless x is an integer, when

$$J_c \to e^{-2a}\sum_{j=x}^{\infty}\frac{a^j}{j!}\binom{j}{\frac{1}{2}(j-x)},$$

where j increases by steps of 2 units.

Thus the p.m.f. is

$$f(x)=e^{-2a}\left(\frac{a^x}{x!0!}+\frac{a^{x+2}}{(x+1)!1!}+\frac{a^{x+4}}{(x+2)!2!}+\dots\right),$$

which holds for $x=0, \pm 1, \pm 2, \dots$; by Exercise 4.1 the distribution is symmetrical since $\phi(t)$ is real.

We may now verify that these probabilities account for the whole of the c.f. and hence that all the terms have been found. (Cf. Irwin, 1937.)

Conditions for a function to be a characteristic function

4.7 Any function that is not negative in its range of definition and is integrable in the Stieltjes sense can be a probability function; and any non-decreasing function that increases from 0 to 1 in its range of definition can be a d.f. There are much more restrictive conditions to be obeyed before a given function can be a c.f.

In the first place, let us note that it is a necessary and sufficient condition for a function $\phi(t)$ to be a c.f. that the integral at (4.4)

$$\frac{1}{2\pi}\int_{-\infty}^{\infty}\frac{1-e^{-itx}}{it}\phi(t)\,dt$$

shall (except for an additive constant $F(0)$) be a d.f. This, however, is not a very helpful criterion in practice.

Looking to the definition of $\phi(t)$ as $\int_{-\infty}^{\infty}e^{itx}dF$, we see that necessary conditions for $\phi(t)$ to be a c.f. are

(a) that $\phi(t)$ be continuous in t,
(b) that $\phi(t)$ is defined in every finite t-interval,
(c) that $\phi(0)=1$,
(d) that $\phi(t)$ and $\phi(-t)$ shall be conjugate quantities,
(e) that $|\phi(t)| \leqslant \int_{-\infty}^{\infty}|e^{itx}|dF = 1 = \phi(0)$.

Sufficient conditions for $\phi(t)$ to be a c.f. are

(a) that $\phi(0)=1$ and $\phi(t)\to 0$, as $|t|\to\infty$,
(b) that $\phi(t)$ is an even function of t,
(c) that $\phi(t)$ is a convex function over $(0,\infty)$.

These conditions were first established by Pólya (cf. Feller, 1971, Chapter 15) and have been extended by Shimizu (1972) to cover non-symmetric real-valued functions.

4.8 A general theorem of Cramér (1937) states that for a bounded and continuous function $\phi(t)$ to be a c.f. it is necessary and sufficient that $\phi(0)=1$ and that

$$\int_0^A \int_0^A \phi(t-u)\exp\{ix(t-u)\}\,dt\,du$$

is real and non-negative for all real x and all $A>0$.

There are also some notable specialized theorems on the subject. For example Marcinkiewicz (1938) proved that $\exp\{P(t)\}$, where $P(t)$ is a polynomial in t, cannot be a c.f. unless $P(t)$ is of the first or second degree—cf. Exercise 4.5 for a simplified version, and Exercise 4.11. Lukacs and Szász (1952) have given conditions under which a *rational* function can be a c.f. See also Exercise 4.10.

Limiting properties of distribution and characteristic functions

4.9 We now proceed to discuss a set of results involving the tendency of d.f.s and c.f.s to limiting forms. Suppose there is given a sequence of d.f.s $F_n(x)$ depending on a parameter n that can increase indefinitely. To each F_n there will correspond a c.f. ϕ_n. The type of question to be discussed is: if F_n tends to a limit G, will ϕ_n tend to a limit ϕ and is ϕ the c.f. of G? Conversely, if ϕ_n tends to a limit ϕ, does F_n tend to a limit G and is G a d.f. having ϕ for its c.f.? The answers to these questions, as will be seen below, are affirmative under certain general conditions.

4.10 Let us first of all consider what is meant by a d.f. tending to another. If both are continuous, $F_n(x)$ is said to tend to $G(x)$ if, given any ε, there is an n_0 such that $|F_n(x)-G(x)|<\varepsilon$ for all $n>n_0$ and for all x. If there are discontinuities present, F_n will be said to tend to G if it does so in every point of continuity of G. Since by definition our functions are taken to be continuous on the right at saltuses, this evidently conforms to the definition for the continuous case and to the common-sense requirements of the situation.

4.11 We require a number of results concerning the convergence of d.f.s. The first is that if a sequence of d.f.s $\{F_n(x)\}$ converges to the d.f. $G(x)$ at all its points of continuity, then the convergence is uniform in every closed interval of continuity of $G(x)$.

Let $G(x)$ be continuous in the range a to b. Divide the range into a finite number of parts, say at $a=\xi_1, \xi_2, \ldots, \xi_h=b$ such that the increase of $G(x)$ in each interval is at most ε. Choose n_0 so large that $|F_n(\xi_j)-G(\xi_j)|<\varepsilon, j=1,\ldots,h$, $n>n_0$. Then for such an n and any x in the interval (a,b) there is an r such that x lies in (ξ_r, ξ_{r+1}) and

$$G(x)-2\varepsilon \leqslant G(\xi_r)-\varepsilon < F_n(\xi_r) \leqslant F_n(x) \leqslant F_n(\xi_{r+1}) < G(\xi_{r+1})+\varepsilon \leqslant G(x)+2\varepsilon \quad (4.19)$$

and hence

$$|F_n(x)-G(x)|<2\varepsilon.$$

This is true for all x in (a,b) and hence the theorem is established.

In particular, if $G(x)$ is continuous the convergence is uniform throughout the whole interval $-\infty < x < \infty$.

4.12 The second theorem we require (the Montel–Helly theorem) is that if the sequence $\{F_n(x)\}$ is monotonic and bounded for all x (which is so for d.f.s) then we can pick out a subsequence $\{F_{n'}(x)\}$ which converges to some monotonic increasing function G (not necessarily a d.f. itself, for it may not vary from 0 to 1).

Consider first of all a series of values $x_1, x_2, \ldots$. It is known that every bounded set of numbers contains a convergent sequence. Hence we can pick out from the sequence $\{F_n(x_1)\}$ a convergent sequence, say $\{F_{n_1}(x_1)\}$. Then from the subsequence $\{F_{n_1}(x_2)\}$ we can pick out a subsequence $\{F_{n_2}(x_2)\}$ and $\{F_{n_2}(x)\}$ is thus convergent at both x_1 and x_2. Continuing in this way we may, by picking out the first function in $\{F_{n_1}(x)\}$, the second in $\{F_{n_2}(x)\}$, and so on, arrive at a sequence of functions $G_1(x)$, $G_2(x)$ ... that converges at each of the values $x_1, x_2, \ldots$, etc. This is the so-called Weierstrassian diagonal process.

It follows that the sequence G_n is convergent at every *rational* point x. Since $G_n(a) \leqslant G_n(x) \leqslant G_n(b)$ for every x between a and b, we see that if $G_n(a)$ and $G_n(b)$ converge, the limiting values of $G_n(x)$ lie between those limits, say $G(a)$ and $G(b)$.

Then the function $u(x) =$ upper bound of $G_n(x)$ (x not necessarily rational) is well defined and non-decreasing and so has no more than a denumerable number of points of discontinuity. If u is continuous at x, we take y and z such that $y<x<z$ and $u(z)-u(y)<\varepsilon$. Then if a and b are rational points such that $y<a<x<b<z$ it follows that $u(y) \leqslant G(a) \leqslant G(b) \leqslant u(z)$. Moreover, as all the limiting values of $G_n(x)$ are between $G(a)$ and $G(b)$, they are between $u(y)$ and $u(z)$. Hence, as ε can be arbitrarily small, we see that $G(x)$ tends to $u(x)$ at every point of continuity of u. Finally, by the diagonal process, we can select a sequence that will also be convergent at the points of discontinuity of $u(x)$. The theorem is established.

4.13 It is not true that if a sequence of functions $\{F_n(x)\}$ all lie between 0 and 1 inclusive then their limit $G(x)$ can attain those values. A counter-example is furnished by the distribution.

$$\left.\begin{aligned} F_n(x) &= 0, && x< -n,\\ &= \tfrac{1}{2}, && -n \leqslant x \leqslant n,\\ &= 1, && x>n. \end{aligned}\right\} \tag{4.20}$$

The limit of $F_n(x)$ here is $\frac{1}{2}$, $-\infty < x < \infty$. A condition that is sufficient (but not necessary) has been given by D. G. Kendall and Rao (1950).

If $g(x)$ is a monotone non-decreasing function such that $\int |g(x)|\, dF_n(x) \leqslant A$ for all $n > n_0$ and $|g(x)| \to \infty$ as $|x| \to \infty$, then $G(x)$ is a d.f. We may take $E(x)=0$ without loss of generality.

From the generalization of the Bienaymé–Chebyshev inequality (3.97)

$$F_n(-u) + 1 - F_n(u) \leqslant A/g(u), \quad n > n_0,$$

and hence, for sufficiently large u

$$F_n(-u)+1-F_n(u) \leqslant \varepsilon,$$

provided that $|g(u)| \to \infty$ with u. Since this is so uniformly in u,

$$G(-u)+1-G(u) \leqslant \varepsilon \tag{4.21}$$

provided that (as we can ensure without loss of generality) $G(x)$ is continuous at $\pm u$. Thus $G(-\infty)=0$ and $G(\infty)=1$.

The First Limit Theorem

4.14 We now prove the theorem: if a sequence of d.f.s $\{F_n\}$ tends to a continuous d.f. G, then the corresponding sequence of c.f.s ϕ_n tends to ϕ uniformly in any finite t-interval, where ϕ is the c.f. of G.

It is required to prove that, given ε, there is an n_0 independent of t such that

$$|\phi(t)-\phi_n(t)| = \left|\int_{-\infty}^{\infty} e^{itx}(dG-dF_n)\right| < \varepsilon, \qquad n > n_0.$$

Select two points of continuity of G, u and $-u$. We can make u as large as we please. We then split the integral

$$\int_{-\infty}^{\infty} e^{itx}(dG-dF_n) \tag{4.22}$$

into two parts, that in the range $-u$ to $+u$ and that in the remaining portion of the range. Now

$$\left|\int_{x<-u}^{x>u} e^{itx}dG\right| \leqslant \int_{x<-u}^{x>u} dG \leqslant 1-G(u)+G(-u),$$

and by taking u large enough we can make this quantity less than $\varepsilon/6$.

Similarly

$$\left|\int_{x<-u}^{x>u} e^{itx}dF_n\right| \leqslant 1-F_n(u)+F_n(-u),$$

and since F_n tends to G (and that uniformly) this, for some large u, will be less than $\varepsilon/3$. Hence for some n_0 the portion of (4.22) outside the range $-u$ to $+u$ will be less in modulus than $\varepsilon/6+\varepsilon/3=\frac{1}{2}\varepsilon$. Consider now the other part

$$\int_{-u}^{u} e^{itx}(dG-dF_n). \tag{4.23}$$

This expression is the limit of the sum

$$\sum e^{itx_j}[\{G(\xi_{j+1})-G(\xi_j)\}-\{F_n(\xi_{j+1})-F_n(\xi_j)\}], \tag{4.24}$$

ξ_j, ξ_{j+1} being the boundaries of the interval into which the range is subdivided and x_j a value in that interval. The difference between this sum and the limiting value can be made less than $\frac{1}{4}\varepsilon$ if the intervals are small enough; for if they are less than η in width the difference of e^{itx_j} and $e^{it\xi_j}$ is less in modulus than $\eta|t|$, by the mean value theorem, and thus in any t-range $\pm T$ the difference of (4.23) and (4.24) is less in modulus than

$$\eta T\left|\sum [\{G(\xi_{j+1})-G(\xi_j)\}-\{F_n(\xi_{j+1})-F_n(\xi_j)\}]\right|<2\eta T,$$

which is less than $\frac{1}{4}\varepsilon$ if $\eta<\varepsilon/(8T)$.

Now the sum of (4.24) will itself be less than $\frac{1}{4}\varepsilon$ for some $n>n_0$, for it is the sum of a finite number of terms each of which tends to zero. Consequently (4.23) is less than $\frac{1}{2}\varepsilon$ and hence

$$|\phi(t)-\phi_n(t)|<\varepsilon, \qquad n>n_0.$$

Converse of the First Limit Theorem

4.15 The converse result is even more important.

Let $\{\phi_n\}$ be a sequence of c.f.s corresponding to the sequence of d.f.s $\{F_n\}$. Then if $\phi_n(t)$ tends to $\phi(t)$ for all real t, and uniformly in a finite t-interval $|t|<a$,(*) $\{F_n\}$ tends to a d.f. G and ϕ is the c.f. of G.

As a preliminary lemma, let us prove that if G is a d.f. with c.f. ϕ, then for all real ξ and all $h>0$

$$\frac{1}{h}\int_{\xi}^{\xi+h}G(u)\mathrm{d}u-\frac{1}{h}\int_{\xi-h}^{\xi}G(u)\mathrm{d}u=\frac{1}{\pi}\int_{-\infty}^{\infty}\left(\frac{\sin t}{t}\right)^2 e^{-2it\xi/h}\phi\left(\frac{2t}{h}\right)\mathrm{d}t. \qquad (4.25)$$

In fact, put

$$H(x)=\frac{1}{h}\int_x^{x+h}G(u)\mathrm{d}u.$$

This is a continuous d.f. and its c.f. is

$$\int_{-\infty}^{\infty}e^{itx}\mathrm{d}H=\frac{1}{h}\int_{-\infty}^{\infty}e^{itx}\{G(x+h)-G(x)\}\,\mathrm{d}x,$$

which by a partial integration becomes

$$\frac{1}{h}\left[\frac{\{G(x+h)-G(x)\}e^{itx}}{it}\right]_{-\infty}^{\infty}-\frac{1}{ith}\int_{-\infty}^{\infty}e^{itx}\{\mathrm{d}G(x+h)-\mathrm{d}G(x)\}$$

$$=-\frac{1}{ith}\int_{-\infty}^{\infty}\{e^{it(x-h)}\mathrm{d}G(x)-e^{itx}\mathrm{d}G(x)\}$$

$$=\phi(t)\frac{1-e^{-ith}}{ith}.$$

(*) Or equivalently, if $\phi(t)$ is continuous at $t=0$ or if $\phi(t)$ is a c.f.

Substituting for $H(x)$ in (4.4) we get

$$\frac{1}{h}\int_{\xi+h}^{\xi+2h} G(u)\mathrm{d}u - \frac{1}{h}\int_{\xi}^{\xi+h} G(u)\mathrm{d}u = \frac{1}{2\pi h}\int_{-\infty}^{\infty}\left(\frac{1-\mathrm{e}^{-ith}}{it}\right)^2 \mathrm{e}^{-it\xi}\phi(t)\mathrm{d}t,$$

whence, writing ξ for $\xi+h$, we find, after a little re-arrangement, equation (4.25).

Reverting now to the theorem required to be proved, note that it is sufficient to establish that if $\phi_n \to \phi$ uniformly in *some* interval $|t| < a$, then $\{F_n\}$ tends to some d.f. G in every point of continuity of G. When this is established it follows from the First Limit theorem that ϕ is the c.f. of G and that ϕ_n converges to ϕ uniformly in *every* finite t-interval.

As shown in **4.12**, given a sequence $\{F_n\}$ we may always choose from it a subsequence $\{F_{n'}\}$ such that $\{F_{n'}\}$ converges to a non-decreasing function G in every continuity point of G.

Let us then choose such a sequence. We have of necessity $0 \leqslant G \leqslant 1$, and G may be supposed everywhere continuous on the right. It is then a d.f. if $G(+\infty) - G(-\infty) = 1$, and this we proceed to prove. From (4.25) with $\xi = 0$ we have

$$\frac{1}{h}\int_0^h F_{n'}(u)\mathrm{d}u - \frac{1}{h}\int_{-h}^0 F_{n'}(u)\mathrm{d}u = \frac{1}{\pi}\int_{-\infty}^{\infty}\left(\frac{\sin t}{t}\right)^2 \phi_{n'}\left(\frac{2t}{h}\right)\mathrm{d}t.$$

By hypothesis ϕ_n tends uniformly to ϕ for $|t| < a$ and hence $\phi_{n'}$ does so, and it is easily seen that the integral on the right is uniformly convergent. Thus, given ε, we can find h_0 such that for $h > h_0$

$$\left(\frac{1}{h}\int_0^h - \frac{1}{h}\int_{-h}^0\right) G(u)\mathrm{d}u = \frac{1}{\pi}\int_{-\frac{1}{2}ah}^{\frac{1}{2}ah}\left(\frac{\sin t}{t}\right)^2 \phi\left(\frac{2t}{h}\right)\mathrm{d}t + \eta,$$

where $|\eta| < \varepsilon$. Now let h tend to infinity. As G is a non-decreasing function the left-hand side tends to $G(+\infty) - G(-\infty)$. The right-hand side tends, in virtue of the uniformity of $\phi_{n'}$ and the consequent continuity of ϕ near $t = 0$, to

$$\frac{1}{\pi}\int_{-\infty}^{\infty}\left(\frac{\sin t}{t}\right)^2 \mathrm{d}t,$$

which is equal to unity.

Hence G, the limit of the subsequence $\{F_{n'}\}$, is a d.f. whose c.f. is ϕ.

But any sequence of ϕ_n tends to ϕ, in virtue of the uniformity of the convergence, and hence any convergent subsequence of $\{F_n\}$ tends to G. Consequently $\{F_n\}$ tends to G in every point of continuity of G and the theorem follows.

Example 4.6 (Limiting distributions for the binomial)

In Example 3.5, it was shown in effect that the binomial distribution has the c.f.

$$\phi(t) = (q + pe^{it})^n.$$

Now the probability at $x = j$ is $\binom{n}{j} q^{n-j} p^j$. This is greater than the probability at

$x = j+1$ if

$$\binom{n}{j} q^{n-j} p^j > \binom{n}{j+1} q^{n-j-1} p^{j+1}$$

or

$$j > pn - q.$$

For large n the maximum probability will then be in the neighbourhood of $j = pn$, and is there

$$\binom{n}{pn} q^{qn} p^{pn}.$$

In virtue of Stirling's approximation to the factorial in (3.64) this approximates to

$$\frac{n^n e^{-n} \sqrt{(2\pi n)} q^{qn} p^{pn}}{(pn)^{pn} e^{-pn} \sqrt{(2\pi pn)} (qn)^{qn} e^{-qn} \sqrt{(2\pi qn)}} \sim \frac{1}{\sqrt{(2\pi pqn)}}$$

and therefore tends to zero.

Thus every probability in the binomial distribution tends to zero and the distribution does not tend to any limiting distribution.

Suppose, however, that we standardize the distribution. Putting $y = (x - \mu_1')/\sigma$ we have

$$\phi_x(t) = \int_{-\infty}^{\infty} e^{itx} dF(x) = \int_{-\infty}^{\infty} \exp\{it(\sigma y + \mu_1')\} dF(y)$$
$$= \exp(it\mu_1')\phi_y(\sigma t).$$

Hence

$$\phi_y(t) = \exp(-it\mu_1'/\sigma)\phi_x(t/\sigma).$$

The effect on $\phi(t)$ of standardizing is then to replace t by t/σ and to multiply by $\exp(-it\mu_1'/\sigma)$.

For the binomial, $\mu_1' = np$, $\mu_2 = npq$, and thus the c.f. of the standardized binomial is

$$\exp\left\{-\frac{itnp}{(npq)^{\frac{1}{2}}}\right\}\left\{q + p\exp\frac{it}{(npq)^{\frac{1}{2}}}\right\}^n.$$

Thus

$$\begin{aligned}
\log\phi &= \frac{-itnp}{(npq)^{\frac{1}{2}}} + n\log\left\{1 + p\left(\exp\frac{it}{(npq)^{\frac{1}{2}}} - 1\right)\right\} \\
&= \frac{-itnp}{(npq)^{\frac{1}{2}}} + n\log\left\{1 + \frac{pit}{(npq)^{\frac{1}{2}}} - \frac{pt^2}{2npq} + \frac{p\theta t^3}{6(npq)^{\frac{3}{2}}} + \cdots\right\}, \qquad 0 \leqslant |\theta| \leqslant 1, \\
&= n\left\{\frac{-pt^2}{2npq} - \frac{1}{2}\frac{(pit)^2}{npq}\right\} + O(t^3 n^{-\frac{1}{2}}) \\
&= -\tfrac{1}{2}t^2 + O(t^3 n^{-\frac{1}{2}}).
\end{aligned}$$

Thus for any finite t, log ϕ tends uniformly to $-\frac{1}{2}t^2$ and hence

$$\phi(t) \to e^{-\frac{1}{2}t^2}.$$

Thus the standardized binomial distribution tends to the distribution whose c.f. is $e^{-\frac{1}{2}t^2}$; from Example 4.1 this is the c.f. of the normal distribution with $\mu = 0$, $\sigma = 1$. It is important to recognize the nature of this convergence. The d.f. of the binomial approaches that of the normal, but as the former is a step function, the p.m.f of the binomial does not approach the normal p.d.f. at any point.

Multivariate characteristic functions

4.16 The joint c.f. of a bivariate distribution $F(x_1, x_2)$ is defined as at (3.73) by

$$\phi(t_1, t_2) = \int_{-\infty}^{\infty}\int_{-\infty}^{\infty} e^{it_1x_1 + it_2x_2} dF(x_1, x_2) \tag{4.26}$$

and generally, that of a multivariate distribution $F(x_1, x_2, \ldots, x_n)$ as

$$\phi(t_1, t_2, \ldots, t_n) = \int_{-\infty}^{\infty}\int_{-\infty}^{\infty} \cdots \int_{-\infty}^{\infty} e^{it_1x_1 + it_2x_2 \ldots + it_nx_n} dF(x_1, x_2, \ldots, x_n). \tag{4.27}$$

If any subset of the t_j are put equal to zero, we obtain the c.f. of the (marginal) distribution of the other variables; e.g., $\phi(t_1, 0)$ is the c.f. of x_1 alone, and so on.

If $x_1, x_2, \ldots, x_n$ are independent we have

$$\begin{aligned}\phi(t_1, t_2, \ldots, t_n) &= \int_{-\infty}^{\infty} e^{it_1x_1} dF_1(x_1) \int_{-\infty}^{\infty} e^{it_2x_2} dF_2(x_2) \ldots \int_{-\infty}^{\infty} e^{it_nx_n} dF_n(x_n) \\ &= \phi_1(t_1)\phi_2(t_2) \ldots \phi_n(t_n)\end{aligned} \tag{4.28}$$

where ϕ_r is the c.f. of F_r. Similarly

$$\psi(t_1, t_2, \ldots, t_n) = \sum_{j=1}^{n} \log \phi_j(t_j). \tag{4.29}$$

Thus the joint c.f. of a number of independent variables is the product of their c.f.s; and the joint c.g.f. is the sum of their c.g.f.s. This is a fundamentally important result in the theory of sampling.

4.17 In generalization of (4.4) we have

$$\begin{aligned}F(x_1, x_2, \ldots, x_n) - F(0, x_2, \ldots x_n) \ldots - F(x_1, x_2, \ldots, 0) + F(0, 0, x_3, \ldots, x_n) + \ldots \\ + F(x_1, x_2, \ldots, 0, 0) - \ldots = \frac{1}{(2\pi)^n} \int_{-\infty}^{\infty} \cdots \int_{-\infty}^{\infty} \frac{1 - e^{ix_1t_1}}{it_1} \cdots \frac{1 - e^{-ix_nt_n}}{it_n} \\ \phi(t_1, t_2, \ldots, t_n) dt_1 \ldots dt_n.\end{aligned} \tag{4.30}$$

There are 2^n terms on the left of (4.30), $\binom{n}{r}$ of which have r zeros among the arguments of F. The integrals are to be understood as principal values $\lim_{c \to \infty} \int_{-c}^{c} \cdots \int_{-c}^{c}$.

The proof is similar to that for the univariate case. For example, with two variates we have, in an obvious generalization of (4.10),

$$I_c=\int_{-\infty}^{\infty}\int_{-\infty}^{\infty}\int_{-c}^{c}\int_{-c}^{c}\left\{\frac{\sin t_1u_1}{t_1}-\frac{\sin t_1(u_1-x_1)}{t_1}\right\}\left\{\frac{\sin t_2u_2}{t_2}-\frac{\sin t_2(u_2-x_2)}{t_2}\right\}\mathrm{d}t_1\mathrm{d}t_2\mathrm{d}F(u_1,u_2)$$

and

$$\begin{aligned}\lim I_c&=\lim\int_{-\infty}^{\infty}\int_{-c}^{c}\left\{\frac{\sin t_1u_1}{t_1}-\frac{\sin t_1(u_1-x_1)}{t_1}\right\}\mathrm{d}t_1(2\pi)\{\mathrm{d}F(u_1,x_2)-\mathrm{d}F(u_1,0)\}\\&=(2\pi)^2\{F(x_1,x_2)-F(0,x_2)-F(x_1,0)+F(0,0)\}.\end{aligned}$$

Similarly, (4.5) generalizes to

$$f(x_1,x_2,\ldots,x_n)=\frac{1}{(2\pi)^n}\int_{-\infty}^{\infty}\cdots\int_{-\infty}^{\infty}\exp\left(-i\sum_{j=1}^{n}t_jx_j\right)\phi(t_1,t_2,\ldots,t_n)\mathrm{d}t_1\ldots\mathrm{d}t_n. \tag{4.31}$$

As in the univariate case, the c.f. defines the distribution uniquely and the corresponding analogues of the limit theorems also remain true. Thus the converse of the result of **4.16** is ensured by the unique correspondence of the c.f. and the d.f.: if a joint c.f. factorizes into the individual c.f.s, the variables are independent.

Sufficient conditions for $\phi(t)$ to be a multivariate c.f. are given by Plachky and Thomsen (1981).

4.18 If we have a distribution $F(x)$ and some function of the variate such as y $(=y(x))$, we define the c.f. of y as

$$\phi_y(t)=\int_{-\infty}^{\infty}\exp(ity)\mathrm{d}F(x). \tag{4.32}$$

The distribution of y is then given by inversion; for instance, the probability function of y, say $g(y)$, is given by

$$g(y)=\frac{1}{2\pi}\int_{-\infty}^{\infty}\exp(-ity)\phi_y(t)\mathrm{d}t, \tag{4.33}$$

and a similar result holds for a function g of several variates.

Conditional characteristic functions

4.19 We may express the c.f. of a conditional distribution in terms of the underlying multivariate c.f. In the bivariate case, e.g. (4.26) may be written, using **1.32–3**, as

$$\begin{aligned}\phi(t_1,t_2)&=\int_{-\infty}^{\infty}\int_{-\infty}^{\infty}\mathrm{e}^{it_1x_1+it_2x_2}p(x_1\,|\,x_2)f_2(x_2)\mathrm{d}x_1\mathrm{d}x_2\\&=\int_{-\infty}^{\infty}\mathrm{e}^{it_2x_2}\phi_{12}(t_1)f_2(x_2)\mathrm{d}x_2,\end{aligned}$$

where ϕ_{12} is the c.f. of the conditional distribution $p(x_1|x_2)$. We see that here $\phi_{12}(t_1)f_2(x_2)$ plays the role usually taken by $f_2(x_2)$ in obtaining the c.f. of the marginal distribution of x_2, $\phi_2(t_2)$. Thus, applying (4.5),

$$2\pi\phi_{12}(t_1)f_2(x_2) = \int_{-\infty}^{\infty} \phi(t_1, t_2)e^{-it_2x_2}dt_2.$$

Dividing this equation by its value at $t_1 = 0$, when $\phi_{12}(0) = 1$, we find the required conditional c.f.

$$\phi_{12}(t_1) = \int_{-\infty}^{\infty} \phi(t_1, t_2)e^{-it_2x_2}dt_2 \Big/ \int_{-\infty}^{\infty} \phi(0, t_2)e^{-it_2x_2}dt_2. \tag{4.34}$$

Essentially the same result holds if x_1, x_2 are sets of variables, now using (4.31)—the general result will be given when we discuss multiple regression in Volume 2A.

The problem of moments

4.20 It is of some interest to consider how far a set of moments (assuming that they all exist) determine a distribution uniquely. To give some point to the discussion let us note that in some circumstances it is possible for two different distributions to have the same set of moments.

Consider the integral

$$\int_0^{\infty} t^{p-1}e^{-qt}dt = \Gamma(p)/q^p, \qquad p > 0, \qquad \mathcal{R}(q) > 0.$$

Put $p = (n+1)/\lambda$, n a non-negative integer; $0 < \lambda < \frac{1}{2}$; $q = \alpha + i\beta$; $\beta/\alpha = \tan\lambda\pi$; $x^\lambda = t$. We find on substitution, since $(1 + i\tan\lambda\pi)^{(n+1)/\lambda}$ is real, that the imaginary part of the integral is zero and thus that

$$\int_0^{\infty} x^n \exp(-\alpha x^\lambda)\sin(\beta x^\lambda)dx = 0.$$

Hence the distributions

$$f(x) = k\exp(-\alpha x^\lambda)\{1 + \varepsilon\sin(\beta x^\lambda)\}, \quad 0 \leqslant x < \infty; \quad \alpha > 0; \quad 0 < \lambda < \tfrac{1}{2}; \quad |\varepsilon| < 1 \tag{4.35}$$

have the same moments for all ε in the range $|\varepsilon| < 1$. This is not a case of practical significance, but Exercise 6.21 below deals with an important distribution (the lognormal) that is not determined by its moments.

We may derive a similar family with range infinite in both directions by putting $p = (2n+1)/\rho$, $q = \alpha + i\beta$, $\beta/\alpha = \tan\frac{1}{2}\rho\pi$, $x^\rho = t$, $\rho = 2s/(s+1)$, s a positive integer. The family then becomes

$$f(x) = k\exp\{-\alpha|x|^\rho\}\{1 + \varepsilon\cos(\alpha|x|^\rho)\}, \qquad -\infty < x < \infty. \tag{4.36}$$

Shantaram and Harkness (1972) show that it is possible to construct uncountably many distributions all having the same moments, provided the d.f. satisfies the functional equation

$$F(x) = c\int_0^{ax} [1 - F(u)]\,\mathrm{d}u,$$

$a > 1$, $x > 0$.

Leipnik (1981) discusses the "strong" non-uniqueness of the moments sequence for the lognormal distribution. Holgate (1989) provides several series representations for the c.f. of the lognormal.

4.21 The problem of moments in its full generality considers a set of constants $c_0, c_1, c_2, \ldots$ and inquires whether they can be the moments of a distribution. For statistical purposes this is not of particular interest.(*) We are more concerned with the problem: given that the set of constants are, in fact, the moments of a distribution, can any other distribution have the same set?

Note in the first instance that this problem need only be considered when absolute moments of all orders exist. It is not difficult to see that more than one distribution can exist having a limited number of given moments finite and the remainder infinite.

We will prove in the first place the theorem that a set of moments determines a distribution uniquely if the series $\sum_{j=0}^{\infty} \nu_j t^j/j!$ converges for some real non-zero t. We write ν and μ without the prime in this and the following sections, but the moments may be about any origin.

The c.f. is continuous in t and by 3.5 its derivatives exist at $t = 0$ if the moments exist. We have then in the neighbourhood of $t = 0$ the finite expansion

$$\phi(t) = \sum_{j=0}^{r} (it)^j \mu_j/j! + R_r, \tag{4.37}$$

where R_r is less in absolute value than $\dfrac{\nu_{r+1}|t|^{r+1}}{(r+1)!}$.

Thus if $\sum_{j=0}^{\infty} \nu_j t^j/j!$ converges, $\nu_j t^j/j!$ tends to zero and hence $\phi(t)$ is equal to the sum of the series $\sum_{j=0}^{\infty} (it)^j \mu_j/j!$ if it exists. Moreover, this series is dominated by $\sum \nu_j t^j/j!$ and hence is absolutely convergent if the latter is convergent. Hence we have the infinite expansion

$$\phi(t) = \sum_{j=0}^{\infty} (it)^j \mu_j/j! \tag{4.38}$$

and thus $\phi(t)$ is uniquely determined in the neighbourhood of $t = 0$. In the neighbourhood of $t = t_0$ we have

$$\phi(t) = \sum_{j=0}^{r} \left\{ \frac{i^j (t - t_0)^j}{j!} \int_{-\infty}^{\infty} x^j e^{it_0 x}\,\mathrm{d}F \right\} + R_r'$$

(*) For some results on the general problem, see Stieltjes (1918) and Hamburger (1920). For a general review, see Shohat and Tamarkin (1943) and chapter III of Widder (1941).

and the modulus of the coefficient of $(t-t_0)^j/j!$ is not greater than ν_j. Consequently $\phi(t)$ can be expanded everywhere as a convergent Taylor series and is equal to the sum of that series. Hence $\phi(t)$ may be extended from the neighbourhood $t=t_0$ by analytic continuation through any finite t-interval. Hence $\phi(t)$ is everywhere uniquely defined. But $\phi(t)$ determines the d.f. and hence the latter is uniquely determined.

It is clear that only when the analytic expansion (4.38) is not permissible can the moments fail to determine the d.f. through the c.f.

4.22 A few simple but effective results follow as corollaries.

(a) The moments uniquely determine the distribution if the upper limit of $(\nu_n^{1/n})/n$ is finite. For the series whose general term is $\nu_n t^n/n!$ is convergent if

$$\limsup\,(\nu_n t^n/n!)^{1/n} < 1.$$

Replacing the factorial by its Stirling approximation we see that this will be true if

$$\limsup\,(\nu_n^{1/n})/n < k/t$$

where k is some constant. If the upper limit is finite the inequality can be satisfied for some non-zero t.

(b) It is also sufficient for the upper limit of $(\mu_{2n}^{1/2n})/2n$ to be finite, a form which enables us to disregard the absolute moments. It is, in fact, easy to see that this upper limit and the upper limit of $(\nu_n^{1/n})/n$ are finite or infinite together.

(c) The moments uniquely determine the distribution if the range is finite. For, taking an origin at the start of the distribution (of range h, say), we see that $\nu_r \leqslant h^r$ and hence $(\nu_n^{1/n})/n \leqslant h/n$, which tends to zero.

4.23 Two further criteria may be mentioned. The first is due to Carleman (1925). A set of moments determines a distribution uniquely if (in the case of limits $-\infty$ to $+\infty$)

$$\sum_{j=0}^{\infty} \frac{1}{(\mu_{2j})^{1/2j}} \tag{4.39}$$

diverges. For the limits 0 to ∞ the corresponding series is

$$\sum_{j=0}^{\infty} \frac{1}{(\mu_j)^{1/2j}}. \tag{4.40}$$

Secondly, if there is a probability function, the moments determine it uniquely if, for limits $-\infty$ to $+\infty$,

$$f(x) < M|x|^{\beta-1}\exp(-\alpha|x|^{\lambda}) \quad \text{for} \quad |x|>x_0, \qquad M,\beta,\alpha>0,\ \lambda\geqslant 1, \tag{4.41}$$

and for limits 0 to ∞,

$$f(x) < M|x|^{\beta-1}\exp(-\alpha|x|^{\lambda}) \quad \text{for} \quad |x|>x_0, \qquad M,\beta,\alpha>0,\ \lambda\geqslant \tfrac{1}{2}. \tag{4.42}$$

This result is due ultimately to Stieltjes. It follows without difficulty from the Carleman criterion.

It is interesting to note that if for some x_0

$$f(x) > \exp(-\alpha|x|^{\lambda}), \qquad \alpha > 0, \quad x > x_0, \tag{4.43}$$

then the problem of moments is necessarily indeterminate (as usual, $\lambda < \frac{1}{2}$ for the range 0 to ∞ and $\lambda < 1$ for the range $-\infty$ to $+\infty$). This follows from the examples in equations (4.35) and (4.36), for we can add to (4.43), *without rendering $f(x)$ negative at any x*, a function all of whose moments are zero.

Example 4.7 (Determination of the normal distribution by its moments)

The moments of the distribution

$$f = \frac{1}{\sigma\sqrt{(2\pi)}} \exp\{(-x^2/(2\sigma^2))\}, \qquad -\infty < x < \infty,$$

are given (cf. Example 3.4) by

$$\mu_{2r+1} = 0, \qquad \mu_{2r} = \frac{(2r)!}{2^r r!}\sigma^{2r}.$$

Thus the upper limit of $(\mu_{2n}^{1/2n})/2n$ is, from the Stirling approximation to the factorial, asymptotically equivalent to

$$\frac{\sigma}{2n\sqrt{2}}\left[\frac{e^{-2n}(2n)^{2n}\sqrt{(4\pi n)}}{e^{-n}n^n\sqrt{(2\pi n)}}\right]^{1/2n} \sim \frac{\sigma}{(2en)^{\frac{1}{2}}}.$$

The upper limit is then zero and the distribution is uniquely determined by its moments.

4.24 Cramér and Wold (1936) extended Carleman's criterion to multivariate distributions. We need only consider the moments of the marginal distributions, $\mu'_{r00}\ldots$, $\mu'_{0r0\ldots}$, $\mu'_{00r\ldots}$, etc. If the sum of these for all variates is λ_r then the criteria of (4.39) and (4.40) remain with λ_r instead of μ_r; for example, a distribution ranging from $-\infty$ to ∞ is completely determined by its moments if

$$\sum_{j=0}^{\infty} \frac{1}{(\lambda_{2j})^{1/2j}} \tag{4.44}$$

diverges.

4.25 If a moment of order r exists it must be given by the rth derivative of the c.f. at $t = 0$. Thus, if $\phi(t)$ can be expanded as an infinite Taylor series, that series can only be $\Sigma\,(it)^j\mu'_j/j!$. And if this series does not converge, $\phi(t)$ cannot be so expanded. But it can always be expanded asymptotically as a finite series with remainder

$$\phi(t) = \sum_{j=0}^{r} (it)^j\mu'_j/j! + R.$$

This illustrates the source of a difficulty in discussing limiting properties when the infinite series does not converge, for it is known that there are an infinite number of functions that have a given set of coefficients in an asymptotic expansion. For instance, if $\alpha(t)$ has an asymptotic expansion in t the functions $\alpha(t)+kt^{-\log t}$ all have the same expansion. It is therefore hardly surprising that when the infinite sum $\sum (it)^j\mu_j'/j!$ fails to converge, there may be more than one p.f. with the same set of moments.

But it does not follow from what has been said that there *must* be more than one probability distribution. There must be more than one function, but those functions may not qualify as probability distributions, e.g. they may be negative in part of the range. In the example just given, $t^{-\log t}$ cannot be a c.f., for it does not satisfy the condition that $\phi(t)$ and $\phi(-t)$ should be conjugate.

4.26 We now proceed to what is known as the Second Limit theorem, which is concerned with the way in which a sequence of d.f.s $\{F_n(x)\}$ tends to a limit if the corresponding sequence of moments of order j, say $\mu_j(n)$, tends to a limit μ_j. Our method of proof is due to D. G. Kendall and Rao (1950).

We require in the first place a result concerning the expansion of a c.f. that is rather more precise than those we have used so far. If μ_{2m} exists then the c.f. can be expressed in the form

$$\phi(t)=\sum_{j=0}^{2m-1}(it)^j\mu_j/j!+\rho(it)^{2m}\mu_{2m}/(2m)! \tag{4.45}$$

for all real t, ρ being such that $|\rho|<1$. Also $\lim_{t\to 0}\rho=1$.

For $m=0$ this is true trivially. For $m>0$ we have for real x, t,

$$e^{ixt}=\sum_{j=1}^{n}\frac{(ixt)^j}{j!}+\rho'\frac{(ixt)^{n+1}}{(n+1)!},\qquad |\rho'|\leqslant 1, \tag{4.46}$$

and on substitution in the integral defining $\phi(t)$ we get, for $n=2m-1$, equation (4.45). For t, μ_{2m} zero we may define $\rho=1$; for other values, considering the difference of (4.46) taken to n and $n+1$ terms we find

$$|\rho'-1|\leqslant |xt|/(n+2). \tag{4.47}$$

Since

$$|\rho-1|\mu_{2m}=\int_{-\infty}^{\infty}(\rho'-1)x^{2m}\,dF$$

we have

$$|\rho-1|\mu_{2m}<\varepsilon+\int_{-u}^{u}|\rho'-1|x^{2m}\,dF,$$

for any given ε, if u is large enough (for otherwise μ_{2m} would not converge). But for fixed u the integral can be made as small as we please by taking t sufficiently small, because of (4.47). Thus we have

$$|\rho-1|\mu_{2m}<2\varepsilon$$

and thus $\rho\to 1$ when $t\to 0$.

4.27 A second result we require (a converse of the previous one) is as follows: If for some positive integer m

$$\phi(t)=\sum_{j=0}^{2m}(it)^j\lambda_j/j!+o(t^{2m}) \tag{4.48}$$

as t tends to zero through real values, then the first $2m$ moments exist and $\mu_r=\lambda_r$, $r=0, 1, \ldots, 2m$.

Let δ^2 denote the operation of taking central differences with interval $2h$, i.e.

$$\delta^2 A(t)=A(t+2h)-2A(t)+A(t-2h).$$

Then if the remainder term in (4.48) is $t^{2m}B(t)$, where $B(t)$ tends to zero as $t\to\infty$, a little calculation shows that for $t=0$

$$\left(\frac{\delta^2}{4h^2}\right)^m t^{2m}B(t)=\sum_{j=0}^{2m}\binom{2m}{j}(-1)^j(m-j)^{2m}B(2mh-2jh).$$

Thus for $|B(t)|<\varepsilon$

$$\left(\frac{\delta^2}{4h^2}\right)^m t^{2m}B(t)\leqslant\sum_{j=0}^{2m}\binom{2m}{j}|m-j|^{2m}\varepsilon \tag{4.49}$$

when $|h|<\eta/2m$ and $|t|\leqslant\eta$. Hence if (4.48) is true

$$\left[\lim_{h\to 0}\left(\frac{\delta^2}{4h^2}\right)^m t^{2m}B(t)\right]_{t=0}=0. \tag{4.50}$$

But $\lim_{h\to 0}\left(\dfrac{\delta^2}{4h^2}\right)^m$, $=D^{2m}$ say, acting on a polynomial is easily seen to be the $(2m)$th derivative. Thus applying the operation to (4.48) we find, remembering (4.50), that when $t=0$

$$D^{2m}\phi(t)=(-1)^m\lambda_{2m}. \tag{4.51}$$

Now

$$\delta^2 e^{ixt}=-(2\sin xh)^2 e^{ixt}$$

and thus

$$\lim_{h\to 0}\int_{-\infty}^{\infty}\left(\frac{\sin xh}{xh}\right)^{2m}x^{2m}\,dF=\lambda_{2m}. \tag{4.52}$$

From the uniform convergence to unity of $(\sin xh/xh)^{2m}$ we see that for any finite interval (a, b)

$$\int_a^b x^{2m}\,dF\leqslant\lambda_{2m},$$

and thus the moment of order $2m$ must be finite. Hence from (4.45), as t tends to zero,

$$\phi(t)=\sum_{j=0}^{2m-1}(it)^j\mu_j/j!+\frac{(it)^{2m}\mu_{2m}}{(2m)!}[1+o(1)]. \tag{4.53}$$

Comparing this with (4.48) we see that

$$\mu_r = \lambda_r, \qquad r = 0, 1, \ldots, 2m.$$

The result of this section implies that if the $(2m)$th derivative of the c.f. exists at $t = 0$, the $(2m)$th moment also exists. We have seen in 3.5 that if the rth moment exists the rth derivative of the c.f. will exist at $t = 0$, a statement of which the above is the converse for even moments. For odd moments the existence of the derivative at $t = 0$ is not sufficient for the existence of the corresponding moment. Pitman (1956) has shown that necessary and sufficient conditions for the existence of the derivative are (1) that the principal value of μ_r' exists and (2) that

$$\lim_{x\to\infty} x^r\{F(-x)+1-F(x)\} = 0.$$

4.28 The results of the two preceding sections are necessary to establish the uniqueness of asymptotic expansions of the c.f. under certain conditions. We now prove some theorems on sequences of distributions and moments.

Let $F_n(x)$, the nth member of a sequence $\{F_n(x)\}$, possess a finite moment of the jth order, say $\mu_j(n)$, for all $n > n_0$. Let

$$\lim_{n\to\infty} \mu_j(n) = \lambda_j$$

for every value of j; and let $\{F_n(x)\}$ converge to a limit function $G(x)$ at all its points of continuity. ($G(x)$ is necessarily bounded, monotonically non-decreasing and may be taken to be continuous on the right.) Then $G(x)$ is a d.f. with all moments finite; $\{\lambda_j\}$ is a sequence of moments; and λ_j is the jth moment of $G(x)$.

That $G(x)$ is a d.f. follows from 4.13, for the second moments $\mu_2(n)$ form a convergent and therefore bounded sequence. Hence, by the First Limit Theorem of 4.14, the sequence of c.f.s $\{\phi_n(t)\}$ converges to the c.f. of $G(x)$, say $\phi(t)$.

Now when $n > n_0$

$$\phi_n(t) = \sum_{j=0}^{2m-1} (it)^j\mu_j(n)/j! + \rho(it)^{2m}\mu_{2m}(n)/(2m)! \tag{4.54}$$

and the last term on the right must approach a limit since all the other terms do so. Thus

$$\phi(t) = \sum_{j=0}^{2m-1} (it)^j\lambda_j/j! + R(it)^{2m}\lambda_{2m}/(2m)!, \qquad |R| \leqslant 1. \tag{4.55}$$

This is true for all m and by comparing each of these formulae with the next following one we have, as in 4.26, $R \to 1$ as $t \to 0$. From the results of 4.27 the theorem now follows.

Example 4.8 (Uniqueness of the characteristic function)

Note that the result of 4.28 follows even if $F_n(x)$ does not have all moments finite. For example, the symmetric distribution

$$f = \frac{k}{(1+t^2/\nu)^{\frac{1}{2}(\nu+1)}}, \qquad -\infty < t < \infty; \quad \nu > 1,$$

has finite moments only up to and including $\mu_{\nu-1}$. In fact (cf. Example 3.3)

$$\mu_{2r} = \frac{\Gamma(\frac{1}{2}\nu - r)\Gamma(r+\frac{1}{2})}{\Gamma(\frac{1}{2}\nu)\Gamma(\frac{1}{2})}\nu^r, \qquad 2r < \nu. \tag{4.56}$$

By use of the Stirling approximation to the Gamma function at (3.64), we find that as $\nu \to \infty$

$$\mu_{2r} \to \frac{(2r)!}{2^r r!}, \tag{4.57}$$

namely to the moments of the standardized normal distribution (Example 4.7), which uniquely determine it. If, then, our limit function converges it must do so to the normal form, i.e. to a form for which all moments exist.

It is, in fact, easy to see directly that the density function tends to $k \exp(-\frac{1}{2}t^2)$.

The Second Limit Theorem

4.29 Let $\{F_n(x)\}$ converge to $G(x)$; let $\mu_j(n)$ exist for $n > n_0$ and for all $j \geqslant 0$; and let $\mu_j(n)$ be bounded above by some constant A_j. Then all the moments λ_j of $G(x)$ exist and $\mu_j(n) \to \lambda_j$ as $n \to \infty$.

From the First Limit Theorem we know that the sequence of c.f.s $\{\phi_n(t)\}$ converges to $\phi(t)$, the c.f. of $G(x)$. Also

$$\phi_n(t) = 1 + it\mu_1(n) + O(t^2), \qquad n > n_0,$$

where the constant implied by $O(t^2)$ can be taken as independent of n and t. Thus if m is some other value $> n_0$,

$$|\mu_1(n) - \mu_1(m)| \leqslant A_1|t| + \delta_{m,n}$$

where $\delta_{m,n} \to 0$ as m, n tend independently to infinity. This, being true for every non-zero t, implies that

$$\lim_{m,n\to\infty} |\mu_1(n) - \mu_1(m)| = 0$$

and hence there exists a constant λ_1 such that

$$\lim \mu_1(n) = \lambda_1. \tag{4.58}$$

We now proceed inductively by the same kind of argument to establish the limit of $\mu_2(n)$. We require

$$\phi_n(t) = 1 + it\mu_1(n) + (it)^2\mu_2(n)/2! + O(t^3),$$

an expression which follows from (4.54) because the sixth moment is bounded. We then find

$$|\mu_2(n) - \mu_2(m)| \leqslant \tfrac{2}{3}A_6^{\frac{1}{6}}|t| + \delta'_{m,n}$$

and, as before, this implies

$$\lim \mu_2(n) = \lambda_2. \tag{4.59}$$

Thus we establish that the λ_j are all finite. It then follows from the theorem of **4.28** that they are the moments of $G(x)$.

4.30 For most purposes we require a converse form of the Second Limit Theorem: let the moments $\mu_j(n)$ exist and the limits $\lim \mu_j(n) = \lambda_j$ be the moments of a distribution $G(x)$ *that is uniquely determined by its moments.* Then $\{F_n(x)\}$ converges to $G(x)$ in all points of continuity.

We can always find a sequence in $\{F_n(x)\}$ that converges to a d.f. It then follows from the theorem of **4.28** that this function must be the unique function having the moments λ_j and thus it is $G(x)$.

Suppose now that at some point of continuity, say a,

$$\alpha \equiv \lim\sup F_n(a) \neq G(a).$$

We can choose a subsequence converging to α at $x = a$; and a further sub-subsequence which converges both to α and also to $G(a)$, because $G(x)$ is continuous at $x = a$. This contradiction can only be avoided if

$$\lim\sup F_n(a) = G(a);$$

and by a similar argument

$$\lim\inf F_n(a) = G(a).$$

The converse theorem is thus established.

Example 4.9 (Distribution uniquely determined by its moments)

The discrete (Poisson) distribution whose probability at $x = j (j = 0, 1, \ldots)$ is $e^{-\lambda}\lambda^j/j!$ has by Example 3.10 the c.f.

$$\phi(t) = \exp\{\lambda(e^{it} - 1)\},$$

and hence all cumulants equal to λ.

The distribution is evidently the only one with such cumulants, for $\sum \kappa_j(it)^j/j!$ is convergent and equals $\lambda(e^{it} - 1)$, so that the c.g.f. and the c.f. are uniquely determined.

Now as λ tends to infinity the probability at x_j, $e^{-\lambda}\lambda^j/j!$, tends to zero and thus the distribution does not tend to a limit. This is consistent with the behaviour of the cumulants, which increase without limit.

Suppose, however, that we standardize the distribution. Then

$$\kappa_1 = 0, \qquad \kappa_r = \lambda/\kappa_2^{\frac{1}{2}r} = \lambda^{-(\frac{1}{2}r-1)}, \qquad r \geqslant 2.$$

Hence as $\lambda \to \infty$ all cumulants higher than the second tend to zero, and by Example 3.11 these are the cumulants of the normal distribution

$$f = \frac{1}{\sqrt{(2\pi)}}\exp(-\tfrac{1}{2}x^2), \qquad -\infty < x < \infty,$$

which is completely determined by its moments (Example 4.7), and hence by its cumulants. Thus the converse of the Second Limit Theorem applies, and the standardized discrete distribution tends to the continuous normal form.

4.31 In connexion with the dual role of the c.f. and the d.f., it is worth remarking that the behaviour of one in the neighbourhood of the origin is related to the behaviour of the other at infinity. In fact, the mth derivative of $\phi(t)$ at $t=0$ is the mth moment, the existence of which depends on the behaviour of $x^m f(x)$ at infinity. Conversely, from (4.5) we see that the mth derivative of $f(x)$, if it exists, is given by $\int_{-\infty}^{\infty}(-it)^m\phi(t)e^{-ixt}\,dt/2\pi$, which in modulus is not greater than $\int_{-\infty}^{\infty}t^m\phi(t)\,dt/2\pi$, which depends for its convergence on the behaviour of $t^m\phi(t)$ at infinity.

4.32 It is natural to inquire whether there is some other transform of the d.f. that enjoys properties similar to those of the c.f. The answer appears to be in the negative. Lukacs (1952) has proved a theorem to the following effect: Let $\kappa(x, t)$ be a complex function defined for all real x, t, bounded and measurable in t. Define

$$\phi(t) = \int_{-\infty}^{\infty}\kappa(x, t)\,dF(x);$$

and let the two following conditions hold:

(1) $\phi_1(t) \equiv \phi_2(t)$ if and only if $F_1(x) = F_2(x)$.
(2) If $F(x) = \int_{-\infty}^{\infty}F_1(x-y)\,dF_2(y)$, then $\phi(t) = \phi_1(t)\phi_2(t)$.

Then $\kappa(x, t)$ must have the form $\exp\{itA(x)\}$ where $A(x)$ is a real-valued function that assumes all values of a set dense on the real line. It is very remarkable that we do not require among the conditions either an inversion theorem or limit theorems.

Infinitely divisible and stable characteristic functions

4.33 We shall see in **8.43** that a c.f. is sometimes the product of other c.f.s, and in particular may be a power of another c.f., so that

$$\phi(t) = \{\phi_j(t)\}^n. \tag{4.60}$$

If a given c.f. $\phi(t)$ can thus be represented as the nth power of some other c.f., $\phi_j(t)$, for every positive integer n, $\phi(t)$ is called an *infinitely divisible* c.f., and the corresponding d.f. is also called infinitely divisible. Since $\phi(t)$ is fixed, $\phi_j(t)$ is a function of n through the relation (4.60), and we relabel it $\phi_n(t)$, so that (4.60) becomes

$$\phi(t) = \{\phi_n(t)\}^n. \tag{4.61}$$

Example 4.10 (The Cauchy as an infinitely divisible distribution)

The Cauchy c.f. of Example 4.2,

$$\phi(t) = \exp(-|t|) = \left\{\exp\left(-\left|\frac{t}{n}\right|\right)\right\}^n$$

for any positive integer n, and

$$\phi_n(t) = \exp\left(-\left|\frac{t}{n}\right|\right) = \phi\left(\frac{t}{n}\right)$$

is a c.f., so $\phi(t)$ is infinitely divisible.

Example 4.11 (The normal and infinite divisibility)

The normal c.f. of Example 4.1,

$$\phi(t) = \exp(-\tfrac{1}{2}t^2\sigma^2) = \{\exp(-\tfrac{1}{2}t\sigma^2/n)\}^n$$

for any positive integer n, and

$$\phi_n(t) = \exp(-\tfrac{1}{2}t^2\sigma^2/n) = \phi\left(\frac{t}{n^{\frac{1}{2}}}\right)$$

is a c.f., so $\phi(t)$ is infinitely divisible.

4.34 Now consider two linear functions of independent variates $x_j (j = 1, 2, \ldots)$ with the same c.f. $\phi(t)$. Let $y_j = a_j x_j + b_j$, where the $a_j > 0$. The c.f. of y_j is

$$\phi_j(t) = \phi(a_j t)e^{itb_j} \tag{4.62}$$

and thus

$$\phi_j(t)\phi_k(t) = \phi(a_j t)\phi(a_k t)\exp\{it(b_j + b_k)\}. \tag{4.63}$$

If the product (4.63) is of the same form as (4.62), i.e.

$$\phi(a_j t)\phi(a_k t)\exp\{it(b_j + b_k)\} = \phi(a_0 t)e^{itb_0}, \quad a_0 > 0, \tag{4.64}$$

$\phi(t)$ is called a *stable* c.f., and the corresponding d.f. a stable d.f. (4.64) simplifies to

$$\phi(a_j t)\phi(a_k t) = \phi(a_0 t)e^{itb}, \tag{4.65}$$

where $b = b_0 - b_j - b_k$. Clearly, if (4.65) holds, we may extend it to

$$\prod_{j=1}^{n} \phi(a_j t) = \phi(at)e^{itb}, \quad a > 0, \tag{4.66}$$

by defining a and b appropriately. Thus, with $a_j \equiv 1$, we have

$$\phi(at) = \{\phi(t)\}^n e^{-itb} = \{\phi(t)e^{-itb/n}\}^n$$

or

$$\phi(t) = \left\{\phi\left(\frac{t}{a}\right)\exp[-(itb)/(na)]\right\}^n. \tag{4.67}$$

The expression in braces on the right of (4.67) is a c.f., the exponential factor merely representing a shift in location. Since (4.67) holds for any positive integer n, it satisfies (4.61). Thus any stable distribution is infinitely divisible.

Bondesson (1979) showed that any distribution on $(0, \infty)$ is infinitely divisible if the density function is of the form $f(x) = cx^{\beta-1}h(x)$ where $\beta > 0$, $h(0) = 1$ and h is a completely monotone function on $(0, \infty)$ which satisfies some additional regularity conditions. A broad class of generalizations of the Gamma distribution have densities of this form. The result demonstrates the infinite divisibility of the *F*, lognormal, inverse Gaussian and Burr distributions among others. These distributions are described in Chapters 5 and 6.

4.35 Infinitely divisible and stable c.f.s are of great importance in probability theory, but we shall here mention only a few of their properties that are useful in Statistics, referring the reader to Lukacs (1970) for a general exposition and a bibliography.

First, we observe that an infinitely divisible c.f. $\phi(t)$ can have no real zeros, for as $n \to \infty$ through the positive integers, we see from (4.61) that $\phi_n(t) = \{\phi(t)\}^{1/n}$ can only tend to limiting values of 1 (if $\phi(t) \neq 0$) and 0 (if $\phi(t) = 0$) because $|\phi_n(t)| \leqslant 1$. Moreover the limit of $\phi_n(t)$ is itself a c.f., and (cf. 4.7) equals 1 at $t = 0$ and is continuous in t. It cannot therefore jump to the value 0. Thus $\phi(t) \neq 0$ for all t.

Borges (1966) showed that the fourth cumulant $\kappa_4 \geqslant 0$ for all infinitely divisible distributions; see also Exercise 4.30.

4.36 Whereas infinitely divisible distributions may be discrete (cf. Exercise 4.25), all stable distributions are continuous and unimodal.

There is an explicit general form for a stable c.f., which we write

$$\log \phi(t) = ait - c|t|^{\alpha}\left\{1 + i\beta \frac{t}{|t|} w(|t|, \alpha)\right\}, \tag{4.68}$$

where

$$w(|t|, \alpha) = \begin{cases} \tan(\frac{1}{2}\pi\alpha), & \alpha \neq 1, \\ \frac{2}{\pi}\log|t|, & \alpha = 1, \end{cases}$$

a, $c \geqslant 0$, $|\beta| \leqslant 1$ and $0 < \alpha \leqslant 2$ are real.

In (4.68), α is called the characteristic exponent and β is a skewness parameter. When $\beta = 0$, or $\alpha = 2$, (4.68) reduces to

$$\log \phi(t) = ait - c|t|^{\alpha}, \tag{4.69}$$

so that if we take a as origin, as we shall, $\phi(t)$ is real and, by Exercise 4.1, the stable distribution is symmetric then and only then, with $c^{1/\alpha}$ as a scale parameter.

In (4.68) with $a = 0$, changing the sign of the variate changes only sgn t. Thus if x has a stable distribution with skewness parameter β, $(-x)$ has the same distribution with skewness parameter $(-\beta)$, c and α remaining unchanged.

Example 4.12 (Examples of stable laws)

In (4.69), putting $\alpha = 1$, $a = 0$, $c = 1$, gives the Cauchy c.f. of Example 4.10, while $\alpha = 2$, $a = 0$, $c = \frac{1}{2}\sigma^2$ gives the normal c.f. of Example 4.11. The location and scale parameters a and $c^{1/\alpha}$ may, of course, be varied.

4.37 Apart from the two distributions in Example 4.12, no other symmetric stable distribution has a known elementary form for its p.d.f. Even in the much wider class (4.68), only one such case is known.

Example 4.13 (An asymmetric stable law)

In (4.68), put $\alpha = \frac{1}{2}$, $\beta = -1$, $a = 0$, $c = 1$. We find

$$\log \phi(t) = -|t|^{\frac{1}{2}}(1 - i \operatorname{sgn} t).$$

Since $(1 - i \operatorname{sgn} t)^2 = -2i \operatorname{sgn} t$, we may write this

$$\log \phi(t) = -|t|^{\frac{1}{2}}(-2i \operatorname{sgn} t)^{\frac{1}{2}} = -(-2it)^{\frac{1}{2}}.$$

This stable c.f. is discussed in Exercise 11.25 below, where it may be seen that its p.d.f. is

$$f(y) = (2\pi)^{-\frac{1}{2}} \exp\left(-\frac{1}{2y}\right) y^{-\frac{3}{2}}, \quad 0 \leqslant y < \infty, \tag{4.70}$$

a distribution none of whose moments exists; it is sometimes called the Lévy distribution. It is evident from (4.68) that the mean of a stable distribution only exists if $\alpha > 1$, and the variance only in the extreme case $\alpha = 2$, which is the normal distribution.

4.38 Series expansions for the p.d.f. of stable laws may be obtained by inversion of the c.f. The results were obtained independently by Bergström (1952) and Feller; see Feller (1971, section 17.6).

Since $\phi(t)$ and $\phi(-t)$ are complex conjugates, the p.d.f. may be written, using (4.5), as

$$f(x) = \frac{1}{\pi} \operatorname{Re} \int_0^\infty e^{-ixt} \phi(t) \, dt. \tag{4.71}$$

For $t > 0$, the coefficient of $|t|^\alpha$ in the exponent of (4.68) may be rewritten as $c\{1 + i\beta \tan \frac{1}{2}\pi\alpha\} = c_0 \exp(\frac{1}{2}\pi\gamma i)$. We may then relocate and rescale x so that $a = 0$, $c_0 = 1$. The integral (4.71) becomes

$$f(x \mid \alpha, \gamma) = \frac{1}{\pi} \operatorname{Re} \int_0^\infty \exp\{-ixt - t^\alpha e^{\frac{1}{2}\pi\gamma i}\} \, dt. \tag{4.72}$$

We now restrict attention to $0 < \alpha < 1$; a similar development is possible for $1 < \alpha < 2$. Also, we take $x > 0$, since it may be shown that

$$f(x \mid \alpha, \gamma) = f(-x \mid \alpha, -\gamma). \tag{4.73}$$

As the imaginary part of $t \to -\infty$, the integral (4.72) tends to zero because of the linear term in the exponent. Thus we may integrate over the negative imaginary axis. In effect, this means that we can make the substitution

$$xt = u \exp(-\tfrac{1}{2}\pi i) \tag{4.74}$$

which reduces the integral to

$$f(x|\alpha, \gamma) = \frac{1}{\pi x} \operatorname{Im} \int_0^\infty \exp(-u - gu^\alpha)\, du \tag{4.75}$$

where $g = x^{-\alpha} \exp\{\frac{1}{2}\pi i(\gamma - \alpha)\}$. We may now treat (4.75) as though all the coefficients were real. Expanding $\exp(-gu^\alpha)$ as a series and integrating term by term, we obtain

$$f(x|\alpha, \gamma) = \frac{1}{\pi x} \sum_{k=0}^{\infty} (-1)^k \frac{\Gamma(k\alpha + 1)}{k!} x^{-\alpha k} \sin \frac{k\pi(\gamma - \alpha)}{2}. \tag{4.76}$$

Similarly, for $1 < \alpha \leqslant 2$, $x > 0$

$$f(x|\alpha, \gamma) = \frac{1}{\gamma\pi} \sum_{k=0}^{\infty} (-1)^k \frac{\Gamma\{(k+1)/\alpha\}}{k!} x^k \cos\left[\frac{k\pi(\gamma - \alpha)}{2\alpha} + \frac{\pi\gamma}{2\alpha}\right]. \tag{4.77}$$

When $\beta = -1$, the random variable is restricted to $(0, \infty)$ and to $(-\infty, 0)$ when $\beta = 1$. Brockwell and Brown (1978) give series expansions for these cases.

The role of the stable laws in limit theorems will be discussed in **8.52**.

4.39 Because the stable distributions' p.d.f.s cannot, except in the three cases discussed above, be expressed simply, they are usually studied through the c.f. However, by numerical evaluation of integrated convergent series expansions for the p.d.f., Fama and Roll (1968) constructed 4 d.p. tables of the symmetric stable d.f. corresponding to (4.69) for $\alpha = 1.0$ (0.1) 1.9, 1.95, 2.0; $a = 0$, $c = 1$ and the variate (their $|u|$) $= 0.05$ (0.05) 1.0 (0.1) 2.0 (0.2) 4.0 (0.4) 6, 7, 8, 10, 15, 20, with an additional table of the $100p$ and $100(1-p)$ percentiles, $p = 0.52$ (0.02) 0.94 (0.01) 0.97 (0.005) 0.995, 0.9995. Holt and Crow (1973) used various approximations to give 4 d.p. tables and charts of the general stable p.d.f. corresponding to (4.68) for $\alpha = 0.25$ (0.25) 1.75; $\beta = -1$ (0.25) 1; $a = 0$, $c = 1$ and the variate $x \geqslant 0$ in steps varying from 0.001 and 0.01 near the origin to 10 and 100 in the extreme tails. (For negative x, the sign of β must be changed—cf. **4.36**.) A curiosity appears because of the discontinuity of the function $w(|t|, \alpha)$ at $\alpha = 1$: for $\alpha = 1$ and $\beta > 0$, the stable distributions have larger positive tails, but for $\alpha \neq 1$, $\beta > 0$ larger negative tails.

4.40 The empirical characteristic function (e.c.f.) is

$$\hat{\phi}(t) = \frac{1}{n} \sum_{j=1}^{n} \exp(itx_j). \tag{4.78}$$

Press (1972) proposed several methods for estimating m parameters $\boldsymbol{\theta}$ in $\phi(t, \boldsymbol{\theta})$ using the e.c.f.; in particular, we may

(i) minimize

$$\int [\hat{\phi}(t) - \phi(t, \boldsymbol{\theta})]^2 W(t)\, dt,$$

where $W(t)$ is a suitable weighting function, or

(ii) evaluate $\hat{\phi}(t_k)$ for selected t_k, $k=1, \ldots, m$, and solve the estimating equations $\hat{\phi}(t_k)=\phi(t_k, \theta)$, $k=1, \ldots, m$.

Feuerverger and McDunnough (1981) and Koutrouvelis (1982) show that e.c.f.-based estimators have high efficiency (cf. 17.28-9, Volume 2A) for the parameters of the stable laws.

Sums of random variables

4.41 We first notice one very simple but powerful result concerning the c.f. of the sum of independent random variables. If we have n random variables $x_1, \ldots, x_n$ distributed as $dF_1, \ldots, dF_n$, the c.f. of their sum, $Z=x_1+\ldots+x_n$ being the integral of e^{itz} over the range of z, is

$$\begin{aligned}\phi(t) &= \int_{-\infty}^{\infty} \cdots \int_{-\infty}^{\infty} e^{itz}\, dF_1 \ldots dF_n \\ &= \int_{-\infty}^{\infty} e^{itx_1} dF_1 \int_{-\infty}^{\infty} e^{itx_2} dF_2 \ldots \int_{-\infty}^{\infty} e^{itx_n} dF_n \\ &= \phi_1(t)\phi_2(t) \ldots \phi_n(t). \end{aligned} \tag{4.79}$$

Thus for independent random variables the c.f. of a sum is the product of the c.f.s; and the c.g.f. of a sum is the sum of the individual c.g.f.s. In particular, if all the ϕ_j are the same, we have simply $\phi(t) = \{\phi_1(t)\}^n$.

If follows that the rth cumulant of a sum of independent random variables is the sum of the rth cumulants of its components, and it is this property that accounts for the name "cumulant"—independent variables' cumulants cumulate.

4.42 The Inversion Theorem of 4.3 ensures that the c.f. uniquely determines the d.f. and that the converse is true. However, when we consider sums of random variables, we have no such guarantees. In particular, the following examples show that:

(i) the random variables $X+Y_1$ and $X+Y_2$ may have the same distribution (where $Y_2 \neq Y_1$) even if they are all independent;
(ii) if the c.f. of $X+Y$ is equal to the product of their individual c.f.s., it does not follow that X and Y are independent.

Example 4.14 (Conditions for the unique factorization of the c.f.)

The essential point in this example, which was originally due to A. Khinchin, is that two independent random variables, X, Y, may have c.f.s that coincide in some interval of t containing the origin. If the c.f. of X is zero outside this interval, (4.79) ensures that the sum of two independent Xs has the same distribution as $X+Y$.

Starting from the integral

$$\int_0^{\infty} \frac{\sin \alpha x}{x} dx = \tfrac{1}{2}\pi, \qquad \alpha > 0,$$

we have, on integration with respect to α,

$$\int_0^\infty \frac{\cos \beta x - \cos \alpha x}{x^2} \mathrm{d}x = \tfrac{1}{2}\pi(\alpha - \beta), \qquad \alpha < \beta.$$

It is then easy to show, by the double use of identities such as

$$\sin \theta \cos m\theta = \tfrac{1}{2}\{\sin (m+1)\theta - \sin (m-1)\theta\}$$

that

$$\frac{2}{\pi}\int_0^\infty \frac{1-\cos x}{x^2} \cos tx \, \mathrm{d}x = 1 - |t| \quad \text{for} \quad |t| \leqslant 1$$
$$= 0 \qquad \text{for} \quad |t| > 1.$$

Thus the distribution

$$\mathrm{d}F = \frac{1}{\pi}\frac{1-\cos x}{x^2}\mathrm{d}x, \qquad -\infty < x < \infty,$$

has the c.f.

$$\left.\begin{aligned} \phi_1(t) &= 1 - |t|, \qquad & |t| \leqslant 1 \\ &= 0, & |t| > 1 \end{aligned}\right\} \tag{4.80}$$

for the integral

$$\frac{1}{\pi}\int_{-\infty}^\infty \frac{1-\cos x}{x^2} \sin tx \, \mathrm{d}x = 0$$

because the function $\sin tx$ is odd. Moreover, since

$$\sum_{n=1}^\infty n^{-2} = \pi^2/6 \quad \text{and} \quad \sum_{r=1}^\infty \cos\{(2r-1)\theta\}/(2r-1)^2 = (\pi^2 - 2\pi\theta \operatorname{sgn}\theta), \qquad -\pi \leqslant \theta \leqslant \pi,$$

we may show that the discrete distribution

$$\left.\begin{aligned} f(0) &= \tfrac{1}{2} \\ f(x) &= \frac{2}{x^2}, \quad x = n\pi, \; n \text{ a positive or negative odd integer,} \end{aligned}\right\} \tag{4.81}$$

has the c.f.

$$\phi_2(t) = \tfrac{1}{2} + \frac{4}{\pi^2}\sum_{r=1}^\infty \cos\{(2r-1)\pi t\}/(2r-1)^2,$$

which reduces to

$$\phi_2(t) = 1 - |t|, \qquad |t| \leqslant 1. \tag{4.82}$$

For $|t| > 1$ the c.f. consists of periodic repetitions of the value inside the range $|t| \leqslant 1$.

It follows from (4.80) and (4.82) that *for all t*, $\phi_1(t)\phi_2(t) = \phi_1(t)\phi_1(t)$. Thus the sum of the independent first and second variables is distributed in the same form as the sum of two independent variables like the first. Thus it is not true, in general, that if $X + Y_1$ and $X + Y_2$ have the same distribution, then $Y_1 = Y_2$.

The factorization of a c.f. into two others is unique if all three are infinitely divisible (cf. **4.33**); they cannot then take zero values, by **4.35**.

Example 4.15 (Independence and products of c.f.s; James, 1952)

The converse of the proposition of (4.79) is untrue. Although we saw in **4.16–17** that, if X and Y are independent, their joint c.f. is the product of their individual c.f.s. *and conversely*, it is not true that if the c.f. of their sum is the product of their individual c.f.s they must be independent.

Consider

$$f(x,y) = \frac{1}{2\pi}x^{-\frac{1}{2}}y^{-\frac{1}{2}}e^{-\frac{1}{2}(x+y)}\{1 + \varepsilon 2\pi(xy)^{\frac{1}{2}}(x-y)(xy-x-y+2)e^{-\frac{1}{2}(x+y)}\},$$

$$0 \leqslant x, y < \infty. \quad (4.83)$$

Within the braces, the multiplier of ε, say M, has a lower bound, so there will be some value of ε, perhaps small, for which (4.83) is non-negative throughout the range of x and y. Moreover, if we interchange x and y in M, we change its sign and hence the integral of the term in ε over the range $0 \leqslant x, y < \infty$ is zero. The integral of the first term is clearly $\{2^{\frac{1}{2}}\Gamma(\frac{1}{2})\}^2/(2\pi) = 1$. Thus (4.83) is a genuine density.

The c.f. of x and y is obtained by integrating $\exp(it_1x + it_2y)$ over (4.83) and is found to reduce to

$$\phi(t_1, t_2) = \frac{1}{(1-2it_1)^{\frac{1}{2}}(1-2it_2)^{\frac{1}{2}}} - \frac{2i\varepsilon t_1 t_2(t_1 - t_2)}{(1-it_1)^3(1-it_2)^3}. \quad (4.84)$$

To obtain the c.f. of x we put $t_2 = 0$, obtaining

$$\phi_1(t_1) = (1-2it_1)^{-\frac{1}{2}}.$$

Similarly for the c.f. of y we put $t_1 = 0$, obtaining

$$\phi_2(t_2) = (1-2it_2)^{-\frac{1}{2}}.$$

For the c.f. of $x+y$ we put $t_1 = t_2 = t$, obtaining $\phi_3(t) = (1-2it)^{-1}$ which is the product of $\phi_1(t)$ and $\phi_2(t)$. The relationship between the c.f.s is the same as if ε were zero, owing to the fact that in (4.84) the term containing ε has factors t_1, t_2 and $(t_1 - t_2)$.

Equating coefficients of t^2 in $\phi_3(t) = \phi_1(t)\phi_2(t)$ shows that $E(xy) = E(x)E(y)$, x and y being uncorrelated as in **2.28**, although not independent.

EXERCISES

4.1. Show that if a p.f. $f(x)$ is symmetrical the c.f., apart from a term involving the mean, is an even function, i.e. $\phi(t) = \phi(-t)$, and that therefore $\phi(t)$ is real; and conversely, if $\phi(t)$ is real the p.f., if any, is symmetrical.

4.2 Show that the c.f.

$$\phi(t) = \left(\frac{e^{it} - 1}{it}\right)^n, \quad n \text{ a positive integer}$$

is that of

$$f(x) = \frac{1}{(n-1)!}\sum_{j=0}^{[x]} (-1)^j \binom{n}{j} (x-j)^{n-1}$$

4.3 Show that the distribution whose c.f. is $1/(1+t^2)$ is the double exponential (or Laplace) distribution $\frac{1}{2}e^{-|x|}$, $-\infty < x < \infty$, and that it is symmetrical about zero with even cumulants $\kappa_{2r} = 2(2r-1)!$, so that it has variance 2 and kurtosis coefficient $\gamma_2 = 3$.

4.4 If for a distribution

$$\kappa_r = \lambda a^r, \quad \lambda, a > 0,$$

show that the distribution is discrete (the Poisson) with variate values $0, a, 2a, \ldots, ra, \ldots$ and the probability at ra equal to $e^{-\lambda}\lambda^r/r!$

4.5 Show that the function $\exp(-t^\alpha)$ cannot be a c.f. unless $\alpha = 2$.

4.6 Show that there is only one distribution with moments given by

$$\mu_r' = \Gamma(\nu + r)/\Gamma(\nu)$$

and that it is

$$dF = \frac{1}{\Gamma(\nu)} e^{-x} x^{\nu-1} dx, \quad 0 \leqslant x < \infty.$$

4.7 Show that the distribution

$$dF = \frac{\sqrt{2}}{\pi}\frac{dx}{1+x^4}, \quad -\infty < x < \infty,$$

has c.f.

$$\phi(t) = \sqrt{2}\exp(-|t|/\sqrt{2})\sin(|t|/\sqrt{2} + \tfrac{1}{4}\pi).$$

Hence verify that $\mu_2 = 1$ as in Exercise 3.8 and that no higher moment is finite.

4.8 Show that the distribution

$$dF = \frac{rb^{2r-1}\sin(\pi/2r)}{\pi(b^{2r} + x^{2r})}dx, \quad -\infty < x < \infty; \quad b > 0; \quad r \text{ a positive integer},$$

has c.f.

$$\phi(t) = \sum_{s=0}^{r-1} \exp\left(-b|t|\sin\frac{2s+1}{2r}\pi\right)\sin\left(\frac{2s+1}{2r}\pi + b|t|\cos\frac{2s+1}{2r}\pi\right)\sin(\pi/2r).$$

(Exercise 4.7 is the special case $b = 1$, $r = 2$.)

4.9 If z has the Cauchy distribution of Example 4.2, show that $x = \pi^{-1} \log |z|$ has the distribution

$$f = \frac{1}{\cosh \pi x}, \qquad -\infty < x < \infty,$$

with c.f.

$$\phi(t) = \operatorname{sech} \tfrac{1}{2} t,$$

that $\mu_2 = \frac{1}{4}$, $\mu_4 = \frac{5}{16}$ and that the variate $y = (2\pi)^{\frac{1}{2}} x$ has p.d.f. and c.f. of the same functional form apart from a constant $(2\pi)^{-\frac{1}{2}}$ in the former.

4.10 Show that $(1 + t^{2k})^{-1}$ cannot be a c.f. for $k > 1$; and also that $(1 - t^2)^{-1}$ cannot be a c.f.

4.11 Using the theorem of Marcinkiewicz referred to in **4.8**, show that, if all cumulants are zero for r greater than some $r_0 > 2$, they must be zero for $r > 2$; and hence that the distribution is normal.

4.12 Show that the distribution

$$F_n(x) = \left(1 - \frac{1}{n}\right) \frac{1}{\sqrt{(2\pi)}} \int_{-\infty}^{x} e^{-\frac{1}{2}u^2} du + \frac{1}{2n} \{1 + \operatorname{sgn}(x - n)\}$$

tends to normality as $n \to \infty$ but that the higher moments tend to infinity.

(D. G. Kendall and Rao, 1950)

4.13 A theorem due to Weierstrass states that any function continuous in the range (a, b) can be represented by a uniformly convergent series of polynomials $\sum_{n=0}^{\infty} P_n(x)$, $P_n(x)$ being of degree n in x. Deduce that if two p.d.f.s, f_1 and f_2, have the same set of moments,

$$\int_a^b (f_1 - f_2)^2 \, dx = 0,$$

and hence that the moments determine a distribution uniquely if it is continuous and of finite range.

4.14 If θ is a non-negative function of the variate x and

$$\alpha(t) = \int_{-\infty}^{\infty} \theta^t \, dF(x),$$

show that the p.f. of θ, if any, is given by

$$f(\theta) = \frac{1}{2\pi i} \int_{-i\infty}^{i\infty} \theta^{-t-1} \alpha(t) \, dt.$$

4.15 Show that if a c.f. $\phi(t)$ has two derivatives, then

$$\left| \left(\frac{d\phi}{dt}\right)^2_{t=0} \right| \leqslant \left| \left(\frac{d^2\phi}{dt^2}\right)_{t=0} \right|$$

and generalize this result.

4.16 *Mixtures of Poisson distributions.* λ is a positive random variable with c.f. equal to $\phi(t)$. Another variable y is conditionally distributed in the (Poisson) form of Exercise 3.1 with parameter λ. Show that the unconditional c.f. of y is $\phi\{(e^{it}-1)/i\}$ and that its mean equals that of λ while its variance equals $\text{var}(\lambda)+E(\lambda)$.

4.17 Show that the distribution

$$f=\frac{1}{2\pi(1-\rho^2)^{\frac{1}{2}}}\exp\left\{\frac{-1}{2(1-\rho^2)}(x_1^2-2\rho x_1x_2+x_2^2)\right\},\qquad -\infty<x_1,x_2<\infty,\quad |\rho|<1$$

is uniquely determined by its moments.

4.18 If a vàriate is distributed in the normal form

$$f=\frac{1}{\sigma\sqrt{(2\pi)}}e^{-\frac{1}{2}x^2/\sigma^2},\qquad -\infty<x<\infty,$$

show that the c.f. of its square is given by $\phi(t)=(1-2it\sigma^2)^{-\frac{1}{2}}$.

4.19 If two variates are distributed as in Exercise 4.17 show that the joint c.f. of x_1^2 and x_2^2 (with variables t_1 and t_2 respectively) is $\{(1-2it_1)(1-2it_2)+4\rho^2t_1t_2\}^{-\frac{1}{2}}$.

4.20 A discrete distribution defined at integral values in the range 0 to $\frac{1}{2}n(n-1)$ has a probability generating function

$$P(t)=\frac{1}{n!}\prod_{j=1}^{n}\frac{t^j-1}{t-1}.$$

By considering the c.g.f., and using the relation (3.61), for Bernoulli numbers, show that the standardized distribution tends to the normal form as $n\to\infty$.

4.21 Show that the c.f. of the *logistic distribution*

$$f=\frac{1}{4\cosh^2(\frac{1}{2}x)}=\frac{\exp(-x)}{\{1+\exp(-x)\}^2},\qquad -\infty<x<\infty,$$

is $\phi(t)=\Gamma(1-it)\Gamma(1+it)=\pi t/\sinh(\pi t)$ using (3.67). Hence show using (3.61) that its c.g.f. is

$$\psi(t)=-\log\left(\frac{\sinh(\pi t)}{\pi t}\right)=\sum_{j=1}^{\infty}\frac{(it)^{2j}}{(2j)!}(-1)^{j-1}\frac{(2\pi)^{2j}}{2j}B_{2j},$$

so that it is symmetrical about zero with even cumulants $\kappa_{2j}=(-1)^{j-1}(2\pi)^{2j}B_{2j}/(2j)$ and in particular the variance $\sigma^2=\pi^2/3$ and the kurtosis $\beta_2=\mu_4/\mu_2^2=4.2$.

Using Exercise 2.9, show that the mean difference $\Delta=2$, and from Exercise 2.24 that the mean deviation $\delta_1=2\log 2$, so that $\Delta/\sigma=1.103$ and $\delta_1/\sigma=0.764$.

4.22 If x is an integer-valued random variable, show that the limits of integration in the inversion formula (4.5) may be replaced by $-\pi$ and π. Use this fact to invert the c.f. of a Poisson variate, given in Example 3.10.

4.23 *Mixtures by location parameter.* x is distributed with p.f. $f(x-\mu)$ conditional upon the value of μ, which is a random variable with p.f. $g(\mu)$. Show using c.f.s that the unconditional distribution of x is that of the sum of independent variables y, z, with p.f.s $f(y)$, $g(z)$.

4.24 *Mixtures by scale parameter.* x is distributed with p.f. $\sigma f(x\sigma)$ $(\sigma>0)$ conditional upon the value of σ, which is a random variable with p.f. $g(\sigma)$. Show using c.f.s that the unconditional distribution of x is that of the ratio of independent variables y, z, with p.f.s $f(y)$, $g(z)$.

4.25 Show that the Poisson c.f. in Example 4.9 is infinitely divisible.

4.26 Using the sufficient conditions given in 4.8, verify that

$$\phi(t) = \exp(-|t|^{\alpha})$$

is a valid c.f. when $0 < \alpha \leqslant 1$, but that the conditions are not satisfied for $1 < \alpha \leqslant 2$.

(Cf. Feller, 1971, chapter 15.)

4.27 Verify that (4.75) reduces to

$$f(x) = \frac{1}{\pi(1+x^2)}$$

when $\gamma = 0$, $\alpha = 1$.

4.28 Verify that (4.76) reduces to

$$f(x) = \frac{1}{2\sqrt{\pi}} \exp(-\tfrac{1}{4}x^2)$$

when $\gamma = 0$, $\alpha = 2$.

4.29 If X is a discrete random variable with positive probability assigned to k distinct values and Y is a random variable with the same first $2k$ moments as X, show that Y must have the same distribution as X. Further, if Y is known to be discrete, show that the equality of the first k moments suffices. (Arnold and Norton, 1985.)

4.30 When a distribution is infinitely divisible and has $\mu_4 < \infty$, show that $\beta_2 - \beta_1 - 3 \geqslant 0$, which strict equality attained for the normal and Poisson distributions.

(Rohatgi and Szekely, 1989.)

CHAPTER 5

STANDARD DISTRIBUTIONS

5.1 There are certain distribution and probability functions that, for both theoretical and practical reasons, occupy a central position in statistical theory. In this and the next chapter we shall consider their properties, leaving their uses to be developed and illustrated later in the book. We shall, however, indicate briefly some of the ways in which they arise, even at the expense of anticipating ideas introduced at a subsequent stage. This will not impair the logical continuity of our development and will give concreteness to a treatment that might otherwise appear somewhat abstract.

Comprehensive accounts of all the major theoretical distributions are given by N. L. Johnson and Kotz (1969, 1970, 1972), N. L. Johnson *et al.* (1993) and Patil *et al.* (1985a, b, c).

The binomial distribution

5.2 A convenient conceptual framework for developing statistical distributions is the *urn scheme*; that is, we consider a population of N individuals (balls in an urn) which are identical save for their colour. We shall assume that Np of the balls are one colour, say red, and that the rest, Nq, are black; $p+q=1$. In a single trial, a ball is selected from the urn and its colour noted; the ball is then returned to the urn. Thus, further trials may be performed under conditions identical to the first. If each ball has the same chance of selection in each trial, the experiments correspond to simple random sampling from the population (cf. Chapter 9).

If a single trial were performed a large number of times, we would expect the proportion of times in which a red ball was selected to approach $(Np)/N=p$. Now consider pairs of trials. Since conditions are identical at each trial, we would expect two red balls to occur with a relative frequency of $(Np)^2/N^2=p^2$; likewise, two black balls would occur with expected relative frequency q^2, and one of each with expected frequency $2pq$, since we may select red on either the first or the second drawing. We now regard the limiting relative frequency as a probability (cf. **8.11**) and henceforth develop the discussion in probabilistic terms.

Generally, if we perform n trials, the probability of j red balls and $(n-j)$ black balls will be

$$f_j=\binom{n}{j}p^jq^{n-j}, \quad j=0,1,2,\ldots,n. \tag{5.1}$$

The term p^jq^{n-j} gives the probability for a specific sequence of j reds and $(n-j)$ blacks and $\binom{n}{j}$ is the number of such sequences. By construction, $\sum f_j=1$; in general,

$$\sum_{j=0}^{n}\binom{n}{j}p^jq^{n-j}=(q+p)^n, \tag{5.2}$$

and the f_js are the terms of the *binomial series* expansion. Thus, we refer to (5.1) as the *binomial distribution*. The number of red balls, j, is often termed the number of "successes".

5.3 Distributions very close to the binomial form occur in practice, particularly in artificial experiments with coin-tossing or dice-throwing. Table 5.1 shows some data due to Weldon, who threw 12 dice 26 306 times and noted the values at each throw. This is equivalent to the drawing of samples of 12 from a large population. The occurrence of a 5 or a 6 on any die was regarded as a success.

Table 5.1 Distribution of the number of successes (throws of 5 or 6) in 26 306 throws of 12 dice
(Weldon's data, cited by F. Y. Edgeworth, *Encyclopaedia Britannica*, 11th edn, **22**, 39)

No. of successes	Observed frequency	Binomial distribution with the same $p=0.3377$	No. of successes	Observed frequency	Binomial distribution with the same $p=0.3377$
0	185	187	6	3067	3043
1	1149	1146	7	1331	1330
2	3265	3215	8	403	424
3	5475	5465	9	105	96
4	6114	6269	10 and over	18	16
5	5194	5115			
			TOTALS	26 306	26 306

If the dice were perfect the proportion p of successes would be $\frac{1}{3}$; and the appropriate binomial would be, in the form (5.2), $(\frac{2}{3}+\frac{1}{3})^{12}$. In practice, as always, the dice were not perfect, the proportion of cases exhibiting a 5 or a 6 being 0.3377. Taking this as the value of p, we get the binomial $(0.6623+0.3377)^{12}$, which when multiplied by the total frequency 26 306 gives the theoretical frequencies shown in the third column of Table 5.1. The agreement with observation is evidently fairly good.

A. W. Kemp and C. D. Kemp (1991) give an account of the history behind this data set and suggest several alternative models.

5.4 We have already found some moments and the factorial moments of the distribution in Examples 3.2 and 3.8. The c.f. of the distribution is, as in Example 3.5,

$$\phi(t) = P(e^{it}) = \sum_{j=0}^{n}\binom{n}{j}q^{n-j}p^{j}e^{ijt}$$
$$= (q+pe^{it})^{n}. \tag{5.3}$$

Taking logarithms and expanding we have for the c.g.f. (writing $\theta = it$)

$$\psi(t) = n\log\{1+p(e^{it}-1)\} = n\log\left(1+p\sum_{r=1}^{\infty}\theta^{r}/r!\right).$$

Expanding the logarithm and identifying powers in θ we then find

$$\kappa_1=\mu_1'=np;\quad \kappa_2=\mu_2=np(1-p)=npq;\quad \kappa_3=\mu_3=np(1-p)(1-2p)=npq(q-p);$$
$$\kappa_4=np(1-p)(1-6p+6p^2)=npq(1-6pq), \tag{5.4}$$

so that

$$\mu_4=3n^2p^2q^2+npq(1-6pq); \tag{5.5}$$

$$\gamma_1=\mu_3/\mu_2^{\frac{3}{2}}=\frac{q-p}{(npq)^{\frac{1}{2}}}=\sqrt{\beta_1}; \tag{5.6}$$

$$\gamma_2=\kappa_4/\kappa_2^2=\frac{1-6pq}{npq}=\beta_2-3. \tag{5.7}$$

The factorial moments about the origin are particularly simple. We have already found in Example 3.8 that

$$\left.\begin{aligned}\mu'_{[r]}&=p^r n^{[r]}, r\leqslant n\\ &=0, r>n.\end{aligned}\right\} \tag{5.8}$$

The mean deviation is given in Exercise 5.4.

5.5 There are some interesting recurrence relations connecting the moments of the binomial, due to Romanovsky (1923).

From (5.3), the c.f. about the mean is

$$\phi(t)=e^{-np\theta}\{1+p(e^\theta-1)\}^n.$$

Differentiating with respect to θ, we find

$$\phi'(t)=\frac{npq(e^\theta-1)}{1+p(e^\theta-1)}\phi(t). \tag{5.9}$$

Taking the denominator in (5.9) to the left side, and identifying coefficients of $\theta^{r-1}/(r-1)!$, we find for $r\geqslant 2$

$$\mu_r=npq\sum_{j=0}^{r-2}\binom{r-1}{j}\mu_j-p\sum_{j=0}^{r-2}\binom{r-1}{j}\mu_{j+1}, \tag{5.10}$$

a recurrence relation for μ_r in terms of lower central moments.

Now, differentiating the c.f. with respect to p, we have

$$\frac{\partial\phi(t)}{\partial p}=n\phi(t)\left\{-\theta+\frac{e^\theta-1}{1+p(e^\theta-1)}\right\}.$$

Identifying coefficients of $\theta^r/r!$, we get simply $\dfrac{\partial\mu_r}{\partial p}$ on the left, while the first term on the right, $-n\theta\phi(t)$, yields the coefficient $-nr\mu_{r-1}$. The other term on the right is seen from (5.9) to be $\phi'(t)/(pq)$, and since the coefficient of $\theta^r/r!$ in $\phi'(t)$ is μ_{r+1}, the coefficient we require is $\mu_{r+1}/(pq)$. Thus we have for $r\geqslant 1$

$$\mu_{r+1}=pq\left(nr\mu_{r-1}+\frac{\partial\mu_r}{\partial p}\right). \tag{5.11}$$

For example, $\mu_1 = 0$, $\mu_2 = npq = np(1-p)$ and hence, as stated in (5.4),

$$\mu_3 = pq(n-2np) = npq(q-p).$$

Exercise 5.1 gives an even simpler recurrence relation for the cumulants.

Exercise 5.2 shows that (5.11) holds for the incomplete moments, and Exercise 5.3 generalizes (5.10) similarly.

5.6 If $p = q = \frac{1}{2}$, the binomial distribution is obviously symmetrical, and if $p \neq q$ the distribution is skew. It will be unimodal unless $p(n+1) \leqslant 1$ (which holds for $p = \frac{1}{2}$ only at $n = 1$) since the probability of r successes exceeds that of $(r-1)$ if and only if

$$\binom{n}{r} p^r q^{n-r} > \binom{n}{r-1} p^{r-1} q^{n-r+1},$$

i.e., $r < (n+1)p$ or

$$r - p < np = \mu_1'. \tag{5.12}$$

Thus the probability of r successes increases steadily so long as $r < \mu_1' + p$, and decreases steadily for $r > \mu_1' + p$ with maximum probability at $r = [p(n+1)]$, the integer part of $p(n+1)$. There are equal maxima at $(r-1)$ and r if $r = p(n+1)$, when $r - 1 < \mu_1' < r$. Thus, the mode is always within one unit of the mean and the median lies in the interval between them. Some typical distributions are shown below.

Terms of the binomial distribution for $n = 20$ and values of p from 0.1 to 0.5 (each term has been multiplied by 10 000)

Number of successes	$p=0.1$ $q=0.9$	$p=0.2$ $q=0.8$	$p=0.3$ $q=0.7$	$p=0.4$ $q=0.6$	$p=0.5$ $q=0.5$
0	1216	115	8	—	—
1	2702	576	68	5	—
2	2852	1369	278	31	2
3	1901	2054	716	123	11
4	898	2182	1304	350	46
5	319	1746	1789	746	148
6	89	1091	1916	1244	370
7	20	545	1643	1659	739
8	4	222	1144	1797	1201
9	1	74	654	1597	1602
10	—	20	308	1171	1762
11	—	5	120	710	1602
12	—	1	39	355	1201
13	—	—	10	146	739
14	—	—	2	49	370
15	—	—	—	13	148
16	—	—	—	3	46
17	—	—	—	—	11
18	—	—	—	—	2
19	—	—	—	—	—
20	—	—	—	—	—

The binomial probabilities may be calculated directly from (5.1) or from the recurrence relation

$$f_j = \frac{(n-j+1)p}{jq} f_{j-1}.$$

For the binomial and, later, for other distributions we outline exact and approximate procedures for the evaluation of the d.f., as well as providing listings of available tables. Most major computer packages now provide subroutines that evaluate the d.f.s of standard distributions; the computational procedures are based upon results such as those outlined here.

Tables of the binomial distribution

(a) The *Biometrika Tables* give the individual terms of the distribution for $p = 0.01$, $0.02(0.02)0.1(0.01)0.5$ and $n = 5(5)30$, to 5 decimal places.

(b) *Tables of the Binomial Probability Distribution* (National Bureau of Standards, Applied Mathematics Series, 6, Washington, 1950) gives the individual terms and the d.f. for $p = 0.01(0.01)0.50$ and $n = 2(1)49$, to 7 d.p.

(c) Romig (1953) extends (b) above to $n = 50(5)100$, to 6 d.p.

(d) *Tables of the Cumulative Binomial Probabilities* (U.S. Army Ordnance Corps Pamphlet ORD P 20–1, Washington, 1952) gives the d.f. for $p = 0.01(0.01)0.50$ and $n = 1(1)150$, to 7 d.p.

(e) *Tables of the Cumulative Binomial Probability Distribution* (Harvard University Press, 1955) gives the d.f. for $p = 0.01(0.01)0.50$, $p = \frac{1}{16}(\frac{1}{16})\frac{1}{2}$, $p = \frac{1}{12}(\frac{1}{12})\frac{1}{2}$ and $n = 1(1)50(2)100(10)200(20)500(50)1000$.

(f) Weintraub (1963) gives the d.f. to 9 d.p. for $p = 0.00001$, $0.0001(0.0001)0.001(0.001)0.1$ and $n = 1(1)100$.

(g) Robertson (1960) gives the d.f. for $p = 0.001(0.001)0.020$, $n = 2(1)100(2)200(10)400(200)1000$, and for $p = 0.021(0.001)0.050$, $n = 2(1)50(2)100(5)200(10)300(20)500$ and $500(50)1000$.

(h) Miller (1954) gives the coefficients $\binom{n}{r}$ for $2 \leqslant r \leqslant \frac{1}{2}n \leqslant 100$; $r = 2(1)12$, $n \leqslant 500$; $r = 2(1)11$, $n = 500(1)1000$; $r = 2(1)5$, $n = 1000(1)2000$; $r = 2, 3$, $n = 2000(1)5000$.

From Example 4.6, the standardized binomial distribution tends to the normal distribution as n increases. Raff (1956) showed that if $np^{\frac{1}{2}} > 1.07$, the error in using the normal d.f. instead of the binomial never exceeds 0.05 for any r, and he compared other approximations to the binomial, as did Gebhardt (1969). Peizer and Pratt (1968) and Pratt (1968) gave an approximation accurate to order $n^{-\frac{3}{2}}$. See also Ling (1978) and Ghosh (1980).

5.7. The d.f. of the binomial distribution may also be evaluated using E. S. Pearson and N. L. Johnson's *Tables of the Incomplete Beta Function* (2nd edn, Cambridge U.P., 1968).

Consider the integral

$$B_{1-p}(n-r+1, r) = \int_0^{1-p} u^{n-r}(1-u)^{r-1}\, du. \tag{5.13}$$

Substituting $u = (1-p)(1-v)$ in the integral, (5.13) becomes

$$= \int_0^1 (1-p)^{n-r+1}(1-v)^{n-r}\{p+(1-p)v\}^{r-1}\,dv$$

$$= (1-p)^{n-r+1}\sum_{j=0}^{r-1}\binom{r-1}{j}p^j(1-p)^{r-1-j}\int_0^1 v^{r-j-1}(1-v)^{n-r}\,dv. \quad (5.14)$$

The integral in (5.14) is $B(r-j, n-r+1) = (r-j-1)!\ (n-r)!/(n-j)!$ so that

$$B_{1-p}(n-r+1, r) = \sum_{j=0}^{r-1}\frac{(r-1)!(n-r)!}{j!(n-j)!}p^j(1-p)^{n-j}. \quad (5.15)$$

Division of (5.15) by $B(n-r+1, r)$, i.e. multiplication by $\dfrac{n!}{(r-1)!(n-r)!}$, gives the Incomplete Beta Function

$$I_q(n-r+1, r) = \frac{B_{1-p}(n-r+1, r)}{B(n-r+1, r)} = \sum_{j=0}^{r-1}\binom{n}{j}p^j(1-p)^{n-j} = F(r-1), \quad (5.16)$$

the sum of the first r terms of the binomial distribution. Since $I_x(a, b) = 1 - I_{1-x}(b, a)$, (5.16) may also be written as $1 - I_p(r, n-r+1)$, so that $I_p(r, n-r+1)$ is the remainder after the first r terms.

The sum of the first $r+1$ terms is, similarly, $I_q(n-r, r+1)$, and hence the $(r+1)$th term is

$$\binom{n}{r}q^{n-r}p^r = I_q(n-r, r+1) - I_q(n-r+1, r). \quad (5.17)$$

For example, when $n=20$, $r=11$, $p=0.4$ we have for the remainder after 11 terms $I_{0.4}(11, 10)$ which from the tables is found to be 0.127 521 2. The value given by summing the last six non-zero terms in the table in 5.6 is 0.127 6, the error in the last place being due to rounding up. The remainder after 12 terms is $I_{0.4}(12, 9) =$ 0.056 526 4. The 12th term (11 successes) is then the difference of these two remainders = 0.0710, as shown in the table for the expected frequency per 10 000 of 11 successes.

The Poisson distribution

5.8 Cases sometimes occur in which the proportion p of successes in the population is very small. We may suppose that the sample size n is large enough to render np itself appreciable though p is small; and we are thus led to consider the limiting form of the binomial distribution as $n\to\infty$, $p\to 0$ subject to the condition that np remains finite, and tends to λ, say.

Under these conditions the term

$$\binom{n}{r}p^r q^{n-r} = \frac{n!}{(n-r)!r!}\frac{\lambda^r}{n^r}\left(1-\frac{\lambda}{n}\right)^{n-r}$$

$$\doteq \frac{\sqrt{(2\pi)}e^{-n}n^{n+\frac{1}{2}}}{\sqrt{(2\pi)}(n-r)^{n-r+\frac{1}{2}}e^{-n+r}n^r}\cdot\frac{\lambda^r}{r!}e^{-\lambda}$$

$$\doteq \frac{1}{\left(1-\dfrac{r}{n}\right)^n e^r}\cdot\frac{\lambda^r}{r!}e^{-\lambda} \doteq \frac{\lambda^r}{r!}e^{-\lambda}.$$

Thus the terms of the binomial approach

$$f_r = e^{-\lambda}\frac{\lambda^r}{r!}, \quad r = 0, 1, 2, \ldots; \quad \lambda > 0. \tag{5.18}$$

The distribution (5.18) is called the Poisson distribution, having been given by S. D. Poisson in 1837.

We may therefore expect (5.18) to describe the distribution of an event that occurs rarely in a short period, but where we observe the frequency of its occurrence in longer periods, each consisting of many short periods. Table 5.2 gives an example where the agreement seems very good.

Table 5.2 Number of major labour strikes in the U.K., 1948–59, commencing in each week
(Source: *Ministry of Labour Gazette*)

No. commencing in week	No. of weeks with this no. commencing	Poisson distribution with the same mean 0.90
0	252	254.5
1	229	229.1
2	109	103.1
3	28	30.9
4 or more	8	8.4
Total no. of weeks	626	626.0

5.9 For the c.f. of (5.18) we have from (5.3) (writing $\theta = it$)

$$\phi(t) = \lim_{n\to\infty}(q + pe^{\theta})^n = \lim_{n\to\infty}\left\{1 + \frac{\lambda}{n}(e^{\theta} - 1)\right\}^n$$
$$= \exp\{\lambda(e^{\theta} - 1)\} \tag{5.19}$$

which has already been shown in Example 3.10 to be the c.f. of the distribution (5.18), with c.g.f.

$$\psi(t) = \lambda(e^{\theta} - 1)$$

and all cumulants equal to λ. We thus find from (3.38)

$$\mu_1' = \lambda, \quad \mu_2 = \lambda, \quad \mu_3 = \lambda, \quad \mu_4 = \lambda + 3\lambda^2. \tag{5.20}$$

If we let $n \to \infty$, $p \to 0$, $np \to \lambda$ in (5.10) and (5.11) we find

$$\mu_r = \lambda \sum_{j=0}^{r-2}\binom{r-1}{j}\mu_j \tag{5.21}$$

and

$$\mu_{r+1} = r\lambda\mu_{r-1} + \lambda\frac{d\mu_r}{d\lambda}, \tag{5.22}$$

both of which may be proved directly as in Exercise 5.28.

The factorial moments are given in Exercise 5.11 and the mean deviation in Exercise 5.6.

Since $f_r/f_{r-1} = \lambda/r$, the probability at r increases steadily so long as $r<\lambda$ (the mean) and declines steadily to zero for $r>\lambda$. The maximum occurs at $r=[\lambda]$, with an equal adjacent maximum at $\lambda-1$ if λ is an integer. Thus, as for the binomial in **5.6**, the mode is always within one unit of the mean, with the median between them. For low values of λ the distribution is very skew, being J-shaped for $\lambda<1$, but it approaches unimodal symmetry as λ increases.

The d.f. $F(x)=\sum_{j=0}^{x} f_j$ may be evaluated in a manner similar to that of **5.7**.

The integral

$$G_\lambda(r-1)=\int_\lambda^\infty e^{-u}u^{r-1}du=\int_0^\infty e^{-(v+\lambda)}(v+\lambda)^{r-1}dv$$

$$=e^{-\lambda}\sum_{s=0}^{r-1}\binom{r-1}{s}\lambda^s\int_0^\infty e^{-v}v^{r-s-1}dv.$$

The integral on the right is $\Gamma(r-s)=(r-s-1)!$, so that

$$\frac{G_\lambda(r-1)}{(r-1)!}=\frac{1}{\Gamma(r)}\int_\lambda^\infty e^{-u}u^{r-1}du=e^{-\lambda}\sum_{s=0}^{r-1}\lambda^s/s!=F(r-1), \tag{5.23}$$

the sum of the first r terms of the Poisson distribution. In the notation of K. Pearson's *Tables of the Incomplete Gamma Function* (1922), (5.23) is $1-I\left(\frac{\lambda}{\sqrt{r}},\ r-1\right)$, so that the argument used in these tables is a difficult one to work with in the present case. Special tables of the d.f. are listed below. It is easier to calculate $e^{-\lambda}\lambda^r/r!$ directly rather than to use an expression analogous to (5.23) for the individual terms.

Tables of the Poisson distribution

(a) *Biometrika Tables for Statisticians* give the terms $e^{-\lambda}\lambda^r/r!$ to 6 d.p. for $\lambda=0.1(0.1)15$.

(b) The d.f. can be obtained using (5.23) from the tables of the χ^2 distribution described in **16.4** below—cf. Exercise 16.7. Thus *Biometrika Tables for Statisticians* gives $\sum_{j=0}^{r-1} e^{-\lambda}\lambda^j/j!$ to 5 d.p. for r up to 35 and $\lambda=0.000\,5(0.000\,5)0.005(0.005)0.05(0.05)1.0(0.1)5.0(0.25)10.0(0.5)20(1)60$. These can be used to supplement the tables of the individual terms, especially in the range $0\leqslant\lambda\leqslant 1$.

(c) Molina (1942) gives the terms and the sums to 6 or 7 decimal places for $\lambda=0.001(0.001)0.010(0.010)0.30(0.10)15(1)100$.

(d) *Tables of the Individual and Cumulative Terms of Poisson Distribution* (Van Nostrand, Princeton, 1962) give 8 d.p. tables of the f_r, the d.f. $F(r)$ and $1-F(r-1)$ for λ-values similar to those in (c) and also $\lambda=100(5)200$.

As $\lambda \to \infty$, we see in **5.8** that this effectively removes the limitation to small p, and we should thus expect the normal approximation to the binomial (cf. Example 4.6) to apply to the Poisson also; it does, as we saw in Example 4.9. Peizer and Pratt (1968) and Pratt (1968) give a much closer normal approximation. See also Ling (1978).

The differences between the binomial probabilities and their Poisson counterparts have been studied extensively. If $b(j, n, p)$ and $p(j, np)$ denote the binomial and Poisson probabilities respectively, $\lambda = np$ and $e_j = b(j, n, p) - p(j, np)$, it may be shown that (Poor, 1991)

$$\max_j |e_j| = e_1 = np[(1-p)^{n-1} - e^{-np}]$$

$$= \frac{\lambda^2 e^{-\lambda}}{2n}|2-\lambda| + O(n^{-2}).$$

Further, using (3.107), Romanowska (1977) showed that

$$D(n,p) = \sum_{j=0}^{\infty} |e_j| \leqslant p\{2/(1-p)\}^{\frac{1}{2}}.$$

For further discussion, see Kennedy and Quine (1989).

Generalizations of the binomial and the Poisson

5.10 We now consider a generalization of the binomial and the Poisson distributions. In **5.2** our approach was based on the drawing of sets of n from the same population. Suppose, however, that each set of n consists of an observation drawn from each of n different populations with proportions $p_1, p_2, \ldots, p_n$. Then the probabilities may be generated by the terms of the product

$$(p_1+q_1)(p_2+q_2)\ldots(p_n+q_n) = \prod_{j=1}^{n}(p_j+q_j) \qquad (5.24)$$

which of course reduces to the binomial distribution if all the ps are equal.

The characteristic function of this distribution is

$$\phi(t) = \prod(q_j + p_j e^{\theta})$$

from which we have

$$\psi(t) = \sum \log(q_j + p_j e^{\theta})$$

$$= \sum \log(1 + p_j\theta + \tfrac{1}{2}p_j\theta^2 + \ldots)$$

$$= \theta\sum p_j + \tfrac{1}{2}\theta^2\sum(p_j - p_j^2) + \ldots, \text{etc.},$$

giving

$$\mu_1' = \sum p_j, \quad \kappa_2 = \mu_2 = \sum p_j q_j. \tag{5.25}$$

Writing now $\bar{p}$ for the mean of the ps in the different populations, we have

$$\mu_1' = n\bar{p}$$

$$\begin{aligned}\mu_2 &= \sum pq = \sum p - \sum p^2 \\ &= \sum p - \frac{1}{n}\left(\sum p\right)^2 - \left\{\sum p^2 - \frac{1}{n}\left(\sum p\right)^2\right\} \\ &= n\bar{p} - n\bar{p}^2 - n\sigma_p^2\end{aligned}$$

(where σ_p^2 is written for the variance of p)

$$= n\bar{p}\bar{q} - n\sigma_p^2. \tag{5.26}$$

A comparison of these results with those for the binomial distribution shows that the variance of the distribution arrayed by (5.24) is *less* than that of the binomial with the same average p by an amount equal to n times the variance of p.

If instead of n binomials with parameters p_j we consider n Poisson populations with parameters λ_j/n, it follows from (5.19) that

$$\begin{aligned}\phi(t) &= \prod_{j=1}^{n} \exp\left\{\frac{1}{n}\lambda_j(e^{\theta} - 1)\right\} \\ &= \exp\left\{\frac{1}{n}\sum \lambda_j(e^{\theta} - 1)\right\} = \exp\{\bar{\lambda}(e^{\theta} - 1)\}.\end{aligned} \tag{5.27}$$

Hence, $\mu_1' = \mu_2 = \bar{\lambda}$, not withstanding the inequality of the λs.

The generalization which we have developed is a special form of *stratified sampling*. It has sometimes been called Poissonian sampling, a usage that can be confused with sampling from a Poisson distribution.

5.11 Consider now a second generalization, where each complete set of n is drawn from one of k different populations, with probabilities $p_1, p_2, \ldots, p_k$, each population having the same chance of selection. (In the previous case we supposed any set of n obtained by taking one observation from each of n populations. We now suppose that the set is drawn from one population only.) The set of probabilities is now given by the terms of

$$\frac{1}{k}\sum_{j=1}^{k} (q_j + p_j)^n \tag{5.28}$$

and evidently the moments about the origin are the means of the moments of the k populations. Thus, from (5.4),

$$\mu_1' = \frac{1}{k}\sum np_j,$$

$$\mu_2' = \frac{1}{k}\left\{\sum np_j + \sum n(n-1)p_j^2\right\}.$$

Writing $\bar{p}$ for the mean of the ps as before, we have

$$\mu_1' = n\bar{p}$$

$$\begin{aligned}\mu_2 &= n\bar{p} + \frac{1}{k}\sum n(n-1)p_j^2 - n^2\bar{p}^2 \\ &= n\bar{p}\bar{q} + \frac{1}{k}\sum n(n-1)p_j^2 - n(n-1)\bar{p}^2 \\ &= n\bar{p}\bar{q} + n(n-1)\left(\frac{1}{k}\sum p_j^2 - \bar{p}^2\right) \\ &= n\bar{p}\bar{q} + n(n-1)\sigma_p^2. \end{aligned} \tag{5.29}$$

In this case the variance is *greater* than it would be if the distribution were of the ordinary binomial type by an amount $n(n-1)$ times the variance of p.

For k Poisson populations we have, on letting $np_j \to \lambda_j$,

$$\mu_1' = \bar{\lambda}, \qquad \mu_2 = \bar{\lambda} + \sigma_\lambda^2, \tag{5.30}$$

and here, in contrast to (5.27), the variance of the distribution is also affected.

This second generalization is a special case of *cluster sampling*. It is sometimes called Lexian sampling, after W. H. R. A. Lexis, who considered it in 1877.

Altham (1978) presents generalizations of the binomial assuming that there are first order interactions between the Bernoulli trials. The distribution may be either more or less dispersed than the binomial, depending on the nature of the interactions.

5.12 It is often found in practice that data derived from the sampling of qualitative variates over an extended area or an extended period of time do not conform to the simple binomial type. For example, suppose we regard the possession of blue eyes as a success, and take a number of samples of n from the population of the United Kingdom in different localities. We should probably find that the proportions in these samples did not follow the simple binomial form. The variance calculated from the known n and average observed p would probably turn out to be smaller than the variance observed among the different samples. If so, we should conclude from (5.29) that the proportion p varied from place to place in the population, the excess in the variance of the proportions observed being due to the variation in p itself in the sections of the population from which the samples were chosen. See Exercise 5.19 for an alternative model of binomial sampling. Such processes will be explored in detail in **5.20–4** below.

Pólya urn scheme: the hypergeometric distributions

5.13 We now return to the urn scheme of **5.2** and revise the rules so that whenever a ball of a certain colour is selected, $c+1$ balls of that type are returned to the urn. Thus, successive trials may cease to be independent and the probability of j successes in n trials becomes

$$f_j = \binom{n}{j}\frac{Np(Np+c)\dots(Np+jc-c)Nq(Nq+c)\dots\{Nq+(n-j-1)c\}}{N(n+c)\dots(N+nc-c)}. \tag{5.31}$$

The structure of (5.31) is similar to that of (5.1); indeed, it reduces to (5.1) when $c = 0$. When $c = -1$, no balls are returned, and we have sampling *without replacement*. The probability function becomes

$$f_j = \frac{1}{N^{[n]}}\binom{n}{j}(Np)^{[j]}(Nq)^{[n-j]}, \tag{5.32}$$

where $N^{[n]} = N(N-1) \ldots (N-n+1)$ and j ranges over the integers max $(0, n-Nq) \leqslant j \leqslant \min(n, Np)$. Since $N^{[n]} = N!/(N-n)!$, (5.32) may also be written as

$$f_j = \binom{Np}{j}\binom{Nq}{n-j}\bigg/\binom{N}{n}. \tag{5.33}$$

This is known as the *hypergeometric distribution*. As $N \to \infty$, (5.32) approaches the binomial form.

When $c = +1$ in the urn scheme, there is an element of *contagion*, since the occurrence of a particular colour makes further occurrences more likely. From (5.31) we obtain

$$\begin{aligned} f_j &= \binom{n}{j}(Np+j-1)^{[j]}(Nq+n-j-1)^{[n-j]}/(N+n-1)^{[n]} \\ &= \binom{Np+j-1}{j}\binom{Nq+n-j-1}{n-j}\bigg/\binom{N+n-1}{n} \end{aligned} \tag{5.34}$$

where $j = 0, 1, \ldots, n$. This is known as the *negative hypergeometric* distribution for reasons that will become evident in **5.16–18**.

Other values of c may be used, but 0, ± 1 are the most common. Likewise, more general patterns of ball replacement may be considered. For fuller details on urn models, see N. L. Johnson and Kotz (1977b).

Properties of the hypergeometric distribution

5.14 The probability generating function for the hypergeometric distribution is

$$\begin{aligned} P(t) &= \frac{1}{N^{[n]}}\sum_{j=0}^{n}\left\{\binom{n}{j}(Np)^{[j]}(Nq)^{[n-j]}t^j\right\} \\ &= \frac{(Nq)^{[n]}}{N^{[n]}}\sum\frac{(Np)^{[j]}n^{[j]}}{(Nq-n+j)^{[j]}}\frac{t^j}{j!}. \end{aligned} \tag{5.35}$$

This is the hypergeometric function

$$\frac{(Nq)^{[n]}}{N^{[n]}}F(\alpha, \beta, \gamma, t)$$

if

$$\alpha = -n, \quad \beta = -Np, \quad \gamma = Nq - n + 1. \tag{5.36}$$

We have (cf. Romanovsky, 1925)

$$F(\alpha, \beta, \gamma, t) = 1 + \frac{\alpha\beta}{\gamma}\frac{t}{1!} + \frac{\alpha(\alpha+1)\beta(\beta+1)}{\gamma(\gamma+1)}\frac{t^2}{2!} + \dots$$

and it is well known that this function satisfies the differential equation

$$t(1-t)\frac{\mathrm{d}^2F}{\mathrm{d}t^2} + \{\gamma - (\alpha+\beta+1)t\}\frac{\mathrm{d}F}{\mathrm{d}t} - \alpha\beta F = 0, \tag{5.37}$$

as may be readily verified from the equation itself.

If in (5.35) we put $t = \mathrm{e}^{\theta}$ (where $\theta = it$) we have the c.f. ϕ of the distribution. On making this substitution in (5.37) we find, after some reduction and replacement of the values of α, β, γ by those of (5.36),

$$(1-\mathrm{e}^{\theta})\left\{\frac{\mathrm{d}^2\phi}{\mathrm{d}\theta^2} - (n+Np)\frac{\mathrm{d}\phi}{\mathrm{d}\theta} + nNp\phi\right\} - Nnp\phi + N\frac{\mathrm{d}\phi}{\mathrm{d}\theta} = 0. \tag{5.38}$$

Since $\phi = \Sigma \mu_j' \theta^j/j!$ we find, from the coefficient of θ^0 in this expression,

$$-Nnp + N\mu_1' = 0$$

$$\mu_1' = np, \tag{5.39}$$

the same result as for the binomial. The mean of the hypergeometric distribution is independent of N.

In terms of the c.f. about the mean, we find substituting $\mathrm{e}^{np\theta}\phi$ for ϕ in (5.38),

$$(1-\mathrm{e}^{\theta})\left[\frac{\mathrm{d}^2\phi}{\mathrm{d}\theta^2} + \frac{\mathrm{d}\phi}{\mathrm{d}\theta}\{n(p-q) - Np\} + (N-n)pqn\phi\right] + N\frac{\mathrm{d}\phi}{\mathrm{d}\theta} = 0 \tag{5.40}$$

whence, identifying coefficients in θ, θ^2, θ^3, we find

$$\left.\begin{aligned} \mu_2 &= \frac{npq(N-n)}{N-1}, \qquad \mu_3 = \frac{npq(q-p)(N-n)(N-2n)}{(N-1)(N-2)}, \\ \mu_4 &= \frac{npq(N-n)}{(N-1)(N-2)(N-3)} \\ &\quad \times [N(N+1) - 6n(N-n) + 3pq\{N^2(n-2) - Nn^2 + 6n(N-n)\}], \end{aligned}\right\} \tag{5.41}$$

and generally, if U denotes the operation of raising the order of a central moment by unity, i.e. $U\mu_r = \mu_{r+1}$, we have, on identifying coefficients of $\theta^r/r!$ in (5.40),

$$N\mu_{r+1} = \{(1+U)^r - U^r\}[\mu_2 - \{Np + n(q-p)\}\mu_1 + \{npq(N-n)\mu_0\}]. \tag{5.42}$$

As we expect, when $N \to \infty$ the moments tend to those of the binomial distribution at (5.4-5).

The factorial moments are given in Exercise 5.11.

5.15 An example of the occurrence of the hypergeometric distribution in practice is given in Table 5.3, giving the frequency of occurrence of playing cards of a certain suit in hands dealt to an individual. Here $N=52$ is the number of cards in the pack, $n=13$ is the number of cards dealt to the individual, and $p=\frac{1}{4}$ since there are 4 suits in a pack. The appropriate probabilities are thus

$$\frac{1}{52^{[13]}}\binom{13}{j}13^{[j]}39^{[n-j]}$$

leading to the figures shown in the third column. The agreement is reasonably good.

Table 5.3 Number of cards of a specified suit in 3400 hands of 13 cards
(K. Pearson (1924))

Number of cards in the hand	Observed frequency	Expected frequency from hypergeometric distribution	Number of cards in the hand	Observed frequency	Expected frequency from hypergeometric distribution
0	35	44	5	444	424
1	290	272	6	115	141
2	696	700	7	21	30
3	937	974	8	11	4
4	851	811	9 and over	0	0
			TOTALS	3400	3400

Lieberman and Owen (1961) give the d.f. and p.f. for $N=1(1)50(10)100$ and all n; and for $N=200(100)2000$ and selected n. Wise (1954) developed a quickly convergent expansion for the d.f. in terms of the Incomplete Beta-function. Sandiford (1960) shows that a close approximation is obtained from a binomial distribution with the same mean and variance. Ord (1968a) compares these and gives other approximations. Ling and Pratt (1984) give exceptionally accurate approximations due to Peizer.

Sequential sampling: the negative binomial distribution

5.16 Instead of drawing a sample of fixed size, n, we may alter the *stopping rule* and elect to continue sampling until the kth success has been recorded. Such a sampling regime is often relevant in medical trials and in industrial quality control where a number of early successes may indicate considerable shifts in recovery rates or percent defectives, respectively, so that the average sampling effort may be reduced.

We now count the number of failures up to the kth success, recognizing that each sequence of trials must end with the kth success. It follows that the probability of j failures is

$$f_j=\binom{k+j-1}{j}p^{k-1}q^j.p$$

$$=\binom{k+j-1}{j}p^kq^j,\quad j=0,1,\ldots. \tag{5.43}$$

This is known as the *negative binomial* distribution, originally because the f_js may be obtained from the negative binomial series

$$p^{-k} = (1-q)^{-k} = \sum \binom{k+j-1}{j} q^j. \tag{5.44}$$

It follows directly from (5.43) and (5.44) that the probability generating function is

$$P(t) = (1-q)^k/(1-qt)^k. \tag{5.45}$$

Expression (5.45) remains a valid p.g.f. for non-integer $k>0$, when the negative binomial is also known as the *Pascal* distribution. When $k=1$, we have the *geometric* distribution, since then $f_j = pq^j$.

Since $f_j/f_{j-1} = (k+j-1)q/j$, we see that $f_j > f_{j-1}$ if and only if $p<(k-1)/(j+k-1)$. Thus, the distribution is J-shaped if $k=1$ (or $k \leqslant 1$) and unimodal otherwise, with its maximum at $j=[(k-1)q/p]$.

The probability with which more than $(j+k)=n$ individuals will be required for k successes is exactly the probability of fewer than k successes in n ordinary binomial observations, i.e., is given by (5.16). Thus, the negative binomial d.f. is the complement of this, namely

$$F(n) = \sum_{s=k}^{n} f_s = 1 - I_{1-p}(n-k+1, k) = I_p(k, n-k+1).$$

Williamson and Bretherton (1963) have given tables of the negative binomial distribution. Peizer and Pratt (1968) and Pratt (1968) give a very good normal approximation. See also Ling (1978).

Exercise 5.30 asks the reader to fit a geometric distribution to observations.

5.17 From (5.45), the c.f. of the negative binomial distribution is (writing θ for it)

$$\phi(t) = p^k(1-qe^\theta)^{-k}. \tag{5.46}$$

Thus, for the c.g.f.

$$\psi(t) = k\log p - k\log(1-qe^\theta) = -k\log\left\{1 - \frac{q}{p}(e^\theta - 1)\right\}. \tag{5.47}$$

Expanding and identifying coefficients, we have

$$\left.\begin{aligned} \kappa_1 &= \frac{kq}{p}; \quad \kappa_2 = \frac{kq}{p^2}; \quad \kappa_3 = \frac{kq(1+q)}{p^3}; \\ \kappa_4 &= \frac{kq(1+4q+q^2)}{p^4}. \end{aligned}\right\} \tag{5.48}$$

(5.46) and (5.48) may be derived from equations (5.3–4) by substituting $-q/p$ for p, $1/p$ for q and $-k$ for n.

The same substitutions made in the recurrence relations (5.10–11) for binomial central moments, and that in Exercise 5.1 for binomial cumulants, yield

$$\left.\begin{aligned} \mu_s &= \frac{q}{p}\left(\frac{k}{p}\sum_{j=0}^{s-2}\binom{s-1}{j}\mu_j + \sum_{j=0}^{s-2}\binom{s-1}{j}\mu_{j+1}\right), \\ \mu_{s+1} &= q\left(\frac{ks}{p^2}\mu_{s-1} - \frac{\partial\mu_s}{\partial p}\right), \\ \kappa_{s+1} &= -q\,\partial\kappa_s/\partial p. \end{aligned}\right\} \quad (5.49)$$

The f.m.g.f. is given by

$$P(1+t) = p^k\{1-q(1+t)\}^{-k} = (1-qt/p)^{-k}.$$

Hence we find, for the factorial moments about the origin,

$$\mu'_{[s]} = \left(\frac{q}{p}\right)^s (k+s-1)^{[s]}. \quad (5.50)$$

From (5.48), $\mu_2 > \mu'_1$, whereas for the ordinary binomial $\mu_2 < \mu'_1$, and for the Poisson $\mu_2 = \mu'_1$. A preliminary calculation μ_2/μ'_1 is often a good guide to which of the three is likely to fit the data best. Exercise 5.10 gives another useful distinguishing feature.

As $k \to \infty$ with μ'_1 fixed, the negative binomial distribution approaches the Poisson with $\lambda = \mu'_1$; the result is readily established using the c.f. (5.46).

5.18 Corresponding to (5.31), the form for f_j in urn models with n fixed, the general expression for sequential sampling is

$$f_j = \binom{k+j-1}{j}\frac{(Np)(Np+c)\dots(Np+kc-c)Nq(Nq+c)\dots(Nq+jc-c)}{N(N+c)\dots\{N+(j+k-1)c\}} \quad (5.51)$$

When $c = -1$, this reduces to

$$f_j = \binom{k+j-1}{j}(Np)^{[k]}(Nq)^{[j]}/N^{[j+k]}, \quad (5.52)$$

$j = 0, 1, \dots, \min(Nq, N-k)$; this is the *negative hypergeometric* distribution and corresponds to (5.34) when (k, N, Nq) in (5.52) are reparametrized as $(Np, N+n-1, n)$. The word "negative" is used here, by extension from the negative binomial, because of the sequential sampling scheme that generates the distribution.

The probability generating function corresponds to the hypergeometric series $F(\alpha, \beta, \gamma, t)$, where $\alpha = k$, $\beta = -Nq$, and $\gamma = -N+k$. In Exercise 5.19, this distribution arises as the *beta-binomial* distribution, when α, β, and γ need not be integers.

C. D. Kemp and A. W. Kemp (1956) carried out a systematic investigation of the various distributions whose probability generating functions may be represented as convergent hypergeometric series.

The other distribution of principal practical interest arises when $c = +1$ in (5.51), yielding

$$f_j = \binom{k+j-1}{j}(Np+k-1)^{[k]}(Nq+j-1)^{[j]}/(N+j+k-1)^{[j+k]}, \quad j \geqslant 0. \tag{5.53}$$

This is a special case of the *Beta-Pascal*; see Exercise 5.19.

The urn schemes described here are capable of considerable extension; see N. L. Johnson and Kotz (1977b, chapters 4 and 5).

5.19 The Poisson, the binomials, and the hypergeometric distributions all satisfy the relationship

$$\frac{f_{j+1}-f_j}{f_j} = \frac{(j-a)}{b_0+b_1j+b_2j^2}. \tag{5.54}$$

This family of distributions was examined systematically in Ord (1967b). If the variable is defined on intervals of length h and the coefficients are suitably rescaled, then the limit of (5.54) as $h \to 0$ yields the differential equation

$$\frac{1}{f}\frac{\mathrm{d}f}{\mathrm{d}x} = \frac{x-a}{b_0+b_1x+b_2x^2}. \tag{5.55}$$

This is the defining equation of the Pearson family, which will be considered in the next chapter.

Bowman *et al.* (1991) use (5.54) as the basis for developing approximations to general discrete distributions whose first three moments are known.

Mixtures of distributions

5.20 Instead of drawing samples from a single urn, we may first select an urn from m possible urns, initially containing N_sp_s red and N_sq_s black balls, respectively, $s = 1, \ldots, m$. We may assume that the sth urn is selected with probability u_s, $\sum u_s = 1$. If the probability function for the sth urn is f_{js}, it follows that the probability function for the overall experiment is

$$f_j = \sum_{s=1}^{m} u_s f_{js}. \tag{5.56}$$

The p.f. (5.56) is said to describe a *finite mixture*; (5.28) gives a special case when $u_s = m^{-1}$ for all s. If μ'_{rs} denotes the rth moment about the origin for the sth population, it follows from (5.56) that

$$\mu'_r = \sum_s u_s \mu'_{rs}$$

so that

$$\mu_2 = \sum u_s \mu_{2s} + \sum u_s(\mu'_{1s} - \mu'_1)^2, \tag{5.57}$$

a generalization of (5.29).

Example 5.1 (Moments of discrete mixture distributions)

With suitable reparameterization, the factorial moments for the binomial, Poisson and negative binomial may be written as

$$\mu'_{[r]} = a_r\theta^r, \qquad r = 1, 2, \ldots,$$

where a_r is a function of r only. Thus, the factorial moments for the finite mixture are

$$\mu'_{[r]} = a_r \sum u_s\theta_s^r$$

the parameters may be estimated by the Method of Moments (cf. **3.42–3**) using the first $(2m-1)$ moment expressions.

Finite mixtures tend to become rather unwieldy for $m \geqslant 3$ because of the number of parameters. For further general discussions, see Douglas (1980), and Everitt and Hand (1981). Lindsay (1989) has developed an efficient computational scheme for the method of moments for univariate normal mixtures; Lindsay and Basak (1993) extend this approach to multivariate mixtures.

5.21 One way of reducing the number of parameters is to specify a distributional form for the probability function $\{u_s\}$. Suppose, for example, we are counting the number of newly-hatched larvae of a certain species of insect caught in a trap. The number of nearby egg-masses from which the larvae emerged is unknown but will certainly influence the number of larvae caught. If M is the variable denoting the number of egg masses and j_s denotes the catch from the sth egg mass, the total catch is

$$j = j_1 + j_2 + \ldots + j_M. \tag{5.58}$$

Expression (5.58) is a sum of a variable number of variables, also known as a *randomly-stopped sum*. Suppose that the $\{j_s\}$ are independent and identically distributed, with probability generating function $g(t) = \Sigma p(r)t^r$. Also, we assume that the p.g.f. of M is $G(t) = \Sigma u_s t^s$ and that of j is $H(t)$.

We have that

$$\begin{aligned} f(j) &= f(j_1 + \ldots + j_M = j) \\ &= \sum u_s f(j_1 + \ldots + j_M = j \mid M = s) \\ &= \sum u_s f(j_1 + \ldots + j_s = j), \end{aligned}$$

using the ideas of conditional distributions in **1.32** and **8.7**. Then

$$\begin{aligned} H(t) &= \sum f(j)t^j \\ &= \sum u_s \left[\sum f(j_1 + \ldots + j_s)t^j \right], \end{aligned} \tag{5.59}$$

where the inner summation is over all j_i such that $\sum_i j_i = j$. Since the j_i are independent, it follows that the term in square brackets decomposes into

$$\prod_{i=1}^{s}\left[\sum_{j_i} p(j_i)t^{j_i}\right]^s = [g(t)]^s. \tag{5.60}$$

Thus (5.59) becomes

$$H(t) = \sum u_s[g(t)]^s = G[g(t)]. \tag{5.61}$$

That is, the p.g.f. for j may be determined from the p.g.f.s of the constituent parts. This result is extended to continuous "catch" variables in Exercise 5.22. The distribution of j is known as a *generalized*, or *randomly-stopped sum*, distribution.

Example 5.2 (The negative binomial distribution as a randomly stopped sum; Quenouille, 1949)

Suppose M has a Poisson distribution with parameter λ, so that by Exercise 1.13

$$G(t) = e^{\lambda(t-1)}. \tag{5.62}$$

Let

$$g(t) = -\alpha \log(1-qt) = \alpha \sum_{i=1}^{\infty} q^i t^i / i, \tag{5.63}$$

where $\alpha = -1/\log(1-q)$. (5.62) is the p.g.f. of the *logarithmic series distribution*, to be discussed in **5.28**. Using (5.62) and (5.63), (5.61) yields

$$\begin{aligned} H(t) &= \exp[\lambda\{g(t)-1\}] \\ &= \exp[-\lambda - \lambda\alpha \log(1-qt)] \\ &= \left(\frac{1-q}{1-qt}\right)^k, \end{aligned} \tag{5.64}$$

where $k = \lambda\alpha$. (5.64) corresponds to (5.46) when $e^{\theta} = t$, the p.g.f. of the negative binomial, save that k is now *any* positive number. Thus a Poisson-stopped sum of logarithmic series variables has a negative binomial distribution.

Other examples of Poisson-stopped sums are the Neyman Type A, the Polya-Aeppli, and the Hermite distributions; see Exercises 5.7, 5.8, and 5.21, respectively. For further examples and properties, see Douglas (1980). Exercise 5.22 gives the form of the mean and variance for any generalized distribution.

5.22 Another form of mixture is to specify a continuous weighting (or mixing) distribution for a parameter θ; that is, given a probability function dependent on a parameter θ, $f(x|\theta)$, we may specify a weighting distribution for θ, $p(\theta)$ say. Integrating with respect to θ, we obtain

$$f(x) = \int_{-\infty}^{\infty} f(x|\theta)p(\theta)\,d\theta. \tag{5.65}$$

Essentially we now have a bivariate distribution of x and θ, and making use of the basic ideas in **1.32-3**, $f(x|\theta)$ is the conditional distribution of x given θ, while $p(\theta)$ is the marginal distribution of θ. Their product, the integrand above, is the bivariate

Table 5.4 Accidents to 647 women in 5 weeks
(Greenwood and Yule (1920))

Numbers of accidents	Observed frequency	Expected frequency	
		Poisson	Negative binomial
0	447	406	442
1	132	189	140
2	42	45	45
3	21	7	14
4	3	1	5
5 and over	2	0.1	2
TOTALS	647	648	648

distribution, as at (1.29). The marginal distribution $f(x)$ is obtained as at (1.30). $f(x)$ is sometimes known as a *compound* distribution.

5.23 This type of mixing is common in distributions that at first sight might be expected to be of the Poisson type. For example, suicide is a rare event and it might be supposed that if we took a series of large samples, say the population of the United Kingdom in successive years, the frequencies of suicides would follow the Poisson distribution. This, however, is not necessarily so, for all members of the population are not equally exposed to risk, and the temptation to suicide may vary from year to year, e.g. being greater in years of economic depression. This inequality of risk is typical of one field in which the Poisson distribution has been freely applied, namely, industrial accidents. Table 5.4 shows, in the second column, the frequency of accidents occurring to women making high-explosive shells during a 5-week period. The Poisson expected frequencies shown in the third column provide a very poor fit. One possible reason is that the liability of individuals to accident varies.

As a working hypothesis (cf. Greenwood and Yule, 1920) suppose that the population is a mixture of individuals with different degrees of accident proneness, represented by different values of λ in a Poisson distribution; and suppose that in the population the distribution of λ is of the Gamma form of Example 3.6,

$$dF = \frac{c^r}{\Gamma(r)} e^{-c\lambda}\lambda^{r-1}\, d\lambda, \qquad 0 \leqslant \lambda < \infty; \quad r > 0, \quad c > 0.$$

The probability of j successes is then

$$\int_0^\infty f(\lambda) e^{-\lambda}\frac{\lambda^j}{j!}\, d\lambda = \int_0^\infty \frac{c^r}{\Gamma(r)} e^{-c\lambda}\lambda^{r-1} e^{-\lambda}\frac{\lambda^j}{j!}\, d\lambda,$$

or the coefficient of t^j in the probability generating function

$$P(t) = \frac{c^r}{\Gamma(r)} \int_0^\infty e^{-c\lambda}\lambda^{r-1} e^{-\lambda+\lambda t}\, d\lambda,$$

which, on substituting $v = (c+1-t)\lambda$, becomes

$$P(t) = \frac{c^r}{(c+1-t)^r} = \left(\frac{c}{c+1}\right)^r \left(1 - \frac{t}{c+1}\right)^{-r}. \tag{5.66}$$

Replacing r by k and $(c+1)^{-1}$ by q, we see that (5.66) corresponds to (5.64), the p.g.f. of the negative binomial distribution. Alternatively, we could have used Exercise 4.16 to deduce the negative binomial c.f. from the Gamma c.f. at (3.19).

If we equate the observed mean and variance in Table 5.4 to κ_1 and κ_2 in (5.48), and solve for r and c, (5.43) gives the expected frequencies in the final column of Table 5.4, which evidently agree much better with the data.

Thus in addition to the sequential sampling model that gave rise to it in **5.16**, the negative binomial has been derived from two quite distinct mechanisms here and in Example 5.2. Given a random sample from the population, the two mechanisms are indistinguishable on the basis of the data alone.

Other compound Poisson distributions include the Poisson-lognormal (c.f. Anscombe, 1950) and the Poisson-Inverse Gaussian (c.f. Ord and Whitmore, 1986).

In Exercise 5.19, a Beta mixing distribution is applied to the binomial and the negative binomial to yield the beta-binomial and beta-Pascal distributions. For further results on mixtures, see Douglas (1980). Other exercises dealing with mixtures include 4.23–4, 5.7–9, and 5.24–5.

5.24 Various relationships between distributions discussed thus far are summarized in Fig. 5.1.

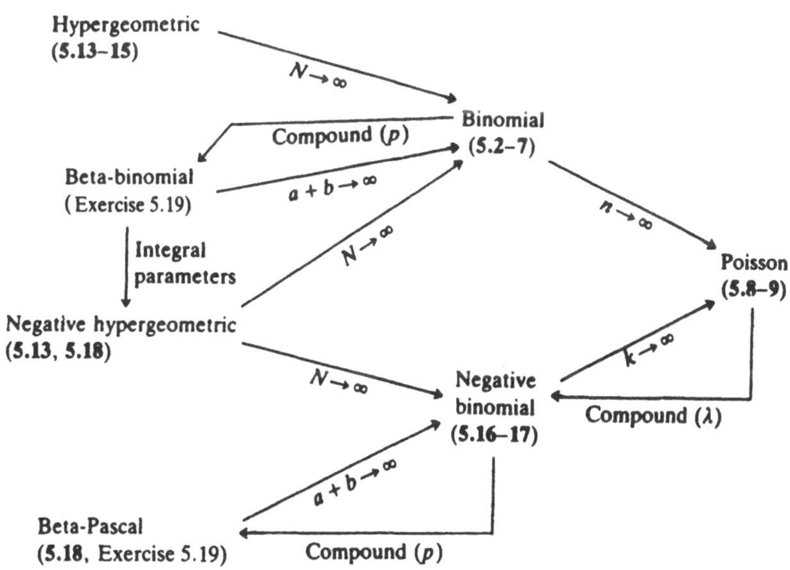

Fig. 5.1 Interrelationships among discrete distributions

Notes: (1) All limits are taken with means fixed.

(2) Compound (θ) means mixing on parameter θ.

Distributions modified by the recording process: censoring and truncation

5.25 The observation process may affect our ability to record data in a variety of ways. For example, suppose we sample telephone calls at a certain time of day in order to determine the distribution of call length. The probability that a particular call is included is proportional to the length of that call, so the observable probability function, $f_w(x)$, is related to the true probability function of call length, $f(x)$, by the relationship

$$f_w(x) \propto xf(x)$$

or

$$f_w(x) = xf(x)/\mu_1', \tag{5.67}$$

where it is assumed that $x \geqslant 0$. We see that (5.67) is the first moment distribution introduced in **2.25**. More generally, we may define a *weighted*, or *size-biased*, distribution as one that has a p.f. given by

$$f_w(x) \propto w(x)f(x). \tag{5.68}$$

When $w(x) = x^\alpha$, a natural class to consider is the log-exponential family introduced by Patil and Ord (1976) with

$$f_w(x) = x^\theta A(x)B(\theta), \quad x \geqslant 0. \tag{5.69}$$

Since f_w is of the same form as f, such form-invariance is a characteristic property of (5.69).

Example 5.3 (Weighted sampling for the Poisson distribution)

If X has a Poisson distribution, (5.67) is

$$\begin{aligned} f_w(x) &= \frac{xe^{-\lambda}\lambda^x}{\lambda \cdot x!} \\ &= \frac{e^{-\lambda}\lambda^{x-1}}{(x-1)!}, \quad x \geqslant 1 \\ &= 0, \quad x = 0. \end{aligned}$$

Thus, the weighted distribution of $X - 1$ is the same as the original distribution of X.

5.26 In survival and life-testing studies, the object being tested is often observed for a certain period, say c units, and the experiment is then terminated. Thus, the distribution of X leads to the probability function:

$$\left.\begin{aligned} f_c(x) &= f(x), \quad x = 0, 1, \ldots, c-1 \\ &= 1 - F(c-1), \quad x = c. \end{aligned}\right\} \tag{5.70}$$

The distribution is said to be *censored* at $x = c$. This is known as right-censoring, since it is the right tail that is not observed; left-censoring is also possible, but less important in practice.

Example 5.4 (Censored geometric distribution)

Suppose X has the geometric distribution

$$f(x) = pq^x, \quad x = 0, 1, \ldots.$$

If the distribution is censored at c, $1 - F(c) = q^c$.

Censored distributions are particularly important for continuous variables; see Exercise 5.29.

5.27 Another modification due to the data-recording process is that of *truncation*. For example, if families are selected from school records, not only will selection be size-biased, but there is no possibility of selecting childless families. Again, in the study of species abundance in entomology, the numbers of species with one, two, or more specimens present is duly recorded, but there is a total lack of information on the number of species with no individuals present. In general, the truncated distribution has p.f.

$$\begin{aligned} f_T(x) &= f(x)/[F(b) - F(a)], \quad x = a+1, \ldots, b \\ &= 0, \quad x \leqslant a \quad \text{or} \quad x > b. \end{aligned}$$

We have discussed a case of particular importance, *zero-truncation*, when the p.m.f. is

$$f_T(x) = f(x)/[1 - F(0)], \quad x = 1, 2, \ldots. \tag{5.71}$$

Example 5.5 (The zero-truncated negative binomial distribution)

From (5.43) and (5.71), the zero-truncated negative binomial has p.m.f.

$$f_T(x) = \binom{k+x-1}{x} p^k q^x/(1-p^k), \quad x = 1, 2, \ldots. \tag{5.72}$$

When $k = 1$, we have the geometric distribution $f(x) = pq^x$, $x = 0,1,2, \ldots$ and its zero-truncated form

$$f_T(x) = pq^x/(1-p) = pq^{x-1}, x = 1, 2, \ldots.$$

Thus, given that $X \geqslant 1$, the remainder $(X - 1)$ has the same distribution as the original X. The same holds true for $(X - a)$, given $X \geqslant a$. This is known as the Markov, or lack of memory, property and holds only for the geometric and exponential distributions (see 5.32).

If the truncation (or censoring) point is a function of the unknown parameters, this poses additional problems in estimation—see Exercises 17.24 and 17.13, Volume 2A. Truncation does not necessarily reduce variance; cf Exercise 5.31. Truncation and censoring in general will be discussed later.

The logarithmic series distribution

5.28 An interesting distribution arises if we consider the zero-truncated negative binomial as $k \to 0$. From (5.72),

$$f_T(x) = \frac{k(k+x-1)^{[x-1]}}{x!} p^k q^x/(1-p^k). \tag{5.73}$$

By L'Hôpital's rule, we have

$$\lim_{k \to 0} \frac{k}{p^{-k} - 1} = \lim \frac{k}{\exp(-k \log p) - 1} = \frac{-1}{\log p}$$

so that

$$\lim_{k\to 0} f_T(x) = \alpha q^x/x, \qquad x = 1, 2, \ldots, \tag{5.74}$$

where $\alpha = -1/\log(1-q)$.

It follows that the probability generating function is

$$P(t) = \frac{1}{\log(1-q)} \log(1-qt). \tag{5.75}$$

This distribution is due to Fisher *et al.* (1943). As we have introduced it, the process of derivation looks rather artificial; for we obtained the negative binomial with index k, and we are now letting k tend to zero. But as we observed in **5.16** and **5.21**, the distribution remains valid for any positive k, however close to zero, and the limiting process is therefore not invalid.

The logarithmic series distribution has been widely used by entomologists to describe species abundance (cf. Example 5.2).

An example of a distribution of the logarithmic series type is given in Table 5.5; but this is to be regarded as illustrative of goodness of fit, not as evidence that the data are generated by the particular limiting process by which we have derived the distribution.

Table 5.5 No. of species of butterflies caught in light-traps in Malaya
(From Fisher *et al.* (1943))

No. of species	Observed frequency	Expected frequency (log. series)
1	118	135.05
2	74	67.33
3	44	44.75
4	24	33.46
5	29	26.69
6	22	22.17
7	20	18.95
8	19	16.53
9	20	14.65
10	15	13.14
11	12	11.91
12	14	10.89
13	6	10.02
14	12	9.28
15	6	8.63
16	9	8.07
17	9	7.57
18	6	7.13
19	10	6.74
20	10	6.38
21	11	6.06
22	5	5.77
23	3	5.50
24	3	5.25

Replacing t by e^θ in (5.75), we find the c.f. of the distribution

$$\phi(t) = 1 - \alpha \log\left\{1 - \frac{q}{p}(e^\theta - 1)\right\}$$

where $\alpha = -1/\log(1-q)$. Noting the formal similarity of this to the negative binomial c.g.f. (5.47), we have for the moments about the origin (not the cumulants—cf. Exercise 5.22) simply (5.48) with k replaced by α,

$$\mu_1' = \frac{\alpha q}{p}; \quad \mu_2' = \frac{\alpha q}{p^2}; \quad \mu_3' = \frac{\alpha q(1+q)}{p^3}; \quad \mu_4' = \frac{\alpha q(1+4q+q^2)}{p^4}; \tag{5.76}$$

whence

$$\left.\begin{aligned} \mu_2 &= \frac{\alpha q(1-\alpha q)}{p^2}, \\ \mu_3 &= \frac{\alpha q(1-3\alpha q+2\alpha^2 q^2)}{p^3}, \\ \mu_4 &= \frac{\alpha q\{1+4q+q^2-4\alpha q(1+q)+6\alpha^2 q^2-3\alpha^3 q^3\}}{p^4}. \end{aligned}\right\} \tag{5.77}$$

Tables of the p.m.f. and d.f. are given by Williamson and Bretherton (1964) for $\mu_1' = 1.1(0.1)2.0(0.5)5.0(1.0)10.0$. G. P. Patil *et al.* (1964) give extensive tables of the p.m.f. and d.f.

5.29 For further discussion of discrete distributions, the reader should consult N. L. Johnson *et al.* (1993) and Patil *et al.* (1985a). We now turn to consider univariate continuous distributions.

The uniform distribution

5.30 The simplest continuous distribution is the *uniform* or *rectangular*, with density

$$\begin{aligned} f(x) &= \frac{1}{b-a}, \quad a < x \leqslant b \\ &= 0, \quad \text{otherwise;} \end{aligned}$$

whence

$$\begin{aligned} F(x) &= 0, \quad x \leqslant a \\ &= \frac{x-a}{b-a}, \quad a < x \leqslant b \\ &= 1, \quad x > b. \end{aligned}$$

If we put $z = \dfrac{x-a}{b-a}$, we have the standard forms

$$f(z) = 1, \quad 0 < z \leqslant 1 \tag{5.78}$$

and

$$F(z) = z, \quad 0 < z \leqslant 1. \tag{5.79}$$

Since $\int_0^1 z^j\,dz = 1/(j+1)$, it follows that (cf. Exercise 2.5)

$$\mu_1' = \tfrac{1}{2}, \quad \mu_2 = \tfrac{1}{12}, \quad \text{while } \beta_1 = 0 \text{ by symmetry and } \beta_2 = 1.8.$$

The main theoretical interest in the uniform distribution derives from the ***probability integral transform***, described in **1.27**. This transform is used extensively in random number generation; see **9.22**.

The Poisson process: exponential and Gamma distributions

5.31 Let $j(t)$ describe the number of events occurring during a time interval $(0, t]$. If T_r denotes the time elapsed until the occurrence of the rth event, we have the identity of the probabilities

$$P[j(t) < r] = P[T_r > t], \tag{5.80}$$

regardless of the distribution of $j(t)$. In particular, consider a *Poisson process* for which $j(t)$ has a Poisson distribution with mean λt. Then

$$\begin{aligned} P[j(t) < 1] &= P[j(t) = 0] \\ &= e^{-\lambda t} = P[T_1 > t]. \end{aligned} \tag{5.81}$$

Since $P(T_1 > t) = 1 - F_1(t)$, where $F_1(t)$ is the d.f. of T_1, we have the density

$$f_1(t) = \lambda e^{-\lambda t}, \quad 0 < t < \infty. \tag{5.82}$$

This is known as the *exponential* distribution. More generally,

$$\begin{aligned} P[j(t) < r] &= e^{-\lambda t} \sum_{i=0}^{r-1} \frac{(\lambda t)^i}{i!} \\ &= P[T_r > t] \end{aligned}$$

from which T_r has the density

$$f_r(t) = \frac{\lambda^r t^{r-1} e^{-\lambda t}}{(r-1)!}, \quad 0 < t < \infty. \tag{5.83}$$

This is the *Gamma* distribution with parameters r and λ, also known as the *Erlang* distribution in the queueing theory literature. The Gamma distribution may be defined for any $r > 0$, when the density is written as

$$f_r(t) = \frac{\lambda^r t^{r-1} e^{-\lambda t}}{\Gamma(r)}, \quad 0 < t < \infty. \tag{5.84}$$

The c.f. of the Gamma distribution, $\phi(u) = \left(1 - \frac{iu}{\lambda}\right)^{-r}$, was derived in (3.19), Example 3.6; in our present notation the first four moments are, using (3.20),

$$\mu_1' = \frac{r}{\lambda}, \quad \mu_2 = \frac{r}{\lambda^2}, \quad \mu_3 = \frac{2r}{\lambda^3}, \quad \text{and} \quad \mu_4 = \frac{3r(r+2)}{\lambda^4}. \tag{5.85}$$

5.32 Because of its link with the Poisson, the exponential distribution is the starting point for many studies in reliability, life-testing, and survival analysis. It possesses the *Markov*, or *lack-of-memory*, property that, using (5.81),

$$\frac{1-F(t+u)}{1-F(t)}=\frac{e^{-\lambda(t+u)}}{e^{-\lambda t}}=e^{-\lambda u}$$
$$=1-F(u), \tag{5.86}$$

so that with origin at any value t, the d.f. is unchanged; truncation on the left makes no difference.

Furthermore, it is the only continuous distribution possessing this property; the geometric is the only discrete distribution (cf. Example 5.5). A weaker version of the lack-of-memory property is the statement that the *expected residual lifetime* does not depend on t, or

$$E[T-t\,|\,T>t]=\frac{1}{\lambda},\quad \text{for all } t. \tag{5.87}$$

Result (5.87) follows directly from (5.85) and (5.86). More surprisingly, (5.87) also serves to characterize the exponential distribution, as we now show. By definition,

$$E[T-t\,|\,T>t]=\int_t^\infty (u-t)f(u)\,du/G(t),$$

where $G(t)=1-F(t)$. Integrating by parts, the right-hand side becomes

$$\frac{1}{G(t)}\left\{[-(u-t)G(u)]_t^\infty+\int_t^\infty G(u)\,du\right\}=\frac{1}{G(t)}\int_t^\infty G(u)\,du.$$

From (5.87), this conditional expectation must be $1/\lambda$ for all t, so that

$$1-F(t)=G(t)=\lambda\int_t^\infty G(u)\,du$$

or

$$f(t)=\lambda G(t). \tag{5.88}$$

(5.88) is a simple differential equation subject to the boundary condition $G(0)=1$. The unique solution is $G(t)=e^{-\lambda t}$.

This characterization is the starting point for a detailed discussion of distributions presented in Kotz and Shanbhag (1980) who demonstrate the value of characterizations in the formulation and selection of statistical models.

The Weibull distribution: hazard rate

5.33 If the Poisson process is not homogeneous in time, but has mean $\lambda g(t)$ rather than λt, we obtain a class of distributions with (5.81)–(5.82) generalized to

$$P[T_1\leqslant t]=F_1(t)=1-\exp[-\lambda g(t)]. \tag{5.89}$$

In particular, when $g(t) = t^\alpha$, $\alpha > 0$, we have the *Weibull distribution*,[(*)] with

$$f(t) = \alpha\lambda t^{\alpha-1}\exp[-\lambda t^\alpha], \quad 0 < t < \infty. \tag{5.90}$$

The moments about the origin of the Weibull distribution are

$$\mu_r' = \lambda^{-r/\alpha}\,\Gamma\left(\frac{r}{\alpha}+1\right), \quad r = 1, 2, \ldots. \tag{5.91}$$

5.34 When testing the survival time of a component, the hazard rate (or failure rate) is defined as the probability that the component fails during the interval $(t, t+\mathrm{d}t]$, given survival up to and including t. In the limit, this yields the *hazard rate*

$$h(t) = \frac{f(t)}{G(t)}. \tag{5.92}$$

From (5.88) we see that $h(t) = \lambda$ for the exponential. Distributions with $h'(t) > 0$ for all t are known as increasing hazard rate (IHR) distributions; similarly, decreasing hazard rate distributions (DHR) have $h'(t) < 0$ for all t. The Weibull is IHR (DHR) when $\alpha > 1\,(<1)$. Distributions specified by hazard rates are considered in greater detail in **6.42–4**.

5.35 A variety of distributions may be derived from the Gamma; some of these are described briefly below.

(i) If X is Gamma distributed, $Y = X^{-1}$ has a reciprocal Gamma distribution; see Exercise 6.6.

(ii) Suppose X follows a Gamma distribution with scale parameter λ and we form a mixture by assuming λ to follow a Gamma distribution also. The resulting distribution for X is a Beta of the second kind (see **6.7** and Exercise 5.32).

(iii) If X_1 and X_2 are independent Gamma variates with the same scale parameter, $Y_1 = X_1/(X_1+X_2)$ follows a Beta distribution of the first kind; see **6.6**. Further, $Y_2 = X_1/X_2$ has a Beta distribution of the second kind; see **6.7** and **16.15**.

(iv) If X_1 and X_2 are independent exponentials with common parameter λ, $Y = X_1 - X_2$ has density

$$f(y) = \frac{\lambda}{2}\mathrm{e}^{-\lambda|y|}, \quad -\infty < y < \infty. \tag{5.93}$$

This is known as the *double exponential* or *Laplace* distribution and also as Laplace's First Law of Error. Laplace's Second Law is the normal distribution, which we now consider in detail.

[(*)] So called after Weibull (1951), who studied it in developing a theory of the strength of materials.

The normal distribution: c.f. and moments

5.36 We have already noted in Examples 4.6 and 4.9 that the binomial and the Poisson distributions tend, when standardized, to the form

$$f(x) = \frac{1}{\sqrt{(2\pi)}} e^{-\frac{1}{2}x^2}, \qquad -\infty < x < \infty. \tag{5.94}$$

The more general form

$$f(x) = \frac{1}{\sigma\sqrt{(2\pi)}} \exp\left\{-\frac{1}{2\sigma^2}(x-\mu_1')^2\right\}, \qquad -\infty < x < \infty; \quad \sigma > 0, \tag{5.95}$$

is known as the normal distribution.(*) It is the most important theoretical distribution in Statistics. We describe (5.95) by saying that x is $N(\mu_1', \sigma^2)$. (5.94) is the standardized normal distribution, $N(0, 1)$. We recall some of its properties which have been given in examples.

The c.f. of (5.95) is easily found to be—cf. Example 3.4—

$$\phi(t) = \exp(it\mu_1' - \tfrac{1}{2}t^2\sigma^2) \tag{5.96}$$

giving

$$\mu_{2r} = \frac{(2r)!}{2^r r!}\sigma^{2r}, \quad \mu_{2r+1} = 0, \quad r \geqslant 1. \tag{5.97}$$

For the c.g.f. we have from Example 3.11

$$\psi(t) = it\mu_1' - \tfrac{1}{2}t^2\sigma^2$$

so that

$$\kappa_2 = \sigma^2, \qquad \kappa_r = 0, \qquad r > 2. \tag{5.98}$$

We also have $\beta_2 = 3$, $\gamma_2 = 0$, which accounts for the standard adopted for mesokurtosis in 3.32.

The normal d.f.

5.37 The density of the standardized normal distribution will occur so often that we shall introduce a special symbol for it. We write

$$\alpha(x) = \frac{1}{\sqrt{(2\pi)}} e^{-\frac{1}{2}x^2}.$$

For the d.f. we have

$$F(x) = \int_{-\infty}^{x} \alpha(y)\mathrm{d}y = \frac{1}{2} + \int_0^x \alpha(y)\mathrm{d}y.$$

(*) The description of the distribution as the "normal", used by Galton, is almost universal in English, although "Gaussian" is a frequent synonym in special contexts. Continental writers refer to it variously by the names of Laplace and/or Gauss. As an approximation to the binomial it was published by Demoivre in 1733 but he did not discuss its properties.

We may expand $\alpha(y)$ as a power series and integrate term by term to obtain

$$F(x) = \frac{1}{2} + \frac{1}{\sqrt{(2\pi)}}\left(x - \frac{x^3}{2\cdot 3} + \frac{x^5}{2!2^2\cdot 5} - \dots\right). \tag{5.99}$$

This, however, is not a very useful expression for the calculation of $F(x)$. It converges too slowly except when x is small. A better expansion may be obtained, following Kerridge and Cook (1976).

We expand $f(z)$ about $z = \frac{1}{2}x$ in an infinite Taylor series and integrate z from 0 to x term by term, obtaining

$$\int_0^x f(z)dz = \sum_{r=0}^{\infty} \frac{f^{(r)}(\frac{1}{2}x)}{r!} \int_0^x (z - \tfrac{1}{2}x)^r dz.$$

The integral is zero by symmetry for odd r, and equals $\dfrac{2}{r+1}(\frac{1}{2}x)^{r+1}$ for even r, so

$$\int_0^x f(z)dz = 2\sum_{s=0}^{\infty} \frac{f^{(2s)}(\frac{1}{2}x)}{(2s+1)!}(\tfrac{1}{2}x)^{2s+1}.$$

Putting $f(z) = \exp(-\frac{1}{2}z^2)$, we have

$$\int_0^x \exp(-\tfrac{1}{2}z^2)dz = 2\exp\left\{-\frac{1}{2}\left(\frac{x}{2}\right)^2\right\}\sum_{s=0}^{\infty}\frac{H_{2s}(\frac{1}{2}x)}{(2s+1)!}(\tfrac{1}{2}x)^{2s+1},$$

where $H_r(x)$ is the Chebyshev–Hermite polynomial defined at (6.21) below. If we write $J_s(x) = H_s(x)x^s/s!$, we have finally for the normal d.f. $F(x)$

$$F(x) = \frac{1}{2} + \frac{x\exp(-x^2/8)}{(2\pi)^{\frac{1}{2}}}\sum_{s=0}^{\infty}\frac{J_{2s}(\frac{1}{2}x)}{2s+1}, \tag{5.100}$$

and we may calculate the successive J_{2s} by using the recurrence relation established at (6.26) below, which in terms of the Js is

$$(r+1)J_{r+1}(x) = x^2\{J_r(x) - J_{r-1}(x)\}.$$

(5.100) converges much faster than (5.99), even for small x. Good as this is, one may very simply obtain an even better expression.

The integral (4.14) for the d.f. in terms of the c.f. reduces in this normal case to

$$F(x) = \frac{1}{2} + \frac{1}{\pi}\int_0^{\infty}\exp(-\tfrac{1}{2}t^2)\sin(xt)t^{-1}dt = \frac{1}{2} + \frac{1}{\pi}\int_0^{\infty} y(t)dt, \quad \text{say.}$$

If we approximate this integral by summation over n intervals of length h, and use the Euler-Maclaurin sum formula (3.51), neglecting the derivatives at the terminals, we have

$$F(x) \doteq \frac{1}{2} + \frac{h}{\pi}\sum_{r=0}^{n-1} y\{(r+\tfrac{1}{2})h\}. \tag{5.101}$$

We may choose n and h so that the range of the discrete approximation $(0, nh)$ covers the effective range of the normal p.d.f. to the accuracy desired. Thus, if we put $F(nh) = 1$ to 10 d.p., the *Biometrika Tables* show that we must have nh equal to about $6\frac{1}{4}$. Moran (1980) conveniently chooses $n = 13$, $h = 2^{\frac{1}{2}}/3$ with product 6.13, and finds the resulting approximation

$$F(x) = \frac{1}{2} + \frac{1}{\pi}\sum_{r=0}^{12} \exp\{-(r+\tfrac{1}{2})^2/9\}\sin\{x(r+\tfrac{1}{2})2^{\frac{1}{2}}/3\}(r+\tfrac{1}{2})^{-1}$$

to be accurate to 9 d.p. for $x \leqslant 7$. Greater accuracy may be obtained by increasing n, decreasing h and increasing nh.

Analogous approximations for the density may be obtained from (4.15).

Mills' ratio

5.38 For large x an asymptotic series may be employed. Let us note that

$$\frac{\exp\{-\frac{1}{2}x^2(1+u^2)\}}{1+u^2} = \int_{\frac{1}{2}x^2}^{\infty} \exp\{-v(1+u^2)\}\,dv.$$

We may integrate both sides for u from $-\infty$ to ∞, and because of the uniform convergence on the right-hand side, may there invert the order of integration. We then have

$$\begin{aligned}\int_{-\infty}^{\infty} \frac{\exp\{-\frac{1}{2}x^2(1+u^2)\}}{1+u^2}\,du &= \int_{\frac{1}{2}x^2}^{\infty}\int_{-\infty}^{\infty} \exp\{-v(1+u^2)\}\,du\,dv \\ &= \sqrt{(2\pi)}\int_{\frac{1}{2}x^2}^{\infty} \frac{1}{\sqrt{(2v)}}\,e^{-v}dv \\ &= \sqrt{(2\pi)}\int_{x}^{\infty} e^{-\frac{1}{2}u^2}du.\end{aligned}$$

Hence

$$\begin{aligned}1 - F(x) &= \frac{1}{\sqrt{(2\pi)}}\int_{x}^{\infty} e^{-\frac{1}{2}u^2}du \\ &= \frac{1}{2\pi}\int_{-\infty}^{\infty} \frac{\exp\{-\frac{1}{2}x^2(1+u^2)\}}{1+u^2}\,du \\ &= \frac{1}{\sqrt{(2\pi)}}\,e^{-\frac{1}{2}x^2}.\frac{1}{\sqrt{(2\pi)}}\int_{-\infty}^{\infty} \frac{\exp(-\frac{1}{2}x^2u^2)}{1+u^2}\,du.\end{aligned} \qquad (5.102)$$

The ratio of the tail area of a distribution, $1 - F(x)$, to its bounding ordinate $F'(x)$, is known as Mills' ratio[(*)] and denoted by $R(x)$. Thus in this normal case

$$\begin{aligned}R(x) &= \{1 - F(x)\}\Big/\frac{1}{\sqrt{(2\pi)}}\,e^{-\frac{1}{2}x^2} \\ &= e^{\frac{1}{2}x^2}\int_{x}^{\infty} e^{-\frac{1}{2}x^2}du.\end{aligned} \qquad (5.103)$$

[(*)] It was tabulated by Mills (1926) for the normal case, but had, in effect, been considered by earlier writers. Its reciprocal $F'(x)/\{1 - F(x)\}$ is the hazard rate defined in (5.92).

From (5.102) we have

$$R(x) \frac{1}{\sqrt{(2\pi)}} \int_{-\infty}^{\infty} \frac{\exp(-\frac{1}{2}x^2u^2)}{1+u^2}\,\mathrm{d}u, \tag{5.104}$$

and on replacing $\frac{1}{2}x^2u^2$ by t,

$$R(x) = \frac{x}{2\sqrt{\pi}} \int_0^{\infty} \frac{\mathrm{e}^{-t}t^{-\frac{1}{2}}}{t+\frac{1}{2}x^2}\,\mathrm{d}t.$$

Expanding the denominator and expressing the resultant integrals in terms of Gamma functions we have

$$\begin{aligned} R(x) &= \frac{x}{2\sqrt{\pi}} \int_0^{\infty} \frac{2}{x^2}\left(1 - \frac{2t}{x^2} + \frac{4t^2}{x^4} - \ldots\right)\mathrm{e}^{-t}t^{-\frac{1}{2}}\,\mathrm{d}t \\ &= \frac{1}{x} - \frac{1}{x^3} + \frac{1\cdot 3}{x^5} - \ldots (-1)^j \frac{1\cdot 3\cdot 5 \ldots (2j-1)}{x^{2j+1}} + R_j(x), \end{aligned} \tag{5.105}$$

where the remainder term

$$|R_j(x)| = \frac{2^j x^{-(2j+1)}}{\sqrt{\pi}} \int_0^{\infty} \frac{t^{j+1}\mathrm{e}^{-t}t^{-\frac{1}{2}}}{t+\frac{1}{2}x^2}\,\mathrm{d}t.$$

This series does not converge but

$$\begin{aligned} |R_j(x)| &< \frac{2^j x^{-(2j+1)}}{\sqrt{\pi}} \int_0^{\infty} t^{j-\frac{1}{2}}\mathrm{e}^{-t}\,\mathrm{d}t \\ &< x^{-(2j+1)} 1\cdot 3\cdot 5 \ldots (2j-1). \end{aligned} \tag{5.106}$$

Thus the remainder at any point in the summation is less in absolute value than the last term taken into account, and the series is asymptotic. For large x the expansion is reasonably effective.

Exercise 5.17 indicates how the leading term in (5.105) may be used to obtain an expansion of the inverse of the d.f.

5.39 Useful forms for the calculation of $F(x)$, or equivalently of $R(x)$, may be expressed as continued fractions. One such, given by Laplace in 1805, is as follows. Define $z(t)$ by

$$\frac{1}{x} z(t) = R\{x(1-t)\} = \frac{1}{\sqrt{(2\pi)}} \int_{-\infty}^{\infty} \frac{\exp\{-\frac{1}{2}x^2u^2(1-t)^2\}\,\mathrm{d}u}{1+u^2}, \tag{5.107}$$

using (5.104). Then

$$\frac{1}{x}\frac{\mathrm{d}z}{\mathrm{d}t} = -x\{(1-t)z - 1\}. \tag{5.108}$$

Hence if z has the expansion in powers of t

$$z = \sum_{j=0}^{\infty} y_j t^j$$

we find, on substitution in (5.107) and identification of coefficients,

$$\frac{1}{x^2}(r+1)y_{r+1}+y_r-y_{r-1}=0. \tag{5.109}$$

Hence

$$\frac{y_r}{y_{r-1}}=\frac{1}{1+\dfrac{r+1}{x^2}\dfrac{y_{r+1}}{y_r}}.$$

Repeated applications of this formula give

$$\frac{y_1}{y_0}=\frac{x}{x+}\,\frac{2}{x+}\,\frac{3}{x+}\cdots\frac{n}{x+}\cdots$$

Also, from (5.107) we find $y_0=xR(x)$, $y_1/x^2=1-y_0$. Substituting for y_1 we have, after a little rearrangement,

$$R(x)=y_0/x=\frac{1}{x+}\,\frac{1}{x+}\,\frac{2}{x+}\,\frac{3}{x+}\cdots\frac{n}{x+}\cdots \tag{5.110}$$

This expression was used by Sheppard (1939) in calculating his tables of the normal distribution, described in **5.41** below. In Appendix Tables 1 and 2 we give some tables that will suffice to illustrate the theory and examples given in this book.

5.40 Another continued fraction, which converges much more rapidly than Laplace's for small x, has been given by Shenton (1954). Defining

$$\bar{R}(x)=\int_0^x\alpha(y)\,dy/\alpha(x)=\frac{1}{2\alpha(x)}-R(x),$$

Shenton finds

$$\bar{R}(x)=\frac{x}{1-}\,\frac{x^2}{3+}\,\frac{2x^2}{5-}\,\frac{3x^2}{7+}\cdots \tag{5.111}$$

Ruben (1962, 1963, 1964), Ray and Pitman (1963) and Rabinowitz (1969) give other asymptotic expressions for $R(x)$. Further related results are given in Exercises 5.12–16 and 5.18.

Divgi (1979) gives a convergent series in orthogonal polynomials for $R(x)$ that approximates the d.f. within 3×10^{-7} using 11 terms, the error tending to zero as $x\to\infty$.

Hastings (1955) gives Chebyshev polynomial approximations to the normal p.d.f. and d.f. and a rational approximation to the *inverse* d.f., i.e. the solution for x of $F(x)=c$. See also Exercise 5.17.

For a discussion of measures of accuracy for normal approximations, see Hodges (1987).

Owen (1980) provides a table of integrals based on the normal law, which facilitates theoretical developments.

Tables of the normal distribution

5.41 (a) Sheppard's (1939) tables give, among other things, Mills' ratio to 12 decimal places by intervals of 0.01 of x; the same to 24 decimal places by intervals of 0.1; and the negative natural logarithm of the tail area to 16 places by intervals of 0.1.

(b) The *Biometrika Tables*, in a table based on Sheppard's work, give the p.d.f. and the d.f. of the standardized distribution, to 7 decimal places for $x=0(0.01)4.50$ and to 10 d.p. for $x=4.50(0.01)6.00$, as well as an auxiliary table which extends to $x=500$ and inverse tables giving x and the p.d.f. from the value of the d.f.

(c) *Tables of Normal Probability Functions* (National Bureau of Standards, Applied Mathematics Series, 23, Washington, 1953) gives the p.d.f. and $\{2F(x)-1\}$, where F is the d.f., for $x=0(0.0001)1(0.001)7.800$ to 15 d.p., and the p.d.f. and $2\{1-F(x)\}$ to 7 *significant figures* for $x=6(0.01)10$, as well as auxiliary tables for large values of x.

(d) *The Kelley Statistical Tables* (Harvard U.P., 1948) give an inverse table of x and the p.d.f., each to 8 d.p., for values of the d.f. 0.000 1(0.0001)0.9999, with a finer tabulation for 10 lower and 10 higher values extending to 0.000 000 001 and 0.999 999 999.

(e) Smirnov (1965) gives 7 d.p. tables of the d.f. and p.d.f. and also a 5 d.p. table of $-\log\{1-F(x)\}$ for $x=5(1)50(10)100(50)500$.

(f) J. S. White (1970) gives inverse tables to 20 d.p. of x for $F(x)=0.5(0.005)0.995$ and for $1-F(x)=10^{-k}$, 2.5×10^{-k} and 5×10^{-k} with $k=1(1)20$.

Properties of the normal distribution

5.42 The shape of the standardized normal distribution

$$\alpha(x)=\frac{1}{\sqrt{(2\pi)}}\,e^{-\frac{1}{2}x^2}$$

is illustrated in Fig. 5.2. It is symmetrical and unlimited in range, falling off to zero very rapidly as the variate increases. There are points of inflection at unit distance on either

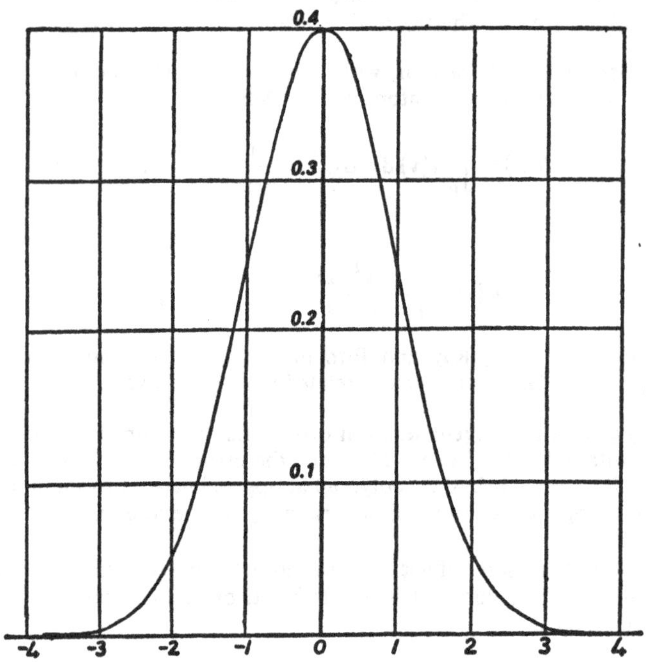

Fig. 5.2 The normal density function $y=\frac{1}{\sqrt{(2\pi)}}\,e^{-\frac{1}{2}x^2}$

side of the mean, and the quartiles are ±0.674 489 75 from the mean, as may be seen from tables.

For the mean deviation (2.15), we have

$$\delta_1 = \frac{1}{\sqrt{(2\pi)}} \int_{-\infty}^{\infty} |x| e^{-\frac{1}{2}x^2} dx = \sqrt{\frac{2}{\pi}} \int_0^{\infty} x e^{-\frac{1}{2}x^2} dx = \sqrt{\frac{2}{\pi}} = 0.797\,88. \qquad (5.112)$$

The mean difference is obtained from **2.25** as

$$\Delta = 2 \int_{-\infty}^{\infty} \int_{-\infty}^{x} (x-y)\alpha(y)\alpha(x)\, dy\, dx$$

and making the polar transformation

$$x = r\cos\theta, \quad y = r\sin\theta$$

with Jacobian r, we find

$$\begin{aligned}\Delta &= \frac{2}{2\pi} \cdot \int_0^{\infty} r^2 e^{-\frac{1}{2}r^2} dr \cdot \int_{-3\pi/4}^{\pi/4} (\cos\theta - \sin\theta)\, d\theta \\ &= \frac{1}{\pi} \cdot \left(\frac{\pi}{2}\right)^{\frac{1}{2}} \cdot 2^{\frac{3}{2}} = \frac{2}{\pi^{\frac{1}{2}}} = 1.1284. \qquad (5.113)\end{aligned}$$

Destandardizing, (5.112–113) are found to be the values of δ_1/σ and Δ/σ.

The resemblance of the formulae $\delta_1 = \left(\frac{2}{\pi}\right)^{\frac{1}{2}}$ and $\Delta = 2/\pi^{\frac{1}{2}}$ in (5.112–113) is not an accident. By the remark below (2.24), Δ is the mean *deviation* δ_1 of $x - y$, where x and y are independently standardized normal, and we shall see (e.g. as a special case of Example 11.2) that $(x - y)$ is therefore normal with zero mean and variance 2, so that its value of $\delta_1/2^{\frac{1}{2}} = \left(\frac{2}{\pi}\right)^{\frac{1}{2}}$ by (5.112), and hence for x, $\Delta = 2/\pi^{\frac{1}{2}}$ as given in (5.113).

5.43 As an illustration of the occurrence in practice of a distribution which is very close to the normal, the height data of Table 1.7 may be taken. Table 5.6 shows the actual frequencies and the expected frequencies for the normal with the same mean and standard deviation (67.46 and 2.56 inches respectively).

The correspondence is evidently fairly good. It must, however, be noted that whereas the theoretical distribution has infinite range, the practical distribution has not, since height must be positive. In this particular case the total probability of the normal distribution outside the range 57–77 inches is so small that the point is unimportant; but when distributions of finite range are represented by those of infinite range it is as well to remember that the fit near the tails may not be very close.

The key role of the normal distribution in Statistics derives from the Central Limit theorem; see **8.50**. Further, many of the sampling distributions for random samples from a normal distribution may be obtained exactly; see Chapters 11, 15 and 16. We conclude our present discussion of the normal with an examination of normal mixtures.

Table 5.6 The height data of Table 1.7 compared with expected frequencies of a normal distribution with the same mean and variance

Height (inches)	Observed frequency	Expected frequency	Height (inches)	Observed frequency	Expected frequency
57–	2	1	68–	1230	1234
58–	4	3	69–	1063	989
59–	14	11	70–	646	682
60–	41	33	71–	392	405
61–	83	88	72–	202	207
62–	169	200	73–	79	91
63–	394	395	74–	32	34
64–	669	669	75–	16	11
65–	990	976	76–	5	3
66–	1223	1227	77–	2	1
67–	1329	1326			
			TOTALS	8585	8586

Mixtures of normal distributions

5.44 The simplest mixture to specify is the finite mixture of two normals with density, $f_m(x)$, given by

$$f_m(x) = (1-\pi)f_1(x|\mu_1, \sigma_1) + \pi f_2(x|\mu_2, \sigma_2),$$

where the f_i are normal densities with parameters μ_i and σ_i and π is the mixing proportion, as first proposed by Karl Pearson (1894) as an alternative to the normal. Pearson suggested fitting by the Method of Moments; cf. **3.42–3** and **6.11**. Parameter estimation for this model is by no means simple, and the problem continues to attract attention; see Gupta and Huang (1981) and Everitt and Hand (1981).

If x is $N(\mu, \sigma^2)$ and the mixing distribution for μ is $N(\lambda, \omega^2)$, it follows that x is $N(\lambda, \sigma^2+\omega^2)$. This result is exploited in a different way when considering conjugate prior distributions in **8.31**. Two other results of interest are now presented.

Example 5.6 (The normal-reciprocal Gamma mixture)

Let x be $N(\mu, \sigma^2)$ with a mixing distribution for σ^{-2} which is Gamma (p, θ). Then

$$\begin{aligned} f_m(x) &= \int_0^\infty \frac{1}{\sigma(2\pi)^{\frac{1}{2}}} \exp\left[-\frac{(x-\mu)^2}{2\sigma^2}\right] \frac{1}{\Gamma(p)} \sigma^{-2(p-1)} \theta^p e^{-\theta/\sigma^2} d(\sigma^{-2}) \\ &= \frac{\Gamma(p+\frac{1}{2})}{\Gamma(p)(2\pi)^{\frac{1}{2}}} \frac{\theta^p}{[\theta+\frac{1}{2}(x-\mu)^2]^{p+\frac{1}{2}}} \end{aligned} \tag{5.114}$$

which the substitution $\frac{1}{2}(x-\mu)^2 = \theta t^2/p$ reduces to *Student's t* distribution; cf. Example 4.8 and (16.15) below.

Example 5.7 (The normal-exponential mixture)

Let x denote the sum of independent normal increments over time (a Wiener process) so that at time t, x is $N(0, \sigma^2 t)$ with c.f. $\phi_t(u) = \exp(-\frac{1}{2}\sigma^2 t u^2)$ by (5.96). If the observed time, t, follows an exponential distribution with parameter λ, the. c.f. of x is

$$\phi(u) = \int_0^\infty \phi_t(u) f(t) \mathrm{d}t = \int_0^\infty \mathrm{e}^{-\frac{1}{2}\sigma^2 u^2 t} \lambda \mathrm{e}^{-\lambda t} \mathrm{d}t$$

$$= \frac{\lambda}{\lambda + \frac{1}{2}\sigma^2 u^2}. \tag{5.115}$$

Thus the c.f. of $\left(\frac{\sigma^2}{2\lambda}\right)^{-\frac{1}{2}} x$ is that of the Double exponential distribution (cf. Exercise 4.3 and 5.35).

5.45 A large number of continuous univariate distributions have been proposed for various purposes; for further details, the reader should consult N. L. Johnson and Kotz (1970) and Patil *et al.* (1985b). A brief summary of the major distributions is presented in Table 5.7.

Generalizations and extensions

5.46 In Chapter 7, we shall consider various multivariate extensions of these distributions. The multivariate normal is so important that we shall treat it at length in Chapters 15 and 16; the emphasis in those chapters will be on sampling distributions.

We now turn to the exponential family, which is important in statistical inference.

The exponential family of distributions

5.47 Consider the general distributional form, dependent on a parameter θ,

$$f(x) = \exp\{A(\theta)B(x) + C(x) + D(\theta)\}, \tag{5.116}$$

where A, B, C, D are arbitrary functions of their arguments. (5.116) is called the *exponential family* of distributions, members of which may be continuous or discrete. The subclass of the family with $B(x) = x$,

$$f(x) = \exp\{xA(\theta) + C(x) + D(\theta)\} \tag{5.117}$$

will be called the *natural exponential family*, and $A(\theta)$ will be called the *natural parameter*—we may replace $A(\theta)$ by θ without changing the notation for the arbitrary function $D(\theta)$.

Example 5.8 (The normal as a member of the exponential family)

The substitutions in (5.117)

$$A(\theta) = \theta, \quad C(x) = -\tfrac{1}{2}\{x^2 + \log(2\pi)\}, \quad D(\theta) = -\tfrac{1}{2}\theta^2,$$

immediately yield, on comparison with (5.95), the normal distribution with mean θ and unit variance which is therefore a member of the natural exponential family and uses the natural parameter.

Table 5.7 The major continuous distributions, their density functions, moments, c.f.s and references in the text

Distribution	Range	$f(x)$	μ_1'	μ_2	β_1	β_2-3	c.f.	Principal references in text
Uniform (rectangular)	(0, 1]	1	$\frac{1}{2}$	$\frac{1}{12}$	0	−1.2	$(e^{it}-1)/it$	Exercise 3.23, 1.27, 5.30
Normal (Gaussian)	$(-\infty, \infty)$	$\frac{1}{\sqrt{2\pi}}e^{-\frac{1}{2}x^2}$	0	1	0	0	$\exp(-\frac{1}{2}t^2)$	3.4, 5.36–44
Exponential (Pearson type X)	$(0, \infty)$	e^{-x}	1	1	4	6	$(1-it)^{-1}$	5.31–2
Laplace (double exponential)	$(-\infty, \infty)$	$\frac{1}{2}e^{-\lvert x\rvert}$	0	2	0	3	$(1+t^2)^{-1}$	Exercise 4.3, 5.35, 5.44
Logistic (cf. Burr family)	$(-\infty, \infty)$	$\frac{e^{-x}}{(1+e^{-x})^2}$	0	$\frac{\pi^2}{3}$	0	1.2	$\Gamma(1-it)\Gamma(1+it)$	Exercise 4.21, 6.37
Cauchy	$(-\infty, \infty)$	$\frac{1}{\pi(1+x^2)}$	—	—	—	—	$\exp(-\lvert t\rvert)$	3.4, 3.16, 4.6
Lognormal (Johnson S_L)	$(0, \infty)$	$\frac{1}{\sigma x\sqrt{2\pi}}\exp\left\{\frac{-(\ln x)^2}{2\sigma^2}\right\}$	$\mu_j'=\exp(\frac{1}{2}j^2\sigma^2)$				n.a., see Exercise 6.21	6.29–31
Gamma (Pearson type III; chi-squared)	$(0, \infty)$	$\frac{x^{p-1}e^{-x}}{\Gamma(p)}$	p	p	$4p^{-1}$	$6p^{-1}$	$(1-it)^{-p}$	5.31, 6.9, 16.2–9
Beta (first kind) (Pearson type I)	(0, 1)	$\frac{x^{p-1}(1-x)^{q-1}}{B(p,q)}$	$\mu_j'=\frac{B(p+j,q)}{B(p,q)}$				n.a.	5.35, 6.5
Beta (second kind) (variance-ratio F; Pearson type VI)	$(0, \infty)$	$\frac{x^{p-1}}{B(p,q)(1+x)^{p+q}}$	$\mu_j'=\frac{B(p+j,q-j)}{B(p,q)}$, $j<q$				cf. 16.17	5.35, 6.7, 16.15–22
Student's t (Pearson type VII)	$(-\infty, \infty)$	$\frac{v^{-\frac{1}{2}}}{B\left(\frac{v}{2},\frac{1}{2}\right)\left(1+\frac{x^2}{v}\right)^{(v+1)/2}}$	0, $v\geqslant 2$	$\frac{v}{v-2}$, $v\geqslant 3$	0, $v\geqslant 4$	$\frac{6}{v-4}$, $v\geqslant 5$	cf. Example 3.15	5.44, 16.10–14
Pareto (Pearson type XI)	$(0, \infty)$	$q/(1+x)^{q+1}$	Beta (second kind) above with $p=1$					Exercises 2.19, 2.29
Weibull	$(0, \infty)$	$cx^{c-1}\exp(-x^c)$	$\mu_j'=\Gamma(1+j/c)$				n.a.	5.33–4
Inverse Gaussian	$(0, \infty)$	$\left(\frac{\lambda}{2\pi x^3}\right)^{\frac{1}{2}}\exp\left\{\frac{-\lambda(x-\mu)^2}{2\mu^2 x}\right\}$	μ	$\frac{\mu^3}{\lambda}$	$\frac{9\mu}{\lambda}$	$\frac{15\mu}{\lambda}$	$\exp\left[\frac{\lambda}{\mu}\left\{1-\left(1-\frac{2\mu^2 it}{\lambda}\right)^{\frac{1}{2}}\right\}\right]$	Exercises 11.28–9
Extreme value	$(-\infty, \infty)$	$\exp(-x-e^{-x})$	0.5572 (Euler's γ)	$\frac{\pi^2}{6}$	1.30	2.4	$\Gamma(1-it)$	14.19, Exercise 14.4

Notes: (i) All distributions have been expressed in standard form; the substitution $y=x\sigma+\mu$ may be made in each case.
(ii) μ_j' has been quoted in place of μ_1', μ_2, β_1, β_2-3 when the form of these expressions is too cumbersome.
(iii) n.a. means that the characteristic function is not expressible in closed form.

Example 5.9 (The binomial as a member of the exponential family)
In (5.117), the substitutions

$$A(\theta) = \log\{\theta/(1-\theta)\}, \quad C(x) = \log\binom{n}{x}, \quad D(\theta) = n\log(1-\theta),$$

yield the binomial distribution (5.1), in another notation. Thus the binomial belongs to the natural exponential family, but p is not the natural parameter. The hypergeometric distribution (5.33) cannot be expressed in the form (5.116), so it is not a member of the exponential family.

Example 5.10 (The Poisson as a member of the exponential family)
If in (5.117) we put

$$A(\theta) = \log\theta, \quad C(x) = -\log(x!), \quad D(\theta) = -\theta,$$

we obtain the Poisson distribution (5.18). Again, the parameter of the Poisson is seen not to be the natural one.

Example 5.11 (The negative binomial as a member of the exponential family)
In (5.117),

$$A(\theta) = \log\{\theta/(1-\theta)\}, \quad C(x) = \log\binom{k+x-1}{x}, \quad D(\theta) = (k-j)\log\theta$$

gives the negative binomial form (5.43).

Example 5.12 (The Gamma as a member of the exponential family)
The Gamma distribution (a special case of (5.84) with $\lambda = 1$)

$$f(x) = e^{-x}x^{\theta-1}/\Gamma(\theta), \qquad \theta > 0, \quad 0 \leqslant x < \infty,$$

is a member of the exponential family (5.116) with

$$A(\theta) = \theta, \quad B(x) = \log x, \quad C(x) = -(x + \log x), \quad D(\theta) = -\log\Gamma(\theta),$$

but not of the natural family (5.117) since $B(x) \neq x$. However, if instead we use the scale parameter λ in (5.84) as θ, so that

$$f(x) = \theta^r x^{r-1} e^{-x\theta}/\Gamma(r),$$

the reader may verify that the distribution of $y = -x$ is of the natural exponential form with natural parameter θ.

Example 5.13 (Location and scale parameters in the exponential family)
Although (5.117) is a wide class of distributions, it is still limited. For example, if θ is a location parameter, so that the distribution is of the form $f(x-\theta)$ as in 2.2, we see that this form implies $\frac{\partial}{\partial x}\log f(x-\theta) = -\frac{\partial}{\partial\theta}\log f(x-\theta)$, so that at $\theta = 0$ we find from (5.117)

$$\frac{\partial}{\partial x}\log f(x) = -\frac{\partial}{\partial \theta}\log f(x) = -\{A'(0)x + D'(0)\} = -(ax+b),$$

say. Integration gives

$$\log f(x) = -(\tfrac{1}{2}ax^2 + bx + c) = -\frac{a}{2}\left(x+\frac{b}{a}\right)^2 + \text{const.}$$

Thus f must be normal with mean $(-b/a)$ and variance $(1/a)$, and no other member of the natural exponential family is of simple location-parameter form, though at the end of **14.19** we find that the extreme value distributions (14.65) are location-parameter distributions in the wider exponential family (5.116).

Similarly for a scale-parameter distribution of form $f(x\,|\,\theta) = \theta^{-1}f(x\,|\,\theta)$, implies that

$$-x\frac{\partial}{\partial x}\log f(x\,|\,\theta) = 1 + \theta\frac{\partial}{\partial \theta}\log f(x\,|\,\theta) \tag{5.117}$$

so that, at $\theta = 1$, (5.117) gives

$$-x\frac{\partial}{\partial x}\log f(x\,|\,\theta) = 1 + \theta\frac{\partial}{\partial \theta}\log f(x) = 1 + A'(1)x + D'(1),$$

and

$$\frac{\partial}{\partial x}\log f(x) = -\left(a+\frac{b}{x}\right).$$

Integrating,

$$\log f(x) = -(ax + b\log x + c)$$

and

$$f(x) \propto e^{-ax}x^b$$

a scale-parameter gamma distribution as in Example 5.12, which is the only class in (5.117) having a simple scale parameter.

5.48 We shall be in a position to understand the importance of the exponential family when we begin to discuss the theory of estimation in Chapter 17, Volume 2A. Here, we confine ourselves to the distribution theory, which is particularly straightforward for the natural exponential family using the natural parameter θ,

$$f(x) = \exp\{\theta x + C(x) + D(\theta)\}. \tag{5.118}$$

Since $\int_{-\infty}^{\infty} f(x)\mathrm{d}x = 1$, we have

$$\int_{-\infty}^{\infty} e^{\theta x + C(x)}\mathrm{d}x = e^{-D(\theta)}$$

and the c.f. of x is therefore

$$\phi(t) = e^{D(\theta)} \int_{-\infty}^{\infty} e^{(\theta+it)x + C(x)} dx$$
$$= \exp\{D(\theta) - D(\theta+it)\},$$

or, writing

$$\exp\{D(\theta)\} = G(\theta)$$
$$\phi(t) = G(\theta)/G(\theta+it). \tag{5.119}$$

Hence the c.g.f. is

$$\psi(t) = \log G(\theta) - \log G(\theta+it) = D(\theta) - D(\theta+it), \tag{5.120}$$

with cumulants

$$\kappa_r = \left[\frac{\partial^r \psi(t)}{\partial (it)^r}\right]_{t=0} = \left[-\frac{\partial^r}{\partial (it)^r} D(\theta+it)\right]_{t=0}$$
$$= -\frac{\partial^r}{\partial \theta^r} D(\theta), \tag{5.121}$$

whence

$$\left.\begin{aligned} \kappa_1 &= E(x) = -\frac{\partial}{\partial\theta} D(\theta), \\ \kappa_r &= \frac{\partial^{r-1}}{\partial\theta^{r-1}} E(x), \quad r \geqslant 2. \end{aligned}\right\} \tag{5.122}$$

In particular, $\dfrac{\partial \kappa_1}{\partial\theta} = \kappa_2 > 0$, so the mean is always strictly increasing in θ.

If $\kappa_1 = \kappa_2$, so that the mean is unchanged by differentiation, it will be similarly unchanged by further differentiation, so we must have all the cumulants equal, and the distribution is Poisson since the c.f. is thus determined. Morris (1982, 1983) studies the six subclasses of the natural exponential family for which the variance is at most a quadratic function of the mean—the normal, Poisson, binomial, negative binomial and Gamma distributions are five of them.

R. A. Johnson *et al.* (1979) generalized (5.122) to give the first two moments in (5.116), and also treat the case of the multivariate exponential family.

Example 5.14 (C.f. of the normal, via the exponential family)

In Example 5.8, $D(\theta) = -\frac{1}{2}\theta^2$, so from (5.119) the c.f. is

$$\phi(t) = e^{-\frac{1}{2}\theta^2 + \frac{1}{2}(\theta+it)^2} = e^{\theta it - \frac{1}{2}t^2},$$

agreeing with (5.96) with $\sigma^2 = 1$. The cumulants are obtained as

$$\kappa_1 = -\frac{\partial}{\partial\theta}(-\tfrac{1}{2}\theta^2) = \theta,$$

$$\kappa_2 = \frac{\partial}{\partial\theta}\kappa_1 \qquad = 1,$$

$$\kappa_r = \frac{\partial^r}{\partial\theta^r}\theta \qquad = 0, \quad r > 2.$$

Example 5.15 (C.f. of the Gamma, via the exponential family)

For the second Gamma distribution in Example 5.12, we see that $D(\theta) = r\log\theta$, so from (5.120) the c.g.f. of $y = -x$ is

$$\psi(t) = r\log\theta - r\log(\theta + it) = -r\log\left(1 + \frac{it}{\theta}\right),$$

and the c.f. is

$$\phi(t) = \left(1 + \frac{it}{\theta}\right)^{-r}.$$

Thus the c.f. of x itself is

$$\phi(-t) = \left(1 - \frac{it}{\theta}\right)^{-r},$$

agreeing with Example 3.6.

Power series distributions

5.49 With appropriate substitutions, (5.118) may be rewritten as

$$f(x) = \theta^x c(x)/\mathrm{d}(\theta); \tag{5.123}$$

when x is discrete, (5.123) defines the class of generalized power series distributions (GPSD). Clearly, the binomial, Poisson and negative binomial fall within the class. It follows from the definitions that the f.m.g.f. is

$$P(1+t) = \mathrm{d}[\theta(1+t)]/\mathrm{d}(\theta). \tag{5.124}$$

It follows that the cumulants satisfy the recurrence relations

$$\kappa_r = \theta\,\mathrm{d}\,\kappa_{r-1}/\mathrm{d}\theta \tag{5.125}$$

whereas the factorial cumulants satisfy

$$\kappa_{[r]} = \theta\frac{\partial\kappa_{[r-1]}}{\partial\theta} - (r-1)\kappa_{[r-1]}. \tag{5.126}$$

For other properties of the GPSD, see Khatri (1959) and Patil (1962a, b).

Perez-Abreu (1991) examines Poisson approximations for the GPSD.

Distributions on the circle and on the sphere

5.50 We have so far discussed only distributions of variates whose possible values form some interval of the real line. Statistical distributions are almost always of this type, but there is a class of situations for which representation on the line is not natural, namely distributions of directions measured from a fixed line as origin. For example, the direction of movement of an animal and the direction of magnetization of a rock are common observations in zoological and geological research. Such observations are naturally recorded as angles over the range $(0, 2\pi)$ or $(0, 360°)$, so that they may be regarded as points on the circumference of the unit circle or of the unit sphere.

One is immediately tempted to ask whether any special treatment of such data on the unit circle is necessary. After all, one can record the observations as angles, and then draw an ordinary histogram on the line from 0 to 360°, and analyse it by familiar linear methods. It is easy to see that this leads to artificiality that may cause trouble in the analysis, if the data are really spread out over the whole circumference of the circle, and not merely a short arc of the circle. First, where should we "cut" the circle? If we cut it at the fixed line origin, and analyse the results linearly, any two observations at $\alpha°$ and $(360° - \alpha°)$ would have an arithmetic mean of 180°, whereas if $\alpha < 180$ they are both closer to 0° (i.e. 360°) than to 180°. The same essential phenomenon occurs if we choose any other cutting-point for the circle. In fact, directional observations have a natural invariance (modulo 2π) on the circle, and no linear representation can in general do them justice.

5.51 For observations on the circle, even when represented circularly, the familiar measures of location and dispersion will not do. For example, a measure of location on the circle should possess the same invariance property (mod 2π) as the observations themselves: and, e.g., the range of the observations must be redefined as the shortest arc covering the observations. Similarly, the d.f. of the variate x on the circle must satisfy

$$F(x+2\pi) - F(x) = 1 \qquad (5.127)$$

identically in x, since a complete tour of the circle covers all the observations, wherever we start from; it follows that the c.f. of x,

$$\phi(t) = E(e^{itx}) = E(e^{it(x+2\pi)}),$$

whence if $|\phi(t)| > 0$,

$$1 = e^{it2\pi} = \cos 2\pi t + i \sin 2\pi t. \qquad (5.128)$$

(5.128) holds, and the c.f. need only be defined, for integer values of t.

5.52 If the direction is uniformly distributed on the circle,

$$f(x) = \frac{1}{2\pi}, \qquad 0 < x \leqslant 2\pi, \qquad (5.129)$$

the simplest form possible. A more general distribution is the *Von Mises distribution*

$$f(x) = \{2\pi I_0(\kappa)\}^{-1} \exp\{\kappa \cos(x-\mu)\}, \qquad (5.130)$$

$$0 < x, \quad \mu \leqslant 2\pi; \quad \kappa > 0,$$

where

$$I_0(\kappa) = \sum_{s=0}^{\infty} \frac{\kappa^{2s}}{2^{2s}(s!)^2}$$

is the modified Bessel function of the first kind and of order zero. In (5.130), μ is the fixed direction from which x is measured and is thus a location parameter, and κ is a dispersion parameter. For a derivation of (5.130), see Exercise 5.33. (5.130) is a unimodal distribution, symmetrical about μ. When $\kappa = 0$, (5.130) reduces to (5.129). When $\kappa \to \infty$, (5.130) concentrates wholly at the point $x = \mu$. In fact, if we write $z = \kappa^{\frac{1}{2}}(x-\mu)$ we see from (5.130) that the distribution of z is

$$g(z) \propto \kappa^{-\frac{1}{2}} \exp\{\kappa \cos(\kappa^{-\frac{1}{2}}z)\}$$

and as $\kappa \to \infty$, $\cos(\kappa^{-\frac{1}{2}}z) \sim 1 - z^2/(2\kappa)$, so

$$g(z) \doteq \text{const.} \exp\{-z^2/2\},$$

so that x is approximately normal with mean μ and variance κ^{-1}. For better normal approximations see G. W. Hill (1976).

Mardia (1972) reproduces 5 d.p. tables, due to E. Batschelet, of the d.f. of (5.130), with $\mu = 180°$, $\kappa = 0(0.2)10$ and $x = 0°(5°)180°$; and also gives the 100δ percent points, $\delta = 0.0005$, 0.005, 0.025, 0.05, for $\kappa = 0(0.1)5(0.2)10.0(0.5)13$, 14, 15, 20(10)50, 100.

5.53 In some circumstances, three-dimensional angles are measured, so that instead of distributions on a circle we have distributions on the surface of a unit sphere. Two angles (x, y) are needed to specify a point on the sphere relative to a fixed direction, and they are customarily assumed to be independently distributed.

To extend the uniform distribution on the circle at (5.129) to that on the sphere, let y be distributed as at (5.129); we need only observe that the second angle being introduced, x, has its cosine uniformly distributed on the interval $(-1, 1)$. Transforming from $\cos x$ to x, the Jacobian of the transformation is $\sin x$, and we thus have

$$f(x, y) = \frac{1}{2\pi} \cdot \tfrac{1}{2} \sin x = \frac{1}{4\pi} \sin x, \qquad 0 < x \leqslant \pi; \quad 0 < y \leqslant 2\pi, \tag{5.131}$$

the uniform distribution on the sphere. Putting $\mu = 0$ in (5.130), it similarly generalizes on the sphere to

$$f(x, y) \propto \exp\{\kappa \cos x\} \sin x$$

which, on evaluating the constant, becomes

$$f(x, y) = \frac{\kappa}{4\pi \sinh \kappa} \exp\{\kappa \cos x\} \sin x, \qquad 0 < x \leqslant \pi; \quad 0 < y \leqslant 2\pi, \tag{5.132}$$

which is *Fisher's (spherical) distribution*. It is symmetrical if rotated about its directional axis. When $\kappa = 0$, (5.132) reduces to the uniform (5.131), and as $\kappa \to \infty$, concentrates about its axis.

(5.132) is the product of the uniform marginal distribution of y on $(0, 2\pi]$ and the independent marginal distribution of x,

$$f(x) = \frac{\frac{1}{2}\kappa}{\sinh \kappa} \exp(\kappa \cos x) \sin x. \qquad 0 < x \leqslant \pi. \tag{5.133}$$

We approximate as before for large κ, first putting $z = \kappa^{\frac{1}{2}}x$. We find

$$g(z) \propto \frac{\kappa^{\frac{1}{2}}}{\sinh \kappa} \exp\{\kappa \cos(\kappa^{-\frac{1}{2}}z)\} \sin(\kappa^{-\frac{1}{2}}z)$$

which since $\sinh \kappa \sim \frac{1}{2}e^{\kappa}$, $\sin(\kappa^{-\frac{1}{2}}z) \sim \kappa^{-\frac{1}{2}}z$, $\cos(\kappa^{-\frac{1}{2}}z) \sim 1 - z^2/(2\kappa)$, becomes

$$g(z) \propto z \exp(-\tfrac{1}{2}z^2). \tag{5.134}$$

Thus $u = \frac{1}{2}z^2$ has the exponential distribution $h(u) = e^{-u}$, i.e. the Gamma (5.83) with $r = \lambda = 1$. We express this by saying that $z^2 = \kappa x^2$ is asymptotically distributed "as χ^2 with 2 degrees of freedom"—we shall explain this terminology in **16.2** below. Essentially, as we shall see, it implies that κx^2 is distributed like the sum of squares of *two* independent normal variates, whereas for the Von Mises distribution (5.130), with $\mu = 0$ as here, κx^2 was asymptotically the square of a single normal variable.

Mardia (1972) gives the 100δ percent points, with parameters as for the Von Mises percentiles in **5.52**.

Mardia's (1972) monograph is a comprehensive discussion of the statistical analysis of directions; see also the review by Mardia (1981).

EXERCISES

5.1 Show that the c.f. about zero of the binomial distribution satisfies the relation

$$\frac{\partial^2 \log \phi(t)}{\partial (it)^2} = pq \frac{\partial^2 \log \phi(t)}{\partial p \partial (it)}$$

and hence that

$$\kappa_{r+1} = pq \frac{\partial \kappa_r}{\partial p}, \qquad r \geqslant 1.$$

(Cf. Frisch (1926), and Haldane (1939) who gives formulae up to κ_{12}.)

5.2 For the *incomplete* moments about the mean of the binomial, starting from some variate value ρ,

$$\mu_r = \sum_{j=\rho}^{n} (j-np)^r \binom{n}{j} p^j q^{n-j},$$

show by differentiating with respect to p that equation (5.11) holds. (Frisch, 1925)

5.3 Writing $T_j = \binom{n}{j} p^j q^{n-j}$, show that the incomplete moments of the binomial are given by

$$\mu_0 = \sum_{j=\rho}^{n} T_j, \qquad \mu_1 = pqT_\rho, \qquad \mu_2 = pqT_\rho\{\rho - (n+1)p\} + npq\mu_0,$$

$$\mu_3 = pqT_\rho[\{\rho - (n+1)p\}^2 + pq(2n-1)] + npq(q-p)\mu_0,$$

and generally

$$\mu_r = pqT_\rho(\rho - np)^{r-1} + npq \sum_{j=0}^{r-2} \binom{r-1}{j} \mu_j - p \sum_{j=0}^{r-2} \binom{r-1}{j} \mu_{j+1}.$$

(Frisch (1926). This is the generalization of equation (5.10) to incomplete moments.)

5.4 Writing $f_{r,n}$ for the binomial probability $\binom{n}{r} p^r q^{n-r}$, show that

$$(np-r)f_{r,n} = npq(f_{r,n-1} - f_{r-1,n-1}), \quad r \geqslant 1$$

$$npf_{0,n} = npqf_{0,n-1},$$

and hence that the mean deviation of the binomial is (cf. 2.18)

$$\delta_1 = 2\sum_{r=0}^{[np]} (np-r)f_{r,n} = 2npqf_{[np],n-1}.$$

Show that when np is an integer, $f_{np,n-1} = f_{np,n}$ (the maximum of the $f_{r,n}$ (cf. 5.6)), so that the relation (A) of Exercise 2.25 holds exactly in this case.

(Cf. Frame (1945); also N. L. Johnson (1957). Ramasubban (1958) obtains the mean deviation and mean difference for many of the discrete distributions in this chapter.)

5.5 From equation (5.42) derive the recurrence formula for the moments of the binomial distribution $\{(1+U)^r - U^r\}(npq\mu_0 - p\mu_1) = \mu_{r+1}$; and that for the Poisson distribution $\{(1+U)^r - U^r\}\lambda\mu_0 = \mu_{r+1}$.

(K. Pearson, 1924)

5.6 Show as in Exercise 5.4 that for the Poisson distribution $f(x) = e^{-\lambda}\lambda^x/x!$, the mean deviation is $\delta_1 = 2\lambda f([\lambda]) = 2\mu_2 f([\mu_1'])$. (Cf. (A) of Exercise 2.25.)

5.7 *Neyman's Type A contagious distribution.* The p.m.f. of a variate r defined at $r = 0, 1, \ldots$ is given by

$$f_r = \frac{\lambda_2^r}{r!} e^{-\lambda_1} \sum_{j=0}^{\infty} \frac{j^r}{j!} (\lambda_1 e^{-\lambda_2})^j, \qquad \lambda_1, \lambda_2 > 0.$$

Show that the factorial cumulant g.f. is $\lambda_1(e^{\lambda_2 t} - 1)$ and that $\mu_1' = \lambda_1\lambda_2$; $\mu_2 = \lambda_1\lambda_2(1+\lambda_2)$. Hence write down the c.f. and invert it, using Exercise 4.22.

(Neyman, 1939; Barton (1957) shows that the distribution may be highly multimodal.)

5.8 The *Pólya-Aeppli distribution*—cf. Anscombe (1950). The p.m.f. of a variate defined at $r = 0, 1, \ldots$ is given by

$$f_0 = e^{-\lambda_1}$$

$$f_r = e^{-\lambda_1} \tau^r \sum_{j=1}^{r} \binom{r-1}{j-1} \frac{1}{j!} \left\{\frac{\lambda_1(1-\tau)}{\tau}\right\}^j, \qquad r \geqslant 1; \quad 0 < \tau < 1, \quad \lambda_1 > 0.$$

Show that this may be written

$$f_r = \exp(-\lambda_1/\tau)\tau^r \sum_{j=0}^{\infty} \binom{j+r-1}{r} \left\{\frac{\lambda_1(1-\tau)}{\tau}\right\}^j \Big/ j!, \quad r \geqslant 0,$$

and hence or otherwise that the factorial c.g.f. is $\lambda_1 t/(1-\tau-\tau t)$, with mean $\lambda_1(1-\tau)^{-1}$ and variance $\lambda_1(1+\tau)(1-\tau)^{-2}$. Show that f_0 denotes a mode if and only if $\lambda_1(1-\tau) < 1$, and that, irrespective of this, there is a mode elsewhere if $\lambda_1 > 2$.

(This and Exercise 5.7 are instances of generalized distributions, introduced in **5.21**. In both Exercises, M of **5.21** has a Poisson distribution with parameter λ_1. In Exercise 5.7, the number of individuals in a cluster is also a Poisson variate with parameter λ_2, but in Exercise 5.8 it has the geometric form $(1-\tau)\tau^{x-1}, x \geqslant 1$.)

5.9 Show that a mixture of Poisson distributions as in **5.22**, using any unimodal continuous mixing distribution for λ, is a unimodal discrete distribution.

(Holgate, 1970; this result does not hold for unimodal discrete mixing, as can be seen by observing that Neyman's distribution in Exercise 5.7 can also be derived as a mixture of Poisson distributions with parameters $\lambda = \lambda_2 x$, where x itself is Poisson with parameter λ_1.)

5.10 Show that the p.m.f. ratio $f(x)/f(x-1)$ is a linear function of $1/x$ if $f(x)$ is (a) the binomial, (b) the Poisson, (c) the negative binomial with origin where the first non-zero probability occurs, and (d) the logarithmic series distribution. Using the slope and intercept of the linear function in each case, show how these four distributions may be distinguished graphically.

(Ord, 1967a)

5.11 Show that for the Poisson p.m.f. (5.18), $\mu_{[r]}' = \lambda^r$; and that for the hypergeometric p.m.f. (5.33), $\mu_{[r]}' = n^{[r]}(Np)^{[r]}/N^{[r]}$.

5.12 Use the Laplace continued fraction (5.110) to show that

$$\frac{x}{x^2+1} < R(x) < \frac{x^2+2x}{x^3+3x}.$$

5.13 Show that Mills' ratio

$$R(x) = x\int_{-\infty}^{\infty} \frac{1}{u^2+x^2}\frac{1}{\sqrt{(2\pi)}}\,e^{-\frac{1}{2}u^2}\,du.$$

Replacing the integral by the sum $h\sum_{j=-\infty}^{\infty} \frac{1}{\sqrt{(2\pi)}}\frac{e^{-\frac{1}{2}j^2h^2}}{x^2+j^2h^2}$, show that for $h=\frac{1}{4}$, $x=1$, the sum to 13 terms gives $R=0.158\,655\,24$. (Das (1956). The true value is 0.158 655 25.)

5.14 For the Laplace continued fraction (5.110) for Mills' ratio, show that if the sth convergent c_s is denoted by a_s/b_s,

$$a_0=0,\quad a_1=1,\quad b_0=1,\quad b_1=x$$

and

$$c_s - c_{s-1} = \frac{(-1)^{s-1}(s-1)!}{b_{s-1}b_s}$$

$$c_s - c_{s-2} = \frac{(-1)^{s-1}x(s-2)!}{b_{s-2}b_s}.$$

Hence show that $c_0 < c_2 < c_4 < \ldots < R(x) < \ldots < c_5 < c_3 < c_1$. (Shenton, 1954)

5.15 In the previous exercise show that

$$R(x) = \int_0^{\infty} \exp(-\tfrac{1}{2}t^2 - tx)\,dt$$

and hence that

$$R(x)b_s - a_s = \frac{(-1)^s\int_x^{\infty}(t-x)^s e^{-\frac{1}{2}t^2}\,dt/\sqrt{(2\pi)}}{\frac{1}{\sqrt{(2\pi)}}e^{-\frac{1}{2}x^2}}.$$

In the Schwarzian inequality

$$\begin{vmatrix} \int f_1^2\,dx & \int f_1f_2\,dx \\ \int f_1f_2\,dx & \int f_2^2\,dx \end{vmatrix} > 0$$

put

$$f_1 = (t-x)^{\frac{1}{2}s}e^{-\frac{1}{4}t^2},\qquad f_2 = (t-x)^{\frac{1}{2}s+1}e^{-\frac{1}{4}t^2}$$

and hence show that

$$R > \frac{\alpha_{2s} + (2s)!\sqrt{(x^2+8s+4)}}{2\beta_{2s}},$$

$$R < \frac{2\gamma_{2s+1}}{\alpha_{2s+1} + (2s+1)!\sqrt{(x^2+8s+8)}},$$

where

$$\alpha_s = b_s a_{s+2} + b_{s+2} a_s - 2b_{s+1} a_{s+1},$$
$$\beta_s = b_s b_{s+2} - b_{s+1}^2, \quad \gamma_s = a_s a_{s+2} - a_{s+1}^2.$$

Hence derive Birnbaum's inequality

$$\tfrac{1}{2}\{-x + \sqrt{(x^2+4)}\} < R(x) < 4/\{3x + \sqrt{(x^2+8)}\}, \quad x > 0,$$

the positive sign of the radical being taken. (Shenton, 1954)

5.16 If $v(x) = 1/R(x)$ and $\lambda(x) = dv/dx = v(v-x)$ show that $0 < \lambda < 1$. If

$$\pi(x) = e^{-x^2} \bigg/ \left\{ \int_{-\infty}^{x} e^{-\frac{1}{2}u^2} du \int_{x}^{\infty} e^{-\frac{1}{2}u^2} du \right\},$$

show that

$$d\pi/dx = 2x\pi(x)\{\lambda(x') - 1\},$$

where $-|x| \leqslant x' \leqslant |x|$. Hence show that $\pi(x)$ is a decreasing function of x^2.

(Sampford, 1953)

5.17 Using $R(x) \sim x^{-1}$ from (5.105), show that for the normal distribution the solution for x of the equation $1 - F(x) = z^{-1}$ has the expansion, as $x(z) \to \infty$,

$$x \sim (2\log z)^{\frac{1}{2}}\{1 - \tfrac{1}{2}(\log 4\pi + \log\log z)/(2\log z)\}.$$

Verify numerically that this gives $x = 2.36$ for $F(x) = 0.99$, against the true value 2.33 obtained from Appendix Table 2.

5.18 Show that the function

$$\bar{R}(x) = y(x) = e^{\frac{1}{2}x^2} \int_0^x e^{-\frac{1}{2}u^2} du$$

satisfies the differential equation

$$\frac{dy}{dx} = xy + 1$$

with initial condition $y(0) = 0$. Hence show that for $r = 0, 1, 2, \ldots$

$$y^{(2r)}(0) = 0, \quad y^{(2r+1)}(0) = 2^r r!$$

and

$$y(x) = \sum_{r=0}^{\infty} \frac{x^{2r+1}}{(2r+1)!} 2^r r! = \sum_{r=0}^{\infty} \frac{x^{2r+1}}{(2r+1)(2r-1)\ldots 3\cdot 1}.$$

(Pólya, 1945)

5.19 *Mixtures of binomials.* If in a binomial distribution (5.1), the parameter p is itself distributed in the form

$$g(p) = \frac{p^{a-1}(1-p)^{b-1}}{B(a,b)}, \quad 0 \leqslant p \leqslant 1; \quad a, b > 0, \tag{A}$$

show that the unconditional distribution of the number of successes is

$$h(x) = \frac{1}{n+1}\frac{B(a+x, b+n-x)}{B(x+1, n-x+1)B(a,b)}, \quad x = 0, 1, 2, \ldots, n,$$

with mean $na/(a+b)$ and variance

$$nab(a+b+n)/\{(a+b)^2(a+b+1)\}.$$

(This distribution is sometimes called the *Beta-binomial*; if (A) is applied to the negative binomial, we similarly obtain the *Beta-Pascal* distribution.)

5.20 x is the number of successes in n binomial trials with parameter p, where n itself is a random variable. Show that if n is the number of successes in another series of N binomial trials with parameter P, x will also be distributed as a binomial based on N trials with parameter Pp. Hence or otherwise show that if, instead, n is a Poisson variate with parameter λ, x is also a Poisson variate with parameter λp, and that in this case x is distributed independently of the number of failures, $n-x$.

(Rao and Rubin (1964) show that the Poisson is the only distribution for which independence holds.)

5.21 Let x have a Poisson distribution with parameter λ. Consider (a) the generalized Poisson distribution where M of 5.21 is binomial with p.g.f. $(q+pt)^2$; (b) the compound Poisson distribution where λ has a normal distribution with mean μ and variance σ^2, $\mu \gg \sigma^2$. Show that both models give rise to a p.g.f. of the form

$$H(t) = \exp\{\gamma_1(t-1) + \gamma_2(t^2-1)\},$$

known as the Hermite distribution. (Kemp and Kemp, 1965)

5.22 *Generalized distributions.* As in 5.21, z is the sum of n independent variates x_i, each with c.f. $\phi_1(t)$, where n is a random variable, taking the values $0, 1, 2, \ldots$, with c.f. $\phi_2(t)$. Show that the unconditional c.f. of z is $\phi_2\left(\frac{\log \phi_1(t)}{i}\right)$, and that $E(z) = E(n)E(x)$, $\text{var}\, z = E(n)\,\text{var}\, x + \text{var}\, n\,\{E(x)\}^2$. Show that the d.f. of z has a saltus at zero even if the x_i are continuous.

If n is a Poisson variate with parameter λ, show that the c.f. of z is $\exp\{\lambda[\phi_1(t)-1]\}$. Show that the cumulants of z are λ times the moments of x, and hence that z has kurtosis $\beta_2 - 3 = c\beta_1$, where $c \geqslant 1$ (cf. the general inequality in Exercise 3.19).

5.23 If $x_1, x_2, \ldots, x_n$ are independent and identically distributed with mean μ and variance σ^2, show that the distributions of $Z_i = (x_i - \mu)/\sigma$ and $\bar{Z}_n = (\bar{x}_n - \mu)\sqrt{n}/\sigma$ are identical if and only if the distribution is normal. (Use characteristic functions.)

5.24 *Mixtures of normal distributions.* x is conditionally normally distributed with mean μ and variance λ. μ and λ are random variables with joint c.f. $\phi(t, u) = E\{\exp(it\mu + iu\lambda)\}$. Show that the unconditional c.f. of x is $\phi(t, \frac{1}{2}it^2)$.

5.25 In Exercise 5.24, show that if μ and λ are distributed independently of each other with respective c.f.s

$$\phi_1(t) = (1 + t^2\sigma^2)^{-1} \qquad \phi_2(t) = (1 - 2iu\sigma^2)^{-m}, \qquad m \text{ a positive integer},$$

(cf. Exercises 4.3, 4.18 for these c.f.s) then x is distributed as the sum of $(m+1)$ independent variates, each distributed exactly like μ, and that if instead μ is constant at zero and $m = 1$, x has the c.f. $\phi_1(t)$.

If μ is constant at zero and $\phi_2(t) = \exp\{-(-2it)^{\frac{1}{2}}\}$, show that x has the Cauchy distribution of Example 4.2, with mean infinite by Example 3.3, even though each component of the mixture has zero mean.

5.26 If $f(x) = pf_1(x) + p_2 f_2(x)$ where $f_j(x)$ is the p.d.f. of a normal distribution with parameters (μ_j, σ_j^2), $j = 1, 2$, show that a sufficient condition for $f(x)$ to be unimodal is that

$$|\mu_2 - \mu_1| \leqslant 2 \min(\sigma_1, \sigma_2).$$

When $\sigma_1 = \sigma_2 = \sigma$ show that this condition can be sharpened to

$$|\mu_2 - \mu_1| \leqslant \sigma\{4 + 2|\log(p_1/p_2)|\}^{\frac{1}{2}}.$$

(Behboodian, 1970.)

5.27 For a variate with the binomial distribution (5.1), show by using a Taylor series expansion that the variate $l_{ab} = \log\left(\frac{j+a}{n+b}\right)$ has mean

$$E(l_{ab}) = \log p + \{(a - \tfrac{1}{2}) + (\tfrac{1}{2} - b)p\}/(np) + O(n^{-2})$$

the term in n^{-1} disappearing when $a = b = \frac{1}{2}$, and that the variance is then

$$V(l_{\frac{1}{2},\frac{1}{2}}) = \frac{q}{np} + \frac{\frac{1}{2}q^2}{(np)^2} + O(n^{-3}),$$

which to the same order is the expectation of $(n-j)/\{(j+\frac{1}{2})(n-j+\frac{1}{2})\}$.
Show that the skewness and kurtosis of $l_{\frac{1}{2},\frac{1}{2}}$ are

$$\gamma_1 = -(1+q)/(npq)^{\frac{1}{2}} + O(n^{-\frac{3}{2}})$$

$$\gamma_2 = (1 + 6q + 2q^2)/(npq) + O(n^{-2}).$$

Taking the limit as $n \to \infty$ with $np = \lambda$, show that for the Poisson distribution

$$E\{\log(x+a)\} = \log\lambda + (a - \tfrac{1}{2})/\lambda + O(\lambda^{-2})$$

so the term in λ^{-1} disappears at $a = \frac{1}{2}$ when

$$V\{\log(x + \tfrac{1}{2})\} = \lambda^{-1} + 2\lambda^{-2} + O(\lambda^{-3}).$$

(cf Pettigrew *et al.* (1986))

5.28 Show that for the c.f. about the mean of the Poisson distribution,

$$\frac{\partial\phi(t)}{\partial(it)} = \lambda(e^{it} - 1)\phi(t)$$

and hence establish (5.21). Show also that

$$\frac{\partial\phi(t)}{\lambda} = \lambda^{-1}\frac{\partial\phi(t)}{\partial(it)} - it\phi(t)$$

and deduce (5.22).

5.29 If x is a non-negative variable with finite mean μ, Exercise 2.24 shows that $\mu = \int_0^b [1 - F(x)]\,dx$. If the distribution is right-censored at c, show that the mean becomes

$$\mu_c = \int_0^c [1 - F(x)]\,dx.$$

If x follows an exponential distribution with parameter λ, right-censored at c, show that $\mu_c = \frac{1}{\lambda}(1 - e^{\lambda c})$.

5.30 The following data show the frequency with which long-term loans of 18 854 books were made from Sussex University Library in 1976–7:

Times borrowed	1	2	3	4	5	6	7	8	9	10	11	≥12
No. of books	9674	4351	2275	1250	663	355	154	72	37	14	6	3

(Data from Burrell and Cane, 1982)

Given that $\bar{x} = 1.50$, use the Method of Moments to find the corresponding expected frequencies for
(a) a zero-truncated Poisson distribution ((5.18), (5.71)) ;
(b) a geometric distribution ((5.44) when $k = 1$).
Show that (b) represents the data better.

5.31 A non-negative distribution, with variance σ^2 and coefficient of variation $V = \sigma/\mu_1'$, is zero-truncated. Show that the resulting variance is $\dfrac{\sigma^2}{1 - F(0)}\left\{1 - \dfrac{F(0)}{V^2[1 - F(0)]}\right\}$, and that this exceeds σ^2 if $V^2 \geqslant [1 - F(0)]^{-1}$. For the negative binomial distribution of (5.43) and (5.72), show that this holds if $k < 1$.

5.32 If the distribution of x, given λ, has density

$$f(x|\lambda) = x^{r-1}\lambda^r e^{-x\lambda}/\Gamma(r),$$

while λ has density

$$f(\lambda) = \lambda^{s-1}\mu^s e^{-\mu\lambda}/\Gamma(s),$$

show that x unconditionally has density

$$f(x) = \frac{x^{r-1}\mu^s}{(\mu + x)^{r+s}} \cdot \frac{1}{B(r, s)}.$$

5.33 Suppose that x_1 and x_2 are independent and normally distributed with means μ_1 and μ_2 and common variance σ^2. Make the transformation to polar coordinates,

$$x_1 = r\cos\theta, \quad x_2 = r\sin\theta.$$

Hence show that the conditional distribution of θ, given r, is the Von Mises distribution (5.148). Show that this reduces to the uniform, (5.147), when $\mu_1 = \mu_2 = 0$.

CHAPTER 6

SYSTEMS OF DISTRIBUTIONS

6.1 In this chapter we continue the account, begun in the last, of the standard distributions of statistical theory. From the variety of forms assumed by the frequency distributions of experience, as exemplified in Chapter 1, it is evident that an elastic system would be required to describe them all in mathematical terms. Three main approaches will be considered: the first, due to Karl Pearson, seeks to ascertain a *family* of distributions that will satisfactorily represent observed data; the second, named after Gram, Charlier and Edgeworth, seeks to represent a given density function as a series in the derivatives of the normal density function; the third, due to Edgeworth and later writers, seeks a transformation of the variate that will throw the distribution at least approximately into a known form.

General accounts of the various approaches are given by Elderton and Johnson (1969) and by Ord (1972), who provide extensive bibliographies.

Pearson distributions

6.2 We noted in **5.19** that several important discrete distributions satisfy a difference equation that can be expressed in the limiting form

$$\frac{df}{dx}=\frac{(x-a)f}{b_0+b_1x+b_2x^2}. \tag{6.1}$$

Equation (6.1) also describes the equilibrium distribution of a birth-death process (Cox and Miller, 1965, pp. 213–5) when the Kolmogorov equations reduce to

$$\tfrac{1}{2}\frac{d^2}{dx^2}[b(x)f(x)]=\frac{d}{dx}[a(x)f(x)],$$

where $a(x)$ and $b(x)$ denote the instantaneous mean and variance, for changes in x; if we take $a(x)$ to be linear and $b(x)$ to be quadratic we arrive at (6.1) under suitable boundary conditions.

This equation may be considered from a different standpoint. The unimodal distributions of Chapter 1 suggest that it might be worth while examining the class of density functions that (a) have a single mode, so that df/dx vanishes at some point $x=a$; (b) have smooth contact with the x-axis at the extremities, so that df/dx vanishes when $f=0$. Evidently these conditions are in general obeyed by any distribution of the family (6.1). As will be seen below, however, there are also solutions of (6.1) in particular cases that are J- or U-shaped.

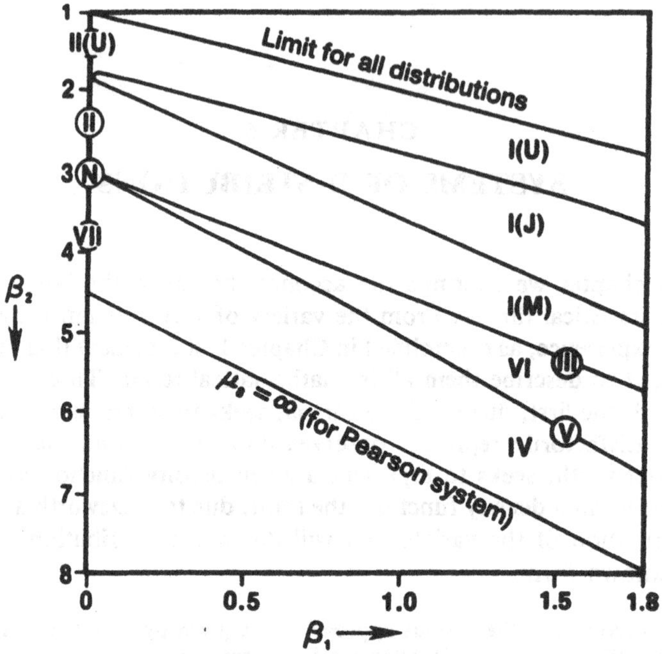

Fig. 6.1 The β_1, β_2 chart for the Pearson system (based upon Table 43, *Biometrika Tables for Statisticians*, vol. I, 1966, p. 234)

Equations of bounding distributions:

Upper limit for all distributions: $\beta_2 - \beta_1 - 1 = 0$ (see Exercise 3.19).
Boundary of I(J) area: $4(4\beta_2 - 3\beta_1)(5\beta_2 - 6\beta_1 - 9)^2 = \beta_1(\beta_2 + 3)^2(8\beta_2 - 9\beta_1 - 12)$.
Type III line: $2\beta_2 - 3\beta_1 - 6 = 0$.
Type V curve: $\beta_1(\beta_2 + 3)^2 = 4(4\beta_2 - 3\beta_1)(2\beta_2 - 3\beta_1 - 6)$.

Other notation:
J = J-shaped, U = U-shaped, M = unimodal, N = normal

Johnson and Rogers (1951) define the region in the (β_1, β_2) plane of Fig. 6.1 above within which no distribution whatever can be unimodal; it lies just below the upper limit for all distributions in Fig. 6.1.

The family of density functions defined by (6.1) are known as Pearson distributions. Before obtaining explicit solutions of the equation, we consider certain general results that hold for all members of the family. We have immediately

$$(b_0 + b_1 x + b_2 x^2)\,\mathrm{d}f = (x - a)f\,\mathrm{d}x$$

or

$$x^n(b_0 + b_1 x + b_2 x^2)\frac{\mathrm{d}f}{\mathrm{d}x}\mathrm{d}x = x^n(x - a)f\,\mathrm{d}x.$$

Integrating the left-hand side by parts over the range of the distribution, we find, assuming that the integrals exist,

$$[x^n(b_0+b_1x+b_2x^2)f]_{-\infty}^{\infty}-\int_{-\infty}^{\infty}\{nb_0x^{n-1}+(n+1)b_1x^n+(n+2)b_2x^{n+1}\}f\,\mathrm{d}x$$
$$=\int_{-\infty}^{\infty}x^{n+1}f\,\mathrm{d}x-a\int_{-\infty}^{\infty}x^nf\,\mathrm{d}x. \tag{6.2}$$

Let us assume that the expression in square brackets vanishes at the extremities of the distribution, i.e. that $\lim_{x\to\pm\infty} x^{n+2}f\to 0$ if the range is infinite. We then have, substituting moments for integrals in (6.2):

$$-nb_0\mu'_{n-1}-(n+1)b_1\mu'_n-(n+2)b_2\mu'_{n+1}=\mu'_{n+1}-a\mu'_n$$

or

$$nb_0\mu'_{n-1}+\{(n+1)b_1-a\}\mu'_n+\{(n+2)b_2+1\}\mu'_{n+1}=0. \tag{6.3}$$

The m.g.f., c.g.f. and cumulants obey similar recurrence relations—cf. Exercise 6.7.

(6.3) permits the determination of any moment from those of lower orders. In fact, all moments can be expressed in terms of a, b_0, b_1, b_2, μ_0 ($\equiv 1$) and μ'_1. Conversely we can express the four constants in terms of the moments μ'_1 to μ'_4, or, if we take the mean μ'_1 as origin, the three central moments μ_2 to μ_4. Putting $\mu'_1=0$ and $n=0, 1, 2, 3$ successively in (6.3), we find equations for a, b_0, b_1, b_2 which result in

$$\left.\begin{aligned} b_1=a&=-\frac{\mu_3(\mu_4+3\mu_2^2)}{A} &&=-\frac{\sqrt{\mu_2}\sqrt{\beta_1}(\beta_2+3)}{A'}\\ b_0&=-\frac{\mu_2(4\mu_2\mu_4-3\mu_3^2)}{A} &&=-\frac{\mu_2(4\beta_2-3\beta_1)}{A'}\\ b_2&=-\frac{(2\mu_2\mu_4-3\mu_3^2-6\mu_2^3)}{A} &&=-\frac{(2\beta_2-3\beta_1-6)}{A'}\end{aligned}\right\} \tag{6.4}$$

where

$$\left.\begin{aligned} A&=10\mu_4\mu_2-12\mu_3^2-18\mu_2^3\\ A'&=10\beta_2-12\beta_1-18.\end{aligned}\right\} \tag{6.5}$$

b_0 in (6.4) can be positive or negative but not zero, since by Exercise 3.19 its numerator is positive. Further, $b_1=a=0$ if $\beta_1=0$, e.g. if the distribution is symmetrical.

From (6.4), we may write (6.3) as

$$-nA'(b_0\mu_{n-1}+b_1\mu_n)=\mu_{n+1}\{6(\beta_2-\beta_1-1)-n(2\beta_2-3\beta_1-6)\}.$$

If μ_n exists and μ_{n+1} does not, the coefficient of μ_{n+1} must be zero. By Exercise 3.19 its first term is positive, so if $2\beta_2-3\beta_1-6\leqslant 0$ (the region on and above the Type III line in Fig. 6.1) the coefficient of μ_{n+1} is positive, so that μ_{n+1} and by extension the moments of all orders must exist. If $2\beta_2 - 3\beta_1 - 6 > 0$, μ_{n+1} will exist only if $n < 6(\beta_2 - \beta_1 - 1)/(2\beta_2 - 3\beta_1 - 6)$. The lowest line in Fig. 6.1 is determined by putting $n = 7$, which is the boundary for μ_8 to exist. As we shall see in **10.4** below, the sampling variance of a sample moment of order r depends on the population moment of order $2r$, and since the Pearson system uses four moments, we require μ_8 to exist if the sample moments are to provide consistent estimators (cf **17.7-8**, Volume 2A).

In equation (6.1) the mode is evidently at the point $x = a$. From (6.4), we have for the Pearson measure of skewness (3.31)

$$\frac{\text{mean} - \text{mode}}{\sigma} = \frac{-a}{\sqrt{\mu_2}} = \frac{\sqrt{\beta_1}(\beta_2 + 3)}{10\beta_2 - 12\beta_1 - 18}, \tag{6.6}$$

the form given at (3.87).

6.3 If, instead of using μ_1' as origin, we use the mode a thus, by putting $X = x - a$, the constants b_0, b_1 and b_2 will take values different from those in (6.4): we shall denote these values by B_0, B_1 and B_2. (6.1) now becomes, putting $a = 0$,

$$\frac{\partial \log f}{\partial X} = \frac{X}{B_0 + B_1 X + B_2 X^2}. \tag{6.7}$$

By equating coefficients in $b_0 + b_1 x + b_2 x^2 \equiv B_0 + B_1(x - a) + B_2(x - a)^2$ we obtain, using $b_1 = a$ from (6.4),

$$\left.\begin{aligned} B_0 &= b_0 + a^2(1 + b_2), \\ B_1 &= a(1 + 2b_2), \\ B_2 &= b_2. \end{aligned}\right\} \tag{6.8}$$

We find from (6.7) that

$$\frac{d^2 f}{dX^2} = \frac{d}{dX}\frac{Xf}{B_0 + B_1 X + B_2 X^2} = \frac{f}{(B_0 + B_1 X + B_2 X^2)^2}(B_0 - (B_2 - 1)X^2).$$

Thus any points of inflection in the density function are given by

$$X^2 = B_0/(B_2 - 1).$$

Hence there cannot be more than two of them, and if two exist, they are equidistant from the mode. It is not to be inferred that a member of the family cannot have a single point of inflection, for one solution may be outside the permissible range of x.

6.4 The explicit expression of the density function f requires the integration of the right-hand side of (6.7). We first assume $B_2 > 0$, if necessary changing the sign of X to achieve this—the case $B_2 = 0$ is discussed in **6.9** below and in Exercise 6.1.

We may distinguish three main types of distribution according as the quadratic in the denominator on the right of (6.7) has real roots of opposite sign, real roots of the same sign, or complex roots. If the roots are real, $B_1^2 \geqslant 4B_0B_2$ and B_0B_2, the product of the roots, is negative if the roots have opposite sign (i.e. lie on opposite sides of the mode, our origin) and is positive if they have the same sign. Thus

(i) if the real roots have opposite sign, $K = B_1^2/(4B_0B_2) \leqslant 0$;
(ii) if the real roots have the same sign, $K \geqslant 1$;
(iii) if the roots are complex, $B_1^2 < 4B_0B_2$ and $0 < K < 1$.

K is thus a criterion for assigning a distribution to a type of the Pearson system. Using (6.8), we see that $B_1^2 - 4B_0B_2 = b_1^2 - 4b_0b_2$, reflecting the fact that the reality of the roots of the quadratic cannot be affected by a change of origin, here from the mode to the mean. However, a new origin can affect the signs of the roots and K must then be replaced by the criterion appropriate to the new origin. Using the mean, we see that

$$K = \frac{B_1^2}{4B_0B_2} = 1 + \frac{\kappa - 1}{1 + 4b_2(1 + b_2)\kappa}, \tag{6.9}$$

where $\kappa = b_1^2/(4b_0b_2)$ is the same function of the bs as K is of the Bs.

From (6.9), we see that κ and K are 0 or 1 together, while since $b_2 > 0$,

(i) $\kappa < 0$ does not imply $K < 0$ unless

$$-\{4b_2(1 + b_2)\}^{-1} < \kappa < 0; \quad \text{otherwise } K > 1.$$

(ii) $\kappa > 1$ implies $K > 1$.
(iii) $0 < \kappa < 1$ implies $0 < K < 1$.

(iii) reflects the fact that the complexity of the roots is independent of the origin; (ii) shows that if both roots are on the same side of the mean, they must be on the same side of the mode, but (i) implies that the converse is not true.

Thus the value of κ gives useful sufficient conditions for the value of K. From (6.4), we see that

$$\kappa = \frac{b_1^2}{4b_0b_2} = \frac{\beta_1(\beta_2 + 3)^2}{4(4\beta_2 - 3\beta_1)(2\beta_2 - 3\beta_1 - 6)}. \tag{6.10}$$

Type I (*Beta distribution of the first kind*)

6.5 If the roots are real and of opposite sign, we have, using the mode as origin, $K \leqslant 0$, with $K = 0$ if and only if $B_1 = 0$, i.e. from (6.8) and (6.4) if $\beta_1 = 0$, as it is for any symmetric distribution. Generally, for $K \leqslant 0$, we have

$$B_0 + B_1X + B_2X^2 = B_2(X + \alpha_1)(X - \alpha_2), \qquad \alpha_1, \alpha_2 > 0.$$

Then

$$\frac{d}{dX}(\log f) = \frac{X}{B_2(X+\alpha_1)(X-\alpha_2)}$$

$$= \frac{\alpha_1}{B_2(\alpha_1+\alpha_2)}\cdot\frac{1}{(X+\alpha_1)} + \frac{\alpha_2}{B_2(\alpha_1+\alpha_2)}\cdot\frac{1}{(X-\alpha_2)},$$

giving

$$f = k(X+\alpha_1)^{\alpha_1/B_2(\alpha_1+\alpha_2)}(X-\alpha_2)^{\alpha_2/B_2(\alpha_1+\alpha_2)}.$$

This may be written in the form

$$f = k\left(1+\frac{X}{a_1}\right)^{m_1}\left(1-\frac{X}{a_2}\right)^{m_2} \tag{6.11}$$

where

$$\frac{m_1}{a_1} = \frac{m_2}{a_2} > 0.$$

The range for which $f \geqslant 0$ is from $-a_1$ to a_2, and by integrating between these values we find

$$f = \frac{a_1^{m_1}a_2^{m_2}}{(a_1+a_2)^{m_1+m_2+1}B(m_1+1, m_2+1)}\left(1+\frac{X}{a_1}\right)^{m_1}\left(1-\frac{X}{a_2}\right)^{m_2}. \tag{6.12}$$

The origin here is the mode. Taking an origin at the start of the distribution and measuring in units (a_1+a_2) times the original, (6.12) reduces to

$$f = \frac{1}{B(m_1+1, m_2+1)}x^{m_1}(1-x)^{m_2}, \quad 0 \leqslant x \leqslant 1. \tag{6.13}$$

This is Pearson's Type I. Type II is the symmetrical case with $K=0$—cf. Exercise 6.1.

6.6 (6.13) is usually written in the standard form

$$f = \frac{1}{B(p,q)}x^{p-1}(1-x)^{q-1}, \qquad 0 \leqslant x \leqslant 1; \quad p, q > 0. \tag{6.14}$$

Since its d.f. is the Incomplete Beta function, (6.14) is called the *Beta distribution of the first kind* with parameters p and q. Its percentage points are tabulated in the *Biometrika Tables*, extended by Amos (1963). If $p, q > 1$, f has a unique mode at $(p-1)/(p+q-2)$ and is zero at $x=0, 1$. If p or $q=1$, f has a corresponding terminal value q or p; if $p=q=1$, (6.14) becomes the uniform distribution

$$f(x) = 1, \qquad 0 \leqslant x \leqslant 1. \tag{6.15}$$

If p or q is between 0 and 1, f is infinite at the corresponding terminal. Thus a variety of U-shaped distributions is obtainable with $0 < p, q \leqslant 1$, and J-shaped distributions with

$0<p\leqslant 1<q$ or $0<q\leqslant 1<p$. Fig. 6.1 indicates the regions in the (β_1, β_2) plane where these occur.

Since the range of (6.14) is finite, all its moments exist as we saw towards the end of 6.2; we obtained the first two in Examples 2.2 and 2.8, and the general expression is similarly seen to be

$$\mu_r' = B(p+r, q)/B(p, q) = \Gamma(p+r)\Gamma(p+q)/\{\Gamma(p)\Gamma(p+q+r)\}.$$

TYPE VI (*Beta distribution of the second kind*)

6.7 If the real roots of the quadratic have the same sign and are not equal, $K>1$ and simple changes in the argument of **6.5** show that we may write the density in the form

$$f = kx^{-q_1}(x-c)^{q_2}, \qquad 0<c\leqslant x<\infty; \quad q_1>q_2+1>0.$$

This was called Type VI by Pearson; the transformation $y=c/x$ reduces it to the Beta form (6.14) with $p=q_1-q_2-1$, $q=q_2+1$. In the standard form

$$f = \frac{1}{B(p,q)}\frac{x^{p-1}}{(1+x)^{p+q}}, \qquad 0\leqslant x<\infty; \quad p, q>0, \tag{6.16}$$

it is called a Beta distribution of the second kind. ((6.14) is of the first kind.) It is discussed in slightly different form as the F-distribution in **16.15–21** below. (6.16) is reducible to (6.14) by putting $x=y/(1-y)$.

The rth moment of (6.16) exists only if $r<q$, and is then found to be

$$\mu_r' = B(p+r, q-r)/B(p, q) = \Gamma(p+r)\Gamma(q-r)/\{\Gamma(p)\Gamma(q)\}.$$

Exercise 6.6 treats the case when the roots are equal.

TYPE IV

6.8 If the roots of $B_0+B_1X+B_2X^2$ are complex we have $0<K<1$ and

$$\frac{\mathrm{d}}{\mathrm{d}X}(\log f) = \frac{X}{B_2\left\{\left(X+\frac{B_1}{2B_2}\right)^2+\frac{B_0}{B_2}-\frac{B_1^2}{4B_2^2}\right\}}$$

$$= \frac{X}{B_2\{(X+\gamma)^2+\delta^2\}},$$

since $B_0B_2>B_1^2$. Integrating, we have

$$\log f = \log k + \frac{1}{2B_2}\log\{(X+\gamma)^2+\delta^2\} - \frac{\gamma}{B_2\delta}\arctan\frac{X+\gamma}{\delta},$$

$$f = k\{(X+\gamma)^2+\delta^2\}^{1/2B_2}\exp\left\{-\frac{\gamma}{B_2\delta}\arctan\frac{X+\gamma}{\delta}\right\}.$$

This is Pearson's Type IV and is usually written in the form

$$f = k\left(1+\frac{x^2}{a^2}\right)^{-m} \exp\{-\nu \arctan(x/a)\}, \qquad m > \tfrac{1}{2}. \tag{6.17}$$

The distribution has unlimited range in both directions and is unimodal. Its first four moments exist if $m > \frac{5}{2}$. The lack of a closed form for its d.f. makes the Type IV difficult to handle in practice, notwithstanding the existence of special tables. As an alternative to numerical integration, Shenton and Carpenter (1965) provide a series expansion for the Mills' ratio (cf. **5.38**); Woodward (1976) gives an approximation for the tail area probabilities.

Type VII (Student's t-distribution) is the symmetrical case $K = 0$, when B_1 and ν are also zero, and the normal distribution is its extreme case $B_1 = B_2 = 0$—cf. Exercise 6.1.

Type III (*Gamma distribution*)

6.9 Pearson distinguished nine other types, some entirely trivial, some no longer of interest. We need mention only one, which has extensive theoretical applications. Some particulars of the others will be found in Exercises 6.1, 6.4 and 6.6.

If, in (6.7), $B_2 = 0$ the distribution becomes

$$f = k\left(1+\frac{x}{a}\right)^{p} e^{-px/a}, \qquad -a \leqslant x < \infty.$$

Here the origin is at the mode. If we transfer the origin to the start of the distribution we arrive at a form in which we have already encountered it (Examples 3.6, 3.12 and 5.31). A scale change then gives the standard form

$$f = \frac{1}{\Gamma(\lambda)} x^{\lambda-1} e^{-x}, \qquad \lambda > 0, \qquad 0 \leqslant x < \infty. \tag{6.18}$$

In this case the criterion κ of (6.10) is infinite—cf. Exercise 6.5. The density is unimodal except for values of λ less than or equal to unity, when it is J-shaped. It is known as Type III or the Gamma distribution, the latter name being due to the fact that its distribution function is an Incomplete Γ-function. Tables of the d.f. and of percentage points of the distribution of $2x$ in (6.18) are described in **16.4** below.

6.10 An extensive table of percentage points for the whole family of Pearson distributions was given by N. L. Johnson *et al.* (1963) and is reproduced, with many enlargements, in Vol. II of the *Biometrika Tables*. Bowman and Shenton (1979a, b) have developed accurate approximations for the percentage points of the Pearson distributions using ratios of polynomials $\beta_1^{\frac{1}{2}}$ and β_2. Davis and Stephens (1983) provide a computer algorithm that implements this approach.

These tables reflect a major use of the Pearson system, that of approximating theoretical distributions whose first four moments are known. E. S. Pearson (1963) and Solomon and Stephens (1978) show that in this case the approximations are remarkably accurate, especially in the longer tail of the distribution, which of course is the dominating contributor to the central moments.

The more traditional use of Pearson distributions is to represent observational data. The increased use and complexity of computer simulation studies makes the summary of (large) empirical distributions by a fitted distribution an attractive option, as random values can then be generated from the fitted distribution (cf. **9.22–7**). For the present, we concentrate upon the technical details of fitting. Our account will be fairly brief, as a systematic account is given in Elderton and Johnson (1969).

6.11 From (6.4), we see that all the Pearson distributions are determined by their first four moments, provided these exist; some of the special cases are determined by fewer moments. The family may be represented by plotting β_1 against β_2, as in Fig. 6.1. Plots of $\beta_1^{\frac{1}{2}}$ against β_2 or β_1 against δ, defined in Exercise 6.22, are also useful. Thus, to select and fit a Pearson distribution we may use the Method of Moments as follows:

(1) calculate the values of the first four moments, β_1 and β_2, from the observed distribution (we shall here use β_1 and β_2 to represent these values whether they are computed from data or from the approximating distribution);
(2) determine the Pearson type to which the observed distribution corresponds, either using the criterion κ of (6.10) or plotting the observed (β_1, β_2) values on Fig. 6.1;
(3) equate the observed moments to the moments of this type of distribution expressed in terms of its parameters; and
(4) solve the resulting equations for those parameters, whereupon the fitted distribution is determined.

The following example will illustrate the process:

Example 6.1 (Fitting a Pearson distribution by the Method of Moments)

In Table 1.15 there are shown, in the column totals, a distribution of 9440 beans according to length. The figures are repeated in Table 6.1 in **6.20** below. It is required to fit a Pearson distribution to these data.

For the moments it is found that, with Sheppard's corrections (in units of 0.5 mm),

$$\mu_1' \text{(origin at the mode 14.5)} = -0.190\,783\,898 \qquad \mu_3 = -5.306\,566\,352$$

$$\mu_2 = 3.238\,424\,951 \qquad \mu_4 = 50.999\,624\,044$$

$$\beta_1 = 0.829\,135\,838, \qquad \gamma_1 = \sqrt{\beta_1} = -0.910\,569$$

$$\beta_2 = 4.862\,944\,362.$$

To determine the Pearson type, we plot the observed values (β_1, β_2) on Fig. 6.1: the point clearly lies in the Type IV region. Alternatively, we find that

$$\kappa = \frac{51.262}{84.040} = 0.61,$$

again suggesting that Type IV is appropriate. Putting $\tan\theta = x/a$ and $2m - 2 = r$ in equation (6.18) we find

$$\mu_n' = k\int_{-\frac{1}{2}\pi}^{\frac{1}{2}\pi} a^{n+1}\cos^{r-n}\theta\sin^n\theta e^{-\nu\theta}\,d\theta,$$

whence, integrating by parts with $\cos^{r-n}\theta\sin\theta$ as one part,

$$\mu_n' = \frac{a}{r-n+1}\{(n-1)a\mu_{n-2}' - \nu\mu_{n-1}'\},$$

a particular case of (6.3). Hence, in terms of moments about the mean,

$$\mu_1' = -\frac{a\nu}{r}, \qquad \mu_2 = \frac{a^2}{r^2(r-1)}(r^2+\nu^2),$$

$$\mu_3 = -\frac{4a^3\nu(r^2+\nu^2)}{r^3(r-1)(r-2)}, \qquad \mu_4 = \frac{3a^4(r^2+\nu^2)\{(r+6)(r^2+\nu^2)-8\nu^2\}}{r^4(r-1)(r-2)(r-3)},$$

whence it is found that

$$r = \frac{6(\beta_2-\beta_1-1)}{2\beta_2-3\beta_1-6},$$

$$\nu = \frac{r(r-2)\sqrt{\beta_1}}{\sqrt{\{16(r-1)-\beta_1(r-2)^2\}}},$$

$$a = \sqrt{\left[\frac{\mu_2}{16}\{16(r-1)-\beta_1(r-2)^2\}\right]}.$$

Substituting for β_2, β_1 and μ_2 we find

$$r = 14.697\,72, \qquad m = 8.348\,86$$
$$\nu = 18.380\,43, \qquad a = 4.159\,49.$$

The signs here want a little watching. r and m present no difficulty; but a is to be taken positive and ν positive since μ_3 is negative.

So far the process is straightforward, but to evaluate the constant k we require tables of the distribution function.[*] On evaluation of k we find

$$f = 0.395\,121\left(1+\frac{x^2}{17.301\,34}\right)^{-8.348\,86}\exp\left(-18.380\,43\arctan\frac{x}{4.159\,49}\right).$$

The d.f. may now be calculated by numerical integration. Table 6.1, column (3), in **6.20** below gives the results for comparison with the observed frequencies.

[*] The tables were given in the first two editions of K. Pearson's *Tables for Statisticians*, Part I, but omitted from later editions: these tables were the forerunners of the Biometrika Tables.

6.12 Several extensions to the Pearson system have been proposed (cf. Ord, 1972, pp. 8–9), including the use of polynomials of general order fitted directly to histogram estimates of $d\log f/dx$; see Dunning and Hanson (1977). However, most of these modifications have seen limited application.

A multimodal generalization of the Pearson system is obtainable if we replace (6.1) by

$$\frac{1}{f}\frac{df}{dx}=\frac{-g(x)}{v(x)},$$

where $g(x)$ is a polynomial of degree k and $v(x)$ is another, of degree $\leqslant 2$. The roots of $g(x)$ are at modes or antimodes of $f(x)$, while the roots of $v(x)$ correspond to zero or infinite values of $f(x)$ unless a value x is a root of both $g(x)$ and $v(x)$, when there is a degenerate boundary value for $f(x)$. Thus if $k\geqslant 3$, multimodal distributions can be obtained. Cobb *et al.* (1983) distinguish four main types, with $v(x)$ respectively equal to $1, x, x^2$ and $x(1-x)$ and corresponding ranges for $f(x)$ $(-\infty,\infty)$, $(0,\infty)$, $(0,\infty)$ and $(0, 1)$. Integration by parts as in **6.2** gives

$$\int_{-\infty}^{\infty} x^n g(x)\,dF(x)=\int_{-\infty}^{\infty}\frac{d}{dx}\{x^n v(x)\}\,dF(x),$$

which yields recurrence relations for the moments, of which (6.3) is a special case.

An important special case arises when the lower end-point of the distribution is known. The location parameter may be adjusted so that x is defined on $(0,\infty)$ and (6.1) becomes

$$\frac{df}{dx}=\frac{(x-a)f}{x(b_1+b_2x)}. \tag{6.19}$$

The reduced system includes the main Types I and VI, as well as Types III and V (cf. Exercise 6.6). Only the first three moments are now required to identify the parameters; see Exercise 6.25. To choose an appropriate type, we may use

$$J=\frac{\mu_3\mu_1'}{2\mu_2^2}, \tag{6.20}$$

where $J \gtreqless 1$ for Types III, VI, and I, respectively; see Exercise 6.26. For further discussion of this subfamily, see Ord (1972, pp. 13–14) and Müller and Vahl (1976). Solomon and Stephens (1978) illustrate the fitting process.

6.13 The fitting of Pearson distributions by the method of moments, as in Example 6.1, may be considered from two rather different standpoints. If our purpose is to obtain a mathematical expression which will approximate a distribution whose first four moments are known, the method is satisfactory. This applies in actuarial work and in the approximation of sampling distributions. However, when the observed data form a random sample drawn from a population, the method of moments yields only estimates of the population moments and these do not, in general, lead to efficient estimates of the population parameters. A more effective approach is to select a particular type (by κ, J, or some other criterion) and then to estimate the parameters of that distribution by Maximum Likelihood (cf. **8.18–23** and Chapter 18, Volume 2A). Parrish (1983) describes a general loss-function approach which selects and then fits the type giving the highest likelihood over all members of the family.

Other systems of distributions have been studied, with a view to representing density functions by series expansions. It is well known in mathematical and physical work that functions can often be usefully expressed as a series of terms such as powers of the variable (Taylor's series) or trigonometrical functions (Fourier's series). Neither of these forms is very suitable for density functions, but we proceed to consider another set of functions with more promising possibilities.

Chebyshev-Hermite polynomials

6.14 Writing, as in **5.37**,

$$\alpha(x)=\frac{1}{\sqrt{(2\pi)}}e^{-\frac{1}{2}x^2}$$

and

$$D=\frac{d}{dx},$$

consider successive derivatives of $\alpha(x)$ with respect to x. We have

$$\begin{aligned} D\alpha(x) &= -x\alpha(x) \\ D^2\alpha(x) &= (x^2-1)\alpha(x) \\ D^3\alpha(x) &= (3x-x^3)\alpha(x), \end{aligned}$$

and so on. The result will obviously be, in general, a polynomial in x multiplied by $\alpha(x)$. We then define the Chebyshev-Hermite polynomial $H_r(x)$ by the identity

$$(-D)^r\alpha(x)=H_r(x)\alpha(x). \tag{6.21}$$

Evidently $H_r(x)$ is of degree r in x and the coefficient of x^r is unity. By convention $H_0=1$. We have

$$\alpha(x-t)=\frac{1}{\sqrt{(2\pi)}}\exp\left(-\tfrac{1}{2}x^2+tx-\tfrac{1}{2}t^2\right)=\alpha(x)\exp\left(tx-\tfrac{1}{2}t^2\right)$$

and also, by Taylor's theorem,

$$\alpha(x-t)=\sum_{j=0}^{\infty}\frac{(-1)^j}{j!}t^jD^j\alpha(x)=\sum_{j=0}^{\infty}\frac{t^j}{j!}H_j(x)\alpha(x).$$

Consequently $H_r(x)$ is the coefficient of $\dfrac{t^r}{r!}$ in $\exp\left(tx-\tfrac{1}{2}t^2\right)$. It follows that

$$H_r(x) = x^r - \frac{r^{[2]}}{2 \cdot 1!}x^{r-2} + \frac{r^{[4]}}{2^2 \cdot 2!}x^{r-4} - \frac{r^{[6]}}{2^3 \cdot 3!}x^{r-6} + \dots \tag{6.22}$$

The first ten polynomials are

$$\left.\begin{aligned}
H_0 &= 1 \\
H_1 &= x \\
H_2 &= x^2 - 1 \\
H_3 &= x^3 - 3x \\
H_4 &= x^4 - 6x^2 + 3 \\
H_5 &= x^5 - 10x^3 + 15x \\
H_6 &= x^6 - 15x^4 + 45x^2 - 15 \\
H_7 &= x^7 - 21x^5 + 105x^3 - 105x \\
H_8 &= x^8 - 28x^6 + 210x^4 - 420x^2 + 105 \\
H_9 &= x^9 - 36x^7 + 378x^5 - 1260x^3 + 945x \\
H_{10} &= x^{10} - 45x^8 + 630x^6 - 3150x^4 + 4725x^2 - 945
\end{aligned}\right\} \tag{6.23}$$

6.15 The polynomials have a number of interesting properties. Differentiating the identity

$$\exp\left(tx - \tfrac{1}{2}t^2\right) = \sum_{j=0}^{\infty} \frac{t^j H_j(x)}{j!} \tag{6.24}$$

with respect to x and identifying coefficients in t^r we have for $r \geqslant 1$

$$\frac{\mathrm{d}}{\mathrm{d}x} H_r(x) = rH_{r-1}(x)$$

and generally for $r \geqslant j$

$$\mathrm{D}^j H_r(x) = r^{[j]} H_{r-j}(x). \tag{6.25}$$

Differentiating (6.24) with respect to t and identifying coefficients in t^{r-1} we have for $r \geqslant 2$

$$H_r(x) - xH_{r-1}(x) + (r-1)H_{r-2}(x) = 0. \tag{6.26}$$

From (6.25) and (6.26) together we find

$$\frac{\mathrm{d}^2 H_r(x)}{\mathrm{d}x^2} - x\frac{\mathrm{d}H_r(x)}{\mathrm{d}x} + rH_r(x) = 0, \qquad r \geqslant 2. \tag{6.27}$$

It is also known that the equation in x, $H_r(x) = 0$, has r real roots, each not greater in absolute value than $\sqrt{\{\frac{1}{2}r(r-1)\}}$. (Cf. Charlier, 1931.)

6.16 The polynomials have an important orthogonal property, namely, that

$$\left.\begin{aligned}\int_{-\infty}^{\infty} H_m(x)H_n(x)\alpha(x)\,\mathrm{d}x &= 0, \qquad m \neq n\\ &= n!, \qquad m = n.\end{aligned}\right\} \tag{6.28}$$

For, integrating by parts, we have, if $m \leqslant n$,

$$\int_{-\infty}^{\infty} H_m H_n \alpha\,\mathrm{d}x = (-1)^n \int_{-\infty}^{\infty} H_m \mathrm{D}^n \alpha\,\mathrm{d}x$$

$$= (-1)^n [H_m \mathrm{D}^{n-1}\alpha]_{-\infty}^{\infty} + (-1)^{n-1}\int_{-\infty}^{\infty} \frac{\mathrm{d}H_m}{\mathrm{d}x}\mathrm{D}^{n-1}\alpha\,\mathrm{d}x.$$

The term in square brackets is zero and, from (6.25), the integral becomes

$$m(-1)^{n-1}\int_{-\infty}^{\infty} H_{m-1}\mathrm{D}^{n-1}\alpha\,\mathrm{d}x.$$

Continuing the process, we obtain (6.28). In particular, $m = 0$ in (6.28) gives $\int_{-\infty}^{\infty} H_n(x)\alpha(x)\,\mathrm{d}x = 0$ for all $n > 0$.

The Gram-Charlier series of Type A

6.17 Suppose now that a density function can be expanded formally in a series of derivatives of $\alpha(x)$. (We shall discuss in **6.22** the conditions under which such an expansion is valid.) We then have

$$f(x) = \sum_{j=0}^{\infty} c_j H_j(x)\alpha(x). \tag{6.29}$$

Multiplying by $H_r(x)$ and integrating from $-\infty$ to ∞ we have, from the orthogonality relation (6.28),

$$c_r = \frac{1}{r!}\int_{-\infty}^{\infty} f(x)H_r(x)\,\mathrm{d}x.$$

The reader familiar with harmonic analysis will recognize the resemblance between this procedure and the evaluation of constants in a Fourier series.

Substituting the explicit value of $H_r(x)$ from (6.22) we find

$$c_r = \frac{1}{r!}\left\{\mu'_r - \frac{r^{[2]}}{2\cdot 1!}\mu'_{r-2} + \frac{r^{[4]}}{2^2\cdot 2!}\mu'_{r-4} - \cdots\right\}. \tag{6.30}$$

In particular, for moments about the mean,

$$\left.\begin{aligned}
c_0 &= 1\\
c_1 &= 0\\
c_2 &= \tfrac{1}{2}(\mu_2 - 1)\\
c_3 &= \tfrac{1}{6}\mu_3\\
c_4 &= \tfrac{1}{24}(\mu_4 - 6\mu_2 + 3)\\
c_5 &= \tfrac{1}{120}(\mu_5 - 10\mu_3)\\
c_6 &= \tfrac{1}{720}(\mu_6 - 15\mu_4 + 45\mu_2 - 15)\\
c_7 &= \tfrac{1}{5040}(\mu_7 - 21\mu_5 + 105\mu_3)\\
c_8 &= \tfrac{1}{40320}(\mu_8 - 28\mu_6 + 210\mu_4 - 420\mu_2 + 105)
\end{aligned}\right\} \qquad (6.31)$$

Thus we find the formal expansion

$$f(x) = \alpha(x)\{1 + \tfrac{1}{2}(\mu_2 - 1)H_2 + \tfrac{1}{6}\mu_3 H_3 + \tfrac{1}{24}(\mu_4 - 6\mu_2 + 3)H_4 + \ldots\}. \qquad (6.32)$$

If $f(x)$ is standardized the series becomes

$$f(x) = \alpha(x)\{(1 + \tfrac{1}{6}\mu_3 H_3 + \tfrac{1}{24}(\mu_4 - 3)H_4 + \ldots\}. \qquad (6.33)$$

This is the so-called Gram–Charlier series of Type A, although it appeared in the work of P. L. Chebyshev (who also developed (6.40) below) and L. H. F. Oppermann before that of J. P. Gram in 1879, Thiele (1903, originally 1889) and C. V. L. Charlier in the early 1900s, cf Charlier (1931). If only the three terms shown in (6.33) are used, it makes allowance for skewness and kurtosis.

Edgeworth's form of the Type A series

6.18 Consider the Fourier transformation of a term $H_r(x)\alpha(x)$. Since

$$\sqrt{(2\pi)}\alpha(t) = e^{-\frac{1}{2}t^2} = \int_{-\infty}^{\infty} e^{itx}\frac{1}{\sqrt{(2\pi)}}e^{-\frac{1}{2}x^2}\,dx$$

we have

$$\sqrt{(2\pi)}\frac{d^r}{dt^r}\alpha(t) = (-1)^r\sqrt{(2\pi)}H_r(t)\alpha(t) = \int_{-\infty}^{\infty} i^r x^r \frac{e^{itx}}{\sqrt{(2\pi)}}e^{-\frac{1}{2}x^2}\,dx$$

and thus the transform of $x^r\alpha(x)$ is $i^r\sqrt{(2\pi)}H_r(t)\alpha(t)$. Conversely

$$x^r\alpha(x) = \frac{1}{2\pi}\int_{-\infty}^{\infty} e^{-itx}i^r\sqrt{(2\pi)}H_r(t)\alpha(t)\,dt.$$

Interchanging x and t, we find

$$\sqrt{(2\pi)}(-i)^r t^r\alpha(t) = \int_{-\infty}^{\infty} e^{-ixt}H_r(x)\alpha(x)\,dx$$

and hence, changing the sign of t, that the transform of $H_r(x)\alpha(x)$ is $\sqrt{(2\pi)}i^r t^r\alpha(t)$.

Consider now the expression

$$\{\exp(\kappa_r D^r)\}\alpha(x). \tag{6.34}$$

Its c.f. is

$$\int_{-\infty}^{\infty} e^{itx}\exp(\kappa_r D^r)\alpha(x)\,dx = \int_{-\infty}^{\infty} e^{itx}\sum\left(\frac{\kappa_r^j D^{rj}}{j!}\right)\alpha(x)\,dx$$
$$= \sum\frac{\kappa_r^j}{j!}\int_{-\infty}^{\infty} e^{itx}D^{rj}\alpha(x)\,dx$$
$$= \sum\frac{\kappa_r^j}{j!}\int_{-\infty}^{\infty} e^{itx}(-1)^{rj}H_{rj}(x)\alpha(x)\,dx$$
$$= \sum\frac{\kappa_r^j}{j!}\sqrt{(2\pi)}(-1)^{rj}t^{rj}\alpha(t)$$
$$= \sqrt{(2\pi)}\alpha(t)\exp\{\kappa_r(-it)^r\}. \tag{6.35}$$

In a similar way it will be seen that the c.f. of

$$\exp\left\{-\frac{\kappa_1-a}{1!}D+\frac{\kappa_2-b}{2!}D^2-\frac{\kappa_3}{3!}D^3+\frac{\kappa_4}{4!}D^4\ldots\right\}\alpha(x) \tag{6.36}$$

is

$$\sqrt{(2\pi)}\alpha(t)\exp\left\{\frac{\kappa_1-a}{1!}it+\frac{\kappa_2-b}{2!}(it)^2+\frac{\kappa_3}{3!}(it)^3+\frac{\kappa_4}{4!}(it)^4+\ldots\right\}. \tag{6.37}$$

More generally, if we replace $\alpha(x)$ by the unstandardized

$$\beta(x)=\frac{1}{\sigma\sqrt{(2\pi)}}e^{-\frac{1}{2}(x-m)^2/\sigma^2}=\alpha\{(x-m)/\sigma\}/\sigma,$$

the c.f. of

$$\exp\left\{-\frac{\kappa_1-a}{1!}D+\frac{\kappa_2-b}{2!}D^2-\frac{\kappa_3}{3!}D^3+\frac{\kappa_4}{4!}D^4\ldots\right\}\beta(x) \tag{6.38}$$

is

$$\sqrt{(2\pi)}\alpha(t\sigma)e^{imt}\exp\left\{\frac{\kappa_1-a}{1!}it+\frac{\kappa_2-b}{2!}(it)^2+\frac{\kappa_3}{3!}(it)^3+\frac{\kappa_4}{4!}(it)^4\ldots\right\} \tag{6.39}$$

as may be seen by the same line of argument.

Now suppose that (6.38) represents a density function. Its c.g.f. is then the logarithm of (6.39), i.e. is equal to

$$\frac{(\kappa_1-a+m)it}{1!}+\frac{\kappa_2-b+\sigma^2}{2!}(it)^2+\frac{\kappa_3}{3!}(it)^3+\frac{\kappa_4}{4!}(it)^4+\ldots$$

and hence its cumulants are κ_1-a+m, $\kappa_2-b+\sigma^2$, κ_3, κ_4, ..., κ_r, ..., etc. We may take $a=m$ and $b=\sigma^2$ and thus we obtain a distribution whose cumulants are κ_1,

$\kappa_2, \ldots$ etc. Now if these are in fact the cumulants of a distribution the series (6.38) must be equal to that distribution, provided that (1) the series converges to a density function, and (2) it is uniquely determined by its moments.

If we take the density function to be standardized, then $\kappa_1 = 0$, $\kappa_2 = 1$ and (6.38) becomes

$$f(x) = \exp\left\{-\kappa_3 \frac{D^3}{3!} + \kappa_4 \frac{D^4}{4!} - \ldots\right\} \alpha(x), \tag{6.40}$$

where we have written $\alpha(x)$ for $\beta(x)$ because now m vanishes and $\sigma^2 = 1$.

6.19 A series of this kind was derived by Edgeworth (1904), though from an entirely different approach through the theory of elementary errors. Equation (6.40) is formally identical with (6.33), and the reader who consults the original memoirs on this subject may be puzzled by the fact that Edgeworth claimed his series to be different from the Type A series and better as a representation of density functions. The explanation is that for practical purposes we use only a finite number of terms in the series and neglect the remainder. If we take the first k terms in (6.33) the result is in general different from that obtained by taking the first $(k-1)$ terms of the operator in the exponential of (6.40). The argument centred on the fact (cf. Example 6.3 below) that the terms in (6.33) do not tend regularly to zero from the point of view of elementary errors, so that in general no term is negligible compared with a preceding term.

6.20 In (6.29), the coefficients c_j evaluated at (6.31) become, when standardized and converted into cumulants using (3.38),

$$\left.\begin{aligned}
c_0 &= 1, \qquad c_1 = c_2 = 0 \\
c_3 &= \frac{\kappa_3}{6} \\
c_4 &= \frac{\kappa_4}{24} \\
c_5 &= \frac{\kappa_5}{120} \\
c_6 &= \frac{1}{720}(\kappa_6 + 10\kappa_3^2) \\
c_7 &= \frac{1}{5040}(\kappa_7 + 35\kappa_4\kappa_3) \\
c_8 &= \frac{1}{40320}(\kappa_8 + 56\kappa_5\kappa_3 + 35\kappa_4^2)
\end{aligned}\right\} \tag{6.41}$$

We see that c_j depends on κ_j, and we shall find in Chapter 10 that for $j > 4$ estimation of κ_j is unreliable owing to sampling fluctuations. When sampling effects are not in question the series may be taken to more terms, usually not higher than the term in

H_6. We should then have to investigate how far the observed distribution can be represented by the series

$$f(x) = \alpha(x)\left(1 + \frac{\kappa_3}{6}H_3 + \frac{\kappa_4}{24}H_4 + \frac{\kappa_5}{120}H_5 + \frac{\kappa_6 + 10\kappa_3^2}{720}H_6\right) \tag{6.42}$$

in the hope that the remainder after these terms could be neglected in comparison. If cumulants above the fourth are neglected, (6.42) differs from (6.33) by the term $\kappa_3^2 H_6/72$ in parentheses. (6.42) is often called the Edgeworth form of the Type A series.

It may be noted in passing that, in contrast to the Pearson family, the d.f. of such a series is easy to obtain. If as in (6.29)

$$f(x) = \sum_{r=0}^{\infty} c_r H_r(x)\alpha(x),$$

then using (6.21),

$$\begin{aligned}\int_{-\infty}^{x} f(u)\,\mathrm{d}u &= \sum_{r=0}^{\infty} c_r \int_{-\infty}^{x} H_r(u)\alpha(u)\,\mathrm{d}u \\ &= \int_{-\infty}^{x} \alpha(u)\,\mathrm{d}u - \sum_{r=1}^{\infty} c_r H_{r-1}(x)\alpha(x).\end{aligned} \tag{6.43}$$

The coefficients in (6.41) are, apart from the factor $(r!)^{-1}$, those in (3.37) when $\kappa_1 = \kappa_2 = 0$—see Davis (1976).

Example 6.2 (Fitting the Edgeworth series by moments)

Consider the fitting of a Type A series to the bean data of Table 6.1.

We have already found the first four moments in Example 6.1. Standardizing them, we have

$$\mu_3 = -0.910\,569$$
$$\mu_4 = 4.862\,944$$

and we also find the standardized

$$\mu_5 = -12.574\,125$$
$$\mu_6 = 53.221\,083.$$

Hence we obtain the standardized cumulants and substitute them into (6.42) to give

$$f(x) = \alpha(x)\{1 - 0.151\,762H_3 + 0.077\,622\,7H_4 - 0.028\,903\,6H_5 + 0.014\,273\,5H_6\}.$$

Columns (4) to (6) of Table 6.1 show the expected frequencies given by taking the first three, the first four and the first five terms of this series. A glance at the figures will show that the four- and five-term series are worse than the three-term, for each gives a negative frequency at a high value and a second mode at a low value, contrary to the data. The representation is clearly not satisfactory and worse than that given by the Pearson Type IV fit in column (3).

Table 6.1 Various systems fitted to the distribution of length of beans (Table 1.15)
(from Pretorius, 1930, Johnson, 1949a)

Length of beans (mm) (1)	Observed frequency (2)	Pearson Type IV (3)	Edgeworth Type A (three terms) (4)	Edgeworth Type A (four terms) (5)	Edgeworth Type A (five terms) (6)	Lognormal (7)	Johnson Type S_u (8)
>17.25	—	{	16.3	−15.2	2.0	—	0.1
17.0	6	1.4	12.8	13.7	−35.3	{	2.0
16.5	55	28.5	25.6	116.6	22.3	10.1	32.2
16.0	275	299.3	241.7	370.4	438.1	280.5	290.1
15.5	1129	1181.6	1012.7	926.2	1214.0	1255.2	1151.5
15.0	2082	2132.6	2155.4	1833.0	1866.9	2163.8	2130.3
14.5	2294	2229.8	2593.0	2506.4	2112.8	2179.6	2240.6
14.0	1787	1638.9	1788.4	2082.6	1916.7	1598.5	1642.5
13.5	929	968.9	713.4	921.3	1183.4	965.0	970.6
13.0	437	503.6	280.7	199.0	371.2	515.3	508.7
12.5	199	243.7	258.7	132.1	66.9	254.5	249.3
12.0	115	113.8	206.2	178.1	101.2	119.5	118.0
11.5	70	52.5	98.7	117.0	107.1	54.4	55.2
11.0	36	24.2	29.6	43.5	54.0	24.3	25.7
10.5	18	11.3	5.9	10.0	15.4	10.7	12.1
10.0	7	5.4	{	{	{	4.8	5.8
9.5	1	2.6	.9	1.7	3.3	2.1	2.7
>9.25	—	1.9	—	—	—	1.7	2.6
TOTALS	9940	9940	9440	9440	9440	9440	9440

(The braces means that the expected frequencies shown are rounded up and include some small values in blank rows covered by the brackets.)

Tetrachoric functions

6.21 The numerical evaluation of $H_r(x)\alpha(x) = (-D)^r\alpha(x)$ may be carried out directly by evaluating the polynomial in (6.23) and multiplying by $\alpha(x)$, or from tables of $H_r(x)$ and $\alpha(x)$, but simpler tables are available.

The *Biometrika Tables*, Vol. II, gives $D^r\alpha(x)$ for $x = 0(0.02)4.00(0.05)6.20$ for $r = 1,2$(6 d.p.); $r = 3,4$(5 d.p.); $r = 5,6$(4 d.p.); $r = 7$(3 d.p.); $r = 8$(2 d.p.) and $r = 9$(1 d.p.).

K. Pearson tabulated

$$\tau_r(x) = \frac{H_{r-1}(x)\alpha(x)}{(r!)^{\frac{1}{2}}}. \tag{6.44}$$

This was known as a tetrachoric function, for reasons that will become apparent in Chapter 26, Volume 2A when we discuss the estimation of correlation coefficients from qualitative data.

Smirnov (1965) gives 7 d.p. values of (6.44) for $r = 2(1)21$, $x = 0(0.002)4$, together with a 10 d.p. table of the coefficients in the recurrence relations derived from (6.27).

6.22 So far, it has been assumed that a density function possesses a convergent Type A series. We shall not here enter into a discussion of the conditions under which this is so, except to warn the reader that mistakes have been made on the subject and to quote some theorems without proof.

(1) Cramér (1925). If $f(x)$ is a function with a continuous derivative such that

$$\int_{-\infty}^{\infty}\left(\frac{\mathrm{d}f}{\mathrm{d}x}\right)^2 e^{\frac{1}{2}x^2}\mathrm{d}x$$

converges and if $f(x)$ tends to zero as $|x|$ tends to infinity, then $f(x)$ may be developed in the series

$$f(x)=\sum_{j=0}^{\infty}\frac{c_j}{j!}D^j\alpha(x), \tag{6.45}$$

where

$$c_j=\int_{-\infty}^{\infty}f(x)H_j(x)\mathrm{d}x.$$

This series is absolutely and uniformly convergent for $-\infty<x<\infty$.

(2) A theorem by Cramér (1925) based on one by Galbrun. If $f(x)$ is of bounded variation in every finite interval and if

$$\int_{-\infty}^{\infty}|f(x)|e^{\frac{1}{2}x^2}\mathrm{d}x$$

exists, then the expansion of $f(x)$ in the series (6.45) converges everywhere to the sum $\frac{1}{2}\{f(x+0)+f(x-0)\}$. The convergence is uniform in every finite interval of continuity.

Cramér has also shown that this last theorem cannot be substantially improved upon as regards the behaviour of $f(x)$ at infinity.

Exercise 6.9 is to show that a convergent Type A expansion requires that the distribution be determined by its moments.

Bhattacharya and Ghosh (1978, 1988) give the general conditions under which an Edgeworth expansion exists; in particular, they show that a valid expansion may be developed using approximate moments, calculated as in **10.5**. These conditions are further weakened by Bai and Rao (1991).

Kolassa and McCullagh (1990) show that an Edgeworth expansion is as accurate for a discrete distribution, provided that a continuity correction is used and that Sheppard's corrections are applied to the cumulants.

6.23 From the statistical viewpoint, however, the important question is not whether an *infinite* series can represent a density function, but whether a finite number of terms can do so to a satisfactory approximation. Two things seem clear:

(a) The sum of a finite number of terms of the series may give negative values, particularly near the tails (as, for instance, in Example 6.2).
(b) The series may behave irregularly in the sense that the sum of k terms may give a worse fit than the sum of $(k-1)$ terms.

Barton and Dennis (1952)—see also Draper and Tierney (1972)—show that if cumulants above the fourth are neglected, the three-term expansion (6.33) and the four-term (6.42) do not give a unimodal, or even a non-negative, density, unless the skewness and kurtosis coefficients $\beta_1 = \gamma_1^2$ and $\beta_2 = \gamma_2 + 3$ are restricted in value. For example, (6.42) is non-unimodal if $|\beta_1| \geqslant 0.25$, and becomes negative if $|\beta_1| \geqslant 0.5$; (6.33) is non-unimodal if $|\beta_1| \geqslant 0.7$ and becomes negative if $|\beta_1| > 1.2$; neither is unimodal if $\beta_2 > 5.4$ and both become negative for $\beta_2 > 7$. Berndt (1957) makes a similar investigation when only κ_3 is used. Balitskaya and Zolotuhina (1988) analytically determine and graphically illustrate the precise region in the (γ_1, γ_2) plane where (6.42) is never negative: it is roughly elliptical with major axis on the γ_2 axis between 0 and 4 and the minor axis between $\gamma_1 = \pm\frac{2}{3}$. Thus we see that no platykurtic distribution can be properly represented by (6.42).

Thus the finite series is useful only in cases of moderate skewness, and in such cases a Pearson distribution, or a transformed distribution of the type we consider later, may be just as good. For many statistical purposes we are most interested in the tails of a distribution, and it is here that the inadequacies of the series approach appear.

Example 6.3 (Series expansions for the Gamma distribution)

As an illustration of the irregular behaviour of terms in the Type A series, consider the Gamma distribution with p.d.f.

$$f = \frac{1}{\Gamma(\lambda)} x^{\lambda-1} e^{-x}, \qquad 0 \leqslant x < \infty; \quad \lambda > 0.$$

Its c.f. is, from Example 3.6,

$$\phi(t) = 1/(1 - it)^{\lambda}$$

and thus $\kappa_r = \lambda(r-1)!$ or, on standardizing,

$$\kappa_r = \frac{(r-1)!}{\lambda^{\frac{1}{2}r-1}}.$$

From the manner of the formation of terms in (6.41) it is evident that the coefficient c_r is the sum of terms $\kappa_r, \kappa_{r-3}\kappa_3, \ldots, (\kappa_{q_1}\kappa_{q_2} \ldots \kappa_{q_m})$, where $(q_1 \ldots q_m)$ is a partition of r such that no q is less than 3. It will then be clear that, since κ_q is of order $\lambda^{1-\frac{1}{2}q}$, the term of greatest order in λ is that with the greatest number of parts in $(\kappa_{q_1} \ldots \kappa_{q_m})$. For example, if $r=9$ it is (3^3), if $r=8$ it is (4^2), and so on.

From these considerations we can find the order in λ of the terms in the Type A series. They are

Term	c_0	c_3	c_4	c_5	c_6	c_7	c_8	c_9	c_{10}	c_{11}	c_{12}
Order in λ	0	$-\frac{1}{2}$	-1	$-1\frac{1}{2}$	-1	$-1\frac{1}{2}$	-2	$-1\frac{1}{2}$	-2	$-2\frac{1}{2}$	-2

The terms decrease in order of λ, but not at all regularly, and it is clear that in general no coefficient will be negligible compared with a preceding one if λ is large. The asymptotic qualities of such series obviously require careful investigation in particular cases.

More generally, when a Type A expansion is used to approximate a sampling distribution for a random sample of size n, it is found that c_r is of the same order in n as the orders of λ given above. By contrast, the coefficients of the Edgeworth expansion do decrease uniformly and it may be shown that $c_r \sim 0(n^{1-\frac{1}{2}r})$, $r \geqslant 3$.

Other expansions

6.24 Hall (1983) has developed a general inverse Edgeworth expansion such that (6.42) may be represented as

$$\int_{-\infty}^{\xi} f(x)\,\mathrm{d}x = \int_{-\infty}^{k} \alpha(x)\,\mathrm{d}x,$$

where $\xi = x - a_1(x) - a_2(x) - \ldots$; typically, $a_r(x)$ is $O(n^{-\frac{1}{2}r})$ when approximating a sampling distribution.

Robinson (1978) gives an Edgeworth expansion for approximating sampling distributions when sampling is without replacement from a finite population.

If the density is a member of the exponential family (5.118), it is possible to derive an indirect Edgeworth expansion which is generally more accurate, particularly in the tails; see **11.13–16**.

Expansions can also be obtained in terms of derivatives of the Gamma and Beta distributions. The appropriate polynomials are those of Laguerre and Jacobi respectively. See G. Szegö, *Orthogonal Polynomials*, New York, 1939, and Durbin and Watson (1951), especially p. 172. Wallace (1958) discusses the validity of asymptotic expansions generally, including non-normal p.d.f.s as generating distributions, and also the problem of normalization discussed in **6.25–6** below. Gray *et al.* (1975) give an expansion of any theoretical $f(x)$ in terms of any $\alpha(x)$ (not necessarily normal) using derivatives but not cumulants, which therefore need not exist. In the Gamma case of Example 6.3, this expansion improves on Edgeworth's in the tails of the distribution.

Charlier proposed a Type B series based on derivatives of the Poisson probabilities $e^{-\lambda}\lambda^x/x!$ with respect to λ, or equivalently, on first differences of those probabilities with respect to x. (Cf. Exercise 6.8.) It has not come into general use. A further form (Type C) proposed by Charlier to avoid negative values for the density has also failed to stand the test of experience.

Polynomial transformations to normality

6.25 Many of the important theoretical distributions occurring in Statistics depend on some variable n in such a way that as n tends to infinity the distribution tends to normality. For large n it is often a sufficient approximation to assume the distribution normal, but for small or moderate n this may be hardly exact enough. In such a case we are nevertheless able to use the normal integral by seeking a polynomial variate transformation

$$\xi = b_0 + b_1 x + b_2 x^2 + b_3 x^3 + \ldots \tag{6.46}$$

By choosing the bs appropriately we can bring the distribution of ξ much nearer to normality than that of x and hence from (6.46) find the d.f. of x from that of ξ, assumed normal.

Consider the Edgeworth form of the Type A expansion (6.38)

$$\exp\left\{-\frac{\kappa_1 - m}{1!}\mathrm{D} + \frac{\kappa_2 - \sigma^2}{2!}\mathrm{D}^2 - \frac{\kappa_3}{3!}\mathrm{D}^3 + \ldots\right\}\beta(x). \tag{6.47}$$

We have retained the terms in D and D^2 because the approximation may perhaps be slightly improved by taking m and σ^2 in the ξ-distribution not quite equal to the mean and variance of x.

We now assume that the cumulant κ_r is of order n^{1-r}, a case of fairly common occurrence; by choice of m and σ^2, we may write

$$\kappa_1 - m = l_1\sigma, \qquad l_1 = O(n^{-\frac{1}{2}}),$$
$$\kappa_2 - \sigma^2 = l_2\sigma^2, \qquad l_2 = O(n^{-1}).$$

Then σ^2 is of order κ_2, i.e. n^{-1}, and thus

$$\frac{\kappa_r}{\sigma^r} = O(n^{1-\frac{1}{2}r}).$$

Thus (6.47) may be written

$$\exp\{-l_1\sigma\mathrm{D} + \tfrac{1}{2}l_2\sigma^2\mathrm{D}^2 - \tfrac{1}{6}l_3\sigma^3\mathrm{D}^3 + \tfrac{1}{24}l_4\sigma^4\mathrm{D}^4 - \tfrac{1}{120}l_5\sigma^5\mathrm{D}^5 + \tfrac{1}{720}l_6\sigma^6\mathrm{D}^6 - \dots\}\beta(x), \tag{6.48}$$

where

$$l_1 \text{ and } l_3 \text{ are } O(n^{-\frac{1}{2}}), \qquad l_5 \text{ is } O(n^{-\frac{3}{2}}),$$
$$l_2 \text{ and } l_4 \text{ are } O(n^{-1}), \qquad l_6 \text{ is } O(n^{-2}), \text{ etc.}$$

Expanding the exponential in (6.48) and retaining only terms up to and including $O(n^{-2})$ in the ls we find that it is equal to

$$\begin{aligned}
&1 - l_1\sigma\mathrm{D} + \tfrac{1}{2}l_2\sigma^2\mathrm{D}^2 - \tfrac{1}{6}l_3\sigma^3\mathrm{D}^3 + \tfrac{1}{24}l_4\sigma^4 D^4 - \tfrac{1}{120}l_5\sigma^5\mathrm{D}^5 + \tfrac{1}{720}l_6\sigma^6\mathrm{D}^6 + \tfrac{1}{2}(l_1^2\sigma^2\mathrm{D}^2\\
&+ \tfrac{1}{4}l_2^2\sigma^4\mathrm{D}^4 + \tfrac{1}{36}l_3^2\sigma^6\mathrm{D}^6 + \tfrac{1}{576}l_4^2\sigma^8\mathrm{D}^8 - l_1l_2\sigma^3\mathrm{D}^3 + \tfrac{1}{3}l_1l_3\sigma^4\mathrm{D}^4 - \tfrac{1}{12}l_1l_4\sigma^5\mathrm{D}^5\\
&+ \tfrac{1}{60}l_1l_5\sigma^6\mathrm{D}^6 - \tfrac{1}{6}l_2l_3\sigma^5\mathrm{D}^5 + \tfrac{1}{24}l_2l_4\sigma^6\mathrm{D}^6 - \tfrac{1}{72}l_3l_4\sigma^7\mathrm{D}^7 + \tfrac{1}{360}l_3l_5\sigma^8\mathrm{D}^8) + \tfrac{1}{6}(-l_1^3\sigma^3\mathrm{D}^3\\
&- \tfrac{1}{216}l_3^3\sigma^9\mathrm{D}^9 + \tfrac{3}{2}l_1^2l_2\sigma^4\mathrm{D}^4 - \tfrac{1}{2}l_1^2l_3\sigma^5\mathrm{D}^5 - \tfrac{1}{8}l_1^2l_4\sigma^6\mathrm{D}^6 + \tfrac{1}{288}l_3^2l_4\sigma^{10}\mathrm{D}^{10} - \tfrac{1}{12}l_1l_3^2\sigma^7\mathrm{D}^7\\
&+ \tfrac{1}{24}l_2l_3^2\sigma^8\mathrm{D}^8 + \tfrac{1}{2}l_1l_2l_3\sigma^6\mathrm{D}^6 + \tfrac{1}{24}l_1l_3l_4\sigma^8\mathrm{D}^8) + \tfrac{1}{24}(l_1^4\sigma^4\mathrm{D}^4 + \tfrac{1}{1296}l_3^4\sigma^{12}\mathrm{D}^{12}\\
&+ \tfrac{2}{3}l_1^3l_3\sigma^6\mathrm{D}^6 + \tfrac{1}{6}l_1^2l_3^2\sigma^8\mathrm{D}^8 + \tfrac{1}{54}l_1l_3^3\sigma^{10}\mathrm{D}^{10}).
\end{aligned} \tag{6.49}$$

The result of applying (6.49) to $\beta(x)$ in (6.48) is a similar expression, which we will not bother to write out at length, with the operator $\sigma^r\mathrm{D}^r$ replaced by

$$(-1)^r H_r\left(\frac{x-m}{\sigma}\right)\beta(x).$$

The d.f. is given by integrating this expression, and we then have for the d.f. at $m + \sigma x$ (arranging the terms in order of magnitude in n)

$$\begin{aligned}
&\int_{-\infty}^{x} \alpha(x)\mathrm{d}x + \alpha(x)[-(l_1 + \tfrac{1}{6}l_3H_2) - (\tfrac{1}{2}l_1^2H_1 + \tfrac{1}{2}l_2H_1 + \tfrac{1}{6}l_1l_3H_3 + \tfrac{1}{24}l_4H_3\\
&\quad + \tfrac{1}{72}l_3^2H_5) - (\tfrac{1}{6}l_1^3H_2 + \tfrac{1}{2}l_1l_2H_2 + \tfrac{1}{12}l_1^2l_3H_4 + \tfrac{1}{12}l_2l_3H_4 + \tfrac{1}{24}l_1l_4H_4\\
&\quad + \tfrac{1}{120}l_5H_4 + \tfrac{1}{72}l_1l_3^2H_6 + \tfrac{1}{144}l_3l_4H_6 + \tfrac{1}{1296}l_3^3H_8) - (\tfrac{1}{24}l_1^4H_3 + \tfrac{1}{8}l_2^2H_3 + \tfrac{1}{4}l_1^2l_2H_3\\
&\quad + \tfrac{1}{36}l_1^3l_3H_5 + \tfrac{1}{12}l_1l_2l_3H_5 + \tfrac{1}{48}l_1^2l_4H_5 + \tfrac{1}{48}l_2l_4H_5 + \tfrac{1}{120}l_1l_5H_5 + \tfrac{1}{720}l_6H_5\\
&\quad + \tfrac{1}{144}l_1^2l_3^2H_7 + \tfrac{1}{144}l_2l_3^2H_7 + \tfrac{1}{1152}l_4^2H_7 + \tfrac{1}{144}l_1l_3l_4H_7 + \tfrac{1}{720}l_3l_5H_7 + \tfrac{1}{1296}l_1l_3^3H_9\\
&\quad + \tfrac{1}{1728}l_3^2l_4H_9 + \tfrac{1}{31104}l_3^4H_{11})].
\end{aligned} \tag{6.50}$$

6.26 Now let ξ be a normal variate. We will determine ξ in terms of x so that

$$G(\xi) = \int_{-\infty}^{\xi} \alpha(y)\mathrm{d}y = F(x). \tag{6.51}$$

$F(x)$ being the distribution function given by the expansion (6.50).

We have, by Taylor's theorem,

$$G(\xi) = G\{x + (\xi - x)\} = G(x) + \sum_{r=1}^{\infty} \frac{(\xi - x)^r}{r!} \frac{\mathrm{d}^r}{\mathrm{d}x^r} G(x)$$

and on using (6.21), this becomes

$$G(\xi) = G(x) - \sum_{r=1}^{\infty} \frac{(x - \xi)^r}{r!} H_{r-1}(x)\alpha(x) \tag{6.52}$$

and this is to be equal to (6.50).

The next step is to invert this series to give $(x - \xi)$ in powers of x. Assuming

$$x - \xi = a_0 + a_1 x + a_2 x^2 + \ldots, \tag{6.53}$$

which is (6.46) slightly re-arranged for the sake of convenience, we see that when $x = 0$, $\xi = -a_0$, $x - \xi$ is of order $n^{-\frac{1}{2}}$; and hence, to order n^{-2}, we have from (6.52) with $x = 0$, on using (6.23),

$$\xi - \frac{\xi^2}{2}(+0) + \frac{\xi^3}{6}(-1) = -\left(a_0 - \frac{a_0^3}{6}\right)$$

and this is equal to the expression in square brackets in (6.50) with $x = 0$.

We then find

$$a_0 = l_1 - \tfrac{1}{6}l_3 - \tfrac{1}{2}l_1 l_2 + \tfrac{1}{6}l_1^2 l_3 + \tfrac{1}{4}l_2 l_3 + \tfrac{1}{8}l_1 l_4 + \tfrac{1}{40}l_5 - \tfrac{7}{36}l_1 l_3^2 - \tfrac{15}{144}l_3 l_4 + \tfrac{52}{648}l_3^3.$$

We can now find a_1 in (6.50) by identifying coefficients in x, and so on. After some algebraic reduction we find, writing the terms in descending order in n,

$$\begin{aligned} x - \xi = {} & l_1 + \tfrac{1}{6}l_3(x^2 - 1) + \tfrac{1}{2}l_2 x - \tfrac{1}{3}l_1 l_3 x + \tfrac{1}{24}l_4(x^3 - 3x) - \tfrac{1}{36}l_3^2(4x^3 - 7x) - \tfrac{1}{2}l_1 l_2 \\ & + \tfrac{1}{6}l_1^2 l_3 - \tfrac{1}{12}l_2 l_3(5x^2 - 3) - \tfrac{1}{8}l_1 l_4(x^2 - 1) + \tfrac{1}{120}l_5(x^4 - 6x^2 + 3) + \tfrac{1}{36}l_1 l_3^2(12x^2 - 7) \\ & - \tfrac{1}{144}l_3 l_4(11x^4 - 42x^2 + 15) + \tfrac{1}{648}l_3^3(69x^4 - 187x^2 + 52) - \tfrac{3}{8}l_2^2 x + \tfrac{5}{6}l_1 l_2 l_3 x \\ & + \tfrac{1}{8}l_1^2 l_4 x - \tfrac{1}{48}l_2 l_4(7x^3 - 15x) - \tfrac{1}{30}l_1 l_5(x^3 - x) + \tfrac{1}{720}l_6(x^5 - 10x^3 + 15x) \\ & - \tfrac{1}{3}l_1^2 l_3^2 x + \tfrac{1}{72}l_2 l_3^2(36x^3 - 49x) - \tfrac{1}{384}l_4^2(5x^5 - 32x^3 + 35x) + \tfrac{1}{36}l_1 l_3 l_4(11x^3 - 21x) \\ & - \tfrac{1}{360}l_3 l_5(7x^5 - 48x^3 + 51x) - \tfrac{1}{324}l_1 l_3^3(138x^3 - 187x) \\ & + \tfrac{1}{864}l_3^2 l_4(111x^5 - 547x^3 + 456x) - \tfrac{1}{7776}l_3^4(948x^5 - 3628x^3 + 2473x). \end{aligned} \tag{6.54}$$

This is our required expression of the variate ξ in terms of the variate x. To order n^{-2} at least, ξ will be normally distributed.

It is often more convenient to express x in terms of ξ. This may be done by noting that

$$\begin{aligned} x - \xi &= g(x) = g(\xi + x - \xi) \\ &= g(\xi) + (x - \xi)g'(\xi) + \ldots \\ &= g(\xi) + g'(\xi)\{g(\xi) + (x - \xi)g'(\xi)\} + \ldots \end{aligned}$$

and by continuing the process

$$x-\xi = g(\xi)+g(\xi)g'(\xi)+g(\xi)g'^2(\xi)+\tfrac{1}{2}g^2(\xi)g''(\xi)+g(\xi)g'^3(\xi)+\tfrac{3}{2}g^2(\xi)g'(\xi)g''(\xi)$$
$$+\tfrac{1}{6}g^3(\xi)g'''(\xi)+\ldots \qquad (6.55)$$

Hence, using the value of ξ given by (6.54), we find, after some reduction,

$$\begin{aligned} x-\xi = {} & l_1+\tfrac{1}{6}l_3(\xi^2-1)+\tfrac{1}{2}l_2\xi+\tfrac{1}{24}l_4(\xi^3-3\xi)-\tfrac{1}{36}l_3^2(2\xi^3-5\xi)-\tfrac{1}{6}l_2l_3(\xi^2-1) \\ & +\tfrac{1}{120}l_5(\xi^4-6\xi^2+3)-\tfrac{1}{24}l_3l_4(\xi^4-5\xi^2+2)+\tfrac{1}{324}l_3^3(12\xi^4-53\xi^2+17) \\ & -\tfrac{1}{8}l_2^2\xi-\tfrac{1}{16}l_2l_4(\xi^3-3\xi)+\tfrac{1}{720}l_6(\xi^5-10\xi^3+15\xi)+\tfrac{1}{72}l_2l_3^2(10\xi^3-25\xi) \\ & -\tfrac{1}{384}l_4^2(3\xi^5-24\xi^3+29\xi)-\tfrac{1}{180}l_3l_5(2\xi^5-17\xi^3+21\xi) \\ & +\tfrac{1}{288}l_3^2l_4(14\xi^5-103\xi^3+107\xi)-\tfrac{1}{7776}l_3^4(252\xi^5-1688\xi^3+1511\xi). \end{aligned} \qquad (6.56)$$

l_1 appears only in the first term on the right of (6.56), since it represents the location of x, and cannot affect the cumulants other than the mean.

The approach of these sections is due to Cornish and Fisher (1937). Fisher and Cornish (1960) give higher-order expansions and tables to facilitate their use.

The reader may have noticed that in (6.54) and (6.56) the coefficient of l_r is $H_{r-1}/r!$ This is one of several formulae for the coefficients proved by G. W. Hill and Davis (1968), who have computerized the expansions. They also generalize the polynomial transformation to any desired regular continuous distribution (e.g. the χ^2 distribution of Chapter 16—cf. A. W. Davis (1971)).

When the cumulants can be expanded in a power series in n^{-1}, Withers (1984) derives expressions for (6.50), (6.54) and (6.56) in terms of the coefficients in the series.

Example 6.4 (Cornish-Fisher expansion for the Gamma distribution)

Consider again the Gamma distribution of Example 6.3,

$$f=\frac{1}{\Gamma(\lambda)}x^{\lambda-1}e^{-x}, \qquad 0\leqslant x<\infty; \quad \lambda>0,$$

whose standardized κ_r is of order $\lambda^{1-\frac{1}{2}r}$. Thus, as $\lambda\to\infty$, $\kappa_r\to 0$ for $r>1$ and the distribution tends to normality by (5.98) and the Inversion theorem.

We will take l_1 and l_2 of (6.48) to be zero, which implies that our normal variate ξ is to have the same mean and variance as that of x. We have

$$l_3=2\lambda^{-\frac{1}{2}}, \quad l_4=6\lambda^{-1}, \quad l_5=24\lambda^{-\frac{3}{2}}, \quad l_6=120\lambda^{-2}.$$

(6.50) then becomes

$$G(x)-\alpha(x)\left\{\frac{1}{3\lambda^{\frac{1}{2}}}H_2+\frac{1}{4\lambda}H_3+\frac{1}{18\lambda}H_5+\frac{1}{5\lambda^{\frac{3}{2}}}H_4+\frac{1}{12\lambda^{\frac{3}{2}}}H_6+\frac{1}{162\lambda^{\frac{3}{2}}}H_8\right.$$
$$\left.+\frac{1}{6\lambda^2}H_5+\frac{1}{32\lambda^2}H_7+\frac{1}{15\lambda^2}H_7+\frac{1}{72\lambda^2}H_9+\frac{1}{1944\lambda^2}H_{11}\right\}.$$

Let us, as a simple illustration, find the d.f. of x for $\lambda=9$, $x=12$. The mean of the distribution is then 9 and its variance 9, so that this corresponds to a deviation

$(12-9)/\sqrt{9}$ in standardized form, equal to unity. It is found from (6.23) and an additional equation for H_{11} that

$$H_2 = 0, \quad H_3 = -2, \quad H_4 = -2, \quad H_5 = 6, \quad H_6 = 16, \quad H_7 = -20,$$
$$H_8 = -132, \quad H_9 = 28, \quad H_{10} = 1216, \quad H_{11} = 936.$$

We then find for the d.f.

$$\int_{-\infty}^{1} \alpha(x)dx + \alpha(1)(0.015\,163\,5).$$

The values for the normal function are obtained from the tables and we get the value

$$0.841\,345 + (0.241\,970\,7)(0.015\,163\,5) = 0.8450,$$

which is exact to four places. The approximation is evidently fairly good, even for values of λ as low as 9.

We could have found the same result by using (6.54). Substituting $x = 1$ in that equation we find $\xi = 1.015\,386$, and the d.f. of the normal distribution at this value of ξ is 0.8450 as before.

Suppose now that we wish to find the value x whose d.f. is $F(x) = 0.99$ when $\lambda = 15$.

The value ξ corresponding to the standardized normal $F(x) = 0.99$ is found from tables to be 2.326 348. We then have from (6.56)

$$x - \xi = \frac{1}{3\lambda^{\frac{1}{2}}}(\xi^2 - 1) + \frac{1}{4\lambda}(\xi^3 - 3\xi) + \dots$$

which will be found to give $x = 2.697\,22$. De-standardizing, this becomes $15 + x\sqrt{15} = 25.45$. This is exact to two places of decimals.

The example shows that, notwithstanding the irregular behaviour of the infinite Type A series, a satisfactory approximation may be obtained from its first few terms, at least in certain cases. We may remark without proof that by an adaptation of a procedure given by Cramér (1928) it may be shown that an asymptotic expansion exists for the distribution of this example.

Haldane (1938) considers power transformations $y = \{x/E(x)\}^h$, with h chosen to minimize skewness and kurtosis, for variates x with all cumulants of order n, the sample size. His results are given in Exercise 28.25 (Volume 2A). 16.7 below treats a special case.

The more general power transformation $y = \{(x+c)^h - 1\}/h$ was suggested by Box and Cox (1964) as a prior "normalizing" transformation for statistical analysis. John and Draper (1980) suggest the use of $y = \{(|x| + c)^h - 1\}\,\mathrm{sgn}(x)/h$ when the range of x is unbounded in both directions and kurtosis is a more critical departure from normality than skewness.

Functional transformations to normality: Johnson distributions

6.27 The idea of transforming from one variate to another with a more convenient density function may be developed further. We need not confine ourselves to a polynomial expression of the type of (6.46); nor is that kind of representation necessarily the best. Several writers have studied more general types. Our treatment follows that of N. L. Johnson (1949a).

We shall consider a transformation of the type

$$\xi = \gamma + \delta g\left(\frac{x-\mu}{\lambda}\right) \tag{6.57}$$

where μ, λ, γ and δ are parameters at choice and g is some convenient function. For practical purposes it is desirable that g should not itself depend on variable parameters and that it should be a monotonic function; otherwise it might be troublesome to match the ranges of ξ and x. Without loss of generality we may suppose that g is monotone increasing. If we write the transformation in the form

$$\frac{\xi-\gamma}{\delta} = g\left(\frac{x-\mu}{\lambda}\right) \tag{6.58}$$

the parameters μ, γ appear as location parameters and λ, δ as scale parameters. We may suppose the latter pair to be positive. We shall try to choose the parameters and the function g so that ξ is normal or approximately so.

6.28 Without loss of generality we write

$$y = \frac{x-\mu}{\lambda}. \tag{6.59}$$

If ξ is normal with density function $\alpha(\xi)$ we have for the density function of y

$$\begin{aligned} p(y) &= \alpha(\xi)\left|\frac{\mathrm{d}\xi}{\mathrm{d}y}\right| \\ &= \frac{1}{\sqrt{(2\pi)}}\exp[-\tfrac{1}{2}\{\gamma+\delta g(y)\}^2]\delta g'(y) \end{aligned} \tag{6.60}$$

where $g'(y) = \mathrm{d}g(y)/\mathrm{d}y$. We shall examine three types of system:

(1) The S_L or lognormal system: $g(y) = \log y$.
(2) The S_B (bounded range) system: $g(y) = \log\{y/(1-y)\}$.
(3) The S_U (unbounded range) system: $g(y) = \sinh^{-1} y = \log\{y + \sqrt{(y^2+1)}\}$.

Other systems, of course, are possible. The variety of shapes given by these three is, however, quite as great as that of the Pearson system, for the S_B and S_U systems occupy non-overlapping regions covering the whole of the (β_1, β_2) plane; the S_L system defines the curve that separates them—see Fig. 6.2.

The lognormal distribution

6.29 We have

$$\xi = \gamma + \delta \log y. \tag{6.61}$$

As ξ goes from $-\infty$ to ∞, y goes from 0 to ∞. We may set $\lambda = 1$ in (6.59) without loss of generality. The distribution has infinitely high-order contact at each terminal

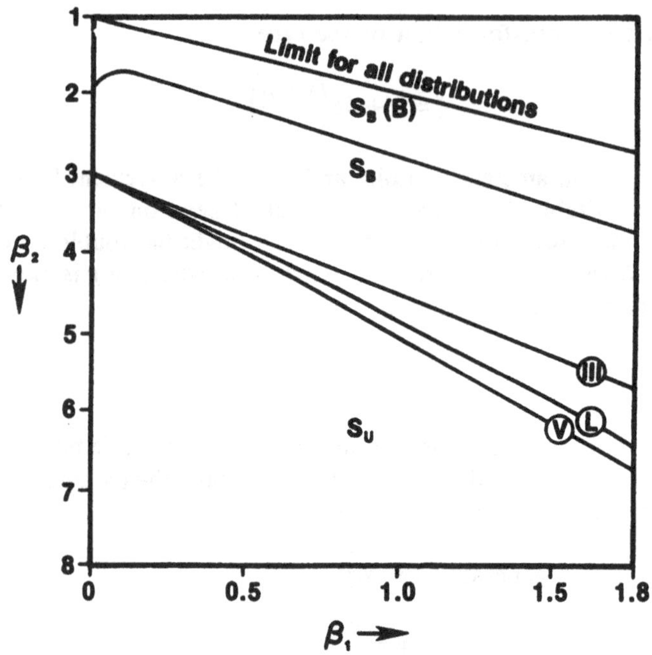

Fig. 6.2 The β_1, β_2 chart for the Johnson systems (adapted from Johnson, 1949a)

Upper limit for all distributions: $\beta_2 - \beta_1 - 1 = 0$ (see Exercise 3.19).
The distributions in the $S_B(B)$ region are bimodal (see (6.71)).
Lognormal curve: denoted by L.
Pearson Type III line and Type V curve: denoted by III, V respectively.

of the range; from (6.60) the density function is

$$p(y) = \frac{\delta}{y\sqrt{(2\pi)}} \exp\{-\tfrac{1}{2}(\gamma + \delta \log y)^2\}, \qquad 0 \leqslant y < \infty.$$

The limit, for any positive r, is

$$\lim_{y \to 0} y^r p(y) = \lim \frac{\delta e^{-\frac{1}{2}\gamma^2}}{\sqrt{(2\pi)}} \frac{y^{r-1-\gamma\delta}}{\exp\{\frac{1}{2}(\delta \log y)^2\}} = 0$$

and

$$\lim_{y \to 0} y^r p(y) = \lim \frac{\delta e^{-\frac{1}{2}\gamma^2}}{\sqrt{(2\pi)}} \frac{y^{r-1-\gamma\delta}}{y^{\frac{1}{2}\delta^2 \log y}} = 0.$$

Also the distribution is unimodal, for

$$\frac{dp(y)}{dy} = \frac{\delta}{\sqrt{(2\pi)}} \exp\{-\tfrac{1}{2}(\gamma + \delta \log y)^2\}\left\{-\frac{1}{y^2} - \frac{\delta}{y^2}(\gamma + \delta \log y)\right\},$$

which, apart from the values $y = 0$ and $y \to \infty$, vanishes where

$$1 + \delta(\gamma + \delta \log y) = 0,$$

i.e. at the single modal value $y=\exp\left\{-\frac{\gamma}{\delta}-\frac{1}{\delta^2}\right\}$.

6.30 For the moments of y about zero we have

$$\begin{aligned}\mu_r'(y) &= \int_0^\infty y^r p(y)\mathrm{d}y \\ &= \frac{1}{\sqrt{(2\pi)}}\int_{-\infty}^{\infty}\exp\left\{\frac{r(\xi-\gamma)}{\delta}\right\}\exp(-\tfrac{1}{2}\xi^2)\mathrm{d}\xi \\ &= \exp\left(\frac{1}{2}\frac{r^2}{\delta^2}-\frac{r\gamma}{\delta}\right), \end{aligned} \tag{6.62}$$

using (3.15). Thus we find, putting $\omega=\exp(1/2\delta^2)$, $\rho=\exp(-\gamma/\delta)$,

$$\left.\begin{aligned}\mu_1' &= \omega\rho \\ \mu_2 &= \omega^2\rho^2(\omega^2-1) \\ \mu_3 &= \omega^3\rho^3(\omega^2-1)^2(\omega^2+2) \\ \mu_4 &= \omega^4\rho^4(\omega^2-1)^2(\omega^8+2\omega^6+3\omega^4-3),\end{aligned}\right\} \tag{6.63}$$

and

$$\beta_1 = (\omega^2-1)(\omega^2+2)^2 \tag{6.64}$$

$$\beta_2-3 = (\omega^2-1)(\omega^6+3\omega^4+6\omega^2+6). \tag{6.65}$$

The equations (6.64–5) determine the curve L in Fig. 6.2, which lies below the Type III line because the equations give $(2\beta_2-3\beta_1-6)=\omega^4(\omega^2-1)(2\omega^2+3)>0$, and also since K at (6.10) >1. L lies above the Type V curve, so that it is always in the Pearson Type VI region in Fig. 6.1.

The median is the value of y corresponding to $\xi=0$, since the transformation is monotonic, i.e. from (6.61) the median is exp $(-\gamma/\delta)$. Thus the lognormal distribution has mode, median and mean in that order, the spacings between them being approximately 2:1 in ratio, as at (2.13). The skewness is always positive and the distribution is always leptokurtic. In general the density of y will rise from zero to a maximum and then tail off more slowly as y tends to infinity. The distribution is not uniquely determined by its moments—cf. Exercise 6.21—since the analytic expansion (4.38) does not hold here, and the c.f. cannot be obtained from them.

6.31 The values of μ_2 and μ_3 determine ω and ρ, and hence δ and γ. Thus, apart from the location and scale parameters μ and σ, we may fit a lognormal distribution from the second and third moments of the x-distribution. The simplest procedure, suggested by Wicksell (1917), is as follows:

Using (6.64) we find $t\equiv(\omega^2-1)^{\frac{1}{2}}$ as the positive root of

$$t^3+3t-\sqrt{\beta_1}=0. \tag{6.66}$$

ρ is found from μ_2 and t, and hence γ and δ are found from the definitions of ω and ρ. Yuan (1933) has given tables to facilitate the solution of (6.66).

The moments of x are identical with those of y, apart from their means, which differ by the location parameter μ determined, using (6.63), by

$$\mu = \mu_1'(x) - \frac{\sqrt{\mu_2}}{t}. \tag{6.67}$$

For the three parameter lognormal (μ, γ, δ), Hill (1963) shows that the likelihood function (cf. **8.18**) becomes infinite as μ approaches $x_{(1)}$, the smallest order statistic, leading to inadmissible estimates. Nevertheless, Cohen and Whitten (1980) show that "local" ML estimators based upon the solution to the three first order derivatives of the log-likelihood function generally perform well, as do estimators based on the smallest order statistic and the first two moments after transformation.

The local ML estimators are given by solving the following expression for μ:

$$\left(\frac{\gamma}{\delta} + \frac{1}{\delta^2}\right) + \frac{\sum z_i u_i}{\sum u_i} = 0$$

where $z_i = \ln\ (x_i - \mu)$ and $u_i = (x_i - \mu)^{-1}$, together with

$$-\gamma/\delta = \left(\sum z_i/n\right) = \bar{z} \quad \text{and} \quad n/\delta^2 = \sum (z_i - \bar{z})^2.$$

For further discussion, see Chapter 2 by Shimizu in Crow and Shimizu (eds., 1988).

For a more extensive study of the history, theory and applications of the distribution, reference may be made to Aitchison and Brown (1957) and to Crow and Shimizu (eds, 1988).

Example 6.5 (Fitting the lognormal distribution)

For the purposes of comparison we will fit a lognormal distribution to the bean data used in Examples 6.1-2. From these, we see that the distribution has negative skewness, whereas (6.64) is positive. We will therefore fit a lognormal distribution with the sign reversed, i.e. so that lower values of ξ correspond to higher values of x.

From Example 6.1 we then have $\sqrt{\beta_1} = +0.910\,569$ and (6.66) becomes

$$t^3 + 3t - 0.910\,569 = 0.$$

The positive root is $t = 0.294\,968$. Also from Example 6.1, $\mu_2 = 3.238\,425$. Then from the second equation of (6.63) we have $\rho = 5.8516$. We notice in passing that from (6.65) we shall then have, for ξ, $\beta_2 = 4.510$ as against the β_2 found for x in Example 6.1 of 4.863. Our transformation brings the values of β_2 roughly into agreement but not completely so.

We then have

$$1/(2\delta^2) = \log_e \omega = \tfrac{1}{2}\log_e(1 + t^2) = 0.041\,718$$
$$-\gamma/\delta = \log_e \rho = 1.7667$$

giving

$$\delta = 3.462, \qquad \gamma = -6.116.$$

The unit (length of intervals) is $\frac{1}{2}$ millimetre and hence, from the data of Example 6.1, the variate x has a mean $14.5 - \frac{1}{2}(0.190\,78) = 14.405$. The location parameter μ is by (6.67) equal to

$$14.405 + \tfrac{1}{2}\sqrt{\mu_2/t} = 17.455,$$

the sign here again being changed because the start of the distribution is above the mean, and the $\frac{1}{2}$ mm unit again being taken into account. Thus the fitted distribution starts at 17.455 mm.

Column (7) of Table 6.1 in **6.20** gives the corresponding expected frequencies. The fit is not particularly good, being rather worse than that of Type IV but better than those of the Edgeworth series.

Exercises 6.13–15 apply the transformation (6.61) to the Gamma and the two kinds of Beta distributions.

Johnson's S_B system

6.32 Distributions of the S_B system with $g(y) = \log\{y/(1-y)\}$ have the variate transformation

$$\xi = \gamma + \mathrm{d}\log\frac{x-\mu}{\lambda+\mu-x}. \tag{6.68}$$

This may itself be regarded, of course, as a simple transformation of the lognormal $\xi = \gamma + \delta\log(x-\mu)$ with $y = 1 - \mu/x$. We may also write

$$\xi = \gamma + \delta\log\{y/(1-y)\} = \gamma + 2\delta\tanh^{-1}(2y-1), \tag{6.69}$$

the latter form enabling numerical transformations to be made with the aid of tables of inverse hyperbolic tangents. The density function of y is given by

$$p(y) = \frac{\delta}{\sqrt{(2\pi)}}\frac{1}{y(1-y)}\exp\left\{-\frac{1}{2}\left(\gamma + \delta\log\frac{y}{1-y}\right)^2\right\}, \qquad 0 \leqslant y \leqslant 1. \tag{6.70}$$

It may be shown without difficulty that this has contact of infinitely high order at both terminals of its range.

6.33 The density function may, however, be bimodal. For the modal values of y we find, on differentiating (6.70),

$$2y - 1 = \delta\left(\gamma + \delta\log\frac{y}{1-y}\right).$$

Putting $y = \frac{1}{2}(y'+1)$ we find

$$y' - \gamma\delta = \delta^2\log\frac{1+y'}{1-y'}.$$

The straight line, in Cartesian co-ordinates (y', u), $u = y' - \gamma\delta$ meets

$$u = \delta^2\log\{(1+y')/(1-y')\}$$

in one or three points. In the first case the distribution is unimodal, in the second bimodal. It may be shown that the necessary and sufficient conditions for bimodality are

$$\delta < 1/\sqrt{2}, \qquad |\gamma| < \delta^{-1}\sqrt{(1-2\delta^2)} - 2\delta \tanh^{-1}\sqrt{(1-2\delta^2)}. \qquad (6.71)$$

J. Draper (1952) showed that the lower boundary of the bimodal region ($S_B(B)$ in Fig. 6.2) is very close to the lower boundary of the $I(U)$ region in Fig. 6.1.

6.34 The moments of the distribution are very difficult to determine in a manageable form. In the following example, the parameters are estimated by exactly fitting the frequencies in the two end-groups.

Example 6.6 (Fitting the S_B distributions; Johnson, 1949a)

Consider the data of Table 1.13 (cloudiness at Greenwich). On the assumption that degree of cloudiness stated as 0, 1, 2, ... can be regarded as grouping −0.5 to 0.5, 0.5 to 1.5, etc., we may fix the location parameter μ at −0.5 and the scale parameter λ at 11. We then choose γ and δ so as to provide exact agreement in the two end-groups.

Table 6.2 Various systems fitted to the cloudiness data of Table 1.13

Degree of cloudiness	Observed frequencies	$S_B(1)$	$S_B(2)$	Pearson Type I
0	320	320.0	320.0	321.7
1	129	100.9	120.9	121.5
2	74	73.9	72.0	75.1
3	68	63.8	57.5	61.4
4	45	59.8	52.1	56.0
5	45	59.9	51.6	55.2
6	55	63.4	54.9	57.8
7	65	72.0	63.9	65.5
8	90	90.0	85.5	83.2
9	148	135.4	160.7	139.6
10	676	676.0	676.0	678.0
TOTALS	1715	1715.1	1714.9	1715.0

Thus, y for the first group is 1/11 and $\log_e\{y/(1-y)\}$ becomes −2.302 585. The normal variate value corresponding to a value for the d.f. of 320/1715 = 0.186 589 is −0.8965 and hence, from (6.69),

$$-0.8965 = \gamma - 2.3026\delta.$$

Likewise for the eleventh group, $y = 10/11$ and the normal variate value corresponding to a d.f. value of 1639/1715 is 0.268 4. Hence

$$0.2684 = \gamma + 2.3026\delta.$$

These two equations yield

$$\gamma = 0.3110, \qquad \delta = 0.25166.$$

The reader may verify that for these values (6.71) is satisfied so that the distribution is bimodal. The large frequencies in the two end-groups include the modes of the distribution, so Table 6.2 shows only the U-shaped section of the distribution between the modes; the fitted distribution is referred to as S_B (1).

The distribution S_B (2) was fitted in the same way except that it was assumed that the groups were 0 to 0.5, 0.5 to 1.5, etc., and μ was taken as 0, λ as 10. The results of fitting a Pearson Type I distribution of **6.5** are also shown. All three distributions fail to give sufficiently small expected frequencies in the antimodal centre of the range.

Johnson's S_U system

6.35 The S_U system of distributions is defined by

$$\xi = \gamma + \delta \sinh^{-1} y \tag{6.72}$$

and for the density function we have

$$p(y) = \frac{\delta}{\sqrt{(2\pi)}} \frac{1}{\sqrt{(y^2+1)}} \exp[-\tfrac{1}{2}[\gamma + \delta \log\{y + \sqrt{(y^2+1)}\}]^2], \quad -\infty < y < \infty. \tag{6.73}$$

Again there is high-order contact at the ends of the distribution. Exercise 6.18 is to verify that the distributions are unimodal and that the mode lies between the median $\sinh(-\gamma/\delta)$ and zero. Thus the distribution is positively or negatively skewed according as γ is negative or positive.

For the moments we have

$$\mu_r' = \frac{1}{\sqrt{(2\pi)}} \int_{-\infty}^{\infty} \exp(-\tfrac{1}{2}\xi^2) \cdot \frac{1}{2^r} \left\{ \exp\left(\frac{\xi-\gamma}{\delta}\right) - \exp\left(-\frac{\xi-\gamma}{\delta}\right) \right\}^r d\xi,$$

leading to

$$\left.\begin{aligned} \mu_1' &= -\omega \sinh \Omega \\ \mu_2 &= \tfrac{1}{2}(\omega^2-1)(\omega^2 \cosh 2\Omega + 1) \\ \mu_3 &= -\tfrac{1}{4}\omega^2(\omega^2-1)^2\{\omega^2(\omega^2+2)\sin 3\Omega + 3\sinh\Omega\} \\ \mu_4 &= \tfrac{1}{8}(\omega^2-1)^2\{\omega^4(\omega^8+2\omega^6+3\omega^4-3)\cosh 4\Omega \\ &\quad + 4\omega^4(\omega^2+2)\cosh 2\Omega + 3(2\omega^2+1)\} \end{aligned}\right\} \tag{6.74}$$

where $\omega = \exp(1/2\delta^2)$, $\Omega = \gamma/\delta$. The mean is ω times the median, and μ_3 has opposite sign to Ω, i.e. to γ, verifying the remark on skewness above. A method of fitting is illustrated in the following example.

Example 6.7 (Fitting the S_U distribution)

Consider once more the bean-length distribution in Examples 6.1-2 and 6.5. We found there (now using mm units)

$$\mu_1' = 14.5 - 0.191/2 = 14.405 \text{ mm}, \qquad \beta_1 = 0.829,$$

$$\mu_2 = 3.238\,425/4 = (0.9036)^2, \qquad \beta_2 = 4.863.$$

Using these values, Johnson (1949a) obtained the estimates $\delta = 2.64$, $\gamma = 2.38$. We find, using the relations

$$\mu_1' = \mu_1'\{(x-\mu)/\lambda\} = 1.1029, \qquad \mu_2 = \mu_2\left(\frac{x-\mu}{\lambda}\right) = (0.5948)^2,$$

that

$$\lambda = 0.9036/0.5948 = 1.5192, \qquad \mu = 14.405 + 1.1029\lambda = 16.074.$$

Column (8) of Table 6.1 in **6.20** shows the expected frequencies given by an S_U distribution with these values of the constants. The fit seems about as good as the Pearson Type IV distribution.

6.36 Tables to facilitate fitting of the S_B distributions are given by Johnson and Kitchen (1971a, b) and, using quantiles, by Bukač (1972); see also Mage (1980). Bowman *et al.* (1981) give approximations that allow explicit evaluation of estimators based on moments.

Johnson (1965) gives tables for the S_U distribution, reproduced in Volume II of the *Biometrika Tables*. Bowman and Shenton (1980) give an approximation that allows the estimates to be computed directly from the observed values of $\sqrt{\beta_1}$ and β_2.

For both systems, I. D. Hill *et al.* (1976) give computer algorithms for fitting by moments. Wheeler (1980) describes equally efficient methods using quantiles.

One way to select from among the three Johnson systems is to use the (β_1, β_2) chart; see Fig. 6.2. The observed pair of values (0.829, 4.863) in Table 6.1 (Example 6.1) lies inside the S_U region; the results of Examples 6.5 and 6.7 appear to support this choice.

Bowman and Shenton (1988, 1989) generalize an earlier estimation procedure developed by Slifker and Shapiro (1980). Bowman and Shenton consider all possible sets of four symmetrically located quantities at proportions $(\beta, \alpha, 1-\alpha, 1-\beta)$ and estimate the parameters by equating the observed values to their expectations, readily available from tables; see **14.21**. They then select the combination that maximizes the "closeness to normality" of the transformed data, as measured by a normal goodness-of-fit statistic; they use the Shapiro–Wilk statistic (cf **30.43**, Volume 2A), although other measures could also be used.

Tadikamalla and Johnson (1982) use the three transformations in **6.28** to develop a system based on the logistic distribution; see Exercise 6.28. Johnson (1954) had done the same using the double exponential distribution.

Burr's distributions

6.37 The three systems we have considered in this chapter all attempt to fit distributions to the density function; and we have seen that in some cases, at least, this involves troublesome summation or integration when we require the d.f., as we usually do, to calculate fitted frequencies for comparison with observation. The question naturally arises whether we could not directly fit to the d.f. and obtain the density function, if we need it, by the relatively simple process of differentiation.

Such an approach has been considered by Burr (1942) and Hatke (1949). We consider the differential equation

$$\mathrm{d}F = F(1-F)g(x)\,\mathrm{d}x, \tag{6.75}$$

where $g(x)$ is some convenient function, which must be non-negative in $0 \leqslant F \leqslant 1$ and the range of x. (6.75) may be written

$$\mathrm{d}F\left(\frac{1}{F} + \frac{1}{1-F}\right) = g(x)\,\mathrm{d}x,$$

and integration gives the solution

$$F(x) = [1 + \exp\{-G(x)\}]^{-1}, \tag{6.76}$$

where

$$G(x) = \int_{-\infty}^{x} g(t)\,\mathrm{d}t.$$

When $g(t) = 1$, we see that (6.76) gives rise to the logistic distribution of Exercise 4.21. The general form of (6.76) most commonly used is

$$F(x) = 1 - \frac{1}{(1+x^{\alpha})^{\beta}}, \qquad 0 \leqslant x < \infty; \quad \alpha, \beta > 0, \tag{6.77}$$

although many others are possible; see Exercise 6.19. (6.77) may be derived from the Weibull distribution as a mixture; see **7.30**. Burr and Cislak (1968) and Rodriguez (1977) study the shape characteristics of (6.77) and show that it covers a wide range of (β_1, β_2) values from just inside the Type I region to well inside the Type IV region in Fig. 6.1.

Advantages of (6.77) include the fact that only a single functional form is involved and that $x = F^{-1}(p)$ is available in closed form. For further discussion of the Burr system, see Tadikamalla (1980).

6.38 Burr's own method of fitting was to consider *cumulative moments* typified by

$$M_j(a) = \int_{a}^{\infty} (x-a)^j\{1-F(x)\}\,\mathrm{d}x - \int_{-\infty}^{a} (x-a)^j F(x)\,\mathrm{d}x \tag{6.78}$$

and to equate these quantities, calculated for the data, to the theoretical values expressed in terms of the parameters of the distribution; in fact, to do with cumulative moments what Pearson did with ordinary moments. This can be tedious, although tables were provided by Burr and Hatke for the purpose. As for the Johnson family, the use of selected quantiles appears to be the most direct method of fitting.

Another approach to fitting distributions is briefly mentioned at the end of **14.10** below.

Fitting by frequency moments

6.39 Fitting distributions by the method of moments has certain disadvantages, especially when the range is unlimited and observed data are only a sample from the possible observations; outlying members and "tail" frequencies exert an undue effect on the calculated moments. Another possibility is to equate observed and theoretical values of the so-called *frequency moments* or *probability moments* typified by

$$\omega_r = \int_{-\infty}^{\infty} \{f(x)\}^r \, dx, \tag{6.79}$$

in which the variate and the density have exchanged roles. This method has been examined at length by Sichel (1949), especially in one of the cases where the Pearson distributions are somewhat inadequate. The point concerning efficiency in estimation, made in **6.13** for Pearson distributions, applies here also. Further, grouping corrections are required, and the method has not been much tried.

Moriguti (1952) has shown that for any continuous distribution with variance σ^2

$$\sigma^{r-1}\omega_r \geqslant \frac{2r}{3r-1}\left[\left(\frac{r-1}{3r-1}\right)^{\frac{1}{2}} \Big/ B\left(\frac{r}{r-1}, \frac{1}{2}\right)\right]^{r-1}. \tag{6.80}$$

Ord (1968b) shows that ω_2 is asymptotically fully efficient as an estimator of θ for distributions with densities of the form $f(x) \propto (\theta^j + x^j)^{-1}$, which include the Cauchy.

6.40 The fitting of distributions to observational data has a certain intrinsic interest that is apt to outrun its statistical usefulness. The possible reasons for fitting may be as follows:

(a) to validate a hypothesized stochastic model;
(b) to prove an approximation to a (sampling) distribution;
(c) to describe an empirical distribution in a concise fashion, often for use in a simulation study.

If none of these elements is present, the value of the exercise may be questionable.

6.41 Although several different families have been discussed in this chapter, we may observe that they cover substantially overlapping regions in the (β_1, β_2) plane. To determine how different these systems of distributions are, E. S. Pearson *et al.* (1979) examined the percentage points of eight different systems fitted by the first four moments; they provide extensive tables of the 25, 10, 5, 2.5, 1, 0.5, and 0.25 upper and lower percentage points of the distributions for a grid of $\beta_1^{\frac{1}{2}}$ and β_2 values. Their results show that the upper and lower 5 percent points are remarkably consistent across all the families. Thus, if the aim is to approximate the tail area of a distribution, the quality of the approximation may be relatively insensitive to the family selected.

Distributions specified by hazard rates

6.42 In **5.34** we defined the hazard rate as

$$h(t) = f(t)/G(t) \tag{6.81}$$

where $G(t) = 1 - F(t)$. A distribution is an increasing hazard rate (IHR) distribution if $h(t)$ is a non-decreasing function of t with $h(t+u) > h(t)$ for at least some values of $u > 0$. This property is clearly of interest in the context of reliability or survival studies, where the properties of the residual lifetime distribution are important.

An average hazard rate may be defined as

$$\mu(x) = \int_0^x h(t)\mathrm{d}t/x, \tag{6.82}$$

and a distribution has an increasing hazard rate on average (IHRA) if $\mu(x)$ is non-decreasing in x. From (6.82), we have that

$$\begin{aligned} x(x+y)\,[\mu(x+y)-\mu(x)] &= x\int_0^{x+y} h(t)\ \mathrm{d}t - (x+y)\int_0^x h(t)\ \mathrm{d}t \\ &= x\int_x^{x+y} h(t)\ \mathrm{d}t - y\int_0^x h(t)\ \mathrm{d}t \\ &= M(x),\ \text{say.} \end{aligned}$$

If the distribution is IHR, $h(x+u) \geqslant h(x-v)$ for all $u, v > 0$, so that

$$M(x) \geqslant x\,[y\,h(x)] - y\,[x\,h(x)] \geqslant 0,$$

which implies that $\mu(x+y) \geqslant \mu(x)$; that is, the distribution must be IHRA, a result first demonstrated by Bryson and Siddiqui (1969).

6.43 A related class of distributions is given by the requirement that a new unit should be better than a used one (NBU). Thus, if a unit survives until time y, the distribution is said to be NBU if and only if

$$P(T \geqslant x+y \,|\, T \geqslant y) \leqslant P(T \geqslant x), \tag{6.83}$$

where T denotes the lifetime of the component; (6.83) may be rewritten as

$$G(x+y)/G(y) \leqslant G(x), \qquad \text{or}$$

$$G(x+y) \leqslant G(x)\,G(y). \tag{6.84}$$

Finally, a distribution is new better than used in expectation (NBUE) when

$$E(T \,|\, T \geqslant y) \leqslant E(T). \tag{6.85}$$

The NBUE class may be expected to be broader than NBU since (6.85) appears to be a weaker condition. We now establish that this is indeed so.

Recall from Exercise 2.24 that

$$E(T) = \int_0^\infty G(x)\ \mathrm{d}x$$

so that

$$\begin{aligned} E(T \,|\, T \geqslant y) &= \int_y^\infty G(u)\ \mathrm{d}u/G(y) \\ &= \int_0^\infty G(x+y)\ \mathrm{d}x/G(y). \end{aligned}$$

Thus

$$E(T) - E(T|T \geqslant y) = \int_0^\infty \left[G(x) - \frac{G(x+y)}{G(y)}\right] dx$$

and, if (6.94) holds, (6.85) must hold.

6.44 The IHR and NBU classes do not have any transparent relationship. However, following Barlow and Proschan (1975, pp. 158–59) we now show that the NBU class includes the IHRA class.

Since $\mu(x)$ is non-decreasing in x, it follows from (6.82) that

$$(x+y)\mu(x+y) - x\mu(x) - y\mu(y) \geqslant 0,$$

or

$$\int_y^{x+y} h(t) \quad dt - \int_0^x h(t) \quad dt \geqslant 0. \tag{6.86}$$

But $\int h(t)\,dt = -\ln G(t)$, so that (6.86) yields

$$-\ln G(x+y) + \ln G(y) + \ln G(x) - \ln G(0) \geqslant 0$$

or

$$G(x)\,G(y) \geqslant G(x+y). \tag{6.87}$$

Collecting the results together, we have established that

$$\text{IHR} \subset \text{IHRA} \subset \text{NBU} \subset \text{NBUE}. \tag{6.88}$$

Likewise, if we consider decreasing hazard rates (DHR) and new worse than used (NWU) we may show that

$$\text{DHR} \subset \text{DHRA} \subset \text{NWU} \subset \text{NWUE}. \tag{6.89}$$

Although these results have been demonstrated for continuous distributions, the extension to discrete distributions is straightforward. In particular, it may be shown that the Poisson and binomial distributions are IHR, whereas the negative binomial is IHR (DHR) if $k > 1\ (< 1)$.

EXERCISES

6.1 Show that the following are members of the Pearson system of distributions and sketch them for a few values of the constants:

(a) the normal distribution: $B_1 = B_2 = 0$, $B_0 < 0$; $\beta_1 = 0$, $\beta_2 = 3$;

$$f = \frac{1}{\sigma\sqrt{(2\pi)}}\exp\left(-\frac{x^2}{2\sigma^2}\right), \qquad -\infty < x < \infty.$$

(b) the Type II distribution: $B_2 > 0$, $B_1 = 0$, $B_0 < 0$; $\beta_1 = 0$, $\beta_2 < 3$;

$$f = \frac{1}{aB(m+1, m+1)}\left(1 - \frac{x^2}{a^2}\right)^m \qquad -a \leqslant x \leqslant a.$$

(c) the Type VII distribution (Student's): $B_2 > 0$, $B_1 = 0$, $B_0 > 0$; $\beta_1 = 0$, $\beta_2 > 3$;

$$f = \frac{1}{aB(\frac{1}{2}, m - \frac{1}{2})}\left(1 + \frac{x^2}{a^2}\right)^{-m}, \qquad -\infty < x < \infty.$$

6.2 With a simple transformation of the variate when necessary, assign each of the following distributions to one of Pearson's types:

$$\mathrm{d}F = k\mathrm{e}^{-\frac{1}{2}\chi^2}\chi^{\nu-1}\,\mathrm{d}\chi, \qquad \mathrm{d}F = k(1 - r^2)^{\frac{1}{2}(n-4)}\,\mathrm{d}r,$$

$$\mathrm{d}F = \frac{k\,\mathrm{d}t}{(1 + t^2/\nu)^{\frac{1}{2}(\nu+1)}}, \qquad \mathrm{d}F = k\eta^{p-2}(1 - \eta^2)^{\frac{1}{2}(N-p-2)}\,\mathrm{d}\eta.$$

(All these distributions are important in the theory of sampling; see Chapter 16)

6.3 Verify that for the Pearson distributions defined by (6.7), the range is unlimited in both directions if $B_0 + B_1X + B_2X^2$ has no real roots; limited in one direction if the roots are real and of the same sign; and limited in both directions if the roots are real and of opposite sign.

6.4 Show that the Pearson Type VI distribution of 6.7 may be written

$$y = y_0\left(1 - \frac{x^2}{a^2}\right)^{-m}\exp\{-\nu\tanh^{-1}(x/a)\}$$

and discuss the relationship with the Type IV distribution of 6.8.

6.5 Show that for a Pearson Type III distribution of 6.9, $2\beta_2 - 3\beta_1 - 6 = 0$ and hence that the criterion κ of (6.10) is infinite.

6.6 In 6.7, if the real roots of $B_0 + B_1X + B_2X^2$ are equal, show that the criterion $K = \kappa = 1$ and that the corresponding Pearson distribution (Type V) may be written

$$f = \frac{\gamma^{p-1}}{\Gamma(p-1)}x^{-p}\mathrm{e}^{-\gamma/x}, \qquad 0 \leqslant x < \infty,$$

of which (4.70) is the special case $p = \frac{3}{2}$, $\gamma = \frac{1}{2}$.

6.7 Show that for a Pearson distribution, the c.f. satisfies the relation

$$b_2\theta\frac{\mathrm{d}^2\phi}{\mathrm{d}\theta^2} + (1 + 2b_2 + b_1\theta)\frac{\mathrm{d}\phi}{\mathrm{d}\theta} + (a + b_1 + b_0\theta)\phi = 0,$$

where $\theta = it$. Deduce the recurrence relation (6.3) between the moments.

Show also that the c.g.f. satisfies the relation

$$b_2\theta\left\{\frac{d^2\psi}{d\theta^2}+\left(\frac{d\psi}{d\theta}\right)^2\right\}+(1+2b_2+b_1\theta)\frac{d\psi}{d\theta}+(a+b_1+b_0\theta)=0.$$

Hence show that the cumulants obey the recurrence relations

$$\{1+(r+2)b_2\}\kappa_{r+1}+rb_1\kappa_r+rb_2\left\{\kappa_1\kappa_r+\binom{r-1}{1}\kappa_2\kappa_{r-1}+\binom{r-1}{2}\kappa_3\kappa_{r-2}+\dots\right.$$
$$\left.+\binom{r-1}{j}\kappa_{j+1}\kappa_{r-j}+\dots+\binom{r-1}{1}\kappa_{r-1}\kappa_2+\kappa_r\kappa_1\right\}=0.$$

6.8 *The Charlier Type B series.* Defining

$$\gamma(\lambda,x)=e^{-\lambda}\lambda^x/x!,\qquad \nabla\gamma(\lambda,x-1)=\gamma(\lambda,x)-\gamma(\lambda,x-1),$$

and

$$G_r(\lambda,x)=\left\{\frac{d^r}{d\lambda^r}\gamma(\lambda,x)\right\}\bigg/\gamma(\lambda,x),$$

show that

$$G_r(\lambda,x)=\{(-1)^r\nabla^r\gamma(\lambda,x-r)\}/\gamma(\lambda,x)$$

and

$$\sum_{x=0}^{\infty}(G_rG_s\gamma)=0,\qquad r\neq s,$$
$$=r!/\lambda^r,\ r=s.$$

Hence, if f can be expressed as a sum

$$f=\sum_j(b_jG_j\gamma),$$

show that

$$b_j=\frac{\lambda^j}{j!}\sum_{x=0}^{\infty}(fG_j).$$

6.9 Show that no distribution that is not completely determined by its moments can be expanded in a convergent Type A series.

6.10 Show that if y is a function of x that it is desired to represent approximately by the form

$$y=\sum_{j=0}^{r}c_jH_j(x)\alpha(x),$$

then the values of the cs appropriate to the expansion of y in this form minimize

$$\int_{-\infty}^{\infty}\left\{\frac{y-\sum_{j=0}^{r}c_jH_j(x)\alpha(x)}{\sqrt{\alpha(x)}}\right\}^2dx.$$

6.11 If (6.42) with $\kappa_r=0$, $r>4$, adequately represents a distribution that is assumed standardized normal, show that the error in the d.f. resulting from the normality assumption is

$$e(x)=\int_{-\infty}^{x}\{f(u)-\alpha(u)\}\,du=-\alpha(x)\left\{\frac{\kappa_3}{6}H_2(x)+\frac{\kappa_4}{24}H_3(x)+\frac{\kappa_3^2}{72}H_5(x)\right\},$$

and that $e(x)$ has local maxima in its absolute magnitude at those roots in x of

$$\frac{\kappa_3}{6}H_3(x)+\frac{\kappa_4}{24}H_4(x)+\frac{\kappa_3^2}{72}H_6(x)=0$$

where

$$\frac{\kappa_3}{6}H_4(x)+\frac{\kappa_4}{24}H_5(x)+\frac{\kappa_3^2}{72}H_7(x)$$

has the same sign as $e(x)$. The largest such local maximum is a bound for the error in the d.f.

6.12 If $f(x)$ and $g(x)$ are two differentiable density functions with cumulants denoted respectively by κ and κ' show that formally

$$f(x)=\exp\left\{\sum_{j=0}^{\infty}\frac{\kappa_j-\kappa_j'}{j!}\left(-\frac{d}{dx}\right)^j\right\}g(x).$$

6.13 If y is distributed in the Gamma form with p.d.f.

$$f=\frac{1}{\Gamma(\lambda)}y^{\lambda-1}e^{-y},\qquad 0\leqslant y<\infty,$$

and a new variate ξ is defined by

$$\xi=\gamma+\delta\log y,$$

show that for the cumulants of ξ we have

$$\kappa_1(\xi)=\gamma+\delta\frac{d}{d\lambda}\log\Gamma(\lambda),$$

$$\kappa_r(\xi)=\delta^r\frac{d^r}{d\lambda^r}\log\Gamma(\lambda),\quad r>1.$$

Hence show that

$$\beta_1(\xi)=\frac{1}{\lambda}+O(\lambda^{-2}),$$

$$\beta_2(\xi)=3+\frac{2}{\lambda}+O(\lambda^{-2})$$

and that the corresponding coefficients for y,

$$\beta_1(y)=\frac{4}{\lambda},$$

$$\beta_2(y)=3+\frac{6}{\lambda}.$$

(Johnson, 1949a)

6.14 If y is distributed in the second Beta form with p.d.f.

$$f=\frac{\Gamma(\tau)}{\Gamma(\tau-\nu)\Gamma(\nu)}y^{\nu-1}(1+y)^{-\tau},\qquad 0\leqslant y<\infty.$$

show that if $\xi=\gamma+\delta\log y$ we have, for large ν, $\tau-\nu$,

$$\beta_1(\xi)=\frac{1}{\nu}-\frac{4}{\tau}+\frac{1}{\tau-\nu},$$

$$\beta_2(\xi)=3+\frac{2}{\nu}-\frac{6}{\tau}+\frac{2}{\tau-\nu},$$

and

$$\beta_1(y) = \frac{4}{\nu} - \frac{4}{\tau} + \frac{16}{\tau - \nu},$$

$$\beta_2(y) = 3 + \frac{6}{\nu} - \frac{6}{\tau} + \frac{30}{\tau - \nu}.$$

(Johnson, 1949a)

6.15 If R is distributed in the symmetrical form

$$f = \frac{\{\Gamma(\frac{1}{2}n-1)\}^2}{\Gamma(n-2)} R^{\frac{1}{2}(n-4)}(1-R)^{\frac{1}{2}(n-4)}, \qquad 0 \leqslant R \leqslant 1,$$

show that for

$$\zeta = \tfrac{1}{2}\sqrt{(n-3)} \log\{R/(1-R)\}$$

$$\beta_1(\zeta) = 0,$$

$$\beta_2(\zeta) = 3 + \frac{\frac{1}{2}\left[\frac{d^2}{d\lambda^2}\log\Gamma(\lambda)\right]_{\lambda=\frac{1}{2}n-1}}{[\log\Gamma(\lambda)]^2_{\lambda=\frac{1}{2}n-1}}.$$

(Johnson, 1949a)

6.16 Show that the parameters of the reduced Pearson family (6.19) are determined by the relationships

$$b_2 = (\alpha_3 - 2\alpha_2 + \alpha_1)/D$$
$$b_1 = (\alpha_2 - \alpha_1) - b_2(3\alpha_2 - 2\alpha_1)$$
$$a = 2\alpha_1 - \alpha_2 + b_2(4\alpha_1 - 3\alpha_2),$$

where $\alpha_r = \mu_r'/\mu_{r-1}'$ and $D = 4\alpha_3 - 6\alpha_2 + 2\alpha_1$.

6.17 In Exercise 6.16, let $J_1 = \mu_2/\mu_1'^2$ and $J_2 = \mu_3/\mu_2\mu_1'$. Show that

$$J_2 - 2J_1 = 0 \quad \text{for Type III},$$
$$> 0 \quad \text{for Type VI},$$
$$< 0 \quad \text{for Type I},$$

and that, for Type V,

$$\frac{4}{J_2} = \frac{1}{J_1} - 1.$$

Also, verify the values of (6.20) for Types I, III, and VI.

6.18 Verify that the S_U distributions of **6.35** are unimodal and show that they are positively or negatively skewed according as γ is negative or positive.

6.19 Show that the following distribution functions are admissible solutions of equation (6.76), the parameters in all cases being positive:

$F(x) = (1 + e^{-x})^{-\alpha}$, $\quad -\infty < x < \infty$; $\quad$ ($\alpha = 1$ is the logistic distribution of Exercise 4.21.)

$F(x) = (1 + \alpha e^{-\tan x})^{-\alpha}$, $\quad -\frac{1}{2}\pi \leqslant x \leqslant \frac{1}{2}\pi$;

$$F(x) = \left(\frac{2}{\pi}\arctan e^{-x}\right)^{\alpha}, \qquad -\infty < x < \infty;$$

$$F(x) = \left(x - \frac{1}{2\pi}\sin 2\pi x\right)^{\alpha}, \qquad 0 \leqslant x \leqslant 1.$$

(Burr, 1942)

6.20 A variate has zero mean, unit variance and the rth cumulant κ_r is of order $n^{1-\frac{1}{2}r}$, $r \geqslant 2$. By considering an Edgeworth expansion show that the mode, to order n^{-1}, is given by $-\frac{1}{2}\kappa_3$ and the median by $-\frac{1}{6}\kappa_3$. Hence derive the approximate relationship (2.13).

(Haldane, 1942)

6.21 In **6.29**, show that for any $r = 0, 1, 2, 3, \ldots$ and $k = 1, 2, 3, \ldots$,

$$\int_0^{\infty} y^r p(y) \sin\{2\pi k\delta(y + \delta \log y)\}\, dy = 0$$

and hence that the lognormal distribution $p(y)$ is not uniquely determined by its moments.

(Heyde, 1963)

6.22 Defining $\delta = (2\beta_2 - 3\beta_1 - 6)/(\beta_2 + 3)$, show using Exercise 3.19 that $-1 < \delta < 2$.

(Cf. C. C. Craig, 1936a)

6.23 Show that $\alpha(x) = -\int_{-\infty}^{x} D\alpha(y)\, dy$ and hence that for any differentiable function of a standardized normal variable x, $E\{g'(x)\} = E\{xg(x)\}$ if the latter is finite. Hence show that for any $r \geqslant 0$, the moments of x satisfy $\mu_{r+2} = (r+1)\mu_r$, from which if r is an integer we find (5.97) with $\sigma = 1$.

6.24 A translation system of distributions may be based on the logistic distribution, using the same transformations as in the Johnson system **6.28**. If $\xi = \lambda + \delta \sinh^{-1} y$ show that the first four moments about the origin of ξ are

$$\mu_1' = -\theta \operatorname{cosec}\theta \sinh\Omega$$
$$\mu_2' = \theta \operatorname{cosec} 2\theta \cosh 2\Omega - 0.5$$
$$\mu_3' = -0.75\theta \operatorname{cosec} 3\theta \sinh 3\Omega + 0.75\mu_1'$$
$$\mu_4' = 0.5\theta \operatorname{cosec} 4\theta \cosh 4\Omega - \mu_2' - 0.125,$$

where $\theta = \pi/\delta$ and $\Omega = \lambda/\delta$ (Tadikamalla and N. L. Johnson, 1982a, b; 1990).

6.25 Show that the Poisson and binomial distributions have increasing hazard rates (IHR), whereas the negative binomial is IHR if $k > 1$ but has a decreasing rate (DHR) if $k < 1$.

CHAPTER 7

MULTIVARIATE DISTRIBUTIONS

7.1 In Chapter 5 and 6 we have assumed that a single variate x, is to be described by the probability or distribution function. However, it is clear that a number of variates might be considered, as illustrated by the bivariate data in Table 1.15 or the bivariate distributions introduced in **1.32–35**. Generally we shall refer to bivariate distributions for variates (x_1, x_2) and multivariate distributions for $x = (x_1, x_2, \ldots, x_k)'$, $k > 2$. In many cases, no new points of principle arise in going from $k = 2$ to $k > 2$ and we shall often focus upon the bivariate case for reasons of notational simplicity. As in Chapters 5 and 6, we consider only the basic properties of the distributions, leaving their uses to be developed later. This approach is especially true in the case of the multinormal distribution, which is examined in detail in Chapter 15; sampling distributions derived from the multinormal are considered in Chapter 16.

Comprehensive accounts of individual multivariate distributions are given in N. L. Johnson and Kotz (1969, 1972) and Patil *et al.* (1985c) for continuous variates; Kocherlakota and Kocherlakota (1992) and Patil *et al.* (1985a) cover discrete variates.

Basic structure

7.2 Following **1.32–5**, we define the bivariate distribution function as

$$F(x_1, x_2) = \int_{-\infty}^{x_1} \int_{-\infty}^{x_2} dF(u_1, u_2) \tag{7.1}$$

where the bivariate p.f. is

$$dF(x_1, x_2) = f(x_1, x_2)\, dx_1\, dx_2, \tag{7.2}$$

to be interpreted as a probability mass function when the variates are discrete and as a density function when they are continuous. The marginal distribution function for x_1 is given by letting $x_2 \to \infty$ in (7.1) when

$$F_1(x_1) = \int_{-\infty}^{x_1} dF_1(u_1) \tag{7.3}$$

and the conditional distribution for x_2 given x_1 has p.f.

$$f(x_2 | x_1) = \frac{f(x_1, x_2)}{\int_{-\infty}^{\infty} f(x_1, x_2)\, dx_1} \tag{7.4}$$

where the integral is to be interpreted as a summation if x_1 is discrete.

7.3 The moments follow directly as

$$E(x_1^r x_2^s) = \mu'_{rs} = \int_{-\infty}^{\infty}\int_{-\infty}^{\infty} x_1^r x_2^s \, dF(x_1, x_2) \tag{7.5}$$

yielding the means

$$E(x_1) = \mu'_{10} \quad \text{and} \quad E(x_2) = \mu'_{01},$$

and variances

$$\text{var}(x_1) = \mu_{20} = \mu'_{20} - (\mu'_{10})^2 \text{ and var } (x_2) = \mu_{02} = \mu'_{02} - (\mu'_{01})^2.$$

In general, the moments about the mean are denoted by

$$\mu_{rs} = E[(x_1 - \mu'_{10})^r (x_2 - \mu'_{01})^s] \tag{7.6}$$

of which the most important is the covariance, μ_{11}. In turn the correlation is

$$\rho_{12} = \mu_{11}/(\mu_{20}\mu_{02})^{\frac{1}{2}}$$

and we shall set $\rho_{12} = \rho$ when only two variates are being considered. In matrix notation, these expectations are assembled as

$$E(\mathbf{x}) = \boldsymbol{\mu} \text{ and cov } (\mathbf{x}) = E[(\mathbf{x} - \boldsymbol{\mu})(\mathbf{x} - \boldsymbol{\mu})'] = \boldsymbol{\Sigma}. \tag{7.7}$$

7.4 Higher order moments are used primarily to determine the shape of the distribution. Univariate skewness and kurtosis measures were discussed in **3.31–33** and clearly may be used to describe the marginal distributions. Multivariate measures were defined by Mardia (1970c; 1974) as

$$\text{skewness: } \beta_1(k) = E\{[\mathbf{x} - \boldsymbol{\mu})' \boldsymbol{\Sigma}^{-1}(\mathbf{y} - \boldsymbol{\mu})]^3\} \tag{7.8}$$

and

$$\text{kurtosis: } \beta_2(k) = E\{[(\mathbf{x} - \boldsymbol{\mu})' \boldsymbol{\Sigma}^{-1}(\mathbf{x} - \boldsymbol{\mu})]^2\} \tag{7.9}$$

where $\mathbf{x}$ and $\mathbf{y}$ are two independent random vectors from the same distribution with an analogous expression for the sample values. By considering the transformation $\mathbf{z} = \mathbf{Tx}$, it follows that the measures are invariant to linear transformations. In particular when $k = 2$ and $\boldsymbol{\Sigma} = \mathbf{I}$ the measures reduce to

$$\beta_1(2) = \frac{\mu_{30}^2}{\mu_{20}^2} + \frac{\mu_{03}^2}{\mu_{02}^2} + 3\frac{\mu_{21}^2}{\mu_{20}^2\mu_{02}} + 3\frac{\mu_{12}^2}{\mu_{20}\mu_{02}^2} \tag{7.10}$$

and

$$\beta_2(2) = \frac{\mu_{40}}{\mu_{40}^2} + \frac{\mu_{04}}{\mu_{02}^2} + \frac{2\mu_{22}}{\mu_{20}\mu_{02}} \tag{7.11}$$

Only when the variates are independent do the $\beta_j(k)$ reduce to the sums of the univariate measures (constants apart). The binormal case is discussed in Exercise 7.2. Alternative measures are given by Oja (1983).

7.5 As defined in **4.16** the joint characteristic function for $\mathbf{x}$ is

$$\phi(\mathbf{t}) = E[\exp(i\mathbf{t}'\mathbf{x})] = \int_{(k)} \exp(i\mathbf{t}'\mathbf{x}) dF(x_1, \ldots, x_k) \tag{7.12}$$

and the c.f. for any subset is obtained by setting the remaining t_i equal to zero. It follows directly that

$$\phi'(\mathbf{0}) = \frac{\partial\phi(\mathbf{t})}{\partial\boldsymbol{\theta}}\bigg|_{\mathbf{t}=\mathbf{0}} = \boldsymbol{\mu} \text{ and } \phi''(\mathbf{0}) = \frac{\partial^2\phi(\mathbf{t})}{\partial\boldsymbol{\theta}\partial\boldsymbol{\theta}'}\bigg|_{\mathbf{t}=\mathbf{0}} = \boldsymbol{\Sigma}, \tag{7.13}$$

where $\boldsymbol{\theta} = i\mathbf{t}$. In the rest of the chapter we shall often dispense with the use of bold type to denote vectors and matrices, provided no ambiguity arises.

7.6 We now turn to consider individual distributions, starting with discrete variates and then continuous. For any set of marginals, a wide variety of possible joint distributions exists and we therefore restrict attention to those that arise naturally from some specific chance mechanism.

When the distribution is discrete, the p.g.f. is defined by $P(t)$ where $P(e^{iz}) = \phi(z)$. The conditional p.g.f. for x_2 given that $x_1 = y$ is

$$P_x(t|y) = \frac{P^{(y,0)}(0, t_2)}{P^{(y,0)}(0, 1)}, \text{ where } P^{(y,0)}(0, t) = \frac{\partial^y P(t_1, t_2)}{\partial t_1^y}\bigg|_{t_1=0,\, t_2=t} \tag{7.14}$$

with straightforward extensions to the multivariate case.

The multinomial distribution

7.7 Suppose that the members of a population, instead of being classified into two classes as in **5.2**, may be classified into one of $k+1$ classes, $B_1, B_2, \ldots, B_k$, or none of these, which we denote by B_0. If the corresponding proportions in the population are $p_1, p_2, \ldots, p_k$ and p_0, it follows by direct extension of the argument in **5.2** that the probabilities for a random sample of n individuals are given by terms of the form

$$f(x_1, \ldots, x_k) = \frac{n!}{x_0!x_1!\ldots x_k!} p_0^{x_0} p_1^{x_1} \ldots p_k^{x_k}. \tag{7.15}$$

These correspond to the terms of the multinomial expansion of

$$(p_0 + p_1 + p_2 + \ldots + p_k)^n. \tag{7.16}$$

Thus, (7.15) is known as the *multinomial* distribution. Attaching a variable t_i to p_i $(i = 1, \ldots, k)$, we find that the c.f. is

$$(p_0 + p_1 e^{it_1} + p_2 e^{it_2} + \ldots + p_k e^{it_k})^n \tag{7.17}$$

and we may obtain the moments and product-moments in the usual way. It will be sufficient if we write down certain of the cumulants up to the fourth order for $k = 3$. The others follow by symmetry.

$$\left.\begin{aligned}
&\kappa_{1000} = np_0;\quad \kappa_{2000} = np_0q_0;\quad \kappa_{1100} = -np_0p_1;\\
&\kappa_{3000} = np_0q_0(q_0 - p_0);\quad \kappa_{2100} = -np_0p_1(q_0 - p_0);\quad \kappa_{1110} = 2np_0p_1p_2;\\
&\kappa_{4000} = np_0q_0(1 - 6p_0q_0);\quad \kappa_{3100} = -np_0p_1(1 - 6p_0q_0);\\
&\kappa_{2200} = -np_0p_1\{(q_0 - p_0)(q_1 - p_1) + 2p_0p_1\};\\
&\kappa_{2110} = 2np_0p_1p_2(q_0 - 2p_0);\quad \kappa_{1111} = -6np_0p_1p_2p_3.
\end{aligned}\right\} \tag{7.18}$$

Here, as usual, we write $q_i = 1 - p_i$ for brevity. For a method of deriving (7.18), see Exercise 7.1.

It follows from (7.14) and (7.17) that all marginal and conditional forms obtained from (7.15) are again multinomial, or binomial as a special case.

The negative multinomial is discussed in Exercise 7.3; it corresponds to a stopping rule whereby sampling terminates when m members of class B_0 are observed. It is evident that a large number of different stopping rules could be devised, each of which yields the same powers of the p_i in (7.15), but with different combinatorial terms. This observation has some interesting ramifications for statistical inference, see Chapter 31, Volume 2A.

The bivariate binomial distribution

7.8 We may also have a population classified by two attributes, e.g. blue-eyed or not-blue-eyed and male or female. More generally, we could have several attributes and several categories in each, but this does not raise any new points of principle.

Suppose the presence or absence of one quality is denoted by A, A^c, and those of the other quality by B, B^c. The proportions of the four possible combinations may be represented thus:

	B	B^c	TOTAL
A	p_{11}	p_{10}	p
A^c	p_{01}	p_{00}	q
TOTAL	p'	q'	1

(7.19)

From some points of view we may regard this distribution as a multinomial arrayed by

$$(p_{00} + p_{10} + p_{01} + p_{11})^n. \tag{7.20}$$

However, we are usually more interested in the numbers of As and Bs than in the counts for each cell in (7.19). Indeed, the individual cell counts may not be observable. The joint c.f. of the numbers of As and Bs is given by

$$\phi(t) = (p_{00} + p_{10}e^{\theta_1} + p_{01}e^{\theta_2} + p_{11}e^{\theta_1+\theta_2})^n, \tag{7.21}$$

where, as usual, $\theta_1 = it_1$, $\theta_2 = it_2$. If we now define

$$p_{(11)} = p_{11} - pp' = p_{11}p_{00} - p_{10}p_{01} \tag{7.22}$$

we find, on expansion of $\log \phi(t)$, the bivariate cumulants

$$\left.\begin{aligned} \kappa_{11} = np_{(11)}; \quad \kappa_{21} = np_{(11)}(q-p); \quad \kappa_{31} = np_{(11)}(1-6pq); \\ \kappa_{22} = np_{(11)}\{(q-p)(q'-p') - 2p_{(11)}\}. \end{aligned}\right\} \tag{7.23}$$

Other bivariate cumulants such as κ_{12} are obtainable by symmetry; and, of course, such cumulants as κ_{20} are the cumulants of the univariate (marginal) binomials generated by $(q+p)^n$ and $(q'+p')^n$.

7.9 When $p_{(11)}$ in equation (7.22) is zero the c.f. (7.21) becomes

$$\phi(t) = (q + pe^{\theta_1})(q' + p'e^{\theta_2}). \tag{7.24}$$

The attributes are then independent.

If we transfer the origin to the means of the two variates (7.21) becomes

$$\log \phi(t) = n\log(p_{00} + p_{10}e^{\theta_1} + p_{01}e^{\theta_2} + p_{11}e^{\theta_1+\theta_2}) - np\theta_1 - np'\theta_2$$
$$= \tfrac{1}{2}n(pq\theta_1^2 + 2p_{(11)}\theta_1\theta_2 + p'q'\theta_2^2) + O(n\theta^3). \qquad (7.25)$$

If we then standardize, the variances of the two variates being npq and $np'q'$, and let n tend to infinity, we find in the limit that

$$\log \phi(t) = \tfrac{1}{2}(\theta_1^2 + 2\rho\theta_1\theta_2 + \theta_2^2), \qquad (7.26)$$

where

$$\rho = \frac{p_{(11)}}{(pqp'q')^{\frac{1}{2}}}. \qquad (7.27)$$

It follows from Example 3.17 that this is the c.f. of the distribution with p.d.f.

$$f(x_1, x_2) = \frac{1}{2\pi(1-\rho^2)^{\frac{1}{2}}} \exp\left\{\frac{-1}{2(1-\rho^2)}(x_1^2 - 2\rho x_1 x_2 + x_2^2)\right\}, \qquad -\infty < x_1, x_2 < \infty. \qquad (7.28)$$

Thus the bivariate binomial distribution tends to the form (7.28) which is the binormal distribution, to be discussed in Chapters 15 and 16 and, more briefly, in **7.20–22** below.

The variates x_1 and x_2 become independent if and only if $\rho = 0$, which implies that $p_{(11)} = 0$, or $p_{11} = pp'$. This is turn means that the proportion of As among the Bs is the same as the proportion among the B^cs, for then $p_{10} = p - p_{11} = pq'$. The table (7.19) then assumes the form

	B	B^c	TOTAL
A	pp'	pq'	p
A^c	qp'	qq'	q
TOTAL	p'	q'	1

in accordance with (7.24).

7.10 Direct expansion of (7.21) gives the joint p.f., from (7.15) as

$$f(x_1, x_2) = \sum_u \binom{n}{x_1 - u, x_2 - u, u} p_{00}^{v} p_{10}^{x_1 - u} p_{01}^{x_2 - u} p_{11}^{u} \qquad (7.29)$$

where $v = n - x_1 - x_2 + u$ and

$$\binom{n}{a, b, c} = n!/[a!\,b!\,c!\,(n-a-b-c)!], \qquad (7.30)$$

and u varies over $(\max[0, x_1 + x_2 - n], \min[x_1, x_2])$. (7.25) may be expressed as

$$f(x_1, x_2) = \binom{n}{x_1, x_2} p_{00}^{n-x_1-x_2} p_{10}^{x_1} p_{01}^{x_2}\left[1 + \frac{\alpha x_1 x_2}{1!(n - x_1 - x_2 + 1)} + \frac{\alpha^2 x_1^{(2)} x_2^{(2)}}{2!(n - x_1 - x_2 + 2)^{(2)}} + \dots\right], \qquad (7.31)$$

where the term in brackets is a finite hypergeometric series. The conditional expectation and variance of x_1 given x_2 are considered in Exercise 7.4.

Wishart (1949) has given general formulae for the cumulants of multivariate multinomial distributions. Teugels (1990) provides a very general framework for the multivariate binomial.

The bivariate Poisson distribution

7.11 If we set $\lambda_{ij} = np_{ij}$ for $(i,j) = [10, 01, 11]$ in (7.21) and take the limit as $n \to \infty$ with the λ_{ij} fixed, we obtain

$$\phi(t) = \exp[\lambda_{10}(e^{\theta_1} - 1) + \lambda_{01}(e^{\theta_2} - 1) + \lambda_{11}(e^{\theta_1+\theta_2} - 1)]; \tag{7.32}$$

which is the c.f. of the bivariate Poisson distribution. Recurrence relations for the p.m.f. are given in Exercise 7.5. It is evident from (7.32) that the marginal distributions are also Poisson with the means equal to the variances, and given by $\lambda_1 = \lambda_{10} + \lambda_{11}$ and $\lambda_2 = \lambda_{01} + \lambda_{11}$ respectively. However, it follows from (7.32) that $\kappa_{11} = \lambda_{11}$ so that the correlation between x_1 and x_2, $\rho = \lambda_{11}/(\lambda_1\lambda_2)^{\frac{1}{2}}$ is positive unless $\lambda_{11} = 0$ when the variates are independent. The p.m.f. may be derived directly from (7.32) or as a limit from (7.31); see Exercise 7.6.

Inspection of (7.32) might suggest that we could change the sign of λ_{11} and still retain Poisson marginals provided $\lambda_{10} > \lambda_{11}$ and $\lambda_{01} > \lambda_{11}$. However, Griffiths *et al.* (1979) show that (7.32) then ceases to be a valid c.f. although bivariate Poisson distributions with negative correlation can be constructed; see Exercise 7.7.

The bivariate Poisson may be generated directly from $x_j = z_j + z_3$, $j = 1, 2$ where the z_j are independent Poissons with parameters λ_{10}, λ_{01}, and λ_{11}. The multinomials and the Poisson are examples of multivariate power series distributions, cf **5.49**; see also Khatri (1959), Patil (1965) and Sichel (1980).

Multivariate urn schemes

7.12 Following **5.13**, we may develop a multivariate urn scheme by attaching multiple attributes to each member of the population. We may suppose that at the commencement of sampling the urn contains M_i balls of type i $(i = 0, 1, \ldots, k)$ where $M_i = Np_i$ and $p_0 + p_1 + \ldots + p_k = 1$. After each trial where a ball of type i is selected $(c + 1)$ balls of that type are placed back in the urn, so that $c = 0$ correspond to sampling with replacement. A considerable number of extensions is clearly possible, by varying c either with regard to the type of ball or the number of the trial, as well as adding or substracting balls of other types. We do not pursue these possibilities here; the interested reader is referred to N. L. Johnson and Kotz (1977b).

Returning to the setup outlined above, let $f(x_1, \ldots, x_k | m)$ denote the probability that the first m trials yield $\mathbf{x} = (x_1, \ldots, x_k)'$ balls of the k types plus $x_0 = m - x_1 - \ldots - x_k$. It follows that

$$f(\mathbf{x} | m + 1) = \sum_{i=0}^{k} \frac{M_i + (x_i - 1)c}{N + mc} f(\mathbf{x} - \boldsymbol{\delta}_i | m) \tag{7.33}$$

where $\mathbf{x}-\boldsymbol{\delta}_i$ denotes $\mathbf{x}$ with $x_i - 1$ in place of x_i and a term is identically zero if $x_i - \delta_i < 0$ or $M_i + (x_i - 1)c \leqslant 0$. When $c = 0$, (7.33) reduces to (7.15) and when $c = -1$, to the multiple hypergeometric:

$$f(x \mid n) = \prod_{i=0}^{k} \binom{M_i}{x_i} \bigg/ \binom{M}{n}; \tag{7.34}$$

whereas $c = +1$ yields the negative multiple hypergeometric:

$$f(x \mid n) = \prod_{i=0}^{k} \binom{M_i + x_i - 1}{x_i} \bigg/ \binom{M+n-1}{n}; \tag{7.35}$$

(7.35) is derived by different means in **7.15** below.

7.13 It is apparent from (7.34) and (7.35) that these distributions, like the multinomials, have rather restricted dependence structures (see Exercise 7.7) and it is for this reason that we have labelled them "multiple" rather than multivariate. Multivariate distributions may be derived from (7.34) and (7.35) by combining classes, as we did for the bivariate binomial in **7.10**. (Also, sequential sampling models may be developed as in **5.16** for the univariate case.)

Steyn (1955, 1957) developed the class of multiple hypergeometric series distributions, which includes the multinomials as limiting cases as well as (7.34) and (7.35). Also, see Sibuya and Shimizu (1981) for extensions to this class.

Marshall and Olkin (1985) provide a unified treatment of multivariate binomial, Poisson and hypergeometric distributions using the bivariate Bernoulli distribution as a starting point.

Mixture distributions

7.14 In **5.22** we developed a mixture, or compound, distribution by means of the specification that x depends upon some parameter θ, which itself has a distribution:

$$f(x) = \int_{-\infty}^{\infty} f(x \mid \theta) p(\theta) \, d\theta. \tag{7.36}$$

Multivariate versions follow by making x a vector, θ may be either a scalar or a vector.

The simplest distributions are derived by taking the variates $\mathbf{x}$ to be conditionally independent; that is, independent given θ. Dependence is then induced through the mixing distribution. For example, two species of insect may depend upon similar food sources or environmental conditions, even though their abundances are unaffected by the number of individuals of the other species present at a given location. Alternatively, the elements of $\mathbf{x}$ may be directly correlated when the mixing distribution serves primarily to increase variability, even though the mixing may also change the dependence structure.

7.15 As in the univariate case, among discrete distributions most attention focuses upon Poisson mixtures; mixtures of continuous distributions will be discussed in **7.30**. In developing the Poisson mixtures we anticipate some of the bivariate continuous distributions developed later in the chapter.

Example 7.1 (Mixture of independent Poissons)
Let x_1 and x_2 be independent Poisson variates with p.m.f. given by (5.18) and parameters $\lambda\alpha_j, j=1,2$. As in **5.23** we assume that λ follows a gamma distribution so that, from (7.36),

$$f(x_1, x_2) = \int_0^\infty \frac{c^r}{\Gamma(r)} \lambda^{r-1} e^{-c\lambda} \frac{(\lambda\alpha_1)^{x_1}}{x_1!} \frac{(\lambda\alpha_2)^{x_2}}{x_2!} e^{-\lambda(\alpha_1+\alpha_2)} \, d\lambda \tag{7.37}$$

which, after some simplification reduces to

$$f(x_1, x_2) = \frac{\Gamma(x_1+x_2+r)}{\Gamma(r)x_1!x_2!} \frac{\alpha_1^{x_1}\alpha_2^{x_2}c^r}{(\alpha_1+\alpha_2+c)^{x_1+x_2+r}}. \tag{7.38}$$

with p.g.f.

$$P(t_1, t_2) = c^r/[c-\alpha_1(t_1-1)-\alpha_2(t_2-1)]^r. \tag{7.39}$$

Expressions (7.38) and (7.39) represent the negative multinomial with $k=2$ save that r is no longer required to be an integer; the extension to $k>2$ is straightforward. The marginal distributions are negative binomials, as given in (5.66).

Example 7.2 (Bivariate Poisson mixture with single Gamma)
The p.g.f. of the usual bivariate Poisson is, from (7.32) with some notational changes

$$P(t_1, t_2 | \lambda) = \exp[\lambda\alpha_1(t_1-1) + \lambda\alpha_2(t_2-1) + \lambda\alpha_3(t_1-1)(t_2-1)];$$

in this version, the marginal Poisson distributions have parameters $\lambda\alpha_j$ and $\alpha_j \geqslant \alpha_3$, $j=1,2$.

Using a univariate Gamma distribution, the p.g.f. becomes

$$\begin{aligned} P(t_1, t_2) &= [\Gamma(r)]^{-1} \int_0^\infty P(t_1, t_2|\lambda) c^r \lambda^{r-1} e^{-c\lambda} \, d\lambda \\ &= c^r/[c-\alpha_1(t_1-1)-\alpha_2(t_2-1)-\alpha_3(t_1-1)(t_2-1)]^r. \end{aligned} \tag{7.40}$$

It follows from (7.40) that

$$\kappa_{10} = r\alpha_1/c, \quad \kappa_{20} = r\alpha_1(\alpha_1+c)/c^2 \text{ and } \kappa_{11} = r(c\alpha_3+\alpha_1\alpha_2)/c^2.$$

This version of the bivariate negative binomial was introduced by Edwards and Gurland (1961) in the context of accident proneness models.

7.16 A feature of the Poisson distribution is that the single variate p.g.f.s are proportional to $\exp(\lambda t)$. Thus the p.g.f. for any mixture of independent Poissons is given by the c.f. of the mixing distribution:

$$\begin{aligned} P(t) &= \int_{(k)} \exp\left[\sum \lambda_j(t_j-1)\right] f(\lambda_1, \ldots, \lambda_k) \, d\lambda_1 \ldots d\lambda_k \\ &= \phi[(t_1-1)/i, \ldots, (t_k-1)/i]. \end{aligned} \tag{7.41}$$

Example 7.3 (Independent Poissons with bivariate Gamma)

Consider the bivariate gamma with c.f.

$$\phi(t) = [1 - i\beta_1 t_1 - i\beta_2 t_2 - \beta_3 t_1 t_2]^{-r},$$

which arises as an extension of the bivariate exponential in Exercise 7.17, with independence when $\beta_3 = \beta_1\beta_2$ and $0 < \beta_3 \leqslant \beta_1\beta_2$. Using (7.41), we have the p.g.f.

$$P(t_1, t_2) = [1 - \beta_1(t_1 - 1) - \beta_2(t_2 - 1) + \beta_3(t_1 - 1)(t_2 - 1)]^{-r} \tag{7.42}$$

which is of similar form to (7.40) save that the coefficient of the final term is reversed. Since $\kappa_{11} = r(\beta_1\beta_2 - \beta_3)$, the correlation is still positive in (7.42), but it is weaker than in (7.40).

7.17 It is apparent from Examples 7.1–7.3 that a considerable variety of multivariate negative binomial distributions (MNBD) may be developed by mixing arguments. Extensions of these ideas are given by Stein and Juritz (1987); Griffiths and Milne (1987) develop an infinitely divisible class of MNBDs. The bivariate logarithmic series distribution may be obtained as a limiting case; see Exercise 7.10.

Other multivariate Poisson mixtures may be developed along similar lines using the approaches considered in **5.20–3**; see, for example, the bivariate Hermite discussed in Exercise 7.11. Holgate (1966) derives three variants of the bivariate Neyman Type A distribution. Aitchison and Ho (1989) consider the bivariate Poisson-Lognormal.

The Dirichlet mixtures for the multinomial and the negative multinomial are considered in Exercises 7.12 and 7.13.

We now complete our survey of multivariate distributions by considering continuous variates.

The bivariate uniform distribution

7.18 We saw in **1.27** that any continuous probability distribution function may be written as

$$y = F(x) \tag{7.43}$$

where y is uniform on (0, 1); (7.43) was termed the probability integral transform. If x_1 and x_2 are independent it follows, in self-evident notation, that we may write

$$y_1 y_2 = F_1(x_1)F_2(x_2). \tag{7.44}$$

Generally, when x_1 and x_2 are dependent we may specify a bivariate uniform distribution with d.f.

$$\begin{aligned} G(y_1, y_2) &= F_1(x_1)F_2(x_2)\,[1 + H(x_1, x_2)] \\ &= y_1 y_2 [1 + H^*(y_1, y_2)] \end{aligned} \tag{7.45}$$

where $H \geqslant 0$ and $H(x_1, \infty) = H(\infty, x_2) = 0$; the most general form possible for H is just $[F(x_1, x_2) - 1]/F_1(x_1)F_2(x_2)$. A particular case of (7.45), introduced by Morgenstern (1956) and later extended by Farlie (1960) is to set $H(x_1, x_2) = \alpha H_1(x_1)H_2(x_2)$; Morgenstern used $H_i = 1 - F_i$. These distributions are considered in greater detail in **7.32**; for the present, we use (7.45) to consider a bivariate uniform distribution.

Example 7.4 (Morgenstern's bivariate uniform)

Let H^* in (7.45) be $\alpha(1-y_1)(1-y_2)$ so that

$$G(y_1, y_2) = y_1 y_2[1+\alpha(1-y_1)(1-y_2)], \tag{7.46}$$

$-1 \leqslant \alpha \leqslant 1$. The marginals are $G_i(y_i) = y_i$ (ie uniform) and the p.d.f. is

$$g(y_1, y_2) = 1 + \alpha(1-2y_1)(1-2y_2)$$

from which it follows that $\text{cov}(y_1, y_2) = \rho = \alpha/3$ and $|\rho| \leqslant 1/3$ so that the degree of dependence is decidedly limited.

Barnett (1980) considers a variety of bivariate uniform distributions based upon different marginal distributions in (7.45).

7.19 A second form of bivariate uniform arises when

$$\begin{aligned} f(x_1, x_2) &= \text{constant for all } (x_1, x_2) \varepsilon C \\ &= 0, \text{ otherwise} \end{aligned} \tag{7.47}$$

where the set C may be defined arbitrarily. The marginals need not be uniform in this case.

Example 7.5 (Bivariate uniform with non-uniform marginals)

Using (7.47), let

$$f(x_1, x_2) = 2, \qquad x_1 + x_2 \leqslant 1, x_1 \geqslant 0, x_2 \geqslant 0$$

so that the p.d.f. defines a right angled isosceles triangle. Since

$$2\int_0^1\int_0^{1-x_1} x_1^a x_2^b \text{d}x = 2\Gamma(a+1)\Gamma(b+1)/\Gamma(a+b+3), \quad \text{it follows that}$$

$$E(x_j) = 1/3, \ \text{var}(x_j) = 1/18, \ \rho = -1/2.$$

The marginal p.d.f. for x_j is $f_j(x_j) = 2(1-x_j)$, $j = 1, 2$.

Bivariate distributions of this kind arise in sampling at random from irregularly shaped areas; such sampling usually employs a rejection selection method, see **9.24**.

The binormal (bivariate normal) distribution

7.20 The multivariate normal, or multinormal, distribution occupies a central place in statistical inference and we shall discuss it in detail in that context in Chapters 15 and 16. However, consistent with our theme in this chapter, we shall now briefly examine the binormal (bivariate normal) distribution and develop some of its basic properties.

Following Example 1.1 in **1.34**, the p.d.f. is

$$\text{d}F = c_0 \exp\left\{-\frac{1}{2(1-\rho^2)}(z_1^2 + z_2^2 - 2\rho z_1 z_2)\right\}\text{d}z_1\,\text{d}z_2 \tag{7.48}$$

where the z_j denote standardized variates with zero means and unit variances and $c_0^{-1} = 2\pi(1-\rho^2)^{\frac{1}{2}}$. Transforming from z_j to $x_j = \mu_j + \sigma_j z_j$ yields the five parameter binormal with p.d.f.

$$f(x_1, x_2) = c_1 \exp\left\{-\frac{1}{2(1-\rho^2)}\left[\left(\frac{x_1-\mu_1}{\sigma_1}\right)^2 + \left(\frac{x_2-\mu_2}{\sigma_2}\right)^2 - 2\rho\left(\frac{x_1-\mu_1}{\sigma_1}\right)\left(\frac{x_2-\mu_2}{\sigma_2}\right)\right]\right\} \tag{7.49}$$

where $c_1 = c_0/\sigma_1\sigma_2$. Expression (7.49) may be rewritten in matrix form as

$$f(\mathbf{x}) = c_1 \exp\{-\tfrac{1}{2}(\mathbf{x}-\boldsymbol{\mu})'\mathbf{A}(\mathbf{x}-\boldsymbol{\mu})\} \tag{7.50}$$

where

$$\mathbf{x} = \begin{pmatrix} x_1 \\ x_2 \end{pmatrix}, \ \boldsymbol{\mu} = \begin{pmatrix} \mu_1 \\ \mu_2 \end{pmatrix} \text{ and } \mathbf{A} = \frac{1}{\sigma_1^2\sigma_2^2(1-\rho^2)}\begin{pmatrix} \sigma_2^2 & -\rho\sigma_1\sigma_2 \\ -\rho\sigma_1\sigma_2 & \sigma_1^2 \end{pmatrix}.$$

As we shall see in **15.2**, the constant may be expressed as

$$c_1 = (2\pi)^{k/2}|A|^{\frac{1}{2}} \tag{7.51}$$

where k denotes the number of variates, so that (7.50) represents the multinormal distribution given the appropriate extensions to $\mathbf{x}$, $\boldsymbol{\mu}$ and $\mathbf{A}$.

7.21 It follows from Example 3.19 in **3.27** that the c.f. is

$$\phi(t_1, t_2) = \exp\{-\tfrac{1}{2}(t_1^2\sigma_1^2 + 2t_1t_2\rho\sigma_1\sigma_2 + t_2^2\sigma_2^2)\} \tag{7.52}$$

from which the moments may be derived, as in (3.71) up to order four. In particular,

$$E(\mathbf{x}) = \boldsymbol{\mu} \text{ and } \operatorname{cov}(\mathbf{x}) = \boldsymbol{\Sigma} = \begin{pmatrix} \sigma_1^2 & \rho\sigma_1\sigma_2 \\ \rho\sigma_1\sigma_2 & \sigma_2^2 \end{pmatrix} = \mathbf{A}^{-1}. \tag{7.53}$$

The exponent in (7.50) is a quadratic form so that any linear transformation of the form $\mathbf{y} = \mathbf{B}\mathbf{x} + \mathbf{b}$, where $\mathbf{B}$ is a non-singular square matrix and $\mathbf{b}$ is a vector again yields a binormal distribution since the Jacobian is the constant $|B^{-1}|$; cf **1.35**. Given the expectation and covariance matrix for $\mathbf{x}$ as stated in (7.53), it follows that $E(\mathbf{y}) = \mathbf{B}\boldsymbol{\mu} + \mathbf{b}$ and $\operatorname{cov}(\mathbf{y}) = \mathbf{B}\boldsymbol{\Sigma}\mathbf{B}'$.

7.22 If we set $t_2 = 0$ in (7.52), the c.f. reveals that the marginal distributions are always normal. Further, from Example 1.1 we have that

$$y_1 = \left(\frac{(x_1-\mu_1)}{\sigma_1} - \frac{\rho(x_2-\mu_2)}{\sigma_2}\right)\Big/(1-\rho^2)^{\frac{1}{2}}$$

is distributed independently of x_2, from which we see that the conditional distribution of x_1 given x_2 is again normal with

$$E(x_1|x_2) = \mu_1 + \frac{\rho\sigma_1}{\sigma_2}(x_2-\mu_2)$$

$$\text{and } \operatorname{var}(x_1|x_2) = \sigma_1^2(1-\rho^2).$$

The normality of all marginal and conditional distributions extends to the multivariate case, as we shall see in Chapter 15.

Generation of multivariate distributions

7.23 A vast array of multivariate distributions exists in the literature and some of these forms owe their existence to elegant mathematics rather than any foundation laid by a statistical model. Therefore, we shall now focus upon several stochastic mechanisms that lead to the creation of multivariate distributions from sets of univariate ones.

7.24 If u_1, u_2, and u_3 are independent variables, a useful mechanism for generating dependent bivariate variables is to set

$$x = u_1 + u_3 \quad \text{and} \quad y = u_2 + u_3. \tag{7.54}$$

For example, if the u_i are normally distributed with zero means and variances σ_i^2, it follows that x and y are bivariate normal as in (7.49) with zero means and

$$V(x) = \sigma_1^2 + \sigma_3^2, \quad V(y) = \sigma_2^2 + \sigma_3^2, \quad \text{cov}(x, y) = \sigma_3^2. \tag{7.55}$$

Negative covariance is achieved simply by rewriting y as $u_2 - u_3$.

Generally, since u_1, u_2 and u_3 are independent, the joint c.f. of x and y is

$$\phi(t_1, t_2) = E(e^{it_1(u_1+u_3)+it_2(u_2+u_3)}) = \phi_1(t_1)\phi_2(t_2)\phi_3(t_1 + t_2), \tag{7.56}$$

which reduces to (7.52) for the binormal.

Thus, moments are simply derived, although the form of the probability function will often involve a summation (or integration). For example, the bivariate Poisson distribution discussed in **7.11** may be derived by this means.

Example 7.6 (Bivariate exponential using sums)

If (u_1, u_2, u_3) are standard exponential variates (cf **5.31**), an application of (7.54) yields a bivariate exponential with c.f.

$$\phi(t_1, t_2) = [(1 - it_1)(1 - it_2)\{1 - i(t_1 + t_2)\}]^{-1} \tag{7.57}$$

If $z = \min(x, y)$, the joint p.d.f. is, after transforming from (u_1, u_2, u_3) to (x, y, u_3) and integrating out:

$$\begin{aligned} f(x, y) &= e^{-x-y}\int_0^z e^v dv \\ &= (e^z - 1)e^{-x-y}, \; x > 0, \; y > 0. \end{aligned} \tag{7.58}$$

If we set $t_2 = 0$ in (7.57), it is evident that the marginal distributions are Gamma(2). The conditional expectation is $E(x|y) = E(u_1 + u_3 | u_2 + u_3)$

$$= E(u_1) + E(u_3 | u_2 + u_3). \tag{7.59}$$

It follows from Exercise 11.10 that $v = u_3/(u_2 + u_3)$ and $(u_2 + u_3)$ are independent and v is uniformly distributed so that (7.59) yields

$$E(x|y) = 1 + (y/2).$$

Likewise,

$$\mathrm{var}(x|y) = 1 + (y^2/12)$$

and $\mathrm{cov}(x, y) = 1/2$, as is evident from symmetry considerations.

The c.f. may be extended to produce a bivariate Gamma distribution; the conditional expectation and the conditional variance remain linear and quadratic, respectively, in y. However, if the exponentials have different parameters, the regression ceases to be linear; see Exercise 7.16.

7.25 Other combinations of independent variables may also be used. Thus, in life-testing, components A and B may be on parallel circuits, both attached to component C. If their individual lifetimes are u_1, u_2 and u_3, respectively, it follows that the lifetimes of the two circuits are

$$x = \min(u_1, u_3) \quad \text{and} \quad y = \min(u_2, u_3).$$

The joint density $f(x, y)$ is determined by integration of the product $f_1(u_1)f_2(u_2)$ $f_3(u_3)$ over each of the six possible regions:

Case	Region	x	y	$x \gtreqless y$	Integrated over
(1)	$u_1 < u_2 < u_3$	u_1	u_2	$x < y$	$y < u_3 < \infty$
(2)	$u_1 < u_3 < u_2$	u_1	u_3	$x < y$	$y < u_2 < \infty$
(3)	$u_2 < u_1 < u_3$	u_1	u_2	$x > y$	$x < u_3 < \infty$
(4)	$u_2 < u_3 < u_1$	u_3	u_2	$x > y$	$x < u_1 < \infty$
(5)	$u_3 < u_1 < u_2$	u_3	u_3	$x = y = z$	$z < u_1 < \infty$ and
(6)	$u_3 < u_2 < u_1$				$z < u_2 < \infty$

Example 7.7 (The Marshall–Olkin bivariate exponential)

Let u_1, u_2 and u_3 be exponentially distributed with common parameter λ. Following Marshall and Olkin (1967), case (1) yields the p.d.f.

$$f(x, y) = \lambda^2 e^{-\lambda x - \lambda y} \int_y^\infty \lambda e^{-\lambda u}\, du$$

$$= \lambda^2 e^{-\lambda x - 2\lambda y}, \quad 0 < x < y < \infty;$$

the same holds for case (2).

By symmetry, cases (3) and (4) yield

$$f(x, y) = \lambda^2 e^{-2\lambda x - \lambda y}, \quad 0 < y < x < \infty.$$

Finally, for cases (5) and (6), we obtain

$$f(z, z) = \lambda e^{-3\lambda z}, \quad 0 < z < \infty.$$

Integrating $e^{\theta_1 x + \theta_2 y}$ separately over the three pairs of different cases, where $\theta_j = it_j$, we find that the joint c.f. of x and y is

$$\phi(t_1, t_2) = \frac{\lambda}{(3\lambda - \theta_1 - \theta_2)}\left[1 + \frac{2\lambda}{2\lambda - \theta_1} + \frac{2\lambda}{2\lambda - \theta_2}\right]. \tag{7.60}$$

The marginal distributions are exponential with parameter 2λ. The extension to different parameter values is straightforward.

Two other forms of bivariate exponential are considered in Exercises 7.17 and 7.18. A review of bivariate exponential distributions appears in Barnett (1985).

The multivariate analogue to the Gamma distribution in the theory of sampling is the Wishart distribution, which are consider in **16.43–48.**

The Dirichlet distributions

7.26 Let x_j, $j=0, 1, \ldots, k$ be independent Gamma variables with densities:

$$f(x_j) = x_j^{\theta_j - 1} e^{-x_j}/\Gamma(\theta_j), \quad 0 < x_j < \infty. \tag{7.61}$$

It follows from Example 3.12 that the c.f. for $x = \sum_{j=0}^{k} x_j$ is $\phi(t) = \prod_{j=0}^{k} (1 - it)^{-\theta_j}$

$$= (1 - it)^{-\theta} \tag{7.62}$$

where $\theta = \sum_{j=0}^{k} \theta_j$. Since the c.f. uniquely determines the distribution, x must be Gamma distributed with parameter θ. The joint distribution of x and the proportions, $y_j = x_j/x$, $j = 1, \ldots, k$ has p.d.f.

$$f(x, y_1, \ldots, y_k) = \left\{ \prod_{j=0}^{k} f(x_j) \right\} \cdot J \tag{7.63}$$

where the Jacobian, J, is given by

$$J^{-1} = \begin{vmatrix} 1 & 1 & 1 & \cdots & 1 \\ \frac{-x_1}{x^2} & \frac{1}{x} - \frac{x_1}{x^2} & \frac{-x_1}{x^2} & \cdots & \frac{-x_1}{x^2} \\ \vdots & \vdots & \vdots & \vdots & \vdots \\ \frac{-x_k}{x^2} & \frac{-x_k}{x^2} & \cdots & \frac{-x_k}{x^2} & \frac{1}{x} - \frac{x_k}{x^2} \end{vmatrix}$$

which reduces to $J = x^{k-1}$ so that (7.63) factorizes into terms in x and the ys. Thus, the ys are independent of x and we arrive at

$$f(y_1, \ldots, y_k) = \Gamma(\theta) \prod_{j=0}^{k} \left[\frac{y_j^{\theta_j - 1}}{\Gamma(\theta_j)} \right], \tag{7.64}$$

defined for $0 \leqslant y_j \leqslant 1$ over the simplex $y_0 + y_1 + \ldots + y_k = 1$. This is known as the *Dirichlet distribution*. It follows by integrating over the regions $0 \leqslant y_j \leqslant 1 - y_1 - \ldots - y_{j-1}$ for $2 \leqslant j \leqslant k$ that all the marginal distributions are also Dirichlet. In the univariate case, the marginal distributions reduce to the Beta distribution:

$$f(y_1) = \frac{\Gamma(\theta)}{\Gamma(\theta_1)\Gamma(\theta-\theta_1)} y_1^{\theta_1-1}(1-y)^{\theta-\theta_1-1}. \tag{7.65}$$

The low order moments are

$$E(y_j) = \theta_j/\theta, \text{ var }(y_j) = \frac{\theta_j(\theta-\theta_j)}{\theta^2(\theta+1)}$$

$$\text{and corr }(y_i, y_j) = -\left\{\frac{\theta_i\theta_j}{(\theta-\theta_i)(\theta-\theta_j)}\right\}^{\frac{1}{2}}.$$

Thus the correlations are all negative, indicating a restricted dependence structure.

The Dirichlet of the second kind, or inverted Dirichlet, is given by considering $y_j = x_j/x_k$ in (7.61). The resulting p.d.f. is

$$f(y_1, \ldots, y_k) = \frac{\Gamma(\theta)}{(1+y)^\theta} \prod_{j=1}^{k} \frac{y_j^{\theta_j-1}}{\Gamma(\theta_j)}, \tag{7.66}$$

defined for $y_j > 0$. The low order moments are

$$E(y_j) = \frac{\theta_j}{(\theta-1)}, \text{ var }(y_j) = \frac{\theta_j(\theta+\theta_j-1)}{(\theta-1)^2(\theta-2)}$$

$$\text{and } \rho_{jk} = \left[\frac{\theta_j\theta_k}{(\theta+\theta_j-1)(\theta+\theta_k-1)}\right]^{\frac{1}{2}}.$$

The marginal distributions are beta of the second kind, see **6.7**.

A more general multivariate beta distribution with beta conditionals is given by James (1975). For extensions in the context of multivariate sampling theory, see **16.63**.

Bivariate families of distributions

7.27 A variety of bivariate Pearson systems have been suggested (cf. Pretorius, 1930; Mardia, 1970a, chapter 2; and Ord, 1972, chapter 3). These systems involve a pair of partial differential equations that express the partial derivatives of $\log f$ as ratios of polynomials. Van Uven (1947–8) considered the form

$$\frac{\partial \log f}{\partial x_i} = \frac{L_i}{Q_i}, \qquad i = 1, 2, \tag{7.67}$$

where $f \equiv f(x_1, x_2)$ and L_i and Q_i represent, respectively, linear and quadratic functions in x_1 and x_2. These polynomials are specified subject to the condition

$$\frac{\partial^2 \log f}{\partial x_1 \partial x_2} = \frac{\partial}{\partial x_2}\left(\frac{L_1}{Q_1}\right) = \frac{\partial}{\partial x_1}\left(\frac{L_2}{Q_2}\right). \tag{7.68}$$

There are several possible solutions to (7.67); Van Uven (1947–8) gives a complete listing. We shall consider four major forms:

(I) Q_1 and Q_2 have no common factor;
(II) Q_1 and Q_2 have a common (linear) factor;
(III) Q_1 and Q_2 are identical;
(IV) Q_2 reduces to a linear function which is a factor of Q_1.

In case (I), the two random variables are independent; see Exercise 7.20. For cases (II)-(IV), we consider only those solutions which yield marginal distributions of the same univariate type. We find (Mardia, 1970a, pp. 8-9):

$$\text{(II)}\ f(x_1, x_2) \propto x_1^{p_1-1} x_2^{p_2-1} (x_1 + x_2 + b)^{p_3-1} \tag{7.69}$$

$$\text{(III)}\ f(x_1, x_2) \propto (x_1^2 + x_2^2 + 2bx_1x_2 + 1)^p \tag{7.70}$$

$$\text{(IV)}\ f(x_1, x_2) \propto x_1^{p_1-1} (x_2 - x_1)^{p_2-1} e^{-qx_2}, \qquad x_2 > x_1. \tag{7.71}$$

The expressions (7.69-71) are expressed in standardized form, and the usual translations $y_i = (x_i - \mu_i)/\sigma_i$ are possible throughout. Table 7.1 lists some of the major forms that arise. The case (III) distributions with $b = 0$ may be written as functions of $\mathbf{x}'\mathbf{x} = \Sigma x_j^2$ and are said to be *spherically* symmetric; Fang *et al.* (1989) provide a detailed description of such distributions.

Steyn (1960) shows that (7.67) gives rise to linear regression functions, i.e., $E(x_1 | x_2)$ is linear in x_2. If both $\text{var}(x_1 | x_2)$ and $\text{var}(x_2 | x_1)$ are constant, the distribution is bivariate normal—we shall discuss this property in Volume 2A. For other properties of the system and particular distributions, see Mardia (1970a).

Parrish (1981) and Bargmann and Parrish (1981) discuss the use of (7.67) to approximate bivariate distributions whose moments up to the fourth order are known.

In general, the dependence structures allowed by (7.67) are rather restrictive, as observed for the Dirichlet distribution in 7.26. Attempts to increase the extent of dependence by using higher order polynomials appear to create more problems than they solve.

7.28 Bivariate Gram-Charlier and Edgeworth series may be developed in a straightforward fashion; see Exercises 7.21 and 7.22. Barndorff-Nielsen and Pedersen

Table 7.1 Parameter specifications for distributions in the bivariate Pearson family

Distribution	Equation	Parameter values	Range
Beta (first kind)	(7.69)	$b = 1, p_i > 0$	$0 \leqslant x_1 + x_2 \leqslant 1$
Beta (second kind)	(7.69)	$b = 1, p_1, p_2 > 0,$ $p_1 + p_2 + p_3 < 0$	$0 \leqslant x_i < \infty$
Pareto	(7.69)	$p_1 = p_2 = 0, b = -1,$ $p_3 < -2$	$1 \leqslant x_i \infty$
Type VII	(7.70)	$b^2 < 1, p < -\frac{1}{2}$	$-\infty < x_i < \infty$
Cauchy	(7.70)	$b^2 < 1, p = -\frac{3}{2}$	$-\infty < x_i < \infty$
Type II	(7.70)	$b^2 < 1, p > 0$	$x_i^2 < 2p + 4$
Gamma	(7.71)	$p_i > 0, q > 0$	$0 < x_1 < x_2 < \infty$

(1979) give the bivariate Hermite polynomials up to order six. The multivariate Edgeworth expansion is described in Chambers (1967); see also McCullagh (1984) and **11.13–16** below.

Skougaard (1986) gives general conditions for the validity of multivariate Edgeworth expansions.

7.29 Starting from the standardized bivariate normal distribution of Example 3.19,

$$f(x_1, x_2) = \frac{1}{2\pi(1-\rho^2)^{\frac{1}{2}}} \exp\left\{\frac{-1}{2(1-\rho^2)}(x_1^2 - 2\rho x_1 x_2 + x_2^2)\right\}, \tag{7.72}$$

the bivariate Johnson system follows directly upon making two transformations:

$$\xi_i = \gamma_i + \delta_i g_i\left(\frac{x_i - \mu_i}{\lambda_i}\right), \qquad i = 1, 2, \tag{7.73}$$

as in (6.57). Since the median is preserved under monotone transformations, the dependence structure may be described by *median* (*regression*) functions such as

$$g_1^*(z_1 | z_2) = \frac{\rho\delta_2}{\delta_1} g_2(z_2) + \frac{(\rho\gamma_2 - \gamma_1)}{\delta_1}, \tag{7.74}$$

where $z_i = (x_i - \mu_i)/\lambda_i$. In (7.74), $g_1^*(z_1 | z_2)$ represents the median of z_1 given z_2.

The complete system comprises nine possible distributions, given three choices for each marginal. For further properties, see Johnson (1949b) and Mardia (1970a, chapter 4).

Systems based upon mixtures

7.30 The bivariate Burr system, and others, may be developed from mixture models, following an approach similar to that in **7.14**.

Suppose we have two random variables, x_1 and x_2, with joint density depending upon θ,

$$f(x_1, x_2 | \theta) = f_1(x_1 | \theta) f_2(x_2 | \theta). \tag{7.75}$$

If we specify the mixing density $g(\theta)$, we arrive at the bivariate distribution (cf. **5.22**):

$$f(x_1, x_2) = \int f_1(x_1 | \theta) f_2(x_2 | \theta) g(\theta)\, d\theta. \tag{7.76}$$

The assumption of independence conditional on the value of θ in (7.75) is not necessary, but it is convenient and often gives rise to bivariate distributions with a fairly simple structure.

If we define

$$\bar{F}_i(x_i | \theta) = P(X_i > x_i | \theta), \qquad \bar{F}(x_1, x_2 | \theta) = P(X_1 > x_1, X_2 > x_2 | \theta)$$

(7.76) becomes, on integrating from x_i to ∞,

$$\bar{F}(x_1, x_2) = \int \bar{F}_1(x_1 | \theta) \bar{F}_2(x_2 | \theta) g(\theta)\, d\theta. \tag{7.77}$$

Example 7.8 (Bivariate Burr distributions)

Suppose $\bar{F}_i(x_i \mid \theta) = \exp(-\theta a_i x_i^c)$, $i = 1, 2$, so that x_i has a Weibull distribution as in 5.33, given θ. Assume that θ has a Gamma mixing distribution, so that

$$g(\theta) = \frac{\theta^{b-1}\lambda^b e^{-\theta\lambda}}{\Gamma(b)}, \quad 0 < \theta < \infty; \tag{7.78}$$

then

$$\bar{F}(x_1, x_2) = \int_0^\infty \exp(-\theta a_1 x_1^c - \theta a_2 x_2^c) \frac{\theta^{b-1}\lambda^b e^{-\theta\lambda}}{\Gamma(b)} d\theta$$

$$= \frac{\lambda^b}{(\lambda + a_1 x_1^c + a_2 x_2^c)^b}. \tag{7.79}$$

We may set $\lambda = 1$ without loss of generality. (7.79) is a bivariate Burr distribution and, since $x_i \geqslant 0$, we have

$$\bar{F}(x_1, 0) = 1 - F_1(x_1)$$
$$= \bar{F}_1(x_1),$$

so that the marginals are of form (6.77) by construction, and the conditional distributions are of the Burr form also; see N. L. Johnson and Kotz (1981). The median regression function is given in Exercise 7.23.

Whitmore and Lee (1991) develop a bivariate universe Gaussian distribution using a mixture of exponentials. Marshall and Olkin (1988) provide a general framework for considering multivariate distributions as mixtures.

7.31 A common feature of all the bivariate systems discussed so far is the emphasis on a flexible choice of marginal distributions with only a single parameter to describe bivariate dependence. This may be viewed as a weakness, yet much statistical practice relies upon such structures. For example, most multivariate analysis uses only variances and covariances among variables. Nevertheless, higher dimensional versions of the distributions we have considered often have restricted dependence structures.

7.32 An approach that allows completely general margins but has only a single parameter to describe dependence is due to Morgenstern (1956), Gumbel (1960) and later extended by Farlie (1960), or FGM distributions, in the interests of brevity. The general form of the distribution is:

$$F(x_1, x_2) = F_1(x_1)F_2(x_2)\,[1 + \alpha G_1(x_1)G_2(x_2)]; \tag{7.80}$$

where $|\alpha| \leqslant 1$ to ensure that $F(x_1, x_2)$ is a valid d.f. Morgenstern and Gumbel took $G_j = 1 - F_j$, $j = 1, 2$, the form considered in Exercise 1.22 above. The resulting bivariate uniform was considered in Example 7.4. Gumbel considered the exponential scheme; see Exercise 7.19; D'Este (1981) develops the bivariate Gamma form.

A general limitation of the FGM system is that the range of possible values of the correlation coefficient is often very restricted. N. L. Johnson and Kotz (1975b, 1977a) both extend the FGM system to $k > 2$ variates and consider more general G-functions. Huang and Kotz (1984) develop an iterative FGM form that allows higher correlations, even with a single iteration.

A second approach that allows completely general margins, due to Plackett (1965), is as follows. If $F(x), G(y)$ are the desired marginal d.f.s, $H(x, y)$ the desired bivariate d.f., and $\psi(\geqslant 0)$ a parameter expressing the dependence between the variates, then $H(x, y)$ is the root of the quadratic in $H(x, y)$

$$(F - H)(G - H)\psi = H(1 - F - G + H) \tag{7.81}$$

that satisfies the necessary condition (due to Fréchet (1951))

$$\max\{F(x) + G(y) - 1, 0\} \leqslant H(x, y) \leqslant \min\{F(x), G(y)\} \tag{7.82}$$

Mardia (1967) shows that if $S = 1 - (F + G)(1 - \psi)$, the required root is

$$H(x, y) = \begin{cases} [\{S^2 + 4\psi(1 - \psi)FG\}^{\frac{1}{2}} - S]/\{2(1 - \psi)\}, & \psi \neq 1 \\ FG, \quad \psi = 1 \end{cases} \tag{7.83}$$

Given F and G, ψ is a monotone increasing function of $H(x, y)$, being equal to zero when H attains its lower bound in (7.82), and to ∞ when H attains its upper bound there. (7.83) or (7.81) shows that $\psi = 1$ corresponds to independence.

Unfortunately, (7.83) does not produce a bivariate normal distribution (Example 3.19, or (16.43) below) when F and G are univariate normal, although H is then fairly close to bivariate normality.

Exercises 7.24–5 give further interesting properties of $H(x, y)$. Mardia (1970b) gives methods of fitting $H(x, y)$ to observed distributions.

Copulas

7.33 A class of distributions with uniform marginals has the joint d.f.

$$\begin{aligned} F(x, y) &= g^{-1}[g(x) + g(y)], \quad \text{if } g(x) + g(y) \leqslant g(0) \\ &= 0, \text{ otherwise;} \end{aligned} \tag{7.84}$$

where $g(1) = 0$, $g'(u) < 0$ and $g''(u) > 0$ for all $0 < u < 1$. The g-functions are known as *copulas*. A bivariate distribution is said to be a copula if its marginal distributions are uniform, even though the bivariate form may contain singular components or exhibit other non-regular behaviour. Making the inverse probability transformation to (u, v) where $u = F_1(x)$ and $y = F_2(v)$ produces a very general class of distributions, see Genest and MacKay (1986a, b).

Rewriting (7.84) as

$$g(F) = g(x) + g(y) \tag{7.85}$$

and differentiating once with respect to each of x and y yields the p.d.f.

$$f(x, y) = -\frac{g''(F)g'(x)g'(y)}{[g'(F)]^3}, \tag{7.86}$$

with a straightforward extension to (u, v). When $g(u) = -c\log(u)$, $c > 0$, it may be shown that x and y are independent. Indeed, this is the only g-function for which (7.84) yields independence; see Exercise 7.26.

Example 7.9 (A bivariate uniform distribution using copulas)

Let $g(u) = (1-u)^{\alpha}$, where $\alpha \geqslant 1$. Then (7.85) yields

$$\begin{aligned} F(x, y) &= 1 - [(1-x)^{\alpha} + (1-y)^{\alpha}]^{1/\alpha}, \\ &\quad \text{provided } (1-x)^{\alpha} + (1-y)^{\alpha} \leqslant 1 \\ &= 0, \text{ otherwise.} \end{aligned} \tag{7.87}$$

The boundary condition is satisfied with probability $-g(0)/g'(0) = \alpha^{-1}$. When $\alpha = 1$, (7.87) satisfies Fréchet's lower bound, given in (7.82). Another example of (7.84) arises when $g(u) = (u^{-\alpha} - 1)/\alpha$, leading to

$$F(x, y) = [x^{-\alpha} + y^{-\alpha} - 1]^{-1/\alpha}. \tag{7.88}$$

This form of copula is used by R. D. Cook and Johnson (1981, 1986) to produce a general class of Burr, Pareto and logistic distributions.

Dall'Aglio *et al* (1991) provide a detailed development of bivariate distributions based upon copulas.

Hazard rate and NBU distributions

7.34 One extension of the hazard rate in **6.42**, due to Basu (1971), is to specify the multivariate rate by the scalar function

$$h(\mathbf{x}) = f(\mathbf{x})/g(\mathbf{x}) \tag{7.89}$$

where $G(\mathbf{x}) = P(X_1 \geqslant x_1, \ldots, X_k \geqslant x_k)$. When the variates are independent, (7.89) reduces to the product of the univariate rates. The scalar nature of (7.89) is, however, a limitation in understanding the behaviour of the distribution for the different variates. To overcome this, N. L. Johnson and Kotz (1975a) defined the vector hazard rate

$$\mathbf{h}(\mathbf{x}) = \left[\frac{\partial}{\partial x_1}, \ldots, \frac{\partial}{\partial x_k}\right] \log G(\mathbf{x}); \tag{7.90}$$

the elements in the vector are equal to the marginal rates when the variates are independent.

Example 7.10 (Hazard rates for the Marshall-Olkin bivariate exponential)

From Example 7.7 it follows that the d.f. may be written as

$$G(x_1, x_2) = \exp[-\lambda_1 x_1 - \lambda_2 x_2 - \lambda_3 \max(x_1, x_2)]$$

so that $\partial \log G/\partial x_1 = \begin{cases} \lambda_1 & \text{, if } x_1 < x_2 \\ \lambda_1 + \lambda_3 & \text{, if } x_1 > x_2 \end{cases}$

with a similar expression for x_2. Thus,

$$\mathbf{h}(\mathbf{x}) = \begin{cases} (\lambda_1, \lambda_2 + \lambda_3), \text{ if } x_1 < x_2 \\ (\lambda_1 + \lambda_3, \lambda_2), \text{ if } x_1 > x_2 \\ (\lambda_1 + \lambda_2 + \lambda_3,\ \lambda_1 + \lambda_2 + \lambda_3), \text{ if } x_1 = x_2. \end{cases}$$

By contrast, $\mathbf{h}(\mathbf{x})$ in (7.89) yields the product of these elements; when $\lambda_1 = \lambda_2$, this rate would be constant for all $x_1 \neq x_2$.

7.35 A multivariate analogue of (6.83) for new better than used (NBU) distributions is that

$$P(\mathbf{T} \geqslant \mathbf{x} + \mathbf{y} \mid \mathbf{T} \geqslant \mathbf{y}) \leqslant P(\mathbf{T} \geqslant \mathbf{x}) \tag{7.91}$$

$$\text{or } G(\mathbf{x} + \mathbf{y}) \leqslant G(\mathbf{x})G(\mathbf{y}) \tag{7.92}$$

for all $\mathbf{x}, \mathbf{y} \geqslant \mathbf{0}$; although other definitions are possible. The properties of multivariate NBU distributions are reviewed in Marshall and Shaked (1986), who also consider results analogous to (6.88–9).

Bivariate distributions with given conditionals

7.36 Thus far, we have concentrated upon joint distributions with prespecified marginals. However, there are circumstances, particularly in the study of spatial and temporal processes, where it is natural to specify the conditional distributions(s) and to derive the joint distribution from there; for a detailed discussion in the spatial context, see Besag (1974). The following results may be developed from Besag's work, although simpler arguments sometimes suffice in the present context.

Suppose that $f_{y|x}(y|x)$ and $f_{x|y}(x|y)$ are given. In order for these conditionals to be compatible with each other, we require the joint p.f. to be

$$h(x, y) = g_{y|x}(y|x)f_x(x) = f_{x|y}(x|y)g_y(y) \tag{7.93}$$

for all values of x and y. These conditions are often rather severe, as the following example from Arnold and Strauss (1988) illustrates.

Example 7.11 (A bivariate distribution with exponential conditions)

Suppose $f_{x|y}(x|y) = a(y)e^{-xa(y)}$, $x > 0$; that is, the conditional distribution for x given y is exponential and that for y given x is also exponential, but with parameter $b(x)$. Then (7.93) requires that

$$h(x, y) = g(y)a(y)e^{-xa(y)} = f(x)b(x)e^{-yb(x)} \tag{7.94}$$

for all $x > 0$ and $y > 0$; the subscripts may now be dropped without causing confusion.

Assume that

$$a(y) = cy + a,\ b(x) = cx + b, \tag{7.95}$$

so that (7.94) becomes

$$f(x)(cx + b)e^{ax} = g(y)(cy + a)e^{by} \tag{7.96}$$

Since (7.96) must hold for all positive values of x and y, it follows that each side of the equation must be constant and

$$f(x) \propto (cx+b)^{-1}e^{-ax} \tag{7.97}$$

Furthermore, if we construct the ratio

$$\frac{h(x+1, y+1)h(x, y)}{h(x+1, y)h(x, y+1)}$$

for any suitable x and y, the left hand side of (7.96) reduces to $\exp[a(y) - a(y+1)]$ whereas the right side is $\exp[b(x) - b(x+1)]$. Since the two exponents must be equal, it follows that the choices in (7.95) are the most general forms possible if (7.93) is to be satisfied. Finally,

$$h(x, y) = h_0 e^{-(ax+by+cxy)}. \tag{7.98}$$

The proportionality constant may be shown to be

$$h_0 = e^{-u}/[-uEi(u)]$$

$$\text{where } -Ei(u) = \int_u^\infty (e^{-w}/w)\,dw \text{ and } u = ab/c.$$

Besag (1974) gives a number of other joint distributions that satisfy the compatibility conditions, including the normal and the Poisson.

Shape distributions

7.37 Consider a set of objects such as the large stones at an historical site. Let the pairs of variates (x_j, y_j) describe the location of the jth object with reference to some 'true' location, (μ_j, ν_j) say. That is, we assume that the objects have been located with a particular purpose in mind, although there may be errors in the exact siting of each object. After we remove information relating to the absolute locations of the objects, the scale on which they are arrayed and the orientation of the site, the objects' relative positions may be summarized by the set of triangles defined by all sets of three distinct objects. It is this residual pattern that may be defined as *shape*; c.f. D.G. Kendall in the discussion on Bookstein (1986).

7.38 Suppose now that (x_j, y_j) are binormally distributed with means (μ_j, ν_j) and covariance matrix $\sigma^2\mathbf{I}, j = 1, 2, 3$. We may summarize the shape of the triangle using Bookstein's (1986) shape variables:

$$\begin{aligned} U &= -\tfrac{1}{2} + \frac{(x_2 - x_1)(x_3 - x_1) + (y_2 - y_1)(y_3 - y_1)}{(x_2 - x_1)^2 + (y_2 - y_1)^2} \\ V &= \frac{(x_2 - x_1)(y_3 - y_1) - (y_2 - y_1)(x_3 - x_1)}{(x_2 - x_1)^2 + (y_2 - y_1)^2} \end{aligned} \tag{7.99}$$

where $(x_2 - x_1)^2 + (y_2 - y_1)^2 > 0$. The geometrical interpretation of these variables is that (u, v) are coordinates of the third point after translating, rotating and rescaling the triangle so that points 1 and 2 are sent to $(-\frac{1}{2}, 0)$ and $(\frac{1}{2}, 0)$. These shape variables are preferred to the two internal angles as shape measures since the information on the reflections is preserved and the description of pathological cases such as collinear points is straightforward. Other suitable shape variables include D. G. Kendall's (1984) spherical coordinates; there is a one-to-one transformation between Bookstein's and Kendall's coordinate systems.

7.39 In order to arrive at the distribution of (u, v), we outline the approach developed by Mardia and Dryden (1989). In keeping with the geometric interpretation outlined above, they transform from (x_3, y_3) to (u, v) while keeping (x_1, y_1) and (x_2, y_2) fixed; note that the Jacobian is $J = (x_2 - x_1)^2 + (y_2 - y_1)^2$. The inverse transform is, from (7.99).

$$\begin{aligned} x_3 &= (u + \tfrac{1}{2})(x_2 - x_1) - v(y_2 - y_1) + x_1, \\ y_3 &= (u + \tfrac{1}{2})(y_2 - y_1) + v(x_2 - x_1) + y_1. \end{aligned} \tag{7.100}$$

We may integrate over (x_1, y_1, x_2, y_2) by identifying their joint multinormal density and thereby reduce the integral to the calculation of the mean of a non-central chi squared distribution with two degrees of freedom (cf **23.4-6**, Volume 2A).

The joint density of (u, v) eventually reduces to

$$f(u, v) = \frac{1}{3\pi(\frac{1}{2} + \frac{2}{3}u^2 + \frac{2}{3}v^2)}\{1 + \kappa(1 + \cos 2\delta)\}\exp\{-\kappa(1 - \cos 2\delta)\}, \tag{7.101}$$

where

$$\kappa = \left\{\sum_1^3 (\mu_j - \bar{\mu})^2 + (\nu_j - \bar{\nu})^2\right\} / (4\sigma^2),$$

$\bar{\mu} = \sum \mu_j/3$, $\bar{\nu} = \sum \nu_j/3$, and δ is the Riemannian distance between the observed shape (u, v) and the population shape, (θ, ϕ) say, of $(\mu_j, \nu_j)'$, $j = 1, 2, 3$; that is,

$$\cos 2\delta = 1 - \frac{24\{(u - \theta)^2 + (v - \phi)^2\}}{(3 + 4u^2 + 4v^2)(3 + 4\theta^2 + 4\phi^2)}$$

Note that the distribution has three parameters: the population shape $(\theta, \phi)'$ and a concentration parameter κ.

As $\kappa \to 0$ the distribution becomes D. G. Kendall's (1984) uniform distribution in the shape space. As $\kappa \to \infty$ the distribution tends to Bookstein's (1986) normal approximation. An alternative derivation is given by D. G. Kendall (1991).

EXERCISES

7.1 For the multinomial (7.15) show that if $a_i = p_i/p_0$ then

$$\kappa(r_1, r_2, \ldots, r_i + 1, \ldots, r_k) = a_i \frac{\partial}{\partial a_i} \kappa(r_1, r_2, \ldots, r_i, \ldots, r_k)$$

where the cumulant on the right is of order r_1 in p_1 etc. (Guldberg, 1935). Hence derive equations (7.18).

7.2 Verify that the multivariate skewness and kurtosis measures, $\beta_1(k)$ and $\beta_2(k)$ in (7.8-9) are invariant to linear transformations. Hence show that, for the multinormal distribution, $\beta_1(k) = 0$ and $\beta_2(k) = k(k+2)$.

(Mardia, 1970c, 1974)

7.3 *The negative multinomial distribution*. Suppose that a population may be partitioned into $(k+1)$ distinct classes as in 7.7, but that sampling is continued until m members of class B_0 are observed. Show that the probabilities are

$$f(x_1, x_2, \ldots, x_k) = \frac{(R+m-1)!}{(m-1)!\,x_1!\ldots x_k!} p_1^{x_1} p_2^{x_2} \ldots p_k^{x_k} p_0^m,$$

where $R = \Sigma x_i$ and each $x_i \geqslant 0$. Show that this distribution has c.f.

$$p_0^m(1 - p_1 e^{\theta_1} - \ldots - p_k e^{\theta_k})^{-m}$$

with cumulants (for $k = 3$)

$$\kappa_{100} = mp_1/p_0, \quad \kappa_{200} = mp_1(p_0 + p_1)/p_0^2, \quad \text{and} \quad \kappa_{110} = mp_1p_2/p_0^2.$$

[Note that $\kappa_{11} > 0$ for all pairs of variates, whereas $\kappa_{11} < 0$ for the multinomial].

7.4 Use (7.10) to show that the conditional p.g.f. of the bivariate binomial for x_1, given that $x_2 = u$, is

$$(p_{01} + p_{11}t)^u (p_{00} + p_{10}t)^{n-u}.$$

Hence show that the conditional expectation for x_2 is linear in x_1, and that its variance is also a function of x_1.

Consider the extended bivariate binomial with p.g.f.

$$(q + pt_1)^{n_1-k}(q' + p't_2)^{n_2-k}(p_{00} + p_{10}t_1 + p_{01}t_2 + p_{11}t_1t_2)^k,$$

k a non-negative integer, and show that the conditional expectation remains linear.

(Hamdan and Jensen, 1976; Kocherlakota, 1989)

7.5 For the bivariate Poisson with c.f. given by (7.32), show that the p.m.f. satisfies the recurrence relations

$$x_1 f(x_1, x_2) = \lambda_{10} f(x_1 - 1, x_2) + \lambda_{11} f(x_1 - 1, x_2 - 1)$$
$$x_2 f(x_1, x_2) = \lambda_{01} f(x_1, x_2 - 1) + \lambda_{11} f(x_1 - 1, x_2 - 1).$$

(Teicher, 1954)

7.6 (*The bivariate Poisson distribution*). Show that when p_{01}, p_{10} and p_{11} in equation (7.20) are small, but $np_{11} = \lambda_3$, $np_{10} = \lambda_1 - \lambda_3$ and $np_{01} = \lambda_2 - \lambda_3$ are finite, the distribution tends to the form whose general term is

$$f(x_1, x_2) = \sum \frac{\lambda_3^k(\lambda_1 - \lambda_3)^i (\lambda_2 - \lambda_3)^j}{i!\,j!\,k!} e^{-\lambda_1 - \lambda_2 + \lambda_3},$$

where the summation is over all i, j, k such that $x_1 = i + k, x_2 = j + k$. Show that the c.f. is given by (7.56), the three Poisson distributions having parameters $\lambda_1 - \lambda_3, \lambda_2 - \lambda_3$ and λ_3 respectively.

7.7 Consider the bivariate Poisson with p.g.f.

$$P(t_1, t_2) = \exp[\lambda(t_1 - 1)] + \exp[\lambda(t_2 - 1)] + (1 - 2p_0)(t_1 - 1)(t_2 - 1) - 1,$$

where $p_0 = e^{-\lambda} \leqslant \frac{1}{2}$ and $(2 + \lambda)e^{-\lambda} \geqslant 1$. Show that the correlation between x_1 and x_2 is always negative.

(Griffiths, Milne and Wood, 1979)

7.8 Derive (7.38) and (7.39) for the negative multinomial.

7.9 Assume that $x_0, x_1, \ldots, x_k$ are independent Poisson variates with parameters $\lambda_0, \lambda_1, \ldots, \lambda_k$. Show that $n = x_0 + x_1 + \ldots + x_k$ is Poisson distributed with parameter $\lambda = \lambda_0 + \lambda_1 + \ldots + \lambda_k$. Hence show that the joint distribution of $(x_1, \ldots, x_k)$ given n is multinomial with $p_j = \lambda_j/\lambda$.

7.10 (*Bivariate logarithmic series distribution*) Consider the zero-truncated version of the negative multinomial given by (7.38); that is, condition upon the event $x_1 + x_2 \geqslant 1$. Show that, as $r \to 0$, the p.m.f. approaches the bivariate LSD:

$$f(x_1, x_2) = \frac{(x_1 + x_2 - 1)!}{x_1! x_2! B} \frac{\alpha_1^{x_1} \alpha_2^{x_2}}{(c + \alpha_1 + \alpha_2)^{x_1 + x_2}} \qquad \text{(A)}$$

where $B = -\log[1 - (\alpha_1 + \alpha_2)/c]$. Further, show that the marginal distributions derived from (A) are not the univariate LSD given by (5.74).

(Patil and Bildikar, 1967)

7.11 (*Bivariate Hermite distribution*) The bivariate Hermite distribution has p.g.f.

$$P(t_1, t_2) = \exp[\lambda_{10}(t_1 - 1) + \lambda_{20}(t_1^2 - 1) + \lambda_{01}(t_2 - 1) + \lambda_{02}(t_2^2 - 1) + \lambda_{11}(t_1 t_2 - 1)].$$

Show that the distribution may be derived from two independent Poissons using either (A) a randomly stopped sum mechanism (cf **5.21**) with the bivariate binomial (7.21) or (B) a binormal mixing distribution suitably truncated (cf Exercise 5.21).

Hence show that $\kappa_{10} = \lambda_{10}, \kappa_{20} = \lambda_{10} + 2\lambda_{20}$ and $\kappa_{11} = \lambda_{11}$.

(C. D. Kemp and Papageorgiou, 1982).

7.12 (*The Dirichlet-multinomial distribution*) Derive the mixture distribution for multinomial (7.15) with the Dirichlet (7.64) as mixing distribution. Show that the p.m.f. is

$$f(x_1, \ldots, x_k) = \frac{n!\Gamma(\alpha)}{\Gamma(n + \alpha)} \prod_{j=1}^{k} \left[\frac{\Gamma(\alpha_j + x_j)}{\Gamma(\alpha_j)}\right] \qquad \text{(A)}$$

where $\alpha = \alpha_0 + \alpha_1 + \ldots + \alpha_k$. Verify that (A) is of the same form as the negative multiple hypergeometric when all the $\{\alpha_j\}$ are integers. Show that the means and variances are

$$\mu_j' = n\alpha_j/\alpha, \quad \sigma_j^2 = n(n + \alpha)\alpha_j(\alpha - \alpha_j)/[\alpha^2(\alpha + 1)]$$

with correlations $\rho_{ij} = -[\alpha_i\alpha_j/(\alpha - \alpha_i)(\alpha - \alpha_j)]$.

(Mosimann, 1962)

7.13 (*The Dirichlet-negative multinomial distribution*) Repeat the process outlined in Exercise 7.12 for the negative multinomial distribution of Exercise 7.3. Show that the p.m.f. is

$$f(x_1, \ldots, x_k) = \frac{\Gamma(m + \alpha_0)\Gamma(x + m)\Gamma(\alpha)}{\Gamma(\alpha_0)\Gamma(m)\Gamma(m + x + \alpha)} \prod_{j=1}^{k} \frac{\Gamma(x_j + \alpha_j)}{\Gamma(\alpha_j) x_j!}$$

where $\alpha = \alpha_0 + \alpha_1 + \ldots + \alpha_k$ and $x = x_1 + \ldots + x_k$.

Show that $\mu_j' = m\alpha_j/m_0$, $\sigma_j^2 = m(m+\alpha_0)\alpha_j(\alpha_0+\alpha_j)/\alpha_0^2(\alpha_0-1)$ and that $\rho_{ij} = [\alpha_i\alpha_j/(\alpha_0+\alpha_i)(\alpha_0+\alpha_j)]^{\frac{1}{2}}$.

(Mosimann, 1963)

7.14 If X and Y are independently normal with means μ_1, μ_2, and common variance σ^2, show that

$$P(X^2 \geqslant Y^2) = P(U \geqslant 0)P(V \geqslant 0) + P(U < 0)P(V < 0),$$

where $U = X - Y$ and $V = X + Y$. Show that the result still holds when X and Y are correlated.

7.15 Verify (7.63). Hence show that

$$\kappa_{10} = \kappa_{01} = \frac{1}{2\lambda}, \qquad \kappa_{20} = \kappa_{02} = \frac{1}{4\lambda^2}, \qquad \text{and} \qquad \kappa_{11} = \frac{1}{12\lambda^2}$$

so that the correlation between x and y is $\frac{1}{3}$.

7.16 Consider the bivariate exponential distribution in Example 7.5 where u_j has parameter λ_j. Show that the joint distribution of (x, y) has p.d.f.

$$f(x, y) = \frac{\lambda_1\lambda_2\lambda_3}{\lambda_0} e^{-\lambda_1 x - \lambda_2 y}(e^{\lambda_0 z} - 1)$$

where $z = \min(x, y)$ and $\lambda_0 = \lambda_1 + \lambda_2 - \lambda_3 \neq 0$. Further, when $\lambda_0 = 0$ show that the p.d.f. reduces to

$$f(x, y) = \lambda_1\lambda_2\lambda_3 z e^{-\lambda_1 x - \lambda_2 y}.$$

By considering the case $\lambda_1 = \lambda_2 = 1$, $\lambda_3 = 2$ show that $E(y|x)$ is no longer linear in x.

7.17 Assume that (u_1, u_2) and (u_3, u_4) are two independent sets of observations from a standard binormal distribution with correlation parameter ρ. Let

$$x = \tfrac{1}{2}(u_1^2 + u_3^2) \quad \text{and} \quad y = \tfrac{1}{2}(u_2^2 + u_4^2).$$

Show that the joint distribution of x and y has c.f.

$$\phi(t_1, t_2) = [(1 - it_1)(1 - it_2) + \rho^2 t_1 t_2]^{-1};$$

show that the marginals are exponential and that $\operatorname{corr}(x_1, x_2) = \rho^2$.

(Moran, 1967)

7.18 An instrument has two components, A and B say, that operate in parallel, whose lifetimes are exponentially distributed with parameters α and β respectively. When A fails, the parameter of B changes to β_0; when B fails, the A parameter changes to α_0.

Show that the joint p.d.f. is

$$f(x, y) = \begin{cases} \alpha\beta_0 \exp[-\beta_0 y - (\alpha + \beta - \beta_0)x], & x < y \\ \alpha_0\beta \exp[-\alpha_0 x - (\alpha + \beta - \alpha_0)y], & x \geqslant y. \end{cases}$$

Show that

$$E(x) = \frac{(\alpha_0 + \beta)}{\alpha_0(\alpha + \beta)},$$

$$\operatorname{var}(x) = \frac{(\alpha_0^2 + 2\alpha\beta + \beta^2)}{\alpha_0^2(\alpha + \beta)^2}$$

and that $-\frac{1}{3} < \operatorname{corr}(x, y) < 1$.

(Freund, 1961; extended to the Gamma distribution by Becker and Roux, 1981)

7.19 Consider the bivariate exponential given by (7.80) with $G_j = 1 - F_j = \exp(-x_j)$, $j = 1, 2$. Show that $\text{corr}(x_1, x_2) = \alpha/4$ and so ρ is less than $\frac{1}{4}$ in absolute value.

(Gumbel, 1960)

7.20 If the quadratics Q_1 and Q_2 in (7.68) possess no common factor, show that this implies that $Q_i = Q_i(x_i)$, $i = 1, 2$; that is, Q_i is a function of x_i only. In turn, show that this implies that $L_i = L_i(x_i)$ and hence that the random variables, x_1 and x_2, are independent.

(Van Uven, 1947–8; Mardia, 1970a)

7.21 If x, y are jointly distributed in the bivariate normal form

$$f(x, y) = \frac{1}{2\pi(1-\rho^2)^{\frac{1}{2}}} \exp\left\{-\frac{1}{2(1-\rho^2)}(x^2 - 2\rho xy + y^2)\right\}, \qquad -\infty < x, y < \infty,$$

and $D_1 = d/dx$, $D_2 = d/dy$, show that the c.f. of

$$\exp\left\{\kappa_{11} D_1 D_2 + \sum_{r+s \geq 3} (-1)^{r+s} \kappa_{rs} \frac{D_1^r D_2^s}{r!\,s!}\right\} \alpha(x)\alpha(y) \tag{A}$$

$$= \exp\left\{-\tfrac{1}{2}(t_1^2 + 2\rho t_1 t_2 + t_2^2) + \sum_{r+s \geq 3} \kappa_{rs} \frac{(it_1)^r}{r!} \frac{(it_2)^s}{s!}\right\}$$

Hence show that the bivariate form of the Edgeworth series may be written either as at (A) or in the form

$$\exp\left\{\sum_{r+s \geq 3} (-1)^{r+s} \kappa_{rs} \frac{D_1^r D_2^s}{r!\,s!}\right\} f(x, y). \tag{B}$$

(Kendall, 1949b)

7.22 From (A) of the previous exercise show that formally

$$\frac{\partial f}{\partial \kappa_{rs}} = (-1)^{r+s} \frac{D_1^r}{r!} \frac{D_2^s}{s!} f, \qquad r + s \geq 3, \text{ or } r = s = 1,$$

and verify this formula directly for the bivariate normal distribution for $r = s = 1$, namely

$$\frac{\partial f}{\partial \rho} = \frac{\partial^2 f}{\partial x\, \partial y}.$$

(Kendall, 1949b)

7.23 Show that the percentile function, $x = F^{-1}(p)$, for the Burr distribution (6.77) is

$$x = \left\{\left(\frac{1}{1-p}\right)^{1/\beta} - 1\right\}^{1/c}$$

and hence that the median is $\mu_M = (2^{1/\beta} - 1)^{1/c}$. If the joint distribution of x_1 and x_2 is given by (7.79), show that the median regression function for x_2 given x_1 is

$$x_2^* = \mu_M\{(1 + a_1 x_1^c)/a_2\}^{1/c},$$

with β replaced by $\beta + 1$.

7.24 In **7.31**, show that if $F(x)$ is replaced by its complement $1 - F(x)$, $H(x, y)$ at (7.83) has ψ replaced by ψ^{-1}.

(Steck, 1968)

7.25 In 7.31, show that if the marginal distribution of x is dichotomized at any point, and that of y at any point, the expected frequencies

$$\frac{a\,|\,b}{c\,|\,d}$$

in the resulting four quadrants satisfy $\psi = bc/(ad)$ whatever the dichotomizing points.

(Plackett, 1965)

7.26 Show that the variates x and y with d.f. (7.82) are independent if and only if $g(u) = -c\log(u), c > 0$.

(Genest and MacKay, 1986a, b)

7.27 Show that the bivariate exponential distribution in Exercise 7.18 is an IHR distribution using the multivariate hazard rate (7.85).

(N. L. Johnson and Kotz, 1975a)

CHAPTER 8

PROBABILITY AND STATISTICAL INFERENCE

8.1 The previous seven chapters have dealt with the theory of statistical distributions from the descriptive point of view. We have seen that the distributions encountered in practice exhibit certain regular features that lead to theoretical descriptions, that they can be characterized by certain quantities such as their moments and cumulants and that certain general theorems about distribution and density functions can be deduced. We now begin a study of a different kind: we ask whether any meaningful and objective statements can be made about a population or a system generating a population when only a sample of the possible observations is available for scrutiny. This, in broad terms, is the problem of statistical inference.

8.2 Except in trivial cases it is not possible to make inferences from sample to population with the certainty of deductive logic. Either our statements must be somewhat vaguely phrased or, if precise, they are surrounded by a band of doubt. In ordinary speech we use various words and phrases to denote the attitude of mind towards propositions of whose truth we are uncertain. We speak of something as being more or less *probable*, or more or less *likely*; we say that we have a certain degree of *confidence* in a proposition or that the *chances* are against it; we refer to it as being surrounded by or open to *doubt* or requiring *confirmation*. Many of these expressions have acquired technical meanings in Statistics, as we shall indicate in due course. Fundamentally they all express aspects of the same thing, the uncertainty with which we are compelled by circumstances to regard particular propositions.

8.3 In its broadest sense, the theory of probability is a branch of scientific methodology; cf Popper (1968). In this chapter, we focus primarily upon the theory, or "calculus", of probability and give only a brief treatment of statistical inference, which is considered in depth in Volumes 2A and 2B.

The English language is to some extent responsible for a confusion between the narrow and broad definitions. In medieval times and at the Renaissance there was a distinction between *probabilitas*, which related to the intensity of degrees of belief, and the *doctrine of chances*, which calculated the number of ways in which certain classes of event could occur. In the seventeenth and eighteenth centuries the doctrine of chances became assimilated to "probability" so that nowadays the same word is used to denote both. A useful distinction has been lost. We may well admit "probability" as including chances; but we shall continue to distinguish between the *theory of probability*, which is concerned with the mathematical development of the consequences of certain axioms and postulates, and *philosophical concepts of probability* which are used when making statements about the external world. In many respects, use of the term "calculus of probabilities" to describe the mathematical theory provides a clearer distinction. Indeed, this term was used in earlier editions of this volume. However, "theory" is the description now commonly applied to the "calculus", and we shall bow to this convention.

8.4 The subject of probability has a rich and controversial history. The interested reader may consult the classic account by Todhunter (1949, first published in 1865) and the volumes by David (1962) and Maistrov (1974). Over the years, *Biometrika* has published a series of papers on the history of probability and statistics. Selections of these papers appear in the volumes edited by Pearson and Kendall (1970) and by Kendall and Plackett (1977). The history of subjective probability is chronicled in the selection of classic papers edited by Kyburg and Smokler (1964).

Hald (1989) provides an historical account of probability and statistics up to 1750, whereas Stigler (1986) covers the period up to 1990. The two volumes edited by Kruger, Daston and Heidelberger (1987) focus more upon the philosophical development of probability. Dale (1991) outlines the history of inverse probability up to the early twentieth century and provides a detailed assessment of all the major contributors to the field from Thomas Bayes through to Karl Pearson.

The rules of probability

8.5 We now proceed to develop the basic rules of probability using a set theoretic framework but using the labels that have become standard in probability. Thus, the set of all possible outcomes of a statistical experiment is called the sample space (universal set) Ω, and the elementary outcomes of the experiment are the elements $\omega_i \in \Omega$. In a single realization of the experiment, exactly one of the elementary outcomes occurs. For the present, we assume that Ω has $N(<\infty)$ elements, although this assumption will be relaxed later.

An event, A, is a set containing one or more elementary outcomes so that $A \subseteq \Omega$. We say that the event A occurs if the statistical experiment results in one of the elementary outcomes contained in A. Other events may be similarly defined. The further notation taken from set theory may be summarized as follows:

ϕ denotes the impossible event (empty set).

$A \cap B$ denotes the composite event "both A and B occur" (A intersection B), or that the recorded outcome, ω_i, is such that $\omega_i \in A$ and $\omega_i \in B$.

$A \cup B$ denotes the composite event "A or B" (A union B) where *or* is the inclusive or, meaning that at least one event occurs. That is, at least one of the statements $\omega_i \in A$ and $\omega_i \in B$ must be true.

A^c denotes the complementary event, not-A. Thus, if $\omega_i \in A$ then $\omega_i \notin A^c$ and vice versa.

Two events are said to be *mutually exclusive* if $A \cap B = \phi$; clearly, A and A^c are mutually exclusive by construction.

In order to specify a probability function, we use the notion of an additive set function, S, which is defined for all sets $A \subset \Omega$ and which, when $A \cap B = \phi$, satisfies the relation

$$S(A \cup B) = S(A) + S(B). \tag{8.1}$$

Armed with these preliminaries, we now assume that probability is measurable on a continuous scale, so that the probability of any event A can be represented as a real number, $P(A)$ say. This assumption implies, among other things, that any two probabilities may be compared. Event A is more or less probable than event B according as $P(A) >$ or $< P(B)$. When $P(A) = P(B)$, they are equiprobable. These requirements add up to the assumption that P is an additive set function. To complete the specification, we must determine the end-points of the probability scale. We shall let the probability of the certain event be one, and that of the impossible event be zero. Thus, for any event A, $0 \leqslant P(A) \leqslant 1$.

8.6 The ideas outlined above lead to the classical axioms of probability formulated by A. N. Kolmogorov (1933). These are

$$(1)\ P(A) \geqslant 0, \tag{8.2}$$

$$(2)\ P(\Omega) = 1, \tag{8.3}$$

and

$$(3)\ P(A \cup B) = P(A) + P(B), \text{ whenever } A \cap B = \phi. \tag{8.4}$$

It is rather remarkable that these simple rules are all that is required in order to develop the mathematical theory of probability.

The following results may be established directly for any events A, $B \subset \Omega$:

(a) $P(A \cup B) = P(A) + P(B) - P(A \cap B)$
(b) $P(\phi) = 0$
(c) $P(A^c) = 1 - P(A)$
(d) $P(A) = P(A \cap B) + P(A \cap B^c)$

These can be seen from the Venn diagrams in Fig. 8.1 , all are readily demonstrated by standard arguments. For example, $(A \cap B)$ and $(A \cap B^c)$ are mutually exclusive since $B \cap B^c = \phi$. Also,

$$(A \cap B) \cup (A \cap B^c) = A,$$

so that (d) follows from the third axiom. An immediate extension from axiom (3) yields

$$P\left(\bigcup_{i=1}^{k} B_i\right) \equiv P(B_1 \cup B_2 \cup \ldots \cup B_k) = \sum_{i=1}^{k} P(B_i), \tag{8.5}$$

where $B_i \cap B_j = \phi$, for all $i \neq j$. More generally, when no restrictions are placed upon the B_i, Boole's inequality states that

$$\sum_{i=1}^{k} P(B_i) \geqslant P\left(\bigcup_{i=1}^{k} B_i\right). \tag{8.6}$$

When $B_i \cap B_j = \phi$, $i \neq j$ and $\bigcup_{i=1}^{k} B_i = \Omega$, the B_i are mutually exclusive and *exhaustive*; that is, they form a partition of Ω. From (8.5) it follows that

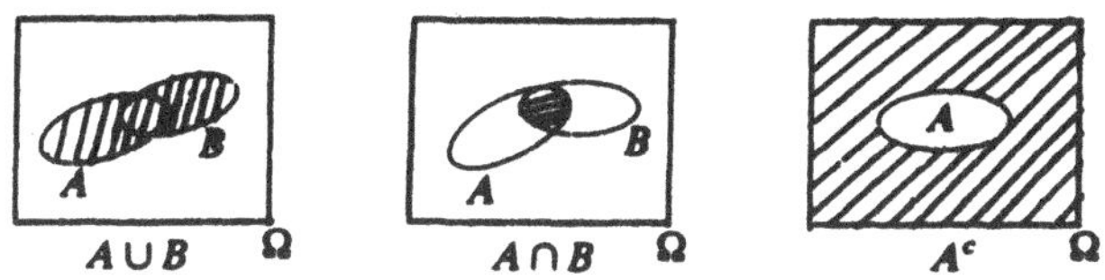

Fig. 8.1 Venn diagrams for different events

$$\sum_{i=1}^{k} P(B_i) = P(\Omega) = 1.$$

For any event A, let

$$C_i = A \cap B_i, \qquad i = 1, \ldots, k; \quad C_{k+1} = A^c,$$

then the C_i form a partition of Ω whenever the B_i do, so that

$$\sum_{i=1}^{k+1} P(C_i) = 1$$

or

$$\begin{aligned}\sum_{i=1}^{k} P(C_i) &= \sum_{i=1}^{k} P(A \cap B_i) \\ &= 1 - P(A^c) = P(A).\end{aligned} \tag{8.7}$$

Result (8.7) is known as the Law of Total Probability.

Conditional probability and Bayes' Theorem

8.7 Conditional distributions were introduced in **1.32–4**. In conditioning, we are restricting the set of possible outcomes of the experiment to some subset $B \subset \Omega$. If the outcome is ω_i and $\omega_i \in A$, it follows that $\omega_i \in A \cap B$. The maximum possible value of $P(A \cap B)$ is $P(B)$, so to obtain a conditional probability function that satisfies the axioms (8.2)–(8.4), we define the conditional probability for A given B as

$$P(A|B) = \frac{P(A \cap B)}{P(B)}, \tag{8.8}$$

provided $P(B) > 0$. In effect, B becomes the new sample space. By symmetry, it is evident that

$$P(A \cap B) = P(A|B)P(B) = P(B|A)P(A). \tag{8.9}$$

The two events A and B are *independent* if and only if

$$P(A \cap B) = P(A)P(B)$$

whence $P(A|B) = P(A)$ and $P(B|A) = P(B)$. This is in accord with our earlier definition of independence in **1.34**.

Combining (8.7) and (8.9), we see that, for all $j = 1, \ldots, k$,

$$P(B_j|A) = P(A|B_j)P(B_j)/P(A)$$
$$= P(A|B_j)P(B_j)\Big/\sum_{i=1}^{k} P(A|B_i)P(B_i), \quad (8.10)$$

which is *Bayes' Theorem*, after Thomas Bayes (1764) who first propounded it. The theorem gives the probabilities of the B_j when A *is known to have occurred*. Thus, the quantities $P(B_j)$ are termed *prior* probabilities and the $P(B_j|A)$ are called *posterior* probabilities. Finally, $P(A|B_j)$ is termed the *likelihood*, so that Bayes' theorem may be restated as

$$\text{posterior} \propto (\text{prior}) \cdot (\text{likelihood}).$$

8.8 Thus far, we have developed the mathematical theory of probability, and the results are unexceptionable. In order to make the theory operational, we must introduce a concept of probability that links the mathematics to an external world of measurable phenomena. The major approaches to probability have used, respectively, the concepts of equally likely events, relative frequencies in repeated trials, and degrees of belief. The interpretation of probability as a degree of belief further subdivides into the logical (or objective) and personal (or subjective) schools of thought.

In principle, any concept leading to a set of probabilities that "obeys the rules" may be admitted as valid. Each individual must decide whether to allow such a definition in his or her scheme of things. In turn, the choice of probabilistic framework will determine the nature of the statistical inferences that can be made.

Equally likely events

8.9 The simplest approach is to assume that all elementary outcomes have the same chance of occurring. That is, given N elements in Ω,

$$P(\{\omega_i\}) = N^{-1}, \quad i = 1, \ldots, N. \quad (8.11)$$

The elementary events are then termed *equally likely*. If the event A contains m such outcomes, it follows that $P(A) = m/N$. As a *definition of* probability, this approach is open to the logical objection of being circular, for it contains the phrase "equally likely". Nevertheless, physical symmetries may exist in a statistical experiment that render the definition usable. Simple examples such as coin-tossing and dice-throwing come to mind. The notion of simple random sampling from a finite population also rests upon the assumption of equally likely events; see Chapter 9.

8.10 We now present a few examples that employ the notion of equally likely outcomes. These examples are intended primarily to illustrate the rules of probability and to emphasize the importance of adhering strictly to those rules.

Example 8.1 (The probability of equally likely events)

Three coins are tossed. What is the probability that they fall either all heads or all tails?

We assume that the probability of a head with any coin is $\frac{1}{2}$ and that the result with one coin is independent of that with the others. Then there are eight possible and equiprobable cases, *HHH*, *HHT*, *HTH*, *HTT*, *THH*, *THT*, *TTH*, *TTT*. Two of these give us all heads or all tails and hence the required probability is $\frac{1}{4}$.

We now consider three alternative, but invalid, arguments.

(i) There are two possibilities; either the three coins all fall alike or two of them are alike and the other different. Of these two possibilities one is of the type required and therefore the probability is $\frac{1}{2}$.

(ii) There are four possibilities; three heads, two heads and a tail, two tails and a head, three tails. Two of these four are of the type required and therefore the probability is $\frac{1}{2}$.

(iii) Of the three coins, two must fall alike. The other must either be the same as these two or different. Thus there are two possibilities and again the chance is $\frac{1}{2}$.

These three arguments are fallacious. For example, in the first case, it is true that there are two possibilities, but they are not equally probable under our assumptions. The reader may care to examine why this is so and how the other two arguments break down on the same point.

Example 8.2 (Enumeration of equally likely outcomes)

In a hand at bridge, what is the probability that a player has at least two aces in his hand? The question as posed is not unambiguous, and there are in fact four distinct probabilities, in any of which we may be interested:

(a) The probability that a specified player (say, South) has at least two aces, irrespective of the other players' hands.

(b) The probability that some player of the four has at least two aces, irrespective of the other three players' hands.

(c) The probability that South and no other player has at least two aces.

(d) The probability that exactly one player of the four has at least two aces.

We proceed to evaluate these in turn.

(a) South may have 0, 1, 2, 3 or 4 aces. The pack of 52 cards may be regarded as consisting of a pack of 4 aces and a pack of 48 other cards. The probability that South gets no ace is

$$p_0 = \binom{4}{0}\binom{48}{13} \bigg/ \binom{52}{13},$$

and the probability of his getting exactly one ace is

$$p_1 = \binom{4}{1}\binom{48}{12} \bigg/ \binom{52}{13}.$$

The probability we require is

$$\begin{aligned} p_a &= 1 - (p_0 + p_1) \\ &= 1 - \frac{38 \cdot 37 \cdot 11}{49 \cdot 25 \cdot 17} = 0.257. \end{aligned}$$

(b) It is tempting to imagine that the second of our probabilities is simply four times the first, but the four constituent events are not mutually exclusive. This is otherwise clear from the fact that $4p_a > 1$. In fact, the probability we now seek is simply the complement of the probability that each player has one ace, and is

$$p_b = 1 - \frac{\binom{4}{1}\binom{48}{12}\cdot\binom{3}{1}\binom{36}{12}\cdot\binom{2}{1}\binom{24}{12}\cdot\binom{1}{1}\binom{12}{12}}{\binom{52}{13}\binom{39}{13}\binom{26}{13}\binom{13}{13}},$$

whence

$$p_b = 1 - \frac{13^3}{49\cdot 25\cdot 17} = 0.895.$$

(c) The third probability is less than p_a by the probabilities that (South, North), (South, East) or (South, West) each have two aces. These three events are mutually exclusive, and each has probability equal to

$$p_{2,2} = \frac{\binom{4}{2}\binom{48}{11}\cdot\binom{2}{2}\binom{37}{11}}{\binom{52}{13}\binom{39}{13}} = 0.022\,48.$$

The probability we require is

$$p_c = p_a - 3p_{2,2} = 0.190.$$

(d) The fourth probability is the sum of the probabilities of four mutually exclusive events, each with probability p_c. Thus

$$p_d = 4p_c = 0.759.$$

Example 8.3 (Inductive methods for enumerating outcomes)

Our first two examples enumerated the basic possibilities directly. A large class of examples for which direct enumeration may be difficult can be solved by inductive or semi-inductive methods. We will illustrate one such method on a very old problem.

n letters, to each of which corresponds an envelope, are placed in the envelopes at random. What is the probability that no letter is placed in the right envelope?

The condition that the letters are put in the envelopes "at random" is to be interpreted as meaning that every possible way of assigning the letters to envelopes is equally probable. The question, using (8.11), then reduces to the purely algebraic one: in what proportion of the possible cases does no letter get into the right envelope?

Suppose that u_n is the number of ways in which all the letters can go wrong. Consider any particular letter. If this occupies another's envelope and vice versa, which can happen in $n-1$ ways, the number of ways in which the remaining $n-2$ letters can go wrong is u_{n-2}. But if the letter occupies another's place and not vice versa, which can happen in $n-1$ ways, there are u_{n-1} ways in which others can go wrong. Hence we have the difference equation

$$u_n = (n-1)(u_{n-1} + u_{n-2}).$$

We may write this as

$$u_n - nu_{n-1} = -\{u_{n-1} - (n-1)u_{n-2}\}$$

and putting

$$v_n = u_n - nu_{n-1},$$

we find

$$\begin{aligned} v_n &= -v_{n-1}, \\ &= (-1)^{n-2}v_2. \end{aligned}$$

Thus

$$u_n - nu_{n-1} = (-1)^{n-2}(u_2 - 2u_1).$$

But $u_1 = 0$ and $u_2 = 1$ and thus

$$u_n - nu_{n-1} = (-1)^n$$

whence

$$u_n = n!\left\{\frac{1}{2!} - \frac{1}{3!} + \ldots + \frac{(-1)^n}{n!}\right\}.$$

The total number of possible ways of arranging the letters is $n!$ and hence the required probability is

$$\frac{1}{2!} - \frac{1}{3!} + \ldots + \frac{(-1)^n}{n!}, \tag{8.12}$$

i.e. (8.12) is the first $n+1$ terms of e^{-1}, the first two cancelling each other and not appearing. It is evident that the limiting value is approached very rapidly.

Frequency theory of probability

8.11 Once we allow the possibility that elementary outcomes are not equally likely, we must modify our definition of probability. The view that has been dominant in the statistical literature in the twentieth century and underlies the approach to statistical inference adopted in Volume 2A is the *frequency* theory.

The frequentist view assumes that it is possible to consider an infinite sequence of independent repetitions of the same statistical experiment. If M denotes the number of experiments performed and M_A the number of times that the event A occurred in those experiments, the probability of event A is the limiting relative frequency

$$P(A) = \lim_{M\to\infty} (M_A/M). \tag{8.13}$$

This definition provides an objective, empirical view whereby the probability of an event is uniquely specified as a number on the scale from zero to one, even though its actual value is unknown. This is in contrast to the assumption of equally likely outcomes where (8.11) allowed a direct statement of the numerical value. It is evident from (8.13) that frequency probabilities satisfy the axioms by construction.

The standard work on the frequency approach is Von Mises (1957). For further discussion, see Barnett (1982, pp. 75–80).

Logical probability

8.12 The principal alternative to the frequentist, or empirical, view is that which considers probability as a *degree of belief*. Consider an essentially unrepeatable phenomenon such as writing Shakespeare's plays. If it is held that probabilities may be specified in such circumstances, these probabilities must represent degrees of belief. That is, the probabilities relate to the strength of evidence in favour of a proposition suggesting a particular author.

There are two approaches to probability as a degree of belief: logical and personal. To the logical school, a probability is uniquely specified, whereas the personal view maintains that each individual may subjectively specify his or her own probabilities. Most sports fans are ardent subjectivists!

In defining probability as a degree of belief, the aim is to extend the application of probabilistic ideas to a broader class of phenomena. However, this is to be achieved without using the frequentist apparatus, even in those circumstances which a frequentist would regard as legitimate. For further discussion on how broad a view one might take and the problems that arise, see Bartlett (1975).

The degree of belief approach to probability is generally termed *Bayesian* after Reverend Thomas Bayes, who first introduced such ideas, see Bayes (1764). Bayes' original essay was reproduced in *Biometrika* in 1958, together with an historical introduction by Barnard. Bayes' ideas are discussed in **8.19–21** below.

8.13 When viewing probability as a degree of belief, all probabilities are conditional. To the logical school, given currently available knowledge H and an event A, the probability $P(A|H)$ expresses the extent to which H implies that A will happen. This probability is unique, so that a probability statement is a logical statement; it is either true or false.

The original development by Keynes (1921) left such probabilities only partially ordered, which is not in accord with the axioms (8.2)–(8.4). Later, Jeffreys (see Jeffreys, 1961) developed a logical framework that produced probabilities which were completely ordered and numerically specified. These probabilities do not rely upon any notion of relative frequency. Rather, as conditional probabilities, they are updated by applications of Bayes' Theorem, given in (8.10).

For the frequentist, the absence of any observational information, H, is a simple matter that does not affect the interpretation of probability. In the construction of logical probabilities, the absence of prior information is a state requiring careful specification and is the source of more than minor inconvenience. We return briefly to this issue in **8.28–9**; see also Jeffreys (1961, pp. 117–25) and Barnett (1982, pp. 80–84, 206–11).

Subjective probability

8.14 As the name implies, the subjective, or personal, view of probability rejects the notion that the probability of an event is uniquely defined for everyone. It is quite feasible for different individuals to assign different values to $P(A|H)$ even given the same information.

One framework for eliciting an individual's probabilities is to present that person with a series of possible wagers concerning the event A. If the individual is indifferent between (a) a certain return of pc units of money and (b) winning c units if A occurs and 0 units if A^c occurs, then $P(A \mid H) = p$ for that individual.

Another way of looking at this, although not the route followed in axiomatic developments, is to say that the individual would win c if A occurred and zero if A^c occurred, so that the *expected* winnings are

$$c \cdot P(A \mid H) + 0 \cdot P(A^c \mid H) = c \cdot P(A \mid H); \tag{8.14}$$

since the individual is indifferent between this amount and pc, it follows that

$$P(A \mid H) = p \tag{8.15}$$

for this individual.

Such a system of wagers is the basis of De Finetti's approach. This approach is subject to the axiom of *coherency* which requires that the individual

(a) will not wish to lay down wagers which he or she must certainly lose;
(b) will not have a preference for a given penalty if there is the option of another penalty which is *certainly* smaller.

For fuller details, see De Finetti (1974, pp. 87–8). (The original work by De Finetti dates back to the late twenties and thirties.) When we add to the system the assumption of *transitivity*:

$$P(A_1 \mid H) > P(A_2 \mid H) \quad \text{and} \quad P(A_2 \mid H) > P(A_3 \mid H) \quad \text{implies} \quad P(A_1 \mid H) > P(A_3 \mid H),$$

it becomes possible to develop an axiomatic system of probability equivalent to (8.2)–(8.4).

8.15 A key assumption in De Finetti's development is that the value of p determined by (8.15) will not vary with the size of the wager. In other words, the utility of money is assumed to be a linear function of the nominal value c. Yet an individual who is willing to wager £1 on the spin of a coin to win £2 if a head shows may refuse such a wager when the stakes are raised to £1000. This "value problem" creates potential problems in attempts to elicit probabilities, an extreme example of which appears in the St. Petersburg paradox initially formulated by Daniel Bernoulli (see Exercise 8.12).

In his development of subjective probability, Ramsey (1931) neatly avoided the value problem by offering two options. The individual is asked whether he or she would prefer to bet in favour of

(i) the event A under examination or
(ii) an event in some standard experiment such as a lottery where the probability of success, p, is known.

By varying the probability of success in (ii), we may determine the value of p for which the individual is indifferent between the two options; then $P(A \mid H) = p$.

It is apparent that Ramsey's approach does not require any specification of utility or value, yet it still leads to a system of probabilities satisfying (8.2)–(8.4).

Further discussion of the assessment of subjective probabilities may be found in Smith (1965), Savage (1971), and Tversky (1974). For the assessment of probabilities when the subject is not *coherent*, see Lindley *et al.* (1979) and French (1980).

In Volume 2B, the theory of statistical inference is developed from the Bayesian perspective.

Countably infinite numbers of events

8.16 Up to this point, we have assumed that the sample space Ω contains only a finite number of outcomes. When Ω contains a countable infinity of possible outcomes, such as the non-negative integers, the axiomatic system may be appropriately extended in the following way. We let the number of outcomes, N, tend to infinity but require that the sums $\Sigma P(B_i) = P(\cup B_i)$ are defined for all, possibly infinite, sequences of mutually exclusive events $\{B_i : i = 1, 2, \ldots\}$.

It is evident that the notion of equally likely events cannot be retained as $N \rightarrow \infty$, since the second axiom would be violated. The frequency approach remains operational, provided the limit is taken in such a way that lim (M_A/M) is well-defined for all A. Further, we shall see that the interpretation of probability as a degree of belief also extends to countably infinite N. However, it should be noted that De Finetti (1974) argues strongly that the study of (subjective) probability should be restricted to finitely additive set functions.

Probability in a continuum

8.17 Suppose now that the sample space Ω is non-countable. We shall restrict attention to Euclidean space and assume that, for a single variable, x takes on values over one or more intervals on the real line. A formal development of probability theory in this case requires the development of the properties of Borel fields (cf. Kingman and Taylor, 1966, Chapter 11). However, we may proceed intuitively by accepting that it is possible to define a probability function on the half-open intervals $a < x \leqslant b$. That is, we may specify

$$P(a < x \leqslant b) \tag{8.16}$$

for all intervals $(a, b]$ within $R = (-\infty, \infty)$. If the $\{B_i\}$ now represent sequences of disjoint intervals, the probability function will satisfy the axioms (8.2)–(8.4), provided $P(\cup B_i)$ and $P(\cap B_i)$ are defined for countable sequences of the B_i; that is, provided P is a σ-additive set function.

Since the probability measure is defined on the half-open intervals in (8.16), both frequentist and degree-of-belief interpretations remain valid. Also, there is a sense in which the notion of equally likely events may be retained. This view is based upon consideration of a set of intervals of equal length, but care must be used in specifying the nature of the limiting process, as the following examples illustrate.

Example 8.4 (Continuous probability measure as a limit)

If a square is inscribed in a circle, what is the probability that a point selected at random within the circle is also inside the square?

Imagine the whole figure divided into small cells of area ε by a rectangular mesh. If we assume that the occurrence of a point in a cell is equally probable for all cells, the probability that a point falls inside both circle and square is the ratio of the number of cells in the latter to those in the former, neglecting the cells at the edges which become of diminishing importance as $\varepsilon \to 0$. In fact, the required probability is the ratio of the area of the square to that of the circle, which is easily seen to be $2/\pi$.

Example 8.5 (Alternative arguments for limiting probabilities)

Consider a straight line OA bisected at B. What is the probability that a point chosen at random on the line falls into the segment OB?

Let us suppose in the first place that the line is divided into n equal segments of length OA/n. If we interpret the choosing of a point at random to mean the choice of one of these intervals, the probability is obviously $\frac{1}{2}$ as $n \to \infty$, for there will be half the intervals in the segment OB.

Now let OP be drawn perpendicular to OA and equal to it in length, and imagine a star of $n+1$ lines drawn through P, including OP and PA, so as to divide the angle OPA $(=\pi/4)$ into equal angles $\pi/(4n)$. These lines cut off segments on OA, and we may, if we regard equal angles as having equal probability, assign to these segments an equal probability, for they subtend equal angles at P. If we make this convention it is evident that as $n \to \infty$ the probability of a point falling into any segment on OA is proportional to the angle subtended at P. For example, the probability that a point falls in the segment OB is $(\arctan \frac{1}{2})/(\frac{1}{4}\pi) = 0.590$.

Now this is not the same answer that we got by assuming all small segments of OA equally probable. There is nothing contradictory in this—the two answers are different because the two limiting processes were different. On a little reflection it will be clear that by moving the point P on the perpendicular to OA and taking a star of lines as before we can make the probability of obtaining a point in OB have any value we like. It is thus abundantly clear that the concept of probability in a continuum depends on the limiting process by which that continuum is reached from a finite subdivision of equiprobable intervals.

8.18 We now give a brief overview of the differences between the frequentist and Bayesian approaches to inference, both to provide motivation for some of the topics discussed in this volume and to prepare the way for Volumes 2A and 2B.

Bayes postulate

8.19 When observational evidence becomes available, we may update prior probabilities using Bayes' Theorem, (8.10). If the prior probabilities are assessed using the approach of **8.14–5**, based upon initial information available to the investigator, we have a set of *informative* prior probabilities, or simply, an informative prior; otherwise the prior is *vague* or *non-informative*. When there is no prior information, Bayes suggested that the prior probabilities should be assumed to be equal. This assumption, known variously as Bayes Postulate, the Principle of the Equidistribution of Ignorance and by one or two other names, furnishes one of the most contentious points in the theory of statistical inference. Before we discuss the point it may be useful to give one or two examples.

Example 8.6 (An application of Bayes' theorem and postulate)

An urn contains four balls, which are known to be either (a) all white, or (b) two white and two black. A ball is drawn at random and found to be white. What is the probability that all the balls are white?

We have here two hypotheses, B_1 and B_2. On B_1 the probability of getting a white ball is 1, on B_2 it is $\frac{1}{2}$. Using (8.10), with H included to represent initial information, we have

$$P(B_1|A,H) = \frac{P(B_1|H)}{P(B_1|H)+\frac{1}{2}P(B_2|H)}$$

$$P(B_2|A,H) = \frac{\frac{1}{2}P(B_2|H)}{P(B_1|H)+\frac{1}{2}P(B_2|H)}.$$

Now, in accordance with Bayes postulate we assume

$$P(B_1|H) = P(B_2|H) = \tfrac{1}{2}$$

and find

$$P(B_1|A,H) = \tfrac{2}{3}$$

$$P(B_2|A,H) = \tfrac{1}{3}.$$

Now suppose that we replace the ball and again draw one at random. If it is found to be black the hypothesis (a) is decisively rejected. But if it turns out to be white we can calculate new posterior probabilities in which our former posterior probabilities become prior. We now have $P(B_1|H) = \frac{2}{3}$, $P(B_2|H) = \frac{1}{3}$, where H includes A, and a renewed application of (8.10) yields the posterior probabilities based on the new event, say A',

$$P(B_1|A',H) = \tfrac{2}{3}/(\tfrac{2}{3}+\tfrac{1}{2}\cdot\tfrac{1}{3}) = \tfrac{4}{5}$$

$$P(B_2|A',H) = \tfrac{1}{2}\cdot\tfrac{1}{3}/(\tfrac{2}{3}+\tfrac{1}{2}\cdot\tfrac{1}{3}) = \tfrac{1}{5}.$$

It will be clear that if we repeat the process and again get a white ball the new posterior probability of (a) will be still higher. This is in agreement with the requirements of common sense; the longer we go on sampling (with replacement) without producing a black ball, the more probable it is that there are no black balls present.

Example 8.7 (Further updating using Bayes' theorem)

Generalizing Example 8.6, suppose we draw balls one at a time, replacing them after each drawing, and obtain n white balls in succession. The probability of this event on hypothesis (a) is unity; that on hypothesis (b) is $1/2^n$. From (8.10) we then have, (A referring to the observation of all n balls as white),

$$P(B_1|A, H) = \frac{\frac{1}{2}}{\frac{1}{2} + \frac{1}{2} \cdot 2^{-n}}$$
$$= 2^n/(2^n + 1).$$

As n becomes larger, $P(B_1|A, H)$ tends to unity and $P(B_2|A, H)$ to zero.

Moreover, this will be true whatever the original prior probabilities may have been. In fact, if that of hypothesis (a) is t and that of (b) is $1 - t$, we find

$$P(B_1|A, H) = \frac{2^n t}{2^n t + (1 - t)},$$

which tends to unity for any non-zero t. This also agrees with common sense. Whatever the original probabilities, the new evidence is so strong as to outweigh them.

Example 8.8 (Laplace's Rule of Succession)

From an urn full of balls of unknown colour a ball is drawn at random, and replaced, m times and a black ball is drawn each time. What is the probability that if a further ball is drawn it will be black?

The question as framed does not admit of a definite answer as there are an infinite number of possible colours and combinations of colours, and we do not know which hypotheses are to be compared. Let us suppose that the balls are either black or white, and consider the hypotheses (1) that all are black, (2) that all but one are black, (3) that all but two are black, and so on. The problem still lacks precision, for the number of balls is not specified. Suppose there are N balls. We shall later let N tend to infinity to get the limiting case.

Consider the hypothesis B_R that there are R black balls and $N - R$ white ones. The probability of choosing a black ball is R/N and that of doing so m times in succession is $(R/N)^m$. If the Bs have equal prior probabilities we have, from (8.10),

$$P(B_R|A, H) = \frac{(R/N)^m}{\sum_{R=0}^{N} (R/N)^m}.$$

Now the probability of getting a further black ball on hypothesis B_R is R/N. Since the hypotheses B_R are mutually exclusive, the probability of getting a further black ball is

$$\sum_{R=0}^{N} \frac{R}{N} P(B_R|A, H) = \frac{\sum (R/N)^{m+1}}{\sum (R/N)^m}.$$

This is the answer to the limited form of the question. As $N \to \infty$ this tends to the ratio

$$\int_0^1 x^{m+1}\, dx \Big/ \int_0^1 x^m\, dx = \frac{m+1}{m+2}.$$

This is a particular case of the so-called Succession Rule of Laplace; see also Exercise 8.19. Enthusiasts have applied it indiscriminately in such unconditioned forms as the statement that if an event is observed to happen m times in succession the chances are $m+1$ to 1 that it will happen again. This is clearly unjustified.

8.20 The principal difficulties arising from Bayes postulate appear from the standpoint of the frequency theory of probability, which would require the states corresponding to the various Bs to be assigned equal probability, if Bayes postulate is to be applied. This has appeared to some statisticians, though not to all, to be asking too much of the universe. However, if we adopt the logical view of probability, it is reasonable to take prior probabilities to be equal when nothing is known to the contrary. Likewise, for adherents of the subjective school, all that is required is that one should not favour any hypothesis over any other when contemplating a series of wagers. Thus, most of those viewing probability as a degree of belief accept Bayes postulate, just as many frequentists explicitly reject it.

L. J. Savage (1954, 1961) argues for the purely subjective use of Bayes prior distributions. For a penetrating exposition and discussion of these views, see Savage (1962).

A difficulty which arises in the continuous case is that such priors may be improper; see **8.28**. When prior information is available, we may use informative priors, a topic we take up in **8.30**.

8.21 There is still so much disagreement on this subject that one cannot put forward any set of viewpoints as orthodox. One thing, however, is clear—anyone who rejects Bayes postulate must put something in its place. The problem that Bayes attempted to solve is supremely important in scientific inference and it scarcely seems possible to have any scientific thought at all without some solution of it, however intuitive and however empirical. We are constantly compelled to assess the degree of credence to be accorded to hypotheses on given data; the struggle for existence, in Thiele's (1903) phrase, compels us to consult the oracles. But, as he added, the oracles do not release us from thinking and from responsibility.

We shall reconsider these issues in depth in Chapter 31, Volume 2A.

Maximum Likelihood

8.22 Various substitutes for Bayes postulate have been proposed. Some of these are put forward as solutions to specific problems, such as the principles of Least Squares and Minimum Chi-squared. However, there is one principle of general application, that of Maximum Likelihood. All these approaches are considered in detail in Chapters 17–19, Volume 2A.

Reverting to (8.10) we may write Bayes' theorem in the form

$$P(B_r|A,H) \propto P(B_r|H)L(A|B_r,H) \tag{8.17}$$

where we now write $L(A|B_r,H)$ for the likelihood. The principle of Maximum Likelihood states that, when confronted with a choice of hypotheses B_r we choose that one (if any) which maximizes L. Whereas Bayes' theorem enjoins the maximization of the joint probability of B_r and A, Maximum Likelihood requires the maximization of the conditional probability of A given B_r.

8.23 It is to be particularly noted that this is not the same thing as choosing the hypothesis with the greatest probability. Some advocates of the Maximum Likelihood principle explicitly deny any meaning to such expressions as "the probability of a hypothesis." We shall see later that in practice the differences between results obtained from Maximum Likelihood and Bayes postulate are not so large as might be expected. There is, however, an important conceptual difference involved.

There is, in fact, a shift of emphasis in the way we regard the likelihood function, reflected by our writing it with an L instead of a P. The ordinary probability function gives the probabilities of A on data B_r and H; A varies, B_r and H are given. From the point of view of likelihood we consider various values of B_r for the observed A and given H; B_r varies, A and H are given. It is this variation of the function for differing values of the Bs that we have in mind in speaking of likelihood.

8.24 Suppose (as is usually the case in statistical work) that the hypotheses with which we are concerned assert something about the numerical value of a parameter θ. For instance, the hypotheses might be $B_1 \equiv \theta < 0$, $B_2 \equiv \theta \geqslant 0$, in which case there are two alternatives. Or we might have $B_1 \equiv \theta = 1$, $B_2 \equiv \theta = 2$, and so on, in which case there is a denumerable infinity of hypotheses.

If now θ can have only discrete values, we may, confronted with an observed event A, require to estimate θ, or to ask what is the "best" value of θ to take, given the evidence A. The method of Bayes would be that in (8.10) we should seek that B_r which made $P(B_r|A, H)$ a maximum. If we know nothing of the prior probabilities $P(B_r|H)$ we should, in accordance with Bayes postulate, assume all such probabilities equal. We then merely have to find that B_r which maximizes $L(A|B_r, H)$. In other words, Bayes postulate and the principle of Maximum Likelihood result in the same numerical answer.

Example 8.9 (Maximum Likelihood estimation for the binomial)

Let us consider sampling with replacement from an urn known to contain N balls, an unknown number R of which are red. We shall assume that R is known to be one of the integers $R_1, \ldots, R_k$; this set may correspond to all the integers $0 \leqslant R_j \leqslant N$.

If we make n selections, the probability of r red balls is given by the binomial

$$p(r|n, R, N) = \binom{n}{r}\pi^r(1-\pi)^{n-r}, \quad r = 0, 1, \ldots, n,$$

where $R = \pi N$. Once the experiment has been performed, the likelihood for R, given $r = t$ say, is

$$L(R|n, t, N) = \binom{n}{t} R^t(N-R)^{n-t}/N^n \tag{8.18}$$

for $R = R_1, \ldots, R_k$. The likelihood may be evaluated for each R_j in turn, and the R_j value giving the largest L is the Maximum Likelihood estimate. Clearly, multiplying (8.18) by a constant prior probability does not affect the results.

8.25 The position does not necessarily hold if the set of permissible values of θ is continuous. We must now replace such expressions as

$$P(B_r|H)$$

by a prior density function $f(\theta|H)$ and in place of (8.17) we have the posterior density function

$$g(\theta|A, H) \propto f(\theta|H)L(A|\theta, H). \tag{8.19}$$

If we now require the "best" value of θ, we may, in accordance with Bayes postulate, take the prior density to be a constant and once again we should have to maximize L for variations of θ.

We might, however, have chosen to represent our hypotheses, not by θ, but by some quantity ϕ that is a function of θ, e.g. the standard deviation instead of the variance. In this case we should have reached equation (8.19) with ϕ written everywhere instead of θ; we should have taken the prior probability as constant; and we should have arrived at the conclusion that we should maximize L for variations of ϕ.

But are being consistent in doing so? By the usual change of variable argument, the prior density for ϕ is

$$f_\phi(\phi|H) = f_\theta(\theta|H)\frac{d\theta}{d\phi},$$

so that if f_θ is constant, f_ϕ cannot be whenever ϕ is a non-linear function of θ. Thus, the use of Bayes postulate appears to involve self-contradiction. However, the principle of Maximum Likelihood is free from this difficulty, for if $L(\theta)$ is maximized at $\hat{\theta}$, and $\phi(\theta)$ is a function of θ, $L(\phi)$ is maximized at $\hat{\phi} = \phi(\hat{\theta})$. Thus it makes no difference to the result which parametrization is used.

8.26 This is one of the grounds on which adherents of the frequency school have rejected Bayes postulate in favour of the principle of Maximum Likelihood; but in our view the matter has been misunderstood. It seems that Bayes postulate and the principle give the same answer in the continuous case as well as in the discrete case when proper regard is had to the limiting processes involved. We saw in **8.17** that in speaking of probability in a continuum it was essential to specify the nature of the process to the limit. If we regard θ (from the frequency viewpoint) as having emanated from a population specified by a uniform density for θ then Bayes postulate applied to this process will clearly give a different answer from that obtained by supposing that θ emanated from a population whose density function is uniform for ϕ. The two are different just as the probabilities in Example 8.5 are different, and for the same reason. Thus the apparent inconsistency is not an inconsistency at all, but a difficulty introduced by ignoring the limiting process in continuous populations.

It remains true, of course, that for many practical purposes we do not know how the actual value of θ arose. If we require a theory of inference that is unaffected by our ignorance on such points, the objection to Bayes postulate remains and does not apply to the principle of Maximum Likelihood. On the other hand we have still not adduced convincing reasons why we should adopt the principle of Maximum Likelihood as a principle of statistical inference. We shall see in Chapter 18 (Volume 2A), that it has properties that give it a kind of posterior justification.

8.27 We now illustrate the foregoing discussion by comparing the results obtained from the Maximum Likelihood and Bayesian arguments for problems concerning the normal distribution.

Example 8.10 (ML estimators for the normal distribution)

Consider are independent sample of size n from the normal distribution

$$f(x) = \frac{1}{\sigma\sqrt{(2\pi)}} \exp\left\{-\frac{1}{2}\left(\frac{x-\mu}{\sigma}\right)^2\right\}$$

If the observations are $x_1, x_2, \ldots, x_n$ the likelihood function may be written

$$L = \frac{1}{\sigma^n(2\pi)^{\frac{1}{2}n}} \exp\left\{-\frac{1}{2}\sum_{j=1}^{n}\left(\frac{x_j-\mu}{\sigma}\right)^2\right\}.$$

We consider a range of possible values of μ that could have generated these observations. To estimate μ we take that value which maximizes L. Since L is a regular function of μ, and $L \to 0$ as $\mu \to \pm\infty$, we choose μ to satisfy

$$\frac{\partial L}{\partial \mu} = 0, \quad \frac{\partial^2 L}{\partial \mu^2} < 0.$$

Since L is positive, we get the same result by maximizing $\log L$, sometimes (as here) a more convenient procedure. We then have

$$\frac{\partial \log L}{\partial \mu} = +\sum\left(\frac{x_j-\mu}{\sigma^2}\right) = 0, \tag{8.20}$$

and thus the estimator of μ, say $\hat{\mu}$, is given by

$$\sum x_j = n\hat{\mu} \tag{8.21}$$

or $\hat{\mu} = \bar{x}$, the mean of the xs.
Since

$$\frac{\partial^2 \log L}{\partial \mu^2} = \frac{-n}{\sigma^2} < 0,$$

this is a unique maximum, and is therefore the Maximum Likelihood solution.

Had we wished to estimate both μ and σ we should find, in addition to (8.20),

$$\frac{\partial \log L}{\partial \sigma} = -\frac{n}{\sigma} + \sum\frac{(x_j-\mu)^2}{\sigma^3} = 0 \tag{8.22}$$

giving

$$\hat{\sigma}^2 = \frac{1}{n}\sum(x-\mu)^2. \tag{8.23}$$

Whereas $\hat{\mu}$ does not depend on σ, $\hat{\sigma}$ does depend on μ. We choose those estimators that maximize the likelihood for simultaneous variations in μ and σ, i.e. solve (8.20) and (8.22) together. This gives us

$$\hat{\sigma}^2 = \frac{1}{n}\sum (x-\bar{x})^2, \tag{8.24}$$

and (8.21) and (8.24) jointly maximize the likelihood.

Non-informative priors

8.28 Prior ignorance about μ may be expressed by the non-informative, uniform prior

$$f(\mu)\,d\mu \propto d\mu, \quad -\infty < \mu < \infty. \tag{8.25}$$

Combining this with the likelihood function in Example 8.10, we see that the posterior probability is maximized at $\hat{\mu} = \bar{x}$, as before. However, it should be noted that (8.25) is an improper prior in that $\int f(\mu)\,d\mu$ does not exist.

Improper priors may lead to paradoxes in multiparameter problems (cf. Dawid *et al.*, 1973), and recently the emphasis has shifted in favour of using De Finetti's idea of *exchangeability* (cf **12.23**) to represent prior ignorance. Further discussion of these ideas will be deferred until Volumes 2A and 2B.

8.29 When σ is unknown, Jeffreys (1961) recommended the use of a uniform prior on the real line for $\log \sigma$, or

$$f(\sigma)\,d\sigma \propto d\sigma/\sigma, \quad 0 < \sigma < \infty. \tag{8.26}$$

Multiplying (8.26) by the likelihood, we obtain the posterior density (8.19), and we see at once that the maximizing value is given by (8.22) with $(n+1)$ in place of n, so that

$$\hat{\sigma}^2 = \frac{1}{(n+1)}\sum (x-\mu)^2. \tag{8.27}$$

Finally, if both μ and σ are unknown, we combine the priors (8.25) and (8.26) and arrive at $\hat{\mu} = \bar{x}$ and (8.27) with μ replaced by $\hat{\mu}$.

Informative priors

8.30 When prior information is available and can be incorporated into the prior probability function, the posterior probability may be determined from (8.19). The estimator of the unknown parameter θ will still be given by maximizing the posterior probability but will, in general, differ from the Maximum Likelihood estimator.

Example 8.11 (The posterior distribution for the normal mean)

Suppose that prior information about the normal mean μ may be represented by

$$f(\mu|\lambda, \omega) \propto \exp\{-\tfrac{1}{2}(\mu-\lambda)^2/\omega^2\}, \quad -\infty < \mu < \infty. \tag{8.28}$$

As in Example 8.10, assume that we are given an independent sample of size n from the normal distribution. Then, from (8.19), the posterior density is

$$f(\mu|\lambda, \omega, \sigma, x) \propto \exp\left\{-\frac{1}{2}\left[\frac{\mu-\lambda}{\omega}\right]^2 - \frac{1}{2}\sum_{j=1}^{n}\left[\frac{x_j-\mu}{\sigma}\right]^2\right\}.$$

Differentiating with respect to μ, we obtain

$$\frac{\partial f}{\partial \mu} = \sum \left(\frac{x_j - \mu}{\sigma^2} \right) + \frac{(\lambda - \mu)}{\omega^2}$$

which yields a maximum at

$$\hat{\mu} = \frac{\bar{x} n \omega^2 + \lambda \sigma^2}{n \omega^2 + \sigma^2}. \tag{8.29}$$

As $n \to \infty$, $\hat{\mu} \to \bar{x}$ regardless of the prior information contained in (λ, ω^2). This reinforces the point made earlier, that sufficiently strong sample information will eventually overwhelm the prior views. Also, we note that $\hat{\mu} \to \bar{x}$ as $\omega^2 \to \infty$. Letting $\omega^2 \to \infty$ is one way of expressing prior ignorance, since (8.28) shows ω to represent the dispersion of μ about λ. Indeed, one way of overcoming the problem of improper priors is to select an informative prior and then to evoke an appropriate limiting argument. As on previous occasions, the proper choice of limiting argument is critical.

Conjugate priors

8.31 The choice of a normal prior density in Example 8.11 is certainly convenient, but is it appropriate? Recalling that such priors express degrees of belief, the final answer for the subjectivist must be an individual one, although the logical theory can hope for a more definitive answer.

In general, arbitrary priors make for intractable mathematics. Since knowledge of the functional form of the prior is itself often vague, this has led to the development of a class of *conjugate* prior distributions, for which the prior and the posterior are of the same functional form. Some of the common forms are summarized below:

Likelihood	*Parameter*	*Prior/Posterior*
Normal	μ	Normal
Normal	σ^2	Gamma (for σ^{-2})
Binomial	π	Beta
Poisson	λ	Gamma

The case of the normal mean was discussed in Example 8.11; the others are considered in Exercises 8.18, 8.20–1.

8.32 The introduction of conjugate priors opens a path whereby prior information can be introduced into a frequentist analysis. Since the likelihood and prior are compatible in form, the frequentist may specify a *prior likelihood* which is considered to be "equivalent" to n_0 observations. When $n_0 = 0$, the prior likelihood would be flat but still proper. When $n_0 > 0$, the Maximum Likelihood estimator is modified in the same fashion as the posterior probability estimator.

Example 8.12 (Use of a prior likelihood)

Let the prior likelihood for the normal mean be

$$L_p(\mu | \lambda, n_0) \propto \exp\{-n_0(\mu - \lambda)^2/(2\sigma^2)\}, \quad -\infty < \mu < \infty,$$

so that in (8.28) we have put $\omega^2 = \sigma^2/n_0$. Following the working of Example 8.11, (8.29) is now

$$\hat{\mu} = \frac{n\bar{x} + n_0\lambda}{n + n_0}. \tag{8.30}$$

Comparing (8.29) and (8.30), we see a difference of emphasis in that the prior likelihood requires specification of λ and ω^2. Once again, the ideas used lead to different formulations, although the end-results may be very similar.

Although conjugate priors have attractive mathematical properties, they may not be very realistic. In recent years, substantial advances have been made in the numerical evaluation of posterior probabilities; see, for example, Naylor and Smith (1988) and Volume 2B.

8.33 Our purpose in this part of the chapter has been to introduce some of the basic issues and to illustrate the probabilistic arguments underlying the several approaches to inference. Full discussions of the frequentist and Bayesian theories of inference appear in Volume 2A and Volume 2B respectively.

8.34 In conclusion, some general comments are in order. It has long seemed to us (Kendall, 1949a) that no one approach is, or ever will be, sufficient for the statistician, whatever its value to the mathematician. The frequentist must incorporate subjective opinions on the choice of sample space and the validity of a sequence of independent repetitions, whereas it is hard for the subjectivist to avoid ideas of relative frequency in eliciting a set of probabilities. The subject matter under discussion and the individual's attitudes will also determine the approach to be followed.

We now return to a discussion of the mathematical aspects of probability theory.

Random variables

8.35 In **1.20** we introduced the concept of a *variate* or *random variable*, with associated distribution function $F(x)$.

If we select an observation at random from this population the probability that it has a value less than or equal to x_0 is $F(x_0)$ or

$$P(X \leqslant x_0) = F(x_0). \tag{8.31}$$

This is the essential link between probabilities and distributions.

When the d.f. $F(x)$ is specified algebraically as a differentiable function, the probability density function for the continuous variate is defined as the derivative of F:

$$f(x) = \mathrm{d}F(x)/\mathrm{d}x, \tag{8.32}$$

or

$$P(X \leqslant x) = F(x) = \int_{-\infty}^{x} f(u)\,\mathrm{d}u. \tag{8.33}$$

8.36 We have spoken above of selection *at random*. It is customary to define randomness in terms of probability itself. A member of a population is said to be chosen at random if it is chosen by a random method; and a random method is one which makes it equally probable that each member of the population will be chosen. Randomness is extremely important in the theory of sampling, and we shall consider it at some length in Chapter 9. At this point, it is sufficient to note that when we speak of random choice, we really mean a method of selection that gives to certain propositions an equal probability and hence allows us to apply the calculus of probability *a priori*. The justification for this is, in the ultimate analysis, empirical. It is found in practice that there exist selective processes that choose members of a population in such a way that the constituent events may be regarded as equiprobable; and the theory of sampling is largely concerned with samples generated by such processes.

Joint and conditional distributions

8.37 Our definition of a random variable assigned a single value to each outcome in the sample space, thereby generating a univariate distribution. By the simple device of assigning a vector of variate values to each outcome, we may define joint distributions for two or more random variables, such as

$$P(X_1 \leqslant x_1, X_2 \leqslant x_2) = F(x_1, x_2), \tag{8.34}$$

and so on. Joint and marginal probability functions are then obtained for both discrete and continuous random variables, as in **1.32–3**. When the random variables are discrete, the conditional p.m.f. is

$$P(X_1 = x_1 \mid X_2 = x_2) = \frac{f(x_1, x_2)}{f_2(x_2)}, \tag{8.35}$$

provided $P(X_2 = x_2) = f_2(x_2) > 0$. When $f_2(x_2) = 0$, (8.35) is taken to be zero. When the random variables are continuous, matters are more intricate.

8.38 The definition of a continuous d.f. $F(x)$ was based on the half-open intervals in (8.16); this is consistent with the definition in (1.7) which implies that the probability that the continuous variate exactly equals x is zero. The same applies for the joint density function of any pair of continuous variates (x, y). Nevertheless, we may define the conditional distribution function as

$$F(x_1 \mid x_2) = \frac{\int_{-\infty}^{x_1} f(t, x_2)\,dt}{f_2(x_2)}, \tag{8.36}$$

where $f_2(x_2) = \int_{-\infty}^{\infty} f(t, x_2)\,dt$, with the convention that $F(x_1 \mid x_2) \equiv 0$ whenever the denominator of (8.36) is zero. It follows that $F(x_1 \mid x_2)$ is defined almost everywhere. This is consistent with the definition used in **1.32**. The conditional density function is

$$f(x_1 \mid x_2) = f(x_1, x_2)/f_2(x_2). \tag{8.37}$$

By the argument of **8.17**, the nature of the limiting process may affect the form of this function, as Exercise 8.32 illustrates.

It should be emphasized that this problem arises only for the density function; the conditional d.f. of any variate is uniquely determined as above.

8.39 We now take the first steps in developing the theory of sampling distributions, which will be the major topic of interest in Chapters 11–16. We first consider the distributions of sums of random variables and then introduce some important limit theorems that are central to our development of large sample results in Chapter 10.

Sums of random variables

8.40 If the joint distribution of X and Y is $F_{12}(x,y)$, the probability that $Z = X + Y$ is less than or equal to z is

$$P(Z \leqslant z) = \iint dF_{12}(x,y), \tag{8.38}$$

the integrals being taken over the region $x + y \leqslant z$. This integral defines the distribution function of Z.

8.41 More generally, suppose that we have random variables $X_1, X_2, \ldots, X_n$ with corresponding values $x_1, x_2, \ldots, x_n$. We may define a random variable Z with values z, by the equation

$$Z = z(X_1, X_2, \ldots, X_n). \tag{8.39}$$

The d.f. of the random variable is the integral of $dF(x_1, x_2, \ldots, x_n)$ over all values of $x_1, x_2, \ldots, x_n$ such that $z(x_1, x_2, \ldots, x_n) \leqslant z$.

In particular we may define the sum of the random variables by

$$Z = X_1 + X_2 + \ldots + X_n. \tag{8.40}$$

Z is also known as the convolution of $X_1, X_2, \ldots, X_n$, sometimes written as $Z = X_1 * X_2 * \ldots * X_n$.

8.42 From **4.41**, if $X_1, \ldots, X_n$ are independent, the c.f. of Z is

$$\phi_z(t) = \prod_{j=1}^{n} \phi_j(t), \tag{8.41}$$

where $\phi_j(t)$ denotes the c.f. of X_j. Furthermore, if $\phi_j(t) = \phi(t)$ for all j, we have immediately that

$$\phi_z(t) = \{\phi(t)\}^n. \tag{8.42}$$

Simple random sampling corresponds exactly to the case covered by (8.42); that is, observations are drawn independently from a population, with d.f. $F(x)$ say.

Finally, if

$$Z = a_1 X_1 + \ldots + a_n X_n \tag{8.43}$$

where the *a*s are constants, it follows directly from (4.79) that

$$\phi_z(t) = \prod_{j=1}^{n} \phi_j(a_j t), \tag{8.44}$$

with ϕ replacing ϕ_j when the Xs are identically distributed.

Example 8.13 (Distribution of the sample mean for the normal)

Suppose we draw a simple random sample of size two from the normal population with p.d.f.

$$f(x) = \frac{1}{\sigma\sqrt{(2\pi)}} e^{-x^2/2\sigma^2} \quad -\infty < x < \infty,$$

and c.f., from Example 4.1,

$$\phi(t) = \exp(\mu it - \tfrac{1}{2}t^2\sigma^2). \tag{8.45}$$

From (8.42), the c.f. of $Z = X_1 + X_2$ is

$$\begin{aligned}\phi_z(t) &= \{\phi(t)\}^2 \\ &= \exp(2\mu it - t^2\sigma^2).\end{aligned} \tag{8.46}$$

Since the c.f. of Z uniquely determines its d.f., it follows from a comparison of (8.46) with (8.45) that Z is normally distributed with mean 2μ and variance $2\sigma^2$, sometimes written as $Z \sim N(2\mu, 2\sigma^2)$.

It follows immediately from (8.42) that, if $Z = X_1 + \ldots + X_n$, where the Xs are independent and identically distributed with c.f. (8.45), that $Z \sim N(n\mu, n\sigma^2)$. Finally, if Z is given by (8.43), it follows from (8.44) that Z remains normally distributed with mean $\mu \sum a_j$ and variance $\sigma^2 \sum a_j^2$. When $a_j = n^{-1}$ for all j so that

$$Z = \sum_{j=1}^{n} X_j/n = \bar{X},$$

$\bar{X}$ is normally distributed with mean μ and variance σ^2/n.

8.43 We now consider some statistically important theorems concerning the limiting distribution of the sum of n random variables as n tends to infinity.

The Weak Law of Large Numbers

8.44 Let $X_1, X_2, \ldots$ be a set of independent random variables distributed in the same form with mean μ. Let $\bar{X}_n$ be the mean of the first n,

$$\bar{X}_n = \frac{1}{n}\sum_{i=1}^{n} X_i.$$

Then we shall see that $\bar{X}_n$ also has mean μ, whatever the value of n. The Weak Law of Large Numbers states that, in effect, $\bar{X}_n$ becomes more and more narrowly dispersed about μ as n increases. More precisely, given any positive ε,

$$\lim_{n\to\infty} P\{|\bar{X}_n - \mu| > \varepsilon\} = 0. \tag{8.47}$$

If we may assume that the variance of any X exists, and is equal to σ^2, say, the proof is very easy. For as we shall see later, $\bar{X}_n$ is a random variable with variance σ^2/n and, by virtue of the Bienaymé–Chebyshev inequality (3.95),

$$P\{|\bar{X}_n-\mu|>\varepsilon\}\leqslant\sigma^2/n\varepsilon^2.$$

For given ε the probability can therefore be made as small as we please by increasing n.

Result (8.47) is often referred to as *convergence in probability*; that is, we say $\bar{X}_n$ converges in probability to μ as n increases, sometimes written as

$$p\lim\bar{X}_n=\mu.$$

8.45 The theorem remains true even if σ^2 is not finite, but a different type of proof is required. (Cf. Exercise 8.30.) It is also easily adapted to the case where the Xs have different means.

The Strong Law of Large Numbers

8.46 The Weak Law of Large Numbers states a limiting property of sums of random variables. The Strong Law states something about the behaviour of the sequence $S_n\equiv\sum_{i=1}^{n}X_i$ for all values of n; something, as it were, about its properties on the way to the limit.

In fact, given any positive numbers ε and δ there is an N such that for every $M>0$

$$P\{|\bar{X}_n-\mu|\geqslant\varepsilon\}\leqslant\delta,\quad n=N,N+1,\ldots,N+M. \tag{8.48}$$

The Weak Law states that $|\bar{X}_n-\mu|$ is ultimately small but not that every value is small; it might be that for some n it was large, although such cases could only occur infrequently. The Strong Law says that the probability of such an event is extremely small. The law is true for independent variables that are identically distributed under the sole condition that μ exists; in other cases further conditions must in general be added. For a proof of the Strong Law, see Feller (1968, Chapter 10).

The Central Limit theorem

8.47 A much more precise result, the Central Limit theorem, is available in cases where the variances exist. In fact, the mean $\bar{X}_n$ tends to be distributed normally about μ with variance σ^2/n. More precisely, if $G(z)$ represents the normal distribution function then, for every z

$$\lim P\left\{\frac{\bar{X}_n-\mu}{\sigma/\sqrt{n}}\leqslant z\right\}=G(z). \tag{8.49}$$

This theorem is also true if the Xs have different distributions, μ then being replaced by the mean of individual μs and $\sigma/\sqrt{n}$ being replaced by the variance of $\bar{X}_n$.

Let us suppose that each X has a finite absolute third moment. Then as in **3.15** we have, for the c.f. of the rth variate (with first and second moments, say, μ'_{1r} and μ'_{2r}),

$$\phi_r(t)=1+\mu'_{1r}(it)+\tfrac{1}{2}\mu'_{2r}(it)^2+R, \tag{8.50}$$

where $|R|$ is not greater than $\frac{1}{6}\nu'_{3r}|t|^3$. Likewise, for the c.g.f. we have

$$\psi_r(t) = \mu'_{1r}(it) + \tfrac{1}{2}\mu_{2r}(it)^2 + S,$$

where $|S|$ is some constant times $|t|^3$, say $k|t|^3$. Thus, for the c.g.f. of the sum of the n variables we have

$$\psi(t) = \sum_{r=1}^{n} \mu'_{1r}(it) + \tfrac{1}{2}\sum_{r=1}^{n} \mu_{2r}(it)^2 + \sum S.$$

Let us now take the mean of this sum as origin and standardize by putting $u = \Sigma X / (\Sigma\mu_{2r})^{\frac{1}{2}}$. This gives us for u

$$\psi(t) = -\tfrac{1}{2}t^2 + \sum S \Big/ \left(\sum \mu_{2r}\right)^{\frac{3}{2}}.$$

If each ν_{3r} is finite the numerator of the remainder term is bounded by $nK|t|^3$, where K is the largest value of the $|\kappa_{3r}|$. The denominator is not less than $n^{\frac{3}{2}}M_{2r}$, where M_{2r} is the smallest variance. Thus the remainder term is of order $n^{-\frac{1}{2}}$ and tends to zero. Thus

$$\psi(t) \to -\tfrac{1}{2}t^2$$

and consequently the distribution of the standardized sum tends to normality. The result is exact for the normal distribution, as we saw in Example 8.13.

General results of this kind go back to Laplace, Demoivre's derivation of the normal limit to the binomial in 1733 being a special case. This form of the result is essentially due to Liapunov (1900, 1901).

8.48 The theorem may be proved under conditions that do not require finite third moments. In fact, it is a necessary and sufficient condition that

$$\lim_{n\to\infty} \frac{1}{M_n} \sum_{r=1}^{n} \int_{|x| > \epsilon\sqrt{M_n}} x^2 \, dF_r = 0, \tag{8.51}$$

where $M_n = \Sigma\mu_{2r}$.

This condition, due to Lindeberg (1922) and Cramér (1937), implies that the total variance M_n tends to infinity and that every μ_{2r}/M_n tends to zero; in fact that no random variable dominates the others.

The theorem may fail to hold for variables whose variance is not finite; for example, we shall see in Example 11.1 that the mean of n variates, each distributed in the Cauchy form

$$f(x) = \frac{1}{\pi(1+x^2)}, \quad -\infty < x < \infty, \tag{8.52}$$

is distributed in precisely the same form. This is easily deduced from the c.f., given in Example 4.2 using (8.44). Exercise 11.25 will give an extreme case where the sample mean has a distribution whose dispersion *increases* with n.

8.49 Nevertheless, central limit theorems may be developed for such distributions using the stable laws introduced in **4.36**. If $X_1, \ldots, X_n$ are independent with common d.f., F say, F belongs to the *domain of attraction* of stable law G if there exist constants $a_n > 0$ and b_n such that the distribution of

$$S_n = \frac{X_1 + \ldots + X_n}{a_n} - b_n \tag{8.53}$$

tends to G as $n \to \infty$. Roughly speaking, if $E(X^c)$ exists for $0 \leqslant c < \alpha < 2$, it turns out that the G here is the symmetric stable law with characteristic exponent α, whose c.f. is given by (4.69). For fuller details, see Feller (1971, section 9.8). Since the Cauchy is a stable law with $\alpha = 1$, the result holds for (8.52) when $a_n = n$, $b_n = 0$.

8.50 The Central Limit theorem occupies an important place in statistical theory. We have already met instances of its operation in the tendency of the binomial and the Poisson distributions to normality. We shall meet many more later. Central Limit theorems are also obtainable for sums of random variables that are not independent. General accounts of these limit theorems will be found in Feller (1968, Chapter 10; 1971, Chapter 7), Ibragimov and Linnik (1971) and Petrov (1974).

For an informative history of the Central Limit Theorem up to 1986, see LeCam (1986).

8.51 A question that naturally arises is how rapidly the distribution of the mean approaches normality. Let

$$Z = (\bar{X}_n - \mu)\sqrt{n}/\sigma,$$

then, following the notation of (8.54), it may be shown that

$$\sup_{-\infty < z < \infty} |P(z \leqslant z) - G(z)| \leqslant c\nu_3/(\sigma^3 n^{\frac{1}{2}}) \tag{8.54}$$

where $\nu_3 = E[\,|(X-\mu)^3|\,]$ and c is an absolute constant, $c \leqslant 0.82$. This result was established by Berry (1941) and Esseen (1945) and is known as the Berry–Esseen theorem.

Example 8.14 (Application of the Berry–Esseen theorem)

Suppose X is uniformly distributed on $[-1, 1]$. Then $\mu = 0$, $\sigma^2 = \frac{1}{3}$ and $\nu_3 = \frac{1}{4}$. The upper bound given by (8.54) with $c = 0.82$ is $1.065\,n^{-\frac{1}{2}}$. As is to be expected, the bounds are rather wide since we have not used any information about the distribution other than ν_3. However, the 2-point distribution with $P(X = -0.368) = 0.711$ and $P(X = 0.906) = 0.289$ has virtually the same values of μ, σ and ν_3 so the same bound applies here, although the approach to normality is clearly much slower. For a full discussion of the rate of convergence to normality in the Central Limit Theorem, see Petrov (1974).

8.52 Under conditions that parallel those for the univariate case, Central Limit theorems can be proved for multivariate random variables. In particular, corresponding to the theorem of **8.47**, the means of n xs and ys tend to bivariate normality as n increases under similar very general conditions, and similarly for more than two variates.

EXERCISES

8.1 (*The Inclusion-Exclusion Law*) Let $A_1, A_2, \ldots A_n$ be n compatible events, i.e. such that any number of them from 0 to n may be simultaneously realized. Let $p_i = P(A_i)$, $p_{ij} = P(A_i \cap A_j)$, and so on. Show that the probability that no event is realized is

$$1 - \sum p_i + \sum p_{ij} - \sum p_{ijk} + \ldots + (-1)^n p_{12\ldots n},$$

where summations take place over all possible different values of the subscripts.

8.2 Show that the probability that, of n compatible events, exactly r unspecified events are realized is

$$\sum p_{(r)} - \binom{r+1}{1} \sum p_{(r+1)} + \binom{r+2}{2} \sum p_{(r+2)} + \ldots + (-1)^{n-r} \binom{n}{n-r} p_{(n)},$$

where $p_{(j)}$ denotes a p with j subscripts.

8.3 In the previous exercise show that the probability that at least r unspecified events are realized is

$$\sum p_{(r)} - \binom{r}{1} \sum p_{(r+1)} + \binom{r+1}{2} \sum p_{(r+2)} - \ldots + (-1)^{n-r} \binom{n-1}{n-r} p_{(n)},$$

reducing to the complement of Exercise 8.1 when $r = 1$.

8.4 Use Exercise 8.2 to solve the envelope problem of Example 8.3. Show also that the probability that exactly r unspecified letters get into the right envelopes is

$$\frac{1}{n!}\left\{\binom{n}{r}(n-r)! - \binom{r+1}{1}\binom{n}{r+1}(n-r-1)! + \text{etc.}\right\} = \frac{1}{r!}\left\{1 - 1 + \frac{1}{2!} - \ldots + \frac{(-1)^{n-r}}{(n-r)!}\right\}.$$

Hence show that, as n increases, the probabilities rapidly approach those of the Poisson distribution with $\lambda = 1$.

8.5 Given n independent but compatible events $A_1 \ldots A_n$, and that the probability that A_i alone happens is α_i, show that the probability p_i that A_i happens, whether the others happen or not, is given by

$$p_i = \alpha_i/(\alpha_i + t),$$

where t is a root of

$$\prod_{i=1}^{n} (\alpha_i + t) = t^{n-1}.$$

Show that in general there are two values of p_i satisfying the conditions.

8.6 From a heap of counters of unknown number N a player takes a handful of n at random. Examine this argument: it is equally probable that N is odd or even. If it is odd, the probability that n is odd is greater than $\frac{1}{2}$, whereas if it is even the probability that n is odd is $\frac{1}{2}$. Thus the probability that n is odd is greater than $\frac{1}{2}$, and the player should bet on getting an odd number.

8.7 An event happens at random on an average once in time t. Regarding occurrences in equal small intervals as equiprobable, show that the probability that it does not happen in a specified interval T is $\exp(-T/t)$; that is, T is exponentially distributed.

8.8 If a is uniformly distributed in the interval $0 < a \leqslant 6$ and b is uniform in the interval $0 < b \leqslant 9$, show that the probability that $x^2 - ax + b = 0$ has two real roots is $\frac{1}{3}$.

8.9 A straight line in a plane is represented by the pedal equation

$$x \cos \theta + y \sin \theta = p.$$

A random line is drawn in such a way that elements $d\theta\, dp$ are equally probable. If it intersects a closed convex curve of length l_1, show that the probability that it will also intersect a convex curve of length l_2, lying inside the first, is l_2/l_1.

Deduce that if a random line intersects a circle, the probability that it also intersects a fixed diameter of the circle is $2/\pi$.

8.10 If, in the previous exercise, the co-ordinates are changed by a translation of the origin and a rotation about it, show that $dp'\, d\theta' = dp\, d\theta$ where the primes relate to the new co-ordinates. Hence show that the "randomness" of lines drawn in the foregoing manner is independent of the co-ordinate system.

8.11 Three points are taken at random on a circle. Show that the probability that they lie on the same semicircle is $\frac{3}{4}$. (Assume that in the limit elementary intervals of arc are equally probable.)

Explain the fallacy in the following argument: one pair of points *must* lie on a semicircle terminating at one of them. The probability that the third point lies on this semicircle is $\frac{1}{2}$, which is therefore the required answer.

Consider also this argument: it does not matter where the first point is chosen. Imagine the circle cut at that point and unrolled into a horizontal straight line. The probability that the second and third points fall into the left half of the line is $\frac{1}{2} \times \frac{1}{2} = \frac{1}{4}$; similarly for the right half; and hence the probability required is $\frac{1}{2}$.

8.12 (*St Petersburg Paradox*) Suppose an invitation is extended to play the following game, just once. A fair coin will be spun until the first head appears. If the first head appears on the kth toss, you win £(2^k). What is the highest entry fee that you, personally, would be willing to pay to play such a game? Show that the expected winnings are infinite.

(Cf. Feller, 1968, section 10.4.)

8.13 An event of constant probability π is observed r times in n independent trials. Show by Bayes' theorem or the method of Maximum Likelihood that the value indicated for π is r/n. (For the application of Bayes' theorem it may be assumed that π is uniformly distributed on the interval $[0, 1]$.)

8.14 In the previous exercise, trials are repeated until r events are observed, the number of trials necessary being n. Show that Bayes' theorem and the method of Maximum Likelihood still give r/n as the appropriate estimate of π.

8.15 Show that r/n, averaged over all samples, equals π in Exercise 8.13, but is not equal to π in Exercise 8.14.

8.16 On the assumptions of the previous exercises show that if n trials are conducted and r is the number of events observed, the probability that the event occurs on the next trial is $(r+1)/(n+2)$. (Use a more general form of the rule of succession of Example 8.8.)

8.17 An urn is known to contain balls of different colours but the number of colours M is unknown. A sample of n balls is drawn with replacement and is found to contain balls of $m(\leqslant n)$ different colours. Show that the estimator of M indicated by the method of Maximum Likelihood is m.

8.18 Consider again Exercise 8.13. Show that the density

$$f(\pi | a, b) \propto \pi^a(1-\pi)^b$$

is a conjugate prior for π (cf. 8.31). Hence, show that the estimator that maximizes the posterior probability is $\hat{\pi} = (a+r)/(n+a+b)$.

(Jeffreys (1961) proposed a non-informative prior with $a = b = -\frac{1}{2}$, corresponding to the assumption that arc $\sin(\pi^{\frac{1}{2}})$ is uniformly distributed.)

8.19 Following Example 8.8, let $\pi = P$(red ball) and use the prior distribution given in Exercise 8.18. If m red balls are observed in succession show that $\hat{\pi} \to 1$ as $m \to \infty$ even though the posterior density remains zero at $\pi = 1$ if $b > 0$.

8.20 Let $x_1, \ldots, x_n$ be n independent observations from a Poisson distribution with parameter λ. Show that the density

$$f(\lambda | \theta, c) \propto \lambda^c \exp(-\lambda\theta), \qquad 0 < \lambda < \infty$$

is a conjugate prior for λ. Hence show that the estimator which maximizes the posterior probability is

$$\hat{\lambda} = (\bar{x} + c)/(n + \theta).$$

Examine the behaviour of the prior distribution as $c \to 0$ with $c/\theta = \alpha$ fixed.

8.21 Show that the prior density

$$f(\sigma^2 | c, \omega^2)\mathrm{d}(\sigma^2) = \frac{(c\omega^2)^c}{\Gamma(c)}\sigma^{-2c-2}\exp(-c\omega^2/\sigma^2)\mathrm{d}(\sigma^2), \qquad 0 < \sigma^2 < \infty$$

is the conjugate prior for the normal distribution parameter σ^2. Hence show that the estimator for σ^2 which maximizes the posterior density is

$$\hat{\sigma}^2 = \frac{\sum (x_i - \mu)^2 + 2c\omega^2}{n + 2c + 2}.$$

8.22 If x and y are independent random variables with c.f.s ϕ_1 and ϕ_2 respectively, show that the c.f. of $x - y$ is $\phi_1(t)\phi_2(-t)$. Hence show (cf. Exercise 4.1) that if $x + y$ is distributed in the same form as $x - y$, then y is symmetrically distributed about zero.

8.23 x and y are distributed in the bivariate normal form of (7.48). Find the c.f. of $x + y$ and hence show that $x + y$ is normally distributed with variance $2(1+\rho)$. Deduce that $|\rho| \leqslant 1$.

8.24 Show that the sum of a number of independent random variables, each distributed in the Poisson form with possibly different parameters, is also distributed in the Poisson form.

8.25 x is a standardized normal variate and $y = x^2$. Show that the joint c.f. of x and y is

$$\phi_{x,y}(t_1, t_2) = (1 - 2it_2)^{-\frac{1}{2}} \exp\left\{-\frac{1}{2}\frac{t_1^2}{1 - 2it_2}\right\}$$

and deduce the c.f.s of x and of y alone. Show that x and y are not independent and that the product-cumulants $\kappa_{rs} = 0$ for any $r \neq 2$ and any s.

8.26 A random sample of n values $x_1, x_2, \ldots, x_n$ is chosen from a population that has mean μ and takes only positive values. Show that the probability that $\sum_{i=1}^{n} x_i$ exceeds some constant λ is not greater than $n\mu/\lambda$.

8.27 x is a variate with mean $\mu > 0$, variance σ^2 and coefficient of variation $c = \sigma/\mu$. Show that

$$P\{x > 0\}E(x^2) \geqslant \mu^2,$$

so that

$$P\{x > 0\} \geqslant \frac{1}{1+c^2}.$$

Hence, taking the median M as origin, show that $\left|\frac{\mu - M}{\sigma}\right| \leqslant 1$ as in Exercise 3.22.

8.28 Find the distribution of the sum of k independent random variables, each distributed in the form

$$f(x) = e^{-x}, \qquad 0 \leqslant x < \infty.$$

Graph the distribution for some values of k and observe its tendency to normality.

8.29 The random variable X has p.m.f. $P(X = x) = A/|x|$ for $x = \pm(p+1)^2$, $p = 1, 2, \ldots$. Show that if A is suitably chosen this gives a properly defined p.f.

Show that the expectation of $|x|$ does not exist. Show also that if an expectation be defined as the principal value $\lim_{n\to\infty} \sum_{-n}^{n} xf(x)$, it does exist. By taking a different origin show that the principal value about some other point does not exist. Hence, by taking one random variable to be a constant, show that the expectation of the sum of two independent random variables is not necessarily equal to the sum of their expectations if the principal value be admitted as the expectation.

(Fréchet, 1937)

8.30 By expanding their c.f.s as

$$\phi_j(t) = 1 + \mu it + o(t),$$

show that the mean $\bar{X}_n$ of n identical independent variates X_j has a c.g.f. for which

$$\lim_{n\to\infty} \psi(t) = \mu it,$$

and hence that $\bar{X}_n$ has the unit distribution of Example 4.3 in the limit. This establishes the Weak Law of Large Numbers of **8.44**.

8.31 Using the joint c.f. of Exercise 8.25, or otherwise, show that if $x_1, x_2, \ldots, x_n$ are independent normal variates with means μ_i and variance 1, the c.f. of $\sum_{i=1}^{n} x_i^2$ is

$$\phi(t) = (1 - 2it)^{-\frac{1}{2}n} \exp\left(\frac{\lambda it}{1 - 2it}\right)$$

where $\lambda = \sum_{i=1}^{n} \mu_1^2$. (Exercise 4.18 is the special case $n = 1, \lambda = 0$ if $\sigma^2 = 1$.)

8.32 In the transformation from (x_1, x_2) to (y_1, y_2) in **1.35**, let x_1, x_2 be non-negative, put

$$y_1 = x_1 + x_2, \qquad y_2 = x_1/x_2$$

and show that

$$g_1(y_1, y_2) = f\left(\frac{y_1 y_2}{1+y_2}, \frac{y_1}{1+y_2}\right)\frac{y_1}{(1+y_2)^2}.$$

Alternatively, transform from (x_1, x_2) to y_1 and $y_3 = x_1 - kx_2$ to obtain

$$g_2(y_1, y_3) = f\left(\frac{y_3 + ky_1}{1+k}, \frac{y_1 - y_3}{1+k}\right)\frac{1}{1+k}.$$

Show that the conditional p.d.f.s $p_1(y_1 | y_2 = k)$ and $p_2(y_1 | y_3 = 0)$ obtained from (1.29) are not identical, even though $y_2 = k$ if and only if $y_3 = 0$. If $f(x_1, x_2) = 1$, $0 \leqslant x_1, x_2 \leqslant 1$, show that

$$\left.\begin{aligned} p_1(y_1 | y_2 = k) &= 2y_1/(1+k)^2, \\ p_2(y_1 | y_3 = 0) &= 1/(1+k), \end{aligned}\right\} \qquad 0 \leqslant y_1 \leqslant 1 + k.$$

(J. S. Williams, 1966)

8.33 $x_1, x_2, \ldots, x_n$ are independently and identically distributed with cumulants κ_s, and γ_1 and γ_2 defined by (3.89–90). If $l = \sum_{j=1}^{n} c_j x_j$, show from **3.13** that the cumulants of l are given by $\lambda_r = \sum_j c_j^r \kappa_r$, and hence that $\gamma_1^* = \lambda_3/\lambda_2^{\frac{3}{2}}$ and $\gamma_2^* = \lambda_4/\lambda_2^2$ satisfy $|\gamma_1^*| \leqslant |\gamma_1|$, $\gamma_2^* \leqslant \gamma_2$. If $c_j = \dfrac{1}{n}$, so that l is the sample mean, show that γ_1^* and $\gamma_2^* \to 0$, as does every $\lambda_p/\lambda_2^{p/2}$ for $p > 2$. Hence verify the Central Limit theorem in this case where all cumulants are finite.

CHAPTER 9

RANDOM SAMPLING

The sampling problem

9.1 In the previous chapter we have referred incidentally to the sampling problem, which can be stated quite simply: given a sample from a population, to infer from it some or all of the properties of that population. We noted that only in exceptional cases is it possible to make assertions about the population with complete certainty, and that consequently it is necessary to fall back on statements of a less categorical kind expressible in terms of probability.

9.2 In order to be able to apply the theory of probability to this problem, it is necessary that the sampling should be random. By a random sample, we mean a sample which has been selected in such a manner that every possible sample has a calculable chance of selection. Two remarks are necessary concerning this definition. First, it contains the word "calculable" rather than "calculated," because the chances of selection often need not be exhaustively calculated for every sample—the specification and control of the sampling procedure is all that is required for the application of probability theory. Second, the definition does not imply that every possible sample must have an *equal* chance of selection. If this is in fact the case, and successive drawings are independent, the sampling procedure is called *simple random sampling*, in accordance with the usage of 8.44. For many theoretical purposes, simple random sampling is the selection process considered although the search for efficiency in the design of sampling inquiries often leads to the deliberate abandonment of simple random sampling. The point here is that such abandonment does not impair the application of the theory of probability, so long as the (unequal) chances of selection are uniquely determined by the sampling process, and are therefore calculable.

In practice, we often meet with samples that are not random, having been chosen purposively in some way. In such circumstances, it is not possible to make precise probability statements about the population, and where a decision has to be taken one is forced to rely on subjective judgments of an unsatisfactory kind. It is for this reason that random sampling is of primary importance in sampling investigations of a population.

Henceforth, we shall discuss only random samples, and to avoid constant repetition we shall understand the unqualified use of the words "sample" or "sampling distribution" to refer to simple random sampling conditions.

9.3 It is useful to begin a discussion of random sampling by considering the nature of the sampling process and the types of population from which samples can be chosen.

In the first place, the population may be finite and *existent*, e.g. the population of human beings in Europe at a fixed point of time, or the population of apples on a given tree. A sampling process that chooses members one at a time from this population will eventually exhaust the supply of members if continued long enough. Thus the sampling, though random, is not simple, for successive drawings are not independent.

We may, however, reduce this process to one of simple sampling by replacing the selected member after each selection. The population then remains the same at each selection. The two cases are distinguished as *sampling without replacement* and *sampling with replacement*.

Furthermore, if the population is large compared with the size of the sample, the removal of relatively few members will not materially affect the constitution of the remaining population, which may thus be regarded as approximately the same for subsequent samplings. That is, sampling with replacement from a finite population may be regarded as sampling from an infinite population, to a reasonable degree of approximation, provided the proportion sampled is small.

When sampling from a finite population, we select particular individuals. However, what is of statistical interest is one or more characteristics of these individuals, such as height or income. Thus, given a population of N individuals with a total of K distinct values for the characteristic of interest, $x_1 < x_2 < \ldots < x_K$, say, we may describe the population by the frequency function f_j, $j = 1, \ldots, K$ corresponding to the values $\{x_j\}$. The sampling process now corresponds to that of selecting one of the values $\{x_1, \ldots, x_K\}$ with probabilities $\{f_1/N, f_2/N, \ldots, f_K/N\}$.

9.4 This representation paves the way for the more general notion of random sampling from a distribution. Given the set of possible x-values and the associated distribution function, we may select a value *at random* by the following device:

(a) select a real number, u, at random on the interval $(0, 1]$; i.e., u has a uniform distribution on $(0, 1]$;

(b) determine x (or x_j) as the unique value satisfying
  (i) $u = F(x)$, when $f(x)$ is continuous, or
  (ii) $F(x_{j-1}) < u \leqslant F(x_j)$, when $f(x_j)$ is discrete.

This definition may seem rather indirect, but it is completely general and it is useful operationally, as we shall see in **9.11-15** below.

It is evident that the set of x-values may be finite, countably infinite, or non-countably infinite. If $f(x)$ is continuous and is derived by some limiting process, the nature of the limiting process must be taken into account; see **8.17**.

Our extended definition allows for the possibility that the population may be purely hypothetical. The concept of the hypothetical population is necessitated by the notion of probability as a limiting relative frequency, as introduced in **8.11**. It is not required (and indeed has been explicitly rejected by Jeffreys) in definitions of probability as a degree of belief. However, in order to consider relative frequencies, we must be able to contemplate a series of replications under identical conditions and to specify the set of possible values that might be realized..

Consider, for example, the act of throwing a die and observing the uppermost face. Successive throws may be regarded as a sampling process, and we might agree that the hypothetical population is described by the set of possible outcomes $\{1, 2, \ldots, 6\}$ with associated probabilities $\pi_j, j = 1, \ldots, 6$. Again, if we grow wheat on a plot of soil, we will achieve a yield x, measured in some standard units. In order to regard this as a sample, we must be willing to specify a set of possible yields from that plot and an associated distribution function. Once given these constructs, we must be willing to assume that our observed value was randomly selected from the set of possibilities.

At first sight, this is a rather baffling conception. However, such a structure is fundamental in a frequency-based theory of inference, and the approach is justified as being empirically useful.

Randomness in sampling practice

9.5 In its colloquial use the word "random" is applied to any method of choice that lacks aim or purpose. We speak of drawing names at random out of a hat, choosing plants at random from a field of corn, selecting family budgets at random from the population, meaning thereby that the selection is completely haphazard.

Now it is found in practice that choice by a human being is not random in the stricter sense that, equally frequently, it produces events that we expect to have equal prior probabilities. Some examples will make this clear.

Example 9.1 (Size-biased sampling)

In the course of work at Rothamsted Experimental Station, sets of eight wheat plants were chosen for measurement. In each set, six were chosen by approved methods, referred to below, and may be taken to be truly random. The other two were chosen haphazardly by eye. If, in any set, the eight plants were numbered in order of height, the two selected by eye could have any number from one to eight; and if they were chosen at random, they should occupy these places with approximately equal frequency in a large number of sets. Table 9.1 shows what actually occurred on two different occasions (a) on May 31st, before the ears of wheat had formed, and (b) on June 28th, after the ears had formed.

Table 9.1 Distribution of plants chosen haphazardly in ranks 1 to 8
(Yates, 1935)

Date	Observation	Numbers bearing specified rank								TOTAL
		1	2	3	4	5	6	7	8	
May 31st	Shoot height . .	9	7	11	8	11	18	21	31	116
June 28th	Ear height . . .	9	19	27	23	15	10	5	4	112

The divergence of actual from expected results is quite striking. On May 31st, before the ears had formed, the observer was strongly biased towards the taller shoots;

whereas in June he was biased strongly towards the central plants and avoided short and tall plants.

Thus it is seen that bias can appear even in a trained observer, and that the bias need not be consistent in over- or under-estimation in different circumstances.

Example 9.2 (Bias in scale reading)

Table 9.2 shows the frequencies of final digits in a number of measurements made by four different observers.

It is hard to suppose that there was any genuine difference which would lead to the appearance of certain digits at the expense of others, and we may confidently suppose that the deviations from approximate equality indicate bias on the part of the observer.

Observer A had a decided preference for 0, 2, 8 and 9, avoiding the centre of the scale. Observer B is quite good, his deviations from expected values being small, though he also showed a preference for 0. Observer C was poor, rounding off nearly one measurement in two to the whole or half unit. Observer D was obviously very bad indeed, nearly 57 percent of his measurements being rounded off to the whole or half unit.

Table 9.2 Bias in scale reading: distribution of final digits in measurements by four observers
(Yule, 1927)

Final digit	Frequency of final digit per 1000			
	A	B	C	D
0	158	122	251	358
1	97	98	37	49
2	125	98	80	90
3	73	90	72	63
4	76	100	55	37
5	71	112	222	211
6	90	98	71	62
7	56	99	75	70
8	126	101	72	44
9	129	81	65	16
TOTAL	1001	999	1000	1000

The observations were all made reading a scale, those under A being on drawings to the nearest tenth of a millimetre, those under B, C and D being measurements on the heads of living subjects to the nearest millimetre. We may conclude from this that different observers may exhibit different degrees of bias even under comparable circumstances, and that even those who are aware of the existence of the possibility of bias and the necessity for taking great care (as observer A was) may nevertheless fail to avoid it.

Example 9.3 (Bias in observation)

An observer was placed before a machine consisting of a circular disc divided into ten equal sections in which were inscribed the digits 0 to 9. The disc rotated at high speed and every now and then a flash occurred from a nearby electric lamp of such short duration that the disc appeared at rest. The observer had to watch the disc and write down the number occurring in the division indicated by a fixed pointer.

This was a machine designed to produce random numbers, and had been found by another observer to do so, but this particular observer produced a definite bias. The frequencies of digits in 10 000 run off by him are shown in Table 9.3.

Table 9.3 Distribution of digits obtained by an observer in using a randomizing machine
(Kendall and Babington Smith, 1939)

Digit . . .	0	1	2	3	4	5	6	7	8	9	TOTAL
Frequency . .	1083	865	1053	884	1057	1007	1081	997	1025	948	10 000

If the observer was unbiased the digits should appear in approximately equal numbers; but there is a bias in favour of all the even numbers and against the odd numbers 1, 3 and 9. The cause of this bias is obscure, for the observer did not have to estimate (as in the previous examples) but merely to write down something that he saw, or thought he saw. The explanation seemed to be that he had a strong number-preference, i.e. that he actually mis-saw the numbers, or that his brain controlled his ocular impressions and censored them. We have here to deal with one of the deadliest forms of bias in psychology.

Example 9.4 (Forecasting bias)

Every year a number of crop reporters in England and Wales estimate the prospective yields of certain crops, forecasts being obtained at different periods of the year and final estimates when the crop is harvested. Table 9.4 shows the average estimated yield of potatoes at the various times for the years 1929–1936.

Table 9.4 Bias in crop forecasting: forecasts of yields of potatoes in England and Wales (tons per acre)
(From the official agricultural statistics)

Year	Sept. 1st		Oct. 1st		Nov. 1st		Final estimate
	Yield	% difference from final	Yield	% difference from final	Yield	% difference from final	
1929	5.7	−17.4	6.2	−10.1	6.5	−5.8	6.9
1930	6.0	−7.7	6.1	−6.2	6.1	−6.2	6.5
1931	5.5	0.0	5.3	−3.6	5.3	−3.6	5.5
1932	6.4	−3.0	6.2	−6.1	6.3	−4.5	6.6
1933	6.4	−4.5	6.2	−7.5	6.4	−4.5	6.7
1934	6.0	−15.5	6.3	−11.3	6.7	−5.6	7.1
1935	5.6	−9.7	5.7	−8.1	6.0	−3.2	6.2
1936	6.0	−3.2	5.9	−4.8	5.8	−6.5	6.2

This table exhibits very clearly an effect that has shown itself in many English crop reports (and appears also in other countries), namely, the chronic pessimism of crop forecasts. In every case but one in the table, the forecasts are below the final estimate. Nor do crop reporters seem able to learn by experience that they are underestimating. Nothing in this table indicates that the differences between forecast and final estimate diminished during the period concerned.

It should also be noticed that these estimates are the weighted averages of a large number of independent observations. One of the commoner misunderstandings in this type of work is based on the supposition that, though individuals may make mistakes, their errors will cancel out in the aggregate. Our present example shows this to be untrue in general. There can appear a systematic bias affecting all the individuals making estimates.

9.6 The foregoing examples are enough to indicate that human bias is prevalent. Trained observers may be biased even when conscious of their own imperfections; different observers may be biased in different ways in similar circumstances; and the same observer may be biased in different ways in different circumstances. It is abundantly clear that we must look for true randomness elsewhere than in mere lack of purpose on the part of human observers. There may be persons whose psychological processes are so finely balanced that they can deliberately select random samples, but few statisticians who have experimented in this interesting field would regard themselves as so gifted.

9.7 When we consider sampling from a finite population of size N, the notion of randomness implies that each individual has the same chance of selection. If a large number of selections is made with replacement, we expect the proportion of times a given individual is selected to be approximately $1/N$.

But this is not enough. Suppose we have a population of two members A and B, and sample with replacement. Then a method that chooses A and B alternately and produces the series $ABAB$ ... selects each member approximately equally frequently; but it is not what we customarily mean by a random method. What we require of a random method is that in such circumstances it should produce a series in which no systematic arrangement is evident. Not only single characteristics, but all possible groups of characteristics should appear equally frequently.

9.8 A further point is to be noted. We may, in drawing the sample, be interested in one particular variate exhibited by the members, and it is possible that a method may give a satisfactory random sample *so far as this variate is concerned* without doing so for other variates. Suppose, for example, we are anxious to take a random sample from the inhabitants of a particular area. If we are concerned with a variate such as eye-colour it might be sufficient to choose a house every so often, say every tenth house, and select one inhabitant of that house as part of the sample. Such a method would not give every inhabitant of the area an equal chance of being chosen; but if we look back to the time when the inhabitants took up residence, we may imagine that the colour of their eyes did not influence their geographical distribution, and thus if we consider the distribution of the inhabitants into houses as independent of eye colour, we may suppose that so far as eye-colour is concerned the sample is random. But the matter would stand differently if we were sampling for income, for we may reasonably expect poorer people to live in more crowded conditions, and a sample of one person from each house chosen would therefore under-represent the poor. Thus our sample would not be random with respect to income.

Thus a method that may properly be deemed to be random for one variate may not be so for another.

The technique of random sampling

9.9 Suppose, then, that we are given a population and a variate is specified. How are we to draw a random sample, i.e. how can we find a method which is random for that population and that variate? The answer lies partly in theory and partly in practice.

(a) In the first place we must require that there is no obvious connection between the method of selection and the variates under consideration. The method and the variates must be independent so far as our prior knowledge is concerned. In sampling a field of wheat for shoot height, for example, we must not use a method that could be influenced by that height, such as skimming a hoop over the field and selecting the plants round which it fell (for the hoop might tend to catch on the taller plants). Again, in sampling the inhabitants of a town by choosing names from a telephone directory, we should undoubtedly tend to get the more well-to-do classes and hence, if the variate under consideration is wealth or any related characteristic such as number of children, political opinion, standard of education and so on, the sample would not be random. If we were concerned with characteristics such as height, hair colour, or blood group the sample might be random, though it is not difficult in many similar cases to think of reasons why the variate might be linked with wealth.

If this matter is viewed from the standpoint of the subjective theory of probability, the absence of knowledge about relationship between the method of selection and the characteristic under consideration may be sufficient to ensure randomness, for the probabilities of elementary propositions then become equal—the probabilities being measures of prior attitudes of mind.[*] But if the frequency viewpoint is adopted, it is not enough that there should be absence of knowledge of this kind, for unknown to the observer there may be some systematic relationship between the variate values and the probability that that individual is selected. The presumption is that if we make as great an effort as possible to ascertain whether any relationship exists and fail to find it, there is no relationship; and hence we can assume the randomness of the method with more or less confidence. But in this approach the assumption of randomness is ultimately part of the general uncertainty of the inference from sample to population.

(b) Secondly, we may rely on previous experience of a random method of selection

[*] At least, this is our interpretation of the position, but we may be putting a gloss on the views of its proponents that they would not accept.

to justify its use on new occasions. This is evidently an extrapolation, and though most people would regard it as reasonable, the fact has to be realized. The subjective theory of probability can embrace this extrapolation within its scope, for the probabilities given by the method are assessable in terms of prior knowledge; but the frequency theory has to take the extrapolation as an additional assumption.

9.10 One method of drawing random samples consists of constructing a model of the population and sampling from the model. We may, for instance, note down the characteristics of each member on a card and sample by choosing cards from the pack corresponding to the whole population. This is the method adopted in lotteries and the process is known as lottery or ticket sampling. It is moderately effective but suffers in practice from two disadvantages: the labour of constructing the card population, and the danger of bias in the drawing of cards. To justify the assumption of randomness, we must ensure a thorough shuffling of the cards, which is difficult to achieve. The same object can be attained much more simply by the use of random sampling numbers, which we now consider.

Random (sampling) numbers

9.11 The easiest way to represent a finite population is to attach a positive integer to each member, most simply by numbering the members from 1 onwards. The set of integers completely describes the population and the problem of drawing a random sample reduces to finding a series of random integers. The advantages of this method are obvious: no physical model population has to be constructed; the numbering can be carried out in any convenient manner; and the series of random integers can be applied to any enumerable population so that any series of random integers has a very wide range of application.

If the numbering of the population is carried out in such a way as to be independent of certain characteristics of the population, any set of integers will serve to draw a sample random with respect to those characteristics. The randomness in such a case lies in the allocation of integers to the population, not in deciding which integers to select for the sample. But in practice a procedure of this kind is of no value, since it only throws us back to the problem of numbering the population "at random". The usual course is to number the population in any convenient way, related to the characteristics or not, and then seek a set of integers that are a random set from the possible integers of the population.

9.12 One of the more obvious ways of drawing random samples from an enumerated population is to use haphazard numbers taken from some totally unrelated source. Suppose, for instance, we wished to take a sample from the stars visible in the sky. We will ignore the small complications due to the existence of double stars and unresolved objects. Since the position of a star on the celestial sphere is defined by latitude and longitude, what is required is a series of random pairs of latitudes and longitudes. At first sight it seems plausible to take an ordinary atlas and choose the figures set out in the index for place-names arranged alphabetically; for there is little reason to expect any relationship between the distribution of stars in the sky and the distribution of places on the Earth's surface. A little reflection, however, will show that the method is unsound. There are large stretches of territory and sea on the Earth that have no place-names on them—the poles, deserts and oceans; consequently no numbers will occur for these regions and the corresponding areas on the celestial sphere would have no chance of being included.

9.13 As a next attempt we might take a book containing a number of digits, e.g. a telephone directory, or a set of statistical tables or mathematical tables, open it at hazard and choose the digits that first strike the eye, or that occur at the top of the page, and so on. This is an improvement, but it is still open to objection.

Example 9.5 (Sampling using published listings)

(a) *Telephone directories.* Table 1.4 in **1.9** shows the distribution of 10 000 digits taken from a London telephone directory. Pages were chosen by opening the directory haphazardly; numbers of less than four digits and numbers in heavy type were ignored; and of the four-digit numbers remaining the last two digits were taken for all numbers on the page. If the numbers were random we should expect about 1000 of each digit in the total of 10 000. Actually there are very considerable deviations from this expectation, and we shall see in Example 15.3 that they cannot be explained as sampling fluctuations. There are significant deficiencies in 5s and especially 9s, due to several causes such as the tendency to avoid these digits because they sound alike, the reservation of numbers ending in 99 for testing purposes by telephone engineers and so on. It is evident that tables of random numbers could not be constructed from directories such as this.

(b) *Mathematical tables.* Evidently care has to be exercised in using mathematical tables in constructing random series. Suppose, for instance, we take a set of logarithm tables. There are clearly relationships between successive logarithms, expressible by the fact that differences are approximately constant if the interval is small. Moreover there is a very curious theorem about digits in certain classes of table that throws theoretical doubt on the method. Consider the logarithms to base 10 of the integers from 1 onwards. Suppose we choose the kth digit in each and so obtain a series of digits 0–9. Then the proportional frequency of any digit in this series does *not* tend to a limit as the length of the series increases, whatever k may be.[*] Just what does happen does not appear to be known, but it seems that certain systematic effects begin to show themselves and these will obviously endanger the randomness of the series. Other tables, such as the successive integers in the decimal representation of π, have also been used. Even when such lists appear random, their use is severely restricted by the limited number of integers available and the cost of generating further values.

(c) *Statistical tables.* If we have a volume of statistics such as populations of towns and rural districts, there are some grounds for supposing that if the numbers are large—say, four figures or more—the final digits will be random. Here again, however, the use

[*] Cf. J. Franel, *Vierteljahrschrift der Naturforschenden Gesellschaft in Zürich* (1917), 62, 286.

of such tables requires care—they may have been compiled by an observer with number preferences, and some rounding may have taken place.

9.14 Historically, the need for random numbers was met by the publication of tables of random numbers. These include:

(a) Tippett's numbers comprise 41 600 digits taken from census reports combined into fours to make 10 400 four-figure numbers (*Tracts for Computers*, No. 15, Cambridge U.P.).

(b) Kendall and Babington Smith's numbers comprise 100 000 digits grouped in twos and fours and in 100 separate thousands (*Tracts for Computers*, No. 24). These numbers were obtained from a machine specially constructed for the purpose on the lines very briefly described in Example 9.3.

(c) Fisher and Yates' numbers comprise 15 000 digits arranged in twos (*Statistical Tables for Biological Agricultural and Medical Research*). These numbers were obtained from 15th–19th digits in A. J. Thompson's tables of logarithms and were subsequently adjusted, it having been found that there were too many sixes.

(d) The Rand Corporation has published *A Million Random Digits* (1955) arranged in groups of five. They were constructed by a kind of electronic roulette wheel; but it is interesting to observe that even after searches for systematization had been made in the circuits, the numbers were still not quite random and had to be adjusted.

The above-mentioned tables consist of randomized digits. Wold (*Tracts for Computers*, No. 25) has compiled a table of 25 000 normal observations, by converting the Kendall–Babington Smith tables with the aid of the normal integral. The Rand Corporation's book also contains 100 000 normal observations. Fieller and others (*Tracts for Computers*, No. 26) gave 3000 bivariate normal observations for each correlation coefficient value 0.1(0.1)0.9. Krishna Iyer and Sinha (1967) give 1050 bivariate normal observations for each correlation value 0.05(0.10)0.95.

Quenouille (1959) gave 1000 random values from each of eight distributions, these being the normal, lognormal, exponential, double exponential, uniform and three Edgeworth expansions. *Tables of Normal and Log-Normal Random Deviates* (Almqvist and Wiksell, 1962) have been prepared from (b) and (d) above at the University of Gothenburg. Barnett (*Tracts for Computers*, No. 27) gives 10 000 random exponential observations and 1000 random standardized normal squares.

9.15 Before considering the basis of these tables it may be helpful to give an example of their use. Here are the first 200 digits in the Kendall–Babington Smith tables:

Table 9.5 Random sampling numbers
(Tracts for Computers, No. 24)

23 15	75 48	59 01	83 72	59 93	76 24	97 08	86 95	23 03	67 44
05 54	55 50	43 10	53 74	35 08	90 61	18 37	44 10	96 22	13 43
14 87	16 03	50 32	40 43	62 23	50 05	10 03	22 11	54 38	08 34
38 97	67 49	51 94	05 17	58 53	78 80	59 01	94 32	42 87	16 95
97 31	26 17	18 99	75 53	08 70	94 25	12 58	41 54	88 21	05 13

Example 9.6 (Sampling using tables of random numbers)

To draw a sample of 10 men from the population of 8585 men of Table 1.7.

The first process is to number the population; and here, as in most similar cases, one numbering has already been provided by the frequency distribution. We take numbers 1 and 2 to be those in the group 57– inches, numbers 3 to 6 those in the group 58–, and so on, those in the group 77– inches being numbers 8584 and 8585.

Now we take 10 four-figure numbers from the tables, e.g. reading across in Table 9.5 we have 2315, 7548, 5901, 8372, 5993, 7624, [9708], [8695], 2303, 6744, 0554, 5550.

The two numbers in square brackets are greater than 8585 and we ignore them. We now select the individuals corresponding to the remaining 10 numbers. They will be found to be in the intervals 65–, 70–, 68–, 72–, 68–, 70–, 65–, 69–, 63–, 68– inches respectively. The mean of these values considered as located at the centres of intervals is 68.24, as against the population value of 67.46 in Example 2.1.

Example 9.7 (Random permutations)

To construct a series of random permutations of the digits 1 to 5.

Here we are not concerned with the digits 0, 6, 7, 8 and 9 and so ignore them in the table of random numbers. We read through the table and note the digits as they occur, e.g. in Table 9.5 we have 2315, 7548, etc. The 7 is to be ignored and also the second 5, for one 5 has already occurred. We then reach the permutation 23 154. Then we start again, the next series being 8, 5901, 8372, 5993, 7624, etc., giving the permutation 51 324; and so on. This method is clearly rather wasteful of random digits—Plackett (1968) discusses more efficient methods.

Tables of 400 random permutations of the integers 1 to 20 are given in Kendall and Gibbons (1990). Permutations of fewer than 20 can be derived by omitting unwanted integers, and there can be derived by obvious procedures 800 permutations of the integers 1 to 10, 1600 of the integers 1 to 5 and so on. Cochran and Cox, in their *Experimental Designs* (Wiley, 2nd edn, 1957), give 1000 permutations of each of 9 and 16 integers.

Moses and Oakford (*Tables of Random Permutations*, Allen and Unwin, 1963) used the Rand's million digits (9.14(d)) to construct 960 random permutations of 9 integers; 850 of 16; 720 of 20; 448 of 30; 400 of 50; 216 of 100; 96 of 200; 35 of 500 and 20 of 1000.

9.16 Random sampling numbers must satisfy certain conditions before they can be used. Any set of numbers whatever is random in the sense that it might arise from random sampling; but such a set might not be suitable as a table of random sampling numbers. From the examples already given it is clear that we desire such a table to have very great flexibility. It should give random results in as many cases as possible, whether used in part or in whole.

Now it is impossible to construct a table of random sampling numbers that will satisfy this requirement entirely. Suppose, to take an extreme case, we constructed a table of $10^{10^{10}}$ digits. The chance of any digit being a zero is $\frac{1}{10}$ and thus the chance that any given block of a million digits are all zeros is 10^{-10^6}. Such a set should therefore arise fairly often in the set of $10^{10^{10}-6}$ blocks of a million. If it did not, the whole set would not be satisfactory for certain sampling experiments. Clearly, however, the set of a million zeros is not suitable for drawing samples in an experiment requiring less than a million digits.

Thus, it is to be expected that in a table of random sampling numbers there will occur patches that are not suitable for use by themselves. The unusual must be given a chance of occurring in its due proportion, however small. Kendall and Babington Smith attempted to deal with this problem by indicating the portions of their table (5 thousands out of 100) that it would be better to avoid in sampling experiments requiring fewer than 1000 digits.

9.17 If a table of random numbers is used to draw members from a population of ten, we expect the members to appear in approximately equal proportions. In other words we expect such a table to contain the ten digits 0–9 in approximately equal proportions. Similarly we expect the hundred pairs 00–99 to appear in approximately equal proportions, and so on. Various tests of this kind, based on a comparison between actual and expected frequencies, can be devised. No table can satisfy them all, but if it satisfies tests that (a) ensure the randomness of the numbers for the commoner types of sampling inquiry for which it is likely to be used and (b) are capable of revealing any particular sort of bias to which the numbers are susceptible because of their mode of formation, it is likely to be of general application.

Other tests of randomness based on runs, changes of sign or on the autocorrelation properties of the series, may also be used; see, for example, Kendall and Ord (1990, Chapter 2). For a systematic account of tests for random numbers, see Fishman and Moore (1982).

9.18 Random sampling numbers offer the best method known at the present time of drawing random samples from an enumerable population and, as we shall see below, may also be used to draw samples from a continuous distribution specified mathematically. But cases sometimes occur in which they cannot be employed. For instance, if we wish to take a sample of a liquid or a powder, we cannot in practice number each particle and extract it from the population for examination. In such cases we are usually compelled to fall back on more intuitively founded procedures. To take a random sample from a milk churn, for instance, we might stir the contents thoroughly and scoop up a sample haphazardly. Sometimes, when the population is of manageable size, we can proceed systematically by dividing it into a number of parcels and selecting parcels by the ordinary technique of random numbers. Most sciences have their own peculiar sampling problems and no attempt is made here to discuss them all.

Pseudorandom numbers

9.19 The methods described thus far were developed for sampling from finite populations when a list of the population members is available. Developments in computer technology have shifted the emphasis in two rather different ways. First of all, many population listings are now available as computer data bases, so that the sampling process can be performed much more efficiently if the random numbers are directly available in the computer. Secondly, the vast increases in computing speed have made simulation an indispensable tool for the statistician. Large-scale studies of sampling distributions and the like are now performed routinely and require ready access to large streams of random numbers. We shall try to resist the picturesque description of these as "Monte Carlo studies"; "sampling experiment" is more informative, if less exciting.

An early response to these developments was to store tables of random numbers in the computer and use them as required. However, to be effective, such tables must be large and, in consequence, must often be stored in slower-access memory: this is generally considered to be too slow. A second method was to generate numbers by some physical device whose construction could reasonably guarantee random selection. This technique fell from favour both because of lack of speed and of the difficulty of checking the randomness of the device (perhaps a rather curious criticism). However, it should be noted that Inoueh *et al.* (1983) report the satisfactory performance of a radiation device in generating random numbers, and they have stored large tracts of random numbers on disk.

By and large, the modern response has been to use a pseudorandom number (PRN) generator. A PRN generator is an arithmetic (deterministic) device that produces a stream of numbers which appear to behave in a random fashion. That is, they satisfy various tests for randomness, both those mentioned in **9.17** and others (cf. Atkinson, 1980).

Once a stream of uniformly distributed random numbers is available, we must find ways of converting these to the distribution required. We now examine in turn the methods used to generate PRN and methods of converting them.

9.20 The most commonly used class of methods are the *congruential* methods. Given the initial PRN x_0 (known as the *seed*) and the constants a, b, and m, successive random numbers are generated by the relationship

$$x_{i+1} = (ax_i + b)(\text{mod}\, m), \tag{9.1}$$

where $0 \leqslant x_i < m$. The expression (mod m) means that $0 \leqslant ax_i + b - km = x_{i+1} < m$, where k is an integer.

Example 9.8 (Generation of pseudo-random numbers)

Let $a = 827$, $b = 0$, $m = 10^3$, and take $x_0 = 179$. We obtain

$$x_1 = 148\,033\,(\text{mod}\,10^3) = 033$$
$$x_2 = 27\,291\,(\text{mod}\,10^3) = 291$$
$$x_3 = 240\,657\,(\text{mod}\,10^3) = 657$$

and so on.

Two features of this example are worth noting. First of all, we set $b = 0$; (9.1) is then said to define a *multiplicative* congruential generator. Much of the theoretical and empirical work done on PRN generators refers to this class. Secondly, we choose m to be a power of 10 (the number base used) so that the modulus operation requires only that we retain the (three) least significant digits. This device reduces the time taken to generate a PRN; $m = 2^c$ is the most common form since most computers use binary arithmetic.

9.21 If the PRN generator is to produce a series of numbers that satisfies the tests for randomness, the choice of a, b and m is clearly critical. If we suppose that $m = 2^c$, where c is the length of a standard work area (c binary digits), it may be shown (Knuth, 1969) that a choice of a such that $a = \pm 3 \pmod 8$ will give a maximal period of 2^{c-2}. That is, 2^{c-2} different numbers will be generated before the generator returns to the original seed, x_0. Since modern computers often have $c = 31$ or $c = 35$, this is a satisfactory property for most purposes. Although various additional guidelines have been suggested (cf. Kennedy and Gentle, 1980, Chapter 6), the final choice of a is guided by experience as much as anything else.

Marsaglia (1972) developed a lattice test based on consideration of successive n-tuples of PRN sequences. An ideal generator should produce n-tuples that are uniformly distributed over the n-dimensional cube. Unfortunately, Marsaglia (1972) found the multiple congruential generator to be seriously deficient in this respect. Three approaches have been suggested to overcome this difficulty. One is to change m; Fishman and Moore (1982) find that several generators based on $m = 2^c - 1$ pass the usual statistical tests and perform well on the lattice test and that these can be coded almost as efficiently as those based on $m = 2^c$.

Secondly, we may use the "coupled" generators first suggested by MacLaren and Marsaglia (1965). The method assumes two different generators yielding sequences $\{x_i\}$ and $\{x_i'\}$. The first k values of $\{x_i\}$ are then entered into a table $\{t_1, \ldots, t_k\}$. The following steps are then performed:

(i) generate the next values x and x';
(ii) use x' to select randomly an integer j, $1 \leqslant j \leqslant k$;
(iii) take t_j as the new random number, and replace t_j in the table by the current x.

Nance and Overstreet (1978) show that this scheme works well with $k = 2$. Coupling generators appears to produce sequences with good randomness properties even when the original generators are not very good.

A third approach is to consider linear combinations of independent PRN streams. For example, Wichmann and Hill (1982) develop a simple and effective generator that can be written in a high-level language and produces consistent results on different machines. They observe that the fractional part of $(x_1 + \ldots + x_n)$ is uniformly distributed, given that the x_i are independent variates defined on (0, 1) and at least one of them is uniform; cf Exercise 9.20. They then recommend that n multiplicative congruential random number generators be used to produce the final value; $n = 3$ appears to suffice in practice. In addition to the benefits noted above, the use of n generators allows long cycles to be achieved without the need for double precision arithmetic which would slow operating time considerably. M. Brown and Solomon (1979) show that such an approach results in improved sequences in the sense that the joint distributions of $k \geqslant 2$ variates are closer to k independent uniforms.

Marsaglia and Zaman (1991) introduce a broad class of lagged Fibonacci generators of the form

$$x_n = x_{n-r} * x_{n-s}$$

where * denotes some binary operation. In particular, the generators

$$x_n = x_{n-s} + x_{n-r} + c \pmod m$$

and

$$x_n = x_{n-s} - x_{n-r} - c \pmod m$$

with $s > r$, appear to satisfy the standard tests. These forms produce PRNs with a period of $m^r + m^s - 2$, but require a set of s startup values. Marsaglia and Zaman (1991) suggest $m = 6$, $s = 21$ and $r = 2$ as a simple choice that yields a period of 21.94×10^{15}.

Another class of PRNs is the set of generalized feedback shift generators; see Lewis and Payne (1973). Although these generators exhibit good behaviour, they suffer from rather elaborate initialization procedures.

A variety of other approaches are reviewed by Kennedy and Gentle (1980, Chapter 6). From our viewpoint, it is sufficient to recognize that fast and reliable PRN generators may be constructed, although the validity of those available in software packages should be checked before use.

Random values from specific distributions

9.22 Once a random number, u, is available which after scaling may be regarded as uniformly distributed on $(0, 1]$, we may generate a random value for any distribution using the probability integral transformation (see **1.27**); that is,

$$x = F^{-1}(u). \tag{9.2}$$

For example, the exponential distribution has $u = F(x) = 1 - e^{-x}$, whence $x = -\log_e(1-u)$. The alternative form $x = -\log_e u$ is commonly used as it is slightly faster. When (9.2) cannot be inverted in closed form, an approximation may be necessary. However, even when an exact inversion is possible, other techniques may be more efficient (Atkinson and Pearce, 1976). We now briefly examine some of the major alternative approaches.

Distributions may be defined directly in terms of (9.2), so that inversion ceases to be a problem—see the discussion in **14.10** below.

9.23 The simplest approach of general application is the *table look-up method*. Given the values $x_1, \ldots, x_m$ and the corresponding values of the d.f. $F(x_i)$, we generate a uniform random number u on $(0, 1]$ and select

$$x = x_i \quad \text{if } F(x_{i-1}) < u \leqslant F(x_i). \tag{9.3}$$

For continuous distributions, such methods involve an element of approximation, and better methods are usually available. For discrete distributions, variations of this approach may be very efficient (Peterson and Kronmal, 1983).

For more specific algorithms for the Poisson distribution, see Atkinson (1979b,c); for the binomial distribution, see Fishman (1979); and for the logarithmic series see A. W. Kemp (1981a). For bivariate discrete distributions, see A. W. Kemp (1981b) and C. D. Kemp and Loukas (1978).

If the objective is to generate a frequency table, C. D. Kemp and A. W. Kemp (1987) provide a very efficient algorithm whereby the categories are first ordered so that their probabilities satisfy $p_{(1)} \geqslant p_{(2)} \geqslant \dots$ and binomial variates are then generated such that x_1 is bin$[n, p_{(1)}]$, x_2 is bin$[n - x_1, p^*_{(2)} = p_{(2)}/(1 - p_{(1)})]$ and so on.

9.24 Suppose that $f(x)$ is bounded above by $cg(x)$, where $c > 0$ and $g(x)$ is a density for all x. Generally, $g(x)$ is an easily computed function. The *rejection method* generates random numbers from $f(x)$ by

(i) generating a pair of independent uniform random numbers, u_1 and u_2;
(ii) generating a random number from $g(x)$ using u_1, x_g say;
(iii) accepting x_g as a random selection from $f(x)$ if $u_2 \leqslant f(x)/\{cg(x)\}$;
(iv) if x_g is rejected, return to step (i).

The constant c is the smallest number such that $f(x) \leqslant cg(x)$ for all x.

Example 9.9 (Normal random numbers)

Suppose we wish to generate random numbers from the standardized normal distribution with

$$f(x) = \frac{1}{\sqrt{2\pi}} e^{-\frac{1}{2}x^2}.$$

If we use the Laplace distribution (see **5.33**)

$$g(x) = \tfrac{1}{2}\alpha e^{-\alpha|x|},$$

the d.f. is

$$G(x) = \begin{cases} \frac{1}{2}e^{\alpha x}, & x \leqslant 0 \\ 1 - \frac{1}{2}e^{-\alpha x}, & x > 0 \end{cases}$$

which is easily inverted. Once x_g is available, the rule in (iii) becomes

$$\log_e u_2 \leqslant c_1 + \alpha|x_g| - \tfrac{1}{2}x_g^2, \tag{9.4}$$

where $c_1 = \frac{1}{2}\log_e c^2\alpha^2\pi/2$. For example, when $\alpha^2 = 2$ so that var $(x) = 1$ for the Laplace variate, the value of c_1 is 1.

Other forms of $g(x)$ may be used. The choice is a trade-off between the closeness of the bound and the speed with which a rejection rule such as (9.4) can be checked. Several approaches may be combined, as illustrated by the following example, due to Wallace (1974).

Example 9.10 (Gamma random numbers)

It is desired to generate a random number from a population described by the Gamma density

$$f(x) = \frac{x^{\alpha-1}}{\Gamma(\alpha)} e^{-x}, \quad 0 < x < \infty; \alpha > 0.$$

If $\alpha = m$, an integer, x is the sum of m independent exponentials. Let $u_1, \ldots, u_m$ be independent and uniformly distributed on $(0, 1]$, then x_m is Gamma (m) if we generate

$$x_m = -\sum_{i=1}^{m} \log u_i = -\log\left(\prod_{i=1}^{m} u_i\right). \tag{9.5}$$

When $m < \alpha < m+1$, let $q = \alpha - m$ and generate

$$x_{m+1} \text{ with probability } q \text{ or } x_m \text{ with probability } 1-q.$$

Finally, given a new uniform random number u, we accept x as Gamma (α) if

$$u \leqslant (x/m)^q (1 + qz),$$

where z is the integral part of $(x/m) - 1$.

Forsythe (1972) introduced a general rejection method that involves first selecting a particular interval in accordance with (9.3) and then sampling within the interval using a rejection method. This appears to be computationally efficient for a variety of distributions (Atkinson and Pearce, 1976); see also Marsaglia (1984).

9.25 A variety of distributional results may be used to assist in the construction of random numbers from other distributions. The following table indicates several such possibilities, whose derivation will be found elsewhere in this volume.

Table 9.6 Transformations useful for converting PRN to desired form

Original variable(s) x	Transformation	Distribution of transformed variate
Uniform (0, 1]	$-\log_e x$	Exponential
Exponential	x^α	Weibull (α)
k independent $N(0, 1)$	$\sum_{i=1}^{k} x_i^2$	Chi-squared (k d.fr.)
2 independent Gammas	$x_1/(x_1+x_2)$	Beta (first kind)
Beta (first kind)	$x/(1-x)$	Beta (second kind)
Uniform (0, 1]	$\log_e[x/(1-x)]$	Logistic
2 independent $N(0, 1)$	x_1/x_2	Cauchy

Such transformations are not necessarily efficient in terms of computational speed, but they are readily programmed and sufficiently quick for many practical purposes. Random numbers from the Johnson and the Burr families (see **6.28** and **6.37**) are readily generated by the appropriate transformations.

More specific algorithms for the Gamma distribution are described in Atkinson and Pearce (1976). For the Beta distribution, see Atkinson (1979a), Atkinson and Whittaker (1979), Schmeiser and Shalaby (1980) and Boswell and De Angelis (1981). For a detailed treatment of multivariate distributions, see M. E. Johnson (1987).

9.26 One further transformation that is well known and easy to program is due to Box and Müller (1958). If x and y are independent $N(0, 1)$ and we transform to polar coordinates

$$x = r\cos\theta, \qquad y = r\sin\theta,$$

the Jacobian is

$$\frac{\partial(x, y)}{\partial(r, \theta)} = \begin{vmatrix} \cos\theta & \sin\theta \\ -r\sin\theta & r\cos\theta \end{vmatrix} = r$$

whence

$$\begin{aligned} f(r, \theta) &= rf(x)f(y) \\ &= r\exp\{-\tfrac{1}{2}(x^2 + y^2)\}/(2\pi) \\ &= r\exp(-\tfrac{1}{2}r^2)/(2\pi), \end{aligned}$$

where $0 < r < \infty$ and $0 < \theta \leqslant 2\pi$; thus r and θ are independent. Moreover, $u_1 = \exp(-\frac{1}{2}r^2)$ and $u_2 = \theta/(2\pi)$ are independently uniform on $(0, 1]$. Inverting our derivation we see that if u_1, u_2 are independently uniform $(0, 1]$,

$$x = (-2\log u_1)^{1/2}\cos 2\pi u_2$$

and (9.6)

$$y = (-2\log u_1)^{1/2}\sin 2\pi u_2$$

give a pair of independent normal random values.

Marsaglia and Bray (1964) improved the operating speed of this approach by the following device.

Let u_1 and u_2 be uniformly distributed on $(-1, 1]$, subject to the condition $w = u_1^2 + u_2^2 \leqslant 1$. That is, $f(u_1, u_2)$ is a disc of uniform height and radius 1, centred at the origin. It follows that w and $z = u_1/u_2$ are independent. The pair of random deviates follows as

$$x = u_1 v, \quad y = u_2 v, \quad v = \left(\frac{-2\log w}{w}\right)^{1/2}. \tag{9.7}$$

Atkinson and Pearce (1976) demonstrate the greater speed of (9.7) over the Box–Müller version (9.6). We may also note that z has a Cauchy distribution.

9.27 A variety of algorithms has been generated by composition, or mixture, methods. The required probability function may be represented as

$$f(x) = \sum_{i=1}^{k} \pi_i g_i(x), \qquad \sum \pi_i = 1, \tag{9.8}$$

where the components g_i are selected so that the more commonly occurring terms have a very simple form (e.g., sums of uniform variates). Although the remaining terms are complex, their probability of selection is very low. For further details of this and other methods, see Newman and Odell (1971), Atkinson and Pearce (1976), Kennedy and Gentle (1980, chapter 6), and Ripley (1983). Extensive bibliographies appear in Sowey (1972, 1978, 1986) and Sahai (1980).

Sampling for attributes

9.28 As an introduction to the statistical problems related to sampling, we shall consider attribute sampling, which raises many of the difficulties of principle, but is not obscured by too much mathematics.

Suppose we sample from a population whose members exhibit either an attribute, A, or its complement, not-A. In a random sample of size n, we find that r individuals exhibit the attribute, giving the sample proportions $p = r/n$ and $q = 1 - p$. We are now using p and q for the proportions in the sample, rather than in the population from which the sample was selected, for which we must define different symbols. We shall here, and henceforth wherever possible, use Greek symbols for population parameters.

We will assume that the population is large, or that sampling is with replacement, so that the probability of obtaining an A at any drawing is not affected by other drawings and is therefore a constant, say π.

The problems we have to consider are of three types:

(a) Suppose we have some reason for supposing that the proportion of As in the population is given by a known π. Does the observed proportion p bear out this hypothesis or is it so divergent from π as to lead us to doubt the hypothesis? For example, in an experiment with plants exhibiting two strains of a quality such as height in pea plants, we may wish to test whether the breeding follows the simple Mendelian law of dominant and recessive. If we begin with two pure strains tall and short, cross-breed a first generation and then produce a second generation by interbreeding, the proportional frequencies of "short" and "tall" in this generation will be $\frac{3}{4}$ and $\frac{1}{4}$ if "short" is dominant and $\frac{1}{4}$ and $\frac{3}{4}$ if "tall" is dominant, provided that the simple Mendelian law holds. Suppose we carry out such an experiment and find that for 400 plants the frequencies are 70 and 330. Can the divergence from the theoretical values 100 and 300 have arisen by chance, or is it large enough to throw doubt on the hypothesis that the simple Mendelian law is operating?

(b) In the foregoing type of problem we have some reason for testing a value of π given *a priori*; but we may know nothing of π, and in such a case our principal problem is to estimate it from the sample.

(c) Then, having estimated it, we wish to know the degree of reliability of the estimate. How far is the estimate likely to deviate from the real value of π?

9.29 Consider the first type of problem, in which π is given *a priori*. If we select repeated samples of size n from the population, the distribution of the number of A-individuals in the sample is given by the binomial distribution of **5.2** with parameter π. The probability of obtaining r or fewer A-individuals is the sum of the first $(r+1)$ terms in this distribution. Call this P. We test the hypothesis by determining beforehand a (generally small) probability of error, say α, that we are prepared to tolerate in rejecting the hypothesis, and compare P with α. If P is the larger, we do not reject the hypothesis, while if α is the larger we do. In hypothesis testing, P is known as the "P-value".

Superficially, this procedure is a reasonable one; we reject the hypothesis if the observed value of the statistic is located in the most "unlikely" portion of its sampling distribution. However, a satisfactory analysis of the logic of test inference must await a more sophisticated discussion in Volume 2A.

It will have been noted that P, determined as above, relates to only one of the "tails" of the sampling distribution. It is, in fact, more common in practice for tests to employ a P calculated from *both* tails of the distribution. That is to say, quite frequently we are interested in departure from the hypothesis under test that would have the effect either of decreasing or of increasing the expected number of A-individuals in the sample, while on other occasions we are interested only in departures in one direction. Once again, full discussion of the rationale of these procedures must be deferred to Volume 2A. Here, it is sufficient to note that in our Mendelian law example we should be following the intuitively sound procedure in regarding only low numbers of "tall" plants as providing evidence against the hypothesis that "tall" is dominant.

Example 9.11 (Test for bias in a coin)

A coin was tossed 20 times and came down "heads" 15 times. Does this conflict with the hypothesis that the coin was unbiased?

Here we have to test the hypothesis that $\pi = \frac{1}{2}$. Bias in the coin can operate in either direction (towards "heads" or "tails"), so we calculate P including both extremes of the sampling distribution. The probability that in 20 tosses we should get 15 or more heads is the sum of the first six terms in $(\frac{1}{2} + \frac{1}{2})^{20}$, and from the table in **5.6** this is 0.0207. The probability of 15 or more tails is, by symmetry, the same. P is therefore 0.0414. If we had only been prepared to tolerate a chance of wrongly rejecting the hypothesis that is lower than this, the hypothesis would not be rejected. In the contrary case, the hypothesis would have been rejected. Commonly, a chance of 0.05 is tolerated, and in this case the hypothesis would be rejected.

9.30 In the example just given we purposely took a fairly low value of n in order that the terms of the binomial could be calculated directly. In practice n is often fairly large—100 or more—and the evaluation and summation of individual terms would be most tedious. We can, if complete accuracy is desired, use the method of summation given in **5.7**, and use the incomplete B-function or the tables listed there. But for most ordinary purposes it is enough to use the normal approximation to the binomial. We saw in Example 4.6 that as $n \to \infty$ the binomial tends to the normal distribution, with mean $n\pi$ and variance $n\pi(1-\pi)$ if we de-standardize. Probabilities can therefore be evaluated from the normal integral. For example, we may set up a decision rule such as "reject the hypothesis if the observed value lies more than k standard deviations from the mean". That is, the rejection region lies outside the interval $n\pi \pm k[n\pi(1-\pi)]^{\frac{1}{2}}$. Values of P corresponding to $k = 1$, 2, 3 and 4 are

0.317 3, 0.045 5, 0.002 7, and 0.000 06, respectively. If the sample value $r = np$ lies outside the interval, the hypothesis is rejected.

It makes no difference whether we compare the actual frequencies or the proportions, since mean and standard deviation are equally affected when the variate is divided by a constant.

Example 9.12 (Test for bias in dice)

In some dice-throwing experiments Weldon threw dice 49 152 times, and of these 25 145 yielded a 4, 5, or 6. Is this consonant with the hypothesis that the dice were unbiased?

If the dice are unbiased the probability of a 4, 5 or 6 is $\frac{1}{2}$. Thus $n\pi$ is 24 576 and the observed np is 569 in excess of this value, while

$$\{n\pi(1-\pi)\}^{\frac{1}{2}} = \sqrt{(49\,152 \times \tfrac{1}{2} \times \tfrac{1}{2})} = 110.9.$$

The observed deviation is more than 5 times this quantity and we accordingly suspect very strongly that the dice were biased.

Standard error

9.31 The quantity $\{n\pi(1-\pi)\}^{\frac{1}{2}}$ is a particular case, appropriate to the binomial, of an important statistical concept known as the standard error. It is the standard deviation of the sampling distribution of the statistic, i.e., the square root of the sampling variance. It is particularly important in the large class of cases in which the sampling distribution can be taken to be normal either exactly or to an adequate degree of approximation.

9.32 Let us now turn to the case (b) in **9.28**, where no value of π is given *a priori*. If the sample gives a proportion of *A*s equal to p, what shall we take as our estimator of π? The most obvious course is to take p itself; and this is the course dictated by the more sophisticated ideas described in Chapter 8.

Consider first of all the method of Maximum Likelihood. The probability of obtaining r *A*s and $n-r$ not-*A*s is

$$\binom{n}{r}\pi^r(1-\pi)^{n-r} \tag{9.9}$$

This is the likelihood, and neglecting constants we have to maximize

$$L = k\pi^r(1-\pi)^{n-r}.$$

for variations in π. It is sufficient to maximize $\log L$. We have

$$\frac{\partial}{\partial\pi}(\log L) = \frac{r}{\pi} - \frac{n-r}{(1-\pi)} = 0,$$

giving, as our estimator $\hat{\pi}$,

$$\hat{\pi} = p, \tag{9.10}$$

the second derivative of $\log L$ being negative at this unique maximum.

If we assume that the prior distribution of π is uniform on (0, 1], the posterior distribution for π given p is, from Example 8.9,

$$P(\pi|p) \propto \pi^r(1-\pi)^{n-r} \tag{9.11}$$

which is also maximized at $\hat{\pi} = p$.

There is a third way of looking at the estimation problem. Suppose we were to consider repeated samples of size n, drawn with replacement from the same population. Then the expected value of the estimator p is

$$\begin{aligned} E(p) &= \sum_{r=0}^{n} (r/n)\binom{n}{r}\pi^r(1-\pi)^{n-r} \\ &= \frac{1}{n}\sum_{r=0}^{n} r\binom{n}{r}\pi^r(1-\pi)^{n-r} \\ &= \frac{1}{n}\cdot n\pi\sum_{r=1}^{n}\binom{n-1}{r-1}\pi^{r-1}(1-\pi)^{n-r} \\ &= \pi. \end{aligned} \tag{9.12}$$

Whenever the mean value of the estimator is equal to the value of the population parameter, the estimator is said to be *unbiased*. This is a somewhat charged term—who, after all, would wish to use "biased" procedures?—but it describes a property of estimators that seems desirable.

It may be argued that the unbiased estimator with the smallest possible variance should be taken as the estimator of π. As we shall see in Chapter 17, Volume 2A, p is that estimator.

9.33 In this case, therefore, all the approaches lead to the same conclusion. This happy state of affairs does not always exist, as we shall see later. Consider now the next stage in **9.28**: (c) what is the reliability of the estimator? In other words, how far does it differ from the true value of the parameter?

When n is large, we know that the distribution of p is approximately normal with mean π and standard error, $\sigma_p = [\pi(1-\pi)/n]^{\frac{1}{2}}$. The probability that $|p-\pi| > z\sigma_p$ decreases as z increases, but an exact probability would require knowledge of σ_p, which depends upon the unknown parameter π.

As stated, the problem may only be resolved approximately. If n is large, σ_p is of order $n^{-\frac{1}{2}}$, so that we may put

$$\pi = p + kn^{-\frac{1}{2}}.$$

Thus,

$$\sigma_p^2 = \frac{\pi(1-\pi)}{n} = \frac{1}{n}(p + kn^{-\frac{1}{2}})(q - kn^{-\frac{1}{2}}).$$

Neglecting terms of order n^{-1}, we obtain

$$\sigma_p^2 = \frac{pq}{n}\left\{1 + \frac{k(q-p)}{pqn^{\frac{1}{2}}}\right\}. \tag{9.13}$$

Hence for large n the standard error of p is approximately equal $\sqrt{(pq/n)}$; and we thus reach the fundamental result that in large samples for attributes the standard error may be calculated by using the estimate p of the parameter instead of the (unknown) value of that parameter.

In Volume 2A we shall consider a different approach, which frees us from the necessity of making this type of approximation.

Example 9.13 (Sampling distribution of proportion)

In a sample of 600, 240 are found to possess the attribute A. Thus, $p = 0.40$ and $\sigma_p \doteq 0.02$. Our result enables us to make probability statements such as

$$P\{|p-\pi| \geqslant 2(0.02)\} \doteq 0.045\,5.$$

This is a statement about sampling variability in p, given π. It is tempting to invert this statement to yield the statement that with probability $1 - 0.0455 = 0.9545$, π lies in the interval $p \pm 2\sigma_p$, or [0.36, 0.44] in this case. Such an inversion requires a theory of interval estimation, which is developed in Chapter 20, Volume 2A.

9.34 We now turn to a general consideration of the problems of sampling that have been exemplified above. In the first place, let us note the role of the sampling distribution in this branch of the subject. We construct from the observations some function of them, which we call a statistic t. The sampling distribution of this statistic will in general (but not always) depend on some parameters of the population. The observed value of t then permits the making of statements, by inverse probability, likelihood or otherwise, about these parameters, and so we are able to draw inferences about the population. The sampling distribution is thus fundamental to the whole subject and several subsequent chapters will be devoted entirely to the methods of finding sampling distributions when the population is specified.

If we wish to test some hypothesis about the population that determines its parameters *a priori*, the problem is simple. Given the values of the parameters, we can determine from the sampling distribution of the statistic the probability that it will deviate from what we expect by as much as, or more than, its observed value actually does, and use this to assess the acceptability of the hypothesis. Complications can arise even here, however, for many statistics can be computed from the same sample, and they need not necessarily all lead to the same conclusion about the hypothesis; for instance, a sample might have a mean that throws doubt on the hypothesis and a variance that does not. We shall discuss this problem in Volume 2A.

9.35 When the parameters of the population are not given *a priori*, we have the extra problem of estimating the parameters from the sample before using these estimates to find a sampling distribution that may be used as in **9.34**. In some cases we can find a statistic whose sampling distribution depends on only one parameter of the population (as for attributes in **9.33**).

9.36 Certain important approximations to sampling distributions are available when the sample size is large. In particular, we saw in **8.50** that under very general conditions the sum of n independent variables, distributed in whatever form, tends to normality as n tends to infinity. Many of the ordinary statistics in current use can be expressed as the sum of variates, e.g. all the moments; and many others may also be shown to tend to normality for large samples. Thus we may approximate—

(a) by taking a statistic, calculated from the sample, to be the estimate of the corresponding parameter in that population, e.g. the variance of the sample may be taken as an estimate of the variance of the population;

(b) by calculating the mean and variance of the sampling distribution using, instead of the unknown parameter values, the statistic values calculated according to (a);

(c) by assuming that the sampling distribution is normal and hence determining probabilities from the normal integral with the aid of the mean and variance of the sampling distribution (the latter being the square of the standard error).

9.37 Just how large n must be for such approximations to be valid it is not always easy to say. For some distributions, particularly that of the mean, quite a satisfactory approximation is given by low values of n, say $n > 30$. For others n has to be much higher before the approximation begins to give satisfactory results, e.g. for the product-moment correlation coefficient, to be discussed in Chapter 16, even values as high as 500 are not good enough in samples from a normal population. In fact, the form of the population, as well as the statistic considered, influences the rapidity of approach of the sampling distribution to the normal form; cf **8.51**.

Example 9.14 (Sampling without replacement)

Returning to the binomial distribution, let us consider how far our procedure needs to be modified when sampling without replacement.

If a random sample of n is drawn without replacement from a population of N, a proportion π of which bear the attribute A, the sampling distribution of the number r of attribute-bearers in the sample is given by the hypergeometric distribution (**5.13**) and a typical term is, writing (5.32) in our present notation,

$$\binom{n}{r}\frac{(N\pi)^{[r]}(N-N\pi)^{[n-r]}}{N^{[n]}}.$$

The mean value of p in this distribution, from (5.39), is known to be π; and thus the sample proportion remains an unbiased estimator of the population proportion. It is not a Bayes estimator or a Maximum Likelihood estimator (see Exercise 9.10)—these do not take simple forms in this case—but it differs from them only by terms of order n^{-1}.

The variance of p, from (5.41), is given by

$$\operatorname{var} p = \frac{N-n}{N-1}\cdot\frac{\pi(1-\pi)}{n}. \tag{9.14}$$

This differs from the ordinary binomial case (sampling with replacement) by the factor $(N-n)/(N-1)$ which is less than unity for $n > 1$. The variance in sampling without replacement is therefore smaller than with replacement, as seems intuitively obvious from the fact that the extreme sampling possibilities are greater in the latter case.

Example 9.15 (Sequential sampling for attributes)

We now consider the sequential sampling model introduced in **5.16**. That is, we continue sampling until the kth success, observing j failures along the way and $n = j + k$ observations in all. The random variable j has a negative binomial distribution, as given in (5.43). In our present notation, this becomes

$$\text{Prob}(j) = \binom{k+j-1}{j}\pi^k(1-\pi)^j, \quad j = 0, 1, \ldots, \tag{9.15}$$

with mean $k(1-\pi)/\pi$ and variance $k(1-\pi)/\pi^2$ by (5.48). The Maximum Likelihood estimator of π is p, as before. This is an example of the general result that the Maximum Likelihood estimator is unaffected by the "stopping rule", i.e., the rule that determines the cessation of sampling. However, p is no longer unbiased. We shall now show that $(k-1)/(n-1)$ is an unbiased estimator of π.

$$\begin{aligned} E\left(\frac{k-1}{n-1}\right) &= \pi^k \sum_{j=0}^{\infty} \left(\frac{k-1}{j+k-1}\right)\binom{k+j-1}{j}(1-\pi)^j \\ &= \pi^k \sum_j \binom{k+j-2}{j}(1-\pi)^j \\ &= \pi^k[1-(1-\pi)]^{-(k-1)} \\ &= \pi. \end{aligned} \tag{9.16}$$

If $k = 1$, the estimator $(k-1)/(n-1) = 0$ if $n > 1$, and if we define it to be equal to 1 if $n = 1$, it remains unbiased, as may easily be confirmed.

The biased character of k/n follows, for it cannot have the same expectation as $(k-1)/(n-1)$, being always greater. However, n/k is an unbiased estimator of $1/\pi$ as $E(n) = E(k+j) = k + E(j) = k + k(1-\pi)/\pi = k/\pi$.

At first sight, this seems rather paradoxical. n/k is an unbiased estimator of $1/\pi$, but k/n is a biased estimator of π. The reason is, of course, our somewhat arbitrary definition of bias as a departure from a mean value. It should not occasion surprise that if $E(x) = \alpha$, $E(1/x)$ is not equal to $1/\alpha$. For a positive variate this is, in fact, inevitable (cf. Exercises 9.13 and 9.15).

The general theory of sequential sampling will be developed in Chapter 24, Volume 2A. Its characteristic feature is that we decide as we go along whether to draw further members for the sample, in the light of the values drawn up to date.

EXERCISES

9.1 Of 10 000 babies born in a particular country 5100 are male. Taking this to be a random sample of the births in that country, show that it throws considerable doubt on the hypothesis that the sexes are born in equal proportions.

Consider how far this conclusion would be modified if the sample consisted of 1000 births, 510 of which were male.

9.2 In a sample of n_1 from a population P_1, the proportion observed with an attribute A is p_1. In another independent sample of n_2 from a population P_2 the proportion is p_2. To test whether the populations have identical proportions of A in them, it is proposed to consider the difference $p_1 - p_2$. Show that on the hypothesis of identical proportions the difference has zero mean and a variance that may be estimated as

$$\bar{p}(1-\bar{p})\Big/\left(\frac{1}{n_1}+\frac{1}{n_2}\right)$$

where $\bar{p} = (n_1p_1 + n_2p_2)/(n_1+n_2)$.

9.3 If a proportion π has to be estimated from a simple random sample with proportion p, and if f is the prior probability of π, then the posterior probability of π is, according to Bayes' theorem, proportional to

$$f\pi^{np}(1-\pi)^{nq}.$$

Show that this is a maximum if

$$\frac{1}{f}\frac{\partial f}{\partial \pi} + n\frac{p-\pi}{\pi(1-\pi)} = 0.$$

Hence, in general, as n increases, the solution tends to $\pi = p$, whatever the prior probability of π.

9.4 By using the Bienaymé–Chebyshev inequality, show that in sampling from a large population for an attribute, the probability that the observed proportion p in a sample of n differs from the population proportion π by more than k is not greater than $1/(4nk^2)$.

(This is an exact result, no assumptions about the limiting normality of the binomial or the use of estimates in calculating standard errors being involved. The limits are, however, much too wide in general.)

9.5 In a large population the proportion π of members bearing an attribute A is small. Show that in samples of size n the proportion of As is the Maximum Likelihood estimator of π; that it is unbiased; and that the variance of the number, say n_1, of As observed may be estimated approximately by n_1; that is, n_1 is the ML estimator for the Poisson distribution with parameter $\lambda = n\pi$, cf. **5.8**.

9.6 In k different populations the proportions of an attribute are $\pi_1, \pi_2, \ldots, \pi_k$. One population is chosen at random (with probability $1/k$) and from it a sample of n members is drawn with replacement at random, yielding a proportion p bearing the attribute. Show that, under repetitions of this process (including the renewed selection of a population), the mean value of p is $\pi(=\sum_{i=1}^{k}\pi_i/k)$ and that its variance is $\pi(1-\pi)/n + (n-1)\Sigma(\pi_i-\pi)^2/(nk)$.

Obtain the corresponding formulae if the probability of selection of the ith population is α_i, where $\sum_{i=1}^{k}\alpha_i = 1$ (cf. **5.11**).

9.7 In the previous exercise, a sample of kn is taken by drawing n members from each population. Show that the proportion p in the kn sample members together has mean π and variance (cf. **5.10**)

$$\pi(1-\pi)/(nk)-\Sigma(\pi_i-\pi)^2/(nk^2).$$

9.8 From a population in which the proportion of individuals possessing a certain attribute is π, a sample of size n is drawn with replacement. From this a subsample of n_1 is drawn with replacement. If p, p_1 are the proportions in the sample and subsample respectively show that $p-p_1$ has zero mean and variance

$$\pi(1-\pi)\frac{n-1}{nn_1}.$$

9.9 In the previous exercise, if the sample of size n is selected from a population of size N without replacement and the subsample of n_1 is selected from the n without replacement, show that $p-p_1$ has zero mean and variance

$$\pi(1-\pi)\frac{N}{N-1}\cdot\frac{n-n_1}{nn_1}.$$

9.10 For the hypergeometric distribution given in Example 9.14, show that the Maximum Likelihood estimator $N\hat{\pi}$ of $N\pi$ is the smallest integer greater than

$$\frac{r(N+1)}{n}-1,$$

so that $Np-q\leqslant N\hat{\pi}\leqslant Np+p$. Hence show that the variance of $\hat{\pi}$ is, to $O(N^{-1})$,

$$\frac{N-n}{N-3}\cdot\frac{\pi(1-\pi)}{n}.$$

9.11 Define a variable x as unity if the individual concerned possesses an attribute A and zero in the contrary case. If p is the proportion in a sample of n bearing the attribute show that, for sampling either with or without replacement, $E(x)=E(x^2)=E(x^k)=\pi$. Hence obtain equation (9.14).

9.12 In Example 9.15, show that

$$E(k/n)=k\pi^k(1-\pi)^{-k}\int_0^{1-\pi}t^{k-1}(1-t)^{-k}dt$$

$$=\pi\sum_{j=0}^{\infty}\frac{k!j!}{(k+j)!}(1-\pi)^j.$$

Hence show that $\pi\leqslant E(k/n)\leqslant(k\pi+1)/(n+1)$ and that, when $k=1$, $E(n^{-1})=-[\pi/(1-\pi)]\log_e\pi$.

9.13 For a variate x which is defined in the range $0<x<\infty$, show that if $E(x)$ and $E(1/x)$ exist, $E(x)E(1/x)\geqslant 1$, the inequality becoming an equality only if the distribution of x is wholly concentrated at a single value—this is the inequality $H<\mu_1'$ given at (2.9). Hence show that if t is an unbiased estimator of θ, $1/t$ cannot be an unbiased estimator of $1/\theta$.

9.14 For the ordinary binomial show that $E(n/r)$ does not exist. Show also that $E\{(n+1)/(r+1)\}=\{1-(1-\pi)^{n+1}\}/\pi$ and hence that, asymptotically in n, $(n+1)/(r+1)$ is an unbiased estimator of $1/\pi$.

9.15 If y is distributed independently of x in Exercise 9.13, show that

$$E\left(\frac{y}{x}\right) \geqslant \frac{E(y)}{E(x)}.$$

Hence, if s_1^2, s_2^2 are independent unbiased estimators of variances σ_1^2, σ_2^2 respectively, show that $E\left(\frac{s_i^2}{s_j^2}\right) \geqslant \frac{\sigma_i^2}{\sigma_j^2}$ for $i \neq j = 1, 2$.

9.16 Show that the method described in Example 9.10 generates values from a Gamma (α) distribution. (Wallace, 1974)

9.17 If $u_1, \ldots, u_{12}$ are independent and uniformly distributed on (0, 1], show that $x = \Sigma u_i - 6$ has mean 0, variance 1, $\beta_1 = 0$ and $\beta_2 = 2.9$, so that x is approximately normally distributed.

(This used to be a popular method for generating normal random numbers but is now little used because of the "light" tails; cf. Example 11.9.)

9.18 Suppose that u_1 and u_2 are independent uniform random variables defined on $(-1, 1]$. If these are generated subject to the condition $w = u_1^2 + u_2^2 \leqslant 1$, verify that (9.7) yields a pair of independent standard normal variates. Also, show that a proportion $\pi/4 = 0.785$ of the original pairs satisfy the condition. (Marsaglia and Bray, 1964)

9.19 A positive variate x has p.f. $f(x)$, mean μ and variance σ^2. If y has the first moment distribution $g(y) = yf(y)/\mu$, show that for any positive constant a and integer $r \geqslant 1$,

$$aE\{(x+a)^{-r}\} + \mu E\{(y+a)^{-r}\} = E\{(x+a)^{-r+1}\}$$

and hence that

$$E\{(x+a)^{-1}\} \leqslant \frac{\sigma^2 + a\mu}{a(\sigma^2 + a\mu + \mu^2)}.$$

(Wooff (1985) deduces the upper bound for any r from the recurrence relation.)

9.20 Suppose that x_1 and x_2 are independently distributed on (0, 1] and that x_1 is a uniform variate. Show that f, the fractional part of $(x_1 + x_2)$, is uniformly distributed on (0, 1], whatever the distribution of x_2.

(Wichmann and Hill, 1982)

CHAPTER 10

STANDARD ERRORS

10.1 In **9.31–37** we discussed the estimation of statistical parameters from large samples and the type of judgement of their reliability that depends on the use of the standard error. It was remarked that an estimate of a parameter may be obtained by calculating the corresponding statistic from the sample and it was established that for samples of size n the standard error, similarly estimated, gives an approximate measure of precision, provided (a) that the sampling distribution of the statistic under discussion tends to normality and (b) that n is large enough for this tendency to be effective.

Since the majority of statistics in current use tend to normality the theory of large samples is, in the main, devoted to the determination of standard errors. In this chapter we describe the principal methods available for the purpose, and incidentally derive formulae for the standard errors of various statistics considered in previous chapters. To avoid the usual square roots associated with the standard error we shall write our results as sampling variances and covariances. Thus, for a statistic t we write the variance of its sampling distribution as var t. The covariance of the joint distribution of two statistics t and u, that is, the first product moment of their joint sampling distribution, is written cov (t, u).

10.2 By definition, the rth moment of a statistic t, that is the rth moment of its sampling distribution, is the mean value of t^r taken over all possible samples, and may be written $E(t^r)$ (cf. **2.27**). If the joint d.f. of the variates $x_1, \ldots, x_n$, from which t is calculated, is $F(x_1, \ldots, x_n)$, then the rth moment of t is the integral of t^r dF over the domain of the xs. In particular, if the sample is simple random we have

$$E(t^r) = \int_{-\infty}^{\infty} \ldots \int_{-\infty}^{\infty} t^r \,\mathrm{d}F(x_1) \ldots \mathrm{d}F(x_n). \tag{10.1}$$

We are particularly interested in this chapter in the first and second moments of t, that is, the mean and variance of its sampling distribution—the latter will be called its sampling variance. It may be recalled from **2.28** that the mean value of a sum is the sum of the mean values and that, if the variables are independent, the mean value of a product is the product of the mean values. These two results will be repeatedly required.

Standard errors of moments

10.3 In the following sections, we shall adopt the usual convention in regard to the distinction of population values and statistics by writing Greek letters to represent the former and Roman letters to represent the latter. We have, then, for the rth moment statistic m_r', corresponding to the rth moment μ_r' of the population about the value zero,

$$m_r' = \frac{1}{n}\sum_{j=1}^{n} x_j^r \tag{10.2}$$

and for the rth central moment

$$m_r = \frac{1}{n}\sum_{j=1}^{n} (x_j - m_1')^r. \tag{10.3}$$

Consider now the expectation of m_r'. We have

$$\begin{aligned} E(m_r') &= \frac{1}{n}\sum E(x^r) = E(x^r) \\ &= \mu_r'. \end{aligned} \tag{10.4}$$

The sampling variance of m_r' is, by definition, $E\{m_r' - E(m_r')\}^2$ and thus assuming, as we do throughout, that the appropriate moments exist,

$$\begin{aligned} \operatorname{var} m_r' &= E\left\{\frac{1}{n}\sum x^r - \mu_r'\right\}^2 \\ &= \frac{1}{n^2} E\left[\left\{\sum x^r\right\}^2 - 2n\mu_r'\sum x^r + n^2\mu_r'^2\right] \\ &= \frac{1}{n^2} E\left[\left\{\sum x^r\right\}^2\right] - \mu_r'^2 \\ &= \frac{1}{n^2} E\left\{\sum x_j^{2r} + \sum (x_j^r x_k^r)\right\} - \mu_r'^2, \end{aligned}$$

the second summation extending over the $n(n-1)$ cases in which $j \neq k$ (permutation of j and k thus being allowed). Since the xs are independent the mean value of the product is the product of the mean values, and thus

$$\begin{aligned} \operatorname{var} m_r' &= \frac{1}{n^2}\{n\mu_{2r}' + n(n-1)\mu_r'^2\} - \mu_r'^2, \\ &= \frac{1}{n}(\mu_{2r}' - \mu_r'^2). \end{aligned} \tag{10.5}$$

This is an exact result.

In a similar way, if we have two moments, m_q', m_r', their sampling covariance is given by

$$\begin{aligned} \operatorname{cov}(m_q', m_r') &= E\{(m_q' - \mu_q')(m_r' - \mu_r')\} \\ &= E\left\{\left(\frac{1}{n}\sum x^q - \mu_q'\right)\left(\frac{1}{n}\sum x^r - \mu_r'\right)\right\} \\ &= \frac{1}{n^2}E\left(\sum x^{q+r}\right) + \frac{1}{n^2}E\sum_{j\neq k}(x_j^q x_k^r) - \frac{1}{n}\mu_q' E\left(\sum x^r\right) \\ &\quad - \frac{1}{n}\mu_r' E\left(\sum x^q\right) + E(\mu_q'\mu_r') \\ &= \frac{1}{n}(\mu_{q+r}' - \mu_q'\mu_r'), \end{aligned} \tag{10.6}$$

which reduces to (10.5) when $q = r$.

The same results hold if μ_r' and m_r' are moments about any fixed point.

10.4 The formulae for moments about the mean are not so simple, for the mean itself is subject to sampling fluctuations. We see from (10.5) with $r = 1$ that

$$\operatorname{var} m_1' = (\mu_2' - \mu_1'^2)/n = \mu_2/n. \tag{10.7}$$

Thus by the Bienaymé–Chebyshev inequality (3.95), deviations of m_1' from μ_1' are of order $n^{-\frac{1}{2}}$. Retaining only terms of order n^{-1}, we have

$$E(m_r) = E\left\{\frac{1}{n}\sum (x - m_1')^r\right\}$$

in which we need to expand the sum only to its first three terms, the remainder being of lower order than n^{-1}. We take our origin at μ_1', so that every unit power of x in the expansion will have zero expectation, and we find

$$\begin{aligned} E(m_r) &= \frac{1}{n}E\left\{x_j^r - \frac{r}{n}\sum x_j^{r-1}\sum x_k + \frac{r(r-1)}{2n^2}\sum x_j^{r-2}\sum x_k \sum x_l\right\} + o(n^{-1}) \\ &= \frac{1}{n}E\left\{\left(1 - \frac{r}{n}\right)\sum x_j^r + \frac{r(r-1)}{2n^2}\left(\sum x_j^r + \sum x_j^{r-2}x_k^2\right)\right\} + o(n^{-1}). \end{aligned}$$

The middle sum on the right has n terms so we may neglect it to order n^{-1}. The final sum has $n(n-1)$ terms, so we have

$$\left.\begin{aligned} E(m_r) &= \left(1 - \frac{r}{n}\right)\mu_r + \frac{r(r-1)}{2n}\mu_{r-2}\mu_2 + o(n^{-1}) \\ &= \mu_r + O(n^{-1}). \end{aligned}\right\} \tag{10.8}$$

Now if, in finding $E(m_r^2)$, we retain only sums containing a number of terms that is at least one order in n greater than that sum's divisor in n within the braces below, the overall expression will be correct to order n^{-1}. We find

$$\begin{aligned} E(m_r^2) &= \frac{1}{n^2}E\left[\left\{\sum x_j^r - \frac{r}{n}\sum x_j^{r-1}\sum x_k + \frac{r(r-1)}{2n^2}\sum x_j^{r-2}\sum x_k\sum x_l\right\}^2\right] + o(n^{-1}) \\ &= \frac{1}{n^2}E\left[\sum x_j^{2r} + \sum x_j^r x_k^r + \frac{r^2}{n^2}\left(\sum x_j^{r-1}\sum x_k\right)^2\right. \\ &\quad + \left\{\frac{r(r-1)}{2n^2}\right\}^2\left\{\sum x_j^{r-2}\left(\sum x_k\right)^2\right\}^2 - \frac{2r}{n}\sum x_j^r \sum x_k^{r-1}\sum x_l \\ &\quad \left. + \frac{r(r-1)}{n^2}\sum x_j^r\sum x_k^{r-2}\left(\sum x_l\right)^2 - \frac{r^2(r-1)}{n^3}\sum x_j^{r-1}\sum x_k^{r-2}\left(\sum x_l\right)^3\right] + o(n^{-1}) \\ &= \frac{1}{n}\{\mu_{2r} + (n-1)\mu_r^2 + r^2\mu_{r-1}^2\mu_2 - 2r(\mu_{r+1}\mu_{r-1} + \mu_r^2) + r(r-1)\mu_r\mu_{r-2}\mu_2\} + o(n^{-1}). \end{aligned}$$

From (10.8),

$$\{E(m_r)\}^2 = \left(1 - \frac{2r}{n}\right)\mu_r^2 + \frac{r(r-1)}{n}\mu_r\mu_{r-2}\mu_2 + o(n^{-1})$$

and upon subtraction we find

$$\operatorname{var} m_r = \frac{1}{n}(\mu_{2r} - \mu_r^2 + r^2\mu_{r-1}^2\mu_2 - 2r\mu_{r+1}\mu_{r-1}) + o(n^{-1}). \tag{10.9}$$

Similarly it may be shown that

$$\operatorname{cov}(m_r, m_q) = \frac{1}{n}(\mu_{r+q} - \mu_r\mu_q + rq\mu_{r-1}\mu_{q-1}\mu_2 - r\mu_{r-1}\mu_{q+1} - q\mu_{r+1}\mu_{q-1}) + o(n^{-1}). \tag{10.10}$$

Finally, it should be noted that the remainder terms in (10.8–10) may all be shown to be of order $n^{-3/2}$.

Example 10.1 (Standard error of mean)

For the height distribution of Table 1.7 we found (Examples 2.1 and 2.7) that $m_1' = 67.46$, $\sqrt{m_2} = 2.57$. Suppose we regard this distribution as a simple random sample from the adult male inhabitants of the United Kingdom living at the time when the data were collected. What can we say about the mean of the population?

The standard error of the mean depends on μ_2. This is an unknown quantity, but we may, in accordance with the general principles of large-sample theory, use m_2 instead. We then find

$$\text{Standard error of } m_1' = \frac{2.57}{\sqrt{8585}} = 0.028 \text{ approximately.}$$

Thus we can say that the population mean probably lies in the range of twice this amount on either side of the sample mean, i.e. in the range, 67.46 ± 0.06, and very probably in 67.46 ± 0.08. Our estimate of the mean would almost certainly be less than a tenth of an inch in error.

Example 10.2 (Covariance between mean and even moments)

To show that in samples from a symmetrical population the sampling covariance between the mean and any central moment of even order is zero to order n^{-1}.

We have, by definition,

$$\operatorname{cov}(m_1', m_r) = \frac{1}{n^2}E\left[\left\{\sum x_j\left\{\sum x_j^r - \frac{r}{n}\sum x_j x_k^{r-1}\right\}\right\}\right] + o(n^{-1})$$

$$= \frac{1}{n^2}E\left\{\sum x_j^{r+1} - \frac{r}{n}\sum x_j^2 x_k^{r-1}\right\} + o(n^{-1}),$$

the other terms vanishing, since they involve a unit power of x, if we take our origin at μ_1' as in the developments above, we find

$$\text{cov}(m_1', m_r) = \frac{1}{n}(\mu_{r+1} - r\mu_2\mu_{r-1}) + o(n^{-1}).$$

Now if r is even, μ_{r+1} and μ_{r-1}, being moments of odd order, will vanish for a symmetrical population and hence

$$\text{cov}(m_1', m_r) \doteq 0.$$

That is, the mean and any even moment about the mean are uncorrelated to order n^{-1}.

Standard errors of functions of random variables

10.5 Suppose that x_i has mean θ_i, and that the variances and covariances of the k variates $x_1, x_2, \ldots, x_k$ are of order n^{-r}, $r>0$. (In practice, r is usually equal to unity.) The Bienaymé–Chebyshev inequality (3.95) then ensures that x_i converges to θ_i as $n\to\infty$, in the sense that the probability that $|x_i-\theta_i|$ exceeds any fixed $\varepsilon>0$ tends to zero.

Consider the function $g(x_1, x_2, \ldots, x_k)$, which we write $g(x)$ for brevity. If $g_i'(\theta)$ denotes $\partial g(x)/\partial x_i$ evaluated at $\theta_1, \theta_2, \ldots, \theta_k$, we have the Taylor series expansion

$$g(x) = g(\theta) + \sum_{i=1}^{k} g_i'(\theta)(x_i-\theta_i) + O(n^{-r}). \tag{10.11}$$

Thus, since $E(x_i)=\theta_i$,

$$E\{g(x)\} = g(\theta) + O(n^{-r}).$$

This formula is taken to a further term in Exercise 10.17.

If not all the $g_i'(\theta)=0$, we have further

$$\begin{aligned}\text{var}\{g(x)\} &= E\left\{\left[\sum_{i=1}^{k} g_i'(\theta)(x_i-\theta_i)\right]^2\right\} + o(n^{-r})\\ &= \sum_{i=1}^{k}\{g_i'(\theta)\}^2\text{var}\,x_i + \sum\sum_{i\neq j=1}^{k} g_i'(\theta)g_j'(\theta)\text{cov}(x_i,x_j)\\ &\quad + o(n^{-r}).\end{aligned} \tag{10.12}$$

Similarly for two functions $g(x)$, $h(x)$, we find

$$\begin{aligned}\text{cov}\{g(x), h(x)\} &= \sum_{i=1}^{k} g_i'(\theta)h_i'(\theta)\text{var}(x_i) + \sum\sum_{i\neq j=1}^{k} g_i'(\theta)h_j'(\theta)\text{cov}(x_i,x_j)\\ &\quad + o(n^{-r}).\end{aligned} \tag{10.13}$$

If all the first derivatives above are zero, we must take further terms in the Taylor expansion, and the variance is more complicated—cf.Exercises 10.17–18. This approach is often known as the *delta method*, because of the reliance upon first derivatives.

10.6 Three particular specializations of (10.12) are of interest:

(a) g is a function of a single random variable. We then find

$$\operatorname{var}\{g(x)\} \doteq \left(\frac{\mathrm{d}g}{\mathrm{d}x}\right)_\theta^2 \operatorname{var} x. \tag{10.14}$$

In particular, if $g(x)$ is simply a linear function of the form $cx+e$ we find that $\operatorname{var} g = c^2 \operatorname{var} x$, as is otherwise obvious, the result in this case being exact.

(b) g is a linear function of the random variables. If

$$g(x_1, \ldots, x_k) = \sum_{i=1}^{k} a_i x_i \tag{10.15}$$

then (10.12) gives

$$\operatorname{var} g = \sum a_i^2 \operatorname{var} x_i + \sum_{i \neq j} a_i a_j \operatorname{cov}(x_i, x_j), \tag{10.16}$$

again an exact result. In particular, if the xs are independent, the variance of g is a weighted sum of their individual variances.

(c) g is a ratio of two random variables, say x_1/x_2. To avoid difficulties to be discussed in **11.9**, we assume here that $x_2 > 0$ if it is discrete and $\geqslant 0$ if it is continuous. Equation (10.12) gives

$$\begin{aligned}\operatorname{var}(x_1/x_2) &\doteq \frac{\operatorname{var} x_1}{\theta_2^2} + \frac{\theta_1^2 \operatorname{var} x_2}{\theta_2^4} - \frac{2\theta_1 \operatorname{cov}(x_1, x_2)}{\theta_2^3} \\ &= \left\{\frac{E(x_1)}{E(x_2)}\right\}^2 \left\{\frac{\operatorname{var} x_1}{E^2(x_1)} + \frac{\operatorname{var} x_2}{E^2(x_2)} - \frac{2\operatorname{cov}(x_1, x_2)}{E(x_1)E(x_2)}\right\}.\end{aligned} \tag{10.17}$$

The second factor in (10.17) is the sum of the squares of the coefficients of variation of the two variables, minus twice the square of what may analogously be called their coefficient of covariation.

Results for the product of two random variables are given in Exercise 10.23. Using (10.17) in a normal approximation may cause problems when n is not large; see Exercise 10.29.

Example 10.3 (Sampling variance of fourth cumulant)

To find the sampling variance of the fourth cumulant, we shall, for reasons which will become clear in Chapter 12, continue to write κ, not k, for the *sample* cumulant. Thus we require the variance of

$$\kappa_4 = m_4 - 3m_2^2.$$

From (10.12) with $x_1 = m_4$, $x_2 = m_2$ and $g = x_1 - 3x_2^2$, we obtain

$$\operatorname{var} \kappa_4 = \operatorname{var} m_4 + 36\mu_2^2 \operatorname{var} m_2 - 12\mu_2 \operatorname{cov}(m_4, m_2).$$

Substituting from (10.9) and (10.10) we find

$$\operatorname{var} \kappa_4 = \frac{1}{n}\{\mu_8 - 12\mu_6\mu_2 - 8\mu_5\mu_3 - \mu_4^2 + 48\mu_4\mu_2^2 + 64\mu_3^2\mu_2 - 36\mu_2^4\}.$$

For a normal population, $\mu_4 = 3\mu_2^2 = 3\sigma^4$, $\mu_6 = 15\sigma^6$, $\mu_8 = 105\sigma^8$ and we find in this case

$$\operatorname{var} \kappa_4 = 24\sigma^8/n.$$

Example 10.4 (Sampling variance of coefficient of variation)

To find the sampling variance of the coefficient of variation defined at (2.28),

$$V = m_2^{\frac{1}{2}}/m_1',$$

where we assume $m_1' > 0$ always in order to use (10.17). Writing $\mathbf{V}$ for the population coefficient $\mu_2^{\frac{1}{2}}/\mu_1'$, where $\mu_1' > 0$ also, we have

$$\text{var}(V) \doteq \mathbf{V}^2\left\{\frac{\text{var}(m_2^{\frac{1}{2}})}{E^2(m_2^{\frac{1}{2}})} + \frac{\text{var}\, m_1'}{E^2(m_1')} - \frac{2\,\text{cov}(m_2^{\frac{1}{2}}, m_1')}{E(m_2^{\frac{1}{2}})E(m_1')}\right\}. \tag{10.18}$$

We saw in (10.7) that

$$\text{var}\, m_1' = \mu_2/n.$$

Also, from (10.14) and (10.9),

$$\text{var}(m_2^{\frac{1}{2}}) \doteq \frac{\text{var}\, m_2}{4\mu_2} = \frac{\mu_4 - \mu_2^2}{4n\mu_2}.$$

Finally, in Example 10.2 we saw that to order n^{-1}

$$\text{cov}(m_2, m_1') \doteq \mu_3/n,$$

and from (10.13) we have

$$\text{cov}(m_2^{\frac{1}{2}}, m_1') \doteq \frac{1}{2\mu_2^{\frac{1}{2}}}\text{cov}(m_2, m_1') = \mu_3/(2n\mu_2^{\frac{1}{2}}).$$

Substituting these various values in (10.18) we find

$$\text{var}\, V \doteq \frac{\mathbf{V}^2}{n}\left\{\frac{\mu_4 - \mu_2^2}{4\mu_2^2} + \frac{\mu_2}{\mu_1'^2} - \frac{\mu_3}{\mu_2\mu_1'}\right\}.$$

In the normal case ($\mu_3 = 0$, $\mu_4 = 3\mu_2^2$) this becomes

$$\text{var}\, V \doteq \frac{\mathbf{V}^2}{n}\left\{\frac{1}{2} + \frac{\mu_2}{\mu_1'^2}\right\}$$

$$= \frac{\mathbf{V}^2}{2n}(1 + 2\mathbf{V}^2). \tag{10.19}$$

10.7 Suppose now that $g(y, x)$ is a function of a sample of observations $y = (y_1, y_2, \ldots, y_N)$ and x (which may itself be a function of y), a statistic with mean θ and variance of order n^{-r} as in **10.5**, so that as $n \to \infty$, x tends to θ. In applications, n is usually a function of N; often $n = N$. A question of general interest is how the variance of $g(y, x)$ compares with that of $g(y, \theta)$, when θ is known. Expanding in a Taylor series about θ, we have as at (10.11)

$$g(y, x) = g(y, \theta) + g_x'(\theta)(x - \theta) + O(n^{-r}),$$

so that asymptotically

$$E\{g(y, x)\} \doteq E\{g(y, \theta)\}$$

and, from (10.16),

$$\operatorname{var}\{g(y,x)\} \doteq \operatorname{var}\{g(y,\theta)\} + \{g_x'(\theta)\}^2 \operatorname{var} x \\ + 2g_x'(\theta)\operatorname{cov}\{g(y,\theta), x-\theta\}. \tag{10.20}$$

(10.20) shows that there is no general relationship between the sizes of the two variances that we wish to compare; either may be the larger.
(a) If $\operatorname{cov}\{g(y,\theta), x-\theta\}=0$, the last term in (10.20) is zero, and $\operatorname{var}\{g(y,x)\}$ cannot be smaller asymptotically, as we might intuitively have expected to be true generally, since it contains extra variation through x;
(b) if $g_x'(\theta)=0$, the last two terms in (10.20) are zero, and the two variances are asymptotically equal;
(c) in general, however, the last term in (10.20) may be negative and outweigh the middle term on the right, so that $\operatorname{var}\{g(y,\theta)\}$ is the larger.

Example 10.5 contains simple instances of both cases (b) and (c).

Example 10.5 (Variances of sample moments)

Let us compare the variance of the rth central moment of a sample,

$$g(y, m_1') = \frac{1}{n}\sum_{i=1}^{n} (y_i - m_1')^r = m_r,$$

with that of the sample rth moment about the population mean μ_1',

$$g(y, \mu_1') = \frac{1}{n}\sum_{i=1}^{n} (y_i - \mu_1')^r = m_r^*,$$

say. Here m_1' is x and μ_1' is θ and

$$g_x'(\theta) = \left[\frac{\partial g(y, m_1')}{\partial m_1'}\right]_{m_1'=\mu_1'} = -\frac{r}{n}\sum_{i=1}^{n} (y_i - \mu_1')^{r-1}.$$

As $n\to\infty$, this tends to $-r\mu_{r-1}$. From (10.7), $\operatorname{var} m_1' = \mu_2/n$, and writing $\operatorname{cov}\left\{\frac{1}{n}\sum (y_i-\mu_1')^r, m_1'-\mu_1'\right\}$ as $\operatorname{cov}(m_r', m_1')$, where these moments are about the population mean μ_1', (10.6) gives this covariance the value μ_{r+1}/n (its second term being zero since $\mu_1'=0$ about the population mean). Substituting into (10.20), we find

$$\operatorname{var} m_r \doteq \operatorname{var} m_r^* + r^2\mu_{r-1}^2\mu_2/n - 2r\mu_{r-1}\mu_{r+1}/n.$$

If r is even and the population is symmetrical, we have

$$\operatorname{var} m_r \doteq \operatorname{var} m_r^*;$$

this is case (b) above. If r is odd, this does not hold; for example, when $r=3$,

$$\operatorname{var} m_3 \doteq \operatorname{var} m_3^* + (9\mu_2^3 - 6\mu_4\mu_2)/n,$$

and from (10.9) we have in the normal case, since $\mu_4 = 3\mu_2^2$, and $\mu_6 = 15\mu_2^3$, $\text{var}\, m_3 \doteq 6\mu_2^3/n$. Thus

$$\begin{aligned}\text{var}\, m_3^* &\doteq \text{var}\, m_3 - (9\mu_2^3 - 6\mu_4\mu_2)/n \\ &= 6\mu_2^3/n + 9\mu_2^3/n = 15\mu_2^3/n\end{aligned}$$

is much the larger in the normal case. This is case (c) above.

Exercise 10.8 establishes these variances differently.

10.8 A few points on the use of standard errors may be noted.

(a) The standard error, strictly speaking, has been justified only in the case where the statistic tends to normality. In other cases it may be used, in conjunction with the Bienaymé–Chebyshev inequality, to make statements in probability, but the more precise statements that are customarily made rely on the normal integral.

(b) Where there is serious doubt, it may be necessary to conduct a separate inquiry into the limiting normality of statistics. Many of those in current use, especially those dependent on sums of variates, like the moments, tend to normality under a Central Limit effect; but certain others, especially those dependent on extreme values, do not.

(c) On a similar point, it should also be remembered that some statistics tend to normality more rapidly than others, and a given n may be large for some purposes but not for others. So far as it is possible to generalize with safety, we can usually (but not always) assume values of n greater than 500 to be "large"; values greater than 100 are often great enough to be "large" for our purposes; values below 100 are suspect in many instances; and values below 30 are very rarely "large". Some measures calculated from the moments tend to normality very slowly. $\sqrt{b_1}$ or b_1 (the sample values of $\sqrt{\beta_1}$ or β_1) are cases in point, and more refined methods which we discuss in Chapter 12 are preferable to the use of the standard error.

(d) The sampling variances must relate to the statistic under consideration. For instance, if the standard deviation of a normal distribution is estimated from (5.112) by taking $(\pi/2)^{\frac{1}{2}}$ times the mean deviation of the sample, instead of the more usual $s = m_2^{\frac{1}{2}}$, the formula $\text{var}\, s \doteq (\mu_4 - \mu_2^2)/(4n\mu_2)$ derivable from (10.9) and (10.14) is not applicable.

(e) From (10.5) and (10.9) it will be seen that the sampling variance of a moment depends on the population moment of twice the order, i.e. becomes very large for higher moments, even when n is large. This is the reason why such moments have very limited practical application.

(f) The order of the approximation makes it necessary to exercise care in the neighbourhood of vanishing values of standard errors. For instance, if the coefficient of variation $V = 0$ in a sample, the formula of Example 10.4 would be estimated from the sample as $\text{var}\, V = 0$. But it does not, of course, follow that there is no variation at all in the population, though none exists in the sample.

(g) It is interesting to compare the sampling fluctuations, as expressed in the sampling variance, with Sheppard's corrections to the moments. Writing temporarily s_1^2 for the uncorrected variance in the sample, s_2^2 for the corrected variance, we have

$$\frac{s_2^2}{s_1^2} = 1 - \frac{1}{12}\frac{h^2}{s_1^2},$$

where h is the interval length. For many practical cases, if d is the number of intervals, dh is about equal to $6s_1$, and thus

$$\frac{s_2^2}{s_1^2} \doteq 1 - \frac{3}{d^2}$$

$$\text{or} \quad \frac{s_2}{s_1} = 1 - \frac{3}{2d^2}, \quad \text{approximately.}$$

Writing s for the square root of the sample variance m_2 we have

$$\text{var}\, s \doteq (\mu_4 - \mu_2^2)/(4n\mu_2).$$

For a normal population this is equal to $\sigma^2/(2n)$.

Thus if n is, say, 1000, the standard error of s is about $0.0224\sigma = 2.24$ percent of σ. Sheppard's correction in a case where $d = 20$ is only 0.375 percent of s_1, i.e. only-about a sixth of the standard error. It is as well to make the corrections, even when n is smaller than 1000, in order to avoid systematic error; but the correction should not be misinterpreted as implying a higher degree of reliability in the corrected value than actually exists.

Similar considerations apply *a fortiori* to the higher moments.

Standard errors of bivariate moments

10.9 Extensions of the above formulae to the bivariate case are made without difficulty, only slightly more complicated algebra being involved. The reader will be able to verify the following formulae:

$$\text{var}(m'_{r,s}) = \frac{1}{n}(\mu'_{2r,2s} - \mu'^2_{r,s}) \tag{10.21}$$

$$\text{cov}(m'_{r,s}, m'_{u,v}) = \frac{1}{n}(\mu'_{r+u,s+v} - \mu'_{r,s}\mu'_{u,v}) \tag{10.22}$$

$$\begin{aligned}\text{var}(m_{r,s}) \doteq \frac{1}{n}(&\mu_{2r,2s} - \mu_{r,s}^2 + r^2\mu_{2,0}\mu_{r-1,s}^2 + s^2\mu_{0,2}\mu_{r,s-1}^2 \\ &+ 2rs\mu_{1,1}\mu_{r-1,s}\mu_{r,s-1} - 2r\mu_{r+1,s}\mu_{r-1,s} - 2s\mu_{r,s+1}\mu_{r,s-1})\end{aligned} \tag{10.23}$$

$$\begin{aligned}\text{cov}(m_{r,s}, m_{u,v}) \doteq \frac{1}{n}(&\mu_{r+u,s+v} - \mu_{r,s}\mu_{u,v} + ru\mu_{2,0}\mu_{r-1,s}\mu_{u-1,v} \\ &+ sv\mu_{0,2}\mu_{r,s-1}\mu_{u,v-1} + rv\mu_{1,1}\mu_{r-1,s}\mu_{u,v-1} \\ &+ su\mu_{1,1}\mu_{r,s-1}\mu_{u-1,v} - u\mu_{r+1,s}\mu_{u-1,v} \\ &- v\mu_{r,s+1}\mu_{u,v-1} - r\mu_{r-1,s}\mu_{u+1,v} - s\mu_{r,s-1}\mu_{u,v+1}).\end{aligned} \tag{10.24}$$

Where no ambiguity is involved we may write $\mu_{r,s}$ as μ_{rs}.

Example 10.6 (Variance of the correlation coefficient)

The correlation coefficient in a sample is defined by $r = m_{11}/(m_{20}m_{02})^{\frac{1}{2}}$ and in the population by $\rho = \mu_{11}/(\mu_{20}\mu_{02})^{\frac{1}{2}}$. From (10.17), we have

$$\operatorname{var} r \doteq \rho^2 \left\{\frac{\operatorname{var} m_{11}}{\mu_{11}^2} + \frac{\operatorname{var}(m_{20}m_{02})^{\frac{1}{2}}}{\mu_{20}\mu_{02}} - 2\frac{\operatorname{cov}\{m_{11}, (m_{20}m_{02})^{\frac{1}{2}}\}}{\mu_{11}(\mu_{20}\mu_{02})^{\frac{1}{2}}}\right\}$$

and evaluating var $(m_{20}m_{02})^{\frac{1}{2}}$ and cov $\{m_{11}, (m_{20}m_{02})^{\frac{1}{2}}\}$ from (10.12) and (10.13), we find

$$\operatorname{var} r \doteq \rho^2 \left\{\frac{\operatorname{var} m_{11}}{\mu_{11}^2} + \frac{1}{4}\left(\frac{\operatorname{var} m_{20}}{\mu_{20}^2} + \frac{\operatorname{var} m_{02}}{\mu_{02}^2} + \frac{2\operatorname{cov}(m_{20}, m_{02})}{\mu_{20}\mu_{02}}\right) - \left(\frac{\operatorname{cov}(m_{11}, m_{02})}{\mu_{11}\mu_{02}} + \frac{\operatorname{cov}(m_{11}, m_{20})}{\mu_{11}\mu_{20}}\right)\right\}$$

from which, on substituting for variances and covariances from (10.23) and (10.24), we have

$$\operatorname{var} r \doteq \frac{\rho^2}{n}\left\{\frac{\mu_{22}}{\mu_{11}^2} + \frac{1}{4}\left(\frac{\mu_{40}}{\mu_{20}^2} + \frac{\mu_{04}}{\mu_{02}^2} + \frac{2\mu_{22}}{\mu_{20}\mu_{02}}\right) - \left(\frac{\mu_{31}}{\mu_{11}\mu_{20}} + \frac{\mu_{13}}{\mu_{11}\mu_{02}}\right)\right\}.$$

For the bivariate normal distribution the substitution of values from Example 3.19 gives

$$\operatorname{var} r \doteq \frac{1}{n}(1-\rho^2)^2.$$

The use of the standard error to test a hypothetical non-zero value of ρ is not, however, to be recommended, since the sampling distribution of r tends to normality very slowly. We return to this point in **16.23–33** below.

Standard errors of quantiles

10.10 Among the various quantities measuring location and dispersion that we considered in Chapter 2 there was one group, namely the quantiles, which are not algebraic functions of the observations and whose sampling variances cannot accordingly be determined by the above methods. We proceed to consider them now.

Suppose the population distribution is $dF(x) = f(x)dx$. The probability that, of a sample of n, $(l-1)$ fall below a value x_1, one falls in the range $x_1 \pm \frac{1}{2}dx_1$ and the remaining $(n-l)$ fall above x_1 is proportional to

$$\{F(x_1)\}^{l-1}f(x_1)dx_1\{1-F(x_1)\}^{n-1} = F_1^{l-1}(1-F_1)^{n-l}dF_1, \qquad (10.25)$$

where $F_1 = F(x_1)$. This expression is accordingly the density function of the value not exceeded by a proportion l/n of the sample, i.e. the lth quantile. Put

$$l = nq$$

so that

$$\begin{aligned} n - l &= n(1-q) \\ &= np, \quad \text{say.} \end{aligned}$$

The distribution (10.25) has a modal value $\check{x}$ given by differentiating it with respect to x_l and equating to zero, i.e. (taking logarithms first) by

$$(l-1)\frac{f_l}{F_l}-(n-l)\frac{f_l}{(1-F_l)}+\frac{f_l'}{f_l}=0.$$

If now $n\to\infty$ with q fixed, l and $n-l$ will also $\to\infty$ and the term f_l'/f_l will be relatively negligible. We then have, to order n^{-1},

$$\frac{q}{F}-\frac{p}{1-F}=0 \tag{10.26}$$

or

$$F(\check{x})\doteq q.$$

This is in accordance with our general assumptions. To order n^{-1} the mode of the quantile of the sample is at the corresponding quantile of the population.

Now let us investigate the distribution (10.25) in the neighbourhood of the modal value. Put

$$F(x_l)=q+\xi. \tag{10.27}$$

Equation (10.25) becomes (neglecting constants)

$$(q+\xi)^{nq}(p-\xi)^{np}.$$

Taking logarithms and expanding we have, except for constants,

$$nq\log\left(1+\frac{\xi}{q}\right)+np\log\left(1-\frac{\xi}{p}\right)=nq\left(+\frac{\xi}{q}-\frac{1}{2}\frac{\xi^2}{q^2}\cdots\right)+np\left(-\frac{\xi}{p}-\frac{1}{2}\frac{\xi^2}{p^2}\cdots\right)$$

$$=-\frac{n\xi^2}{2pq}+O(\xi^3).$$

Now for large samples ξ will be small compared with q, and we neglect the terms of higher order. Thus the distribution of ξ is

$$\mathrm{d}F\propto\exp\left(\frac{-n\xi^2}{2pq}\right)\mathrm{d}\xi$$

or, evaluating the necessary constant by integration,

$$\mathrm{d}F=\frac{1}{\surd(2\pi)\surd(pq/n)}\exp\left(\frac{-n\xi^2}{2pq}\right)\mathrm{d}\xi,$$

showing that ξ is in the limit distributed normally with variance

$$\operatorname{var}\xi\doteq\frac{pq}{n}. \tag{10.28}$$

To find the variance of x_l we use (10.14) in (10.27) and find

$$\frac{pq}{n} \doteq \text{var}\,\xi \doteq \{f(x_1)\}^2 \text{var}\,x_1$$

so that

$$\text{var}\,x_1 \doteq \frac{pq}{nf_1^2}, \tag{10.29}$$

where we must have $f_1 > 0$ to satisfy the argument leading to (10.12) and (10.14).

If this formula is applied to grouped distributions, it is to be remembered that f_1 is to be taken as the probability *per unit interval* at x_1, this being the best estimate of the ordinate.

Example 10.7 (Variance of the normal median)

If x_1 is the median, $p = q = \frac{1}{2}$ and (10.29) gives var (median) $= 1/(4nf_1^2)$, where f_1 is the population median ordinate. For instance, if the population is normal with variance σ^2, the median ordinate is (from Appendix Table 1) $0.398\,94/\sigma$. Hence the standard error of the median is

$$\frac{\sigma/n^{\frac{1}{2}}}{2 \times 0.398\,94} = 1.2533\sigma/n^{\frac{1}{2}}.$$

The standard error of the mean in samples of n from any population is $\sigma/n^{\frac{1}{2}}$, which is considerably smaller in the normal case than the standard error of the median.

10.11 To find the covariance of two quantiles we generalize equation (10.25). If we have a random sample of n individuals the probability that $(l-1)$ lie below x_1, one lies at $x_1 \pm \frac{1}{2}dx_1$, $(n-l-m)$ lie between x_1 and x_2, one at $x_2 \pm \frac{1}{2}dx_2$, and the remaining $(m-1)$ above x_2 is

$$dF \propto F_1^{l-1}(F_2 - F_1)^{n-l-m}(1-F_2)^{m-1}dF_1\,dF_2, \tag{10.30}$$

where $F_1 = F(x_1)$, $F_2 = F(x_2)$.

We put

$$l = nq_1,$$
$$m = np_2,$$

and find, for the equations giving the modal values corresponding to (10.26),

$$\frac{q_1}{F_1} - \frac{(q_2 - q_1)}{F_2 - F_1} \doteq 0,$$

$$\frac{q_2 - q_1}{F_2 - F_1} - \frac{p_2}{1 - F_2} \doteq 0,$$

giving, for the limiting modal values,

$$\left.\begin{aligned} F(\check{x}_1) &= q_1 \\ F(\check{x}_2) &= q_2. \end{aligned}\right\}$$

Now put

$$\left.\begin{aligned} F(x_1) &= q_1 + \xi_1 \\ F(x_2) &= q_2 + \xi_2. \end{aligned}\right\} \tag{10.31}$$

The joint distribution of ξ_1 and ξ_2 then becomes

$$\mathrm{d}F \propto (q_1+\xi_1)^{q_1 n}(q_2-q_1+\xi_2-\xi_1)^{(q_2-q_1)n}(p_2-\xi_2)^{p_2 n}\,\mathrm{d}\xi_1\,\mathrm{d}\xi_2.$$

On proceeding as in the previous section, taking logarithms, expanding and neglecting terms in ξ^3 and higher, we find ultimately

$$\mathrm{d}F \propto \exp\left\{-\frac{n}{2(q_2-q_1)}\left(\frac{q_2}{q_1}\xi_1^2 - 2\xi_1\xi_2 + \frac{p_1}{p_2}\xi_2^2\right)\right\}\mathrm{d}\xi_1\,\mathrm{d}\xi_2. \tag{10.32}$$

Thus the joint distribution of ξ_1 and ξ_2 tends to the bivariate normal form, and on comparing (10.32) with Example 3.19, we see that

$$\frac{1}{(1-\rho^2)\operatorname{var}\xi_1} \doteq \frac{nq_2}{(q_2-q_1)q_1}$$

$$\frac{1}{(1-\rho^2)\operatorname{var}\xi_2} \doteq \frac{np_1}{(q_2-q_1)p_2}$$

$$\frac{\rho^2}{(1-\rho^2)\operatorname{cov}(\xi_1,\xi_2)} = \frac{\rho}{(1-\rho^2)\{\operatorname{var}\xi_1\operatorname{var}\xi_2\}^{\frac{1}{2}}} \doteq \frac{n}{(q_2-q_1)}.$$

To order n^{-1}, it follows that,

$$\left.\begin{aligned} \operatorname{var}\xi_1 &= \frac{p_1q_1}{n} \\ \operatorname{var}\xi_2 &= \frac{p_2q_2}{n} \\ \operatorname{cov}(\xi_1,\xi_2) &= \frac{p_2q_1}{n} \end{aligned}\right\} \tag{10.33}$$

The asymmetry of the result for the covariance is due to the fact that p_2 relates necessarily to the *upper* quantile. For the corresponding expression in x_1 and x_2 we have from (10.13) and (10.31)

$$\operatorname{cov}(x_1,x_2) \doteq \frac{p_2q_1}{nf_2f_1}, \tag{10.34}$$

where again f_1 and f_2 must be positive to satisfy the argument leading to (10.13). With equations (10.33) and (10.34) we can find expressions for the variances of the interquartile range and similar statistics.

Example 10.8 (Variance of the interquartile range)

The variance of the difference d of two quantiles at x_1 and x_2 is given, from (10.16), by

$$\text{var}\, d = \text{var}\, x_1 + \text{var}\, x_2 - 2\,\text{cov}(x_1, x_2)$$
$$= \frac{1}{n}\left\{\frac{p_1 q_1}{f_1^2} + \frac{p_2 q_2}{f_2^2} - \frac{2p_2 q_1}{f_1 f_2}\right\}.$$

When the quantiles are the two quartiles, $p_2 = q_1 = \frac{1}{4}$, $p_1 = q_2 = \frac{3}{4}$, and for the variance of the interquartile range we have

$$\text{var}\, d = \frac{1}{16n}\left(\frac{3}{f_1^2} + \frac{3}{f_2^2} - \frac{2}{f_1 f_2}\right),$$

where f_1, f_2 denote the density functions at the two quartiles, f_2 relating to the upper quartile.

For instance, if the distribution is symmetric about zero, $f_1 = f_2$ and we find

$$\text{var}\, d = \frac{1}{4nf_1^2}.$$

In the normal case, we find from the tables that the deviate corresponding to the quartile is $0.674\,5\sigma$ and the ordinate at this point, $f_1 = 0.3178/\sigma$, so that the interquartile range is 1.349σ and its standard error is approximately

$$\frac{\sigma/n^{\frac{1}{2}}}{2 \times 0.3178} = 1.573\sigma/n^{\frac{1}{2}}.$$

10.12 It is worth reiterating the point mentioned in **10.8**(d), that a standard error is related to the way in which a parameter is estimated. For instance, the standard deviation of a normal distribution can be estimated from a sample in several ways: from the second moment; by taking $(\pi/2)^{\frac{1}{2}}$ times the mean deviation; from Example 10.8 by taking $(1.349)^{-1}$ times the interquartile range; and so on. Each method will have its appropriate standard error, that for the first (**10.8**) being $\sigma/(2n)^{\frac{1}{2}}$, and that for the third $1.649\sigma/(2n)^{\frac{1}{2}}$, as is easily deduced from Example 10.8. At a later stage considerations such as this will lead us to inquire: what is the estimate, if any, with the *minimum* sampling variance? For present purposes it is enough to note the importance of not using a quoted formula without reference to the method of estimation of the parameter.

Standard error of the mean deviation

10.13 Suppose that we are sampling from a population with mean μ and variance σ^2. Let d_α be the sample mean deviation measured about the *fixed* point α, i.e.

$$d_\alpha = \frac{1}{n}\sum_{i=1}^{n} |x_i - \alpha|.$$

Then

$$E(d_\alpha) = E|x_i - \alpha| = \delta_\alpha, \tag{10.35}$$

where δ_α is the corresponding population mean deviation, and from (10.16)

$$\text{var}\, d_\alpha = \frac{1}{n^2}\left\{\sum_{i=1}^{n} \text{var}(|x_i - \alpha|) + \sum\sum_{i \neq j} \text{cov}(|x_i - \alpha|, |x_j - \alpha|)\right\}.$$

Since the $|x_i - \alpha|$ are completely independent, their covariances are zero and we have

$$\operatorname{var} d_\alpha = \frac{1}{n}\{E(|x-\alpha|^2) - E^2(|x-\alpha|)\}$$

$$= \frac{1}{n}\{\sigma^2 + (\alpha-\mu)^2 - \delta_\alpha^2\}. \tag{10.36}$$

If $\alpha = \mu$, the population mean, this becomes

$$\operatorname{var} d_\mu = \frac{1}{n}(\sigma^2 - \delta_\mu^2). \tag{10.37}$$

We denoted δ_μ by δ_1 in Chapter 2.

Incidentally, (10.37) establishes that $\delta_\mu \leqslant \sigma$—cf. the stronger inequality given at (2.22).

(10.35-7) are exact results. Consider now the sample mean deviation d_a about a value a, itself subject to sampling variation, which has the properties

$$E(a) = \alpha, \qquad \operatorname{var} a = O(n^{-1}).$$

The $|x_i - a|$ are not now independent, since a is common to them all, and their covariances will not disappear as above. Their sum may, however, be asymptotically negligible compared with the sum of the variances. Since $d_a = d_\alpha\{1 + O(n^{-\frac{1}{2}})\}$, we then find as in **10.5** that *approximately*

$$E(d_a) = \delta_\alpha, \qquad \operatorname{var} d_a = \operatorname{var} d_\alpha.$$

Exercise 10.14 gives the general asymptotic expression for var $d_{\bar{x}}$ and shows that these approximations hold for any symmetrical population when $a = \bar{x}$, the sample mean. For $\bar{x}$ converges to μ as n increases, and for a symmetrical population each deviation that is increased when taken about $\bar{x}$ rather than μ will be offset by one that is decreased by the same amount, unless the deviation itself is very small. We now write simply d for $d_{\bar{x}}$, so that approximately, for symmetrical populations,

$$\operatorname{var} d = \frac{1}{n}(\sigma^2 - \delta_\mu^2). \tag{10.38}$$

For a normal population we have seen at (5.112) that $\delta_\mu = \sigma\sqrt{(2/\pi)}$ and hence

$$\operatorname{var} d \doteq \sigma^2(1 - 2/\pi)/n. \tag{10.39}$$

In this normal case, the mean and variance of d can be obtained exactly. Since

$$E(d) = E|x_i - \bar{x}|$$

and $(x_i - \bar{x})$ has zero mean and exact variance $\sigma^2(n-1)/n$ by (10.16), (5.112) gives

$$E(d) = \left\{\frac{2\sigma^2(n-1)}{\pi n}\right\}^{\frac{1}{2}}.$$

For the sampling variance, Helmert (1876) and Fisher (1920) found

$$\text{var}\, d = \frac{2\sigma^2(n-1)}{\pi n^2}[\tfrac{1}{2}\pi + \{n(n-2)\}^{\frac{1}{2}} - n + \arc\sin\{1/(n-1)\}], \tag{10.40}$$

a proof of which is indicated in Exercise 15.6 below. For large n, this expression tends to (10.39). Godwin (1945, 1948) gives a recurrence relation for the moments, an approximation to the distribution, and an exact expression for the d.f. of d, which is tabulated for $n = 2(1)10$ by Hartley (1945) and reproduced in Vol. II of the *Biometrika Tables*, Vol. I of which gives small-sample tables of the percentage points and the moments of d.

Geary (1936b) derived the asymptotic distribution of d/s and Shenton *et al.* (1979) gave series expansions to $O(n^{-5})$ for its moments.

Standard error of the mean difference

10.14 Let us write g for the sample value of the mean difference without repetition, for which the population value is Δ in **2.22**. We write μ and σ^2 for the population mean and variance and assume $n \geqslant 3$. We have

$$g = \frac{1}{n(n-1)} \sum_{i,j=1}^{n} |x_i - x_j|, \tag{10.41}$$

and

$$E(g) = E|x_i - x_j| = \Delta. \tag{10.42}$$

Also

$$E(g^2) = \frac{1}{n^2(n-1)^2} E\left\{\sum |x_i - x_j|\right\}^2$$
$$= \frac{1}{n^2(n-1)^2} E\left\{\sum (x_i - x_j)^2 + \sum{}' |x_i - x_j||x_i - x_k| + \sum{}'' |x_i - x_j||x_k - x_l|\right\},$$
$$i \neq j \neq k \neq l,$$

where there are $2n(n-1)$ terms in the summation $\sum$, $4n(n-1)(n-2)$ in $\sum'$ and $n(n-1)(n-2)(n-3)$ in $\sum''$. We then find

$$E(g^2) = \frac{1}{n(n-1)}\{2E(x_i - x_j)^2 + 4(n-2)E(|x_i - x_j||x_i - x_k|)$$
$$+ (n-2)(n-3)E(|x_i - x_j||x_k - x_l|)\}.$$

The first expected value is $2\sigma^2$, by **2.22**, and the third is Δ^2 because the xs are all independent there. Writing $\mathscr{J}$ for the second, we have

$$E(g^2) = \frac{1}{n(n-1)}\{4\sigma^2 + 4(n-2)\mathscr{J} + (n-2)(n-3)\Delta^2\}$$

and hence

$$\text{var}\, g = \frac{2}{n(n-1)}\{2\sigma^2 + 2(n-2)\mathscr{J} - (2n-3)\Delta^2\}, \tag{10.43}$$

where

$$\mathscr{J} = \int_{-\infty}^{\infty}\int_{-\infty}^{\infty}\int_{-\infty}^{\infty} |x-y|\,|x-z| f(x) f(y) f(z) \mathrm{d}x\, \mathrm{d}y\, \mathrm{d}z$$

$$= \int_{-\infty}^{\infty}\left\{\int_{-\infty}^{x}\int_{-\infty}^{x}(x-y)(x-z) + \int_{-\infty}^{x}\int_{x}^{\infty}(x-y)(z-x) + \int_{x}^{\infty}\int_{-\infty}^{x}(y-x)(x-z)\right.$$

$$\left. + \int_{x}^{\infty}\int_{x}^{\infty}(y-x)(z-x)\right\} \mathrm{d}F(y)\,\mathrm{d}F(z)\,\mathrm{d}F(x), \tag{10.44}$$

and if we write $G(x) = \int_{-\infty}^{x} tf(t)\,\mathrm{d}t$, integration with respect to y and z gives

$$\mathscr{J} = 4\int_{-\infty}^{\infty}\{(xF-G)+\tfrac{1}{2}(\mu-x)\}^2\,\mathrm{d}F(x).$$

Writing I for the integral on the right, we finally have from (10.43)

$$\operatorname{var} g = \frac{2}{n(n-1)}\{2\sigma^2 + 8(n-2)I - (2n-3)\Delta^2\}, \tag{10.45}$$

and it may be seen from (2.33) that

$$\Delta = 2\int_{-\infty}^{\infty}(xF-G)\,\mathrm{d}F(x).$$

This derivation is due to Lomnicki (1952), a more complicated proof having been first given by Nair (1936).

Sillitto (1969) gives a method of estimating var g from the sample.

Three special cases are of interest:

Normal distribution: any mean, variance σ^2.

$$E(g) = 2\sigma/\sqrt{\pi} = 1.128\,4\sigma,$$

as was shown at (5.113), while

$$\operatorname{var} g = \frac{4\sigma^2}{n(n-1)}\{\tfrac{1}{3}(n+1) + 2(n-2)3^{\frac{1}{2}}/\pi - 2(2n-3)/\pi\} \doteq \sigma^2(0.8068)^2/n. \tag{10.46}$$

Third and fourth moments and an approximation to the distribution of g in the normal case are given by Kamat (1959).

A derivation of (10.46) from (10.43) is indicated in Exercise 15.6 below.

Exponential distribution: $\mathrm{d}F = \exp(-x/\sigma)\,\mathrm{d}x/\sigma$. $(0 \leqslant x < \infty)$

$$E(g) = \sigma,$$

$$\operatorname{var} g = \sigma^2\frac{2(2n-1)}{3n(n-1)} \doteq 4\sigma^2/(3n). \tag{10.47}$$

Uniform distribution: $dF = dx/k$. $(0 < x \leqslant k)$

$$E(g) = \tfrac{1}{3}k,$$

$$\text{var}\, g = \frac{k^2(n+3)}{45n(n-1)} \doteq k^2/(45n). \qquad (10.48)$$

Values of Δ for the Gamma distribution and the logistic distribution were given in Exercises 2.7 and 4.21.

Some commonly required standard errors

10.15 It may be useful to list here for reference some of this standard errors commonly required. For convenience, the variance is given, not the standard error. It should be noted that all these results are approximations, except that for the mean; the omitted terms in the variance are always $o(n^{-1})$.

Statistic	Variance multiplied by n	Normal case
Mean, m_1'	$\mu_2(=\sigma^2)$	μ_2
Sample variance, m_2	$\mu_4-\mu_2^2$	$2\sigma^4$
Sample s.d., s	$(\mu_4-\mu_2^2)/(4\mu_2)$	$\sigma^2/2$
Third moment, m_3	$\mu_6-\mu_3^2-6\mu_4\mu_2+9\mu_2^3$	$6\sigma^6$
Fourth moment, m_4	$\mu_8-\mu_4^2-8\mu_5\mu_3+16\mu_3^2\mu_2$	$96\sigma^8$
$\sqrt{b_1}=m_3/m_2^{\frac{3}{2}}$	See Exercise 10.26.	6. See **12.18** and Exercise 12.9.
$b_2=m_4/m_2^2$	See Exercise 10.27.	24. See Exercise 12.10.
Coefficient of variation, V	See Example 10.4.	$V^2(1+2V^2)/2$
Mean deviation	See **10.13** and Exercise 10.14.	$\sigma^2(1-2/\pi)$
Mean difference	See **10.14**.	$(0.806\,8)^2\sigma^2$
Median	$1/(4f^2)$ where f is population ordinate at its median.	$(1.253\,3)^2\sigma^2$.
Quartile	$3/(16f^2)$ where f is population ordinate at the quartile.	$(1.362\,6)^2\sigma^2$
Deciles	See (10.29).	(deciles 4, 6) $=(1.268\,0)^2\sigma^2$ (deciles 3, 7) $=(1.318\,0)^2\sigma^2$ (deciles 2, 8) $=(1.428\,8)^2\sigma^2$ (deciles 1, 9) $=(1.709\,4)^2\sigma^2$.
Interquartile range	$\frac{1}{16}\left(\frac{3}{f_1^2}+\frac{3}{f_2^2}-\frac{2}{f_1f_2}\right)$ where f_1, f_2 are population quartile ordinates.	$(1.573)^2\sigma^2$
Correlation coefficient r (product moment)	See Example 10.6.	$(1-\rho^2)^2$. Not to be recommended except for very large samples. See **16.23–33**.

10.16 The standard errors derived in this chapter rely upon large sample arguments; in the next chapter, we obtain exact results in those cases where sampling distributions may be obtained explicitly. However, before leaving the topic of approximate standard errors, we briefly examine two empirical methods that are computer intensive.

In order to use the results obtained in this chapter, the standard approach is to replace unknown parameters by appropriate estimates. The use of these estimates, combined with the restriction to first-order terms in the Taylor series expansion, may lead to a bias in the estimate of the standard error, at least in smaller samples. The methods we now describe are attempts to redress that problem.

The jackknife and the bootstrap

10.17 Following early work by Quenouille (1956) (see **17.10** in Volume 2A), Tukey (1958) suggested that an estimate of the sampling variance of a statistic be obtained by the following device. Let t_n denote the sample statistic based on a random sample of n observations: $x_1, \ldots, x_n$. Then let $t_{n-1,i}$ denote the statistic based on the $(n-1)$ values excluding x_i. Finally, let $t_n' = nt_n - (n-1)\bar{t}_{n-1}$, where $\bar{t}_{n-1} = \Sigma\, t_{n-1,i}/n$. Tukey recommended that the variance of t_n be estimated by

$$\hat{V}_J = \frac{n-1}{n} \sum_{i=1}^{n} [t_{n-1,i} - \bar{t}_{n-1}]^2. \tag{10.49}$$

$\hat{V}_J$ reduces to $\dfrac{1}{n(n-1)} \sum (x_i - \bar{x})^2$ when $t_n = \bar{x}$. $\hat{V}_J$ is known as the *jackknife* estimator of the variance of t_n and has produced somewhat mixed results when used for more general functions of the observations (R. G. Miller, 1974; Efron, 1982, Chapters 3 and 4).

10.18 A computationally more intensive approach is the *bootstrap* method (Efron, 1979, 1981, 1982). Given the n sample values $x_1, \ldots, x_n$, we may define the empirical frequency function $\hat{f}$ which has values $1/n$ at each x_i and zero elsewhere. The bootstrap estimate of the sampling variance of a statistic is obtained by selecting a sample of size n with replacement from the population described by $\hat{f}$ and computing the estimate θ^* from this sample; this is done B times, giving θ_j^* for $j = 1, \ldots, B$.

The bootstrap estimator is

$$\theta^* = \frac{1}{B} \sum_{j=1}^{B} \theta_j^*,$$

and the corresponding estimator of the sampling variance is

$$\hat{V}_{BS}(\theta^*) = \sum_{j=1}^{B} (\theta_j^* - \theta^*)^2/(B-1). \tag{10.50}$$

10.19 In order to understand the workings of the bootstrap, we present the following argument, due to Efron (1982, pp. 34–5).

Consider the random variable X, with finite sample space $\mathscr{X} = \{x_1, \ldots, x_M\}$ say, for which

$$P(X=x_j)=f_j, \quad j=1,\ldots,M, \tag{10.51}$$

and $\mathbf{f}=(f_1,\ldots,f_M)$. A sample of size n yields the relative frequencies $\hat{\mathbf{f}}=(\hat{f}_1,\ldots,\hat{f}_M)$.

The statistic of interest may be expressed as the random function $Q=Q(\hat{\mathbf{f}},\mathbf{f})$; the distribution of $\hat{\mathbf{f}}$, given $\mathbf{f}$, is multinomial, which we write as

$$\hat{\mathbf{f}}\,|\,\mathbf{f}\sim Mn(n,\mathbf{f}) \tag{10.52}$$

We now consider the function $Q^*=Q(\hat{\mathbf{f}}^*,\hat{\mathbf{f}})$, where $\hat{\mathbf{f}}^*$ represents the relative frequencies resulting from multinomial sampling based upon $\hat{\mathbf{f}}$, or

$$\hat{\mathbf{f}}^*|\hat{\mathbf{f}}\sim Mn(n,\hat{\mathbf{f}}). \tag{10.53}$$

The bootstrap procedure estimates the unobservable sampling distribution of Q by the B replicates that each generate a value of Q^*, or by complete enumeration in the rare cases where that is possible.

Anticipating Example 15.3, we know that as $n\to\infty$, the distribution of $\hat{\mathbf{f}}$ approaches multivariate normality, which we may express as

$$\hat{\mathbf{f}}-\mathbf{f}\xrightarrow[p]{}N(\mathbf{0},n^{-1}\boldsymbol{\Sigma}) \tag{10.54}$$

where $\boldsymbol{\Sigma}$ has elements

$$\sigma_{jj}=f_j(1-f_j), \qquad \sigma_{jk}=-f_jf_k, \quad j\neq k.$$

Further, the limiting distribution of $\hat{\mathbf{f}}^*-\hat{\mathbf{f}}$ is also given by (10.54). Thus, provided Q is a well-behaved function (cf Efron, 1979), the distribution of Q^* is asymptotically the same as that of Q.

10.20 The argument for general sample spaces is more difficult, but the same type of result holds; see, for example, Babu and Singh (1983). Although our present interest lies in the estimation of the standard error, it is clear that the estimated sampling distribution generated by the bootstrap method allows a variety of other tasks to be performed, such as interval estimation (Hall, 1988) and the estimation of tail area probabilities (Hall, 1990). Kaigh and Cheng (1991) describe bootstrap procedures for the quantiles.

The relationship between the bootstrap and the Edgeworth series expansion is developed in Hall (1992).

10.21 For estimation of the standard error, $B \doteq 200$ appears to suffice but larger values are needed when interval estimates are to be constructed; cf Efron and Tibshirani (1986). One benefit of the bootstrap is that interval estimates may be based upon the complete bootstrap distribution so that monotone transformations of the parameters and their estimators produce consistent results. By contrast, the delta method produces intervals that agree only to terms of order $n^{-1/2}$; see Exercise 10.29.

The bootstrap as described above is nonparametric in nature since no assumptions are made about the form of the population. It is also possible to use a parametric bootstrap; for example, we may resample from a normal distribution with mean and variance estimated by $\bar{x}$ and s^2 respectively. As is to be expected, the parametric versions perform better when the distributional assumptions are satisfied, at least to a close approximation.

For some numerical studies, see Efron (1981) and Wu (1986) among others.

10.22 Jackknife methods work in those circumstances where the statistic $\hat{\theta}$ is representable, at least approximately, as an average. They do not work when $\hat{\theta}$ is a simple quantile or an extreme value.

Example 10.9 (Jackknife for the median)

Consider the sample of n ordered values $x_{(1)} < x_{(2)} < \ldots < x_{(n)}$, where $n = 2m$ is even. Then, by convention

$$t_n = \tfrac{1}{2}(x_{(m)} + x_{(m+1)}),$$

whereas

$$\begin{aligned} t_{n-1,i} &= x_{(m)}, \quad i = m+1, \ldots, 2m \\ &= x_{(m+1)}, \quad i = 1, \ldots, m \end{aligned}$$

so that $\bar{t}_{n-1} = t_n$. Thus, (10.49) reduces to

$$\hat{V}_J = \left(\frac{n-1}{4}\right)(x_{(m+1)} - x_{(m)})^2; \tag{10.55}$$

Efron (1982, p. 16) shows that (10.55) is not a consistent estimator for the sampling variance given in Example 10.7. By contrast, the bootstrap estimator is more plausible, and may be shown to be reasonably efficient; see Efron (1982, chapter 10).

Example 10.10 (Bootstrap for the median)

A random sample of size $n = 25$ was drawn from a normal distribution with $\mu = 5$, $\sigma = 1$. The sample values were:

$$\bar{x} = 5.111, \quad \text{median} = M = 5.248, \quad \text{and} \quad S = 1.253.$$

Bootstrap samples of size 25 were then generated from the original sample for $B = 25$, 50, 100, 200 and 400, where the smaller samples are contained within the larger ones. The estimated mean, median and standard error are given for each value of B in Table 10.1.

By way of comparison, we also generated B samples from a normal with the parameters set at $\bar{x}$ and s^2 (a parametric bootstrap) and from the original normal distribution (a genuine simulation). Finally, the *SE*s obtained by the delta method, based upon Example 10.7, are included for comparison; the value listed under *PB* simply uses S in place of σ.

Too much should not be read into a single example, but it is clear from Table 10.1 that $\widehat{SE}$ is seriously biased downwards when B is too small. To assist comparisons, a total of five runs, each with $B = 400$, is summarized in Table 10.2. The simulation and delta (σ) methods would not be available to us in practice, since they rely upon knowledge of the unknown parameters.

The three usable methods, *B*, *PB* and *DS* all produce average values for the *SE* at around the same level, but the performance of *B* is somewhat more variable. However, both *PB* and *DS* rely upon the assumption that the underlying distribution is normal, whereas *B* is free of any distributional assumptions.

Table 10.1 Summary statistics for the bootstrap analysis of the median in a normal sample of size $n = 25$; (B = bootstrap; PB = parametric bootstrap; S = simulation)

No. of		$\widehat{\text{mean}}$			$\widehat{\text{median}}$			$\widehat{\text{SE}}$	
replicates (B)	B	PB	S	B	PB	S	B	PB	S
25	5.16	5.13	4.97	5.25	5.14	4.95	0.124	0.341	0.270
50	5.17	5.10	4.97	5.25	5.12	4.95	0.121	0.321	0.266
100	5.16	5.10	4.97	5.25	5.11	4.96	0.160	0.311	0.270
200	5.15	5.12	5.00	5.25	5.12	4.99	0.176	0.318	0.265
400	5.16	5.12	4.98	5.25	5.13	4.97	0.169	0.316	0.257
delta	-	-	-	-	-	-	-	0.314	0.251

Table 10.2 Estimates of SE (Median) for five bootstrap runs, each with B = 400

			Run			Average
Estimate	1	2	3	4	5	
bootstrap, B	0.169	0.370	0.406	0.209	0.174	0.266
parametric bootstrap, PB	0.316	0.282	0.337	0.218	0.213	0.273
simulation	0.257	0.251	0.226	0.242	0.233	0.242
delta (s), DS	0.314	0.290	0.327	0.218	0.211	0.272
delta (σ)	0.251	0.251	0.251	0.251	0.251	0.251

EXERCISES

10.1 Regarding the distribution of Table 1.7 as a random sample from a population that is approximately normal, verify that $m_3 = -0.208$ and $m_4 = 137.7$ and show, using the results of Example 2.7, that m_3 does not differ significantly from zero and that m_4 has a standard error of only 4 percent of its value.

10.2 A multinomial distribution has k classes with probabilities $\pi_1, \pi_2, \ldots, \pi_k$. In a sample of size n the observed frequencies in these classes are $f_1, f_2, \ldots, f_k$, and $p_i = f_i/n$. Show that a function $T(np_1, np_2, \ldots, np_k)$, where T does not depend on n except as shown, has asymptotic variance

$$n \operatorname{var} T = \sum_{i=1}^{k} \pi_i \left(\frac{\partial T}{\partial p_i}\right)^2 - \left(n \frac{\partial T}{\partial n}\right)^2,$$

the derivatives being taken at the values $p_i = \pi_i$.

(Fisher, 1970)

10.3 With a random sample of 100 it is found that the odds are 99 to 1 against a certain statistic differing from its expected value by more than two units. Find how large the sample would have to be to increase these odds to 199 to 1; and find the odds that with the sample of 100 the statistic would differ from its expected value by one unit.

10.4 If k random samples of different sizes are drawn from a population in which the proportion of members bearing an attribute A is π, show that the standard error of the mean proportion of A in the samples is $\sqrt{\dfrac{\pi(1-\pi)}{kH}}$ where H is the harmonic mean of the sizes of the samples.

10.5 Using the results of **10.5–6**, show that the sampling correlation in large samples (i.e. the sampling covariance divided by the product of the standard errors) between any regular functions $y(x_1)$ and $z(x_2)$ is the same as that between x_1 and x_2.

10.6 Show from (10.10) that for large samples from a symmetrical distribution any sample central moment of odd order has zero covariance to order n^{-1} with any sample central moment of even order and that the covariance of two sample central moments of even order cannot be negative to order n^{-1}.

10.7 In the notation of this chapter show that, to order n^{-1},

$$\operatorname{cov}(m_r', m_q') = (\mu_{q+r} - \mu_q\mu_r - r\mu_{q+1}\mu_{r-1})/n.$$

10.8 If r is even, show that for a symmetrical distribution, var m_r and cov (m_r, m_q) at (10.9–10) reduce to the exact results (10.5–6) when the sample moment m_r' is calculated about the population mean μ_1'. Verify that this is not so if r is odd by showing that in the normal case, $n \operatorname{var} m_3' = 15\sigma^6$ and $n \operatorname{var} m_3 \doteq 6\sigma^6$.

10.9 The variance of m_2 in large samples is $(\mu_4 - \mu_2^2)/n$. If this is estimated as $(m_4 - m_2^2)/n$ show that the variance of the estimated var m_2 in samples from a normal population is $56\sigma^8/n^3$. Compare with the value of var m_2 in normal variation, $2\sigma^4/n$.

10.10 $x_1, x_2, \ldots, x_p, \ldots, x_n$ are independent, identically distributed variates with zero mean, and $\bar{x}_p = \sum_{i=1}^{p} x_i/p$, $\bar{x} = \sum_{i=1}^{n} x_i/n$. Show that the correlation coefficient (i.e. the covariance divided by the product of the standard errors)

(a) between $\bar{x}_p$ and $\bar{x}_n$ is $(p/n)^{\frac{1}{2}}$;

(b) between $\bar{x}_p - \bar{x}_n$ and $\bar{x}_n$ is zero; and

(c) between $\bar{x}_p$ and $\bar{x}_p - \bar{x}_n$ is $\left(1 - \dfrac{p}{n}\right)^{\frac{1}{2}}$.

10.11 If x has a standardized normal distribution, show that the m.g.f. of $|x|$ is

$$M(t) = e^{\frac{1}{2}t^2}\left(\frac{2}{\pi}\right)^{\frac{1}{2}}\int_{-\infty}^{t} e^{-\frac{1}{2}y^2}\,dy,$$

so that

$$\frac{\partial M(t)}{\partial t} = tM(t) + \left(\frac{2}{\pi}\right)^{\frac{1}{2}}$$

and the moments about zero are

$$\left.\begin{aligned}\mu'_{2r} &= (2r)!/(2^r r!),\\ \mu'_{2r+1} &= 2^r r!(2/\pi)^{\frac{1}{2}},\end{aligned}\right\} r = 0, 1, 2, \ldots$$

Hence or otherwise show that

$$\begin{aligned}\kappa_1 &= (2/\pi)^{\frac{1}{2}},\\ \kappa_2 &= 1 - 2/\pi,\\ \kappa_3 &= (4/\pi - 1)(2/\pi)^{\frac{1}{2}},\\ \kappa_4 &= (1 - 3/\pi)8/\pi,\end{aligned}$$

and use these to evaluate the right-hand side of (6.6). Compare with the value of the left-hand side of (6.6), $\{2/(\pi - 2)\}^{\frac{1}{2}} = 1.32$, given in Exercise 3.27. Explain the result.

Verify that (10.37) is exactly equal to (10.39) in the normal case.

10.12 Show that the sampling variances of the first four cumulants, as calculated from the moments, are given to order n^{-1} by

$$\operatorname{var}\kappa_1 = \frac{1}{n}\kappa_2, \quad \operatorname{var}\kappa_2 = \frac{1}{n}(\kappa_4 + 2\kappa_2^2), \quad \operatorname{var}\kappa_3 = \frac{1}{n}(\kappa_6 + 9\kappa_4\kappa_2 + 9\kappa_3^2 + 6\kappa_2^3),$$

$$\operatorname{var}\kappa_4 = \frac{1}{n}(\kappa_8 + 16\kappa_6\kappa_2 + 48\kappa_5\kappa_3 + 34\kappa_4^2 + 72\kappa_4\kappa_2^2 + 144\kappa_3^2\kappa_2 + 24\kappa_2^4).$$

(*Note.* The sample values of the κs on the left have been written with Greek letters; to replace them in the customary way by ks might produce confusion with the k-statistics to be introduced in Chapter 12.)

10.13 Show that the mean value of the variance is given exactly by

$$E(m_2) = \frac{n-1}{n}\mu_2$$

and that its sampling variance is given exactly by

$$\operatorname{var} m_2 = \left(\frac{n-1}{n}\right)^2\frac{\mu_4 - \mu_2^2}{n} + \frac{2(n-1)}{n^3}\mu_2^2.$$

Hence verify that (10.9) is accurate to order n^{-1}.

10.14 In **10.13**, show as in **2.18** that the mean deviation

$$d_{\bar{x}} = \frac{2}{n}\sum_{x_i<\bar{x}}(\bar{x} - x_i) = \frac{2}{n}\left\{p\sum_{x_i>\bar{x}}x_i - (1-p)\sum_{x_i<\bar{x}}x_i\right\}$$

where there are np observations $\leqslant \bar{x}$ in value. Hence show that

$$\text{var}\, d_{\bar{x}} \doteq \frac{4}{n}\left\{P^2\int_{\mu}^{\infty}(x-\mu)^2\,\mathrm{d}F + (1-P)^2\int_{-\infty}^{\mu}(x-\mu)^2\,\mathrm{d}F\right\} - \delta_{\mu}^2,$$

where $P = F(\mu)$. Show that when the population is symmetrical, this reduces to (10.38).

(Gastwirth, 1974)

10.15 Show that in odd samples of n from a population with uniform density over a unit range, the sampling variance of the distribution of the median is given exactly by $1/\{4(n+2)\}$.

10.16 If two variates are independent, if the distribution of each is symmetric, and if $r+s$ is an even number, show that for their product moments,

$$n\,\text{cov}\,(m_{r,s}, m_{u,v}) = \mu_{r+u,0}\mu_{0,s+v} - \mu_{r0}\mu_{u0}\mu_{0s}\mu_{0v}.$$

10.17 In **10.5** show that to the next order of approximation

$$E\{g(x)\} = g(\theta) + \tfrac{1}{2}\left[\sum_{i=1}^{k} g_i''(\theta)\,\text{var}\,x_i + \sum\sum_{i \neq j} g_{ij}''(\theta)\text{cov}(x_i, x_j)\right] + o(n^{-r})$$

and that if all the $g_i'(\theta) = 0$, we also have to the next order

$$E\{[g(x) - g(\theta)]^2\} = \tfrac{1}{4}E\left\{\left[\sum_{i=1}^{k} g_i''(\theta)(x_i-\theta_i)^2 + \sum\sum_{i \neq j} g_{ij}''(\theta)(x_i-\theta_i)(x_j-\theta_j)\right]^2\right\}$$

and

$$\text{var}\,\{g(x)\} = E\{[g(x) - g(\theta)]^2\} - [E\{g(x) - g(\theta)\}]^2.$$

10.18 In binomial sampling for attributes (see **9.21**) with population parameter π and sample proportion p, use the results of (10.14) and Exercise 10.17 to show that to the first orders

$$\begin{aligned} \text{var}\,\{p(1-p)\} &= \pi(1-\pi)(1-2\pi)^2/n, \qquad & \pi \neq \tfrac{1}{2}, \\ &= (n-1)/(8n^3), & \pi = \tfrac{1}{2}. \end{aligned}$$

Verify directly from the moments of the binomial that

$$\text{var}\,\{p(1-p)\} = \frac{\pi(1-\pi)(n-1)^2}{n^3}\left\{(1-2\pi)^2 + \frac{2\pi(1-\pi)}{n-1}\right\}$$

exactly, agreeing with the above first-order results. (The first-order result for $\pi = \frac{1}{2}$ is exact here, because $p(1-p)$ is exactly represented by the second-order Taylor expansion.)

10.19 Show that in samples of size n from the uniform distribution $f(x) = k$, $0 \leqslant x \leqslant 1/k$, the variance of the mean deviation is approximately $1/(48k^2n)$.

10.20 Using the first result of Exercise 10.17, and the results of Exercise 10.13, show that, writing $s^2 = m_2 n/(n-1)$, the standard deviation of a population is estimated with bias of order less than that of s itself by

$$s\left(1 + \frac{m_4 - s^4}{8ns^4}\right),$$

reducing to $s\{1 + (4n)^{-1}\}$ in the normal case.

10.21 Two lists of M and N items have D items in common. Independent random samples of sizes m and n respectively ($m \geqslant n \geqslant 2$) are taken without replacement and found to have d items in common. Show that an unbiased estimate of D is given by

$$\hat{D} = \frac{MN}{mn} d$$

and that

$$\text{var}\, \hat{D} = \frac{MND}{mn}\left\{1 + \frac{m-1}{M-1} \cdot \frac{n-1}{N-1}(D-1)\right\} - D^2.$$

Show that as d approaches n, $\hat{D}$ exceeds N and thus that the estimate takes "impossible" values.
(Cf. Goodman (1952); also Deming and Glasser (1959))

10.22 In Exercise 10.21, show that an unbiased estimate of var $\hat{D}$ is given by

$$\hat{V} = \hat{D}\left(\frac{M-1}{m-1} \cdot \frac{N-1}{n-1} - 1\right) + \hat{D}^2\left(1 - \frac{mn}{MN} \cdot \frac{M-1}{m-1} \cdot \frac{N-1}{n-1}\right)$$

and hence that $\hat{V}$ may be negative, and thus an "impossible" estimate of var $\hat{D}$, when m, n, M and N are small.

10.23 x and y are jointly distributed with moments μ_{rs}. Show that

$$\text{var}\,(xy) = \mu_{10}'^2\mu_{02} + \mu_{20}\mu_{01}'^2 + \mu_{20}\mu_{02} + \text{cov}(x^2, y^2) - \mu_{11}^2 - 2\mu_{10}'\mu_{11}\mu_{01}' \qquad \text{(A)}$$

exactly, and that when x is independent of y, this reduces to

$$\text{var}\,(xy) = \mu_{10}'^2\mu_{02} + \mu_{20}\mu_{01}'^2 + \mu_{20}\mu_{02}. \qquad \text{(B)}$$

Show that both (A) and (B) agree to order n^{-1} with the large-sample result obtained from (10.12) if $(x - \mu_{10}')$ and $(y - \mu_{01}')$ are both of order $n^{-\frac{1}{2}}$.

If m_{20}, m_{02} are unbiased estimates of μ_{20}, μ_{02}, show that an unbiased estimate of (B) is

$$x^2 m_{02} + y^2 m_{20} - m_{20}m_{02}.$$

(Cf. Goodman (1960); Goodman (1962) generalizes to the product of more than two variates, and Bohrnstedt and Goldberger (1969) give the covariance of two products xy, uv.)

10.24 Using the results of **10.10–11**, show that if M, Q_1 and Q_3 are the median and quartiles of a large sample of size n from a symmetrical population $f(x)$,

$$\text{cov}\,(Q_3 - M, M - Q_1) \doteq -\frac{1}{4n}\left(\frac{1}{f_M} - \frac{1}{2f_Q}\right)^2 < 0,$$

and hence use (10.12) to show that the skewness measure in Exercise 3.27,

$$W = \frac{(Q_3 - M) - (M - Q_1)}{(Q_3 - M) + (M - Q_1)},$$

has large-sample variance

$$\text{var}\, W \doteq \frac{1}{8n(Q_3 - M)^2}\left(\frac{1}{f_Q^2} + \frac{2}{f_M^2} - \frac{2}{f_M f_Q}\right),$$

reducing to $1.839/n = (1.356)^2/n$ in the normal case.

10.25 Generalize the result of Exercise 10.24 to show that for any continuous distribution the large-sample variance of W is

$$\frac{1}{n(Q_3-Q_1)^4}\left\{\left[(M-Q_1)\left(\frac{1}{f_M}-\frac{1}{2f_{Q_3}}\right)+(Q_3-M)\left(\frac{1}{f_M}-\frac{1}{2f_{Q_1}}\right)\right]^2+\frac{1}{2}\left[\left(\frac{M-Q_1}{f_{Q_3}}\right)^2+\left(\frac{Q_3-M}{f_{Q_1}}\right)^2\right]\right\}.$$

10.26 Use (10.12), (10.9) and (10.10) to show that the large-sample variance of the skewness measure $b_1^{\frac{1}{2}}=m_3/m_2^{3/2}$ is

$$n^{-1}\left\{\frac{\mu_6}{\mu_2^3}-6\beta_2+9+\frac{\beta_1}{4}(9\beta_2+35)-\frac{3\mu_5\mu_3}{\mu_2^4}\right\},$$

reducing to $6/n$ in the normal case.

10.27 As in Exercise 10.26, show that the large-sample variance of the kurtosis measure $b_2=m_4/m_2^2$ is

$$n^{-1}\left\{\frac{\mu_8}{\mu_2^4}-\frac{4\mu_6\mu_4}{\mu_2^5}+4\beta_2^3-\beta_2^2+16\beta_2\beta_1-\frac{8\mu_5\mu_3}{\mu_2^4}+16\beta_1\right\},$$

reducing to $24/n$ in the normal case.

10.28 *Capture–Recapture* An initial set of n_1 individuals is selected (captured) from a population of (unknown) size N. These individuals are marked and returned to the population. A further group of n_2 individuals is selected, including r marked individuals. Assuming simple random sampling, show that the estimator for N, $\hat{N}_1=n_1n_2/r$, has infinite expectation when $n_1+n_2<N$ but that

$$\hat{N}_2=\frac{(n_1+1)(n_2+1)}{(r+1)}-1$$

has expected value $N-(N+1)p_0$, where $p_0=P(r=0)$. When n_1n_2/N is large, show that the approximate variance for $\hat{N}_2$ reduces to

$$\frac{N(N-n_1)(N-n_2)}{n_1n_2}.$$

10.29 Let x_1 and x_2 be two independent variates with means μ_1, μ_2 and variances that are $O(n^{-1})$. Use a normal approximation to evaluate

$$P(x_1/x_2\leqslant(\mu_1/\mu_2)+bn^{-1/2})$$

with mean μ_1/μ_2 and variance given by (10.17). Show that the probability statements for x_1/x_2 and x_2/x_1 agree only as far as terms of $O(n^{-1/2})$.

CHAPTER 11

EXACT SAMPLING DISTRIBUTIONS

11.1 The role of the sampling distribution in statistical inference has been indicated in Chapter 8. In the present chapter we propose to give an account of the main methods of finding such distributions when the population from which the sample was derived is specified. It will, as usual, be assumed that the sampling is simple random. Thus, for a distribution, $F(x)$, the joint distribution of n values $x_1, \ldots, x_n$ is $F(x_1)F(x_2) \ldots F(x_n)$; and if z is a statistic

$$z = z(x_1, \ldots, x_n) \tag{11.1}$$

the d.f. of z is given by

$$G(z_0) = \int \ldots \int dF(x_1) \ldots dF(x_n), \tag{11.2}$$

the integration being taken over the domain of the xs such that $z(x_1, \ldots, x_n) \leqslant z_0$.

Formally, (11.2) is the solution of our problem, which thus reduces to the purely mathematical one of evaluating certain multiple integrals or sums. The methods with which we are here concerned are fundamentally devices of various kinds to facilitate the integrative process. They fall into three main groups:

(a) straightforward evaluation of the integral (11.2) by ordinary analytical processes such as a convenient change of variables;
(b) the use of geometrical terminology to effect the same object and to avoid cumbrous analytical formulae; and
(c) the use of characteristic functions.

The analytical method

11.2 As an illustration of the straightforward analytical approach, let us find the distribution of the sum of squares of $n(>2)$ independent variables, each of which is distributed normally with zero mean and unit variance. The joint density of the n variables is then the product of n quantities of type $\frac{1}{\sqrt{(2\pi)}} e^{-\frac{1}{2}x^2}$, that is to say

$$f = \frac{1}{(2\pi)^{\frac{1}{2}n}} \exp\{-\tfrac{1}{2}(x_1^2 + x_2^2 + \ldots + x_n^2)\}. \tag{11.3}$$

We require the sampling distribution of

$$z = x_1^2 + x_2^2 + \ldots + x_n^2. \tag{11.4}$$

We have thus to evaluate the multiple integral

$$G(z_0) = \int \cdots \int \frac{1}{(2\pi)^{\frac{1}{2}n}} \exp\left(-\tfrac{1}{2}\sum x^2\right) dx_1 \ldots dx_n$$

over the domain of xs conditioned by $z \leqslant z_0$.

Make the polar transformation to variables $z, \theta_1, \theta_2, \ldots, \theta_{n-1}$

$$\left.\begin{aligned} x_1 &= z^{\frac{1}{2}} \cos\theta_1 \cos\theta_2 \ldots \cos\theta_{n-1}, \\ x_j &= z^{\frac{1}{2}} \cos\theta_1 \cos\theta_2 \ldots \cos\theta_{n-j} \sin\theta_{n-j+1}, \\ & \qquad j = 2, 3, \ldots, n-1; \quad n > 2, \\ x_n &= z^{\frac{1}{2}} \sin\theta_1. \end{aligned}\right\} \tag{11.5}$$

The Jacobian of this transformation is given by

$$\frac{\partial(x_1, \ldots, x_n)}{\partial(z, \theta_1, \ldots, \theta_{n-1})},$$

which is equal to $\frac{1}{2}z^{\frac{1}{2}n-1}$ times the determinant

$$\begin{vmatrix} \cos\theta_1 \cos\theta_2 \ldots \cos\theta_{n-1} & \cos\theta_1 \cos\theta_2 \ldots \cos\theta_{n-2} \sin\theta_{n-1} & \ldots & \sin\theta_1 \\ -\sin\theta_1 \cos\theta_2 \ldots \cos\theta_{n-1} & -\sin\theta_1 \cos\theta_2 \ldots \cos\theta_{n-2} \sin\theta_{n-1} & \ldots & \cos\theta_1 \\ -\cos\theta_1 \sin\theta_2 \ldots \cos\theta_{n-1} & -\cos\theta_1 \sin\theta_2 \ldots \cos\theta_{n-2} \sin\theta_{n-1} & \ldots & 0 \\ \ldots & \ldots & \ldots & \ldots \\ -\cos\theta_1 \cos\theta_2 \ldots \sin\theta_{n-1} & +\cos\theta_1 \cos\theta_2 \ldots \cos\theta_{n-1} & \ldots & 0 \end{vmatrix}$$

Taking out common factors in columns, we find that this determinant is equal to $\cos^{n-1}\theta_1 \cos^{n-2}\theta_2 \ldots \cos\theta_{n-1} \sin\theta_1 \sin\theta_2 \ldots \sin\theta_{n-1}$ multiplied by

$$\begin{vmatrix} 1 & 1 & 1 & \ldots & 1 \\ -\tan\theta_1 & -\tan\theta_1 & -\tan\theta_1 & \ldots & \cot\theta_1 \\ -\tan\theta_2 & -\tan\theta_2 & -\tan\theta_2 & \ldots & 0 \\ \ldots & \ldots & \ldots & \ldots & \ldots \\ -\tan\theta_{n-2} & -\tan\theta_{n-2} & \cot\theta_{n-2} & \ldots & 0 \\ -\tan\theta_{n-1} & \cot\theta_{n-1} & 0 & \ldots & 0 \end{vmatrix}$$

and, on subtracting each column from the preceding one, the first determinant is found to reduce to $\cos^{n-2}\theta_1 \cos^{n-3}\theta_2 \ldots \cos\theta_{n-2}$.

Thus our integral becomes

$$\int \cdots \int \frac{1}{(2\pi)^{\frac{1}{2}n}} e^{-\frac{1}{2}z} \tfrac{1}{2} z^{\frac{1}{2}n-1} \cos^{n-2}\theta_1 \ldots \cos\theta_{n-2}\, dz\, d\theta_1 \ldots d\theta_{n-1}. \tag{11.6}$$

The advantage of the transformation is that the limits of the variables are now much simpler, z itself can vary from 0 upwards, θ_{n-1} from 0 to 2π and the other θs from

$-\frac{1}{2}\pi$ to $\frac{1}{2}\pi$. Thus the integral (11.6) divides into a product of integrals, those in θ_j being constants, and we find for the distribution function of z

$$G(z_0) = k\int_0^{z_0} e^{-\frac{1}{2}z} z^{\frac{1}{2}n-1} dz. \tag{11.7}$$

The constant k may be evaluated by integration between 0 and ∞ and hence the distribution sought is

$$dG = \frac{1}{2^{\frac{1}{2}n}\Gamma(\frac{1}{2}n)} e^{-\frac{1}{2}z} z^{\frac{1}{2}n-1} dz, \quad 0 \leqslant z < \infty, \tag{11.8}$$

a Pearson Type III distribution, usually known as the χ^2 distribution, of which the parameter n in (11.8) is called the number of degrees of freedom.

Exercise 11.26 shows that (11.8) holds for $n=2$ also, as does Example 11.6 below; Example 11.16 covers all $n \geqslant 1$.

11.3 The essential feature of the change of variables is the simplification of the domain of integration as defined by the limits of the new variables. In general, we take the statistic whose sampling distribution is being sought as one of the new variables and choose $n-1$ others in any way convenient to the particular problem. Then, if J is the Jacobian of the transformation, namely

$$J = \frac{\partial(x_1, \ldots, x_n)}{\partial(z, \theta_1, \ldots, \theta_{n-1})},$$

the integral (11.2) becomes

$$G(z) = \int \ldots \int f(x_1) \ldots f(x_n) \frac{\partial(x_1, \ldots, x_n)}{\partial(z, \theta_1, \ldots, \theta_{n-1})} dz d\theta_1 \ldots d\theta_{n-1}, \tag{11.9}$$

$f(x_j)$ being the p.d.f. of the population and x_j being expressed in terms of z and the θs. The integration now takes place with respect to the θs, which can usually be chosen so as to vary between limits independent of z; and thus the integral (11.2) is replaced by more easily calculable definite integrals.

As always in such cases J is subject to an ambiguity of sign, which must be determined so as to make the transformed integral positive. The validity of the variate transformation depends on the familiar conditions governing the change of variable in a multiple integral. For example, it is a sufficient condition that the new variables and their first derivatives shall be continuous in the xs and that J does not change sign in the domain of integration. Some further examples will make the general procedure clear.

Example 11.1 (Sampling distribution of the Cauchy)

To find the distribution of the mean of a sample of n values $x_1, \ldots, x_n$ from the Cauchy distribution with p.d.f.

$$f(x) = \pi^{-1}(1+x^2)^{-1}, \quad -\infty < x < \infty.$$

The joint density is

$$\pi^{-n}\prod_{j=1}^{n}(1+x_j^2)^{-1} \tag{11.10}$$

and the statistic z is given by

$$nz=\sum_{j=1}^{n}x_j. \tag{11.11}$$

We have to integrate (11.10) over a domain of xs subject to $\Sigma x \leqslant nz$. Let us take new variables $u_1=x_1$, $u_2=x_2$, ..., $u_{n-1}=x_{n-1}$, and $u_n=\frac{1}{n}\sum_1^n x=z$. Here J is evidently equal to the constant n. Our new variables $u_1, \ldots, u_{n-1}$ may extend from $-\infty$ to $+\infty$ and the new variable u_n from $-\infty$ to z. We then have

$$G(z)=\int_{-\infty}^{z}du_n\int_{-\infty}^{\infty}\cdots\int_{-\infty}^{\infty}\frac{n}{\pi^n}\prod_{j=1}^{n-1}\frac{du_j}{(1+u_j^2)\{1+(nu_n-u_1-\ldots-u_{n-1})^2\}} \tag{11.12}$$

and the density function of z is given by the $(n-1)$-fold multiple integral in $u_1, \ldots, u_{n-1}$ in (11.12). This integral may be evaluated by step-by-step integration. We have

$$\frac{1}{(1+x^2)\{r^2+(a-x)^2\}}=\frac{1}{\{a^2+(r+1)^2\}\{a^2+(r-1)^2\}}\times$$
$$\left[\frac{2ax}{x^2+1}+\frac{a^2+r^2-1}{x^2+1}+\frac{2a^2-2ax}{r^2+(a-x)^2}+\frac{a^2-r^2+1}{r^2+(a-x)^2}\right]$$

whence, integrating with respect to x from $-\infty$ to $+\infty$, we find on the right

$$\frac{1}{\{a^2+(r-1)^2\}\{a^2+(r-1)^2\}}\left[a\log(x^2+1)-a\log\{r^2+(a-x)^2\}\right.$$
$$\left.+(a^2+r^2-1)\arctan x+\frac{a^2-r^2+1}{r}\arctan\frac{x-a}{r}\right]_{-\infty}^{\infty}$$

reducing to

$$\pi\left(\frac{r+1}{r}\right)\left\{\frac{1}{a^2+(r+1)^2}\right\}. \tag{11.13}$$

Thus in (11.12), taking $x=u_{n-1}$, $r=1$, $a=nu_n-u_1-\ldots-u_{n-2}$, we find that the $(n-1)$-fold integral reduces to

$$\int_{-\infty}^{\infty}\cdots\int_{-\infty}^{\infty}\frac{n}{\pi^{n-1}}\prod_{j=1}^{n-2}\frac{du_j}{(1+u_j^2)}\frac{2}{\{2^2+(nu_n-u_1-\ldots-u_{n-2})^2\}}.$$

Integrating with respect to $u_{n-2}, u_{n-3}, \ldots$ successively, we reduce this eventually to

$$\frac{n^2}{\pi\{n^2+(nu_n)^2\}}=\frac{1}{\pi(1+u_n^2)}. \tag{11.14}$$

Thus the density for z is given by

$$g(z)=\pi^{-1}(1+z^2)^{-1}, \qquad -\infty<z<\infty, \tag{11.15}$$

and is the same as that of a single observation.

This is an interesting example of the inapplicability of the Central Limit theorem, the mean of samples of n observations failing to tend to normality for large n since the moments of the distribution do not exist.

Exercise 11.25 gives an even more extreme instance of the failure of the tendency to normality for the stable distribution discussed in Example 4.13.

Example 11.2 (Sampling distribution for a linear function of normal variates)

To find the distribution of a linear function of n independent variables $x_1 \ldots, x_n$ where x_j is distributed normally with zero mean and variance σ_j^2.

Let the linear function be

$$z=a_1x_1+\ldots+a_nx_n. \tag{11.16}$$

Then by a transformation $u_j=x_j/\sigma_j$ we have

$$z=\sum a_j\sigma_ju_j \tag{11.17}$$

and u_j is now distributed with zero mean and unit variance. Our problem is thus equivalent to finding the distribution of a linear function of independent variables each of which is normally distributed with zero mean and unit variance.

Consider a linear transformation of type

$$v_i=\sum_{j=1}^{n} c_{ij}u_j \tag{11.18}$$

and let us determine the cs so that they are orthogonal, i.e. so that

$$\left.\begin{aligned}\sum_{i=1}^{n} c_{ij}c_{ik}&=0, \qquad j\neq k\\ &=1, \qquad j=k\end{aligned}\right\}. \tag{11.19}$$

This can be done in an infinity of ways, for (11.19) imposes $\frac{1}{2}n(n-1)+n$ conditions on n^2 constants. We then find

$$\sum_{i=1}^{n} v_i^2=\sum\left(\sum c_{ij}u_j\right)^2=\sum_{j=1}^{n} u_j^2.$$

Also, for the Jacobian of the transformation

$$J=\left|\frac{\partial v}{\partial u}\right|=|c_{ij}|.$$

If we multiply $|c_{ij}|$ by its transpose we find, in virtue of (11.19), a determinant whose elements are all zero except for units in the diagonals, namely a determinant that is equal to unity. Thus $J^2 = 1$ and $J = 1$. Since the us are independent standardized normal variates with distribution

$$dF = \frac{1}{(2\pi)^{\frac{1}{2}n}} \exp\left(-\tfrac{1}{2}\sum_{j=1}^{n} u_j^2\right) \prod_j du_j, \tag{11.20}$$

it follows that the vs are also, with

$$dF = \frac{1}{(2\pi)^{\frac{1}{2}n}} \exp\left(-\tfrac{1}{2}\sum_{j=1}^{n} v_j^2\right) \prod_j dv_j. \tag{11.21}$$

The distribution of the other variables v is then, from (11.21), seen to be independent of v_1 and we have, for the (marginal) distribution of v_1,

$$dF_1 = \frac{1}{\sqrt{(2\pi)}} \exp(-\tfrac{1}{2}v_1^2)\, dv_1. \tag{11.22}$$

Thus v_1 is distributed normally with unit variance and zero mean, independently of the other v_i. By symmetry, the same is true of any v_j. Now v_1 is an arbitrary linear function of the x_j. If we identify it with z of (11.16), we have the result that z is normally distributed with zero mean and variance obtained from (11.17) as

$$\operatorname{var} z = \sum a_j^2 \sigma_j^2. \tag{11.23}$$

Example 11.3 (Helmert's transformation)

The orthogonal transformation used in the previous example is especially useful for samples from a normal population, for the Jacobian is unity and as at (11.21) the density function remains of the same form. There is one such orthogonal transformation that has a particular interest. Let us put

$$\left.\begin{aligned} u_1 &= (x_1 - x_2)/\sqrt{2} \\ u_2 &= (x_1 + x_2 - 2x_3)/\sqrt{6} \\ u_3 &= (x_1 + x_2 + x_3 - 3x_4)/\sqrt{12} \\ \ldots & \quad \ldots \quad \ldots \\ u_{n-1} &= \{x_1 + x_2 + \ldots + x_{n-1} - (n-1)x_n\}/\sqrt{\{n(n-1)\}} \\ u_n &= (x_1 + x_2 + \ \ldots \ + x_n)/\sqrt{n}. \end{aligned}\right\} \tag{11.24}$$

It is readily verified that (11.19) is satisfied and the Jacobian of the transformation is unity. Thus, if the xs are all independent and normally distributed about zero mean with unit variance, so also are the us.

Now, for *any* orthogonal transformation with u_n as in (11.24),

$$\sum (x-\bar{x})^2 = \sum x^2 - n\bar{x}^2$$
$$= \sum_{i=1}^{n} u_i^2 - u_n^2$$
$$= \sum_{i=1}^{n-1} u_i^2.$$

Thus we see that the sum of squares of n standardized variates *measured from their mean* is distributed like the sum of squares of $n-1$ normal variates with zero mean. It follows from (11.8) that the distribution of

$$w = \sum (x-\bar{x})^2$$

is given by

$$dG = \frac{1}{2^{\frac{1}{2}(n-1)}\Gamma\{\frac{1}{2}(n-1)\}} e^{-\frac{1}{2}w} w^{\frac{1}{2}(n-3)} dw. \tag{11.25}$$

Moreover, this quantity w (a simple multiple of the sample variance) is independent of $\bar{x}$, for the distribution of the us is $k \exp\{-\frac{1}{2}(w+u_n^2)\}$, where $u_n = n^{-\frac{1}{2}}\bar{x}$. Thus in normal samples the mean is distributed independently of the variance. This method of proof was used by F. R. Helmert in 1876; an alternative proof using c.f.s is indicated in Exercise 11.18.

It makes no essential difference if the xs have common mean μ and variance σ^2; $\bar{x}$ is then replaced by $(\bar{x}-\mu)/\sigma$ and w by w/σ^2.

This interesting and important result characterizes the normal distribution; that is to say, if the mean and variance of random samples from a population are independent, the population must be normal. (Cf. Exercise 11.19.)

Order statistics

11.4 It may not always be possible to find variate transformations in which z, the statistic of interest, is one new variate and there are $(n-1)$ others whose domains of integration are not dependent on z. A case of particular interest is provided by the *order statistics*, the theory of which we shall consider in detail in Chapter 14. If we have a sample of n observations, $y_1, y_2, \ldots, y_n$, and rearrange them in ascending order of magnitude, renumbering them so that the suffix 1 is attached to the smallest, y_r is called the rth order statistic. The median and the quantiles are order statistics (the median, for example, being the $\frac{1}{2}(n+1)$th observation or midway between the $\frac{1}{2}n$th and the $\frac{1}{2}(n+2)$th); so also are the smallest value in the sample $(r=1)$ or the largest value $(r=n)$.

If we have a sample of n values $x_1, x_2, \ldots, x_n$ from a d.f. $F(x)$, their distribution is given by

$$dG = dF(x_1)dF(x_2) \ldots dF(x_n).$$

A transformation to

$$y_r = \int_{-\infty}^{x_r} \mathrm{d}F(t) = F(x_r) \qquad (11.26)$$

gives for the distribution of the ys

$$\mathrm{d}H = \mathrm{d}y_1 \mathrm{d}y_2 \dots \mathrm{d}y_n, \qquad (0 \leqslant y_i \leqslant 1, \quad \text{all } i). \qquad (11.27)$$

The ys and the xs are in the same order since y is a non-decreasing function of x.

If we now renumber the ys so that y_1 is the smallest, y_2 the next smallest and so on, the domain of variation is given by

$$0 \leqslant y_1 \leqslant y_2 \leqslant \dots \leqslant y_n \leqslant 1. \qquad (11.28)$$

If the d.f. is continuous, as we now assume, there is probability zero that any two y_i are equal.

To integrate (11.27) over this domain we require

$$\int_0^1 \dots \left[\int_0^{y_4} \left\{\int_0^{y_3} \left(\int_0^{y_2} \mathrm{d}y_1\right) \mathrm{d}y_2\right\} \mathrm{d}y_3\right] \dots \mathrm{d}y_n. \qquad (11.29)$$

This reduces, on step-by-step integration, to $1/n!$. For the *ordered* ys, then, the joint distribution is

$$\mathrm{d}K = n! \prod_{i=1}^{n} \mathrm{d}y_i, \qquad (11.30)$$

the constant obviously being the number of ways of ordering n observations.

Suppose now that we are interested in the (marginal) distribution of y_r, the rth smallest of the ys. To simplify the derivation slightly we make the further transformation

$$\left.\begin{aligned} z_i &= y_i, & i &= 1, 2, \dots, r \\ &= 1 - y_i, & i &= r+1, r+2, \dots, n \end{aligned}\right\} \qquad (11.31)$$

The Jacobian is unity and the distribution of the zs is accordingly

$$\mathrm{d}L = n! \prod_{i=1}^{n} \mathrm{d}z_i \qquad (11.32)$$

over the domain

$$0 \leqslant z_1 \leqslant z_2 \leqslant \dots \leqslant z_r \leqslant 1 - z_{r+1} \leqslant \dots \leqslant 1 - z_n \leqslant 1$$

which may be rewritten

$$0 \leqslant z_1 \leqslant z_2 \leqslant \dots \leqslant z_r; \quad 0 \leqslant z_n \leqslant z_{n-1} \leqslant \dots \leqslant z_{r+1} \leqslant 1 - z_r.$$

To find the distribution of z_r we integrate out the other variables in two separate blocks, as follows:

$$dM = dz_r n! \int_0^{z_r} \dots \int_0^{z_3} \int_0^{z_2} dz_1 dz_2 \dots dz_{r-1} \times \int_0^{1-z_r} \int_0^{z_{r+1}} \dots \int_0^{z_{n-1}} dz_n dz_{n-1} \dots dz_{r+1}$$

$$= dz_r n! \frac{z_r^{r-1}}{(r-1)!} \frac{(1-z_r)^{n-r}}{(n-r)!}$$

$$= \frac{n!}{(r-1)!(n-r)!} z_r^{r-1}(1-z_r)^{n-r} dz_r, \qquad 0 \leqslant z_r \leqslant 1. \tag{11.33}$$

Thus the distribution of z_r is a Beta distribution of the first kind as at (6.14). To obtain that of the rth order statistic x_r we simply put $F(x_r) = z_r$, obtaining

$$dN = \frac{n!}{(r-1)!(n-r)!} \{F(x_r)\}^{r-1}\{1-F(x_r)\}^{n-r} f(x_r) dx_r. \tag{11.34}$$

In the special case of the uniform distribution on (0, 1] with $F(x) = x$, $z_r = x_r$ and (11.33) and (11.34) are identical.

This expression could have been obtained more directly (cf. (10.25)), but we have given a full derivation to illustrate the processes involved.

Example 11.4 (Variance of order statistics)

From (11.33) and Example 2.8, we have the exact result

$$\text{var}(y_r) = r(n-r+1)/\{(n+1)^2(n+2)\},$$

since $y_r = z_r$ by (11.31). Writing np for r, this becomes

$$\text{var}(y_{np}) = np\{n(1-p)+1\}/\{(n+1)^2(n+2)\}.$$

From (11.26) and (10.14) we see that, as $n \to \infty$ with p fixed,

$$\text{var}(y_{np}) \doteq \{F'(x_{np})\}^2 \text{var}(x_{np}),$$

where $F'(x_{np})$ ($\neq 0$) is to be evaluated at the expectation of x_{np}, which is the corresponding population quantile ξ_p defined by $F(\xi_p) = p$. Thus

$$\text{var}(x_{np}) \doteq \{F'(\xi_p)\}^{-2} p(1-p)/n,$$

agreeing with (10.29). The median is the special case $p = \frac{1}{2}$. For the standardized normal distribution, $\xi_{0.5} = 0$ and $F'(0) = (2\pi)^{-\frac{1}{2}}$, so the variance of the sample median is $\pi/(2n)$, as we saw in effect in Example 10.7.

For the uniform distribution on (0, 1], $y_r = x_r$ and we see that the variance of the rth order statistic is a multiple of $r(n-r+1)$, and is therefore maximized at the median if n is odd and at the two central order statistics if n is even.

The geometrical method

11.5 A considerable amount of cumbrous analysis may often be avoided by the use of geometrical representation of the domain of integration. We may imagine the values $x_1, \ldots, x_n$ of any given sample as the coordinates of a point in an n-dimensional Euclidean hyperspace. The function $dF(x) \ldots dF(x_n)$ may then be regarded as the density at the point and the total probability between z_1 and z_2 will be the integral of this density in a region lying between the two loci $z(x_1, \ldots, x_n) = z_1$ and $z(x_1, \ldots, x_n) = z_2$, which in general will be hypersurfaces in the n-fold space, i.e. will themselves be spaces of $(n-1)$ dimensions. The d.f. of z will be the total probability between the hypersurface corresponding to $z = -\infty$ and that corresponding to a value z.

Example 11.5 (Distribution of the sample mean, geometric derivation)

Consider again the problem of Example 11.2. In the n-fold u-space the density is given by

$$\frac{1}{(2\pi)^{\frac{1}{2}n}} \exp\left(-\tfrac{1}{2} \sum u^2\right).$$

The statistic $z(= \Sigma a_j x_j)$ determines a hyperplane

$$z = \sum a_j \sigma_j u_j \tag{11.35}$$

and we have to find the total weight between this hyperplane and the corresponding hyperplane at $-\infty$, i.e. the weight on one side—the "lower" side—of the hyperplane (11.35).

Now $\sum u^2$ is the square of the distance of the point $u_1, \ldots, u_n$ from the origin and is therefore unchanged by any rotation of the coordinate axes. Choose a rotation that brings the axis of one variable perpendicular to the hyperplane (11.35), meeting it in Q. Let P be the sample point $u_1, \ldots, u_n$ and O the origin. Then

$$\sum u^2 = OP^2 = OQ^2 + QP^2,$$

so that the density at P is

$$\frac{1}{(2\pi)^{\frac{1}{2}n}} \exp(-\tfrac{1}{2} OQ^2) \exp(-\tfrac{1}{2} QP^2).$$

For variation over the hyperplane, OQ^2 is constant and the integral of $\exp(-\frac{1}{2}QP^2)$ is thus a constant independent of OQ. Hence the p.d.f. of z is given by

$$f(z) = k \exp(-\tfrac{1}{2} OQ^2),$$

k being some constant.

But OQ is the distance from O to the hyperplane and is given by

$$OQ^2 = \frac{z^2}{\sum a_j^2 \sigma_j^2}.$$

Hence

$$f(z) = k\exp\left\{-\frac{1}{2}\frac{z^2}{\sum a_j^2\sigma_j^2}\right\},$$

i.e. z is distributed normally with variance $a_j^2\sigma_j^2$ about zero mean.

The reader will find it instructive to compare this example with Example 11.2. They are, in effect, the same thing expressed in different language.

Example 11.6 (Distribution of the sample variance, geometric derivation)

Consider again the illustration of **11.2**. The elegance of the geometrical approach is well brought out by the analogous derivation of the result there obtained.

The density function, as before, is given by $k\exp(-\frac{1}{2}OP^2)$. We require the distribution of the statistic $z = OP^2$, and the density is obviously constant over the surface $z =$ constant, that is to say the surface of an n-dimensional hypersphere, where $n \geqslant 2$. The density function of z is then the integral of this constant density between the hyperspheres z and $z + dz$, i.e. is proportional to $\exp(-\frac{1}{2}OP^2)$ times the element of the volume of the hypersphere, which itself is proportional to the nth power of the radius OP. Thus we have

$$dF = k\exp(-\tfrac{1}{2}OP^2)\frac{d}{dz}OP^n dz$$

$$\propto e^{-\frac{1}{2}z}z^{\frac{1}{2}(n-2)}dz$$

giving, on evaluation of the constant,

$$dF = \frac{1}{2^{\frac{1}{2}n}\Gamma(\frac{1}{2}n)}e^{-\frac{1}{2}z}z^{\frac{1}{2}(n-2)}dz$$

as at (11.8).

Now suppose that the variates $x_1, \ldots, x_n$, while still being normally distributed with unit variance, are subjected to p independent homogeneous linear restrictions of type

$$a_1x_1 + a_2x_2 + \ldots + a_nx_n = 0.$$

In the n-space the variables x will then be constrained to lie on p hyperplanes. The first will cut the hypersphere of constant density in a hypersphere of one lower dimension, also, of course, of constant density; the second will cut this in a hypersphere of one lower dimension still, and so on. The result of the p linear restrictions will be to constrain the variables to a hypersphere of p fewer dimensions, and thus the distribution of z in these circumstances will be as before, but with $n - p$ instead of n, i.e.

$$dF = \frac{1}{2^{\frac{1}{2}(n-p)}\Gamma\{\frac{1}{2}(n-p)\}}e^{-\frac{1}{2}z}z^{\frac{1}{2}(n-p-2)}dz. \qquad (11.36)$$

Exercise 11.20 indicates a proof of (11.36) using c.f.s.

Example 11.7 (The sampling distribution of the mean and variance in normal samples)

Writing $\bar{x}$ for the mean of a sample, we have, for the variance s^2,

$$s^2 = \frac{1}{n}\sum(x-\bar{x})^2$$
$$= \frac{1}{n}\sum x^2 - \bar{x}^2.$$

In samples from a normal population with zero mean and unit variance the density at the point $x_1, \ldots, x_n$ is proportional to

$$\exp\left(-\tfrac{1}{2}\sum x^2\right) = \exp\{-\tfrac{1}{2}(ns^2 + n\bar{x}^2)\}. \tag{11.37}$$

Let us find the sampling distributions of s and $\bar{x}$. From (11.37) it is seen that the density function can be expressed simply in terms of those quantities, and we then have to find some transformation of the volume element $dx_1, \ldots, dx_n$. We have considered this from the analytical viewpoint in Example 11.3. Let us now consider it geometrically.

In the n-space consider the unit vector whose direction cosines are $(n^{-\frac{1}{2}}, n^{-\frac{1}{2}}, \ldots, n^{-\frac{1}{2}})$, say OQ where O is the origin. If P is the sample point, let PM be the perpendicular from P on to OQ. Then the length of OM is

$$\frac{x_1}{\sqrt{n}} + \frac{x_2}{\sqrt{n}} + \ldots + \frac{x_n}{\sqrt{n}} = \bar{x}\sqrt{n}.$$

The length of OP is $\sqrt{\sum x^2}$. Thus the length of PM is $(\Sigma x^2 - n\bar{x}^2)^{\frac{1}{2}} = s\sqrt{n}$.

The element of volume at P may be regarded as the product of an elemental increment in OM, equal to $d\bar{x}\sqrt{n}$, and the elemental volume in the perpendicular hyperplane through M. In the hyperplane the contours of equal density, as in the last example, are hyperspheres of radius $s\sqrt{n}$ centred at M, and consequently the element of volume is equal to $k\,d\bar{x}s^{n-2}ds$ multiplied by other elements that need not concern us since they are independent of $\bar{x}$ and s. We then have

$$dF \propto \exp\{-\tfrac{1}{2}(ns^2 + n\bar{x}^2)\}s^{n-2}d\bar{x}ds, \tag{11.38}$$

and this splits into two factors

$$dF_1 \propto \exp(-\tfrac{1}{2}n\bar{x}^2)d\bar{x}, \tag{11.39}$$
$$dF_2 \propto \exp(-\tfrac{1}{2}ns^2)s^{n-2}ds. \tag{11.40}$$

Thus, as we have seen, in samples from a normal population the distributions of mean and variance are independent. We have from (11.40), for the distribution of s^2,

$$dF \propto \exp(-\tfrac{1}{2}ns^2)s^{n-3}ds^2$$

and, on evaluation of the constant,

$$dF = \frac{n^{\frac{1}{2}(n-1)}}{2^{\frac{1}{2}(n-1)}\Gamma\{\frac{1}{2}(n-1)\}}\exp(-\tfrac{1}{2}ns^2)(s^2)^{\frac{1}{2}(n-3)}d(s^2), \qquad 0 \leqslant s^2 < \infty, \tag{11.41}$$

which is (11.25) again, with $w = ns^2$.

It is interesting to compare this with the distribution of the previous example, in which we found the distribution of the sum of squares of the variables *measured from the population mean*. In this case we have found the distribution of $1/n$ times the sum of the squares *measured from the sample mean*. In its equivalent form (11.25), (11.41) is simply (11.36) with a single linear constraint ($p=1$), as we observed in Example 11.3.

Example 11.8 (Student's t-distribution)

In the previous example we have

$$\frac{\bar{x}\sqrt{n}}{s\sqrt{N}} = \frac{OM}{PM} = \cot\phi,$$

where ϕ is the angle POM.

If, then, we define a statistic $z=\bar{x}/s$, z will be constant over the cone obtained by rotating OP about the unit vector, keeping the angle ϕ constant. The distribution of z will then be given by determining the weight between the cones defined by ϕ and $\phi+d\phi$.

Consider the intersection of these cones with the hypersphere of radius OP. They will cut off an annulus on the sphere whose "content" (the n-dimensional analogue of volume) will be proportional to

$$OP\, d\phi.\, PM^{n-2} = OP^{n-1}\sin^{n-2}\phi\, d\phi.$$

The density function is constant and proportional to $\exp\left(-\tfrac{1}{2}OP^2\right)$ on the hypersphere, and thus the total density between the cones is proportional to

$$\int_0^\infty e^{-\frac{1}{2}OP^2}OP^{n-1}\sin^{n-2}\phi\, d\phi\, d(OP)$$
$$\propto \sin^{n-2}\phi\, d\phi, \qquad 0 \leqslant \phi \leqslant \pi.$$

The distribution of $z(=\cot\phi)$ is then given by

$$dF \propto \frac{k\,dz}{(1+z^2)^{\frac{1}{2}n}},$$

or, on evaluation of the constant,

$$dF = \frac{1}{B\{\frac{1}{2}(n-1), \frac{1}{2}\}}\,\frac{dz}{(1+z^2)^{\frac{1}{2}n}}.$$

Since z is the ratio of two functions of unit dimension in the variables, this distribution holds for samples from a normal population irrespective of the scale, that is to say, irrespective of the variance of the population.

The distribution is usually put in a slightly different form. Put

$$t = \frac{\bar{x}n^{\frac{1}{2}}}{\left\{\frac{1}{n-1}\sum(x-\bar{x})^2\right\}^{\frac{1}{2}}} = (n-1)^{\frac{1}{2}}z. \qquad (11.42)$$

In terms of t, we have

$$dF = \frac{1}{(n-1)^{\frac{1}{2}}B\{\frac{1}{2}(n-1), \frac{1}{2}\}} \frac{dt}{\left(1+\frac{t^2}{n-1}\right)^{\frac{1}{2}n}} \tag{11.43}$$

$$= \frac{\Gamma\{\frac{1}{2}(\nu+1)\}}{(\nu\pi)^{\frac{1}{2}}\Gamma(\frac{1}{2}\nu)} \frac{dt}{(1+t^2/\nu)^{\frac{1}{2}(\nu+1)}} \tag{11.44}$$

where $\nu = n-1$.

This celebrated expression is known as Student's t-distribution after the pen-name of its discoverer (1908a). It will be discussed in detail in **16.10–14**.

Example 11.9 (Distribution of the mean of samples from a uniform distribution (Hall, 1927))

Consider now a sample of n values from the uniform distribution

$$dF = dx, \quad 0 < x \leqslant 1.$$

In the n-space the density function will be a constant everywhere inside a hypercube

$$0 \leqslant x_j \leqslant 1, \quad j = 1, \ldots, n \tag{11.45}$$

and zero elsewhere. The unit vector will be along the diagonal of this cube. If P is the sample point $(x_1, \ldots, x_n)$ and PM the perpendicular on to this diagonal, then, as shown in Example 11.7, $OM = \bar{x}\sqrt{n}$. Thus, for the distribution of $\bar{x}$ we require the element of weight (which in this case is proportional to the element of volume) between the hyperplanes $\bar{x}$ and $\bar{x} + d\bar{x}$; and this is equivalent to finding the content of the hyperplane (its "area") cut off by the various faces of the hypercube. The complication of the problem arises from the fact that as $\bar{x}$ increases this region changes its shape according to the number of edges of the hypercube cut by the hyperplane.

Consider the "quadrants"

$$\left.\begin{array}{l} x_j \geqslant r_j \\ r_j = 0 \text{ or } 1 \end{array}\right\} j = 1, 2, \ldots, n, \tag{11.46}$$

whose corners are the corners of the hypercube. Any one of the corners may have 0 or 1 or 2 ... or n of its coordinates equal to unity and the rest zero. We divide the quadrants into $(n+1)$ sets according as the corner has 0, 1, ..., n of its coordinates equal to unity, that is, according as

$$r = \sum_{j=1}^{n} r_j$$

is equal to 0, 1, ..., n. A quadrant of the tth set may be called Q_t. There will be $\binom{n}{t}$ different Q_ts. Let S be any point of Q_0, i.e. any point whose coordinates are all $\geqslant 0$, and let just s of its coordinates be $\geqslant 1$. Then S will belong to just $\binom{s}{0}$, $= 1$, Q_0; $\binom{s}{1}$ Q_1s; $\binom{s}{2}$ Q_2s, and so on. Now if $s > 0$,

$$\sum_{t=0}^{s} (-1)^t \binom{s}{t} = (1-1)^s = 0. \tag{11.47}$$

Hence, if whenever a point belongs to a Q_t we give it a density $(-1)^t$ and then sum over all Q, the resultant density will be 1 or 0 according as the point belongs to the hypercube or not.

Let the segment of the hyperplane

$$z = \sum x \tag{11.48}$$

lying in Q_0 have content $V_n(z)$. Then the segment lying in a member of (11.46) will have content $V_n(z-r)$ which is zero if $r \geqslant z$. Further, the segment of (11.48) lying in any member of (11.46) will have the content

$$\sum_{r=0}^{k} (-1)^r \binom{n}{r} V_n(z-r) \tag{11.49}$$

where $k = [z]$, = the greatest integer less than z.

To find $V_n(z)$, let $\bar{V}_{n-1}(z)$ be the projection of $V_n(z)$ perpendicular to one of the axes, so that

$$V_n(z) = n^{\frac{1}{2}} \bar{V}_{n-1}(z).$$

Now $\bar{V}_n(z)$ is the content of the n-dimensional region bounded by (11.48) and the coordinate hyperplanes—a region whose base is therefore of content $V_n(z)$. The perpendicular from O to this base is $z/n^{\frac{1}{2}}$. Hence

$$\bar{V}_n(z) = \frac{1}{n}\left(\frac{z}{n^{\frac{1}{2}}}\right) V_n(z)$$

and

$$\bar{V}_{n-1}(z) = \frac{1}{n-1}\left(\frac{z}{(n-1)^{\frac{1}{2}}}\right) V_{n-1}(z)$$

or

$$V_n(z) = \frac{1}{n-1}\left(\frac{n}{n-1}\right)^{\frac{1}{2}} z V_{n-1}(z). \tag{11.50}$$

Since $V_2(z) = z\sqrt{2}$ repeated applications of this formula give

$$V_n(z) = \frac{n^{\frac{1}{2}}}{(n-1)!} z^{n-1}.$$

Substituting in (11.49) we find for the content of the region common to the hypercube and the hyperplane

$$f(z) = \frac{n^{\frac{1}{2}}}{(n-1)!} \sum_{r=0}^{k} (-1)^r \binom{n}{r} (z-r)^{n-1} \tag{11.51}$$

for values of z between k and $k+1$.

Since

$$\int_0^n f(z) \frac{dz}{\sqrt{n}} = 1$$

the distribution of the mean $m = z/n$ is given by

$$g(m) = \frac{n^n}{(n-1)!} \sum_{r=0}^{k} (-1)^r \binom{n}{r} \left(m - \frac{r}{n}\right)^{n-1}, \quad \frac{k}{n} \leqslant m \leqslant \frac{k+1}{n}. \tag{11.52}$$

This is the required distribution. It is unusual in consisting of n arcs of degree $(n-1)$ in m, having $(n-1)$-point contact at their joins, that is at the points k/n $(k = 1, 2, \ldots, n)$.

The distribution is symmetrical since the hyperplane $z =$ constant is perpendicular to the unit vector, which itself is an axis of symmetry of the hypercube.

For particular values $n = 2, 3, 4$, (11.52) gives the following results:

$$
\begin{aligned}
n=2: &\quad 4m, && 0 \leqslant m \leqslant \tfrac{1}{2},\\
&\quad 4(1-m), && \tfrac{1}{2} \leqslant m \leqslant 1.\\
n=3: &\quad \frac{27m^2}{2}, && 0 \leqslant m \leqslant \tfrac{1}{3},\\
&\quad \frac{27\{m^2-3(m-\frac{1}{3})^2\}}{2}, && \tfrac{1}{3} \leqslant m \leqslant \tfrac{2}{3},\\
&\quad \frac{27(1-m)^2}{2}, && \tfrac{2}{3} \leqslant m \leqslant 1.\\
n=4: &\quad \frac{128m^3}{3}, && 0 \leqslant m \leqslant \tfrac{1}{4},\\
&\quad \frac{128\{m^3-4(m-\frac{1}{4})^3\}}{3}, && \tfrac{1}{4} \leqslant m \leqslant \tfrac{1}{2},\\
&\quad \frac{128\{(1-m)^3-4(\frac{3}{4}-m)^3\}}{3}, && \tfrac{1}{2} \leqslant m \leqslant \tfrac{3}{4},\\
&\quad \frac{128(1-m)^3}{3}, && \tfrac{3}{4} \leqslant m \leqslant 1.
\end{aligned}
$$

For $n = 2$ the distribution is "triangular." For larger n it approaches normality in virtue of the Central Limit theorem, and the approximation is good even at $n = 3$, as the reader should verify graphically. Stephens (1966) and Buckle *et al.* (1969) give percentiles of the distribution (11.52) for $n \leqslant 30$.

The method of characteristic functions

11.6 It has already been seen in **7.28** that the c.f. of the sum of n independent variables is the product of their c.f.s. This simple property enables us to find the sampling distribution of a wide class of statistics that are expressible as sums, and particularly of the mean.

If we have a sample of n independent values from a population whose c.f. is $\phi(t)$, the c.f. of their sum is $\{\phi(t)\}^n$. Thus the d.f. of their sum z is given by $F(z)$ where (cf. (4.4) and (4.33))

$$F(z) - F(0) = \frac{1}{2\pi}\int_{-\infty}^{\infty} \frac{1-e^{-itz}}{it}\phi^n(t)\,dt \tag{11.53}$$

and the density function is

$$f(z) = \frac{1}{2\pi}\int_{-\infty}^{\infty} e^{-itz}\phi^n(t)\,dt. \tag{11.54}$$

The c.f. of the mean is, by (4.32),

$$E\left\{\exp\left[it\sum_{j=1}^{n} x_j/n\right]\right\} = E\left\{\exp\left[(it/n)\sum_{j=1}^{n} x_j\right]\right\} = \left\{\phi\left(\frac{t}{n}\right)\right\}^n,$$

and this may be inverted as at (11.53–4). By the Inversion Theorem, it is sufficient to recognize the form of the c.f. and write down the corresponding f.f.

The following examples will illustrate the power of these results.

Example 11.10 (Distribution of the sample mean for the binomial distribution)

The c.f. of the binomial distribution is, by (5.3),

$$\phi(t) = (q + pe^{it})^n.$$

The c.f. of the sampling distribution of the sum of m such binomial variables is then

$$\phi^m(t) = (q + pe^{it})^{mn} \tag{11.55}$$

and that of the distribution of their mean is $(q + pe^{it/m})^{mn}$. But this is the c.f. of the binomial distribution with mn replacing n, the interval of the variate being $1/m$ instead of unity; and hence this is the distribution of the mean.

Example 11.11 (Distribution of the sample mean for the Poisson distribution)

The c.f. of the Poisson distribution whose general term is $e^{-\lambda}\lambda^r/r!$ is, by (5.19),

$$\phi(t) = \exp\{\lambda(e^{it} - 1)\}.$$

The c.f. of the mean of n such variates is then

$$\phi^n\left(\frac{t}{n}\right) = \exp\{n\lambda(e^{it/n} - 1)\}$$

and hence the distribution of the mean is the Poisson distribution whose general term is

$$e^{-n\lambda}\frac{(n\lambda)^r}{r!}, \tag{11.56}$$

the interval of the variate being $1/n$ instead of unity.

Example 11.12 (Distribution of the sample mean for the normal distribution)

The c.f. of the normal distribution

$$dF = \frac{1}{\sigma\sqrt{(2\pi)}}\exp\left\{-\frac{1}{2}\left(\frac{x-\mu}{\sigma}\right)^2\right\}dx$$

is, by (5.96),

$$\phi(t) = \exp(it\mu - \tfrac{1}{2}t^2\sigma^2).$$

The c.f. of the distribution of the mean of n values is then

$$\phi^n\left(\frac{t}{n}\right)=\exp\left\{n\left(\frac{it\mu}{n}-\frac{1}{2}\frac{t^2\sigma^2}{n^2}\right)\right\}=\exp\left(it\mu-\frac{1}{2}\frac{t^2\sigma^2}{n}\right). \tag{11.57}$$

This is the c.f. of a normal distribution with mean μ and variance σ^2/n, which is therefore the distribution required.

Example 11.13 (Distribution of the sample mean for the Gamma distribution)
The c.f. of the distribution

$$dF=\frac{1}{\Gamma(\gamma)}e^{-x/a}\left(\frac{x}{a}\right)^{\gamma-1}\frac{dx}{a}, \qquad a>0,$$

is (Example 3.6)

$$\phi(t)=(1-ita)^{-\gamma}.$$

The c.f. of the distribution of the mean n values is then

$$\phi^n\left(\frac{t}{n}\right)=(1-ita/n)^{-n\gamma}.$$

This is the c.f. of the distribution

$$dF=\frac{1}{\Gamma(\gamma n)}e^{-nx/a}\left(\frac{nx}{a}\right)^{n\gamma-1}\frac{n\,dx}{a}. \tag{11.58}$$

Example 11.14 (Distribution of the sample mean for the uniform distribution)
The c.f. of the distribution $dF=dx$ $(0<x\leqslant 1)$ is

$$\phi(t)=\int_0^1 e^{itx}\,dx=\frac{e^{it}-1}{it}.$$

The c.f. of the mean of n values $\phi^n\left(\frac{t}{n}\right)=\left(\frac{e^{it/n}-1}{it/n}\right)^n$, and the density function is thus

$$f(x)=\frac{1}{2\pi}\int_{-\infty}^{\infty}e^{-itx}\left(\frac{e^{it/n}-1}{it/n}\right)^n dt. \tag{11.59}$$

This integral is everywhere analytic and the range of integration may be changed to the contour Γ consisting of the real axis from $-\infty$ to $-c$, the small semicircle of radius c and centre at the origin, and the real axis from c to ∞. Thus

$$f(x)=\frac{1}{2\pi}\int_\Gamma e^{-itx}\left(\frac{e^{it/n}-1}{it/n}\right)^n dt=\frac{(-1)^n}{2\pi}\int_\Gamma e^{-itx}\sum_{j=0}^{n}(-1)^j\binom{n}{j}\frac{e^{itj/n}}{(it/n)^n}dt. \tag{11.60}$$

Now

$$\int_{\Gamma}\frac{e^{igz}}{z}dz=0 \quad \text{if} \quad g>0$$

$$= -2\pi i^{n}\frac{g^{n-1}}{(n-1)!} \quad \text{if} \quad g\leqslant 0.$$

This may be seen by integrating along a contour consisting of Γ and the infinite semicircle *above* the real axis if $g>0$ and *below* it if $g\leqslant 0$.

Substituting in (11.60) we find

$$f(x)=\frac{(-1)^{n}}{2\pi}\int_{\Gamma}\sum(-1)^{j}\binom{n}{j}\frac{e^{itj/n-itx}}{(it/n)^{n}}dt$$

$$=\frac{(-1)^{n-1}}{(n-1)!}\sum_{j\leqslant nx}(-1)^{j}\binom{n}{j}\frac{\left(\frac{j}{n}-x\right)^{n-1}}{\left(\frac{1}{n}\right)^{n}}$$

$$=\frac{n^{n}}{(n-1)!}\sum_{j=0}^{[nx]}(-1)^{j}\binom{n}{j}\left(x-\frac{j}{n}\right)^{n-1}.$$

This, with a few changes of notation, is the same as (11.52). The result was given by Irwin (1927) in this form, but is traceable as far back as Lagrange.

Gray and Odell (1966) obtain the distribution of any linear function of uniform variates with different ranges. Marsaglia (1965) gives the d.f. of the ratio of sums of uniform variates.

11.7 General expressions may also be derived for the distributions of geometric means and of the moments about fixed points.

In fact, if $y=\log x$, the c.f. of y is

$$\phi(t)=\int_{-\infty}^{\infty}e^{it\log x}\,dF=\int_{-\infty}^{\infty}x^{it}\,dF.$$

The d.f. of the sum of n independent values of y, say nz, is then given by (11.53) as

$$F(nz)-F(0)=\frac{1}{2\pi}\int_{-\infty}^{\infty}\frac{1-e^{-itnz}}{it}\phi^{n}(t)\,dt, \tag{11.61}$$

and the distribution of the mean is that of z. But $z=\log u$, where u is the geometric mean, and hence the distribution of u may be found.

The density function, when it exists, is, by (11.54),

$$f(nz)=\frac{1}{2\pi}\int_{-\infty}^{\infty}e^{-itnz}\phi^{n}(t)\,dt.$$

Similarly the c.f. of a power of the variate, say x^r, is given by

$$\phi(t)=\int_{-\infty}^{\infty}\exp(itx^r)\,dF$$

and thus the d.f. of the rth moment about zero, say z, by

$$F(nz)-F(0)=\frac{1}{2\pi}\int_{-\infty}^{\infty}\frac{1-e^{itnz}}{it}\phi^n(t)\,dt. \tag{11.62}$$

Example 11.15 (Distribution of the geometric mean in samples from a uniform distribution)

For the distribution

$$dF=dx/a,\qquad 0<x\leqslant a,$$

the c.f. of $\log x$ is

$$\int_0^a x^{it}\frac{dx}{a}=\frac{a^{it}}{1+it}.$$

The density function of $u=\Sigma\log x$ is then given by (11.54) as

$$\begin{aligned}f(u)&=\frac{1}{2\pi}\int_{-\infty}^{\infty}\frac{e^{-itu}a^{nit}}{(1+it)^n}dt\\&=\frac{1}{2\pi}\int_{-\infty}^{\infty}\frac{e^{it(n\log a-u)}}{(1+it)^n}dt,\qquad n\log a-u\geqslant 0.\end{aligned}$$

This integral may be evaluated by contour integration and we find

$$f(u)=\frac{(n\log a-u)^{n-1}e^{-(n\log a-u)}}{\Gamma(n)},$$

whence, putting $z=e^{u/n}$, we find for the distribution of the geometric mean z

$$f(z)=\frac{n^n z^{n-1}}{a^n\Gamma(n)}\left(\log\frac{a}{z}\right)^{n-1}. \tag{11.63}$$

The c.f. of u is at once seen to be $a^{nit}(1+it)^{-n}$, so that of $(n\log a-u)$ is simply $(1-it)^{-n}$, the c.f. of the Gamma distribution (6.18) with $\lambda=n$, which we found more generally at (3.19).

Example 11.16 (Distribution of the second moment about the population mean in samples from a normal distribution)

If the distribution is

$$dF=\frac{1}{\sigma\sqrt{(2\pi)}}\exp(-\tfrac{1}{2}x^2/\sigma^2)\,dx$$

the c.f. of x^2 is

$$\phi(t) = \frac{1}{\sigma\sqrt{(2\pi)}} \int_{-\infty}^{\infty} \exp(itx^2)\exp(-\tfrac{1}{2}x^2/\sigma^2)\,dx$$
$$= (1-2\sigma^2 it)^{-\frac{1}{2}}.$$

The c.f. of the mean of n such values of x^2, say m_2, is then

$$\phi^n\left(\frac{t}{n}\right) = (1-2\sigma^2 it/n)^{-\frac{1}{2}n} \tag{11.64}$$

and comparison of this with (3.19) in Example 3.6 shows that (11.64) is the c.f. of the distribution

$$dF = \frac{n^{\frac{1}{2}n}}{(2\sigma^2)^{\frac{1}{2}n}\Gamma(\frac{1}{2}n)} \exp(-nm_2/2\sigma^2) m_2^{\frac{1}{2}n-1}\,dm_2, \tag{11.65}$$

a result which may be compared with that of (11.8), to which it reduces on writing $z = nm_2/\sigma^2$.

Our present proof holds for any $n \geqslant 1$, whereas (11.8) was deduced analytically for $n > 2$ and the geometrical derivation of Example 11.6 was valid for $n \geqslant 2$.

The distribution of a sum or difference of independent random variables

11.8 The distribution of the sum of two independent variates may be obtained directly as follows. Their joint distribution is

$$dF(x_1, x_2) = dF_1(x_1)\,dF_2(x_2) = f_1(x_1)f_2(x_2)\,dx_1\,dx_2.$$

The transformation $z = x_1 + x_2$, $y = x_2$, has Jacobian equal to 1, and gives

$$dG(z, y) = f_1(z-y)f_2(y)\,dz\,dy. \tag{11.66}$$

Integrating out y, we obtain, for the marginal distribution of z,

$$dH(z) = \left\{\int_{-\infty}^{\infty} f_1(z-y)f_2(y)\,dy\right\} dz$$

or for the p.d.f of z,

$$h(z) = \int_{-\infty}^{\infty} f_1(z-y)f_2(y)\,dy. \tag{11.67}$$

The d.f. of z is obtained by integrating $h(t)$ from $-\infty$ to z, giving,

$$H(z) = \int_{-\infty}^{\infty} F_1(z-y)f_2(y)\,dy. \tag{11.68}$$

Note that if $x_1 \geqslant 0$, y is to be integrated from $-\infty$ to z; if x_2 also is $\geqslant 0$, y is integrated from 0 to z.

If we are interested in the difference $z = x_1 - x_2$, rather than their sum, the reader will see by retracing the argument that (11.67) is changed only in that $(z-y)$ is replaced by $(z+y)$. In this case, even if x_1 and x_2 are non-negative, z is not.

(11.67) holds for discrete variates (with integration replaced by summation as usual) although no Jacobian enters into the transformation to (11.66), as Example 11.17 shows.

Example 11.17 (Distribution of the sum of two Poisson variates)

Consider the distribution of the sum of two independent Poisson variates. Their joint distribution is

$$f(x_1, x_2) = \mathrm{e}^{-\lambda_1}\frac{\lambda_1^{x_1}}{x_1!}\,.\,\mathrm{e}^{-\lambda_2}\frac{\lambda_2^{x_2}}{x_2!}, \quad x_1, x_2 = 0, 1, 2, \ldots.$$

(11.67) becomes

$$\begin{aligned} h(z) &= \sum_{y=0}^{z} \mathrm{e}^{-\lambda_1}\frac{\lambda_1^{z-y}}{(z-y)!}\,.\,\frac{\mathrm{e}^{-\lambda_2}\lambda_2^{y}}{y!} \\ &= \mathrm{e}^{-(\lambda_1+\lambda_2)}\lambda_1^{z}\sum_{y=0}^{z}\frac{(\lambda_2/\lambda_1)^y}{(z-y)!\,y!} \\ &= \frac{\mathrm{e}^{-(\lambda_1+\lambda_2)}\lambda_1^{z}}{z!}\sum_{y=0}^{z}\binom{z}{y}\left(\frac{\lambda_2}{\lambda_1}\right)^y \\ &= \mathrm{e}^{-(\lambda_1+\lambda_2)}\lambda_1^{z}\left(1+\frac{\lambda_2}{\lambda_1}\right)^z\Big/ z! = \mathrm{e}^{-(\lambda_1+\lambda_2)}(\lambda_1+\lambda_2)^z/z!. \end{aligned}$$

Thus the sum is also distributed in the Poisson form, with parameter equal to the sum of the two components' parameters. It follows at once that the sum of n independent Poisson variates is a Poisson, parameters being additive. Cf. the less general result of Example 11.11.

The difference between two independent Poisson variates cannot be a Poisson, since it can take negative values. If the two Poissons have the same parameter $\lambda = a$, say, their difference will have all odd cumulants zero and even cumulants equal to $2a$. Its distribution was given in Example 4.5.

(11.67) can be used to obtain successively the distribution of the sum of any number of variables whose individual distributions are known. If all the variables have the same distribution, the general form may be suggested when the results for two or three variates have been worked out. Its correctness can then be verified by induction. The following example illustrates the method.

Example 11.18 (Distribution of the sum of two sums of squares)

In Example 11.6 we found that the distribution of the sum of squares of n independent standardized normal variates is given by

$$\mathrm{d}F = \frac{1}{2^{\frac{1}{2}n}\Gamma(\frac{1}{2}n)}\mathrm{e}^{-\frac{1}{2}z}z^{\frac{1}{2}n-1}\,\mathrm{d}z. \tag{11.69}$$

Suppose we had surmised this form from an examination of a few cases for low n. Let x be another variate independently normally distributed about zero mean with unit variance, and $x^2 = v$. Then v has the distribution

$$dF = \frac{1}{2^{\frac{1}{2}}\Gamma(\frac{1}{2})} e^{-\frac{1}{2}v} v^{-\frac{1}{2}} dv, \qquad 0 \leqslant v < \infty.$$

Then, from (11.67) the density function of the distribution of $u = z + v$ is given, since z and v are both non-negative, by

$$\int_0^u \frac{1}{2^{\frac{1}{2}}\Gamma(\frac{1}{2})} e^{-\frac{1}{2}(u-y)} (u-y)^{-\frac{1}{2}} \frac{1}{2^{\frac{1}{2}n}\Gamma(\frac{1}{2}n)} e^{-\frac{1}{2}y} y^{\frac{1}{2}(n-2)} dy$$

$$= \frac{e^{-\frac{1}{2}u}}{2^{\frac{1}{2}(n+1)}\Gamma(\frac{1}{2})\Gamma(\frac{1}{2}n)} \int_0^u (u-y)^{-\frac{1}{2}} y^{\frac{1}{2}(n-2)} dy$$

$$= \frac{B(\frac{1}{2}, \frac{1}{2}n)}{2^{\frac{1}{2}(n+1)}\Gamma(\frac{1}{2})\Gamma(\frac{1}{2}n)} e^{-\frac{1}{2}u} u^{\frac{1}{2}(n-1)}$$

$$= \frac{1}{2^{\frac{1}{2}(n+1)}\Gamma\{\frac{1}{2}(n+1)\}} e^{-\frac{1}{2}u} u^{\frac{1}{2}(n-1)}$$

which is the same as (11.69) with $n+1$ for n. Hence the distribution holds generally.

The distribution of a ratio or product of independent random variables

11.9 Cases sometimes arise in which we require the sampling distribution of the ratio of two independent variates, x_1 and x_2. Their joint distribution is, as before,

$$dF(x_1, x_2) = dF_1(x_1) dF_2(x_2) = f_1(x_1) f_2(x_2)\, dx_1\, dx_2. \tag{11.70}$$

We now transform to new variables

$$u = x_1/x_2,$$
$$v = x_2,$$

and the Jacobian of the transformation is found to be

$$J = v. \tag{11.71}$$

Care is necessary here to ensure that the change of variable obeys the validity conditions outlined in **11.3**. Essentially, we must ensure that $J \neq 0$ within the domain of integration, although zero values on the boundary of the domain do not affect the validity of the transformation. From (11.71) we see that this implies

$$x_2 = v \neq 0 \tag{11.72}$$

except on the boundary. Thus we must in the first place restrict our discussion to ratios of random variables such that the *denominator* can range *either* from 0 to ∞, *or* from $-\infty$ to 0, but not both. In this case we have, from (11.70) and (11.71),

$$dG(u, v) = f_1(uv) f_2(v) v\, du\, dv, \tag{11.73}$$

and thus the marginal distribution of u is

$$dH(u) = \left\{\int_{-\infty}^{\infty} f_1(uv)f_2(v)v\,dv\right\}du, \tag{11.74}$$

and its d.f.

$$H(u) = \int_{-\infty}^{\infty} F_1(uv)f_2(v)\,dv. \tag{11.75}$$

Example 11.19 (Distribution of the ratio of the sample mean and standard deviation)

Consider again the distribution of the ratio $\bar{x}/s$ discussed in Example 11.8. Here $\bar{x}$ is the mean of samples of n from a normal population and is thus distributed as

$$dF_1 \propto \exp(-n\bar{x}^2/2\sigma^2)\,d\bar{x},$$

and s is distributed as

$$dF_2 \propto \exp(-ns^2/2\sigma^2)s^{n-2}\,ds,$$

as we have seen at (11.39) and (11.40), where we had $\sigma^2 = 1$.

Then the distribution of $z(=\bar{x}/s)$ is, from (11.74), a constant times

$$\int_0^{\infty} \exp(-nz^2s^2/2\sigma^2)\exp(-ns^2/2\sigma^2)s^{n-2}\,.\,s\,.\,ds$$

$$\propto \frac{1}{(1+z^2)^{\frac{1}{2}n}},$$

as we found in Example 11.8.

11.10 The distribution of the ratio u of **11.9** can also be obtained in terms of c.f.s if x_2 is non-negative with finite mean. From (11.75), we have for non-negative v

$$H(u) = \int_0^{\infty} F_1(uv)\,dF_2(v),$$

which on using (4.4) becomes, writing $\phi_1(t)$, $\phi_2(t)$ for the c.f.s of x_1, x_2 respectively,

$$\begin{aligned} H(u) &= \int_0^{\infty}\left\{F_1(0) + \frac{1}{2\pi}\int_{-\infty}^{\infty}\frac{1-e^{ituv}}{it}\phi_1(t)\,dt\right\}dF_2(v) \\ &= F_1(0) + \frac{1}{2\pi i}\int_{-\infty}^{\infty}\frac{\phi_1(t)}{t}\left\{\int_0^{\infty}(1-e^{-ituv})\,dF_2(v)\right\}dt \\ &= F_1(0) + \frac{1}{2\pi i}\int_{-\infty}^{\infty}\frac{\phi_1(t)}{t}\{1-\phi_2(-tu)\}\,dt. \end{aligned} \tag{11.76}$$

Differentiating with respect to u, we find that the density of u, if it exists, is given by

$$h(u) = \frac{1}{2\pi i}\int_{-\infty}^{\infty}\phi_1(t)\phi_2'(-tu)\,dt, \tag{11.77}$$

provided that the integral converges, a result originally due to Cramér (1937). Daniels (1954) pointed out that the finiteness of the mean of x_2 is a sufficient condition for the integral on the right of (11.77) to converge absolutely for all u.

Geary (1944) generalized (11.77) to the case of non-independent x_1, x_2 with joint c.f. $\phi(t_1, t_2)$ where x_2 is non-negative with finite mean as before. His result is

$$h(u) = \frac{1}{2\pi i}\int_{-\infty}^{\infty}\left[\frac{\partial\phi(t_1, t_2)}{\partial t_2}\right]_{t_2=-t_1 u} dt_1. \tag{11.78}$$

The proof of (11.78) sketched in Exercise 11.24 below indicates the importance of the condition that $E(x_2)$ be finite. Exercise 11.32 gives the moments of u in terms of ϕ.

The c.f. of u, $\phi_u(t)$, may sometimes more simply be obtained from

$$\phi_u(t) = \int_{-\infty}^{\infty}\left\{\int_{-\infty}^{\infty} e^{itx_1/x_2} dF_1(x_1)\right\} dF_2(x_2) = \int_{-\infty}^{\infty}\phi_1\left(\frac{t}{x_2}\right) dF_2(x_2). \tag{11.79}$$

The same method holds for a *product* of variables—see Example 11.22.

Example 11.20 (Fisher's variance-ratio distribution)

Suppose that we have two independent samples of n_1, n_2 members from normal distributions with variances σ_1^2 and σ_2^2 respectively. We define the sample variances $\sum(x-\bar{x}_j)^2/n_j$ and call them s_1^2 and s_2^2. Their distributions are, by (11.40),

$$dF_1 \propto \exp\left(-\frac{n_1 s_1^2}{2\sigma_1^2}\right) s_1^{n_1-3} ds_1^2, \qquad dF_2 \propto \exp\left(-\frac{n_2 s_2^2}{2\sigma_2^2}\right) s_2^{n_2-3} ds_2^2.$$

The distribution of the ratio $t^2 = s_1^2/s_2^2$ is then given by (11.74) as

$$\begin{aligned} f(t^2) &\propto \int_0^{\infty} \exp\left(-\frac{n_1 t^2 s_2^2}{2\sigma_1^2}\right)(s_2 t)^{n_1-3} \exp\left(-\frac{n_2 s_2^2}{2\sigma_2^2}\right) s_2^{n_2-3}\cdot s_2^2 \cdot ds_2^2 \\ &\propto \int_0^{\infty} \exp\left\{-\frac{s_2^2}{2}\left(\frac{n_1 t^2}{\sigma_1^2}+\frac{n_2}{\sigma_2^2}\right)\right\} s_2^{n_1+n_2-4} t^{n_1-3} ds_2^2 \\ &\propto \frac{(t^2)^{\frac{1}{2}(n_1-3)}}{\left(\frac{n_1 t^2}{\sigma_1^2}+\frac{n_2}{\sigma_2^2}\right)^{\frac{1}{2}(n_1+n_2-2)}}, \qquad 0 \leqslant t^2 < \infty. \end{aligned} \tag{11.80}$$

If we write

$$x = \frac{n_1 t^2}{\sigma_1^2}\Big/\frac{n_2}{\sigma_2^2} = \frac{\sum_i (x_{1i}-\bar{x}_1)^2}{\sigma_1^2}\Big/\frac{\sum_r (x_{2r}-\bar{x}_2)^2}{\sigma_2^2}$$

in (11.80), we reduce it to the second kind of Beta distribution (6.16) with $p = \frac{1}{2}(n_1-1)$, $q = \frac{1}{2}(n_2-1)$. Writing $\nu_1 = n_1 - 1$, $\nu_2 = n_2 - 1$ and $F = \nu_2 x/\nu_1$, we obtain from (6.16) for the distribution of F

$$dG = \frac{\nu_1^{\nu_1/2}\nu_2^{\nu_2/2} F^{\frac{1}{2}(\nu_1-2)} dF}{B(\nu_1/2, \nu_2/2)(\nu_2+\nu_1 F)^{(\nu_1+\nu_2)/2}}. \tag{11.81}$$

R. A. Fisher used the further transformation $z = \frac{1}{2}\log_e F$, whose distribution is at once found from (11.81) to be

$$f(z) = \frac{2\nu_1^{\frac{1}{2}\nu_1}\nu_2^{\frac{1}{2}\nu_2}}{B(\frac{1}{2}\nu_1, \frac{1}{2}\nu_2)} \frac{e^{\nu_1 z}}{(\nu_1 e^{2z} + \nu_2)^{\frac{1}{2}(\nu_1+\nu_2)}}. \tag{11.82}$$

This form was chosen by Fisher because interpolation in tables of z is easier than in tables of F. The distribution is the basis of many tests of hypotheses, and we shall study its properties at some length in **16.15–21.**

Let us obtain the distribution by Cramér's result at (11.77). For the respective c.f.s of s_1^2 and s_2^2 we have (cf. Examples 11.3 and 11.13)

$$\phi_1(t) = \left(1 - \frac{2\sigma_1^2 it_1}{n_1}\right)^{-\frac{1}{2}(n_1-1)}, \qquad \phi_2(t) = \left(1 - \frac{2\sigma_2^2 it_2}{n_2}\right)^{-\frac{1}{2}(n_2-1)}$$

Formula (11.77) then gives us for the ratio $u = \dfrac{s_1^2/\sigma_1^2}{s_2^2/\sigma_2^2}$

$$f(u) \propto \frac{1}{2\pi i}\int_{-\infty}^{\infty} \left(1 - \frac{2it}{n_1}\right)^{-\frac{1}{2}(n_1-1)} \left(1 + \frac{2itu}{n_2}\right)^{-\frac{1}{2}(n_2+1)} dt.$$

A contour integration over the imaginary axis and the infinite semicircle to the left of it then gives us

$$f(u) \propto \left[\frac{d^\alpha}{dt^\alpha}\left(1 - \frac{2tu}{n_2}\right)^{-\frac{1}{2}(n_2+1)}\right]_{t=-\frac{1}{2}n_1}$$

where $\alpha = \frac{1}{2}(n_1 - 3)$, giving

$$f(u) \propto \frac{u^{\frac{1}{2}(n_1-3)}}{\left(1 + \dfrac{n_1 u}{n_2}\right)^{\frac{1}{2}(n_1+n_2-2)}}.$$

This, on the substitution $u = \dfrac{t^2}{\sigma_1^2/\sigma_2^2}$, reduces to the form for t^2 in (11.80). It will be noticed that this proof applies only for odd n_1, since α must be integral.

There are other (not necessarily independent) random variables whose ratio is distributed in this form—cf. Kotlarski (1964).

11.11 In **11.9** we referred to the problem of the distribution of a ratio when the denominator has a range overlapping zero. One way of handling the difficulty is to split the range into two parts at zero. We assume the denominator random variable to be continuous at zero, since otherwise there is a positive probability that the ratio is infinite. We obtain from (11.73), separately for the two parts of the range,

$$dG(u, v) \begin{cases} = f_1(uv)f_2(v)|v|\,du\,dv, & v < 0, \\ = f_1(uv)f_2(v)v\,du\,dv, & v \geqslant 0, \end{cases} \tag{11.83}$$

so that (11.74) is replaced by

$$dH(u) = \left\{\int_0^{\infty} f_1(uv)f_2(v)v\,dv - \int_{-\infty}^{0} f_1(uv)f_2(v)v\,dv\right\}du, \qquad (11.84)$$

and the distribution function follows as before.

In practice, we are usually interested in a ratio whose denominator is non-negative, so that the methods of **11.9–10** apply, as in Examples 11.19 and 11.20. But even when the denominator can take values on either side of zero, the problem can sometimes be simplified by considering the square of the ratio, thus getting back to **11.10**.

Example 11.21 (Distribution of the ratio of two normal variates)

Consider the ratio of two independent normal variables with zero mean and unit variance. From (11.84), its density function is

$$dH(u) = \frac{du}{2\pi}\left\{\int_0^{\infty} e^{-\frac{1}{2}(uv)^2}e^{-\frac{1}{2}v^2}v\,dv - \int_{-\infty}^{0} e^{-\frac{1}{2}(uv)^2}e^{-\frac{1}{2}v^2}v\,dv\right\}.$$

Simplifying, and making use of the symmetry of the normal distribution, this becomes

$$dH(u) = \frac{du}{\pi}\int_0^{\infty} \exp\{-\tfrac{1}{2}v^2(u^2+1)\}v\,dv.$$

This integral is immediately seen to be (cf. the mean deviation of a normal distribution at (5.112)) equal to $1/(1+u^2)$. Thus

$$dH(u) = \frac{du}{\pi(1+u^2)}, \qquad -\infty < u < \infty, \qquad (11.85)$$

which is a Cauchy distribution.

Here, as suggested immediately above Example 11.21, consideration of the square of the ratio is fruitful, leading to the distribution of the ratio of two squares of standardized normal variables, since each of these is distributed in the Type III form of (11.8) with $n = 1$. Thus the distribution of the squared ratio is equivalent to Fisher's distribution of F in Example 11.20. It is easily confirmed that the latter reduces to our result above if we put $n_1 = n_2 = 2$,[(*)] $\sigma_1 = \sigma_2 = 1$.

There are other identical (but not necessarily independent) variables whose ratio has the Cauchy distribution—cf. Laha (1959) and Kotlarski (1964) and a specific instance in Exercise 11.23.

[(*)] Not 1, because we are squaring about the sample means there, and thus lose a degree of freedom. Cf. Example 11.7 on this point.

Example 11.22 (Distribution of the product of two normal variates)

Now consider the product, instead of the ratio, of the variables in Example 11.21. As at (11.79), the c.f. of $z = x_1 x_2$ is

$$\begin{aligned}\phi(t) &= \int_{-\infty}^{\infty} \phi_1(tx_2)\mathrm{d}F_2(x_2)\\ &= \int_{-\infty}^{\infty} \mathrm{e}^{-\frac{1}{2}(tx_2)^2}(2\pi)^{-\frac{1}{2}}\mathrm{e}^{-\frac{1}{2}x_2^2}\,\mathrm{d}x_2\\ &= (2\pi)^{-\frac{1}{2}}\int_{-\infty}^{\infty} \mathrm{e}^{-\frac{1}{2}x_2^2(1+t^2)}\,\mathrm{d}x_2\\ &= (1+t^2)^{-\frac{1}{2}}.\end{aligned}$$

Inverting this c.f., using (4.5), we have

$$\begin{aligned}2\pi f(z) &= \int_{-\infty}^{\infty} (1+t^2)^{-\frac{1}{2}}\mathrm{e}^{-\mathrm{i}zt}\,\mathrm{d}t\\ &= 2\int_0^{\infty} (1+t^2)^{-\frac{1}{2}}\cos zt\,\mathrm{d}t\end{aligned}$$

using symmetries. Thus

$$f(z) = \frac{1}{\pi}\int_0^{\infty} (1+t^2)^{-\frac{1}{2}}\cos zt\,\mathrm{d}t = \frac{1}{\pi}K_0(|z|),$$

where K_0 is the modified Bessel function of the third kind. Thus $f(z)$ is symmetrical about zero, as is intuitively obvious.

Exercise 11.21 contains related results. Lomnicki (1967) gives results for products of n independent identical variates, which need not be normal.

It follows at once from the independence of x_1 and x_2 that $E(z^r) = E(x_1^r)E(x_2^r)$, whatever the distributions of x_1 and of x_2. Here, in the normal case, we find for the skewness and kurtosis of z, from (5.98), that $\gamma_1 = 0$ and $\gamma_2 = 6$, just as for Student's t distribution (11.44) with $\nu = 5$, as can be verified from Example 3.3.

11.12 Up to this point we have been mainly concerned with the distribution of a single statistic calculated from the members of a simple random sample. The methods may, however, readily be generalized to obtain the joint distribution of several statistics. For example, if there are several statistics $z_1, z_2, \ldots, z_p$, and the joint d.f. of the sample values $x_1, \ldots, x_n$ is $F(x_1, \ldots, x_n)$, the c.f. of the zs is given by

$$\phi(t_1, \ldots, t_p) = \int_{-\infty}^{\infty} \cdots \int_{-\infty}^{\infty} \exp(\mathrm{i}t_1 z_1 + \ldots + \mathrm{i}t_p z_p)\,\mathrm{d}F(x_1, \ldots, x_n) \quad (11.86)$$

and the density function of the zs (if it exists) by

$$f(z_1, \ldots, z_p) = \frac{1}{(2\pi)^p} \int_{-\infty}^{\infty} \cdots \int_{-\infty}^{\infty} \exp(-it_1 z_1 - \ldots - it_p z_p)\phi(t_1, \ldots, t_p)\mathrm{d}t_1 \ldots \mathrm{d}t_p. \tag{11.87}$$

Examples of the use of these results will occur later.

Edgeworth and saddlepoint approximations for sums

11.13 Although we have been concerned in this chapter with exact results, it is convenient to discuss here a class of approximations that differs radically from the methods to be discussed in Chapters 12-13. Our treatment is based on that of Barndorff-Nielsen and Cox (1979).

Suppose that we are investigating the distribution of X_n, the sum of n identical independent variables each with cumulants λ_r. X_n will therefore have cumulants $n\lambda_r$ and when standardized will have zero mean, unit variance and higher cumulants $\kappa_r = n\lambda_r/(n\lambda_2)^{\frac{1}{2}r} = n^{1-\frac{1}{2}r}\rho_r$, where $\rho_r = \lambda_r/\lambda_2^{\frac{1}{2}r}$. The exact distribution of X_n is given by (11.54), but even if we know the c.f. of X_n, we may find its inversion too difficult an integration. If we now use the Edgeworth expansion (6.42) to approximate the distribution of X_n, we find

$$f(x) = \alpha(x)\left\{1 + \frac{\rho_3}{6n^{\frac{1}{2}}}H_3 + \frac{1}{24n}(\rho_4 H_4 + \tfrac{1}{3}\rho_3^2 H_6)\right\} + o(n^{-1}). \tag{11.88}$$

We observe from (6.22) that $H_r(x)$ only has a constant term for r even, so that the odd-order $H_r(0) = 0$. Thus, although (11.88) is an expansion in powers of $n^{-\frac{1}{2}}$, it will actually be in powers of n^{-1} near $x = 0$, which since we have standardized is when X_n is near its expectation. Conversely, as we have seen in **6.23**, the Edgeworth expansion may be very poor in the tails of the distribution.

11.14 One may also obtain an Edgeworth expansion indirectly if the distribution of the standardized X_n tends to a member of the natural exponential family with natural parameter, as in **5.48**, which we now write

$$f(x \mid \theta) = G(\theta)H(x)e^{x\theta}. \tag{11.89}$$

As we saw in **5.47-8**, (11.89) covers the normal and the chi-squared distributions, the commonest limiting forms, and has m.g.f.

$$M_x(t) = G(\theta)/G(\theta + t)$$

and c.g.f.

$$\psi_x(t) = \log M_x(t) = \log G(\theta) - \log G(\theta + t),$$

whence it is obvious that the sum of n such independent variables is also a member of the family with $G(\theta)$ replaced by $G^n(\theta)$. We have

$$E(x) = \left[-\frac{\mathrm{d}}{\mathrm{d}\theta} \log G(\theta + t) \right]_{t=0} = -G'(\theta)/G(\theta)$$

$$= -\frac{\mathrm{d}}{\mathrm{d}\theta} \log G(\theta), \tag{11.90}$$

and higher cumulants

$$\kappa_r = -\frac{\mathrm{d}^r}{\mathrm{d}\theta^r} \log G(\theta), \quad r \geqslant 2. \tag{11.91}$$

From the form of (11.89), we see that for any specified value θ_0, we have

$$f(x|\theta_0) = \exp\{(\theta_0 - \theta)x + \log G(\theta_0) - \log G(\theta)\} f(x|\theta). \tag{11.92}$$

11.15 We may now expand $f(x|\theta)$ on the right of (11.92) in the usual (direct) Edgeworth series (11.88), to obtain an (indirect) Edgeworth expansion of $f(x|\theta_0)$ on the left. We may choose the value θ to facilitate the approximation by putting $\theta = \hat{\theta}$, the value for which x is equal to the mean of the distribution of X_n. Then the exponent in (11.92) becomes, using (11.90),

$$\log G(\theta_0) - \left[\log G(\hat{\theta}) + (\theta_0 - \hat{\theta}) \frac{\mathrm{d}}{\mathrm{d}\theta} \log G(\hat{\theta}) \right], \tag{11.93}$$

the terms in square brackets being the first-order Taylor expansion of log $G(\theta_0)$ about $\hat{\theta}$. If we were to expand log $G(\theta_0)$ in (11.93) in an infinite series about $\hat{\theta}$, (11.93) would become

$$\sum_{r=2}^{\infty} \frac{(\theta_0 - \hat{\theta})^r}{r!} \frac{\mathrm{d}^r}{\mathrm{d}\theta^r} \log G(\hat{\theta}) = -\sum_{r=2}^{\infty} \frac{(\theta_0 - \hat{\theta})^r}{r!} \hat{\kappa}_r \tag{11.94}$$

on using (11.91). Here the symbol ^ means that $\theta = \hat{\theta}$ is substituted into the function over which it appears.

If we measure from $x = \hat{\kappa}_1$ as origin, we see from (6.23) that $H_3(0) = 0$, $H_4(0) = 3$ and $H_6(0) = -15$, so that in (11.92) the Edgeworth expansion of $f(x|\hat{\theta})$ on the right is, from (11.88),

$$f(x|\hat{\theta}) = \alpha(x)\left\{1 + \frac{1}{24n}(3\hat{\rho}_4 - 5\hat{\rho}_3^2)\right\} + o(n^{-1}). \tag{11.95}$$

Substituting (11.93) and (11.95) into (11.92), we find

$$f(x|\theta_0) = \frac{G(\theta_0)}{G(\hat{\theta})} \exp\left\{-(\theta_0 - \hat{\theta})\frac{\mathrm{d}}{\mathrm{d}\theta}\log G(\hat{\theta})\right\} \alpha(x)\{1 + O(n^{-1})\}. \tag{11.96}$$

If we now destandardize the distribution, remembering that we must in any case measure from its mean to preserve the form (11.93), the variance alone is affected, and at $x=0$ (11.96) becomes

$$f(x|\theta_0)=\frac{G(\theta_0)}{G(\hat{\theta})}\frac{\exp\{(\theta_0-\hat{\theta})x\}}{(2\pi\hat{\kappa}_2)^{\frac{1}{2}}}\{1+O(n^{-1})\}, \tag{11.97}$$

where we have re-substituted x for its mean for simplicity.

11.16 The leading term in (11.97) is the saddlepoint approximation, developed differently by Daniels (1954). The approximation can be further improved if we multiply it by a constant to ensure that it integrates to unity; its error is then often of order $n^{-\frac{3}{2}}$ rather than n^{-1}. Indeed, it is then sometimes exact, although Daniels (1980) shows that the only exact continuous cases are the normal, the Inverse Gaussian of Exercise 11.28, and the Gamma distribution, which includes the exponential distribution in Example 11.23 below. Exercises 11.30–1 deal with exact discrete cases.

(11.97) is usually more accurate than the direct Edgeworth expansion (11.88), particularly in the tails, but it requires knowledge of the c.g.f. and is not so easily used to obtain the d.f. from the p.d.f. as the Edgeworth expansion was seen to be at (6.43).

Barndorff-Nielsen and Cox (1979) generalize to the multivariate case, and give a number of theoretical applications. See also a quite different and more general approach in Exercise 17.33, Volume 2A, and the papers by Daniels (1983, 1987) which deal with approximations of tail probabilities.

Example 11.23 (Saddlepoint approximation for the sum of exponential variates)

The exponential distribution e^{-x}, $x\geqslant 0$, is a member of the exponential family (11.89) with $\theta=0$, and $G(0)=1$. Its m.g.f., from Example 3.12, is $(1-t)^{-1}$. Thus $G(t)=1-t$. For X_n, the sum of n such independent variables,

$$\log G(t)=n\log(1-t),\qquad \frac{d}{dt}\log G(t)=-n(1-t)^{-1}.$$

and

$$\frac{d^2}{dt^2}\log G(t)=-n(1-t)^{-2}.$$

We now write θ for t, and use (11.90) to determine $\hat{\theta}$ in **11.15**, obtaining from $n(1-\hat{\theta})^{-1}=x$ the solution $\hat{\theta}=1-\dfrac{n}{x}$. The variance at $\hat{\theta}$, from (11.91), is $\hat{\kappa}_2=n(1-\hat{\theta})^{-2}=x^2/n$, so the saddlepoint approximation (11.97) (at $\theta_0=0$, in accordance with the exponential family value) is

$$f(x|0)=\left(\frac{x}{n}\right)^n\frac{\exp\left\{-\left(1-\dfrac{n}{x}\right)x\right\}}{\left(2\pi\dfrac{x^2}{n}\right)^{\frac{1}{2}}}=\frac{e^{-x}x^{n-1}}{(2\pi)^{\frac{1}{2}}e^{-n}n^{n-\frac{1}{2}}}.$$

The denominator is the leading term in the Stirling series for $\Gamma(n)$ at (3.64). Apart from this, the approximation is exact, since we know from Example 11.13 that X_n has a Gamma distribution with parameter n. Since the error is merely a constant multiple, it will be removed if we make the approximation integrate to unity. Finally, since the exponential distribution has cumulants $\lambda_r = (r-1)!$ by Example 3.12, we have $\rho_3 = 2$, $\rho_4 = 6$ and the $O(n^{-1})$ correction term in (11.95) is $\{1-(12n)^{-1}\}$, equivalent to putting the next-order term $\{1+(12n)^{-1}\}$ into the Stirling approximation (cf. (3.64)) in the denominator.

The saddlepoint approximation method may be used for any statistic by using the first four cumulants to estimate the c.f. from its leading term, but the direct approach of **11.14-15** must be replaced by the complex integration technique used by Daniels (1954)—c.f. Easton and Ronchetti (1986).

11.17 Lugannani and Rice (1980) develop a different saddlepoint approach that makes due allowance for the existence of a pole at the origin in the Fourier integral. The successive terms of their expansion decrease by a factor of order n^{-1} over the whole range of $\bar{x}$, rather than $n^{-\frac{1}{2}}$. Daniels (1987) shows how the Lugannani-Rice expansion relates to previous work on saddlepoint approximations and also extends their results to lattice variables.

A general review of saddlepoint methods is given by Reid (1988). A bivariate saddlepoint approximation is derived in Wang (1987).

EXERCISES

11.1 Find the distribution of means of samples of n independent observations from

$$dF = e^{-x}\,dx, \quad 0 \leqslant x < \infty,$$

(a) by induction; (b) by the use of c.f.s.

11.2 Derive by using c.f.s the sampling distribution of Example 11.1 for the mean of samples from the Cauchy distribution

$$dF = \frac{dx}{\pi(1+x^2)}, \quad -\infty < x < \infty.$$

11.3 Show that if $\bar{x}$ is the arithmetic mean and g the geometric mean in samples of n from the Gamma distribution

$$dF = \frac{e^{-x}x^{p-1}}{\Gamma(p)}\,dx, \quad 0 \leqslant x < \infty,$$

the c.f. of $S = \log(\bar{x}/g)$ is

$$\phi(t) = \{\Gamma(p - it/n)/\Gamma(p)\}^n \Gamma(np)/\{\Gamma(np - it)n^{it}\},$$

with mean

$$\kappa_1 = \frac{\Gamma'(np)}{\Gamma(np)} - \frac{\Gamma'(p)}{\Gamma(p)} - \log n.$$

Show further that S is distributed independently of $\bar{x}$; cf Exercise 15.22.

(Bain and Engelhardt (1975) approximate the d.f. of S. Glaser (1976) obtains the exact distribution of $g/\bar{x}$.)

11.4 $x_1, \ldots, x_n$ are n independent standardized normal variates. Show that $x_i^2\Big/\left(\sum_{j=1}^{n} x_j^2\right)$ is distributed independently of $\sum_{j=1}^{n} x_j^2$. Show also that if $\bar{x}$ is the mean of the xs, $(x_i - \bar{x})^2/\sum(x_j - \bar{x})^2$ is independent of $\sum(x_j - \bar{x})^2$; cf. **15.27**.

11.5 Use the result of the previous exercise to show that if m_2, m_3 and m_4 are the sample central moments in a normal sample

$$E\left(\frac{m_3}{m_2^{\frac{3}{2}}}\right)^k = \frac{Em_3^k}{Em_2^{3k/2}},$$

$$E\left(\frac{m_4}{m_2^2}\right)^k = \frac{Em_4^k}{Em_2^{2k}}.$$

11.6 Show that the conditional distribution of Student's t (defined at (11.42)) for fixed s^2 is exactly normal with mean equal to zero (the assumed population mean) and variance

$$(n-1)\sigma^2/(ns^2).$$

11.7 Show that the joint distribution of two order statistics x_r, $x_s\,(r < s)$ is given by

$$dF(x_r, x_s) = \frac{n!}{(r-1)!(s-r-1)!(n-s)!} F_r^{r-1}(F_s - F_r)^{s-r-1}(1-F_s)^{n-s} f(x_r) f(x_s)\, dx_r\, dx_s$$

where F_j is the d.f. of x with $x = x_j$, $j = r, s$. Integrate out x_s to obtain the distribution of x_r at (11.34).

11.8 $x_1, x_2, \ldots, x_n$ are independent Beta variables, distributed in the form (6.14), x_i having parameters (p_i, q_i). Show that if $p_2 = p_1 + q_1$, the product $x_1 x_2$ is distributed in the same form with parameters $(p_1, q_1 + q_2)$; and hence that if $p_r = p_1 + \sum_{s=1}^{r-1} q_s$ for $r = 2, 3, \ldots, n$, the product $x_1 x_2 \ldots x_r$ has the same distribution with parameters $\left(p_1, \sum_{s=1}^{r} q_s\right)$.

(The product of other sets of independent variables can have this distribution—cf. Kotlarski (1962) and also I. R. James (1972).)

11.9 Show that the distribution of the geometric mean of n independent variables, one from each of the distributions

$$\frac{x^{p-1}e^{-x}}{\Gamma(p)}, \frac{x^{p+1/n-1}e^{-x}}{\Gamma(p+1/n)}, \ldots, \frac{x^{p+(n-1)/n-1}e^{-x}}{\Gamma\{p+(n-1)/n\}},$$

is the same as the distribution of the arithmetic mean of n independent variables distributed in the first of these forms. Show that for $n = 2$, $p = \frac{1}{4}$, this gives a pair of independent random variables whose product is exactly normally distributed.

(cf. Kullback, 1934)

11.10 If $x_1, x_2, \ldots, x_n$ are independent Gamma variables

$$f(x_i) = \frac{1}{\Gamma(p_i)} e^{-x_i} x_i^{p_i - 1}, \quad p_i > 0, \quad 0 \leqslant x < \infty,$$

show that the variables

$$y_1 = \sum_{i=1}^{n} x_i,$$

$$y_r = x_r \Big/ \sum_{s=1}^{r-1} x_s, \quad r = 2, 3, \ldots, n$$

are also independent, y_1 being a Gamma variable with parameter $\sum_{i=1}^{n} p_i$ and y_r $(r > 1)$ being distributed in the second kind of Beta distribution (6.16) with parameters p_r and $\sum_{s=1}^{r-1} p_s$. Show that the converse result also holds: if the ys are thus independently distributed, then the xs are. Show that $z_r = x_r \Big/ \sum_{i=1}^{r} x_i$, $r = 2, 3, \ldots, n$ are also a set of independent variables, with parameters as for y_r, but that they have Beta distributions (6.14) of the first kind.

(For more general results, cf. Aitchison (1963).)

11.11 Show that the ratio $v = x/y$ of two independent normal variables has density function

$$f(v) = \frac{1}{\sqrt{(2\pi)}} \frac{\mu_y \sigma_x^2 + \mu_x \sigma_y^2 v}{(\sigma_x^2 + \sigma_y^2 v^2)^{\frac{3}{2}}} \exp\left\{-\frac{1}{2} \frac{(\mu_x - \mu_y v)^2}{\sigma_x^2 + \sigma_y^2 v^2}\right\}$$

where the μ_j and σ_j^2 are the means and variances of the variables, and it is assumed that μ_y is so large compared with σ_y that the range of y is effectively positive.

Hence show that $(\mu_x - \mu_y v)(\sigma_x^2 + \sigma_y^2 v^2)^{-\frac{1}{2}}$ is a standardized normal variate.

(Geary, 1930)

11.12 Generalizing 11.2, suppose that the x_i have unit variances but means μ_i, which may differ. Writing $\lambda = \sum_{i=1}^{n} \mu_i^2$, show that the distribution of z at (11.4) is now

$$f(z) = \frac{e^{-\frac{1}{2}(z+\lambda)} z^{\frac{1}{2}n-1}}{2^{\frac{1}{2}n} \Gamma\{\frac{1}{2}(n-1)\} \Gamma(\frac{1}{2})} \sum_{r=0}^{\infty} \frac{\lambda^r z^r}{(2r)!} B\{\tfrac{1}{2}(n-1), \tfrac{1}{2}+r\},$$

reducing to (11.8) when $\lambda = 0$. (The c.f. of this density was given in Exercise 8.31.)

11.13 Let z_1, z_2 be independently distributed as in Exercise 11.12, with constants (n, λ) equal to (ν_1, λ) for z_1 and to $(\nu_2, 0)$ for z_2. Show that the distribution of $\dfrac{(z_1/\nu_1)}{(z_2/\nu_2)} = w$ is

$$g(w) = e^{-\frac{1}{2}\lambda} \sum_{r=0}^{\infty} \frac{(\frac{1}{2}\lambda)^r}{r!} \left(\frac{\nu_1}{\nu_2}\right)^{\frac{1}{2}\nu_1 + r} \frac{1}{B(\frac{1}{2}\nu_1 + r, \frac{1}{2}\nu_2)} \frac{w^{\frac{1}{2}\nu_1 - 1 + r}}{\left(1 + \dfrac{\nu_1}{\nu_2} w\right)^{\frac{1}{2}(\nu_1 + \nu_2) + r}}$$

which reduces to Example 11.20 when $\lambda = 0$, $\nu_1 = n - 1$, $\nu_2 = n_2 - 1$.

(The distributions of this and the previous exercises are called *non-central*. We shall study them in Volume 2A.)

11.14 x_1 and x_2 are independently distributed in the form (11.8) with $2n$ and $2m$ degrees of freedom respectively. Show that the c.f. of $y = cx_1 - bx_2$ $(c, b > 0)$ is

$$\phi(t) = (1 - 2ict)^{-n}(1 + 2ibt)^{-m}.$$

Show from the negative binomial distribution (5.43) that, with $k + j = n$.

$$p^m \sum_{u=0}^{n-1} \binom{m+u-1}{m-1} q^u + q^n \sum_{v=0}^{m-1} \binom{n+v-1}{n-1} p^v \equiv 1$$

and hence, substituting $p = 1 - q = (1 + 2ibt)c/(c+b)$, that

$$\phi(t) = \sum_{j=1}^{n} \binom{n+m-j-1}{m-1} \left(\frac{c}{c+b}\right)^m \left(\frac{b}{c+b}\right)^{n-j} (1 - 2ict)^{-j}$$

$$+ \sum_{k=1}^{m} \binom{n+m-k-1}{n-1} \left(\frac{b}{c+b}\right)^n \left(\frac{c}{c+b}\right)^{m-k} (1 + 2ibt)^{-k},$$

expressing the c.f. as that of a mixture of $(n+m)$ distributions of the same forms as cx_1 and $(-bx_2)$, with negative binomial probabilities as the weights.

(Jayachandran and Barr, 1970)

11.15 z_1, z_2 are independent variates distributed in the generalized Gamma form

$$dF(z_i) = \frac{\exp\left(-\frac{z_i}{\theta_i}\right)\left(\frac{z_i}{\theta_i}\right)^{a_i}}{\Gamma(a_i+1)} d\left(\frac{z_i}{\theta_i}\right), \quad z_i > 0, \quad \theta_i > 0$$

where a_i is zero or a positive integer. Writing $a_1 + a_2 = a$, $\phi = \frac{1}{\theta_1} + \frac{1}{\theta_2}$, and using the identity

$$\int_0^\infty t^{a_1}(t+s)^{a_2} e^{-\lambda t} dt = \lambda^{-(a+1)} \sum_{r=0}^{a_2} \binom{a_2}{r} (a-r)!(\lambda s)^r,$$

show that the distribution of $u = z_1 - z_2$ is

$$h(u) \propto \exp\left(\frac{u}{\theta_2}\right) A(u, \phi), \quad -\infty < u < \infty,$$

where

$$A(u, \phi) = \begin{cases} \exp(-\phi u)\phi^{-(a+1)} \sum_{r=0}^{a_1} \binom{a_1}{r} (a-r)!(\phi u)^r, & u \geqslant 0, \\ \phi^{-(a+1)} \sum_{r=0}^{a_2} \binom{a_2}{r} (a-r)!(\phi u)^r, & u < 0. \end{cases}$$

Show that if $\theta_1 = 2c$, $\theta_2 = 2b$, $a_1 = n-1$ and $a_2 = m-1$, $h(u)$ is the density corresponding to the c.f. in Exercise 11.14.

The special case $\theta_1 = \theta_2 = 1$, $\phi = 2$, $a_1 = a_2 = \frac{1}{2}a$ yields the distribution of the difference between two identical Gamma variates with parameter an integer.

(Cf. Lentner and Buehler (1963). For the case where $a_1 = a_2$ is non-integral, cf. K. Pearson *et al.* (1932).)

11.16 In Exercise 11.15, show that the distribution of $v = z_1 + z_2$ is (for any, not necessarily integral, $a_i > 0$)

$$g(v) = \sum_{r=0}^{\infty} \binom{a_1+r}{r} \left(\frac{\theta_2}{\theta_1}\right)^{a_1+1} \left(1 - \frac{\theta_2}{\theta_1}\right)^r \frac{e^{-v/\theta_2}\left(\frac{v}{\theta_2}\right)^{a+r+1}}{\Gamma(a+r+2)} d\left(\frac{v}{\theta_2}\right).$$

Show also that the conditional distribution of z_1 given u is, for integral a_i,

$$g(z_1|u) = z_1^{a_1}(z_1-u)^{a_2} \exp(-\phi z_1)/A(u,\phi), \begin{cases} z_1 > 0, \\ z_1 > u. \end{cases}$$

(Lentner and Buehler, 1963)

11.17 x is a standardized normal variate and, independently, z has the density (11.8) with n an even integer $n = 2p > 0$. Use the method of Example 11.22 to show that $w = xz^{\frac{1}{2}}$ is distributed like the sum of p independent variates w_j with density $\frac{1}{2}\exp(-|w_j|)$.

11.18 Derive the joint c.f. of mean and variance in samples of n observations from a normal distribution, and show that it factorizes into their marginal c.f.s, and that therefore they are independently distributed.

11.19 If the c.f. of sample mean and variance in samples of size n is $\phi(t_1, t_2)$ and that of the distribution with p.d.f. $f(x)$ is $\alpha(t)$, show that

$$\left[\frac{\partial}{\partial t_2}\phi(t_1, t_2)\right]_{t_2=0} = \frac{n-1}{n}i\left[\alpha^{n-1}(t_1/n)\int_{-\infty}^{\infty} x^2 e^{it_1x/n}f(x)\,dx - \alpha^{n-2}(t_1/n)\left\{\int_{-\infty}^{\infty} x\, e^{it_1x/n}f(x)\,dx\right\}^2\right].$$

Hence, if mean and variance are independent, show that

$$-\alpha\frac{d^2\alpha}{dt^2} + \left(\frac{d\alpha}{dt}\right)^2 = \sigma^2\alpha^2$$

where σ^2 is the population variance. Hence show that the distribution must be normal.

(Lukacs (1942). Geary (1936a) established the result for distributions possessing finite cumulants of all orders—cf. Example 12.7 below.)

11.20 From Exercise 8.25, show that if there are n independent standardized normal variates x_j, the joint c.f. of

$$x = \sum_{j=1}^{n} a_j x_j \quad \text{and} \quad y = \sum_{j=1}^{n} x_j^2 \quad \text{is}$$

$$\phi_{x,y}(t_1, t_2) = (1 - 2it_2)^{-\frac{1}{2}n}\exp\left\{\frac{-\frac{1}{2}a^2}{1 - 2it_2}\right\}, \quad \text{where} \quad a^2 = t_1^2 \sum_{j=1}^{n} a_j^2.$$

Use this result and (4.34) to show that the conditional c.f. of y given that $x = 0$ is

$$\phi_{21}(t_2) = (1 - 2it_2)^{-\frac{1}{2}(n-1)}$$

corresponding to (11.36) with $p = 1$. If there are p linearly independent functions $x^{(r)} = \sum_{j=1}^{n} a_{jr}x_j$, $r = 1, 2, \ldots, p$, show that the $(p+1)$-variate c.f. $\phi_{x,y}(t_{11}, \ldots, t_{1p}, t_2)$ is as above with $a^2 = \sum_{j=1}^{n}\left(\sum_{r=1}^{p} a_{jr}t_{1r}\right)^2$, and use the generalization of (4.34) to verify (11.36) for any p.

11.21 x_r, y_r, $r = 1, 2, \ldots$, are independent standardized normal variates. Show that $z = x_1y_1 + x_2y_2$ is distributed exactly as $w_1 - w_2$, where the w_j are independent with p.d.f. e^{-w_j} and hence or otherwise that z is distributed in the Laplace form $g(z) = \frac{1}{2}\exp(-|z|)$, and that $|z|$ is again exponentially distributed. Using Exercise 2.7 with $p = 1$ and the remark below (2.24), show that the mean deviation δ_1 of the Laplace distribution is unity and from Exercise 4.3 that $\delta_1/\sigma = 2^{-\frac{1}{2}} = 0.707$.

More generally, show that $\sum_{r=1}^{2n} x_r y_r$ is distributed as the difference between two Gamma variables each with parameter n, i.e. as in Exercise 11.15 with $a_1 = a_2 = n - 1$ and $\theta_1 = \theta_2 = 1$. (The property that $|w_1 - w_2|$ is distributed exactly as w_j characterizes the exponential distribution with $f = \theta\exp(-\theta w)$ among continuous distributions—cf. Puri and Rubin (1970).)

11.22 Generalizing Example 11.21, show that if x and y are bivariate normally distributed with correlation ρ, the statistic

$$z = \frac{x - \mu_x}{\sigma_x}\bigg/\frac{y - \mu_y}{\sigma_y}$$

has the density function

$$g(z) = \frac{(1-\rho^2)^{\frac{1}{2}}}{\pi(1-2\rho z+z^2)}, \qquad -\infty < z < \infty.$$

Generalizing Exercise 11.11, show that if $v = x/y$ and $\mu_y/\sigma_y \to \infty$ so that Prob $(y > 0) \to 1$, the variable $(\mu_x - \mu_y v)/(\sigma_x^2 - 2\rho\sigma_x\sigma_y v + \sigma_y^2 v^2)^{\frac{1}{2}}$ is approximately standardized normal.

(Cf. Geary (1930), Hinkley (1969) and Marsaglia (1965) for the distribution of x/y).

11.23 In Example 11.21, show that if instead of normal variates we consider two independent variates distributed as in Exercise 2.12, their ratio is still distributed in the Cauchy form (11.85).

(Steck, 1958a)

11.24 If x_1, $x_2 (x_2 \geqslant 0,\ E(x_2) = \mu)$ have the joint density $f(x_1, x_2)$ and c.f. $\phi(t_1, t_2)$, show that the density of $u = x_1/x_2$ is $h(u) = \int_0^\infty f(uv, v)v\,dv$, generating (11.74). Now consider the joint p.d.f.

$$g(x_1, x_2) = \frac{x_2}{\mu} f(x_1, x_2).$$

Show that its c.f. is

$$\psi(t_1, t_2) = \frac{1}{i\mu}\frac{\partial\phi(t_1, t_2)}{\partial t_2},$$

and that the c.f. of the variate $w = x_1 - ux_2$ is $[\psi(t_1, t_2)]_{t_2 = -ut_1}$. By showing that the density of w at $w = 0$ is equal to $h(u)/\mu$, use the Inversion Theorem to establish (11.78).

(Daniels, 1954)

11.25 If $x_1, x_2, \ldots, x_n$ are independent standardized normal variates, and $y_j = 1/x_j^2$, with c.f. $\phi_y(t)$, show using (11.79) that the c.f. of x_2/x_1 is $\phi_y(\frac{1}{2}it^2)$, so that from Examples 11.21 and 4.2

$$\phi_y(t) = \exp\{-(-2it)^{\frac{1}{2}}\}.$$

Verify that the density of y is given by (4.70).

Hence show that the c.f. of the mean $\bar{y}$ of the ys is

$$\phi_{\bar{y}}(t) = \phi_y(nt) = \phi_{ny}(t),$$

so that the distribution of the mean of n observations is that of a multiple n of a single observation, becoming more widely dispersed as n increases.

By putting $n = 2$, show that $v = 2x_1x_2/(x_1^2 + x_2^2)^{\frac{1}{2}}$ is a standardized normal variate.

11.26 In **11.2**, show that when $n = 2$, the polar variables z and θ_1 are independently distributed, the former as χ^2 with 2 d.fr. and the latter uniformly on the interval $(0, 2\pi)$. Hence show that $w = z^{\frac{1}{2}}\cos(p\theta_1)$ and $v = z^{\frac{1}{2}}\sin(p\theta_1)$ are independent standardized normal variables for any positive integer p. Putting $p = 2$ in v, derive the final result of Exercise 11.25.

11.27 If x_1 has the symmetric stable c.f. (4.69) with $a = 0$ and exponent α_1, and the non-negative variate x_2 has the stable c.f. (4.68) with $a = 0$ and exponent $\alpha_2 < 1$, show using the method of Example 11.22 that $z = x_1 x_2^{1/\alpha_1}$ is symmetric stable with exponent $\alpha_1\alpha_2$.

If $y_1, y_2, \ldots, y_n$ are independent standardized normal variates, show, using Exercise 11.25, that

$$z = y_1/\{y_2 y_3^2 y_4^{2^2} \ldots y_n^{2^{n-2}}\}$$

is symmetric stable with exponent $\alpha = 1/2^{n-2}$.

(Example 11.21 gives the case $n = 2$; cf. G. W. Brown and Tukey, 1946.)

11.28 Show that the *Inverse Gaussian distribution*,

$$f(x) = (2\pi x^3/\lambda)^{-\frac{1}{2}} \exp\left\{\frac{-\lambda}{2\mu^2}(x-\mu)^2/x\right\}, \quad x, \lambda, \mu > 0 \tag{A}$$

has the c.f.

$$\phi(t) = \exp\left\{\frac{\lambda}{\mu}\left[1 - \left(1 - \frac{2\mu^2}{\lambda} it\right)^{\frac{1}{2}}\right]\right\}$$

and cumulants

$$\kappa_1 = \mu,$$
$$\kappa_r = \mu^{2r-1}(2r-3)!/\{\lambda^{r-1}(r-2)!2^{r-2}\}, \quad r \geqslant 2,$$

so that $\kappa_2 = \mu^3/\lambda$, $\kappa_3^2/\kappa_2^3 = 9\mu/\lambda$ and $\kappa_4/\kappa_2^2 = 15\mu/\lambda$, skewness and kurtosis increasing together.

Show that when $\lambda = 1$ (A) is a member of the natural exponential family (5.117) with natural parameter $\theta = -1/(2\mu^2)$ and $D(\theta) = 1/\mu$. Hence use (5.119) and (5.122) to verify the c.f. and cumulants above when $\lambda = 1$.

Show that the mean of n independent observations from $f(x)$ has the same density with λ replaced by $(n\lambda)$, so that its cumulants are

$$\kappa_1^* = \kappa_1,$$
$$\kappa_2^* = \kappa_2/n.$$

(Tweedie, 1957)

11.29 Writing the exponent of (A) in Exercise 11.28 as $-\frac{1}{2}S$, show that S is distributed in the χ^2 distribution (11.8) with $n = 1$ degree of freedom, exactly like the square of a standardized normal variable, but that $S^{\frac{1}{2}}$ is not standardized normal.

Show that as $\mu \to \infty$, $y = x/\lambda$ in (A) has the limiting distribution (4.70) with c.f. as in Exercise 11.25.

(Cf. Folks and Chhikara, 1978)

11.30 Show that the Poisson distribution is a member of the exponential family (11.89) with $\lambda = e^{\theta}$, and that the sum X_n of n such independent variates has $\log G(\theta) = -ne^{\theta}$. Hence show that the saddlepoint approximation (11.97) for the distribution of X_n is exact (cf. Example 11.17) apart from the Stirling approximation for $x!$ in its denominator. By putting $\lambda = 1$, use the exact probability that $X_n = n$ and the Central Limit theorem to establish the Stirling approximation to the factorial in (3.64).

11.31 Show that the binomial distribution is a member of the exponential family (11.89) with $p = e^{\theta}/(1+e^{\theta})$ and that, regarding this as the sum of n independent variates with $f(1) = p$, $f(0) = 1-p$, we have $G(\theta) = (1-p)^n$. Hence show that the saddlepoint approximation (11.97) for the distribution is exact, apart from the Stirling approximation for each of the factorials in $\binom{n}{x}$.

11.32 In (11.78), show that the mean of u is

$$E(u) = \int_0^\infty \left[\frac{\partial}{\partial(it_1)}\phi(t_1, it_2)\right]_{t_1=0} dt_2,$$

and more generally that for $r \geqslant 0$, $s > 0$,

$$E(x_1^r/x_2^s) = \frac{1}{\Gamma(s)}\int_0^\infty t_2^{s-1}\left[\frac{\partial^r}{\partial(it_1)^r}\phi(t_1, it_2)\right]_{t_1=0} dt_2.$$

Hence, when $r = 0$, obtain the negative moments of x_2 in terms of its c.f. $\phi(0, t_2)$.

(Cf. Cressie *et al.*, 1981)

CHAPTER 12

CUMULANTS OF SAMPLING DISTRIBUTIONS – (1)

12.1 In the previous chapter we have considered methods of deriving exact sampling distributions when the population is completely specified. Those methods are not applicable when the population is not completely known, and they may in any case lead to results that are difficult to apply in practice, e.g. by yielding an integral that has not been tabulated. In such cases we can frequently deal with the problem by finding an approximate form for the sampling distribution, particularly by ascertaining its lower moments, and then fitting a tractable type of distribution such as one of the Pearson system.

A procedure of this kind has already been considered in Chapter 10, where it was seen that approximate expressions could be derived for the first and second moments of sampling distributions in terms of the population moments. When the sampling distribution tends to normality this, in effect, solves our problem, for its first and second moments determine a normal distribution. The methods of this chapter are really developments of this idea. We shall discuss exact methods of finding the moments of sampling distributions in terms of population moments. Our results are important not only on their own account, but in giving an accurate method of judging the degree of approximation of the expressions for large n discussed in Chapter 10. In particular, we shall be able to take up some points that had to be left on one side in that chapter – e.g. the rapidity with which some functions of the moments approach normality.

12.2 The statistics with whose distributions we are customarily concerned may be usefully classified into three groups. The largest group comprises statistics which are symmetric functions of the observations; that is to say, if the observations are $x_1, x_2, \ldots, x_n$, the statistic $t(x_1, x_2, \ldots, x_n)$ is unchanged if we permute the xs. Moreover, our statistics are nearly always algebraic. This class includes the arithmetic mean, moments, quantities such as b_1 and b_2 that are functions of moments and, in the bivariate case, product moments. The second group comprises statistics based on order properties of the sample values – the median, the quantiles, the range, extreme values and so forth. The third group is residual, including those statistics, such as the sample mode, that are not included in the first two groups.

The methods we shall develop in this chapter and the next apply to the first group only. We may recall that there are three different types of moment concerned in the investigation: (a) the moments of the population, (b) the moments of the sample and (c) the moments of the sampling distribution. They will be referred to as population moments, sample moments and sampling moments respectively.

Similarly we shall consider population cumulants, sample cumulants and sampling cumulants.

12.3 In **10.3** we obtained the exact results

$$\left.\begin{aligned} Em'_r &= \mu'_r \\ \operatorname{var} m'_r = E(m'_r - \mu'_r)^2 &= \frac{1}{n}(\mu'_{2r} - \mu'^2_r) \end{aligned}\right\} \tag{12.1}$$

and saw in **10.4** that formulae for sampling moments about the mean were more difficult to obtain. Although we shall later reject this approach in favour of another, it is instructive to consider what happens if we try to generalize the procedure of that chapter to our present problem. Suppose, for example, that we are interested in the sampling distribution of the variance. The above equations give us the first two sampling moments of the second moment about an arbitrary point. For the first sampling moment of the variance we have

$$\begin{aligned} Em_2 &= E\left\{\sum x^2/n - \left(\sum x/n\right)^2\right\} \\ &= E\left\{\frac{n-1}{n^2}\sum x^2 - \frac{1}{n^2}\sum x_i x_j\right\}, \qquad i \neq j \\ &= \frac{n-1}{n}Ex^2 - \frac{n-1}{n}Ex_i x_j. \end{aligned} \tag{12.2}$$

Since x_i and x_j are independent, $Ex_i x_j$ is equal to μ'^2_1. Equation (12.2) then gives us

$$Em_2 = \frac{n-1}{n}(\mu'_2 - \mu'^2_1) = \frac{n-1}{n}\mu_2. \tag{12.3}$$

This may be compared with the approximate expression given by (10.8), namely

$$Em_2 = \mu_2. \tag{12.4}$$

Symmetric functions

12.4 It will be clear that the same method can be used to derive the sampling moment of any order of a statistic that can be expressed as a symmetric function (rational and integral) of the observations. For instance, to find the fourth moment of the sample variance we expand $\{\sum x^2/n - (\sum x/n)^2\}^4$ in terms of sums of products of type $\sum x_i^\alpha x_j^\beta \ldots x_k^\gamma$ and, on taking expectations, replace them by $n(n-1)\ldots(n-t+1)$ $\mu'_\alpha \mu'_\beta \ldots \mu'_\gamma$ where t is the number of different suffixes $i, j \ldots k$ in the sum. This gives us the desired sampling moment in terms of population moments.

The method is straightforward enough. Its execution, however, leads to some tedious algebra and some cumbrous expressions, except in very simple cases. We shall therefore find it convenient to systematize the working in terms of the notation of symmetric functions.

12.5 Suppose that we have a set of xs, $x_1, x_2, \ldots, x_n$. When we write an expression such as $\sum x_i^2 x_j x_l^3 x_k$ we shall suppose that all the suffixes are different and that, subject to this, the summation takes place over all values of x. Thus, in the expression just given, there are $n(n-1)(n-2)(n-3)$ terms in the summation. It is to be noted that in an expression such as $\sum x_i x_j$ every pair occurs twice, for example $x_1 x_2$ and $x_2 x_1$; whereas in $\sum x_i^2 x_j$, $x_1^2 x_2$ and $x_2^2 x_1$ occur, but each does not occur twice because the powers differ.

The *augmented* symmetric functions are defined by

$$[p_1^{\pi_1} p_2^{\pi_2} \ldots p_s^{\pi_s}] = \sum x_i^{p_1} x_j^{p_1} \ldots x_q^{p_2} x_r^{p_2} \ldots x_u^{p_s} x_v^{p_s} \tag{12.5}$$

where there are π_1 powers p_1, π_2 powers p_2, and so on on the right of (12.5). For example

$$\sum x_i^2 x_j x_l^3 x_k = [1^2 23]; \quad \sum x_i^2 x_j^2 x_k^2 = [2^3].$$

More usual functions are the *monomial* symmetric functions defined by

$$(p_1^{\pi_1} p_2^{\pi_2} \ldots p_s^{\pi_s}) = [p_1^{\pi_1} p_2^{\pi_2} \ldots p_s^{\pi_s}]/\{\pi_1!\pi_2! \ldots \pi_s!\}. \tag{12.6}$$

Two particular cases are of special importance: the *unitary* functions

$$a_r = (1^r) = \sum x_i x_j \ldots x_l / r! \tag{12.7}$$

and the one-part functions or power sums

$$s_r = (r) = \sum x_i^r = [r]. \tag{12.8}$$

Tables exist giving these functions in terms of one another. The most useful are the tables giving the power sums in terms of the augmented symmetrics and vice versa (David and Kendall, 1949).

From (12.5) we have the fundamental result

$$E[p_1^{\pi_1} p_2^{\pi_2} \ldots p_s^{\pi_s}] = n(n-1) \ldots (n-\rho+1)(\mu'_{p_1})^{\pi_1}(\mu'_{p_2})^{\pi_2} \ldots (\mu'_{p_s})^{\pi_s} \tag{12.9}$$

where $\rho = \sum_{i=1}^{s} \pi_i$ and $p = \sum_{i=1}^{s} p_i \pi_i$ is the *weight* of the symmetric function. Appendix Table 10 gives the relationships up to weight 6.

Example 12.1 (Sample variance expressed in terms of power sums)

For the sample variance we have, in terms of the power sums,

$$m_2 = \frac{(2)}{n} - \frac{(1)^2}{n^2}.$$

From Appendix Table 10 or directly we have

$$(2) = [2]; \quad (1)^2 = [2] + [1^2].$$

Hence

$$m_2 = \frac{[2]}{n} - \frac{[2]+[1^2]}{n^2}$$

$$= \frac{n-1}{n^2}[2] - \frac{1}{n^2}[1^2]. \tag{12.10}$$

Taking expectations and using (12.9) we have

$$Em_2 = \frac{n-1}{n^2} n\mu_2' - \frac{1}{n^2} n(n-1)\mu_1'^2$$

$$= \frac{n-1}{n}\mu_2. \tag{12.11}$$

Let us note here a very useful abbreviation of the working. The statistic m_2 is independent of the origin of calculation and hence its sampling moments cannot depend on μ_1'. Without loss of generality, therefore, we can take the population mean to be zero, its other moments then becoming central. This implies that we can ignore any augmented symmetric function containing a unit. Thus from (12.10) we have at once

$$Em_2 = \frac{n-1}{n}\mu_2.$$

In a similar way we find

$$m_2^2 = \frac{(2)^2}{n^2} - \frac{2(2)(1)^2}{n^3} + \frac{(1)^4}{n^4}. \tag{12.12}$$

From Appendix Table 10

$$\begin{aligned} (2)^2 &= [4]+[2^2] \\ (2)(1)^2 &= [4]+2[31]+[2^2]+[21^2] \\ (1)^4 &= [4]+4[31]+3[2^2]+6[21^2]+[1^4]. \end{aligned}$$

We can ignore those augmented symmetrics containing a unit, and on substitution in (12.12) we find

$$m_2^2 = \frac{[4]+[2^2]}{n^2} - 2\frac{[4]+[2^2]}{n^3} + \frac{[4]+3[2^2]}{n^4}$$

$$= \frac{(n-1)^2}{n^4}[4] + \frac{n^2-2n+3}{n^4}[2^2],$$

whence, immediately,

$$Em_2^2 = \frac{(n-1)^2}{n^3}\mu_4 + \frac{(n-1)(n^2-2n+3)}{n^3}\mu_2^2. \tag{12.13}$$

We then find, using (12.11),

$$\operatorname{var} m_2 = Em_2^2 - (Em_2)^2$$
$$= \frac{(n-1)^2}{n^3}\mu_4 - \frac{(n-1)(n-3)}{n^3}\mu_2^2. \qquad (12.14)$$

For large n this becomes approximately

$$\operatorname{var} m_2 = \frac{\mu_4 - \mu_2^2}{n},$$

confirming the result we reached at (10.9) with $r = 2$.

If, in (12.14), we put $\kappa_4 = \mu_4 - 3\mu_2^2$, $\kappa_2 = \mu_2$ we find

$$\operatorname{var} m_2 = \left(\frac{n-1}{n}\right)^2 \left\{\frac{\kappa_4}{n} + \frac{2\kappa_2^2}{n-1}\right\}. \qquad (12.15)$$

***k*-statistics**

12.6 The algebraic complexity of the results obtained by this straightforward approach and the amount of work required to reach them, especially before the tables of symmetric functions were available, led to a search for simpler methods. Fisher (1929) revolutionized the subject in two ways: by proposing new symmetric functions of the observations, the so-called k-statistics; and by showing how their sampling cumulants could be obtained by combinatorial methods.

We consider a family of statistics $k_1, k_2, \ldots, k_p, \ldots$ that are symmetric functions of the observations and are such that the mean value of k_p is the pth cumulant κ_p:

$$Ek_p = \kappa_p. \qquad (12.16)$$

There is a possible source of confusion here due to notation. Whereas the moment-statistic m_p is the same function of the sample values as μ_p is of the population values, the same relation does not hold between k_p and κ_p. Conversely, Em_p is not equal to μ_p.

12.7 k_p is uniquely determined by this definition; for if there were two functions k_p and k_p' obeying (12.16) their difference $k_p - k_p'$ would have a zero mean value. But this difference is itself a symmetric function and can therefore be expressed as the sum of terms $\sum x_j^p$, $\sum x_j x_k^{p-1}$, etc., and hence its mean value is a series of terms each of which is a product of moments. The vanishing of this series would imply a relationship among the moments, which is impossible except perhaps for particular populations. Hence $k_p - k_p'$ must vanish identically and $k_p \equiv k_p'$.

Secondly, the ks are, like the central moments, independent of the origin of measurement, except for k_1, which is equal to the mean itself. In fact, we have by Taylor's theorem

$$k_p(x_1+h, x_2+h, \ldots, x_n+h)$$
$$= k_p(x_1, x_2, \ldots, x_n) + \sum_{r=1}^{\infty} \mathrm{D}^r k_p(x_1, x_2, \ldots, x_n) h^r/r!, \qquad (12.17)$$

where

$$\mathrm{D} = \frac{\partial}{\partial x_1} + \frac{\partial}{\partial x_2} + \dots + \frac{\partial}{\partial x_n}.$$

Taking expectations, and recalling from **3.13** that for $p > 1$ κ_p itself is independent of the origin, we have

$$\kappa_p = \kappa_p + \sum_{r=1}^{\infty} E[\mathrm{D}^r k_p(x) h^r / r!], \tag{12.18}$$

Thus each of the terms on the right vanishes separately, for (12.18) is an identity in h. This implies that $E\,\mathrm{D}^r k_p(x) = 0$, and thus we must have $\mathrm{D}^r k_p(x) = 0$ by the argument above, for $p > 1$, and hence, from (12.17), k_p is independent of h. For $p = 1$, k_1 has mean value $\kappa_1 = \mu_1'$ and thus

$$k_1 = \frac{1}{n}\sum x. \tag{12.19}$$

12.8 We now proceed to find explicit expressions for the k-statistics in terms of the observations $x_1, \dots, x_n$. By definition k_p is of degree p in these observations (for k_p is of order p in the moments, that is, the sum of the orders of the moments comprising any term in κ_p is p). We may then write

$$k_p = \sum\sum (x_1^{p_1} x_2^{p_1} \dots x_{\pi_1}^{p_1} x_{\pi_1+1}^{p_2} \dots x_{\pi_1+\pi_2}^{p_2} \dots x_{\pi_1+\dots+\pi_s}^{p_s}) A(p_1^{\pi_1} \dots p_s^{\pi_s}) \tag{12.20}$$

where the second summation extends over all the ways of assigning the $\pi_1 + \pi_2 + \dots + \pi_s = \rho$ subscripts (including permutations) from the n available and the first summation extends over all partitions of the number p, $(p_1^{\pi_1} p_2^{\pi_2} \dots p_s^{\pi_s})$. $A(p_1^{\pi_1} \dots p_s^{\pi_s})$ is a number depending on the partition.

We have

$$p_1\pi_1 + p_2\pi_2 + \dots + p_s\pi_s = p \tag{12.21}$$

and

$$\pi_1 + \pi_2 + \dots + \pi_s = \rho. \tag{12.22}$$

We assume $n \geqslant p$. On taking expectations of (12.20) we have, since the xs are independent,

$$\kappa_p = \sum\{(\mu_{p_1}^{\pi_1} \mu_{p_2}^{\pi_2} \dots \mu_{p_s}^{\pi_s}) AB\}, \tag{12.23}$$

where B is the number of ways of picking out the ρ subscripts from n, permutations allowed, and is therefore equal to $n(n-1)\dots(n-\rho+1) = n^{[\rho]}$.

Now from equation (3.40), we have

$$\kappa_p = p! \sum\sum \left(\frac{\mu_{p_1}}{p_1!}\right)^{\pi_1} \dots \left(\frac{\mu_{p_s}}{p_s!}\right)^{\pi_s} \frac{(-1)^{\rho-1}(\rho-1)!}{\pi_1! \dots \pi_s!}, \tag{12.24}$$

the summation extending over all partitions subject to (12.21) and (12.22). On identifying corresponding terms in (12.23) and (12.24) we find the values of the As and on substituting in (12.20) obtain finally

$$k_p = p! \sum \frac{(-1)^{\rho-1}(\rho-1)!}{n^{[\rho]}} \sum \frac{x_1^{p_1} \dots x_\rho^{p_s}}{(p_1!)^{\pi_1} \dots (p_s!)^{\pi_s} \pi_1! \dots \pi_s!}, \tag{12.25}$$

the explicit expression of k_p in terms of the xs.

We may notice an important simplification of this expression which is crucial in a discussion of the sampling properties of the ks. Apart from factors in ρ and n a typical term in (12.25) may be written

$$p! \left(\frac{x_1^{p_1}}{p_1!} \frac{x_2^{p_1}}{p_1!} \dots \frac{x_{\pi_1}^{p_1}}{p_1!} \dots \frac{x_\rho^{p_s}}{p_s!} \right) \cdot \frac{1}{\pi_1! \dots \pi_s!}$$

where, it is to be remembered, permutations of the subscripts are allowed. There will be a term of this type corresponding to every partition of ρ into πs and of p into p_js. Consequently we may write

$$k_p = \sum \frac{(-1)^{\rho-1}(\rho-1)!}{n^{[\rho]}} \sum x_{\gamma_1} x_{\gamma_2} \dots x_{\gamma_p}, \tag{12.26}$$

where there is a term in the second summation corresponding to every possible way of assigning the subscripts. In this assignment, subscripts are regarded as distinct entities. For example, if we choose p_1 of the subscripts to be 1, p_1 to be 2, ..., p_1 to be π_1, p_2 to be $\pi_1 + 1$, and so on, there will be as many different terms as there are ways of choosing them, i.e.

$$\frac{p!}{(p_1!)^{\pi_1} \dots (p_s!)^{\pi_s} \pi_1! \dots \pi_s!}. \tag{12.27}$$

In fact, (12.26) is a condensed form of (12.25) in which all the terms leading to the same x-product are added together, their number being given by (12.27).

Expression of k-statistics in terms of symmetric products and sums

12.9 We can write down the ks in terms of the augmented symmetric functions at once from the expressions for cumulants in terms of moments. For example, we have

$$\kappa_3 = \mu_3' - 3\mu_2'\mu_1' + 2\mu_1'^3.$$

Hence, comparing (12.24) and (12.25), we have

$$k_3 = \frac{[3]}{n} - \frac{3[21]}{n(n-1)} + \frac{2[1^3]}{n(n-1)(n-2)}.$$

Substituting for the augmented symmetrics in terms of power sums,

$$[3] = (3), \quad [21] = -(3) + (2)(1), \quad [1^3] = 2(3) - 3(2)(1) + (1)^3$$

we find

$$k_3 = \frac{1}{n(n-1)(n-2)}\{n^2(3) - 3n(2)(1) + 2(1)^3\}\ .$$

Writing $s_r = (r)$ we may put this in the form

$$k_3 = \frac{1}{n(n-1)(n-2)}(n^2 s_3 - 3ns_2 s_1 + 2s_1^3).$$

12.10 The first eight k-statistics in terms of the power sums are as follows:

$$\left.\begin{aligned}
k_1 &= \frac{1}{n} s_1 \\
k_2 &= \frac{1}{n^{[2]}}(ns_2 - s_1^2) \\
k_3 &= \frac{1}{n^{[3]}}(n^2 s_3 - 3ns_2 s_1 + 2s_1^3) \\
k_4 &= \frac{1}{n^{[4]}}\{(n^3+n^2)s_4 - 4(n^2+n)s_3 s_1 - 3(n^2-n)s_2^2 + 12ns_2 s_1^2 - 6s_1^4\} \\
k_5 &= \frac{1}{n^{[5]}}\{(n^4+5n^3)s_5 - 5(n^3+5n^2)s_4 s_1 - 10(n^3-n^2)s_3 s_2 + 20(n^2+2n)s_3 s_1^2 \\
&\quad + 30(n^2-n)s_2^2 s_1 - 60ns_2 s_1^3 + 24s_1^5\} \\
k_6 &= \frac{1}{n^{[6]}}\{(n^5+16n^4+11n^3-4n^2)s_6 - 6(n^4+16n^3+11n^2-4n)s_5 s_1 \\
&\quad - 15n(n-1)^2(n+4)s_4 s_2 - 10(n^4-2n^3+5n^2-4n)s_3^2 \\
&\quad + 30(n^3+9n^2+2n)s_4 s_1^2 + 120(n^3-n)s_3 s_2 s_1 + 30(n^3-3n^2+2n)s_2^3 \\
&\quad - 120(n^2+3n)s_3 s_1^3 - 270(n^2-n)s_2^2 s_1^2 + 360ns_2 s_1^4 - 120s_1^6\} \\
k_7 &= \frac{1}{n^{[7]}}\{(n^6+42n^5+119n^4-42n^3)s_7 - 7(n^5+42n^4+119n^3-42n^2)s_6 s_1 \\
&\quad - 21(n^5+12n^4-31n^3+18n^2)s_5 s_2 - 35(n^5+5n^3-6n^2)s_4 s_3 \\
&\quad + 42(n^4+27n^3+44n^2-12n)s_5 s_1^2 + 210(n^4+6n^3-13n^2+6n)s_4 s_2 s_1 \\
&\quad + 140(n^4+5n^2-6n)s_3^2 s_1 + 210(n^4-3n^3+2n^2)s_3 s_2^2 \\
&\quad - 210(n^3+13n^2+6n)s_4 s_1^3 - 1260(n^3+n^2-2n)s_3 s_2 s_1^2 \\
&\quad - 630(n^3-3n^2+2n)s_2^3 s_1 + 840(n^2+4n)s_3 s_1^4 + 2520(n^2-n)s_2^2 s_1^3 \\
&\quad - 2520ns_2 s_1^5 + 720s_1^7\}
\end{aligned}\right\} \quad (12.28)$$

$$\left.\begin{aligned}k_8 = \frac{1}{n^{[8]}}\{&(n^7+99n^6+757n^5+141n^4-398n^3+120n^2)s_8-8(n^6+99n^5+757n^4\\&+141n^3-398n^2+120n)s_7s_1-28(n^6+37n^5-39n^4-157n^3\\&+278n^2-120n)s_6s_2-56(n^6+9n^5-23n^4+111n^3-218n^2+120n)s_5s_3\\&-35(n^6+n^5+33n^4-121n^3+206n^2-120n)s_4^2+56(n^5+68n^4\\&+359n^3-8n^2-60n)s_6s_1^2+336(n^5+23n^4-31n^3-23n^2+30n)s_5s_2s_1\\&+560(n^5+5n^4+5n^3-5n^2-6n)s_4s_3s_1+420(n^5+2n^4-25n^3\\&+46n^2-24n)s_4s_2^2+560(n^5-4n^4+11n^3-20n^2+12n)s_3^2s_2\\&-336(n^4+38n^3+99n^2-18n)s_5s_1^3-2520(n^4+10n^3\\&-17n^2+6n)s_4s_2s_1^2-1680(n^4+2n^3+7n^2-10n)s_3^2s_1^2-5040(n^4-2n^3\\&-n^2+2n)s_3s_2^2s_1-630(n^4-6n^3+11n^2-6n)s_2^4+1680(n^3+17n^2\\&+12n)s_4s_1^4+13\,440(n^3+2n^2-3n)s_3s_2s_1^3+10\,080(n^3-3n^2+2n)s_2^3s_1^2\\&-6720(n^2+5n)s_3s_1^5-25\,200(n^2-n)s_2^2s_1^4+20\,160ns_2s_1^6-5040s_1^8\}\end{aligned}\right\} \quad (12.28)$$

In particular, taking the origin at zero $(s_1 = 0)$, we have

$$\left.\begin{aligned}k_1 &= m_1'\\k_2 &= \frac{n}{n-1}m_2\\k_3 &= \frac{n^2}{(n-1)(n-2)}m_3\\k_4 &= \frac{n^2}{(n-1)(n-2)(n-3)}\{(n+1)m_4-3(n-1)m_2^2\}\end{aligned}\right\} \quad (12.29)$$

expressing the ks in terms of the moment-statistics.

Formulae for k_9, k_{10} and k_{11} are given explicitly by Zia Ud-Din (1954, 1959). The last of these occupies two pages.

12.11 A well-known theorem of symmetric functions states that any rational integral algebraic symmetric function of $x_1, \ldots, x_n$ can be expressed uniquely, rationally, integrally and algebraically in terms of the symmetric sums s_r. It can thus be so expressed in terms of the ks, for from equations such as (12.28) the s_r can be so expressed in terms of the ks. Thus an investigation of the sampling constants of any symmetric function expressible in terms of rational integral algebraic symmetric functions can be translated into an investigation concerning the ks.

The reader who is prepared to take the algebra for granted may prefer to pass over the rest of this chapter and the next rather lightly, noting the main results without following the proofs. We are about to embark on a combinatorial method for deriving systematically the sampling cumulants of k-statistics in terms of population cumulants. We shall find that the results are substantially simpler than equivalent results stated in terms of moments.

Sampling cumulants of *k*-statistics: the combinatorial rules

12.12 The problem of determining the sampling moments or the sampling cumulants of *k*-statistics is that of finding mean values of powers and products of those statistics. To any number a with partition $(a_1^{\alpha_1} a_2^{\alpha_2} \ldots a_s^{\alpha_s})$ there will correspond a moment

$$\mu'(a_1^{\alpha_1} \ldots a_s^{\alpha_s}) = E(k_{a_1}^{\alpha_1} \ldots k_{a_s}^{\alpha_s}) \tag{12.30}$$

and a cumulant $\kappa(a_1^{\alpha_1} \ldots a_s^{\alpha_s})$ related to the moments by the identity (cf. (3.30))

$$\sum \left\{\kappa(a_1^{\alpha_1} \ldots a_s^{\alpha_s}) \frac{t_{a_1}^{\alpha_1}}{\alpha_1!} \ldots \frac{t_{a_s}^{\alpha_s}}{\alpha_s!}\right\} = \log \left\{\sum \mu'(b_1^{\beta_1} \ldots b_m^{\beta_m}) \frac{t_{b_1}^{\beta_1}}{\beta_1!} \ldots \frac{t_{b_m}^{\beta_m}}{\beta_m!}\right\}. \tag{12.31}$$

For example, the fourth cumulant of k_2 will be expressible in terms of the fourth moment of k_2 about the origin and moments of lower order. These quantities will be written $\kappa(2^4)$ and $\mu'(2^4)$, in accordance with (12.30). Again, the cumulant $\kappa(32)$ corresponds to the moment $\mu'(32)$, the mean value of $k_3 k_2$. Generally, in the simultaneous distribution of the *k*s there will be a separate formula of degree a for every partition of a.

Now the product $k_{a_1}^{\alpha_1} \ldots k_{a_s}^{\alpha_s}$ is homogeneous and of total degree a in the *x*s. Hence, when mean values are taken $\mu'(a_1^{\alpha_1} \ldots a_s^{\alpha_s})$ will be homogeneous and of total order a in the μs. Since the κs themselves are of homogeneous order in the μs it follows that $\kappa(a_1^{\alpha_1} \ldots a_s^{\alpha_s})$ is of homogeneous order in the κs. Hence we get the first rule for the sampling cumulants of *k*-statistics:

Rule 1. $\kappa(a_1^{\alpha_1} \ldots a_s^{\alpha_s})$ consists of the sum of terms each of which, except for constants, is a product of population κs of order a.

For instance, $\kappa(2^4)$ is of total order 8 and is therefore the sum of terms in κ_8, $\kappa_6\kappa_2$, $\kappa_5\kappa_3$, κ_4^2, $\kappa_4\kappa_2^2$, $\kappa_3^2\kappa_2$ and κ_2^4. Similarly $\kappa(32)$ will contain a term in κ_5 and one in $\kappa_3\kappa_2$ and no others. As seen in the next rule, no terms in κ_1 appear.

Rule 2. No term in $\kappa(a_1^{\alpha_1} \ldots a_s^{\alpha_s})$ contains κ_1, except $\kappa(1)$ itself.

This follows as in Example 12.1. The *k*-statistics are independent of the origin and hence their sampling distribution cannot depend on the variable quantity κ_1. The exception occurs when we are dealing with the only statistic that is dependent on the origin, namely k_1, and here $\kappa(1) = \kappa_1$ as is evident from the definitions.

12.13 We now enunciate and illustrate the rules by which the terms in $\kappa(a_1^{\alpha_1} \ldots a_s^{\alpha_s})$ can be found. As the proof of the validity of the rules is difficult to grasp until their nature has been comprehended we defer a proof until the next chapter.

To find the term in $\kappa_{b_1}^{\beta_1} \ldots \kappa_{b_m}^{\beta_m}$ in $\kappa(a_1^{\alpha_1} .. a_s^{\alpha_s})$ consider the two-way table

$$\begin{array}{cccc|c} & & & & b_1 \\ & & & & b_1 \\ & & & & \vdots \\ & & & & b_2 \\ & & & & \vdots \\ \hline a_1 & a_1 & \ldots a_2 & \ldots & a \end{array} \tag{12.32}$$

where there is a row corresponding to every κ in the term $\kappa_{b_1}^{\beta_1} \ldots \kappa_{b_m}^{\beta_m}$ and a column corresponding to every part in $\kappa(a_1^{\alpha_1} \ldots a_s^{\alpha_s})$. Consider the various possible arrays that can complete the body of the table by the insertion of positive integers whose row and column sums are the respective b and a numbers; e.g. if we are seeking the coefficient if $\kappa_6\kappa_2^2$ in $\kappa(4^2 2)$ we shall consider such arrays as

$$\begin{array}{ccc|c} 2 & 2 & 2 & 6 \\ 1 & 1 & . & 2 \\ 1 & 1 & . & 2 \\ \hline 4 & 4 & 2 & 10 \end{array} \qquad \begin{array}{ccc|c} 2 & 3 & 1 & 6 \\ 1 & 1 & . & 2 \\ 1 & . & 1 & 2 \\ \hline 4 & 4 & 2 & 10 \end{array} \qquad \begin{array}{ccc|c} 3 & 3 & . & 6 \\ 1 & . & 1 & 2 \\ . & 1 & 1 & 2 \\ \hline 4 & 4 & 2 & 10 \end{array} \qquad (12.33)$$

Then the rules by which these arrays give the coefficients of $\kappa_{b_1}^{\beta_1} \ldots \kappa_{b_m}^{\beta_m}$ are as follows:

Rule 3. Every array in which the numbers in the body of the array fall into two or more blocks, each confined to separate rows and columns, is to be ignored.

For instance, in the foregoing example

$$\begin{array}{ccc|c} 4 & 2 & . & 6 \\ . & 2 & . & 2 \\ . & . & 2 & 2 \\ \hline 4 & 4 & 2 & 10 \end{array}$$

is to be ignored, since the 2×2 block in the top left-hand corner has no row or column member in common with the entry in the bottom right-hand corner.

Rule 4. There will be a contribution to the coefficient of $\kappa_{b_1}^{\beta_1} \ldots \kappa_{b_m}^{\beta_m}$ in $\kappa(a_1^{\alpha_1} \ldots a_s^{\alpha_s})$ corresponding to each array that can complete (12.32) without becoming subject to Rule 3. Each contribution consists of a numerical coefficient multiplied by a function of n called the pattern function.

Rule 5. The numerical coefficient is the number of ways in which the column totals, considered as composed of distinct individuals, can be allocated to form the array concerned, divided by $\beta_1!\beta_2! \ldots \beta_m!$.

Rule 6. The pattern function depends only on the configuration of zeros in the array, not on the actual numbers composing it or on the row and column totals, and is given by considering the separations of the rows into distinct groups or separates.

(i) With a separation into one separate, there is associated the number n; with a separation into two separates, $n(n-1)$; ...; with a separation into q separates, $n(n-1) \ldots (n-q+1)$.

(ii) In each separation we count the number of separates in which a particular column is represented by a non-zero entry. If in ρ separates, we assign the factor

$$\frac{(-1)^{\rho-1}(\rho-1)!}{n(n-1)\dots(n-\rho+1)} \text{ to that column.}$$

This is done for each column and the factors multiplied together.

(iii) For each separation, the number in (i) is multiplied by the product in (ii).

(iv) The result, summed over all separations, gives the pattern function.

Rule 7. Any array containing a row that consists of a single non-zero entry has a vanishing pattern function and is to be ignored.

Rule 8. Any array containing a column that consists of a single non-zero entry has a pattern function $1/n$ times that of the array obtained by omitting that column.

Rule 9. Any array the non-zero elements of which consist of two groups connected only by a single column has a vanishing pattern function and is to be ignored.

Example 12.2 (Combinatorial rules for sampling cumulants)

As an illustration of these rules (which are not as difficult as they look), suppose that we seek the coefficient of $\kappa_6\kappa_2^2$ in $\kappa(4^2 2)$. If the reader will write down the thirty or so possible arrays with column totals 4, 4, 2 and row totals 6, 2, 2, it will be found that the only ones which do not vanish are those of (12.33) and permutations of rows and columns with the same sums, namely

$$\begin{array}{ccc|c} 2 & 2 & 2 & 6 \\ 1 & 1 & . & 2 \\ 1 & 1 & . & 2 \\ \hline 4 & 4 & 2 & 10 \end{array} \quad \begin{array}{ccc|c} 2 & 3 & 1 & 6 \\ 1 & 1 & . & 2 \\ 1 & . & 1 & 2 \\ \hline 4 & 4 & 2 & 10 \end{array} \quad \begin{array}{ccc|c} 3 & 2 & 1 & 6 \\ 1 & 1 & . & 2 \\ . & 1 & 1 & 2 \\ \hline 4 & 4 & 2 & 10 \end{array} \quad \begin{array}{ccc|c} 2 & 3 & 1 & 6 \\ 1 & . & 1 & 2 \\ 1 & 1 & . & 2 \\ \hline 4 & 4 & 2 & 10 \end{array}$$

(a) (b) (c) (d)

$$\begin{array}{ccc|c} 3 & 2 & 1 & 6 \\ . & 1 & 1 & 2 \\ 1 & 1 & . & 2 \\ \hline 4 & 4 & 2 & 10 \end{array} \quad \begin{array}{ccc|c} 3 & 3 & . & 6 \\ 1 & . & 1 & 2 \\ . & 1 & 1 & 2 \\ \hline 4 & 4 & 2 & 10 \end{array} \quad \begin{array}{ccc|c} 3 & 3 & . & 6 \\ . & 1 & 1 & 2 \\ 1 & . & 1 & 2 \\ \hline 4 & 4 & 2 & 10 \end{array} \qquad (12.34)$$

(e) (f) (g)

With practice the reader will find it unnecessary to write down arrays such as (c), (d) and (e), which are merely obtained from (b) by permuting rows and columns, but for clarity at this stage they have been set out in full. There is one trap here to be particularly noticed. In array (b) the two columns summing to 4 and the two rows summing to 2 are different, and their permutations result in 4 different arrays. But in array (f), though the rows and columns are different, permutations produce only 2 different arrays.

Each of these arrays contributed to the coefficient required. Consider first of all that from (a). The numerical coefficient is $\left(\frac{4!}{2!1!1!}\right)\left(\frac{4!}{2!1!1!}\right)\cdot\frac{1}{2} = 72$. The first factor in brackets is the number of ways of allocating 4 individuals in the partition 2, 1, 1, similarly for the second, and we divide by 2! since two of the row totals are equal, this being the only β not equal to unity.

Under Rule 8, the pattern function is $1/n$ times that of

$$\begin{array}{cc} \times & \times \\ \times & \times \\ \times & \times \end{array}$$

The separations, five in number, are as follows:

$$\begin{array}{cc} \times & \times \\ & \\ \times & \times \\ & \\ \times & \times \end{array} \qquad \begin{array}{cc} \times & \times \\ \hdashline \times & \times \\ & \\ \times & \times \end{array} \qquad \begin{array}{cc} \times & \times \\ \hdashline \times & \times \\ \hdashline \times & \times \end{array}\Big\} \qquad \begin{array}{cc} \times & \times \\ & \\ \times & \times \\ \hdashline \times & \times \end{array} \qquad \begin{array}{cc} \times & \times \\ \hdashline \times & \times \\ \hdashline \times & \times \end{array}$$

The first is, of course, the original pattern array (one separate). The next three consist each of two separates, obtained by taking alone the first, second and third row. The last consists of three separates, one for each row.

The contributions respectively under Rule 6 will be found to be

$$n.\left(\frac{1}{n}\right)\left(\frac{1}{n}\right) = \frac{1}{n}$$

$$3n(n-1)\left\{\frac{-1}{n(n-1)}\right\}\left\{\frac{-1}{n(n-1)}\right\} = \frac{3}{n(n-1)}$$

$$n(n-1)(n-2)\left\{\frac{(-1)^2 2!}{n(n-1)(n-2)}\right\}\left\{\frac{(-1)^2 2!}{n(n-1)(n-2)}\right\} = \frac{4}{n(n-1)(n-2)}.$$

The sum of these is $\dfrac{n}{(n-1)(n-2)}$ and hence the contribution from array (a) in (12.34) is $\dfrac{72}{(n-1)(n-2)}$.

Now for arrays (b) to (e), which all have the same numerical factor and the same pattern function and can therefore be considered together. For any one the numerical factor is

$$\left(\frac{4!}{2!1!1!}\right)\left(\frac{4!}{3!1!}\right)\left(\frac{2!}{1!1!}\right)\cdot\frac{1}{2!} = 48$$

and that of the four together is thus 192.

Under Rule 6 the pattern function will depend on the configuration

$$\begin{array}{ccc} \times & \times & \times \\ \times & \times & . \\ \times & . & \times \end{array}$$

where × stands for a non-zero entry and a period for a zero entry. There are five separations of this, one of one separate, three of two separates, and one of three separates. The contribution from the first is

$$n\frac{1}{n}\frac{1}{n}\frac{1}{n}=\frac{1}{n^2}$$

for each column has a non-zero entry in the separate. The contribution from the three separations given respectively by isolating the first, second and third row will be found to be

$$n(n-1)\left[\frac{-1}{n^3(n-1)^3}+\frac{1}{n^3(n-1)^2}+\frac{1}{n^3(n-1)^2}\right]=\frac{2n-3}{n^2(n-1)^2}.$$

The contribution from the separation of three separates is

$$n(n-1)(n-2)\left[\frac{2!}{n(n-1)(n-2)}\cdot\frac{-1}{n(n-1)}\cdot\frac{-1}{n(n-1)}\right]=\frac{2}{n^2(n-1)^2}.$$

The pattern function is the sum of these three contributions and is thus $1/(n-1)^2$.

The contribution from arrays (f) and (g) in (12.34) will be found to be $32/(n-1)^2$.

Hence, adding all the contributions together, we find that the coefficient of $\kappa_6\kappa_2^2$ in $\kappa(4^2 2)$ is

$$\frac{72}{(n-1)(n-2)}+\frac{192}{(n-1)^2}+\frac{32}{(n-1)^2}=\frac{8(37n-65)}{(n-1)^2(n-2)},$$

as shown in equation (12.66) below.

Rule 10. The expression for any $\kappa(a_1^{\alpha_1}\ldots)$ which contains a unit part may be obtained from that without the part by (1) dividing throughout by n and (2) increasing the suffix of one of the κs by unity in every possible way.

For example, it may be seen from (12.15) that

$$\kappa(2^2)=\frac{\kappa_4}{n}+\frac{2\kappa_2^2}{n-1}.$$

Hence

$$\kappa(2^2 1)=\frac{\kappa_5}{n^2}+\frac{4\kappa_3\kappa_2}{n(n-1)}$$

$$\kappa(2^2 1^2)=\frac{\kappa_6}{n^3}+\frac{4\kappa_3^2}{n^2(n-1)}+\frac{4\kappa_4\kappa_2}{n^2(n-1)},$$

and so on.

12.14 The reader cannot be blamed for doubting whether this rather elaborate combinatorial procedure represents much of an advance on the straightforward algebraical approach considered earlier in the chapter. A few of the more complicated cases should be tried both ways. In point of fact, it is so easy to make algebraical slips with either method that most formulae in current use have been checked by being calculated by both methods. The combinatorial method, however, is something more than an algebraical short cut. We shall find later that several results of importance are derivable directly from it, whereas the straightforward evaluation of expectations yields them only by non-straightforward manoeuvres.

Example 12.3 (Variance of sample variance, using sampling cumulants)

For comparison with Example 12.1 let us find the variance of m_2 by the combinatorial method.

From (12.29) we have

$$k_2 = \frac{n}{n-1} m_2.$$

Hence

$$\begin{aligned} \text{var}\, m_2 &= \left(\frac{n-1}{n}\right)^2 \text{var}\, k_2 \\ &= \left(\frac{n-1}{n}\right)^2 \kappa(2^2). \end{aligned}$$

$\kappa(2^2)$ consists of two terms, one in κ_4 and one in κ_2^2. The only array contributing to the first is

$$\begin{array}{cc|c} 2 & 2 & 4 \\ \hline 2 & 2 & 4 \end{array}$$

with a numerical factor unity and a pattern function $1/n$. The arrays giving the second are of type

$$\begin{array}{cc|c} & & 2 \\ & & 2 \\ \hline 2 & 2 & 4 \end{array}$$

If any entry in this were a 2 the row in which it appeared would contain only a single non-zero entry and hence the pattern function of the array would vanish by Rule 7. The only contributing array is therefore

$$\begin{array}{cc|c} 1 & 1 & 2 \\ 1 & 1 & 2 \\ \hline 2 & 2 & 4 \end{array}$$

The numerical coefficient is $\left(\frac{2!}{1!1!}\right)^2 \cdot \frac{1}{2!} = 2$. The pattern function will be found to be

$1/(n-1)$. Hence

$$\kappa(2^2) = \frac{\kappa_4}{n} + \frac{2\kappa_2^2}{n-1}$$

as given in (12.35), and

$$\begin{aligned}\operatorname{var} m_2 &= \left(\frac{n-1}{n}\right)^2 \kappa(2^2)\\ &= \left(\frac{n-1}{n}\right)^2 \left\{\frac{1}{n}(\mu_4 - 3\mu_2^2) + \frac{2}{n-1}\mu_2^2\right\}\\ &= \frac{(n-1)^2}{n^3}\mu_4 + \frac{(3-n)(n-1)}{n^3}\mu_2^2,\end{aligned}$$

agreeing with (12.14).

Example 12.4 (Third moment of sample variance, using sampling cumulants)

To find the third moment of k_2, which is equal to its third cumulant $\kappa(2^3)$. This will be the sum of factors in κ_6, $\kappa_4\kappa_2$, κ_3^2 and κ_2^3.

The coefficient of the first is $1/n^2$. For the second we have to consider the array

$$\begin{array}{ccc|c} 1 & 1 & 2 & 4 \\ 1 & 1 & . & 2 \\ \hline 2 & 2 & 2 & 6 \end{array}$$

all others vanishing except the two equivalent partitions obtained when the column with the single entry appears in the first or second place. The numerical factor is then

$$3.\left(\frac{2!}{1!1!}\right)^2 = 12.$$

The pattern function is $1/n$ times that of

$$\begin{array}{cc} \times & \times \\ \times & \times \end{array}$$

i.e. is $1/\{n(n-1)\}$. The coefficient of $\kappa_4\kappa_2$ is then $12/\{n(n-1)\}$.

For the term in κ_3^2 the only contributory array is

$$\begin{array}{ccc|c} 1 & 1 & 1 & 3 \\ 1 & 1 & 1 & 3 \\ \hline 2 & 2 & 2 & 6 \end{array}$$

with a numerical factor $\left(\frac{2!}{1!1!}\right)^3 \cdot \frac{1}{2!} = 4$ and pattern function $\frac{n-2}{n(n-1)^2}$.

For the last term we have to consider the array

$$\begin{array}{ccc|c} 1 & 1 & . & 2 \\ 1 & . & 1 & 2 \\ . & 1 & 1 & 2 \\ \hline 2 & 2 & 2 & 6 \end{array}$$

with a numerical coefficient 8 and a pattern function $\frac{1}{(n-1)^2}$. Collecting terms together we get

$$\kappa(2^3) = \frac{\kappa_6}{n^2} + \frac{12\kappa_4\kappa_2}{n(n-1)} + \frac{4(n-2)}{n(n-1)^2}\kappa_3^2 + \frac{8}{(n-1)^2}\kappa_2^3.$$

We thus see that if the population is normal the third moment of k_2 reduces to $\frac{8\kappa_2^3}{(n-1)^2}$, i.e. is of order n^{-2}.

12.15 In consequence of Rules 1 and 2 and Exercise 3.5, any $\kappa(a_1^{\alpha_1} \dots a_s^{\alpha_s}) = 0$ if a is odd and the population is symmetrical.

Few things better illustrate the usefulness of expressing the formulae in terms of cumulants and the power of the combinatorial method than the simplification resulting when the population is normal. In this case only terms in κ_2 survive, all higher cumulants vanishing.

Example 12.5 (Sampling cumulants for the normal distribution)

As an illustration let us prove that $\kappa(pq) = 0$ for normal samples unless $p = q$.

The only term which can appear in $\kappa(pq)$ is $\kappa_2^{\frac{1}{2}(p+q)}$ and evidently, if $p+q$ is odd, this itself is impossible and $\kappa(pq)$ must vanish. If $p+q$ is even we have to consider the array

$$\begin{array}{cc|c} & & 2 \\ & & 2 \\ & & \cdot \\ & & \cdot \\ & & \cdot \\ & & 2 \\ \hline p & q & p+q \end{array}$$

Now if any entry in this array is 2 the pattern function of the array vanishes by Rule 7 since the row concerned will contain only one non-zero entry. The reverse can only happen if all the entries are unity, in which case the sums p and q must be equal. This establishes the result.

Example 12.6 (Order in n of sampling cumulants)

If $\kappa(a_1^{\pi_1} \dots a_s^{\pi_s})$ contains r parts, every term in it is of order $n^{-(r-1)}$. For example every term in $\kappa(3^2 2^2)$ is of order n^{-3}.

By Rule 7, every array with a non-vanishing pattern function has at least two non-zero entries (n.z.e.) in each row. By Rule 8, we may confine ourselves to arrays with at least two n.z.e. in each column. We have to show that the pattern function of every such r-column array is of order $n^{-(r-1)}$. Every array has a single-separate separation; by Rule 6, this contributes $n\left(\frac{1}{n}\right)^r = n^{-(r-1)}$. We now need only show that no separation with more than one separate has a contribution of higher order. Consider any 2-separate separation, S_2; if the pattern function is not to vanish by Rule 9, at least 2 columns must have a n.z.e. in each separate, and by Rule 6 the contribution of S_2 is of order $n^2\left(\frac{1}{n^2}\right)^2\left(\frac{1}{n}\right)^{r-2}$ or less, i.e. it is $o(n^{-(r-1)})$. Any 3-separate separation, S_3, may be derived by a sub-separation of an S_2, and using Rule 6 it follows that the contribution of S_3 is of order $n^3\left(\frac{1}{n^3}\right)^3\left(\frac{1}{n}\right)^{r-3}$, i.e. $o(n^{-(r-1)})$. The same argument, step-by-step, shows that no separation can contribute a term of order as high as $n^{-(r-1)}$. This completes the proof.

Example 12.7 (Characterization of the normal distribution by independence of the sample mean and variance)

We may use the properties of sampling cumulants to prove a characterizing property of the normal distribution that was discussed analytically in Example 11.3 and geometrically in Example 11.7: namely, that the mean and variance are independent in normal samples.

First, we recall from **4.16–17** that two variates are independent if and only if their joint c.f. factorizes into their univariate c.f.s, i.e.,

$$\phi(t_1, t_2) = \phi_1(t_1)\phi_2(t_2).$$

As in the univariate case, any finite product moment may be obtained by differentiating the joint c.f. and putting $t_1 = t_2 = 0$–c.f. (3.74), while if the joint distribution is uniquely determined by its moments we may expand the c.f. as an infinite series–c.f. **4.21**. Expanding all three c.f.s and taking logarithms, we have on using (3.74)

$$\sum_{r,s=0}^{\infty} \kappa_{rs}\frac{(it_1)^r}{r!}\frac{(it_2)^s}{s!} = \sum_{r=1}^{\infty} \kappa_{r0}\frac{(it_1)^r}{r!} + \sum_{s=1}^{\infty} \kappa_{0s}\frac{(it_2)^s}{s!}$$

which implies that $\kappa_{rs} = 0$ if the product $rs \neq 0$. Even if the infinite series expansion is impermissible, it is clear that any finite $\kappa_{rs} = 0$ if $rs \neq 0$.

Conversely, if all $\kappa_{rs} = 0$, $rs \neq 0$, and the c.f. may be thus expanded, independence of the variates follows, so that this is then a necessary and sufficient cordition for independence. Its stringency may be gauged from Exercise 8.25, which gave a bivariate distribution for which $\kappa_{rs} = 0$ for any $r \neq 2$ and any s, with the variables clearly dependent.

Now consider the joint distribution in normal samples of mean and variance, or equivalently of k_1 and k_2. Since $\kappa_p = 0$, $p > 2$, we have from Rule 10 that $\kappa(2^s 1^r) = 0$ (any r, $s \neq 0$). Hence k_1 and k_2 are independent, as we have already seen.

We may now establish the converse, due to Geary (1936a), that only in normal variation will mean and variance be independent. For, by Rule 10,

$$\kappa(21^r) = \kappa_{r+2}/n^r, \qquad r > 0.$$

If k_1 and k_2 are independent this is zero for all $r > 0$. This implies normality on the part of the population.

It is remarkable that we have not had to use the relations $\kappa(2^s1^r) = 0$ for $s > 1$ to establish the converse; and that the converse requires independence for only one value of n.

Later writers, e.g. Lukacs (1942), have proved Geary's theorem under less restrictive conditions on the existence of cumulants—cf. Exercise 11.19.

Sampling cumulants of *k*-statistics: the results

12.16 The following formulae give results for statistics of degree not greater than 10, with some of the 12th degree. They hold for n not less than the order of the k-statistic considered.

No separate formulae are needed for the first k-statistic, since Rule 10 gives the general result $\kappa(1^{r+1}) = \kappa_{r+1}/n^r$.

Second *k*-statistic

$$\kappa(2^2) = \frac{\kappa_4}{n} + \frac{2\kappa_2^2}{n-1} \tag{12.35}$$

$$\kappa(2^3) = \frac{\kappa_6}{n^2} + \frac{12\kappa_4\kappa_2}{n(n-1)} + \frac{4(n-2)}{n(n-1)^2}\kappa_3^2 + \frac{8}{(n-1)^2}\kappa_2^3 \tag{12.36}$$

$$\begin{aligned}\kappa(2^4) = {} & \frac{\kappa_8}{n^3} + \frac{24}{n^2(n-1)}\kappa_6\kappa_2 + \frac{32(n-2)}{n^2(n-1)^2}\kappa_5\kappa_3 + \frac{8(4n^2-9n+6)}{n^2(n-1)^3}\kappa_4^2 \\ & + \frac{144}{n(n-1)^2}\kappa_4\kappa_2^2 + \frac{96(n-2)}{n(n-1)^3}\kappa_3^2\kappa_2 + \frac{48}{(n-1)^3}\kappa_2^4\end{aligned} \tag{12.37}$$

$$\begin{aligned}\kappa(2^5) = {} & \frac{\kappa_{10}}{n^4} + \frac{40\kappa_8\kappa_2}{n^3(n-1)} + \frac{80(n-2)}{n^3(n-1)^2}\kappa_7\kappa_3 + \frac{40(5n^2-12n+9)}{n^3(n-1)^3}\kappa_6\kappa_4 \\ & + \frac{16(n-2)(6n^2-12n+7)}{n^3(n-1)^4}\kappa_5^2 + \frac{480}{n^2(n-1)^2}\kappa_6\kappa_2^2 + \frac{1280(n-2)}{n^2(n-1)^3}\kappa_5\kappa_3\kappa_2 \\ & + \frac{320(4n^2-9n+6)}{n^2(n-1)^4}\kappa_4^2\kappa_2 + \frac{480(2n^2-7n+6)}{n^2(n-1)^4}\kappa_4\kappa_3^2 + \frac{1920}{n(n-1)^3}\kappa_4\kappa_2^3 \\ & + \frac{1920(n-2)}{n(n-1)^4}\kappa_3^2\kappa_2^2 + \frac{384}{(n-1)^4}\kappa_2^5\end{aligned} \tag{12.38}$$

$$\kappa(2^6) = \frac{1}{n^5}\kappa_{12} + \frac{60}{n^4(n-1)}\kappa_{10}\kappa_2 + \frac{160(n-2)}{n^4(n-1)^2}\kappa_9\kappa_3 + \frac{240(2n^2-5n+4)}{n^4(n-1)^3}\kappa_8\kappa_4$$
$$+ \frac{96(n-2)(7n^2-14n+9)}{n^4(n-1)^4}\kappa_7\kappa_5 + \frac{4(113n^4-520n^3+950n^2-800n+265)}{n^4(n-1)^5}\kappa_6^2$$
$$+ \frac{1200}{n^3(n-1)^2}\kappa_8\kappa_2^2 + \frac{4800(n-2)}{n^3(n-1)^3}\kappa_7\kappa_3\kappa_2 + \frac{2400(5n^2-12n+9)}{n^3(n-1)^4}\kappa_6\kappa_4\kappa_2$$
$$+ \frac{160(n-2)(31n-53)}{n^3(n-1)^4}\kappa_6\kappa_3^2 + \frac{960(n-2)(6n^2-12n+7)}{n^3(n-1)^5}\kappa_5^2\kappa_2$$
$$+ \frac{1920(n-2)(9n^2-23n+16)}{n^3(n-1)^5}\kappa_5\kappa_4\kappa_3 + \frac{480(11n^3-41n^2+59n-31)}{n^3(n-1)^5}\kappa_4^3$$
$$+ \frac{9600}{n^2(n-1)^3}\kappa_6\kappa_2^3 + \frac{38\,400(n-2)}{n^2(n-1)^4}\kappa_5\kappa_3\kappa_2^2 + \frac{9600(4n^2-9n+6)}{n^2(n-1)^5}\kappa_4^2\kappa_2^2$$
$$+ \frac{28\,800(2n^2-7n+6)}{n^2(n-1)^5}\kappa_4\kappa_3^2\kappa_2 + \frac{960(n-2)(5n-12)}{n^2(n-1)^5}\kappa_3^4 + \frac{28\,800}{n(n-1)^4}\kappa_4\kappa_2^4$$
$$+ \frac{38\,400(n-2)}{n(n-1)^5}\kappa_3^2\kappa_2^3 + \frac{3840}{(n-1)^5}\kappa_2^6 \qquad (12.39)$$

Third k-statistic

$$\kappa(3^2) = \frac{1}{n}\kappa_6 + \frac{9}{n-1}\kappa_4\kappa_2 + \frac{9}{n-1}\kappa_3^2 + \frac{6n}{(n-1)(n-2)}\kappa_2^3 \qquad (12.40)$$

$$\kappa(3^3) = \frac{1}{n^2}\kappa_9 + \frac{27}{n(n-1)}\kappa_7\kappa_2 + \frac{27(3n-4)}{n(n-1)^2}\kappa_6\kappa_3 + \frac{27(4n-7)}{n(n-1)^2}\kappa_5\kappa_4$$
$$+ \frac{54(4n-7)}{(n-1)^2(n-2)}\kappa_5\kappa_2^2 + \frac{162(5n-12)}{(n-1)^2(n-2)}\kappa_4\kappa_3\kappa_2 + \frac{36(7n^2-30n+34)}{(n-1)^2(n-2)^2}\kappa_3^3$$
$$+ \frac{108n(5n-12)}{(n-1)^2(n-2)^2}\kappa_3\kappa_2^3 \qquad (12.41)$$

$$\kappa(3^4) = \frac{1}{n^3}\kappa_{12} + \frac{54}{n^2(n-1)}\kappa_{10}\kappa_2 + \frac{108(2n-3)}{n^2(n-1)^2}\kappa_9\kappa_3 + \frac{27(17n^2-49n+35)}{n^2(n-1)^3}\kappa_8\kappa_4$$
$$+ \frac{108(7n^2-20n+16)}{n^2(n-1)^3}\kappa_7\kappa_5 + \frac{27(17n^2-47n+39)}{n^2(n-1)^3}\kappa_6^2 + \frac{27(37n-70)}{n(n-1)^2(n-2)}\kappa_8\kappa_2^2$$
$$+ \frac{324(19n^2-67n+54)}{n(n-1)^3(n-2)}\kappa_7\kappa_3\kappa_2 + \frac{162(65n^2-245n+234)}{n(n-1)^3(n-2)}\kappa_6\kappa_4\kappa_2$$
$$+ \frac{108(82n^3-481n^2+958n-640)}{n(n-1)^3(n-2)^2}\kappa_6\kappa_3^2 + \frac{108(59n^2-220n+224)}{n(n-1)^3(n-2)}\kappa_5^2\kappa_2$$

$$+\frac{324(75n^3-473n^2+1016n-756)}{n(n-1)^3(n-2)^2}\kappa_5\kappa_4\kappa_3$$
$$+\frac{27(173n^4-1503n^3+4962n^2-7380n+4200)}{n(n-1)^3(n-2)^3}\kappa_4^3$$
$$+\frac{108(71n^2-263n+234)}{(n-1)^3(n-2)^2}\kappa_6\kappa_2^3+\frac{648(79n^2-343n+378)}{(n-1)^3(n-2)^2}\kappa_5\kappa_3\kappa_2^2$$
$$+\frac{486(63n^2-290n+352)}{(n-1)^3(n-2)^2}\kappa_4^2\kappa_2^2+\frac{972(99n^3-688n^2+1612n-1280)}{(n-1)^3(n-2)^3}\kappa_4\kappa_3^2\kappa_2$$
$$+\frac{162(87n^3-594n^2+1420n-1176)}{(n-1)^3(n-2)^3}\kappa_3^4+\frac{972n(23n^2-103n+118)}{(n-1)^3(n-2)^3}\kappa_4\kappa_2^4$$
$$+\frac{648n(103n^2-510n+640)}{(n-1)^3(n-2)^3}\kappa_3^2\kappa_2^3+\frac{648n^2(5n-12)}{(n-1)^3(n-2)^3}\kappa_2^6 \qquad (12.42)$$

Fourth *k*-statistic

$$\kappa(4^2)=\frac{1}{n}\kappa_8+\frac{16}{n-1}\kappa_6\kappa_2+\frac{48}{n-1}\kappa_5\kappa_3+\frac{34}{n-1}\kappa_4^2+\frac{72n}{(n-1)(n-2)}\kappa_4\kappa_2^2$$
$$+\frac{144n}{(n-1)(n-2)}\kappa_3^2\kappa_2+\frac{24n(n+1)}{(n-1)(n-2)(n-3)}\kappa_2^4 \qquad (12.43)$$

$$\kappa(4^3)=\frac{1}{n^2}\kappa_{12}+\frac{48}{n(n-1)}\kappa_{10}\kappa_2+\frac{16(13n-17)}{n(n-1)^2}\kappa_9\kappa_3+\frac{12(41n-65)}{n(n-1)^2}\kappa_8\kappa_4$$
$$+\frac{48(16n-29)}{n(n-1)^2}\kappa_7\kappa_5+\frac{12(37n-70)}{n(n-1)^2}\kappa_6^2+\frac{72(11n-19)}{(n-1)^2(n-2)}\kappa_8\kappa_2^2$$
$$+\frac{288(19n-41)}{(n-1)^2(n-2)}\kappa_7\kappa_3\kappa_2+\frac{48(203n-523)}{(n-1)^2(n-2)}\kappa_6\kappa_4\kappa_2$$
$$+\frac{144(56n^2-257n+302)}{(n-1)^2(n-2)^2}\kappa_6\kappa_3^2+\frac{1440(4n-11)}{(n-1)^2(n-2)}\kappa_5^2\kappa_2$$
$$+\frac{1152(22n^2-106n+133)}{(n-1)^2(n-2)^2}\kappa_5\kappa_4\kappa_3+\frac{8(709n^2-3430n+4456)}{(n-1)^2(n-2)^2}\kappa_4^3$$
$$+\frac{288(19n^3-98n^2+125n+2)}{(n-1)^2(n-2)^2(n-3)}\kappa_6\kappa_2^3+\frac{1728(24n^3-140n^2+200n+4)}{(n-1)^2(n-2)^2(n-3)}\kappa_5\kappa_3\kappa_2^2$$
$$+\frac{432(61n^3-371n^2+552n+12)}{(n-1)^2(n-2)^2(n-3)}\kappa_4^2\kappa_2^2+\frac{864(103n^3-629n^2+948n+24)}{(n-1)^2(n-2)^2(n-3)}\kappa_4\kappa_3^2\kappa_2$$
$$+\frac{288(41n^4-384n^3+1209n^2-1282n-36)}{(n-1)^2(n-2)^2(n-3)^2}\kappa_3^4+\frac{288n(53n^2-179n-52)}{(n-1)^2(n-2)^2(n-3)}\kappa_4\kappa_2^4$$
$$+\frac{1728n(29n^3-196n^2+317n+62)}{(n-1)^2(n-2)^2(n-3)^2}\kappa_3^2\kappa_2^3+\frac{1728n(n+1)(n^2-5n+2)}{(n-1)^2(n-2)^2(n-3)^2}\kappa_2^6 \qquad (12.44)$$

Fifth k-statistic

$$\kappa(5^2) = \frac{1}{n}\kappa_{10} + \frac{25}{n-1}\kappa_8\kappa_2 + \frac{100}{n-1}\kappa_7\kappa_3 + \frac{200}{n-1}\kappa_6\kappa_4 + \frac{125}{n-1}\kappa_5^2$$
$$+ \frac{200n}{(n-1)(n-2)}\kappa_6\kappa_2^2 + \frac{1200n}{(n-1)(n-2)}\kappa_5\kappa_3\kappa_2 + \frac{850n}{(n-1)(n-2)}\kappa_4^2\kappa_2$$
$$+ \frac{1500n}{(n-1)(n-2)}\kappa_4\kappa_3^2 + \frac{600n(n+1)}{(n-1)(n-2)(n-3)}\kappa_4\kappa_2^3$$
$$+ \frac{1800n(n+1)}{(n-1)(n-2)(n-3)}\kappa_3^2\kappa_2^2 + \frac{120n^2(n+5)}{(n-1)(n-2)(n-3)(n-4)}\kappa_2^5 \tag{12.45}$$

Sixth k-statistic

$$\kappa(6^2) = \frac{1}{n}\kappa_{12} + \frac{1}{n-1}(36\kappa_{10}\kappa_2 + 180\kappa_9\kappa_3 + 465\kappa_8\kappa_4 + 780\kappa_7\kappa_5 + 461\kappa_6^2)$$
$$+ \frac{n}{(n-1)(n-2)}(450\kappa_8\kappa_2^2 + 3600\kappa_7\kappa_3\kappa_2 + 7200\kappa_6\kappa_4\kappa_2 + 6300\kappa_6\kappa_3^2$$
$$+ 4500\kappa_5^2\kappa_2 + 21\,600\kappa_5\kappa_4\kappa_3 + 4950\kappa_4^3)$$
$$+ \frac{n(n+1)}{(n-1)(n-2)(n-3)}(2400\kappa_6\kappa_2^3 + 21\,600\kappa_5\kappa_3\kappa_2^2 + 15\,300\kappa_4^2\kappa_2^2$$
$$+ 54\,000\kappa_4\kappa_3^2\kappa_2 + 8100\kappa_3^4)$$
$$+ \frac{n^2(n+5)}{(n-1)(n-2)(n-3)(n-4)}(5400\kappa_4\kappa_2^4 + 21\,600\kappa_3^2\kappa_2^3)$$
$$+ \frac{n(n+1)(n^2+15n-4)}{(n-1)(n-2)(n-3)(n-4)(n-5)}720\,\kappa_2^6 \tag{12.46}$$

Product cumulant formulae

$$\kappa(32) = \frac{1}{n}\kappa_5 + \frac{6}{n-1}\kappa_3\kappa_2 \tag{12.47}$$

$$\kappa(42) = \frac{1}{n}\kappa_6 + \frac{8}{n-1}\kappa_4\kappa_2 + \frac{6}{n-1}\kappa_3^2 \tag{12.48}$$

$$\kappa(52) = \frac{1}{n}\kappa_7 + \frac{10}{n-1}\kappa_5\kappa_2 + \frac{20}{n-1}\kappa_4\kappa_3 \tag{12.49}$$

$$\kappa(62) = \frac{1}{n}\kappa_8 + \frac{12}{n-1}\kappa_6\kappa_2 + \frac{30}{n-1}\kappa_5\kappa_3 + \frac{20}{n-1}\kappa_4^2 \tag{12.50}$$

$$\kappa(72) = \frac{1}{n}\kappa_9 + \frac{14}{n-1}\kappa_7\kappa_2 + \frac{42}{n-1}\kappa_6\kappa_3 + \frac{70}{n-1}\kappa_5\kappa_4 \tag{12.51}$$

$$\kappa(82) = \frac{1}{n}\kappa_{10} + \frac{16}{n-1}\kappa_8\kappa_2 + \frac{56}{n-1}\kappa_7\kappa_3 + \frac{112}{n-1}\kappa_6\kappa_4 + \frac{70}{n-1}\kappa_5^2 \tag{12.52}$$

$$\kappa(43) = \frac{1}{n}\kappa_7 + \frac{12}{n-1}\kappa_5\kappa_2 + \frac{30}{n-1}\kappa_4\kappa_3 + \frac{36n}{(n-1)(n-2)}\kappa_3\kappa_2^2 \tag{12.53}$$

$$\kappa(53) = \frac{1}{n}\kappa_8 + \frac{1}{n-1}(15\kappa_6\kappa_2 + 45\kappa_5\kappa_3 + 30\kappa_4^2) + \frac{n}{(n-1)(n-2)}(60\kappa_4\kappa_2^2 + 90\kappa_3^2\kappa_2) \tag{12.54}$$

$$\kappa(63) = \frac{1}{n}\kappa_9 + \frac{1}{n-1}(18\kappa_7\kappa_2 + 63\kappa_6\kappa_3 + 105\kappa_5\kappa_4) + \frac{n}{(n-1)(n-2)}(90\kappa_5\kappa_2^2 + 360\kappa_4\kappa_3\kappa_2 + 90\kappa_3^3) \tag{12.55}$$

$$\kappa(73) = \frac{1}{n}\kappa_{10} + \frac{1}{n-1}(21\kappa_8\kappa_2 + 84\kappa_7\kappa_3 + 168\kappa_6\kappa_4 + 105\kappa_5^2) + \frac{n}{(n-1)(n-2)}(126\kappa_6\kappa_2^2 + 630\kappa_5\kappa_3\kappa_2 + 420\kappa_4^2\kappa_2 + 630\kappa_4\kappa_3^2) \tag{12.56}$$

$$\kappa(54) = \frac{1}{n}\kappa_9 + \frac{1}{n-1}(20\kappa_7\kappa_2 + 70\kappa_6\kappa_3 + 120\kappa_5\kappa_4) + \frac{n}{(n-1)(n-2)}(120\kappa_5\kappa_2^2 + 600\kappa_4\kappa_3\kappa_2 + 180\kappa_3^3) + \frac{n(n+1)}{(n-1)(n-2)(n-3)}240\kappa_3\kappa_2^3 \tag{12.57}$$

$$\kappa(64) = \frac{1}{n}\kappa_{10} + \frac{1}{n-1}(24\kappa_8\kappa_2 + 96\kappa_7\kappa_3 + 194\kappa_6\kappa_4 + 120\kappa_5^2) + \frac{n}{(n-1)(n-2)}(180\kappa_6\kappa_2^2 + 1080\kappa_5\kappa_3\kappa_2 + 720\kappa_4^2\kappa_2 + 1260\kappa_4\kappa_3^2) + \frac{n(n+1)}{(n-1)(n-2)(n-3)}(480\kappa_4\kappa_2^3 + 1080\kappa_3^2\kappa_2^2) \tag{12.58}$$

$$\kappa(32^2) = \frac{1}{n^2}\kappa_7 + \frac{16}{n(n-1)}\kappa_5\kappa_2 + \frac{12(2n-3)}{n(n-1)^2}\kappa_4\kappa_3 + \frac{48}{(n-1)^2}\kappa_3\kappa_2^2 \tag{12.59}$$

$$\kappa(42^2) = \frac{1}{n^2}\kappa_8 + \frac{20}{n(n-1)}\kappa_6\kappa_2 + \frac{8(5n-7)}{n(n-1)^2}\kappa_5\kappa_3 + \frac{4(7n-10)}{n(n-1)^2}\kappa_4^2 + \frac{80}{(n-1)^2}\kappa_4\kappa_2^2 + \frac{120}{(n-1)^2}\kappa_3^2\kappa_2 \tag{12.60}$$

$$\kappa(52^2) = \frac{1}{n^2}\kappa_9 + \frac{24}{n(n-1)}\kappa_7\kappa_2 + \frac{20(3n-4)}{n(n-1)^2}\kappa_6\kappa_3 + \frac{20(5n-7)}{n(n-1)^2}\kappa_5\kappa_4 + \frac{120}{(n-1)^2}\kappa_5\kappa_2^2 + \frac{480}{(n-1)^2}\kappa_4\kappa_3\kappa_2 + \frac{120}{(n-1)^2}\kappa_3^3 \tag{12.61}$$

$$\kappa(62^2) = \frac{1}{n^2}\kappa_{10} + \frac{28}{n(n-1)}\kappa_8\kappa_2 + \frac{12(7n-9)}{n(n-1)^2}\kappa_7\kappa_3 + \frac{4(41n-56)}{n(n-1)^2}\kappa_6\kappa_4 + \frac{20(5n-7)}{n(n-1)^2}\kappa_5^2 + \frac{168}{(n-1)^2}\kappa_6\kappa_2^2 + \frac{840}{(n-1)^2}\kappa_5\kappa_3\kappa_2 + \frac{560}{(n-1)^2}\kappa_4^2\kappa_2 + \frac{840}{(n-1)^2}\kappa_4\kappa_3^2 \tag{12.62}$$

$$\kappa(3^2 2) = \frac{1}{n^2}\kappa_8 + \frac{21}{n(n-1)}\kappa_6\kappa_2 + \frac{6(8n-11)}{n(n-1)^2}\kappa_5\kappa_3 + \frac{9(3n-5)}{n(n-1)^2}\kappa_4^2 + \frac{18(6n-11)}{(n-1)^2(n-2)}\kappa_4\kappa_2^2 + \frac{18(9n-20)}{(n-1)^2(n-2)}\kappa_3^2\kappa_2 + \frac{36n}{(n-1)^2(n-2)}\kappa_2^4 \tag{12.63}$$

$$\kappa(432) = \frac{1}{n^2}\kappa_9 + \frac{26}{n(n-1)}\kappa_7\kappa_2 + \frac{24(3n-4)}{n(n-1)^2}\kappa_6\kappa_3 + \frac{10(11n-17)}{n(n-1)^2}\kappa_5\kappa_4 + \frac{36(5n-9)}{(n-1)^2(n-2)}\kappa_5\kappa_2^2 + \frac{12(61n-128)}{(n-1)^2(n-2)}\kappa_4\kappa_3\kappa_2 + \frac{36(5n-12)}{(n-1)^2(n-2)}\kappa_3^3 + \frac{360n}{(n-1)^2(n-2)}\kappa_3\kappa_2^3 \tag{12.64}$$

$$\kappa(532) = \frac{1}{n^2}\kappa_{10} + \frac{31}{n(n-1)}\kappa_8\kappa_2 + \frac{101n-131}{n(n-1)^2}\kappa_7\kappa_3 + \frac{5(37n-55)}{n(n-1)^2}\kappa_6\kappa_4 + \frac{5(23n-35)}{n(n-1)^2}\kappa_5^2 + \frac{30(9n-16)}{(n-1)^2(n-2)}\kappa_6\kappa_2^2 + \frac{30(45n-92)}{(n-1)^2(n-2)}\kappa_5\kappa_3\kappa_2 + \frac{60(15n-31)}{(n-1)^2(n-2)}\kappa_4^2\kappa_2 + \frac{30(45n-103)}{(n-1)^2(n-2)}\kappa_4\kappa_3^2 + \frac{720n}{(n-1)^2(n-2)}\kappa_4\kappa_2^3 + \frac{1620n}{(n-1)^2(n-2)}\kappa_3^2\kappa_2^2 \tag{12.65}$$

$$\kappa(4^2 2) = \frac{1}{n^2}\kappa_{10} + \frac{32}{n(n-1)}\kappa_8\kappa_2 + \frac{8(13n-37)}{n(n-1)^2}\kappa_7\kappa_3 + \frac{4(49n-73)}{n(n-1)^2}\kappa_6\kappa_4 + \frac{4(29n-46)}{n(n-1)^2}\kappa_5^2 + \frac{8(37n-65)}{(n-1)^2(n-2)}\kappa_6\kappa_2^2 + \frac{1536}{(n-1)^2}\kappa_5\kappa_3\kappa_2 + \frac{144(7n-15)}{(n-1)^2(n-2)}\kappa_4^2\kappa_2 + \frac{72(21n-50)}{(n-1)^2(n-2)}\kappa_4\kappa_3^2 + \frac{96(10n^2-27n-1)}{(n-1)^2(n-2)(n-3)}\kappa_4\kappa_2^3$$

$$+\frac{144(17n^2-53n-2)}{(n-1)^2(n-2)(n-3)}\kappa_3^2\kappa_2^2+\frac{192n(n+1)}{(n-1)^2(n-2)(n-3)}\kappa_2^5 \tag{12.66}$$

$$\begin{aligned}\kappa(43^2)=&\frac{1}{n^2}\kappa_{10}+\frac{33}{n(n-1)}\kappa_8\kappa_2+\frac{6(19n-25)}{n(n-1)^2}\kappa_7\kappa_3+\frac{3(65n-107)}{n(n-1)^2}\kappa_6\kappa_4\\&+\frac{6(19n-34)}{n(n-1)^2}\kappa_5^2+\frac{18(19n-33)}{(n-1)^2(n-2)}\kappa_6\kappa_2^2+\frac{72(23n-52)}{(n-1)^2(n-2)}\kappa_5\kappa_3\kappa_2\\&+\frac{54(19n-48)}{(n-1)^2(n-2)}\kappa_4^2\kappa_2+\frac{54(33n^2-148n+172)}{(n-1)^2(n-2)^2}\kappa_4\kappa_3^2\\&+\frac{72n(17n-40)}{(n-1)^2(n-2)^2}\kappa_4\kappa_2^3+\frac{108n(27n-70)}{(n-1)^2(n-2)^2}\kappa_3^2\kappa_2^2+\frac{216n^2}{(n-1)^2(n-2)^2}\kappa_2^5\end{aligned} \tag{12.67}$$

$$\begin{aligned}\kappa(32^3)=&\frac{1}{n^3}\kappa_9+\frac{30}{n^2(n-1)}\kappa_7\kappa_2+\frac{2(31n-53)}{n^2(n-1)^2}\kappa_6\kappa_3+\frac{12(9n^2-23n+16)}{n^2(n-1)^3}\kappa_5\kappa_4\\&+\frac{240}{n(n-1)^2}\kappa_5\kappa_2^2+\frac{360(2n-3)}{n(n-1)^3}\kappa_4\kappa_3\kappa_2+\frac{24(5n-12)}{n(n-1)^3}\kappa_3^3+\frac{480}{(n-1)^3}\kappa_3\kappa_2^3\end{aligned} \tag{12.68}$$

$$\begin{aligned}\kappa(42^3)=&\frac{1}{n^3}\kappa_{10}+\frac{36}{n^2(n-1)}\kappa_8\kappa_2+\frac{4(23n-37)}{n^2(n-1)^2}\kappa_7\kappa_3+\frac{4(47n^2-120n+81)}{n^2(n-1)^3}\kappa_6\kappa_4\\&+\frac{12(9n^2-24n+17)}{n^2(n-1)^3}\kappa_5^2+\frac{360}{n(n-1)^2}\kappa_6\kappa_2^2+\frac{288(5n-7)}{n(n-1)^3}\kappa_5\kappa_3\kappa_2\\&+\frac{144(7n-10)}{n(n-1)^3}\kappa_4^2\kappa_2+\frac{24(49n-95)}{n(n-1)^3}\kappa_4\kappa_3^2+\frac{960}{(n-1)^3}\kappa_4\kappa_2^3+\frac{2160}{(n-1)^3}\kappa_3^2\kappa_2^2\end{aligned} \tag{12.69}$$

$$\begin{aligned}\kappa(3^22^2)=&\frac{1}{n^3}\kappa_{10}+\frac{37}{n^2(n-1)}\kappa_8\kappa_2+\frac{6(17n-27)}{n^2(n-1)^2}\kappa_7\kappa_3+\frac{3(61n^2-166n+117)}{n^2(n-1)^3}\kappa_6\kappa_4\\&+\frac{2(59n^2-154n+113)}{n^2(n-1)^3}\kappa_5^2+\frac{6(67n-131)}{n(n-1)^2(n-2)}\kappa_6\kappa_2^2\\&+\frac{24(71n^2-246n+202)}{n(n-1)^3(n-2)}\kappa_5\kappa_3\kappa_2+\frac{36(29n^2-103n+93)}{n(n-1)^3(n-2)}\kappa_4^2\kappa_2\\&+\frac{36(38n^2-155n+160)}{n(n-1)^3(n-2)}\kappa_4\kappa_3^2+\frac{72(14n-23)}{(n-1)^3(n-2)}\kappa_4\kappa_2^3\\&+\frac{144(19n-44)}{(n-1)^3(n-2)}\kappa_3^2\kappa_2^2+\frac{288n}{(n-1)^3(n-2)}\kappa_2^5\end{aligned} \tag{12.70}$$

12.17 Additional formulae for the case of a normal population are as follows.

There are two general formulae,

$$\kappa(2^r)=\frac{2^{r-1}(r-1)!}{(n-1)^{r-1}}\kappa_2^r \tag{12.71}$$

and

$$\kappa(p^q2^r)=\frac{2^r(r+\frac{1}{2}pq-1)!}{(\frac{1}{2}pq-1)!(n-1)^r}\kappa_2^r\kappa(p^q), \tag{12.72}$$

the first being the special case of the second with $p=2$, $q=1$. The following specific formulae are of degree 12 and upwards (those of lower degree, of course, being derivable from equations (12.34) to (12.70) by putting all κs higher than the second equal to zero).

$$\kappa(3^22^4)=\frac{34\,560n}{(n-1)^5(n-2)}\kappa_2^7 \tag{12.73}$$

$$\kappa(3^42)=\frac{7776n^2(5n-12)}{(n-1)^4(n-2)^3}\kappa_2^7 \tag{12.74}$$

$$\kappa(3^42^2)=\frac{108\,864n^2(5n-12)}{(n-1)^5(n-2)^3}\kappa_2^8 \tag{12.75}$$

$$\kappa(3^42^3)=\frac{1\,741\,824n^2(5n-12)}{(n-1)^6(n-2)^3}\kappa_2^9 \tag{12.76}$$

$$\kappa(3^6)=\frac{466\,560n^3(22n^2-111n+142)}{(n-1)^5(n-2)^5}\kappa_2^9 \tag{12.77}$$

$$\kappa(3^62)=\frac{18}{n-1}\kappa_2\kappa(3^6) \tag{12.78}$$

$$\kappa(3^62^2)=\frac{360}{(n-1)^2}\kappa_2^2\kappa(3^6) \tag{12.79}$$

$$\kappa(4^22^2)=\frac{1920n(n+1)}{(n-1)^3(n-2)(n-3)}\kappa_2^6 \tag{12.80}$$

$$\kappa(4^22^3)=\frac{23\,040n(n+1)}{(n-1)^4(n-2)(n-3)}\kappa_2^7 \tag{12.81}$$

$$\kappa(4^22^4)=\frac{322\,560n(n+1)}{(n-1)^5(n-2)(n-3)}\kappa_2^8 \tag{12.82}$$

$$\kappa(4^32)=\frac{20\,736n(n+1)(n^2-5n+2)}{(n-1)^3(n-2)^2(n-3)^2}\kappa_2^7 \tag{12.83}$$

$$\kappa(4^32^2)=\frac{290\,304n(n+1)(n^2-5n+2)}{(n-1)^4(n-2)^2(n-3)^2}\kappa_2^8 \tag{12.84}$$

$$\kappa(4^3 2^3) = \frac{4\,644\,864n(n+1)(n^2-5n+2)}{(n-1)^5(n-2)^2(n-3)^2}\kappa_2^9 \tag{12.85}$$

$$\kappa(4^4) = \frac{6912n(n+1)}{(n-1)^3(n-2)^3(n-3)^3}\{53n^4-428n^3+1025n^2-474n+180\}\kappa_2^8 \tag{12.86}$$

$$\kappa(4^4 2) = \frac{16}{n-1}\kappa_2\kappa(4^4) \tag{12.87}$$

$$\kappa(4^4 2^2) = \frac{288}{(n-1)^2}\kappa_2^2\kappa(4^4) \tag{12.88}$$

$$\kappa(4^5) = \frac{484\times 12^5}{n^4}\kappa_2^{10}\text{ approximately} \tag{12.89}$$

It should be noted that (12.73) to (12.89) are all expressions of even degree. Those of odd degree are all zero, because $\kappa_r = 0$ for $r > 2$ in normal variation. In fact, (12.72) implies that $\kappa(p^q 2^r) = 0$ if pq is odd or if $q = 1$ and $p > 2$. Methods of proof of (12.71) and (12.72) are suggested in Exercises 13.10 and 13.11. Hsu and Lawley (1939) gave exact results for $\kappa(4^5)$ and $\kappa(4^6)$.

Sampling cumulants of a ratio

12.18 As an illustration of the way in which the sampling formulae can be used to approximate to a sampling distribution, let us consider the distribution of $b_1^{\frac{1}{2}}$, the sample value of the measure of skewness at (3.89), in samples from a normal population. We have, in terms of the sample moments,

$$b_1^{\frac{1}{2}} = \frac{m_3}{m_2^{\frac{3}{2}}} = \frac{n-2}{\{n(n-1)\}^{\frac{1}{2}}}\frac{k_3}{k_2^{\frac{3}{2}}}.$$

For a normal distribution the variance of k_3, $\kappa(3^2)$ is, by (12.40), equal to

$$\frac{6n\kappa_2^3}{(n-1)(n-2)}.$$

We therefore consider the statistic

$$x = \sqrt{\frac{(n-1)(n-2)}{6n}}\,k_3 k_2^{-\frac{3}{2}} \tag{12.90}$$

$$= \frac{n-1}{\sqrt{\{6(n-2)\}}}\sqrt{b_1} \tag{12.91}$$

which will, to order n^{-1}, have unit variance. We have

$$x = \sqrt{\frac{(n-1)(n-2)}{6n}}\,\frac{k_3}{\kappa_2^{\frac{3}{2}}}\left\{1+\frac{k_2-\kappa_2}{\kappa_2}\right\}^{-\frac{3}{2}}. \tag{12.92}$$

Since the population is symmetrical, the mean value of x is zero. We then have, expanding (12.92),

$$x^2 = \frac{(n-1)(n-2)}{6n} \cdot \frac{1}{\kappa_2^3}\left\{k_3^2 - \frac{3}{\kappa_2} k_3^2(k_2-\kappa_2) + \frac{6}{\kappa_2^2} k_3^2(k_2-\kappa_2)^2 - \frac{10}{\kappa_2^3} k_3^2(k_2-\kappa_2)^3 \right.$$

$$\left. + \frac{15}{\kappa_2^4} k_3^2(k_2-\kappa_2)^4 - \frac{21}{\kappa_2^5} k_3^2(k_2-\kappa_2)^5 + \frac{28}{\kappa_2^6} k_3^2(k_2-\kappa_2)^6 + \dots\right\}. \quad (12.93)$$

The variance may be obtained by taking expectations on both sides, and since κ_2 is the mean value of k_2 we have

$$\operatorname{var} x = \frac{(n-1)(n-2)}{6n} \cdot \frac{1}{\kappa_2^3}\left\{\mu(3^2) - \frac{3}{\kappa_2}\mu(3^2 2) + \frac{6}{\kappa_2^2}\mu(3^2 2^2) - \frac{10}{\kappa_2^3}\mu(3^2 2^3)\right.$$

$$\left. + \frac{15}{\kappa_2^4}\mu(3^2 2^4) - \frac{21}{\kappa_2^5}\mu(3^2 2^5) + \frac{28}{\kappa_2^6}\mu(3^2 2^6) + \dots\right\}. \quad (12.94)$$

We now express the product μs in terms of product κs by using equation (12.31) and identifying coefficients. For a normal distribution $\kappa(32^r) = 0$ because it is of odd degree and $\kappa_r = 0$ for $r > 2$. We will take our approximation to order n^{-4}, so that κs of five parts or more may be neglected by Example 12.6. We then find

$$\operatorname{var} x = \frac{(n-1)(n-2)}{6n} \cdot \frac{1}{\kappa_2^3}\left[\kappa(3^2) - \frac{3}{\kappa_2}\kappa(3^2 2) + \frac{6}{\kappa_2^2}\{\kappa(3^2 2^2) + \kappa(3^2)\kappa(2^2)\}\right.$$

$$- \frac{10}{\kappa_2^3}\{\kappa(3^2 2^3) + 3\kappa(3^2 2)\kappa(2^2) + \kappa(3^2)\kappa(2^3)\} + \frac{15}{\kappa_2^4}\{6\kappa(3^2 2^2)\kappa(2^2)$$

$$+ 4\kappa(3^2 2)\kappa(2^3) + \kappa(3^2)\kappa(2^4) + 3\kappa(3^2)\kappa^2(2^2)\} - \frac{21}{\kappa_2^5}\{15\kappa(3^2 2)\kappa^2(2^2)$$

$$\left. + 10\kappa(3^2)\kappa(2^3)\kappa(2^2)\} + \frac{28}{\kappa_2^6} \cdot 15\kappa(3^2)\kappa^3(2^2)\right]. \quad (12.95)$$

Substituting the values of equations (12.35) to (12.89) we find, after some algebraic reduction,

$$\operatorname{var} x = 1 - \frac{6}{n-1} + \frac{28}{(n-1)^2} - \frac{120}{(n-1)^3} + \dots$$

$$= 1 - \frac{6}{n} + \frac{22}{n^2} - \frac{70}{n^3} + \dots \quad (12.96)$$

In a similar way it may be shown that

$$\mu_4(x) = 3 - \frac{1056}{n^2} + \frac{24\,132}{n^3} - \dots \quad (12.97)$$

$\mu_3(x)$ is zero, for the distribution is symmetrical.

The exact results for this case are known (cf. Exercise 12.9) and it follows from them that

$$\left\{\frac{(n-2)(n+1)(n+3)}{6n(n-1)}\right\}^{\frac{1}{2}} k_3/k_2^{\frac{3}{2}} = \frac{\{(n+1)(n+3)\}^{\frac{1}{2}}}{n-1}\, x = \left\{\frac{(n+1)(n+3)}{6(n-2)}\right\}^{\frac{1}{2}} b_1^{\frac{1}{2}}$$

has unit variance exactly and fourth moment equal to $3\left(1+\frac{12}{n}-\ldots\right)$, indicating a fairly rapid approach to normality.

There are two ways of improving on the first approximation that x is normally distributed with unit variance. In the first place we may consider a transformation to a polynomial in x, (c.f. **6.25–6**) chosen so that ξ is normally distributed to order n^{-2}. Secondly, we may fit a Pearson distribution to the distribution of x, using the values of moments given by (12.96) and (12.97). The appropriate curve is the Type VII

$$f \propto \left(1+\frac{x^2}{a^2}\right)^{-m}. \tag{12.98}$$

The first line was adopted by Fisher (1929), who obtained the following transformation:

$$\xi = x\left(1+\frac{3}{n}+\frac{91}{4n^2}\right) - \frac{3}{2n}\left(1-\frac{111}{2n}\right)(x^3-3x) - \frac{33}{8n^2}(x^5-10x^3+15x). \tag{12.99}$$

The second was adopted by E. S. Pearson (1930), who tabulated the percentiles of the fitted distribution. The *Biometrika Tables* give the 95% and 99% points (which with changed sign are also the 5% and 1% points, by symmetry) of the distribution of $b_1^{\frac{1}{2}}$ for $n \geqslant 25$, and D'Agostino and Pearson (1973) give the S_U approximation (cf. **6.35**) to the distribution. Mulholland (1977) gives six upper percentage points for $n = 4(1)\,250$.

12.19 By such methods approximations can be obtained to the sampling distributions of any statistics that are expressible as symmetric functions. The reader may refer to investigations on these lines by F. N. David (1949a, b) into the sampling moments of the coefficient of variation, the logarithm of the variance ratio (Fisher's z) and the variance ratio itself. Difficulties associated with the expansion of denominators in ratios may sometimes be overcome by the use of the following device.

Consider a statistic $t = a/b$, where b is always positive. Then

$$\begin{aligned} F(t_0) &= P\{t \leqslant t_0\} = P\{a/b \leqslant t_0\} \\ &= P\{a \leqslant bt_0\} = P\{a - bt_0 \leqslant 0\}. \end{aligned} \tag{12.100}$$

We then consider the statistic $a - bt_0$, which does not suffer from the disadvantage of being a ratio. For example, suppose our statistic is Student's t of Example 11.8, given by

$$t = \frac{(\bar{x}-\mu)(n-1)^{\frac{1}{2}}}{s},$$

where μ is the population mean and $s^2 = \sum (x-\bar{x})^2/n$. Now

$$P\{|t| \leqslant |t_0|\} = P\{t^2 \leqslant t_0^2\}$$
$$= P\{(n-1)(\bar{x}-\mu)^2 - t_0^2 s^2 \leqslant 0\}. \quad (12.101)$$

We then consider

$$u = (n-1)(\bar{x}-\mu)^2 - t_0^2 s^2 \quad (12.102)$$

and can obtain as many moments or cumulants as we like for given μ and t_0^2.

In this particular instance, for normal variation we know that $\bar{x}$ and s are independent and can find the cumulants of u directly. For the c.f. of u, writing it as θ, we have, from (11.39–40) and Example 3.6,

$$\phi(t) \propto \iint \exp[\{(n-1)(\bar{x}-\mu)^2 - t_0^2 s^2\}\theta] \exp\{-\tfrac{1}{2}n(\bar{x}-\mu)^2\}\, d\bar{x} \exp\{-\tfrac{1}{2}ns^2\}s^{n-2}\, ds$$
$$= \left\{1 - \frac{2(n-1)}{n}\theta\right\}^{-\frac{1}{2}} \left\{1 + \frac{2t_0^2\theta}{n}\right\}^{-\frac{1}{2}(n-1)}, \quad (12.103)$$

where we have assumed the population variance to be unity. Hence we find, for u,

$$\kappa_1 = \frac{n-1}{n}(1 - t_0^2) \quad (12.104)$$

$$\kappa_2 = \frac{2(n-1)}{n^2}(n-1+t_0^4). \quad (12.105)$$

As $n \to \infty$, the distribution will tend to normality with these moments. For $n = 20$, $t_0 = 2$ we have $\kappa_1 = -2.85$, $\kappa_2 = 3.42$. This gives a standardized value of $-2.85/\sqrt{3.42} = -1.563$. The probability that this is exceeded in a standardized normal distribution is 0.941, from Appendix Table 2, and this is accordingly our approximation to the probability that t will not exceed $t_0 = 2$ in absolute value. The true value is 0.940.

Although we have considered a normal case, the method is particularly useful in judging the effect of departures from normality on certain statistics. Reference may be made to David and Johnson (1951) for some work on these lines concerning the variance ratio.

Sampling from finite populations

12.20 In sampling without replacement from finite populations the algebra of expectations becomes much more complex because successive drawings are no longer independent. Let us follow through the work of **12.3** in the finite case to illustrate the point. We require (12.2)

$$Em_2 = \frac{n-1}{n}Ex^2 - \frac{n-1}{n}Ex_i x_j, \qquad i \neq j. \quad (12.106)$$

If capital Ms are written to denote the moments of the finite population of N

members, we now have, by definition,

$$Ex^2 = M_2'$$

$$Ex_ix_j = \sum_{i,j=1}^{N} x_ix_j/\{N(N-1)\}, \qquad i \neq j$$

$$= \frac{\left(\sum_1^N x_i\right)^2 - \sum_1^N x_i^2}{N(N-1)} = \frac{N^2M_1'^2 - NM_2'}{N(N-1)}.$$

Substituting in (12.106) we then find

$$Em_2 = \frac{n-1}{n}\left\{\frac{N}{N-1}(M_2' - M_1'^2)\right\}$$

$$= \frac{n-1}{n}\cdot\frac{N}{N-1}M_2 \qquad (12.107)$$

reducing to (12.3) when $N \to \infty$.

Again, with sufficient pertinacity, we can find all the moments of symmetric-function statistics by this approach, but the algebra rapidly becomes unmanageable. A simpler method is as follows.

12.21 Let us first of all note that the expectation of any augmented symmetric function for a sample bears a very simple relation to the corresponding function for the population. Denoting the two functions by the subscripts n and N respectively, we have from (12.9)

$$E[p_1^{\pi_1}\dots p_s^{\pi_s}]_n/n^{[\rho]} = [p_1^{\pi_1}\dots p_s^{\pi_s}]_N/N^{[\rho]} \qquad (12.108)$$

where $\rho = \sum_i \pi_i$. It follows that if K_p is the pth k-statistic for the population, we have

$$E_N k_p = K_p \qquad (12.109)$$

where E_N means an expectation in the set of N values forming the population. This attractive property of k-statistics is one of their most useful features. We have, for example, $k_2 = nm_2/(n-1)$ and hence

$$E_N\frac{nm_2}{n-1} = K_2 = \frac{NM_2}{N-1}$$

from which (12.107) follows immediately.

Secondly, let us imagine the population of N as itself a sample from an infinite population with cumulants κ_p. If E refers to expectation in this infinite population we shall have

$$Ek_p = EK_p = \kappa_p. \qquad (12.110)$$

The sample of n observations may now be considered as having been drawn from the infinite population in two stages, the first of which consists in drawing the population of N, and the second in selecting the sample from the latter. Thus we may regard the expectation operation as consisting of two stages, which we may write symbolically as $E(E_N)$.

Now if a symmetric function f has expectation linear in the cumulants,

$$E(f) = \sum a_j \kappa_j,$$

we may write down the analogous relation

$$E_N(f) = \sum a_j K_j.$$

For if this were not true, $E_N(f)$ could be expressed as some other function of Ks whose expectation E would be the same as that of $\sum a_j K_j$, and this would imply a relationship among the Ks. This is sometimes known as the Irwin-Kendall principle from the authors who introduced it (1944).

From (12.109), we see that

$$E(K_p^s k_p) = E\{K_p^s E_N(k_p)\} = E(K_p^{s+1}) \tag{12.111}$$

and using this with $s = 1$, we find that

$$\left.\begin{aligned} E(k_p - K_p)^2 &= E(k_p^2) + E(K_p^2) - 2E(K_p k_p) = E(k_p^2) - E(K_p^2) \\ \text{or} \qquad \operatorname{var}_N(k_p) &= \operatorname{var}(k_p) - \operatorname{var}(K_p). \end{aligned}\right\} \tag{12.112}$$

Example 12.8 (Moments of the mean in the finite population case)

The mean statistic m_1 is equal to k_1. Thus (12.109) gives

$$E_N(k_1) = K_1.$$

From the known result for the variance of the mean we have

$$E(k_1^2) = \frac{\kappa_2}{n} + \kappa_1^2, \quad E(K_1^2) = \frac{\kappa_2}{N} + \kappa_1^2.$$

Hence, using (12.112),

$$E(k_1 - K_1)^2 = Ek_1^2 - EK_1^2 = \kappa_2\left(\frac{1}{n} - \frac{1}{N}\right).$$

It then follows at once that

$$E_N(k_1 - K_1)^2 = K_2\left(\frac{1}{n} - \frac{1}{N}\right), \tag{12.113}$$

giving us the variance of the mean. If we prefer to express this in terms of moments we have $K_2 = NM_2/(N-1)$ and

$$E_N(k_1 - K_1)^2 = \frac{N-n}{(N-1)n} M_2. \tag{12.114}$$

For the third moment we have

$$\begin{aligned} E(k_1 - \kappa_1)^3 &= E(k_1 - K_1 + K_1 - \kappa_1)^3 \\ &= E(k_1 - K_1)^3 + 3E(k_1 - K_1)^2(K_1 - \kappa_1) \\ &\quad + 3E(k_1 - K_1)(K_1 - \kappa_1)^2 + E(K_1 - \kappa_1)^3. \end{aligned} \tag{12.115}$$

The third term on the right vanishes, since $E_N(k_1 - K_1) = 0$. Now by Rule 10

$$E(k_1 - \kappa_1)^3 = \frac{\kappa_3}{n^2}, \qquad E(K_1 - \kappa_1)^3 = \frac{\kappa_3}{N^2}.$$

Hence, from (12.115) we find

$$E(k_1 - K_1)^3 = \kappa_3\left(\frac{1}{n^2} - \frac{1}{N^2}\right) - 3E(k_1 - K_1)^2(K_1 - \kappa_1).$$

Writing

$$\alpha_r = \frac{1}{n^r} - \frac{1}{N^r} \tag{12.116}$$

we then find

$$\begin{aligned} E(k_1 - K_1)^3 &= \alpha_2\kappa_3 - 3E(K_1 - \kappa_1)E_N(k_1 - K_1)^2 \\ &= \alpha_2\kappa_3 - 3E(K_1 - \kappa_1)\alpha_1 K_2 \\ &= \alpha_2\kappa_3 - 3\alpha_1 E(K_1 - \kappa_1)(K_2 - \kappa_2) \\ &= \alpha_2\kappa_3 - 3\alpha_1\kappa_3/N, \end{aligned}$$

using Rule 10,

$$= \left(\alpha_2 - \frac{3\alpha_1}{N}\right)\kappa_3.$$

Hence, for the third moment,

$$E_N(k_1 - K_1)^3 = \left(\alpha_2 - \frac{3\alpha_1}{N}\right)K_3. \tag{12.117}$$

Substituting for α_2, α_1 and

$$K_3 = \frac{N^2}{(N-1)(N-2)} M_3$$

we find

$$E_N(m_1 - M_1)^3 = \frac{(N-n)(N-2n)}{n^2(N-1)(N-2)} M_3. \tag{12.118}$$

Likewise we find

$$E_N(k_1 - K_1)^4 = K_4\left\{\alpha_3 - 4\alpha_2/N + 6\alpha_1/N^2 - \frac{3(N-1)}{N(N+1)}\alpha_1^2\right\} + 3K_2^2\frac{N-1}{N+1}\alpha_1^2 \qquad (12.119)$$

which is equivalent to

$$E_N(m_1 - M_1)^4 = \frac{N-n}{n^3(N-1)(N-2)(N-3)}\{(N^2 - 6nN + N + 6n^2)M_4 + 3N(n-1)(N-n-1)M_2^2\}. \qquad (12.120)$$

The reader should check that the results (5.41) that we obtained for the hypergeometric distribution are special cases of (12.114), (12.118) and (12.120), given by putting M_r equal to the corresponding binomial moments in (5.4–5) with $n = 1$, and remembering that there the variable is a sum and not a mean as here.

Example 12.9 (k-statistics for finite populations)

We have, generalizing (12.112),

$$\left.\begin{aligned} E(k_r - K_r)(k_p - K_p) &= E(k_r k_p) - E(K_r K_p), \\ \text{or} \qquad & \\ \text{cov}_N(k_r, k_p) &= \text{cov}(k_r, k_p) - \text{cov}(K_r, K_p). \end{aligned}\right\} \qquad (12.121)$$

From (12.121) with $p = 1$ and Rule 10, we find

$$E(k_r - K_r)(k_1 - K_1) = \frac{1}{n}\kappa_{r+1} - \frac{1}{N}\kappa_{r+1} = \alpha_1\kappa_{r+1}$$

from which

$$E_N(k_r - K_r)(k_1 - K_1) = \alpha_1 K_{r+1}.$$

Thus, if the finite population is symmetrical, any k-statistic of even order is uncorrelated with k_1, for K_{r+1} then vanishes since it is homogeneous of odd degree in the central moments, by **12.7–8**. Cf. the large-sample result for even-order central moments in Example 10.2.

Multiple *k*-statistics

12.22 In developing our next point, we encounter a difficulty of notation. We have already used the symbols κ, k and K. In the next chapter, when we proceed to multivariate extensions, we shall require corresponding quantities with several suffixes, e.g. κ_{rs}, k_{rs}, K_{rs}. We now require a notation for a quantity whose expectation is the product of κs, e.g. $\kappa_r\kappa_s$. Previous writers (Tukey (1950), Wishart (1952), Kendall (1952)) notwithstanding, we shall write these quantities with the letter l and thus we define them by

$$El_{rs\ldots u} = \kappa_r\kappa_s\ldots\kappa_u. \qquad (12.122)$$

The l-statistics are called "polykays" by some authors, but we feel that there are limits to linguistic miscegenation that should not be exceeded.

It will, as before, follow that if $L_{rs...u}$ is the corresponding statistic for a finite population

$$E_N l_{rs...u} = L_{rs...u}. \tag{12.123}$$

The usefulness of these functions derives from the fact that we can express them in terms of augmented symmetrics once and for all and hence apply the argument of **12.21** to non-linear functions of cumulants. The necessary table is given by Abdel-Aty (1954) up to weight 12. Wishart (1952) gave expressions up to order 8 for these multiple k-statistics in terms of products of simple k-statistics. For example we have, from his table,

$$k_2 k_1 = l_2 l_1 = l_3/n + l_{21} = k_3/n + l_{21} \tag{12.124}$$

and it follows at once that

$$E(k_2 k_1) = \kappa_3/n + \kappa_2\kappa_1, \tag{12.125}$$

and also that

$$E_N(k_2 k_1) = K_3/n + L_{21}. \tag{12.126}$$

Similarly we have

$$k_1^4 = k_4/n^3 + 4l_{31}/n^2 + 3l_{22}/n^2 + 6l_{211}/n + l_{1111} \tag{12.127}$$

whence

$$E(k_1^4) = \kappa_4/n^3 + 4\kappa_3\kappa_1/n^2 + 3\kappa_2^2/n^2 + 6\kappa_2\kappa_1^2/n + \kappa_1^4, \tag{12.128}$$

again with a parallel expression for $E_N(k_1^4)$.

Example 12.10 (Unbiased estimators for cumulants)

One of the most useful practical applications of these results is that they enable us to write down unbiased estimators of products of cumulants.

We found in Examples 12.1 and 12.3 that

$$\text{var}\, m_2 = \frac{(n-1)^2}{n^2}\left\{\frac{\kappa_4}{n} + \frac{2\kappa_2^2}{n-1}\right\}. \tag{12.129}$$

In some practical applications of this result we may require to calculate the quantity on the right, but if we have only a sample available we are ignorant of κ_4 and κ_2 and have to estimate them. Thus we require a quantity whose expectation is the function on the right in (12.129). In terms of the multiple k-statistics this is written down at sight as

$$\frac{(n-1)^2}{n^2}\left\{\frac{k_4}{n} + \frac{2l_{22}}{n-1}\right\}. \tag{12.130}$$

From Wishart's table we then read off

$$l_{22} = \frac{n-1}{n+1}\left(k_2^2 - \frac{k_4}{n}\right)$$

and, on substitution in (12.130), for the unbiased estimator we find

$$\frac{(n-1)^2}{n^2(n+1)}\left\{\frac{(n-1)k_4}{n}+2k_2^2\right\}. \tag{12.131}$$

Example 12.11 (Unbiased estimator of a product of moments)

To find an unbiased estimator of $\mu_3\mu_1'^2$.

This is equal to $\kappa_3\kappa_1^2$ and the estimator is l_{311}. From Appendix Table 11 we may write down the estimator directly in terms of augmented symmetric functions as

$$l_{311}=\frac{2}{n^{[5]}}[1^5]-\frac{3}{n^{[4]}}[21^3]+\frac{[31^2]}{n^{[3]}}.$$

From Appendix Table 10 we can then substitute for the augmented symmetrics in terms of the power sums (r). We find

$$\begin{aligned}l_{311}=&\frac{2}{n^{[5]}}\{24(5)-30(4)(1)-20(3)(2)+20(3)(1)^2+15(2)^2(1)-10(2)(1)^3+(1)^5\}\\&-\frac{3}{n^{[4]}}\{-6(5)+6(4)(1)+5(3)(2)-3(3)(1)^2-3(2)^2(1)+(2)(1)^3\}\\&+\frac{1}{n^{[3]}}\{2(5)-2(4)(1)-(3)(2)+(3)(1)^2\},\end{aligned}$$

which reduces, on putting $(r)=s_r$, to

$$\begin{aligned}\frac{1}{n^{[5]}}\{&2n(n+2)s_5-2(n^2+2n+6)s_4s_1-(n^2+8n-8)s_3s_2+(n^2+2n+16)s_3s_1^2\\&+(9n-6)s_2^2s_1-(3n+8)s_2s_1^3+2s_1^5\}.\end{aligned}$$

Dwyer (1964) considers the l-statistics of deviations from the sample mean, leading to general moment formulae. Cf. also Dwyer and Tracy (1964), Tracy (1968, 1969), Nagambal and Tracy (1970) and Tracy and Gupta (1973, 1974). See also the collection of papers edited by Tracy (1972).

Dwyer and Tracy (1980) consider product moments for deviations from the sample mean; they give unbiased estimators for these moments when sampling from a finite population. See also Ali and Mikhail (1979) and Mikhail and Ali (1980).

Exchangeability

12.23 A set of random variables is said to be *exchangeable* if, for all $n \geqslant 1$ and for every permutation of the n subscripts, the joint distributions of $(x_{j_1}, \ldots, x_{j_n})$ are identical. The notion of exchangeability is central to the work of De Finetti (1974) and others on Bayesian inference. Clearly, exchangeability is a weaker assumption than the requirement that the random variables be independent and identically distributed. For example, successive observations drawn without replacement from a finite population are dependent, yet exchangeable. Whenever the random variables are exchangeable, even non-symmetric functions of the observations will satisfy

$$E\{g(x_1, \ldots, x_n)\} = E\{g(x_{j_1}, \ldots, x_{j_n})\} \tag{12.132}$$

The nature of the symmetry in (12.132) differs from that employed elsewhere in this chapter, but the nature of the results is often rather similar.

Example 12.12 (Expectations of symmetric functions of exchangeable variates)

Suppose $x_1, \ldots, x_n$ are exchangeable and let

$$Q = \sum_{i=1}^{k} \sum_{j=1}^{k} w_{ij} x_i x_j,$$

where the ws are a set of known constants. It follows that

$$E(Q) = E(x_i^2) \sum_i w_{ii} + E(x_i x_j) \sum\sum_{i \neq j} w_{ij}.$$

If $\sum_i x_i^2 = nm_2$ and $\sum\sum_{i \neq j} x_i x_j = n(n-1)m_{11}$, it follows that $m_2 \sum w_{ii} + m_{11} \sum\sum w_{ij}$ is an unbiased estimator of $E(Q)$, provided that the latter exists.

This approach has been used, for example by Ord (1980), to develop unbiased estimators of the moments of test statistics that are asymmetric functions of the observations.

12.24 In a series of papers, Speed (1983, 1986a,b,c) presents an elegant treatment of generalized k-statistics using an extended form of exchangeability. The results are extended to the multivariate case (1986b) and conditions are given for the almost sure convergence of k-statistics (1986c). McCullagh and Pregibon (1987) consider k-statistics in regression analysis; Speed and Silcock (1988) develop them in the context of random effects models.

EXERCISES

12.1 By using (12.25) directly, write down k_3 and k_4 in terms of the augmented symmetric functions and hence verify (12.29).

12.2 Assuming that

$$k_3 = a_0 s_3 + a_1 s_2 s_1 + a_2 s_1^3$$

obtain the coefficients a by taking expectations of both sides and hence derive k_3 of (12.28).

12.3 Use the approach of Example 12.1 to obtain $\kappa(2^3)$ in terms of the population cumulants.

12.4 Show that the pattern functions of the following patterns:

$$\begin{array}{ccc} \times & \times & \times \\ \times & \cdot & \times \\ \times & \times & \cdot \end{array} \qquad\qquad \begin{array}{cccc} \times & \times & \times & \times \\ \cdot & \cdot & \times & \times \\ \times & \times & \cdot & \cdot \end{array}$$

are $\dfrac{1}{(n-1)^2}$ and $\dfrac{1}{n(n-1)^2}$ respectively.

(Fisher, 1929)

12.5 Show that the pattern function of the pattern

$$\begin{array}{cccc} \times & \times & \times & \times \\ \times & \times & \times & \times \end{array} \quad \text{etc.}$$

with p columns is

$$\frac{1}{n^{p-1}}\left\{1-\frac{(-1)^{p-1}}{(n-1)^{p-1}}\right\}.$$

(Fisher, 1929)

12.6 Verify the formulae of equations (12.37) and (12.43).

12.7 If a pattern contains a column with three entries; if the patterns obtained by suppressing this column and (1) amalgamating the three rows, (2) amalgamating the pairs of three rows, and (3) leaving the rows unamalgamated are A, B_1, B_2, B_3 and C respectively, show that the pattern function for the original pattern is

$$\frac{n}{(n-1)(n-2)}A-\frac{1}{(n-1)(n-2)}(B_1+B_2+B_3)+\frac{2}{n(n-1)(n-2)}C.$$

Deduce that the function of the pattern

$$\begin{array}{ccc} \times & \times & \times \\ \times & \times & \times \\ \times & \times & \times \\ \times & \times & \cdot \end{array}$$

is

$$\frac{n^3-8n^2+17n+2}{(n-1)^2(n-2)^2(n-3)}.$$

(Fisher and Wishart, 1931)

12.8 Use the combinatorial method to verify that, to order n^{-1},

$$\text{var}\, m_4 = (\mu_8 - \mu_4^2 + 16\mu_3^2\mu_2 - 8\mu_5\mu_3)/n,$$

agreeing with (10.9).

12.9 Show that in normal samples k_2 is independent of $k_p/k_2^{\frac{1}{2}p}$ for $p=3, 4, \ldots$ (cf. Exercises 11.4-5). Hence show that this ratio has mean zero for all p, and all odd-order moments zero for p odd. For $p=3$, show that

$$\text{var}(k_3/k_2^{3/2}) = \frac{6n(n-1)}{(n-2)(n+1)(n+3)}$$

$$\mu_4(k_3/k_2^{3/2}) = \frac{108n^2(n-1)^2(n^2+27n-70)}{(n-2)^3(n+1)(n+3)(n+5)(n+7)(n+9)}$$

and verify equations (12.96) and (12.97).

(Cf. Fisher, 1930; μ_6 and μ_8 are listed by P. Williams (1935).)

12.10 Defining y by the relation

$$y = \left\{\frac{(n-1)(n-2)(n-3)}{24n(n+1)}\right\}^{\frac{1}{2}} \frac{k_4}{k_2^2},$$

show by the method of 12.18 that the moments of the distribution of y in samples from a normal population are

$$\mu_1 = 0,$$

$$\mu_2 = 1 - \frac{12}{n} + \frac{88}{n^2} - \frac{532}{n^3} \ldots,$$

$$\mu_3 = 6\left(\frac{6}{n}\right)^{\frac{1}{2}}\left(1 - \frac{65}{2n} + \frac{4811}{8n^2} - \frac{136\,605}{16n^3} \ldots\right),$$

$$\mu_4 = 3 + \frac{468}{n} - \frac{32\,196}{n^2} + \frac{1\,118\,388}{n^3} \ldots.$$

Show as in Exercise 12.9 that

$$\text{var}(k_4/k_2^2) = 24n(n-1)^2/\{(n-3)(n-2)(n+3)(n+5)\}$$

exactly, and hence verify $\mu_2(y)$.

(E. S. Pearson, 1930; Fisher, 1930; the *Biometrika Tables* give percentage points of $b_2 = \dfrac{m_4}{m_2^2} = \dfrac{(n-2)(n-3)}{(n^2-1)}\dfrac{k_4}{k_2^2} + \dfrac{3(n-1)}{(n+1)}$ for $n \geqslant 200$ and D'Agostino and Pearson (1973) give them for $20 \leqslant n \leqslant 200$. Anscombe and Glynn (1983) give a useful approximation to the d.f. of b_2.)

12.11 Show that, in samples from a finite population,

$$E_N(k_2 - K_2)^2 = \frac{(N-n)(Nn-n-N-1)}{n(n-1)N(N+1)}K_4 + \frac{2(N-n)}{(n-1)(N+1)}K_2^2$$

which is equivalent to

$$E_N\left(m_2 - \frac{n-1}{n}\frac{N}{N-1}M_2\right)^2 = \frac{(n-1)N(N-n)}{n^3(N-1)^2(N-2)(N-3)} \times$$
$$\{(nN-N-n-1)(N-1)M_4 - (nN^2-3N^2+6N-3n-3)M_2^2\}.$$

12.12 Show that

$$E_N(k_3-K_3)(k_2-K_2) = \frac{(N-n)\{(n-1)(N-1)-6\}}{n(n-1)N(N+5)}K_5 + \frac{6(N-n)}{(n-1)(N+5)}K_3K_2.$$

12.13 Show that in samples from a symmetrical finite population k_{2r} is uncorrelated with any k-statistic of odd order.

12.14 Show that if $M(z)$ is the m.g.f. of x, the m.g.f. of a function $y=f(x)$ may be written

$$M(\xi) = \left[\exp\left\{\xi f\left(\frac{\mathrm{d}}{\mathrm{d}z}\right)\right\} M(z)\right]_{z=0}.$$

Hence show that the m.g.f. of the k-statistics,

$$\sum\left\{\mu(p_1^{\pi_1}\dots p_s^{\pi_s})\frac{t_{p_1}^{\pi_1}}{\pi_1!}\dots\frac{t_{p_s}^{\pi_s}}{\pi_s!}\right\}$$

is given by

$$\left[\exp\{t_1\mathbf{K}_1+t_2\mathbf{K}_2+t_3\mathbf{K}_3+\dots\}\exp\left\{\kappa_1 s_1+\frac{\kappa_2 s_2}{2!}+\dots\right\}\right]_{t=0}$$

where $\mathbf{K}_r$ is the same function of the operators $\partial/\partial x$ as k_r is of the observations x and $s_r = \sum x^r$.
Deduce that

$$\mathbf{K}_p(s_p) = p!$$

$$\mathbf{K}_p(s_{p_1}s_{p_2}\dots) = 0,$$

where $(p_1, p_2, \dots)$ is any partition of p.

(Fisher, 1929)

12.15 Show that

$$E\{k_p(x_1+h, x_2, \dots, x_n)\} = \kappa_p + \frac{1}{n}h^p$$

and hence that

$$\mathbf{S}_p k_p = p!,$$

$$\mathbf{S}_q k_p = 0, \qquad q \neq p,$$

where $\mathbf{S}_p$ is the same function of the operators $\partial/\partial x$ as s_p is of the observations x.

(Kendall, 1940a)

12.16 Show that the m.g.f. of the k-statistics is given by

$$\left[\exp\left\{\frac{\mathbf{S}_1\kappa_1}{1!}+\frac{\mathbf{S}_2\kappa_2}{2!}+\dots\right\}\exp\{k_1t_1+k_2t_2+\dots\}\right]_{t=0}$$

and hence derive the result $\mathbf{S}_p k_p = p!$ of the previous exercise.

(Kendall, 1942)

12.17 Use Exercise 12.16 to show that, in the expression of k_p in terms of the power sums at (12.28), the sum of the coefficients is $1/n$.

12.18 Show that an unbiased estimator of $\mu_1'^4$ is

$$\frac{1}{n^{[4]}}\{s_1^4 - 6s_2s_1^2 + 3s_2^2 + 8s_3s_1 - 6s_4\}.$$

12.19 Show that

$$l_{222} = -\frac{[1^6]}{n^{[6]}} + \frac{3[21^4]}{n^{[5]}} - \frac{3[2^21^2]}{n^{[4]}} + \frac{[2^3]}{n^{[3]}}$$

as given in Appendix Table 11.

12.20 By the method of Example 12.7 show that if k_1 and k_p are uncorrelated for all $p > 1$, the distribution is normal; and that if k_1 is independent of any $k_p (p > 1)$ the distribution is likewise normal. (Use Exercise 4.5.)

12.21 Show that the joint distribution of k_1 and k_2 in samples of size n has correlation coefficient

$$\rho = \frac{\text{cov}(k_1, k_2)}{\{\text{var}(k_1)\text{var}(k_2)\}^{\frac{1}{2}}} = \frac{\gamma_1}{\left(\gamma_2 + \dfrac{2n}{n-1}\right)^{\frac{1}{2}}},$$

where γ_1, γ_2 are the measures of skewness and kurtosis (cf. (3.89–90)) of the population, so that $\rho = 0$ for any symmetrical population, although Example 12.7 shows that k_1 and k_2 are independent only in the normal case. Use the fact that $\rho^2 < 1$ to verify the result in Exercise 3.19 that $\beta_2 > 1 + \beta_1$.

Show using (5.20) that for the Poisson distribution (5.18) as $n \to \infty$, $\rho \to (1 + 2\lambda)^{-\frac{1}{2}}$, arbitrarily close to 1 as $\lambda \to 0$.

12.22 Generalizing Exercises 12.9–10, show that in normal samples

$$E(k_p^a k_r^b / k_2^s) = E(k_p^a k_r^b)/E(k_2^s) = \frac{(n-1)^{s-1}}{(n+1)(n+3)\dots(n+2s-3)} \frac{E(k_p^a k_r^b)}{\kappa_2^s}$$

for any non-negative integers p, r with $2s = ap + br$ a positive integer. Derive var $(k_3/k_2^{3/2})$ in Exercise 12.9 from (12.40) and var (k_4/k_2^2) in Exercise 12.12 from (12.43) and show that these variables are uncorrelated.

Show that the third moment of k_4/k_2^2 is exactly

$$1728n(n-1)^3(n^2-5n+2)/\{(n-3)^2(n-2)^2(n+3)(n+5)(n+7)(n+9)\}$$

and hence verify $\mu_3(y)$ in Exercise 12.10. Show from (12.67) that

$$E(k_4k_3^2/k_2^5) = 216n^2(n-1)^2/\{(n-2)^2(n+1)(n+3)(n+5)(n+7)\}$$

and hence show that the correlation coefficient between k_4/k_2^2 and k_3^2/k_2^3 in normal samples is asymptotically $(27/n)^{\frac{1}{2}}$.

(Cf. Fisher, 1930)

CHAPTER 13

CUMULANTS OF SAMPLING DISTRIBUTIONS – (2)

13.1 In this chapter we shall discuss two main topics. The first is the extension to bivariate and multivariate statistics of the methods developed in the last chapter for the univariate case. The second is the structure of k-statistics and cumulants, with particular reference to the proof of the combinatorial rules for developing their sampling properties.

Bivariate k-statistics

13.2 Given any bipartite number pp' we shall have for any partition $\{(p_1p_1')^{\pi_1}(p_2p_2')^{\pi_2}\dots\}$ and the bivariate cumulant $\kappa_{pp'}$ a k-statistic $k_{pp'}$ whose mean value is $\kappa_{pp'}$. Explicitly, (12.25) is generalized to

$$k_{pp'} = p!p'!\sum\frac{(-1)^{\rho-1}(\rho-1)!}{n^{[\rho]}}\sum\frac{(x_1^{p_1}y_1^{p_1'}x_2^{p_1}y_2^{p_1'}\dots x_\rho^{p_s}y_\rho^{p_s'})}{(p_1!)^{\pi_1}(p_1'!)^{\pi_1}\dots\pi_1!\pi_2!\dots}. \tag{13.1}$$

In particular, corresponding to (12.28) we have

$$\left.\begin{aligned}
k_{11} &= \frac{1}{n^{[2]}}(ns_{11}-s_{10}s_{01})\\
k_{21} &= \frac{1}{n^{[3]}}(n^2s_{21}-2ns_{10}s_{11}-ns_{20}s_{01}+2s_{10}^2s_{01})\\
k_{31} &= \frac{1}{n^{[4]}}\{n^2(n+1)s_{31}-n(n+1)s_{30}s_{01}-3n(n-1)s_{11}s_{20}\\
&\quad -3n(n+1)s_{21}s_{10}+6ns_{11}s_{10}^2+6ns_{20}s_{10}s_{01}-6s_{01}s_{10}^3\}\\
k_{22} &= \frac{n}{(n-1)(n-2)(n-3)}\left\{(n+1)s_{22}-\frac{2(n+1)}{n}s_{21}s_{01}-\frac{2(n+1)}{n}s_{12}s_{10}\right.\\
&\quad -\frac{(n-1)}{n}s_{20}s_{02}-\frac{2(n-1)}{n}s_{11}^2+\frac{8}{n}s_{11}s_{10}s_{01}+\frac{2}{n}s_{02}s_{10}^2+\frac{2}{n}s_{20}s_{01}^2\\
&\quad \left.-\frac{6}{n^2}s_{10}^2s_{01}^2\right\}
\end{aligned}\right\} \tag{13.2}$$

Generalizing the notation of (12.30), we write, e.g.,

$$E(k_{20}k_{11}^2) = \mu'\begin{pmatrix}2 & 1 & 1\\ 0 & 1 & 1\end{pmatrix}$$

$$E(k_{20}k_{11}k_{02}) = \mu'\begin{pmatrix}2 & 1 & 0\\ 0 & 1 & 2\end{pmatrix}$$

with corresponding κs. The latter may be expressed in terms of the cumulants of the bivariate distribution as in the univariate case; and the coefficients will now depend on partitions of bipartite numbers. Our rules still apply (and in particular the pattern functions corresponding to particular arrays are unchanged); but the numerical coefficients associated with arrays are modified, for we now have to consider the number of ways of allocating two different sorts of individuals in a two-way partition of a bipartite number. An example will make the modification clear.

Suppose we wish to find the coefficient of $\kappa_{33}\kappa_{11}^2$ in $\kappa\begin{pmatrix}2 & 2 & 1\\ 2 & 2 & 1\end{pmatrix}$. The total degree is 10 and we have to consider arrays that complete the table

$$\begin{array}{ccc|c} & & & 6\\ & & & 2\\ & & & 2\\ \hline 4 & 4 & 2 & 10\end{array}$$

i.e. those we discussed in Example 12.2. The pattern functions are those we have already found. For the numerical coefficients we have to regard the column totals as consisting of the pairs (2, 2), (2, 2), and (1, 1) and the row totals as consisting of (3, 3), (1, 1) and (1, 1). For instance, the array

$$\begin{array}{ccc|c} 2 & 2 & 2 & 6\\ 1 & 1 & . & 2\\ 1 & 1 & . & 2\\ \hline 4 & 4 & 2 & 10\end{array}$$

may be written either as

$$\begin{array}{ccc|c} (1,1) & (1,1) & (1,1) & (3,3)\\ (0,1) & (1,0) & . & (1,1)\\ (1,0) & (0,1) & . & (1,1)\\ \hline (2,2) & (2,2) & (1,1) & (5,5)\end{array} \tag{13.3}$$

or as

$$\begin{array}{ccc|c} (2,0) & (0,2) & (1,1) & (3,3)\\ (0,1) & (1,0) & . & (1,1)\\ (0,1) & (1,0) & . & (1,1)\\ \hline (2,2) & (2,2) & (1,1) & (5,5)\end{array} \tag{13.4}$$

each of which will make a contribution to the numerical coefficient. It will be found that no other arrays are possible except those obtained by permuting the first two columns.

The numerical coefficient in (13.3) and the permuted array together is

$$2\left(\frac{2!}{1!\,1!}\right)\left(\frac{2!}{1!\,1!}\right)\left(\frac{2!}{1!\,1!}\right)\left(\frac{2!}{1!\,1!}\right)\cdot\frac{1}{2!}=16.$$

That in (13.4) and the permuted array is

$$2\left(\frac{2!}{2!}\right)\left(\frac{2!}{1!\,1!}\right)\left(\frac{2!}{1!\,1!}\right)\left(\frac{2!}{2!}\right)\cdot\frac{1}{2!}=4.$$

The total contribution is thus 20. The pattern function is $\dfrac{1}{(n-1)(n-2)}$ as in Example 12.2.

In the same way it will be found that for the arrays

$$\begin{array}{ccc|c} 2 & 3 & 1 & 6 \\ 1 & 1 & . & 2 \\ 1 & . & 1 & 2 \\ \hline 4 & 4 & 2 & 10 \end{array} \qquad \begin{array}{ccc|c} 3 & 3 & . & 6 \\ 1 & . & 1 & 2 \\ . & 1 & 1 & 2 \\ \hline 4 & 4 & 2 & 10 \end{array}$$

the coefficients are 48 and 8. Thus, using the pattern functions in Example 12.2, the desired coefficient of $\kappa_{33}\kappa_{11}^2$ is

$$\frac{20}{(n-1)(n-2)}+\frac{48}{(n-1)^2}+\frac{8}{(n-1)^2}=\frac{4(19n-33)}{(n-1)^2(n-2)}.$$

Example 13.1 (Bivariate sampling cumulants)

To find an exact expression for the covariance of the estimates of variance of two variables, not necessarily independent,

$$\kappa\begin{pmatrix}2 & 0\\ 0 & 2\end{pmatrix}.$$

This will clearly consist of three terms, in κ_{22}, $\kappa_{20}\kappa_{02}$ and κ_{11}^2. For the first, we have the array

$$\begin{array}{cc|c} (2,0) & (0,2) & (2,2) \\ \hline (2,0) & (0,2) & (2,2) \end{array}$$

with pattern function $1/n$ and numerical coefficient unity. For the second, the contribution is zero, the only array being

$$\begin{array}{cc|c} (2,0) & . & (2,0) \\ . & (0,2) & (0,2) \\ \hline (2,0) & (0,2) & (2,2) \end{array}$$

which has a vanishing pattern function by Rule 3. For the third term we have

$$\begin{array}{cc|c} (1,0) & (0,1) & (1,1) \\ (1,0) & (0,1) & (1,1) \\ \hline (2,0) & (0,2) & (2,2) \end{array}$$

the pattern function for which is $\frac{1}{(n-1)}$ and numerical coefficient 2. Hence

$$\kappa\begin{pmatrix}2 & 0\\ 0 & 2\end{pmatrix} = \frac{1}{n}\kappa_{22} + \frac{2}{n-1}\kappa_{11}^2.$$

We note the analogy between this expression and the variance of k_{20}, which may be written (cf. (12.35))

$$\kappa\begin{pmatrix}2 & 2\\ 0 & 0\end{pmatrix} = \frac{\kappa_{40}}{n} + \frac{2\kappa_{20}^2}{n-1}.$$

In particular, if the population is bivariate normal all κs except those of the second order vanish by Exercise 3.14. We then find, using the moments given in Example 3.19, for the ratio

$$\kappa\begin{pmatrix}2 & 0\\ 0 & 2\end{pmatrix} \Big/ \left\{\kappa\begin{pmatrix}2 & 2\\ 0 & 0\end{pmatrix}\kappa\begin{pmatrix}0 & 0\\ 2 & 2\end{pmatrix}\right\}^{\frac{1}{2}},$$

$$\text{correlation } (k_{20}, k_{02}) = \frac{\kappa_{11}^2}{\kappa_{20}\kappa_{02}} = \rho^2,$$

where ρ is the correlation parameter of the bivariate normal form.

13.3 It is possible to derive the bivariate formulae from the univariate formulae by the symbolic process, given by Kendall (1940c), which has been illustrated in 3.29 above. This was used by Cook (1951) to derive bivariate formulae up to those of order 4 in each variate. We quote here some of the simpler results.

$$\kappa\begin{pmatrix}2 & 1\\ 0 & 1\end{pmatrix} = \frac{1}{n}\kappa_{31} + \frac{2}{n-1}\kappa_{20}\kappa_{11} \tag{13.5}$$

$$\kappa\begin{pmatrix}2 & 0\\ 0 & 2\end{pmatrix} = \frac{1}{n}\kappa_{22} + \frac{2}{n-1}\kappa_{11}^2 \tag{13.6}$$

$$\kappa\begin{pmatrix}1 & 1\\ 1 & 1\end{pmatrix} = \frac{1}{n}\kappa_{22} + \frac{1}{n-1}\kappa_{20}\kappa_{02} + \frac{1}{n-1}\kappa_{11}^2 \tag{13.7}$$

$$\kappa\begin{pmatrix}3 & 3\\ 0 & 0\end{pmatrix} = \frac{1}{n}\kappa_{60} + \frac{9}{n-1}\kappa_{40}\kappa_{20} + \frac{9}{n-1}\kappa_{30}^2 + \frac{6n}{(n-1)(n-2)}\kappa_{20}^3 \tag{13.8}$$

(which is, of course, a univariate result)

$$\kappa\begin{pmatrix}3 & 2\\ 0 & 1\end{pmatrix}=\frac{1}{n}\kappa_{51}+\frac{6}{n-1}\kappa_{31}\kappa_{20}+\frac{3}{n-1}\kappa_{40}\kappa_{11}+\frac{9}{n-1}\kappa_{30}\kappa_{21}+\frac{6n}{(n-1)(n-2)}\kappa_{20}^2\kappa_{11} \tag{13.9}$$

$$\kappa\begin{pmatrix}3 & 1\\ 0 & 2\end{pmatrix}=\frac{1}{n}\kappa_{42}+\frac{3}{n-1}\kappa_{22}\kappa_{20}+\frac{6}{n-1}\kappa_{31}\kappa_{11}+\frac{3}{n-1}\kappa_{30}\kappa_{12}+\frac{6n}{n-1}\kappa_{21}^2 + \frac{6n}{(n-1)(n-2)}\kappa_{20}\kappa_{11}^2 \tag{13.10}$$

$$\kappa\begin{pmatrix}3 & 0\\ 0 & 3\end{pmatrix}=\frac{1}{n}\kappa_{33}+\frac{9}{n-1}\kappa_{22}\kappa_{11}+\frac{9}{n-1}\kappa_{12}\kappa_{21}+\frac{6n}{(n-1)(n-2)}\kappa_{11}^3 \tag{13.11}$$

$$\kappa\begin{pmatrix}2 & 2\\ 1 & 1\end{pmatrix}=\frac{1}{n}\kappa_{42}+\frac{4}{n-1}\kappa_{22}\kappa_{20}+\frac{4}{n-1}\kappa_{31}\kappa_{11}+\frac{1}{n-1}\kappa_{40}\kappa_{02}+\frac{4}{n-1}\kappa_{30}\kappa_{12} + \frac{5}{n-1}\kappa_{21}^2+\frac{4n}{(n-1)(n-2)}\kappa_{20}\kappa_{11}^2+\frac{2n}{(n-1)(n-2)}\kappa_{20}^2\kappa_{02} \tag{13.12}$$

$$\kappa\begin{pmatrix}2 & 1\\ 1 & 2\end{pmatrix}=\frac{1}{n}\kappa_{33}+\frac{5}{n-1}\kappa_{22}\kappa_{11}+\frac{2}{n-1}\kappa_{13}\kappa_{20}+\frac{2}{n-1}\kappa_{31}\kappa_{02}+\frac{1}{n-1}\kappa_{30}\kappa_{03} + \frac{8}{n-1}\kappa_{21}\kappa_{12}+\frac{4n}{(n-1)(n-2)}\kappa_{11}\kappa_{20}\kappa_{02}+\frac{2n}{(n-1)(n-2)}\kappa_{11}^3 \tag{13.13}$$

There is obviously a strong family resemblance among these formulae and we shall see later how they can be written in a concise form.

Example 13.2 (Bivariate sampling cumulants for finite populations)

The methods may similarly be extended to deal with bivariate finite populations. Formulae in this field are not required often enough to make it worth while listing them, but it may be useful to illustrate the method of deriving them. Let us then derive the variance of k_{11} in samples of n from a finite population of size N. We have, as in **12.21**,

$$E(k_{11}-K_{11})^2=E(k_{11}^2)-E(K_{11}^2). \tag{13.14}$$

From (13.7) we have

$$E(k_{11}^2)=\mu'\begin{pmatrix}1 & 1\\ 1 & 1\end{pmatrix}=\kappa\begin{pmatrix}1 & 1\\ 1 & 1\end{pmatrix}+\kappa_{11}^2$$

$$=\frac{1}{n}\kappa_{22}+\frac{1}{n-1}\kappa_{20}\kappa_{02}+\frac{1}{n-1}\kappa_{11}^2+\kappa_{11}^2, \tag{13.15}$$

$$E(K_{11}^2)=\frac{1}{N}\kappa_{22}+\frac{1}{N-1}\kappa_{20}\kappa_{02}+\frac{1}{N-1}\kappa_{11}^2+\kappa_{11}^2. \tag{13.16}$$

Hence (13.14) becomes

$$\text{var}\, k_{11} = \left(\frac{1}{n}-\frac{1}{N}\right)\kappa_{22}+\left(\frac{1}{n-1}-\frac{1}{N-1}\right)\kappa_{20}\kappa_{02}+\left(\frac{1}{n-1}-\frac{1}{N-1}\right)\kappa_{11}^2,$$

which reduces to

$$\frac{\text{var}\, k_{11}}{N-n}-\frac{1}{nN}\kappa_{22} = \frac{1}{(n-1)(N-1)}\kappa_{20}\kappa_{02}+\frac{1}{(n-1)(N-1)}\kappa_{11}^2. \tag{13.17}$$

We now require to eliminate the non-linear terms on the right. From (13.16) we have

$$E(K_{11}^2)-\frac{1}{N}\kappa_{22} = \frac{1}{N-1}\kappa_{20}\kappa_{02}+\frac{N}{N-1}\kappa_{11}^2. \tag{13.18}$$

From (13.6) in terms of Ks, we likewise find

$$E(K_{20}K_{02})-\frac{1}{N}\kappa_{22} = \kappa_{20}\kappa_{02}+\frac{2}{N-1}\kappa_{11}^2. \tag{13.19}$$

Hence, from (13.17), (13.18) and (13.19),

$$\begin{vmatrix} \dfrac{\text{var}\, k_{11}}{N-n}-\dfrac{1}{nN}\kappa_{22} & \dfrac{1}{(n-1)(N-1)} & \dfrac{1}{(n-1)(N-1)} \\ E(K_{11}^2)-\dfrac{1}{N}\kappa_{22} & \dfrac{1}{N-1} & \dfrac{N}{N-1} \\ E(K_{20}K_{02})-\dfrac{1}{N}\kappa_{22} & 1 & \dfrac{2}{N-1} \end{vmatrix} = 0. \tag{13.20}$$

This reduces to the expression

$$\text{var}\, k_{11} = \frac{N-n}{(n-1)(N+1)(N-2)}\left\{\frac{N-2}{Nn}(Nn-N-n-1)K_{22} + (N-1)K_{20}K_{02}+(N-3)K_{11}^2\right\}. \tag{13.21}$$

In deriving (13.21) from (13.20), we have, in accordance with the Irwin-Kendall principle of **12.21**, replaced any function f in the infinite population by one in the finite population with expectation f, so that κ_{22} is replaced by K_{22}, $E(K_{11}^2)$ by K_{11}^2 and $E(K_{20}K_{02})$ by $K_{20}K_{02}$. Equation (13.21) should be compared with the result of Exercise 12.11.

Proof of the combinatorial rules

13.4 We now proceed to prove the validity of the rules enunciated and exemplified in **12.12–14**. Rules 1 and 2 have already been proved in **12.12**.

As a preliminary let us define an operator ∂_p such that

$$\left.\begin{aligned} \partial_p\mu_r' &= r(r-1)\ldots(r-p+1)\mu_{r-p}', \quad r>p \\ \partial_p\mu_p' &= p! \\ \partial_p\mu_r' &= 0, \quad r<p, \end{aligned}\right\} \tag{13.22}$$

and

$$\partial_p(AB) = (\partial_p A)B + A(\partial_p B), \tag{13.23}$$

so that ∂ acting on a product is distributive.

In virtue of (13.23) we have

$$\begin{aligned} \partial_p(\mu_r')^m &= m(\mu_r')^{m-1}\partial_p\mu_r' \\ &= \frac{\partial}{\partial\mu_r'}(\mu_r')^m\partial_p\mu_r'. \end{aligned}$$

It follows that if f is a polynomial function in the μs

$$\partial_p f = \frac{\partial f}{\partial\mu_1'}\partial_p\mu_1' + \frac{\partial f}{\partial\mu_2'}\partial_p\mu_2' + \ldots \tag{13.24}$$

and this also holds if f can be expanded in a series of polynomials in the μs.

Now consider the expression defining the cumulants in terms of the moments, (3.30),

$$\exp\left(\kappa_1 t + \ldots + \kappa_p\frac{t^p}{p!} + \ldots\right) = 1 + \mu_1' t + \ldots + \frac{\mu_p' t^p}{p!} + \ldots$$

On operating on both sides by ∂_p there results

$$\begin{aligned} &\exp\left(\kappa_1 t + \ldots + \kappa_p\frac{t^p}{p!} + \ldots\right)\left(\partial_p\kappa_1 t + \ldots + \partial_p\kappa_p\frac{t^p}{p!} + \ldots\right) \\ &\qquad = t^p + \mu_1' t^{p+1} + \ldots = t^p\left(1 + \mu_1' t + \ldots + \mu_p'\frac{t^p}{p!} + \ldots\right), \end{aligned}$$

and hence

$$\partial_p\kappa_1 t + \ldots + \partial_p\kappa_p\frac{t^p}{p!} + \ldots = t^p.$$

This is an identity in t and hence

$$\left.\begin{aligned} \partial_p\kappa_p &= p! \\ \partial_q\kappa_p &= 0, \quad q \neq p \end{aligned}\right\}. \tag{13.25}$$

For example,

$$\kappa_4 = \mu_4' - 4\mu_3'\mu_1' - 3\mu_2'^2 + 12\mu_2'\mu_1'^2 - 6\mu_1'^4.$$

$$\partial_1\kappa_4 = 4\mu_3' - 4\mu_3' - 12\mu_2'\mu_1' - 12\mu_2'\mu_1' + 24\mu_1'^3 + 24\mu_2'\mu_1' - 24\mu_1'^3$$
$$= 0.$$

$$\partial_2\kappa_4 = 12\mu_2' - 24\mu_1'^2 - 12\mu_2' + 24\mu_1'^2$$
$$= 0.$$

$$\partial_3\kappa_4 = 24\mu_1' - 24\mu_1'$$
$$= 0.$$

$$\partial_4\kappa_4 = 4!.$$

13.5 Now in accordance with Rule 1, which we established in **12.12**, $\kappa(a_1^{\alpha_1}\ldots a_s^{\alpha_s})$ and hence $\mu(a_1^{\alpha_1}\ldots a_s^{\alpha_s})$ may be expressed in terms of population κs by an equation of the form

$$\mu(a_1^{\alpha_1}a_2^{\alpha_2}\ldots) = \sum\{A(\kappa_{b_1}^{\beta_1}\kappa_{b_2}^{\beta_2}\ldots)\} \tag{13.26}$$

where A is a factor to be found. To save printing we omit the primes on the μs for the rest of this proof. Operate on both sides of (13.26) by $(\partial_{b_1}^{\beta_1}\partial_{b_2}^{\beta_2}\ldots)$. Every term on the right is annihilated except that in $(\kappa_{b_1}^{\beta_1}\kappa_{b_2}^{\beta_2}\ldots)$ and we have

$$A.(b_1!)^{\beta_1}(b_2!)^{\beta_2}\ldots\beta_1!\beta_2!\ldots = (\partial_{b_1}^{\beta_1}\partial_{b_2}^{\beta_2}\ldots)\mu(a_1^{\alpha_1}a_2^{\alpha_2}\ldots). \tag{13.27}$$

We now consider an operator θ_p, analogous to ∂_p, which, when acting on a power of x (of any suffix), reduces the exponent by p and multiplies by $r(r-1)\ldots(r-p+1)$; and we will suppose the operator to be distributive.(*) Regarding $\mu(a_1^{\alpha_1}a_2^{\alpha_2}\ldots)$ as the mean value of $(k_{a_1}^{\alpha_1}k_{a_2}^{\alpha_2}\ldots)$ we see that the result of operating by the ∂s on the mean value expressed in terms of μs is the same as that given by taking the mean value of the operation of the θs. But this latter operation results in a constant, which is equal to its mean value; and we thus have

$$A = \frac{(\theta_{b_1}^{\beta_1}\theta_{b_2}^{\beta_2}\ldots)}{(b_1!)^{\beta_1}(b_2!)^{\beta_2}\ldots\beta_1!\beta_2!\ldots}(k_{a_1}^{\alpha_1}k_{a_2}^{\alpha_2}\ldots). \tag{13.28}$$

Our rules are concerned with the evaluation of this operation.

13.6 Consider now a completed array of type (12.32). A little reflection will show that there is one such array for every term in (13.28) which does not vanish by operation, and that every term in (13.28) will have its corresponding completed array. The numbers in the body of the array are the powers of x occurring in the k-product;

(*) θ_p may be regarded as equivalent to $\left(\frac{\partial^p}{\partial x_1^p} + \frac{\partial^p}{\partial x_2^p} + \ldots + \frac{\partial^p}{\partial x_n^p}\right)$.

added horizontally they compose the orders of the operators; added vertically they compose the orders of the corresponding ks. A completed array is, so to speak, a chart of part of the operation; and the whole operation is the sum of all possible completed arrays.

The operation (13.28) gives us the coefficients in $\mu(a_1^{\alpha_1}a_2^{\alpha_2}\ldots)$, but we wish to find those in the corresponding $\kappa(a_1^{\alpha_1}a_2^{\alpha_2}\ldots)$. The necessary allowance is made by Rule 3, which we now prove; that is, the coefficient of $(\kappa_{b_1}^{\beta_1}\kappa_{b_2}^{\beta_2}\ldots)$ in $\kappa(a_1^{\alpha_1}a_2^{\alpha_2}\ldots)$ is given by all completed arrays, *ignoring those which are resolvable into separate blocks each confined to separate rows and columns.*

Referring to equation (12.31), expressing the relation between multivariate moments and cumulants, we see that $\kappa(a_1^{\alpha_1}a_2^{\alpha_2}\ldots)$ is the sum of terms composed of products of one, two, three ... multivariate moments. The first term is $\mu(a_1^{\alpha_1}a_2^{\alpha_2}\ldots)$ itself. Consider a two-part term such as $\mu(a_1^{\alpha_1'}a_2^{\alpha_2'}\ldots)\mu(a_1^{\alpha_1''}a_2^{\alpha_2''}\ldots)$, where $\alpha_1'+\alpha_1''=\alpha_1$, etc. Its coefficient in the expansion on the right-hand side of (12.31) is

$$-\frac{1}{2}\cdot\frac{2!}{1!\,1!}\,\frac{t_{a_1}^{\alpha_1'}}{\alpha_1'!\,\alpha_1''!}\,\frac{t_{a_2}^{\alpha_2'}}{\alpha_2'!\,\alpha_2''!}\cdots$$

and hence the coefficient with which it appears in the formula for $\kappa(a_1^{\alpha_1}a_2^{\alpha_2}\ldots)$ is

$$-\frac{\alpha_1!}{\alpha_1'!\,\alpha_1''!}\,\frac{\alpha_2!}{\alpha_2'!\,\alpha_2''!}\cdots=-\binom{\alpha_1}{\alpha_1'}\binom{\alpha_2}{\alpha_2'}\cdots \tag{13.29}$$

Now $\mu(a_1^{\alpha_1'}a_2^{\alpha_2'}\ldots)$ will itself have an array of type (12.32) with column totals $(a_1^{\alpha_1'}a_2^{\alpha_2'}\ldots)$ and row totals, say $(b_1^{\beta_1'}b_2^{\beta_2'}\ldots)$; and similarly for $\mu(a_1^{\alpha_1''}a_2^{\alpha_2''}\ldots)$. Provided that $\beta_1'+\beta_1''=\beta_1$ these arrays will correspond to terms in the κs which, when multiplied, will give a term in $(\kappa_{b_1}^{\beta_1}\kappa_{b_2}^{\beta_2}\ldots)$. Thus the product of these terms may be considered as an array of type (12.32) with column totals $(a_1^{\alpha_1}a_2^{\alpha_2}\ldots)$ and row totals $(b_1^{\beta_1}b_2^{\beta_2}\ldots)$ and with the body of the table resolvable into two separate blocks. Since there are α_1 columns of total a_1, there will be $\binom{\alpha_1}{\alpha_1'}\binom{\alpha_2}{\alpha_2'}$ products of this type in the expression that gives $\mu(a_1^{\alpha_1}a_2^{\alpha_2}\ldots)$. This factor is the same as (13.29) but of opposite sign. Hence, if we ignore the separate two-part blocks in the array for μ we shall have allowed for the products of two moments that must be subtracted from μ to give κ.

Now some of these separate blocks will themselves be separable into two blocks, and in subtracting them all from $\mu(a_1^{\alpha_1}a_2^{\alpha_2}\ldots)$ we subtract too much. For example, if there are three separate blocks, L, M, N, we shall, by considering L and $(M+N)$ as two blocks, have subtracted L, M, N. We shall have done the same by considering M and $(L+N)$, and N and $(L+M)$ as two blocks. That is, we have subtracted $2L$, $2M$, $2N$ too much. We must restore these blocks to the array for μ. Such additions, summed over all blocks of three, will be found to equal the terms in the expansion of (12.31) that result from the product of three moments.

In restoring these blocks we restore too many of the cases where there are four separate blocks. These must be subtracted again, and correspond to the negative term in (12.31) involving the product of four moments. Proceeding in this way we establish Rule 3.[*]

13.7 Now we proceed to Rules 4, 5 and 6, which are the fundamental rules of the whole process. Consider again the array of type (12.32), to fix the ideas, say,

$$\begin{array}{ccc|c} 2 & 3 & 1 & 6 \\ 1 & 1 & . & 2 \\ 1 & . & 1 & 2 \\ \hline 4 & 4 & 2 & 10 \end{array} \qquad (13.30)$$

This array will represent a number of terms in the operation, each of which consists of the operation of θ_6 on a term $x^2 \cdot x^3 \cdot x$ (the first row), θ_2 on $x \cdot x$ (the second row), and so on. Provided that the suffixes of the xs in any row are alike, every suffix of the xs will provide a term, for k_p contains terms with every distribution of powers (adding to p) and suffixes. There will, for instance, be terms of the following kind:

$$\begin{array}{ccc} x_1^2 & x_1^3 & x_1 \\ x_2 & x_2 & . \\ x_3 & . & x_3 \end{array}, \quad \begin{array}{ccc} x_1^2 & x_1^3 & x_1 \\ x_1 & x_1 & . \\ x_2 & . & x_2 \end{array}, \quad \begin{array}{ccc} x_1^2 & x_1^3 & x_1 \\ x_1 & x_1 & . \\ x_1 & . & x_1 \end{array}.$$

In fact, for any completed array, we have terms in which

(i) all the xs have the same suffix (n terms, one for each suffix),
(ii) all the xs but one row have the same suffix ($n(n-1)$ terms),
(iii) all the xs but two rows have the same suffix and the remaining two are the same ($n(n-1)$ terms),

and so on. These cases correspond to the various separations dealt with in Rule 5.

Now in case (i) the term in any column arises from the term in x^p in k_p and (apart from numerical factors which are considered presently) is n^{-1}, from equation (12.25). Hence any column that contains an entry contributes a factor n^{-1} and the total function of n arising from case (i) is the product of n and of (n^{-1}) to the power of the number of columns containing a non-zero entry.

Similarly in cases (ii) and (iii) the n-function for each separation is the product of $n(n-1)$ and, for each column, a factor in n^{-1} or $\dfrac{-1}{n(n-1)}$ according as the column contains non-zero entries in one or in both parts of the separation; and so on.

This explains the origin of the pattern function as described in Rule 6. But in order to establish that rule completely (and incidentally to establish Rules 4 and 5) we have to show that the numerical coefficients arising from each separation are the same. When this is done the validity of Rule 6 is demonstrated, for the separate contributions in

[*] Kaplan (1952) has adduced a statistical argument in support of Rule 3.

n may be added together to give the pattern function and the whole multiplied by the numerical coefficient.

θ_1 may be considered as the operation of picking out an x from the operand in all possible ways and replacing it by unity. Similarly $\frac{\theta_p}{p!}$ may be regarded as picking out p xs with the same suffix and replacing them by unity. It is thus evident that operating on a k-product by a θ-product $\frac{\theta_{b_1}\theta_{b_2}\ldots}{b_1!b_2!\ldots}$ of the same degree will yield the number of ways in which sets of xs can be picked out of the k-product so that each set contains b_1 of one suffix, b_2 of a second suffix (which may be the same as the first), and so on.

Now consider the operation (13.28) in which the ks are expressed in the simplified form (12.26). The operations θ being distributive, we shall emerge from the operation with a sum of terms comprising all the possible ways in which the individual xs can be picked out of the k-product such that the row and column totals of the two-way array are satisfied. Consider the sets corresponding to a particular array, such as (13.30). The contribution to the total will consist of the ways of picking out individuals such that

(i) from the individuals in the first k_4 are chosen four in the partition (2, 1, 1),
(ii) from those in the second k_4 are chosen four in the partition (3, 1),
(iii) from those in the k_2 are chosen two in the partition (1, 1),
(iv) these are associated in all possible ways such that individuals in a row arise from the same suffix.

On consideration it will be seen that the total number of ways of doing this is the number of ways of allocating the individuals from column totals as required by Rule 5; *and this is true whether sets of rows have the same suffix or not.*

Rules 5 and 6, and hence Rule 4, follow at once.

13.8 The remaining rules are ancillary.

Rule 7 follows from Rule 2. In fact, the pattern function is independent of the numbers composing the array, and the pattern with a row containing one element can therefore form the skeleton of an array in which that element is unity; and this would entail the appearance of κ_1, which by Rule 2 is impossible.

Rule 8 follows from Rule 6. The column containing the single element appears in just one separate of all the separations, and the contributions to the pattern function are thus all multiplied by n^{-1} owing to its presence.

Rule 10 follows from Rule 8. The addition of a unit part is equivalent to the addition of an extra column containing unity. This multiplies all pattern functions by n^{-1}, leaves numerical coefficients unchanged and increases the suffix of every κ according to the row in which the unit appears.

13.9 It only remains to prove Rule 9. Note that any pattern function can be evaluated linearly in terms of the functions of the pattern obtained by omitting one of the columns. For example, consider the right-hand column of

$$\begin{array}{cccc} . & \times & \times & \times \\ \times & . & \times & \times \\ \times & \times & . & . \\ \times & \times & . & . \end{array} \qquad (13.31)$$

and the contributions to the pattern function from it. The 15 separations which are possible with four rows can be divided into two classes, that in which the two rows in the fourth column lie in the same separate and that in which they do not. In separations of the first type the contributions from the first three columns will be the contributions of all separations of

$$\begin{array}{ccc} \times & \times & \times \\ \times & \times & . \\ \times & \times & . \end{array} \qquad (13.32)$$

in which the first two rows are amalgamated. Considering the function of the first three rows

$$\begin{array}{ccc} . & \times & \times \\ \times & . & \times \\ \times & \times & . \\ \times & \times & . \end{array} \qquad (13.33)$$

in which amalgamation has not taken place, we see that the contribution consists of all contributions which do not occur in the first. Calling the first A and the second B, we see that this contribution is

$$\frac{1}{n}A - \frac{1}{n(n-1)}(B-A) = \frac{1}{n-1}A - \frac{1}{n(n-1)}B,$$

i.e. a linear function of the derived patterns A and B. The proof of the general result follows exactly the same lines.

Now if a pattern may be divided into two groups connected only by a single column we can reduce it step by step by omitting the other columns. We end up with this single column, and the pattern function of this column must vanish; for the column total a corresponds to k_a, whose mean value the one-column array expresses, and since by definition this mean value is κ_a no composite terms such as would be given by two rows or more can appear.

13.10 The adaptation of the argument to the bivariate case is fairly straightforward. The only rule requiring examination is the one giving the numerical coefficients attaching to the pattern function; and on examination it should be clear, from the structure of the bivariate k-statistics, that the procedure sketched in **13.2** applies. The use of an annihilating operator

$$\partial_{pq}\mu_{rs} = r^{[p]}s^{[q]}\mu_{r-p,s-q} \tag{13.34}$$

gives the required result.

We may also state without proof that similar methods apply to multivariate statistics of higher order than the bivariate. For trivariate statistics, for example, we should have to consider partitions into three-part members. Only the simpler results in such cases are required in practice and they can usually be written down from the univariate or bivariate cases from considerations of the type developed in the following sections.

The above proofs of the rules are due to Kendall (1940b). James (1958) gives alternative proofs of the combinatorial rules, covering the general multivariate case, and also including results on moments as well as k-statistics. McCullagh (1984) illuminates the rules, using a tensor notation. Speed (1986b) develops multivariate k-statistics under exchangeability

James and Mayne (1962) give a combinatorial method of obtaining the joint cumulants of any number of functions of any number of variables with finite cumulants. If each variable has its κ_r of order $n^{-(r-1)}$ they give the joint cumulants of order $\leqslant 6$ explicitly to terms of order n^{-4} or better. McCullagh (1984) treats the case of a polynomial function of a number of variables.

The essential structure of the results

13.11 Let us review the results of this and the previous chapter. We have seen that, by the use of simple algebraic methods, we can obtain as many moments or cumulants as we please of any statistic which is expressible as a symmetric function of the observations. We have also noted that the working is greatly simplified by the use of k-statistics and cumulants.

It is reasonable to inquire how this simplification comes about. What are the structural properties of k-statistics and cumulants that give them these advantages, and do they represent the limit of achievement in sampling simplicity? The answer to the last question seems to be affirmative and it is instructive to consider why this should be so.

13.12 The definition of univariate cumulants in terms of moments by the identity in t

$$\exp\left\{\sum \kappa_p t^p/p!\right\} = \sum \mu_p' t^p/p! \tag{13.35}$$

tends to obscure the essential structure of the cumulant–moment relationship. Let us, instead, consider the relation between multivariate cumulants and moments, as defined formally by (12.31). To fix the ideas, suppose we have four variates and require their product moment μ_{1111}'. From (12.31) we find, on examining the coefficient in

$t_1 t_2 t_3 t_4$,

$$\begin{aligned}\mu'_{1111} = \kappa_{1111} &+ (\kappa_{1110}\kappa_{0001} + \kappa_{1101}\kappa_{0010} + \kappa_{1011}\kappa_{0100} + \kappa_{0111}\kappa_{1000}) \\ &+ (\kappa_{1100}\kappa_{0011} + \kappa_{1010}\kappa_{0101} + \kappa_{0110}\kappa_{1001}) \\ &+ (\kappa_{1100}\kappa_{0010}\kappa_{0001} + \kappa_{1010}\kappa_{0100}\kappa_{0001} + \kappa_{1001}\kappa_{0100}\kappa_{0010} + \kappa_{0110}\kappa_{1000}\kappa_{0001} \\ &+ \kappa_{0101}\kappa_{1000}\kappa_{0010} + \kappa_{0011}\kappa_{1000}\kappa_{0100}) \\ &+ \kappa_{1000}\kappa_{0100}\kappa_{0010}\kappa_{0001}. \qquad (13.36)\end{aligned}$$

Such formulae exhibit the essential simplicity of the relationship. In the expression of μs in terms of κs every possibility on the right occurs exactly once. When we invert the relationship to give κs in terms of μs we have, analogously to (13.36),

$$\begin{aligned}\kappa_{1111} = \mu'_{1111} &- (\mu'_{1110}\mu'_{0001} + \mu'_{1101}\mu'_{0010} + \mu'_{1011}\mu'_{0100} + \mu'_{0111}\mu'_{1000}) \\ &- (\mu'_{1100}\mu'_{0011} + \mu'_{1010}\mu'_{0101} + \mu'_{0110}\mu'_{1001}) \\ &+ 2(\mu'_{1100}\mu'_{0010}\mu'_{0001} + \text{etc.}) - 6\mu'_{1000}\mu'_{0100}\mu'_{0010}\mu'_{0001}. \qquad (13.37)\end{aligned}$$

This type of formula differs from (13.36) only in the factor $(\rho-1)!\,(-1)^{\rho-1}$ attached to sets that contain ρ parts.

13.13 **In particular, these formulae remain valid if two or more of the four variates are identical. If, for example, the first and second are identical, expressions such as κ_{1111} are written κ_{211} and become trivariate; and if in addition the third and fourth are identical the cumulant becomes κ_{22} and is bivariate. In the extreme case when all are identical it becomes κ_4. Thus we can obtain all the relations of order 4 for any number of variates by condensing (13.36) or (13.37) by amalgamating suffixes. Certain terms that are distinct in the originals then become the same and can be added. For example, from (13.36) we find**

$$\mu'_{22} = \kappa_{22} + 2\kappa_{21}\kappa_{01} + 2\kappa_{12}\kappa_{10} + \kappa_{20}\kappa_{02} + 2\kappa_{11}^2 + \kappa_{20}\kappa_{01}^2 + \kappa_{02}\kappa_{10}^2 + 4\kappa_{11}\kappa_{10}\kappa_{01} + \kappa_{10}^2\kappa_{01}^2. \qquad (13.38)$$

If we now add the suffixes we reach the known univariate form

$$\mu'_4 = \kappa_4 + 4\kappa_3\kappa_1 + 3\kappa_2^2 + 6\kappa_2\kappa_1^2 + \kappa_1^4. \qquad (13.39)$$

Similarly, in the finite population case, (13.21), on adding its suffixes, reduces to the result of Exercise 12.11. The origin of the numerical coefficients in (13.39) is now clear. The symbolic methods of deriving multivariate from univariate formulae (cf. 3.29) are merely convenient shorthand ways of working the process of amalgamation in reverse, a procedure that is possible because of the symmetry of the terms in (13.36) among the suffixes.

13.14 **It thus appears that the formal definitions of type (12.31) and (13.35) involving the exponential function derive their utility from the enumerative algebraic property of the exponential as illustrated in (13.36); not, for instance, from the analytical properties of the exponential such as its being equal to its own derivative or its being the inverse of the logarithmic function. The numerical coefficients appearing in our relations**

between cumulants and moments are convenient summaries of relations that are structurally much simpler than they appear. Similar considerations apply to k-statistics. Apart from factors in $(-1)^{\rho-1}(\rho-1)!/n^{[\rho]}$ they are, as we have noted in **12.8**, sums of products of the contributory variates x taken in all possible ways once and only once.

Summary of results

13.15 We can now summarize our formulae. For instance (13.36) can be written

$$\mu'_{ijkl} = \kappa_{ijkl} + \sum^{4} \kappa_i\kappa_{jkl} + \sum^{3} \kappa_{ij}\kappa_{kl} + \sum^{6} \kappa_i\kappa_j\kappa_{kl} + \kappa_i\kappa_j\kappa_k\kappa_l, \tag{13.40}$$

where the summations occur over the number of ways of grouping the subscripts, which are shown in the numbers above the summation signs. (There is a change of notation here: μ_i' refers to the mean value of the ith variate, not the mean of the ith power of some variate x.)

Likewise, if $s_{ij\ldots l}$ refers to the products $x_ix_j\ldots x_l$ summed over the sample, we have for the k-statistics

$$\left.\begin{aligned} k_i &= s_i/n \\ k_{ij} &= (ns_{ij} - s_is_j)/n^{[2]} \\ k_{ijk} &= \left(n^2s_{ijk} - n\sum^{3} s_is_{jk} + 2s_is_js_k\right)\Big/n^{[3]} \end{aligned}\right\} \tag{13.41}$$

and so on.

For example, if the ith and jth variates are identical we find, in the more familiar notation,

$$k_2 = (ns_2 - s_1^2)/\{n(n-1)\} \tag{13.42}$$

whereas the bivariate k_{11} is given by

$$k_{11} = (ns_{11} - s_{10}s_{01})/\{n(n-1)\}. \tag{13.43}$$

If we require the bivariate k_{21} we amalgamate i and j and obtain from the third formula of (13.41)

$$k_{21} = \{n^2s_{21} - 2ns_{10}s_{11} - ns_{20}s_{01} + 2s_{10}^2s_{01}\}/n^{[3]}$$

as given in (13.2).

13.16 This notation, which was suggested by Kaplan (1952) as a tensor notation, enables us to summarize the multivariate formulae in little more space than is required for the univariate results. For instance, for four variates we have for the covariance of two pairs

$$\begin{aligned} \kappa(ab, ij) &= E\{(k_{ab} - \kappa_{ab})(k_{ij} - \kappa_{ij})\} \\ &= \frac{1}{n}\kappa_{abij} + \frac{1}{n-1}(\kappa_{ai}\kappa_{bj} + \kappa_{aj}\kappa_{bi}). \end{aligned} \tag{13.44}$$

This embodies seven types of formulae. If all variates are the same (a, b, i, j referring to one variate) we have

$$\operatorname{var} k_2 = \kappa(2^2) = \frac{1}{n}\kappa_4 + \frac{2}{n-1}\kappa_2^2. \tag{13.45}$$

If variate a is the same as i and variate b the same as j, we have

$$\operatorname{var} k_{11} = \kappa\begin{pmatrix}1 & 1\\ 1 & 1\end{pmatrix} = \frac{1}{n}\kappa_{22} + \frac{1}{n-1}\kappa_{20}\kappa_{02} + \frac{1}{n-1}\kappa_{11}^2. \tag{13.46}$$

If a and b are the same, i and j the same, we find

$$\operatorname{cov}(k_{20}, k_{02}) = \kappa\begin{pmatrix}2 & 0\\ 0 & 2\end{pmatrix} = \frac{1}{n}\kappa_{22} + \frac{2}{n-1}\kappa_{11}^2. \tag{13.47}$$

If a, b and i are the same and j is different,

$$\operatorname{cov}(k_{20}, k_{11}) = \kappa\begin{pmatrix}2 & 1\\ 0 & 1\end{pmatrix} = \frac{1}{n}\kappa_{31} + \frac{2}{n-1}\kappa_{11}\kappa_{20}. \tag{13.48}$$

If a and b are the same, i, j are different,

$$\operatorname{cov}(k_{200}, k_{011}) = \kappa\begin{pmatrix}2 & 0\\ 0 & 1\\ 0 & 1\end{pmatrix} = \frac{1}{n}\kappa_{211} + \frac{2}{n-1}\kappa_{110}\kappa_{101}. \tag{13.49}$$

If a and i are the same, b, j are different,

$$\operatorname{cov}(k_{110}, k_{101}) = \kappa\begin{pmatrix}1 & 1\\ 1 & 0\\ 0 & 1\end{pmatrix} = \frac{1}{n}\kappa_{211} + \frac{1}{n-1}(\kappa_{200}\kappa_{011} + \kappa_{110}\kappa_{101}). \tag{13.50}$$

Finally, if all four are different,

$$\operatorname{cov}(k_{1100}, k_{0011}) = \kappa\begin{pmatrix}1 & 0\\ 1 & 0\\ 0 & 1\\ 0 & 1\end{pmatrix} = \frac{1}{n}\kappa_{1111} + \frac{1}{n-1}(\kappa_{1010}\kappa_{0101} + \kappa_{1001}\kappa_{0110}). \tag{13.51}$$

13.17 The following formulae summarize further multivariate results.

$$\kappa(ab, ijk) = \kappa_{abijk}/n + \sum^{6} \kappa_{ai}\kappa_{bjk}/(n-1). \tag{13.52}$$

$$\kappa(ab, ijkl) = \kappa_{abijkl}/n + \left(\sum^{8} \kappa_{ai}\kappa_{bjkl} + \sum^{6} \kappa_{aij}\kappa_{bkl}\right)\Big/(n-1). \tag{13.53}$$

$$\kappa(ab, ij, pq) = \kappa_{abijpq}/n^2 + \sum^{12} \kappa_{abip}\kappa_{jq}/\{n(n-1)\} + \sum^{4} \kappa_{aip}\kappa_{bjq}(n-2)/\{n(n-1)^2\}$$
$$+ \sum^{8} \kappa_{ai}\kappa_{bp}\kappa_{jq}/(n-1)^2. \quad (13.54)$$

$$\kappa(abc, ijk) = \kappa_{abcijk}/n + \left(\sum^{9} \kappa_{ai}\kappa_{bcjk} + \sum^{9} \kappa_{abi}\kappa_{cjk}\right)\Big/(n-1)$$
$$+ \sum^{6} \kappa_{ai}\kappa_{bj}\kappa_{ck}n/\{(n-1)(n-2)\}. \quad (13.55)$$

$$\kappa(abc, ijkl) = \kappa_{abcijkl}/n + \sum^{12} \kappa_{ai}\kappa_{bcjkl}/(n-1) + \left(\sum^{12} \kappa_{abi}\kappa_{cjkl} + \sum^{18} \kappa_{aij}\kappa_{bckl}\right)\Big/(n-1)$$
$$+ \sum^{36} \kappa_{ai}\kappa_{bj}\kappa_{ckl}n/\{(n-1)(n-2)\}. \quad (13.56)$$

$$\kappa(ab, ij, pq, uv) =$$
$$\kappa_{abijpquv}/n^3 + \sum^{24} \kappa_{ai}\kappa_{bjpquv}/\{n^2(n-1)\} + \sum^{32} \kappa_{aip}\kappa_{bjquv}(n-2)/\{n^2(n-1)^2\}$$
$$+ \sum^{8} \kappa_{aipu}\kappa_{bjqv}(n^2 - 3n + 3)/\{n^2(n-1)^3\} + 3\sum^{24} \kappa_{abpu}\kappa_{ijqv}/\{n^2(n-1)\}$$
$$+ \left(\sum^{96} \kappa_{ai}\kappa_{bp}\kappa_{jquv} + \sum^{48} \kappa_{ai}\kappa_{pu}\kappa_{bjqv}\right)\Big/\{n(n-1)^2\}$$
$$+ \sum^{96} \kappa_{ai}\kappa_{bpu}\kappa_{jqv}(n-2)/\{n(n-1)^3\} + \sum^{48} \kappa_{bi}\kappa_{jp}\kappa_{qu}\kappa_{va}/(n-1)^3. \quad (13.57)$$

$$\kappa(abcd, ijkl) = \kappa_{abcdijkl}/n + \sum^{16} \kappa_{ai}\kappa_{bcdjkl}/(n-1) + \sum^{48} \kappa_{abi}\kappa_{cdjkl}/(n-1)$$
$$+ \sum^{72} \kappa_{ai}\kappa_{bj}\kappa_{cdkl}n/\{(n-1)(n-2)\} + \left(\sum^{16} \kappa_{aijk}\kappa_{bcdl}\right.$$
$$\left. + \sum^{18} \kappa_{abij}\kappa_{cdkl}\right)\Big/(n-1) + \sum^{144} \kappa_{ai}\kappa_{bcj}\kappa_{dkl}n/\{(n-1)(n-2)\}$$
$$+ \sum^{24} \kappa_{ai}\kappa_{bj}\kappa_{ck}\kappa_{dl}n(n+1)/\{(n-1)(n-2)(n-3)\}. \quad (13.58)$$

These formulae embrace 41 formulae given by Cook (1951) and a number of others besides, but as Kaplan remarks, some care is required in deriving particular cases. McCullagh (1984) gives similar tensor formulae.

Example 13.3 (Sampling cumulants of the covariance)

For the variance of k_{11} from an infinite population we have, as in Example 13.2,

$$\text{var}\, k_{11} = \frac{1}{n}\kappa_{22} + \frac{1}{n-1}\kappa_{20}\kappa_{02} + \frac{1}{n-1}\kappa_{11}^2.$$

Let us specialize this to the case of a standardized normal population. We then find

$$\text{var}\, k_{11} = \frac{1}{n-1}(1+\rho^2). \tag{13.59}$$

Let us find the third and fourth cumulants of k_{11}.

The third cumulant is given by (13.54) with a, i, p the same and b, j, q the same. We find

$$\kappa\begin{pmatrix}1 & 1 & 1\\ 1 & 1 & 1\end{pmatrix} = \frac{1}{n^2}\kappa_{33} + \frac{1}{n(n-1)}\{6\kappa_{22}\kappa_{11} + 3\kappa_{20}\kappa_{13} + 3\kappa_{31}\kappa_{02}\}$$
$$+ \frac{n-2}{n(n-1)^2}\{3\kappa_{21}\kappa_{12} + \kappa_{30}\kappa_{03}\} + \frac{1}{(n-1)^2}\{2\kappa_{11}^3 + 6\kappa_{11}\kappa_{20}\kappa_{02}\}. \tag{13.60}$$

For the normal case the only surviving terms yield

$$\kappa_3(k_{11}) = \frac{1}{(n-1)^2}(2\rho^3 + 6\rho). \tag{13.61}$$

Likewise we find from (13.57), taking a, i, p, u the same and b, j, q, v the same, the surviving terms, which can only arise from the concluding term,

$$\kappa_4(k_{11}) = \frac{1}{(n-1)^3}\{36\rho^2 + 6 + 6\rho^4\}. \tag{13.62}$$

Thus, for instance, the kurtosis of the distribution of k_{11} is given by (13.59) and (13.62) as

$$\beta_2 - 3 = \frac{6}{n-1}\left(1 + \frac{4\rho^2}{(1+\rho^2)^2}\right). \tag{13.63}$$

Tables of symmetric functions

13.18 *Symmetric Function and Allied Tables*, by F. N. David, M. G. Kendall, and D. E. Barton (C.U.P., 1966), contains full versions of the univariate and bivariate tables discussed in this and the previous chapter, with many others. Mikhail (1968) gives trivariate tables up to weight 6.

EXERCISES

13.1 Show that the correlation between k_{30} and k_{03} in samples of n from a bivariate normal population is ρ^3, where ρ is the correlation parameter of the population.

13.2 Show generally in Exercise 13.1 that the correlation between k_{r0} and k_{0r} is ρ^r.

13.3 A four-variate normal population has variates x_1, x_2, x_3, x_4 with unit variances and the correlation of x_i and x_j is ρ_{ij}. Show that the correlation between the sample covariance of x_1, x_2 and that of x_3, x_4 is given by

$$\frac{\rho_{13}\rho_{24}+\rho_{14}\rho_{23}}{\{(1+\rho_{12}^2)(1+\rho_{34}^2)\}^{\frac{1}{2}}}.$$

13.4 Derive the formulae (13.5–13) from the general formulae in **13.16–17**.

13.5 Use the methods of this chapter to derive the large-sample result for the variance of the correlation coefficient in bivariate normal samples

$$\operatorname{var} r = (1-\rho^2)^2/n.$$

13.6 If x is a random variable with finite cumulants of all orders satisfying $\kappa_r = O(n^{-r+1})$ show that the transformed variate $y=f(x)$ has cumulants which also satisfy this condition. (This does not preclude the cumulants from being of lower order, i.e. $o(n^{-r+1})$.)

(James, 1955)

13.7 Show that for a bivariate normal population k_{tu} and k_{vw} have zero covariance unless $t+u=v+w$.

(Wishart, 1929)

13.8 Show that for a bivariate normal distribution with p.d.f.

$$f=\frac{1}{2\pi\sigma_1\sigma_2(1-\rho^2)^{\frac{1}{2}}}\exp\left\{\frac{-1}{2(1-\rho^2)}\left(\frac{x_1^2}{\sigma_1^2}-\frac{2\rho x_1x_2}{\sigma_1\sigma_2}+\frac{x_2^2}{\sigma_2^2}\right)\right\},$$

$$\operatorname{var}(k_{tu})=\kappa\begin{pmatrix} t & t\\ u & u\end{pmatrix}=t!\,u!\sum_{j=1}^{t+u}\frac{(j-1)!}{j}\frac{\Delta^j 0^{t+u}}{n^{[j]}}\sigma_1^{2t}\sigma_2^{2u}F(-t,-u,1,\rho^2),$$

where $\Delta^j 0^k$ is the jth difference of the kth power of zero and F refers to the hypergeometric function.

(Wishart, 1929)

13.9 In the notation of **13.15**, show that

$$k_{ijkl}=\{n^2(n+1)s_{ijkl}-n(n+1)\sum^4 s_is_{jkl}-n(n-1)\sum^3 s_{ij}s_{kl}+2n\sum^6 s_is_js_{kl}-6s_is_js_ks_l\}/n^{[4]},$$

and hence verify the formulae for k_{31} and k_{22} of (13.2).

(Kaplan, 1952)

13.10 Referring to the result of **13.9**, show that for a normal population the effect of adding a new part 2 to $\kappa(a_1^{\alpha_1}\ldots a_s^{\alpha_s})$ is to give pattern functions $1/(n-1)$ times those of the original. Show also that the effect on the numerical coefficient in an array is to multiply by twice the number of rows in the array. Deduce that the effect of adding a new part 2 is equivalent to operating by

$$\frac{2\kappa_2^2}{n-1}\frac{d}{d\kappa_2}.$$

(Fisher and Wishart, 1931)

13.11 Use the previous exercise to establish the equations (12.71) and (12.72).

13.12 In generalization of Exercise 13.10 show that for a multivariate normal population the effect of adding a covariance k_{pq} (p, q referring to the pth and qth variates) is equivalent to operating by

$$\sum_{rs}\frac{1}{(n-1)}(\kappa_{pr}\kappa_{qs}+\kappa_{ps}\kappa_{qr})\frac{d}{d\kappa_{rs}},$$

where κ_{pq} is the covariance of the variates p and q.

(Fisher and Wishart, 1931)

13.13 Show that in normal samples

$$\text{var}\,(k_r/k_2^{\frac{1}{2}r}) \doteq r!/n, \qquad r>2.$$

(Cf. the exact results for $r=3$, 4 in Exercises 12.9-10.)

CHAPTER 14

ORDER STATISTICS

14.1 In **10.10–11** and **11.4** we discussed some of the sampling properties of medians, quantiles and, more generally, the order statistics of a sample of observations. It will be remembered that the rth order statistic of a sample of n is simply the rth smallest variate value in the sample. We shall denote it by a suffix in parentheses: $x_{(r)}$. Unlike statistics based on symmetric functions of the observations, the order statistics do not lend themselves to investigation by the methods we have been using in Chapters 12 and 13. They have the distinctive property that if we take an ordinary random sample of values $x_1, x_2, \ldots, x_n$ and rearrange them in order, as $x_{(1)}, x_{(2)}, \ldots, x_{(n)}$, the values so derived are no longer independent nor identically distributed even though the original xs were so.

An expository account of many branches of the theory of order statistics, with extensive tables, is given in the book edited by Sarhan and Greenberg (1962) and in the monograph by H. A. David (1981), the appendix to which gives an extensive listing of available tables. Harter (1978, 1983, 1991a,b) gives a comprehensive bibliography up to 1963, to be extended in future. The volumes of tables by Harter (1969) should also be consulted.

Distributions in the continuous case

14.2 Notwithstanding lack of symmetry and independence, some remarkably simple results can be obtained. We have seen in **10.10** and **11.4** that, in samples of size n from any d.f. $F(x)$ with a continuous density $f(x)$, the sampling distribution of the transformed order statistic $F(x_{(r)})$ is given by $G_r(x)$, say, where

$$dG_r = \frac{\{F(x_{(r)})\}^{r-1}\{1-F(x_{(r)})\}^{n-r}\,dF(x_{(r)})}{B(r, n-r+1)} \tag{14.1}$$

is a Beta distribution. Likewise, the joint distribution of $F(x_{(r)})$ and $F(x_{(s)})$, $r<s$, is given (cf. **10.11** and Exercise 11.7) by

$$dG_{r,s} = \frac{\{F(x_{(r)})\}^{r-1}\{F(x_{(s)})-F(x_{(r)})\}^{s-r-1}\{1-F(x_{(s)})\}^{n-s}\,dF(x_{(r)})\,dF(x_{(s)})}{B(r, s-r)B(s, n-s+1)}. \tag{14.2}$$

From (14.1–2) we may obtain the conditional distribution of $x_{(r)}$ given $x_{(s)}$, and that of $x_{(s)}$ given $x_{(r)}$. Exercise 14.9 shows that each of these is of the form (14.1) applied to the original distribution truncated at the value of the given order statistic.

From (14.1), it follows at once that if μ'_p exists for $F(x)$, it must also exist for $G_r(x)$; in the contrary case, the moment of $G_r(x)$ may or may not exist—Exercise 14.22 treats the Cauchy case.

Chan (1967) shows that, provided μ'_1 is finite, the set $E(x_{(n)})$ for all $n \geqslant 1$ characterizes $F(x)$, as does the set $E(x_{(1)})$.

In (14.2), $x_{(r)}$ and $x_{(s)}$ are always positively correlated, as is intuitively obvious—for a proof, see Bickel (1967) or Exercise 14.9.

The generalization to the joint distribution of several order statistics is immediate. For instance, for $x_{(r)}$, $x_{(s)}$, $x_{(t)}$, $x_{(u)}$, $r<s<t<u$, we have, writing F_r for $F(x_{(r)})$, etc.,

$$dG_{r,s,t,u} = \frac{F_r^{r-1}(F_s-F_r)^{s-r-1}(F_t-F_s)^{t-s-1}(F_u-F_t)^{u-t-1}(1-F_u)^{n-u}dF_r\,dF_s\,dF_t\,dF_u}{B(r,s-r)B(s,t-s)B(t,u-t)B(u,n-u+1)}. \quad (14.3)$$

Distributions of the type (14.1) have been studied for various given forms of the population d.f. $F(x)$. When the population is uniformly distributed in the range 0 to 1, $F(x)=x$ and the distribution of $x_{(r)}$ is particularly simple, itself reducing to the Beta distribution. A second type of population leading to simple results is the exponential $f(x)=e^{-x}$, which in several respects occupies in the theory of order statistics the position that the normal distribution holds for symmetric functions of the observations—see especially Exercise 14.16, which shows that certain linear functions of independent exponential variates are also independently exponential, just as in Examples 11.2–3 we saw that orthogonal linear functions of independent normal variates are also independently normal. Nevertheless, the normal distribution has such general importance in statistical theory that even its order statistics, whose distributions often require heavy computations, have been extensively investigated, as we shall see.

The computational task may be simplified by the use of recurrence relations, some of which are given in Exercises 14.5, 14.10, 14.23 and 14.28. For further coverage of such results, see Govindarajulu (1963), Malik *et al.* (1988) and Balakrishnan *et al.* (1988).

Some important and interesting related results are given in Exercises 14.9 and 14.14.

The distribution function of an order statistic

14.3 Even if $F(x)$ is discrete, the d.f. of $x_{(r)}$ is easily obtained, as H. A. David and Mishriky (1968) point out. We now suppose that $x_{(1)} \leqslant x_{(2)} \leqslant \ldots \leqslant x_{(n)}$, with equalities possible. The probability is $\binom{n}{j}\{F(x)\}^j\{1-F(x)\}^{n-j}$ that j of the n observations do not exceed a fixed value x. Thus

$$G_r(x) = \sum_{j=r}^{n}\binom{n}{j}\{F(x)\}^j\{1-F(x)\}^{n-j} \quad (14.4)$$

is the probability that at least r observations in the sample do not exceed x, i.e. that $x_{(r)} \leqslant x$. This is the d.f. of $x_{(r)}$. From (5.16), we have

$$G_r(x) = 1 - I_{1-F(x)}(n-r+1, r) = I_{F(x)}(r, n-r+1). \quad (14.5)$$

This result is completely general: we require differentiability of $F(x)$ only in order to obtain the density of $x_{(r)}$, given at (14.1), from $G_r(x)$, as may be seen by writing out the Incomplete Beta Function as an integral.

For discrete variables taking non-negative integer values, Young (1970) gives recurrence relations for the probability generating function and the moments of the order statistics—see also Gupta and Panchapakesan (1974) and Balakrishnan (1986). Vaughan and Venables (1972) generalize (14.1) to the case where the observations come from different d.f.s.

Median and quartiles in the standardized normal case

14.4 Suppose that the sample size is odd, say $n=2r+1$, so that the median is the $(r+1)$th order statistic, $x_{(r+1)}$. Let us write

$$A_x=\frac{1}{\sqrt{(2\pi)}}\int_{-\infty}^{x}e^{-\frac{1}{2}t^2}\,dt.$$

Then the distribution of the median, from (14.1), is seen to be (writing A for A_x)

$$dG=n!\,A^r(1-A)^r\,dA/(r!)^2.$$

From symmetry, the expected value of the median is zero—cf. Exercise 14.7—and hence its variance is given by

$$\operatorname{var}x_{(r+1)}=\int x^2\,dG=\frac{n!}{(r!)^2}\int_{-\infty}^{\infty}x^2A^r(1-A)^r\cdot\frac{1}{\sqrt{(2\pi)}}e^{-\frac{1}{2}x^2}\,dx. \qquad (14.6)$$

We expand the term $(1-A)^r$ binomially to obtain

$$\operatorname{var}x_{(r+1)}=\frac{n!}{(r!)^2}\int_{-\infty}^{\infty}\sum_{j=0}^{r}(-1)^j\binom{r}{j}A^{r+j}\frac{1}{\sqrt{(2\pi)}}x^2e^{-\frac{1}{2}x^2}\,dx. \qquad (14.7)$$

Each term may be integrated by parts, taking $xe^{-\frac{1}{2}x^2}$ for one part. We have

$$\int_{-\infty}^{\infty}xA^{r+j}\cdot xe^{-\frac{1}{2}x^2}\,dx=[-xA^{r+j}e^{-\frac{1}{2}x^2}]_{-\infty}^{\infty}$$
$$+\int_{-\infty}^{\infty}e^{-\frac{1}{2}x^2}\left\{A^{r+j}+(r+j)A^{r+j-1}\frac{1}{\sqrt{(2\pi)}}xe^{-\frac{1}{2}x^2}\right\}dx. \qquad (14.8)$$

The first term on the right vanishes. With a further integration by parts on the third term, (14.8) becomes

$$\int_{-\infty}^{\infty}e^{-\frac{1}{2}x^2}A^{r+j}\,dx+\frac{(r+j)(r+j-1)}{4\pi}\int_{-\infty}^{\infty}A^{r+j-2}\exp(-\tfrac{3}{2}x^2)\,dx. \qquad (14.9)$$

In (14.9), the first integral is simply $(2\pi)^{\frac{1}{2}}/(r+j+1)$ and on substitution in (14.7) we find

$$\operatorname{var}x_{(r+1)}=\frac{n!}{(r!)^2}\sum_{j=0}^{r}(-1)^j\binom{r}{j}\frac{1}{r+j+1}$$
$$+\frac{n!}{(r!)^2 4\pi}\sum_{j=0}^{r}(-1)^j\binom{r}{j}(r+j)(r+j-1)\frac{1}{\sqrt{(2\pi)}}\int_{-\infty}^{\infty}A^{r+j-2}\exp(-\tfrac{3}{2}x^2)\,dx. \qquad (14.10)$$

The first summation on the right of (14.10) yields unity, for

$$\sum_{j=0}^{r}(-1)^j\binom{r}{j}\frac{1}{r+j+1}=\int_0^1\sum(-1)^j\binom{r}{j}t^{r+j}\,dt$$
$$=\int_0^1 t^r(1-t)^r\,dt=B(r+1,r+1) \qquad (14.11)$$

Thus we have, from (14.10) and (14.11),

$$\operatorname{var} x_{(r+1)} = 1 + \frac{n!}{4\pi(r!)^2}\sum(-1)^j\binom{r}{j}(r+j)(r+j-1)T_{r+j-2}, \tag{14.12}$$

where

$$T_{r+j-2} = \frac{1}{\sqrt{(2\pi)}}\int_{-\infty}^{\infty} A^{r+j-2}\exp\left(-\tfrac{3}{2}x^2\right)\mathrm{d}x. \tag{14.13}$$

14.5 In this kind of way we can ascertain as many moments as we wish of any order statistic. The results depend on integrals of type

$$\int_{-\infty}^{\infty} A^p \exp\left(-\tfrac{1}{2}qx^2\right)\mathrm{d}x, \tag{14.14}$$

of which (14.13) is a particular case. These are sometimes known as Hojo's integrals, from the name of the author who first (1931) studied them in detail and tabulated them for a range of values of p and q. By an easy extension of the method, we may ascertain product moments of two order statistics, and hence the moments of such quantities as the interquartile range and the mid-range. The quadratures involved are extensive and for ordinary purposes the methods of approximation given below are adequate. We may, however, quote a few of Hojo's exact results by way of illustration.

14.6 In normal samples the standard error of the mean is $\sigma/\sqrt{n}$. The standard error of the median is always larger than this, say $c_n\sigma/\sqrt{n}$ where $c_n > 1$ and $n > 2$. Defining the median for even $n = 2r$ as $\frac{1}{2}(x_{(r)} + x_{(r+1)})$ we have the following values for c_n:

n	c_n	n	c_n	n	c_n
2	1.000	6	1.135	10	1.177
3	1.161	7	1.214	15	1.235
4	1.092	8	1.160	20	1.214
5	1.120	9	1.223	∞	1.253

Observe that the convergence of c_n to its limiting value is not monotone because of the changing definitions of the median.

This will be discussed further in 17.12, Volume 2A; the limiting value is $(\pi/2)^{\frac{1}{2}}$ – see Example 10.7.

Likewise, for the mid-range $\frac{1}{2}(x_{(1)} + x_{(n)})$ we have a standard error, say $d_n\sigma/\sqrt{n}$, but d_n tends to infinity with n, as we shall see below (Example 14.4). For increasing sample size the mid-range is an increasingly inferior estimator of the mean of the normal population. The following are some values of d_n:

n	d_n	n	d_n	n	d_n
2	1.000	5	1.142	15	1.540
3	1.042	6	1.190	20	1.691
4	1.092	10	1.362	∞	∞

14.7 The interquartile range, divided by twice 0.674 49 (the distance between the quartiles in a standardized normal population), is sometimes used as an estimator of the population standard deviation. It is interesting to compare the standard error of this quantity with the standard error of the sample s.d., namely $\sigma/\sqrt{(2n)}$. If the standard error is $e_n\sigma\sqrt{(2n)}$, the following are some values of e_n:

n	e_n	n	e_n
2	1.000	10	1.497
4	1.047	12	1.419
6	1.421	∞	1.649
8	1.313		

The values proceed somewhat irregularly owing to the arbitrary element in the definition of quartiles for sample sizes not of the form $4r+1$.

14.8 Hojo (1931, 1933) and K. Pearson (1931) give a number of further results of this character. For other numerical work reference may be made to the book by Sarhan and Greenberg (1962). See also **14.21** below.

Pearson's expansion

14.9 In Chapter 6 we have considered the expansion of a d.f. $F(x)$ or a density function $f(x)$ as a power series in the variate x. We now consider an inverse expansion of x in terms of F, due to Karl Pearson. Let X_r be the value such that

$$F(X_r) = \frac{r}{n+1}.$$

We expand $x_{(r)}$ about X_r in a Taylor series

$$x_{(r)} = X_r + h_r X_r' + \frac{1}{2!}h_r^2 X_r'' + \dots \tag{14.15}$$

where

$$h_r = F(x_{(r)}) - F(X_r) = F_r - r/(n+1),$$

$$X_r' = \frac{\mathrm{d}X_r}{\mathrm{d}F} = \left.\frac{\mathrm{d}x}{\mathrm{d}F}\right|_{x=X_r}, \quad X_r'' = \left.\frac{\mathrm{d}^2x}{\mathrm{d}F^2}\right|_{x=X_r}$$

and so on.

From (14.15) we can express powers of x in series of powers of h; and the expectation of any power of h is easily derivable from (14.1). Provided, then, that our series converge in a suitable manner (or, more generally, give good approximations of an asymptotic kind) we may derive approximations to as many moments as we wish of any order statistic.

Example 14.1 (Variance of the median for the normal distribution)

Consider again the distribution of the median $x_{(r+1)}$ in samples of $n=2r+1$ observations from the standardized normal distribution.

We find, since $(r+1)/(n+1)=\frac{1}{2}$, $X_{r+1}=0$,

$$X_{r+1}' = \left.\frac{\mathrm{d}x}{\mathrm{d}F}\right|_{x=0} = 1\Big/\left.\frac{\mathrm{d}F}{\mathrm{d}x}\right|_{x=0} = \sqrt{(2\pi)},$$

$$X_{r+1}'' = \left.\frac{\mathrm{d}}{\mathrm{d}F}\left(1\Big/\frac{\mathrm{d}F}{\mathrm{d}x}\right)\right|_{x=0} = \left.\frac{\mathrm{d}x}{\mathrm{d}F}\cdot\frac{\mathrm{d}}{\mathrm{d}x}\left(1\Big/\frac{\mathrm{d}F}{\mathrm{d}x}\right)\right|_{x=0} = \sqrt{(2\pi)}\cdot 0 = 0.$$

Similarly $X'''_{r+1} = (2\pi)^{3/2}$. Thus, from (14.15) and (14.4) with $h_{r+1} = A - \frac{1}{2}$, we have

$$x_{(r+1)} = (2\pi)^{\frac{1}{2}}(A - \tfrac{1}{2}) + \tfrac{1}{6}(2\pi)^{3/2}(A - \tfrac{1}{2})^3 + \ldots$$

and

$$x^2_{(r+1)} = 2\pi(A - \tfrac{1}{2})^2 + \tfrac{1}{3}(2\pi)^2(A - \tfrac{1}{2})^4 + \ldots$$

Substituting the expansion to this order for $x^2_{(r+1)}$ in (14.6), we find

$$\text{var}\, x_{(r+1)} = \frac{n!}{(r!)^2}\int_{-\infty}^{\infty} A^r(1-A)^r\{2\pi(A - \tfrac{1}{2})^2 + \tfrac{1}{3}(2\pi)^2(A - \tfrac{1}{2})^4 + \ldots\}\, dA,$$

which reduces to

$$\text{var}\, x_{(r+1)} = \frac{\pi}{2(n+2)} + \frac{\pi^2}{4(n+2)(n+4)} + o(n^{-2}). \tag{14.16}$$

This formula is accurate enough for ordinary purposes. For $n = 11$ it gives a variance of 0.137 agreeing to 3 d.p. with the known true value of 0.137 227.

Example 14.2 (Variance of the median for the uniform distribution)

For the uniform distribution

$$dF = dx, \qquad 0 < x \leqslant 1,$$

we find, for the median $x_{(r+1)}$ in samples of $n = 2r + 1$ observations,

$$X' = 1, \qquad X^{(r)} = 0, \qquad r > 1.$$

The variance of the median is then the expectation of $(x - \frac{1}{2})^2$ in

$$dG = \frac{n!}{(r!)^2}x^r(1-x)^r\, dx, \qquad 0 \leqslant x \leqslant 1,$$

and this is obviously an exact result. We find

$$\text{var}\, x_{(r+1)} = 1/\{4(n+2)\}. \tag{14.17}$$

David and Johnson (1954) have pursued this subject systematically and give expansions for cumulants and product cumulants of order statistics up to and including the fourth order. They choose expansions in terms of $(n+2)^{-1}$ rather than n^{-1} because of the natural appearance of the former quantity in elementary cases—cf. equations (14.16) and (14.17). It is not clear how far these expansions have the required asymptotic properties, but they appear to work well in practice. Exercise 14.3 gives general results for the median, of which Examples 14.1 and 14.2 are special cases. See, also Exercise 14.7.

Hall (1978) gives general series expansions for the means, variances, and covariances.

14.10 Tukey (in Hastings *et al.*, 1947) introduced the class of *lambda distributions*, defined in terms of the inverse of the d.f. as

$$x = \mu + \gamma'[F^\lambda - (1-F)^\lambda], \quad 0 \leqslant F \leqslant 1$$

or, in standardized form, as

$$z = R(F) = \gamma[F^{\lambda} - (1-F)^{\lambda}]. \tag{14.18}$$

$R(F)$ is known as the percentile function and $|z| \leqslant \gamma$. The density of z is

$$f(z) = \frac{1}{\mathrm{d}R/\mathrm{d}F} = \frac{1}{\gamma\lambda[F^{\lambda-1} + (1-F)^{\lambda-1}]}.$$

An attraction of the lambda distributions is that Pearson's expansion may be applied directly to (14.18); see Exercise 14.27. Further, the moments of the order statistics may be evaluated exactly. For example, when $n = 2r+1$, the jth moment of the median is, from (14.1),

$$E(z^j) = \gamma^j \int_0^1 [F^{\lambda} - (1-F)^{\lambda}]^j F^r (1-F)^r \,\mathrm{d}F / B(r+1, r+1).$$

Hence, $E(z) = 0$ and

$$\operatorname{var}(z) = \frac{2\gamma^2[B(r+2\lambda+1, r+1) - B(r+\lambda+1, r+\lambda+1)]}{B(r+1, r+1)}. \tag{14.19}$$

As expected, this reduces to (14.17) when $\gamma = \frac{1}{2}$ and $\lambda = 1$. Results for other order statistics follow in similar fashion.

When $\lambda = 0.135$, the kurtosis measure $\beta_2 = 3$, and the resulting lambda distribution closely approximates the normal ($\sigma^2 = 1$ when $\gamma = 5.063$).

If we rewrite (14.18) as

$$z = \gamma^* \left[\frac{F^{\lambda} - 1}{\lambda} - \frac{(1-F)^{\lambda} - 1}{\lambda}\right]$$

and allow $\lambda \to 0$, we obtain

$$z = \gamma^*[\log F - \log(1-F)]$$

or

$$F = [1 + \mathrm{e}^{-z/\gamma^*}]^{-1},$$

the d.f. of the logistic distribution (cf. Exercise 4.21) with scale parameter γ^*.

Ramberg *et al.* (1979) generalize (14.18) by allowing the two powers λ to be different. The resulting family does not cover the whole (β_1, β_2) plane in Figs. 6.1-2.

Quantiles and extreme values: asymptotic distributions

14.11 In considering the asymptotic distributions of the order statistics we may have two types of limiting process. In the first type we let r and n tend to infinity with the ratio r/n held constant: we are then considering the asymptotic distributions of *quantiles* of the sample. In the second type, r remains fixed as n tends to infinity: this leads to results of a radically different nature from those of the quantiles, and this branch of the subject is often designated as the theory of *extreme values*, the name deriving from the fact that the cases of particular interest are those in which r is close to 1 or to n.

14.12 We have already (10.10–11 and Example 11.4) derived limiting results for quantiles. We saw there that the sample quantile corresponding to $x_{(r)}$ is asymptotically normal. Its mean is X_r (given by $F(X_r) = r/n$) and its variance is

$$p_r q_r/(nf_{r,n}^2), \tag{14.20}$$

where $p_r = r/n$, $q_r = 1 - p_r$ and $f_{r,n}$ is the value of the density at X_r.

More rigorous proofs are given by Rényi (1953) and A. M. Walker (1968). Exercise 14.6 studies (14.20) as a function of p.

Similarly for two quantiles based on $x_{(r)}$, $x_{(s)}$ $(r < s)$ we have bivariate normality, variances given by (14.20) and covariance

$$p_r q_s/(nf_{r,n}f_{s,n}). \tag{14.21}$$

The result generalizes readily to more than two quantiles, the variances and covariances being of type (14.20) and (14.21), and the limiting distribution multivariate normal.

Chernoff *et al.* (1967) give general conditions under which functions of the order statistics are asymptotically normal. See also Shorack (1969, 1972).

Asymptotic distributions of extreme values

14.13 We now turn to the asymptotic theory of extreme values, which we have not so far encountered. Gumbel (1958) treated the subject fully; for a more rigorous discussion, see Galambos (1987).

In (14.1), we make the transformation $y = nF(x_{(r)})$, obtaining as the distribution of y

$$\mathrm{d}H_r(y) = \left(\frac{y}{n}\right)^{r-1}\left(1 - \frac{y}{n}\right)^{n-r} d\left(\frac{y}{n}\right)\bigg/B(r, n-r+1), \tag{14.22}$$

so that y/n is a Beta variate of the first kind. As n tends to infinity, for any fixed r,

$$\lim_{n\to\infty} \mathrm{d}H_r(y) = y^{r-1}\mathrm{e}^{-y}\,\mathrm{d}y/\Gamma(r), \tag{14.23}$$

a Gamma variate with parameter r. When $r = 1$, (14.23) becomes, for the transformed smallest value in the sample, an exponential distribution.

If we transform (14.1) by $y = n(1 - F(x_{(r)}))$, we get a parallel result to (14.23), with $n - r$ written for $r - 1$, and this also reduces to an exponential distribution for the case $r = n$.

These results apply to the transformed variable y. We now inquire directly into the distribution of the original variable $x_{(r)}$. If $F(x)$ is known explicitly, the transformation leading to (14.23) can, of course, be reversed to give the limiting distribution of $x_{(r)}$ itself.

14.14 In the case of the extreme values ($r = 1$ or n), direct progress can be made by the use of an argument due to Fisher and Tippett (1928). Consider the largest value, and suppose that $n = km$. The largest value $x_{(n)}$ may then be regarded as the largest of k largest values of samples of size m. If $x_{(n)}$ has a limiting distribution, it will be identical with that of the largest values in samples of size m, as m tends to infinity with k fixed, except possibly for location and scale factors due to the factor k. Thus, if $G(x)$ is the limiting d.f., it satisfies the functional equation

$$G^k(x) = G(a_k x + b_k). \tag{14.24}$$

This therefore characterizes the possible forms of limiting distribution. To solve (14.24), we use a method due to Jenkinson (1955). Put

$$q(x) = -\log\{-\log G(x)\}, \tag{14.25}$$

so that on taking logarithms twice, (14.24) becomes

$$\log k = q(x) - q(a_k x + b_k). \tag{14.26}$$

We now expand $q(x)$ and $q(a_k x + b_k)$ in a Taylor series about the point x_{on} for which

$$q(x_{on}) = 0. \tag{14.27}$$

(14.26) becomes

$$\log k = -q(a_k x_{on} + b_k) + \sum_{r=1}^{\infty} \{q^{(r)}(x_{on}) - a_k^r q^{(r)}(a_k x_{on} + b_k)\} \frac{(x - x_{on})^r}{r!}. \tag{14.28}$$

This is an identity in x. Equating coefficients, we have

$$\log k = -q(a_k x_{on} + b_k) \tag{14.29}$$

and

$$q^{(r)}(x_{on}) = a_k^r q^{(r)}(a_k x_{on} + b_k), \qquad r \geqslant 1. \tag{14.30}$$

(14.30) gives

$$\frac{q^{(r)}(a_k x_{on} + b_k)}{\{q'(a_k x_{on} + b_k)\}^r} = \frac{q^{(r)}(x_{on})}{\{q'(x_{on})\}^r} = c_r, \tag{14.31}$$

where c_r is independent of k. Since a_k and b_k are constants depending on the value of k, which is arbitrary, we may rewrite (14.31) generally as

$$q^{(r)}(x) = c_r\{q'(x)\}^r. \tag{14.32}$$

Differentiating (14.32) with respect to x, we obtain

$$\begin{aligned} q^{(r+1)}(x) &= rc_r\{q'(x)\}^{r-1} \cdot q''(x) \\ &= rc_r c_2\{q'(x)\}^{r+1}, \end{aligned} \tag{14.33}$$

on using (14.32) with $r = 2$. Comparing (14.32) and (14.33), we have

$$c_{r+1} = rc_r c_2 = r!\, c_2^r, \qquad r \geqslant 1. \tag{14.34}$$

Putting (14.34) back into (14.32) gives

$$q^{(r)}(x) = (r-1)!\, c_2^{r-1}\{q'(x)\}^r, \qquad r \geqslant 1. \tag{14.35}$$

If (14.35) is used in the Taylor expansion of $q(x)$ about x_{on}, we obtain, using (14.27),

$$q(x) = \sum_{r=1}^{\infty} c_2^{r-1}\{q'(x)\}^r (x-x_{on})^r/r$$

$$= -\frac{1}{c_2}\log\{1-c_2 q'(x_{on})(x-x_{on})\}. \quad (14.36)$$

(14.25) and (14.36) give for the characteristic limiting form

$$G(x) = \exp[-\{1-c_2 q'(x_{on})(x-x_{on})\}^{1/c_2}]. \quad (14.37)$$

The three asymptotes

14.15 From **14.13** it follows that since $y = n(1-F(x_{(n)}))$ asymptotically has the d.f. $1-e^{-y}$, $G(x)$ may alternatively be written

$$G(x) = \exp[-n\{1-F(x)\}], \quad (14.38)$$

so that, from (14.25),

$$q(x) = -\log[n\{1-F(x)\}]. \quad (14.39)$$

At x_{on}, by virtue of (14.27) and (14.39), $q(x_{on}) = -\log[n\{1-F(x_{on})\}] = 0$, so that

$$n\{1-F(x_{on})\} = 1. \quad (14.40)$$

The left side of (14.40) is the expected number of values exceeding x_{on} in a sample of size n, so that x_{on} is that value of the variate which has expectation of being exceeded just once in such a sample.[*]

From (14.39) and (14.40)

$$q'(x) = F'(x)/\{1-F(x)\}, \quad (14.41)$$

$$q'(x_{on}) = nF'(x_{on}), \quad (14.42)$$

while, from (14.35),

$$c_2 = q''(x)/\{q'(x)\}^2 = \frac{d}{dx}\{-1/q'(x)\}$$

$$= -\frac{d}{dx}\left\{\frac{1-F(x)}{F'(x)}\right\}, \quad (14.43)$$

by (14.41). If

$$\lim_{x\to\infty} c_2 = 0, \quad (14.44)$$

the limiting form (14.37) becomes, on inserting (14.42) and (14.44),

$$G(x) = \exp[-\exp\{-nF'(x_{on})(x-x_{on})\}], \quad (14.45)$$

---

[*] Gumbel (e.g., 1954) called x_{on} "the expected largest value" though, as he pointed out, this is confusing since it is not the expected value of the largest order statistic.

a limiting form put forward originally by Fisher and Tippett (1928), the explicit condition (14.44) being due to Von Mises (1936). For the exponential distribution

$$dF(x) = e^{-x}\,dx, \qquad 0 \leqslant x < \infty,$$

c_2 defined at (14.43) is zero identically in x, and by analogy distributions for which (14.44), and consequently (14.45), hold have been termed *exponential-type*. Since $G(-\infty) = 0$, $G(+\infty) = 1$, the distribution covers the whole real line.

Exercise 14.2 deduces (14.45) directly for the exponential and logistic distributions.

14.16 If $F(x)$ is not exponential-type, (14.45) is replaced by a different limiting form. From (14.43), we see that

$$c_2 = 1 + F''(x)\{1 - F(x)\}/\{F'(x)\}^2. \tag{14.46}$$

If now

$$\lim_{x \to \infty} xF'(x)/\{1 - F(x)\} = k > 0, \tag{14.47}$$

(14.46) and (14.47) give

$$\lim_{x \to \infty} c_2 = \lim[1 + xF''(x)/\{kF'(x)\}]. \tag{14.48}$$

On using L'Hôpital's rule, (14.47) becomes

$$\lim\{1 + xF''(x)/F'(x)\} = -k, \tag{14.49}$$

and (14.48) and (14.49) imply that

$$\lim_{x \to \infty} c_2 = -1/k. \tag{14.50}$$

The use of (14.42), (14.47) and (14.50) in (14.37) gives us the limiting form

$$G(x) = \exp[-\{1 + n(1 - F(x_{on}))(x - x_{on})/x_{on}\}^{-k}],$$

which finally becomes, on using (14.40),

$$G(x) = \exp\left[-\left(\frac{x}{x_0}\right)^{-k}\right], \tag{14.51}$$

the second type of limiting form, due to Fréchet (1927) and to Fisher and Tippett (1928). The condition (14.47) is implied by

$$\lim_{x \to \infty} x^k\{1 - F(x)\} = A > 0, \tag{14.52}$$

as may be verified by applying L'Hôpital's rule to (14.52). Distributions satisfying (14.52) (and therefore (14.47)) have no moments of order k or more, and hence are termed *Cauchy-type*, after the Cauchy distribution of Example 3.3 for which $k = 1$. Exercise 14.2 deduces (14.51) directly in this case. In (14.51), $G(0) = 0$, $G(+\infty) = 1$, so the distribution is confined to the positive half-line.

14.17 Finally, we consider an original distribution $F(x)$ with a finite terminal x_u

where

$$F(x_u) = 1.$$

If, in addition

$$\left.\begin{array}{ll} F^{(r)}(x_u) = 0, & r = 1, 2, \ldots, k-1 \\ F^{(k)}(x_u) \neq 0, & \end{array}\right\} \tag{14.53}$$

we may expand the exponent of (14.38) in a Taylor series about x_u whose first non-vanishing term is that in $F^{(k)}(x_u)$. If $F^{(k+1)}(x)$ is bounded near x_u, the remainder term is negligible since $x - x_u$ is small, and we obtain

$$G(x) = \exp\left[-\{(-nF^{(k)}(x_u)/k!)^{1/k}(x - x_u)\}^k\right], \tag{14.54}$$

the third limiting form for extreme value distributions given by Fisher and Tippett (1928). With changed notation, this is the complement of the Weibull distribution of **5.33**; here $G(-\infty) = 0$, $G(x_u) = 1$.

The reader should verify that the three limiting forms (14.45), (14.51) and (14.54) each satisfy the functional equation (14.24).

$G(x)$ may be represented in a standard form as

$$G(x) = \exp[-(1 + cx)^{-1/c}],$$

yielding the exponential, Cauchy and Weibull types respectively as $c = 0$, $c < 0$ or $c > 0$. It was shown by Gnedenko (1943) that these are the only possible solutions to (14.24). The c values corresponding to each type are termed the domains of attraction; for a detailed discussion see Pickands (1986).

De Haan (1970) provides an alternate set of necessary and sufficient conditions on $F(x)$ for the limiting distribution to approach one of these three forms. Marshall and Olkin (1983) also discuss the domains of attraction for each of the three limiting cases and extend their results to the multivariate case.

However, a limiting form may not exist at all. Galambos (1984) shows that, when x is a discrete variate defined on the integers, the limiting distribution exists only when

$$\lim_{n\to\infty} \frac{P(x \geqslant n)}{P(x \geqslant n+1)} = 1.$$

For example, both the geometric and the Poisson fail to satisfy this condition.

Gnedenko (1943) and Smirnov (1949) give full analyses of the three asymptotic distributions.

We have been discussing the possible limiting distributions of $x_{(n)}$. If we reverse the sign of the variable, $-x_{(1)}$ becomes $x_{(n)}$ and $G(x)$ becomes $1 - G(x)$. Thus the distribution of the smallest value $x_{(1)}$ is simply

$$H(x) = 1 - G(-x) \tag{14.55}$$

and the results for $x_{(n)}$ apply with only this modification.

Example 14.3 (Extreme value distribution for the normal)

For the normal distribution

$$f(x) = e^{-\frac{1}{2}x^2}/\sqrt{(2\pi)},$$

we have, on using (5.105) in (14.43), for large x,

$$c_2 = -\frac{d}{dx}\{(1-F(x))/F'(x)\} \doteq -\frac{d}{dx}\left(\frac{1}{x}\right) = \frac{1}{x^2}$$

so that

$$\lim_{x\to\infty} c_2 = 0.$$

(14.44) is satisfied in this case, giving (14.45) as the limiting form of the distribution. To evaluate its exponent, we use the fact that at x_{on}, (14.40) holds, so that

$$nF'(x_{on})/x_{on} \doteq n\{1-F(x_{on})\} = 1.$$

Thus

$$nF'(x_{on}) \doteq x_{on},$$

and the exponent of (14.45) is simply $-x_{on}(x-x_{on})$. Taking logarithms,

$$\log n + \log F'(x_{on}) - \log x_{on} = 0,$$

i.e.

$$\log n - \tfrac{1}{2}\log(2\pi) - \tfrac{1}{2}x_{on}^2 = \log x_{on}.$$

As a first asymptotic approximation, this equation has the solution $x_{on} \doteq (2\log n)^{\frac{1}{2}}$. For a better approximation, we may now solve the equation with $(2\log n)^{\frac{1}{2}}$ replacing x_{on} on its right-hand side. We find $x_{on} \doteq (2\log n)^{\frac{1}{2}}\{1-\tfrac{1}{2}(\log 4\pi + \log\log n)/(2\log n)\}$.

McCord (1964) studies the asymptotic moments of the largest value for Cauchy-type, finite terminal, and some exponential-type distributions (excluding the normal and Gamma, but including the exponential itself and the logistic).

Bivariate and multivariate extremes

14.18 If a bivariate density $f(x,y)$ has marginal densities $f_1(x)$, $f_2(y)$, Campbell and Tsokos (1973) show that if $E\left\{\dfrac{f(x,y)}{f_1(x)f_2(y)}\right\}$ is finite, the limiting form of the joint d.f. of $x_{(n)}$ and $y_{(n)}$ can only be

$$G(x,y) = G_1(x)G_2(y)\exp\{W(x,y)\},$$

where G_1, G_2 are the marginal limiting d.f.s for $x_{(n)}$ and $y_{(n)}$. $W(x,y)$ reflects the dependence of $x_{(n)}$ and $y_{(n)}$, being zero when they are asymptotically independent. Remarkably, this is so for the bivariate normal distribution of Examples 3.19 and 15.1 (below), as well as for others—even though the initial variables x, y can be arbitrarily dependent, their extreme values tend to independence. See also Tiago de Oliveira (1962-3, 1980) and Posner *et al.* (1969). Tawn (1988) proposes models for the dependence between the variables and (1990) extends his results to the multivariate case. H. A. David *et al.* (1977) discuss the distribution of the rank of y corresponding to a given order statistic of x. Galambos (1975) considers the joint distribution of extreme values (not necessarily the largest) for a multivariate initial distribution and obtains conditions for their independence—see also Galambos (1987, Chapter 3). Barnett (1976) reviews methods of ordering multivariate data, with an extensive bibliography.

Asymptotic distributions of mth extreme values

14.19 Reverting to (14.1), we now turn from the extreme values to a more general investigation of limiting distributions for other extreme order statistics. We confine ourselves to the case of exponential-type distributions, which are the most important, and consider the mth largest order statistic, $x_{(n-m+1)}$.

If we put $r = n-m+1$ in (14.1) and differentiate its logarithm, as in **10.10**, we obtain

$$(n-m)F'/F-(m-1)F'/(1-F)+F''/F' = 0 \tag{14.56}$$

at the mode of $x_{(n-m+1)}$. From (14.41) and (14.43) it follows that if (14.44) holds

$$q''(x) = F''/(1-F)+(F')^2/(1-F)^2 \to 0$$

so that since $(1-F)\to 0$, F'' must also do so. Applying L'Hôpital's rule to the last term on the left of (14.56), we obtain

$$\lim F''/F' = \lim F'/\{-(1-F)\}, \tag{14.57}$$

so that (14.56) becomes

$$(n-m)F'/F-mF'/(1-F) = 0, \tag{14.58}$$

whence, at the mode $\hat{x}$ of the distribution of $x_{(n-m+1)}$,

$$F(\hat{x}) = 1-m/n. \tag{14.59}$$

Comparison of (14.59) with (14.40) shows that, in the exponential case, when $m=1$ we have $x_{on} = \hat{x}$, i.e. the mode of the distribution of $x_{(n)}$ coincides with the value that is exceeded on the average by one member of the sample.

Exercise 14.11 gives exact results for the mode in the normal case.

We now expand $F(x)$ in a Taylor series about $\hat{x}$,

$$F(x) = F(\hat{x})+(x-\hat{x})F'(\hat{x})+(x-\hat{x})^2F''(\hat{x})/2+\ldots \tag{14.60}$$

Since, by (14.57) and (14.59)

$$F'' \doteq -(F')^2/(1-F) = -n(F')^2/m, \tag{14.61}$$

we rewrite (14.60), using (14.59) and (14.61), as

$$F(x) = (1-m/n)+(x-\hat{x})F'(\hat{x})-n\{F'(\hat{x})\}^2(x-\hat{x})^2/(2m)+\ldots \tag{14.62}$$

$$= 1-(m/n)\left[1-\frac{(x-\hat{x})F'(\hat{x})}{m/n}+\left\{\frac{(x-\hat{x})F'(\hat{x})}{m/n}\right\}^2\cdot\frac{1}{2!}+\ldots\right]$$

$$\doteq 1-(m/n)\exp\{-(x-\hat{x})F'(\hat{x})/(m/n)\} \tag{14.63}$$

approximately. Now (14.1) may be rewritten

$$dG_{n-m+1}(x) \propto \left(\frac{1-F}{F}\right)^{m-1} d(F^n). \tag{14.64}$$

If we write $-y_m$ for the exponent in (14.63) and substitute it into (14.64), we obtain

$$\left(\frac{1}{F}-1\right)^{m-1} = \left(\frac{1}{1-\frac{m}{n}e^{-y_m}}-1\right)^{m-1} \doteq \left(\frac{m}{n}\right)^{m-1} \exp\{-(m-1)y_m\},$$

and

$$d(F^n) = n\left(1-\frac{m}{n}e^{-y_m}\right)^{n-1}\cdot\frac{m}{n}e^{-y_m}\,dy_m.$$

Thus, from (14.64), we have for large n,

$$dG_{n-m+1}(x) \propto \exp[-my_m - me^{-y_m}]\,dy_m$$

and, on evaluating the constants by integration, we get

$$p(y_m) = \frac{m^m}{(m-1)!}\exp[-my_m - me^{-y_m}], \tag{14.65}$$

so that me^{-y_m} is Gamma distributed with index m.

In particular, when $m = 1$, we have

$$p(y) = \exp[-y - e^{-y}], \tag{14.66}$$

which is the density function corresponding to the d.f. (14.45).

The location family $p(y_m - \theta)$ is a member of the exponential family (5.116).

14.20 The limiting form (14.65), due to Gumbel (1934), is far from normal for small m. From its c.f., which the reader is asked to find in Exercise 14.4, its c.g.f. is, with $\theta = it$,

$$\psi = \theta\log m + \log\{\Gamma(m-\theta)/\Gamma(m)\},$$

whence the mean is

$$\begin{aligned}\kappa_{1,m} = \left[\frac{\partial\psi}{\partial\theta}\right]_{\theta=0} &= \log m + \left[\frac{\partial}{\partial\theta}\log\{\Gamma(m-\theta)/\Gamma(m)\}\right]_{\theta=0}\\ &= \log m - \partial\log\Gamma(m)/\partial m,\end{aligned}$$

which by a well-known property of the digamma function $\partial\log\Gamma(m)/\partial m = \Gamma'(m)/\Gamma(m)$ becomes

$$\kappa_{1,m} = \log m + \gamma - \sum_{r=1}^{m-1} r^{-1}, \tag{14.67}$$

where γ is Euler's constant 0.577 2 . . . . For the higher cumulants we have similarly

$$\kappa_{r,m} = \left[\frac{\partial^r\psi}{\partial\theta^r}\right]_{\theta=0} = (-1)^r\frac{\partial^r}{\partial m^r}\log\Gamma(m), \qquad r \geq 2, \tag{14.68}$$

so that the variance is

$$\kappa_{2,m} = \partial^2\log\Gamma(m)/\partial m^2.$$

As $m \to \infty$ we see from (3.63) that $\partial\log\Gamma(m)/\partial m \doteq \log m$ and $\partial^2\log\Gamma(m)/\partial m^2 \doteq m^{-1}$, so that the mean (14.67) tends to zero and the variance to m^{-1}. Similarly, the skewness (γ_1) tends to $m^{-\frac{1}{2}}$ and the kurtosis (γ_2) to $2m^{-1}$. For small m, some numerical values are given in the following table, derived from Gumbel's calculations with corrections by Farebrother (1990):

m	Mean	Variance	Skewness (γ_1)	Kurtosis (γ_2)
1	0.577	1.645	1.140	2.400
3	0.176	0.395	0.621	0.763
5	0.104	0.221	0.469	0.438
10	0.051	0.105	0.324	0.210

For fixed m, the limiting distributions are thus non-normal. They tend to normality as $m \to \infty$, for the distinction between extreme values and quantiles, made in **14.11**, is then lost. We thus see that the mean (14.67) tended to zero as $m \to \infty$ because the limiting normality drives the mean towards the mode which is the origin for y_m.

The asymptotic distributions of the extreme values themselves are obtained through the linear relation used in (14.63),

$$y_m = (n/m)F'(\hat{x})(x-\hat{x}). \tag{14.69}$$

The distribution of y corresponding to the mth smallest order statistic, $x_{(m)}$, is similar by (14.55).

The c.f. and cumulants of (14.66) are given in more detail in Exercise 14.4.

J. J. Dronkers (1958) studied the asymptotic distributions of mth values in detail, deriving limiting forms for all three types of initial distribution. Transitional approximations are also given.

(14.65) is very slowly approached for the normal distribution. For $m = 1$, the approach may be accelerated by using $x_{(n)}^2$ instead of $x_{(n)}$ itself, for as the reader should verify, Exercise 14.4 and Example 14.3 imply that $E\{x_{(n)}^2\} \doteq 2\log n$, var $(x_{(n)}^2) \doteq 2\pi^2/3$, so that just as for $x_{(n)}$ in the exponential and logistic distributions in Exercise 14.2, $x_{(n)}^2$ in normal samples has mean proportional to $\log n$ and constant variance asymptotically. Cf. Haldane and Jayakar (1963).

The order statistics in the normal case

14.21 Many authors have considered the exact distributions of the order statistics in samples of n observations from a standardized normal population. *Biometrika Tables for Statisticians*, Part 1, gives the means of all normal order statistics for $n = 2(1)26(2)50$ to at least two decimal places. Harter (1961) gives them to 5 d.p. for $n = 2(1)100(25)250(50)400$; his table is reproduced by F. N. David *et al.* (1968), in the *Biometrika Tables*, Vol. II, up to $n = 200$, and in Harter (1969), where all intermediate values of n with no prime factor exceeding 7 are also tabulated. Teichroew (1956) gives the means, and Sarhan and Greenberg (1956, 1958, 1962) give the variances and covariances, of all normal order statistics to 10 decimal places for $n = 2(1)20$—the *Biometrika Tables*, Vol. II, reproduce the latter to 6 d.p. Tietjen *et al.* (1977) have tabulated the variances and covariances for $n = 2(1)50$. Davis and Stephens (1978) provide a computer algorithm for generating the variances and covariances of normal order statistics using an improved version of the David–Johnson approximation of **14.9** (cf. Exercise 14.3); further algorithms for the means and for var($x_{(n)}$) are given by Royston (1982) and by Shea and Scallon (1988) respectively. S. S. Gupta (1961) tabulates the 50, 75, 90 and 95 percentage points of the distributions of all normal order statistics for $n = 1(1)10$, and for the extreme and central order statistics for $n = 11(1)20$. Ruben (1954) gives recurrence relations for the moments of $x_{(r)}$ for sample size n in terms of those for lower sample sizes, tabulates the first ten moments for the largest order statistic for $n = 1(1)50$, and from these calculates its first four central moments, and the measures of skewness and kurtosis to a minimum of 7 significant figures—his table is reproduced in Vol. II of the *Biometrika Tables*. For the largest order statistic, Tippett (1925) obtained the results:

n	Mean	Standard error	β_1	β_2
2	0.564	0.826	0.019	3.062
5	1.163	0.669	0.092	3.202
10	1.539	0.587	0.168	3.331
100	2.508	0.429	0.429	3.765
500	3.037	0.370	0.570	4.003
1000	3.241	0.351	0.618	4.088
Asymptote ($n = 1000$)	3.872	0.345	1.300	5.400

The asymptotic values as $n \to \infty$ are obtainable from those in the table in **14.20** for $m = 1$ through (14.69), which here becomes, on using Example 14.3,

$$y = (2\log n)^{\frac{1}{2}}\{x - (2\log n)^{\frac{1}{2}}\}.$$

Thus $E(x) = (2\log n)^{-\frac{1}{2}}0.577 + (2\log n)^{\frac{1}{2}}$, $\operatorname{var} x = 1.645(2\log n)^{-1}$, while $\beta_1 = \gamma_1^2$ and $\beta_2 = \gamma_2 + 3$ are unchanged by the linear transformation. These asymptotic values are given for $n = 1000$ in the final row of the table above, and indicate that only the s.e. is close to its asymptotic value even for so large a sample. The values of β_1 and β_2 illustrate the point that, as n increases, the distribution of the extreme value diverges more and more from the normal form. It is tabulated for $n = 3(1)25(5)60, 100(100)1000$ to 7 d.p. in the *Biometrika Tables*, Vol. II, with percentage points in Vol. I for $n = 1(1)30$.

The general case

14.22 It is easy to show (by assuming the contrary and finding a contradiction) that in any sample $|x_{(r)} - \bar{x}| \leqslant (n-1)^{\frac{1}{2}}s$, where $\bar{x}$, s are the observed mean and s.d.; in particular, this holds for $x_{(1)}$ and $x_{(n)}$—the reader is asked to prove and extend this result in Exercise 14.25.

For the mean value of $x_{(n)}$, Hartley and David (1954) have established an upper bound as follows. By the Cauchy-Schwarz inequality,

$$\int_0^1 n\left(x + \frac{1}{a}\right)F^{n-1}\,\mathrm{d}F \leqslant \left[\left\{\int_0^1 (nF^{n-1})^2\,\mathrm{d}F\right\}\left\{\int_0^1 \left(x + \frac{1}{a}\right)^2\mathrm{d}F\right\}\right]^{\frac{1}{2}},$$

where F is a standardized d.f., and the constant $a = (n-1)/(2n-1)^{\frac{1}{2}}$. Thus

$$\int_0^1 nxF^{n-1}\,\mathrm{d}F + \frac{1}{a} \leqslant \left[\frac{n^2}{2n-1}\cdot\left(1 + \frac{1}{a^2}\right)\right]^{\frac{1}{2}}.$$

The integral on the left is simply $E\{x_{(n)}\}$, by (14.1) with $r = n$. Thus, on substituting for a, we have

$$E(x_{(n)}) \leqslant (n-1)/(2n-1)^{\frac{1}{2}} = a.$$

If the population is symmetrical about zero, we may use the fact that $F(x) = 1 - F(-x)$ in the original inequality to obtain the sharper bound

$$E(x_{(n)}) \leqslant n\left[\left\{1 - 1\Big/\binom{2n-2}{n-1}\right\}\Big/\{2(2n-1)\}\right]^{\frac{1}{2}}. \tag{14.70}$$

This result is due to Moriguti (1951), who gives similar but more complicated results for upper and lower bounds of the variance and the coefficient of variation of $x_{(n)}$ in the symmetrical case.

The obvious lower bound of zero for $E(x_{(n)})$ cannot in general be improved. $E(x_{(n)})$ attains its general upper bound when $nF^{n-1} \propto \left(x + \frac{1}{a}\right)$, i.e.

$$F(x) = \left(\frac{ax+1}{n}\right)^{1/(n-1)}, \qquad -1/a \leqslant x \leqslant (n-1)/a. \tag{14.71}$$

For the smallest order statistic, analogously, the same results hold with $E(x_{(n)})$ replaced by $|E(x_{(1)})|$. Hartley and David (1954) give the general result for any order statistic

$$|E(x_{(m)})| \leqslant \left[\frac{B(2m-1, 2n-2m+1)}{\{B(m, n-m+1)\}^2} - 1\right]^{\frac{1}{2}},$$

but this can only be attained by a d.f. if $m = 1$ or n, the special cases considered above.

C. Singh (1967) uses the Edgeworth expansion (6.42) with $\kappa_r = 0$, $r > 5$ to investigate the distribution of $x_{(1)}$ and $x_{(n)}$ from moderately non-normal populations. Their d.f.s and mean values are more affected by skewness than kurtosis, especially as n increases, but their variances increase with kurtosis. Joshi (1969) gives bounds and approximations for any moment of any order statistic, and C. Singh (1972) uses his (1967) results to find corrections for moderate non-normality to these moments.

The *Biometrika Tables*, Vol. II, give means of the order statistics of the exponential distribution to 2 d.p. for $n = 1(1)60$, and of the the Gamma distribution (6.18) with $\lambda = 0.5$, 1.5, 2, 2.5, 3.5 to 3 d.p. for $n = 1(1)20$. Prescott (1974) gives 4 d.p. variances and covariances of the Gamma order statistics with $\lambda = 2(1)5$ for $n = 2(1)10$. Walter and Stitt (1988) give the means and variances of $x_{(1)}$ and $x_{(n)}$ for the Gamma for $\lambda = 1(1)10, 15, 20$ and $n = 1(1)25$. They also give the means and variances for all order statistics for $\lambda = 1, 5(5)20$ and $n = 1(1)5(5)25$.

Gupta and Shah (1965) give the moments of logistic order statistics for $n = 1(1)10$ and six percentiles of their distributions for $n = 1(1)25$.

Joint distribution of two order statistics

14.23 If in (14.2) we make the transformation

$$y = nF(x_{(r)}) \qquad z = n\{1 - F(x_{(s)})\}, \tag{14.72}$$

we obtain

$$\mathrm{d}H_{r,s}(y,z)=\frac{\left(\frac{y}{n}\right)^{r-1}\left(1-\frac{y}{n}-\frac{z}{n}\right)^{s-r-1}\left(\frac{z}{n}\right)^{n-s}\mathrm{d}y\,\mathrm{d}z}{B(r,s-r)B(s,n-s+1)\cdot n^2}.$$

We keep r and $n-s$ fixed and let n tend to infinity. We find for the limit

$$\mathrm{d}H=\frac{y^{r-1}\mathrm{e}^{-y}\mathrm{d}y}{\Gamma(r)}\cdot\frac{z^{n-s}\mathrm{e}^{-z}\mathrm{d}z}{\Gamma(n-s+1)}, \tag{14.73}$$

so that the order statistics when transformed by (14.72) are asymptotically independently distributed Gamma variables. It follows from (14.73) that $x_{(r)}$ and $x_{(s)}$ themselves are also asymptotically independent. When $s=n-r+1$, (14.73) becomes

$$\mathrm{d}H=\frac{y^{r-1}\mathrm{e}^{-y}}{\Gamma(r)}\mathrm{d}y\cdot\frac{z^{r-1}\mathrm{e}^{-z}}{\Gamma(r)}\mathrm{d}z, \tag{14.74}$$

the marginal distributions of the variables being identical if the order statistics are symmetrically chosen. If $r=1$, (14.74) is the product of two exponential distributions.

Spacings

14.24 When the distribution is uniform on (0, 1], it follows directly that the "spacings" $y_r=x_{(r)}-x_{(r-1)}$ have the joint density

$$f(y_1,\ldots,y_n)=n!,\quad y_r\geqslant 0,\quad \sum_{r=1}^{n}y_r\leqslant 1.$$

By convention, $x_{(0)}=0$ and $x_{(n+1)}=1$. Also,

$$f(y_1,\ldots,y_{n+1})=n!, \tag{14.75}$$

where the distribution is still symmetric but is now degenerate, being confined to the $(n+1)$-simplex $\sum_{r=1}^{n+1}y_r=1$. It follows that if any k of the $\{y_r\}$ are selected,

$$w_k=y_{i_1}+y_{i_2}+\ldots+y_{i_k}$$

follows a Beta distribution of the first kind with parameters k and $n+1-k$.

A general review of the theory of spacings appears in Pyke (1965). Exercise 14.1 gives the expectations of the y_r and Exercise 14.26 studies them as a function of r. The transformed spacings $y_{rs}=F(x_{(s)})-F(x_{(r)})$ are considered in Exercise 14.14; the results follow from considerations similar to those given above.

If a random sample of n observations from the exponential distribution is conditioned upon its sum, the resulting joint distribution is essentially (14.75) with $(n-1)$ in place of n (Exercise 14.8). This argument is sometimes used in ecology to justify simple models of the allocation of resources between species. The spacings model is then known as the random partition or "broken-stick" model.

Exercises 14.16–18 also consider exponential order statistics. Holst (1979) gives a central limit theorem for some functions of spacings. See also Hall (1982).

Ranges and mid-ranges

14.25 The range of a sample has been briefly mentioned in **2.17**. We now generalize the idea by defining the mth range as the difference R_m given by

$$R_m = x_{(n-m+1)} - x_{(m)}, \tag{14.76}$$

namely the distance between the mth values from either end of the sample. The range itself is the special case $m = 1$. We also define the mth mid-range analogously by

$$M_m = \tfrac{1}{2}(x_{(n-m+1)} + x_{(m)}). \tag{14.77}$$

The particular value M_1 is called the mid-range of the sample.

From (14.75) it is clear that the asymptotic distributions of ranges and mid-ranges are distributions of the differences and sums of transformed independent variates. We may, however, derive the exact distributions directly from (14.2) by putting $r = m = n - s + 1$ and then transforming by (14.76) and (14.77). We obtain for the joint distribution of R_m and M_m,

$$\begin{aligned} \mathrm{d}H_m \propto \{F(M - \tfrac{1}{2}R)\}^{m-1}\{F(M + \tfrac{1}{2}R) - F(M - \tfrac{1}{2}R)\}^{n-2m} \\ \times \{1 - F(M + \tfrac{1}{2}R)\}^{m-1} f(M - \tfrac{1}{2}R) f(M + \tfrac{1}{2}R)\, \mathrm{d}R\, \mathrm{d}M, \end{aligned} \tag{14.78}$$

where the suffix m has been omitted on the right for convenience.

The marginal distributions of R_m and of M_m are obtained by integrating out the other variate from (14.78). A little care is necessary in determining the limits of integration, since these depend in each case on the range of the original distribution F. Equivalently, we may proceed directly from (14.2).

The exact distribution of the range

14.26 Putting $m = 1$ in (14.78), and integrating out M, we obtain for the density function of the range

$$\mathrm{d}G(R) = \mathrm{d}R \cdot n(n-1) \int_{-\infty}^{\infty} \{F(x+R) - F(x)\}^{n-2} f(x+R)\, \mathrm{d}F(x). \tag{14.79}$$

The d.f. is at once seen to be

$$G(R) = n \int_{-\infty}^{\infty} \{F(x+R) - F(x)\}^{n-1}\, \mathrm{d}F(x). \tag{14.80}$$

For some particular cases, the distribution of the range is analytically obtainable. More generally, as in the normal case, quadrature is necessary.

For the mean value of the range, we have

$$E(R) = E(x_{(n)}) - E(x_{(1)}),$$

and using (14.1), this is

$$= \int_{-\infty}^{\infty} xn[\{F(x)\}^{n-1} - \{1 - F(x)\}^{n-1}] f(x)\, \mathrm{d}x. \tag{14.81}$$

Since $\{F(x)\}^{n-1} - \{1-F(x)\}^{n-1}$ is monotone non-decreasing, with value -1 as $x \to -\infty$ and $+1$ as $x \to +\infty$, (14.81) converges if and only if $\int_{-\infty}^{\infty} xf(x)\,dx$ does. Thus the mean range exists when the population mean exists. Integrating (14.81) by parts, we have

$$E(R) = (x[\{F(x)\}^n + \{1-F(x)\}^n])_{-\infty}^{\infty} - \int_{-\infty}^{\infty} [\{F(x)\}^n + \{1-F(x)\}^n]\,dx,$$

and since $F^n + (1-F)^n = 1$ at $x = \pm\infty$, we may write this as

$$E(R) = \int_{-\infty}^{\infty} [1 - \{F(x)\}^n - \{1-F(x)\}^n]\,dx. \tag{14.82}$$

Recurrence relations for $E(R)$ like those in Exercise 14.19 are sometimes useful.

(14.82) is due to Tippett (1925); he also established the result, whose proof is simplified by Mardia (1965),

$$E\{R-E(R)\}^m = m(m-1)\int_{-\infty}^{\infty}\int_{-\infty}^{x_{(n)}} \{1 - F_n^n - (1-F_1)^n + (F_n - F_1)^n\}$$

$$\times \{R - E(R)\}^{m-2}\,dx_{(1)}\,dx_{(n)} - (m-1)\{-E(R)\}^m$$

for $m \geqslant 2$. Here, F_s means $F\{x_{(s)}\}$ as before. As below (14.81) we may see from (14.2) with $r=1$, $s=n$ that the mth moment of R exists if the population μ'_m exists.

The standardized normal case has been studied by Tippett (1925) and E. S. Pearson (1926, 1932). Tippett found the first four moments of the distribution of the range, tabulated the mean values for values of n up to 1000 and gave a diagram for determining standard errors. His table of mean values to 5 d.p. is reproduced in the *Biometrika Tables*. The following values illustrate the general behaviour of the distribution:

n	Standard error	β_1 (approximate)	β_2 (approximate)
2	0.853	0.99	3.87
10	0.797	0.16	3.20
100	0.605	0.22	3.39
500	0.524	0.29	3.50
1000	0.497	0.31	3.54
Asymptote ($n = 1000$)	0.488	0.650	4.200

This table should be compared with that for $x_{(n)}$ in **14.21**. Since $x_{(1)}$ and $x_{(n)}$ are asymptotically independent by **14.23**, the cumulants of the distribution of R are asymptotically twice those of $x_{(n)}$ for any symmetrical population. In particular, $E(R) = 2E(x_{(n)})$ is an exact result, and the mean is omitted from our table because the entries would be exactly double those in the table in **14.21**. The s.e. is increased by a factor $\sqrt{2}$; the tables show that even for $n = 100$, this approximation is little in error. Similarly, $\beta_1 = \kappa_3/\kappa_2^{3/2}$ is halved asymptotically and for practical purposes at $n = 100$, and $\beta_2 - 3 = \kappa_4/\kappa_2^2$ is halved, so that β_2 here $= \frac{1}{2}(\beta_2 + 3)$ in **14.21**, again effectively at $n = 100$. Thus we may approximate the range in normal samples as the sum of two independent $x_{(n)}$ variables—see **14.29** below. As in **14.21**, we see that even at $n = 1000$, the distribution is still a long way from its asymptotic form, and that as n increases, it diverges more and more from the normal form.

E. S. Pearson and Hartley (1942) have tabulated to 4 d.p. the d.f. of R/σ in samples of size $n=2$ to 20 from a normal population. Their results, too, are reproduced in the *Biometrika Tables*. Harter (1960) gives to 6 d.p. for $n=2(1)20(2)40(10)100$ the values of R/σ for which the d.f. is equal to 0.0001, 0.0005, 0.001, 0.005, 0.01, 0.025, 0.05, 0.1(0.1)0.9, 0.95, 0.975, 0.99, 0.995, 0.999, 0.9995, 0.9999. He also gives the mean, variance, skewness and kurtosis to at least 8 significant figures for $n=2(1)100$. These tables are reproduced in Harter (1969), which also contains an 8 d.p. table of the d.f. for the same range of n, an 8 d.p. table of the p.d.f. for $n=2(1)16$ and many other related tables.

Subrahmaniam (1969) obtains the distribution and moments of $x_{(r)}$, and jointly of $x_{(r)}$, $x_{(s)}$, in samples from a population represented by an Edgeworth expansion. C. Singh (1967, 1970, 1976) similarly investigates the effect of moderate non-normality on the mean, variance, percentiles and density of the range. The variance is more sensitive than the mean is to population skewness and kurtosis, but skewness always reduces both. High skewness and high kurtosis tend to have opposite, cancelling effects.

Gupta and Shah (1965) obtain the distribution of the range for a logistic population, and compare it with the normal case for $n=2$ and 3—they agree closely.

Burr (1955) and Connor (1969) give the distribution of R in samples from a discrete population, respectively with and without replacement.

Exercise 14.24 gives the exact c.f. of R in the exponential case.

As in **14.22**, it may be seen by assuming the contrary and finding a contradiction that for any sample $R \leqslant (2n)^{\frac{1}{2}}s$—Exercise 14.25 gives a more general result for $x_{(t)}-x_{(r)}$. For the mean value of R, it follows from (14.70) that for symmetrical standardized populations

$$0 \leqslant E(R) \leqslant n\left[2\left\{1-1\Big/\binom{2n-2}{n-1}\right\}\Big/(2n-1)\right]^{\frac{1}{2}} \doteq (n+\tfrac{1}{2})^{\frac{1}{2}},$$

a result due originally to Plackett (1947), who showed that it also holds for non-symmetrical populations. Hartley and David (1954) confirmed this and also derived upper and lower bounds for $E(R)$ in samples from populations with finite range. The following table (extracted from Plackett (1947)) gives values for the upper bound, its asymptotic value and the exact values for the normal and uniform populations.

n	Upper bound for $E(R)$	$(n+\frac{1}{2})^{\frac{1}{2}}$	Exact values Normal	Exact values Uniform
2	1.155	1.581	1.128	1.155
3	1.732	1.871	1.693	1.732
4	2.084	2.121	2.059	2.078
8	2.921	2.915	2.847	2.694
12	3.539	3.536	3.258	2.931

For a general discussion of inequalities for order statistics, see Arnold and Balakrishnan (1989).

The asymptotic distribution of the range

14.27 We now turn to the limiting distribution of R as n becomes large. From (14.2) with $r=n-s+1=1$, we have, for the joint distribution of the smallest and largest observations,

$$\mathrm{d}G_{1,n}(x_1, x_n) = n(n-1)(F_n - F_1)^{n-2}\,\mathrm{d}F_1\,\mathrm{d}F_n.$$

We now transform by the relations

$$u = 2n\{F_1(1-F_n)\}^{\frac{1}{2}}, \qquad v = \tfrac{1}{2}\log\{F_1/(1-F_n)\}, \tag{14.83}$$

the Jacobian of the transformation being $u/(2n^2 f_1 f_n)$. The joint distribution of u and v is then

$$\mathrm{d}F(u, v) = \left(\frac{n-1}{2n}\right)u\left(1 - \frac{u\cosh v}{n}\right)^{n-2}\mathrm{d}u\,\mathrm{d}v. \tag{14.84}$$

As $n \to \infty$, (14.84) tends to

$$\lim_{n\to\infty} \mathrm{d}F(u, v) = \tfrac{1}{2}u\exp(-u\cosh v)\,\mathrm{d}u\,\mathrm{d}v. \tag{14.85}$$

The domain of variation in (14.84) is bounded by $u \geqslant 0$, $\cosh v \leqslant n/u$, so that in (14.85) $\cosh v$, and therefore v, can range to infinity. Hence we obtain from (14.85) the limiting marginal distribution of u:

$$\left.\begin{aligned}\mathrm{d}F(u) &= u\,\mathrm{d}u\int_0^\infty \exp(-u\cosh v)\,\mathrm{d}v\\ &= u\,\mathrm{d}u\int_1^\infty \exp(-ut)(t^2-1)^{-\frac{1}{2}}\,\mathrm{d}t.\end{aligned}\right\} \tag{14.86}$$

Now u may be expressed as

$$u = 2n\{F(M - \tfrac{1}{2}R)[1 - F(M + \tfrac{1}{2}R)]\}^{\frac{1}{2}}. \tag{14.87}$$

If now we assume that the mid-range M converges in probability to the population mid-range μ, u will behave like

$$u' = 2n\{F(\mu - \tfrac{1}{2}R)[1 - F(\mu + \tfrac{1}{2}R)]\}^{\frac{1}{2}}. \tag{14.88}$$

If we may take $\mu = 0$ and if the population is symmetrical, (14.88) reduces to

$$u' = 2nF(-\tfrac{1}{2}R), \tag{14.89}$$

so that u will tend to the probability integral transform (**1.27**) of the range, apart from constant multipliers.

The mean and variance of u are given in Exercise 14.20.

14.28 The value of these results, due to Elfving (1947), is restricted by the condition, imposed in order to reach (14.88), that the mid-range M converges to the population mid-range μ in probability. Even when the population is symmetrical (so that μ is the mean also) and exponential-type, this condition is not necessarily fulfilled, despite the fact that the symmetry implies $E(M) = \mu$ if it exists. To show this, we find from Exercise 14.4 that the variance of the exponential-type extreme distribution (14.66) is $\pi^2/6$. By (14.69) or (14.45), the variate here is

$$y = nF'(x_{on})(x - x_{on}), \tag{14.90}$$

where x is written for $x_{(n)}$. Thus $\operatorname{var} y = \pi^2/6$ and hence, for large samples,

$$\operatorname{var} x = \pi^2/\{6[nF'(x_{on})]^2\}. \tag{14.91}$$

Since $x_{(n)}$ and $x_{(1)}$ are asymptotically independent, the mid-range, which is their mean, has asymptotic variance

$$\operatorname{var} M = \pi^2/\{12[nF'(x_{on})]^2\}. \tag{14.92}$$

Example 14.4 (Variance of the mid-range for the normal distribution)

In the normal case with zero mean, we have seen (Example 14.3) that

$$nF'(x_{on}) \doteq (2 \log n)^{\frac{1}{2}}$$

so that, from (14.92), $\operatorname{var} M = \pi^2/(24 \log n)$. Thus, M converges to the population mid-range since $E(M) = 0$ here and $\operatorname{var} M \to 0$ as $n \to \infty$.

Example 14.5 (Variance of the mid-range for the double exponential distribution)

The double exponential (or Laplace) distribution

$$dF(x) = \tfrac{1}{2} \exp(-|x|)\, dx, \qquad -\infty < x < \infty,$$

has all moments finite. Its d.f. is

$$F(x) = 1 - \tfrac{1}{2} e^{-x}, \qquad x \geqslant 0. \tag{14.93}$$

It is easily verified that (14.44) is satisfied. By (14.40) and (14.93), $x_{on} = \log(\frac{1}{2}n)$ and therefore

$$nF'(x_{on}) = \tfrac{1}{2} n \exp(-\log \tfrac{1}{2} n) = 1.$$

Thus for this case, from (14.92), $\operatorname{var} M = \pi^2/12$ asymptotically and since $\operatorname{var} M$ does not $\to 0$ as $n \to \infty$, M does not converge in probability to its expectation μ.

Exercises 14.12–13 give a number of results concerning the range and mid-range of the uniform distribution.

The reduced range and mid-range

14.29 In view of the result of the last section, we return to the problem of **14.27**, and seek a more general solution of the asymptotic distribution problem for the range and mid-range, or some simple functions of them. We confine our attention to symmetrical exponential-type distributions.

If we equate (14.38) and (14.45), we find

$$1 - F(x_{(n)}) = \frac{1}{n} \exp\{-nF'(x_{on})(x_{(n)} - x_{on})\},$$

and similarly, using (14.55) also,

$$F(x_{(1)}) = \frac{1}{n} \exp\{nF'(x_{on})(x_{(1)} + x_{on})\}.$$

Thus (14.83) becomes

$$\left.\begin{aligned} u &= 2\exp\{-\tfrac{1}{2}nF'(x_{on})(x_{(n)}-x_{(1)}-2x_{on})\}, \\ v &= \tfrac{1}{2}nF'(x_{on})(x_{(n)}+x_{(1)}). \end{aligned}\right\} \tag{14.94}$$

We now define the *reduced range* R^* and the *reduced mid-range* M^* by

$$R^* = nF'(x_{on})(R-2x_{on}), \qquad M^* = nF'(x_{on})M. \tag{14.95}$$

(14.94) then becomes

$$u = 2\exp(-\tfrac{1}{2}R^*), \tag{14.96}$$

$$v = M^*. \tag{14.97}$$

If we apply (14.96) as a transformation of the distribution of u at (14.86), we obtain, for the distribution of R^*,

$$\begin{aligned} dF(R^*) &= 2\exp(-R^*)\,dR^*\cdot\int_0^\infty \exp\{-2\exp(-\tfrac{1}{2}R^*)\cosh v\}\,dv \\ &= 2\exp(-R^*)K_0\{2\exp(-\tfrac{1}{2}R^*)\}\,dR^*, \end{aligned} \tag{14.98}$$

where K_0 is a Bessel function of zero order and imaginary argument. Gumbel (1949), who had earlier obtained (14.98) by a different method, has given tables of the density and the d.f. of R^*, and tabulated its percentage points. Cox (1948), who also obtained this result, used the method of steepest descents to obtain a second approximation, which he compared with (14.98) and Elfving's result at (14.86). The latter is the most accurate in the normal case, and the steepest descents approximation, as might be expected, is also better than (14.98). But as will be seen below, even (14.98) gives a good approximation for small n.

The range proper, R, is a linear function of R^*, by (14.95). As in **14.28**, we know from Exercise 14.4 that, asymptotically,

$$\left.\begin{aligned} E(R^*) &= E(y_{(n)}) - E(y_{(1)}) = 2\gamma, \\ \operatorname{var}(R^*) &= \operatorname{var}(y_{(n)}) + \operatorname{var}(y_{(1)}) = \pi^2/3. \end{aligned}\right\} \tag{14.99}$$

Using (14.95), we have from (14.99), asymptotically,

$$E(R) = 2\gamma/\{nF'(x_{on})\} + 2x_{on}, \qquad \operatorname{var}(R) = \pi^2/[3\{nF'(x_{on})\}^2]. \tag{14.100}$$

By using this and the exact normal population values of $E(R)$ and var (R) (tabulated by Tippett (1925) and E. S. Pearson (1926)) for small n, Gumbel (1947) estimated x_{on} and $F'(x_{on})$ and investigated the closeness of the approximation to the distribution of R obtained from (14.98) by comparing the results it gave with those obtained from the exact tables given by Pearson and Hartley (1942). He found that the distribution function of R was closely approximated by (14.98) for n as low as 6. Thus in the normal case, at least, (14.98) solves the problem of the distribution of the range.

14.30 Turning now to the reduced mid-range M^*, we see from (14.97) that its asymptotic distribution, obtained by integrating u out of (14.85), is

$$dG(M^*) = \tfrac{1}{2}\, dM^* \int_u^\infty u \exp(-u \cosh M^*)\, du. \tag{14.101}$$

The substitution $z = u \cosh M^*$ reduces this to

$$dG(M^*) = dM^*/(2\cosh^2 M^*) = 2\exp(-2M^*)\, dM^*/\{1+\exp(-2M^*)\}^2, \tag{14.102}$$

so that $2M^*$ has the logistic distribution of Exercise 4.21. (14.95) immediately gives the distribution of the mid-range proper, M, from (14.102). It is symmetrical, with zero mean, so that M is, for symmetrical exponential-type populations, an unbiased estimator of the population mid-range (mean), which is zero. But Example 14.5 shows that the even moments of M are heavily dependent on the form of the population.

Exercise 14.21 indicates another proof of (14.102).

Tables for extreme value distributions

14.31 *Probability Tables for the Analysis of Extreme-Value Data* (National Bureau of Standards, Applied Mathematics Series, 22, Washington, 1953) gives:

(1) The d.f. and density of the exponential-type extreme value distribution (14.45) and (14.66) to 7 d.p. over the interval -3.0 to 17.0, including $-2.40(0.05)0(0.1)4.0$, with auxiliary inverse tables giving the extreme value to 5 d.p. for values of its distribution function 0(0.0001)0.005(0.001)0.988(0.0001)0.999 4(0.000 01)1, and giving the density to 5 d.p. for d.f. values (0.0001)0.010(0.001)0.999.

(2) 12 percentage points between 0.005 and 0.995 for the distributions of mth values (14.65) for $m = 1(1)15(5)50$.

(3) The d.f. and density of the reduced range (14.98) for symmetrical exponential-type distributions, for $R^* = 1.0(0.05)11.0(0.5)20.0$, to 7 d.p., with inverse tables giving the reduced range to 3 or 4 d.p. for selected values of its d.f.

14.32 In Volume 2A, when we consider problems of estimation and testing hypotheses, we shall turn to some additional sampling properties of the order statistics.

EXERCISES

14.1 Show, using (14.4), that in samples of size n from a continuous d.f. $F(x)$, the expected value of the spacings

$$E_r = E\{x_{(r+1)} - x_{(r)}\} = \binom{n}{r}\int_{-\infty}^{\infty}\{F(x)\}^r\{1-F(x)\}^{n-r}\mathrm{d}x.$$

Hence establish (14.82) for the mean of the range. Verify that when $f(x)$ is the uniform distribution on (0, 1], $E_r = (n+1)^{-1}$, and that if $f(x) = \mathrm{e}^{-x}$, $x \geqslant 0$, it is $(n-r)^{-1}$.

Show that

$$E_r - E_{r-1} = \binom{n+1}{r}\int_{-\infty}^{\infty}\{F(x)\}^{r-1}\{1-F(x)\}^{n-r}\left\{F(x) - \frac{r}{n+1}\right\}\mathrm{d}x.$$

(The integral form is due to K. Pearson in 1902—see E. S. Pearson (1926) and Sillitto (1951); Exercise 14.26 studies E_r as a function of r.)

14.2 For the exponential d.f. $F(x) = 1 - \mathrm{e}^{-x}$, $x \geqslant 0$ and also for the logistic d.f. $F(x) = (1+\mathrm{e}^{-x})^{-1}$ show directly that the limiting d.f. of $z = x_{(n)} - \log n$ is $G(z) = \exp(-\mathrm{e}^{-z})$ and verify that this is equivalent to (14.45) in each case.

Similarly, for the Cauchy d.f. $F(x) = 0.5 + \pi^{-1}\arctan x$, show directly that the limiting d.f. of $z = \pi x_{(n)}/n$ is $G(z) = \exp(-z^{-1})$, equivalent to (14.51).

14.3 In the notation of 14.9 show that for the median $x_{(r+1)}$ in samples of $n = 2r+1$ observations, to order $(n+2)^{-2}$,

$$E(x_{(r+1)}) = X_{r+1} + \frac{1}{8(n+2)}X''_{r+1} + \frac{1}{128(n+2)^2}X^{(4)}_{r+1},$$

$$\mathrm{var}(x_{(r+1)}) = \frac{1}{4(n+2)}(X'_{r+1})^2 + \frac{1}{32(n+2)^2}\{2X'_{r+1}X'''_{r+1} + (X''_{r+1})^2\},$$

$$\gamma_1(x_{(r+1)}) = \frac{3}{2(n+2)^{\frac{1}{2}}}\cdot\frac{X''_{r+1}}{X'_{r+1}}, \qquad \gamma_2(x_{(r+1)}) = \frac{1}{n+2}\left\{\frac{X'''_{r+1}}{X'_{r+1}} + \frac{3(X''_{r+1})^2}{(X'_{r+1})^2} - 6\right\}.$$

(David and Johnson, 1954)

14.4 Show that the distribution (14.65) has c.f. $\phi(t) = m^{it}\Gamma(m-it)/\Gamma(m)$ and hence that when $m = 1$, (14.66) has cumulants

$$\kappa_1 = \gamma \text{ (Euler's constant, } = 0.5772),$$

$$\kappa_r = (r-1)!\sum_{n=1}^{\infty} n^{-r}, \quad r \geqslant 2,$$

so that

$$\kappa_2 = \sum n^{-2} = \pi^2/6 = 1.645;$$

$$\kappa_3 = 2\sum n^{-3} = 2.404; \quad \kappa_4 = 6\sum n^{-4} = \pi^4/15 = 6.494.$$

Hence derive

$$\beta_1 = 1.30, \qquad \beta_2 = 5.4.$$

Show further that (14.66) has mode zero and median $-\log\log 2 = 0.3665$, so that (2.13) is approximately satisfied.

14.5 Show that if $G_{r,n}$ refers to the distribution (14.1) for the rth order statistic in a sample of n,

$$(n+1)\,\mathrm{d}G_{r,n} = r\,\mathrm{d}G_{r+1,n+1} + (n-r+1)\,\mathrm{d}G_{r,n+1}$$

and hence that the c.f. and moments of $x_{(r)}$ obey a similar recurrence relation. Similarly, if $G_{r,s,n}$ refers to (14.2), show that

$$(n+1)\mathrm{d}G_{r,s,n} = r\,\mathrm{d}G_{r+1,s+1,n+1} + (s-r)\,\mathrm{d}G_{r,s+1,n+1} + (n-s+1)\,\mathrm{d}G_{r,s,n+1}.$$

(David and Joshi (1968) show that these relations also hold for exchangeable variates; Sathe and Dixit (1990) give the general case when the variates are neither independent nor identically distributed.)

14.6 Rewriting (14.20) as $V_p = p(1-p)/(nf_p^2)$, $f_p > 0$, show that if f_p' denotes $\left(\dfrac{\partial f}{\partial x}\right)_p$,

$$\frac{\partial \log V_p}{\partial p} = \frac{1-2p}{p(1-p)} - 2\frac{f_p'}{f_p^2}, \quad \text{and} \quad \frac{\partial^2 \log V_p}{\partial p^2} = -\left(\frac{1}{p^2} + \frac{1}{(1-p)^2}\right) + 4\left(\frac{f_p'}{f_p^2}\right)^2 - 2\frac{f_p''}{f_p^3}.$$

Hence show that V_p has a stationary value at the sample median ($p = 0.5$) if and only if f_p has a stationary value there; and that V_p is a local maximum there if $f_{0.5}' = 0$, $f_{0.5}'' > -4f_{0.5}^3$ and a local minimum if $f_{0.5}' = 0$, $f_{0.5}'' < -4f_{0.5}^3$. Show that the normal distribution falls into the latter category, and the uniform distribution into the former—cf. Example 11.4. Show also that if f_p has a stationary value at $p_0 \neq 0.5$, V_p has no stationary value at p_0.

(Cf. Sen, 1961; H. A. David and Groeneveld (1982) analyse the exact variance of $x_{(r)}$ as a function of r.)

14.7 Show from (14.1) that the distribution, g_{r+1}, of the sample median $x_{(r+1)}$ in samples of $n = 2r+1$ observations from $f(x)$ has the same median as $f(x)$ has, and that if $f(x)$ is symmetric about a value ξ, so is g_{r+1} which then has the same mean as $f(x)$. If $f(x)$ has an absolute maximum at ξ, show that

$$\mathrm{var}\,(x_{(r+1)}) \geqslant [4\{f(\xi)\}^2 \cdot (n+2)]^{-1},$$

and verify that the equality holds for the uniform distribution. Hence show, using Example 11.4, that the asymptotic value for the variance effectively provides a lower bound for the small-sample variance.

(Cf. Chu, 1955)

14.8 Show that in samples of n independent observations from $\mathrm{d}F = \mathrm{e}^{-x}\,\mathrm{d}x$, $0 \leqslant x < \infty$, the conditional distribution of the n observations, x_i, given that their sum is X, is $(n-1)!\,X^{-(n-1)}$, and that when $X = 1$ this is precisely the distribution of the spacings $x_{(r)} - x_{(r-1)}$ $(r = 1, 2, \ldots, n, x_{(0)} \equiv 0, x_{(n)} \equiv 1)$ between the order statistics in a sample of size $(n-1)$ from the uniform distribution on (0, 1], whence the conditional distribution of the partial sums $X_j = \sum_{i=1}^{j} x_i$, $j = 1, 2, \ldots, n$, is that of the uniform order statistics themselves, as in the unconditional distribution of the X_j/X.

14.9 Using (14.1) and (14.2), show that the conditional distribution of $x_{(s)}$ given the value of $x_{(r)}$ is exactly the distribution (14.1) of $x_{(s-r)}$ in samples of size $(n-r)$ from the same distribution truncated at $x_{(r)}$, i.e. with d.f.

$$\{F(x)-F(x_{(r)})\}/\{1-F(x_{(r)})\} \quad \text{for } x \geqslant x_{(r)}.$$

Hence show that $E(x_{(s)}|x_{(r)})$ is a monotone non-decreasing function of the value of $x_{(r)}$. Use this to show that cov $(x_{(s)}, x_{(r)}) > 0$.

Similarly, show that the conditional distribution of $x_{(r)}$ given $x_{(s)}$ is (14.1) for samples of size $(s-1)$ from the d.f. $F(x)/F(x_{(s)})$, $x \leqslant x_{(s)}$.

14.10 In (14.1), expand $\{1-F(x_{(r)})\}^p$ as a binomial series and show that for $0 < p \leqslant n-r$,

$$\mathrm{d}G_{r,n} = \frac{1}{B(r, n-r+1)} \sum_{q=0}^{p} (-1)^q \binom{p}{q} B(r+q, n-r-p+1)\, \mathrm{d}G_{r+q, n-p+q}.$$

Similarly, expanding $\{F(x_{(r)})\}^p$, show that for $0 < p \leqslant r-1$,

$$\mathrm{d}G_{r,n} = \frac{1}{B(r, n-r+1)} \sum_{q=0}^{p} (-1)^q \binom{p}{q} B(r-p, n-r+q+1)\, \mathrm{d}G_{r-p, n-p+q}.$$

Hence derive corresponding recurrence relations for the c.f. and moments of order statistics. Derive the first result of Exercise 14.5 as a special case.

14.11 Show from (14.1) that the modal value of the distribution of $x_{(r)}$ in samples of size n from $f(x) = (2\pi)^{-\frac{1}{2}} \exp(-\frac{1}{2}x^2)$ is the root of

$$(r-1)f(x)\{1-F(x)\} - (n-r)f(x)F(x) = xF(x)\{1-F(x)\},$$

reducing when $r = n$ to $(n-1)f(x) = xF(x)$.

Use this and (5.105) to verify the asymptotic exponential-type result (14.59) in the normal case.

(S. S. Gupta (1961) tabulated the mode of $x_{(n)}$ for $n = 1(1)25(5)50(10)100$.)

14.12 Show that the joint distribution of the sample range and mid-range in samples from a distribution uniform on the interval $(-\frac{1}{2}, \frac{1}{2}]$ is

$$\mathrm{d}F(R, M) = n(n-1)R^{n-2}\,\mathrm{d}R\,\mathrm{d}M, \quad 0 \leqslant R \leqslant 1-2|M| \leqslant 1,$$

that the marginal distribution of M is

$$\mathrm{d}G(M) = n(1-2|M|)^{n-1}\,\mathrm{d}M, \quad |M| \leqslant \tfrac{1}{2},$$

with moments

$$\left.\begin{aligned} \mu_{2k-1} &= 0 \\ \mu_{2k} &= \left\{2^{2k}\binom{2k+n}{2k}\right\}^{-1} \end{aligned}\right\} k = 1, 2, \ldots$$

and that if $t = nM$, $\lim_{n\to\infty} \mathrm{d}P(t) = \exp(-2|t|)\,\mathrm{d}t$.

Show further that the marginal distribution of R is

$$\mathrm{d}H(R) = n(n-1)R^{n-2}(1-R)\,\mathrm{d}R, \quad 0 \leqslant R \leqslant 1.$$

with moments

$$\mu_k' = n(n-1)/\{(n+k)(n+k-1)\}.$$

In particular

$$\begin{aligned} \mu_1' &= (n-1)/(n+1) \\ \mu_2 &= 2(n-1)/\{(n+1)^2(n+2)\}. \end{aligned}$$

Show that the conditional distribution of M given R is uniform on the interval $(-\frac{1}{2}(1-R), \frac{1}{2}(1-R))$ and that the limiting p.d.f. of $u = n(1-R)$ is $u \exp(-u)$.

14.13 Continuing Exercise 14.12, show that the distribution of the ratio $y = M/R$ is given by

$$dF(y) = (n-1)\{1+2|y|\}^{-n}\,dy, \quad -\infty < y < \infty$$

with moments

$$\left.\begin{aligned} \mu_{2k-1} &= 0 \\ \mu_{2k} &= 1\bigg/\left\{2^{2k}\binom{n-2}{2k}\right\}^{-1} \end{aligned}\right\} k = 1, 2, \ldots; \quad 2k \leqslant n-2.$$

Show that the density of $W = 2(n-1)|y|$ is exactly $\left(1+\dfrac{W}{n-1}\right)^{-n}$, a special case of the variance ratio (F) distribution in Example 11.20 or more conveniently at (16.24) below, with $\nu_1 = 2$, $\nu_2 = 2(n-1)$.

When $n = 2$, show that $2y$ is identical with Student's t statistic, defined at (11.42). Compare the distribution of t in this uniform case with the normal form (11.44) with $n = 2$, which is the Cauchy distribution. Show that in the normal case, t^2 has the F-distribution with $\nu_1 = \nu_2 = 1$, rather than $W = |t|$ in the uniform case, with $\nu_1 = \nu_2 = 2$.

14.14 From (14.2), show that the joint distribution of $y = F(x_{(s)}) - F(x_{(r)})$ and $z = F(x_{(r)})$ is

$$dH_{y,z} = \frac{z^{r-1}y^{s-r-1}(1-y-z)^{n-s}\,dy\,dz}{B(r, s-r)B(s, n-s+1)}, \quad 0 \leqslant y+z \leqslant 1,$$

and hence that y is distributed in the first Beta form (6.14) with parameters $p = s-r$, $q = n-s+r+1$, depending only on $(s-r)$. In particular, when $s = r+1$, $p = 1$ and $q = n$ for any r, so that the transformed spacings are identically distributed. Show from (14.1) that $F(x_{(1)})$ and $1 - F(x_{(n)})$ also have this distribution.

Show further that $w = y/z$ is distributed in the second kind of Beta distribution (6.16) with parameters $p = s-r$, $q = r$.

14.15 Show that in a sample of size $n \geqslant 2$ from a standardized normal distribution, $(x_{(j)} - \bar{x})$ is independent of $\bar{x}$. Hence show that

$$\begin{aligned} \kappa_r(x_{(j)} - \bar{x}) &= \kappa_r(x_{(j)}), \quad r > 2 \\ &= \kappa_2(x_{(j)}) - n^{-1}, \quad r = 2, \end{aligned}$$

and that $\sum_{k=1}^{n} \operatorname{cov}(x_{(j)}, x_{(k)}) = 1$ for any j.

(Cf. McKay, 1935)

14.16 If $z_1, \ldots, z_n$ are independently exponentially distributed, so that

$$dF(z_i) = \exp(-z_i)\,dz_i, \quad 0 \leqslant z_i < \infty,$$

show that

$$\begin{aligned} y_1 &= nz_{(1)}, \\ y_r &= (n+1-r)\{z_{(r)} - z_{(r-1)}\}, \quad r = 2, 3, \ldots, n \end{aligned}$$

are also independently exponentially distributed.

For $n=2$, show that, if $f(z_i)=\theta_i\exp(-\theta_i z_i)$, $\theta_i>0$, $y_2=z_{(2)}-z_{(1)}$ is still distributed independently of $(\theta_1+\theta_2)z_{(1)}$, which is exponentially distributed, but that y_2 is not exponential.

(Cf. P. V. Sukhatme, 1937. These are the only n linear functions that are independent of each other—see Tanis (1964).)

14.17 By making the transformation $z_i=-\log F\{x_i\}$ for a random sample $x_1,\ldots,x_n$ from any continuous $F(x)$, use Exercise 14.16 to show that

$$y_n=-n\log F\{x_{(n)}\}$$

$$y_r=-r\log\frac{F\{x_{(r)}\}}{F\{x_{(r+1)}\}},\quad r=1,2,\ldots,n-1$$

are independently exponentially distributed, and hence that

$$[F\{x_{(n)}\}]^n,\ \left[\frac{F\{x_{(r)}\}}{F\{x_{(r+1)}\}}\right]^r,\quad r=1,2,\ldots,n-1$$

are independently uniformly distributed on (0, 1). (Rényi, 1953)

14.18 In Exercise 14.16, show that $z_{(r)}=\sum_{s=1}^{r}\frac{y_s}{n+1-s}$, and that as $n\to\infty$ with r fixed, $nz_{(r)}\doteq\sum_{s=1}^{r}y_s$ is a Gamma variable with parameter r. Hence show that in Exercise 14.17, $-n\log F\{x_{(n-r+1)}\}$ has this limiting distribution and that this implies the limiting distributions obtained for y and z in 14.23.

(Rényi, 1953)

14.19 From (14.82), show that if n is odd, and R_n is the range in samples of size n,

$$2E(R_n)=\sum_{s=2}^{n-1}(-1)^s\binom{n}{s}E(R_s)\quad\text{and}\quad(n+1)E(R_n)=\sum_{s=2}^{n-1}(-1)^s\binom{n+1}{s}E(R_s),$$

and hence that also

$$(n-1)E(R_n)=\sum_{s=2}^{n-1}(-1)^s\binom{n}{s-1}E(R_s).$$

In particular, $2E(R_3)=3E(R_2)$.

14.20 Show that u in (14.86) has mean $\frac{1}{2}\pi$ and variance $(4-\frac{1}{4}\pi^2)$. (Elfving, 1947)

14.21 Use the c.f.s of Exercises 4.21 and 14.4 to establish the result (14.102) for the limiting distribution of the reduced mid-range. Using the remark below (2.24), show that (14.66) has mean difference equal to the logistic mean deviation given in Exercise 4.21, so that from Exercise 14.4, $\Delta/\sigma=1.081$.

14.22 Show from (14.1) that for the Cauchy distribution $dF(x)=dx/\{\pi(1+x^2)\}$, the pth moment of $x_{(r)}$ exists if and only if $p<\min(r,n-r+1)$, i.e., if at least p order statistics lie on each side of $x_{(r)}$. (Example 3.3 with $m=1$ is here the special case $n=r=1$.)

14.23 Writing (14.5) as $G_{r,n}(x)=I_{F(x)}(r,n-r+1)$, show by inverting the order of summation and integration that

$$\sum_{r=1}^{n} r^{-1}G_{r,n}(x) = \sum_{r=1}^{n} r^{-1}G_{1,r}(x)$$

and

$$\sum_{r=1}^{n} (n-r+1)^{-1}G_{r,n}(x) = \sum_{r=1}^{n} r^{-1}G_{r,r}(x),$$

with corresponding relations between the c.f.s, moments and densities in the continuous case.

(Joshi, 1973; Balakrishnan (1987) shows that these identities also hold for exchangeable variates.).

14.24 In Exercise 14.18, show that $z_{(r)}$ in a sample of size n is distributed exactly as $z_{(r-p)}$ in a sample of size $n-p$ from the exponential distribution truncated at $z_{(p)}$—cf. Exercise 14.9.

Show further that the range $R = z_{(n)} - z_{(1)}$ has exact c.f. $E(e^{itR}) = \Gamma(n)\Gamma(1-it)/\Gamma(n-it)$, with $E(R) = \sum_{s=2}^{n} (n+1-s)^{-1}$ and $\text{var}\, R = \sum_{s=2}^{n} (n+1-s)^{-2}$, and that R itself is exponentially distributed when $n=2$—cf. Exercise 11.21.

14.25 Establish the result stated at the beginning of 14.22,

$$|x_{(r)} - \bar{x}| \leqslant (n-1)^{\frac{1}{2}}s.$$

Show that

$$n^2(x_{(n)} - \bar{x})^2 \geqslant n\{(x_{(n)} - \bar{x})^2 + s^2\},$$

and hence that

$$(n-1)^{-\frac{1}{2}}s \leqslant x_{(n)} - \bar{x} \leqslant (n-1)^{\frac{1}{2}}s, \tag{A}$$

and similarly

$$(n-1)^{-\frac{1}{2}}s \leqslant \bar{x} - x_{(1)} \leqslant (n-1)^{\frac{1}{2}}s. \tag{B}$$

Show further that

$$-\left(\frac{n-r}{r}\right)^{\frac{1}{2}} s \leqslant x_{(r)} - \bar{x} \leqslant \left(\frac{r-1}{n-r+1}\right)^{\frac{1}{2}} s,$$

which includes the upper bounds in (A) and (B), and that for $r < t$, if we fix $x_{(r)}$ and $x_{(t)}$ and then minimize s^2 we find

$$x_{(t)} - x_{(r)} \leqslant s\left\{\frac{n(r+n+1-t)}{r(n+1-t)}\right\}^{\frac{1}{2}} = sA(r,t), \text{ say.}$$

Show that this reduces to the result stated in 14.26,

$$R \leqslant s(2n)^{\frac{1}{2}}$$

when $t=n$, $r=1$. Finally, show that $E(x_{(t)} - x_{(r)}) \leqslant \sigma A(r,t)$.

(Cf. Wolkowicz and Styan (1979) and Fahmy and Proschan (1981). Arnold and Groeneveld (1981) give maximal deviations between sample and population means in terms of the population mean, s.d., mean deviation and range.)

14.26 If $f(x)$ is a continuous density with median taken as origin, and E_r is the expected spacing in Exercise 14.1, show that

(i) if $f'(x) \leqslant 0$ for all x, $E_r > E_{r-1}$, $r \geqslant 2$;

(ii) if $f(x)$ is symmetric and unimodal and $|f'(x)| > 0$ for $x \neq 0$, $E_r > E_{r-1}$ for $r > \frac{1}{2}(n+1)$ and $E_r < E_{r-1}$ for $r < \frac{1}{2}(n+1)$;

(iii) if $f(x)$ is unimodal, $|f'(x)| > 0$, $x \neq 0$, and $f(x)$ is skewed to the right in the sense that

$$f[F^{-1}(u)] \geqslant f(F^{-1}(1-u)] \quad \text{for} \quad 0 < u < \tfrac{1}{2},$$

then $E_r > E_{r-1}$ for $r > \frac{1}{2}(n+1)$.

(David and Groeneveld, 1982).

14.27 For Tukey's lambda distribution in **14.10**, show that Pearson's expansion for the variance of the median yields

$$\operatorname{var}(x_{(r+1)}) = \frac{d_1^2}{4(n+2)} + \frac{d_1 d_3}{16(n+2)(n+4)} + o(n^{-2}),$$

where $d_1 = 2\gamma\lambda(\frac{1}{2})^{\lambda-1}$ and $d_3 = 2\gamma\lambda^{[3]}(\frac{1}{2})^{\lambda-3}$ are the first and third derivatives evaluated at $u = \frac{1}{2}$. When $\lambda = 0.135$, $\gamma = 5.063$ (the standardized normal approximation—see **14.10**), show that this reduces to

$$\frac{1.55}{(n+2)} + \frac{2.50}{(n+2)(n+4)} + o(n^{-2}),$$

in close agreement with (14.16).

14.28 Given a sample of n exchangeable variates for which $\mu_{r,n} = E(x_{(r)}|n)$ show that

$$(r-1)\mu_{r,n}^m + (n-r+1)\mu_{r-1,n}^m = n\mu_{r-1,n-1}^m$$

for $m = 1, 2, \ldots$. Hence derive the recurrence relations

$$\mu_{r,n}^m = \sum_{j=n-r+1}^{n} (-1)^{j-n+r-1} \binom{n}{j}\binom{j-1}{n-r} \mu_{1,j}^m$$

$$= \sum_{j=r}^{n} (-1)^{j-r} \binom{n}{j}\binom{j-1}{r-1} \mu_{j,j}^m.$$

(David and Joshi, 1968; extended results are given in Balakrishnan, 1987)

CHAPTER 15

THE MULTINORMAL DISTRIBUTION AND QUADRATIC FORMS

15.1 The univariate normal distribution with p.d.f.

$$f(x) = \frac{1}{\sigma\sqrt{2\pi}} \exp\left\{-\frac{1}{2}\left(\frac{x-\mu}{\sigma}\right)^2\right\}, \qquad -\infty < x < \infty,$$

has been discussed in many contexts in earlier chapters. In the standardized case, the exponent of the density function is simply $-\frac{1}{2}x^2$ as we saw in **5.37**. The joint distribution of two or more variables introduced in **7.20** has a density function that has as its exponent a quadratic form in p variables,

$$f(x_1, \ldots, x_p) \propto \exp\left\{-\frac{1}{2}\sum_{j=1}^{p}\sum_{k=1}^{p} \alpha_{jk}\left(\frac{x_k-\mu_j}{\sigma_j}\right)\left(\frac{x_k-\mu_k}{\sigma_k}\right)\right\}, \qquad -\infty < x_j < \infty. \quad (15.1)$$

Here, as in the univariate case, the μ_j and σ_j are simply location and scale parameters, which may be removed by a standardizing transformation.

In this chapter, and generally when dealing with multivariate problems, we shall find it convenient to use vector and matrix notation for economy and clarity in representation. A bold-face capital letter will represent a matrix, a bold-face small letter a column vector, and a prime will denote transposition, so that a row vector will be represented by a bold-face primed small letter. With this notation, (15.1) becomes

$$f(\mathbf{x}) \propto \exp\{-\tfrac{1}{2}(\mathbf{x}-\boldsymbol{\mu})'\mathbf{A}(\mathbf{x}-\boldsymbol{\mu})\}, \quad (15.2)$$

where $\mathbf{A}$ is a $(p \times p)$ matrix which may be assumed to be symmetric.

Characteristic function and moments

15.2 The condition that (15.2) be a properly defined density function is simply that the matrix $\mathbf{A}$ is non-negative definite. For in such a case (and only in such a case) we can find a real linear transformation that transforms the exponent into a negative sum of squares, and the integral of $f(\mathbf{x})$ will converge. In particular there is an *orthogonal* transformation

$$\mathbf{x} = \mathbf{B}\mathbf{y} \quad (15.3)$$

that transforms the exponent of (15.2) into a sum of squares, the coefficients in this sum being the latent roots of $\mathbf{A}$. Confining ourselves for the present to the non-singular case, where the rank of $\mathbf{A}$ is p,[(*)] and measuring from the vector $\boldsymbol{\mu}$, we now apply this transformation to obtain the c.f. of the distribution, and incidentally to evaluate the proportionality constant that makes its integral equal to one.

In our notation, the c.f. is (cf. (4.27))

$$\phi(\mathbf{t}) \propto \int_{-\infty}^{\infty} \cdots \int_{-\infty}^{\infty} \exp\left(i\mathbf{t}'\mathbf{x} - \tfrac{1}{2}\mathbf{x}'\mathbf{A}\mathbf{x}\right) \prod \mathrm{d}x_j \tag{15.4}$$

which, on applying (15.3), becomes

$$\phi(\mathbf{t}) \propto \int_{-\infty}^{\infty} \cdots \int_{-\infty}^{\infty} \exp\left(i\mathbf{t}'\mathbf{B}\mathbf{y} - \tfrac{1}{2}\mathbf{y}'\mathbf{C}\mathbf{y}\right) \prod_j \mathrm{d}y_j, \tag{15.5}$$

the Jacobian being unity since $\mathbf{B}$ is orthogonal. We have written

$$\mathbf{C} = \mathbf{B}'\mathbf{A}\mathbf{B} \tag{15.6}$$

in (15.5). Since $\mathbf{C}$ is a diagonal matrix, the right side of (15.5) factorizes into p single integrals of type

$$I_j = \int_{-\infty}^{\infty} \exp\left(iu_j y_j - \tfrac{1}{2}c_{jj}y_j^2\right) \mathrm{d}y_j,$$

where u_j is linear in the t_j since $\mathbf{u}' = \mathbf{t}'\mathbf{B}'$. Then I_j is obtained directly from the c.f. of the univariate normal distribution (Example 3.11) as

$$I_j = \left(\frac{2\pi}{c_{jj}}\right)^{\frac{1}{2}} \exp\left(-\frac{1}{2}\frac{u_j^2}{c_{jj}}\right). \tag{15.7}$$

Applying (15.7) to each term in (15.5) gives

$$\phi(\mathbf{t}) \propto (2\pi)^{\frac{1}{2}p} \left(\prod_j c_{jj}\right)^{-\frac{1}{2}} \exp\left(-\frac{1}{2}\sum \frac{u_j^2}{c_{jj}}\right). \tag{15.8}$$

Since $\mathbf{C}$ is diagonal, its determinant is

$$|C| = \prod c_{jj}$$

and that of its inverse matrix $\mathbf{C}^{-1}$ is

$$|C^{-1}| = |C|^{-1} = \left(\prod_j c_{jj}\right)^{-1}. \tag{15.9}$$

Using (15.9) in (15.8) and rewriting its exponent in matrix form, we have

$$\phi(\mathbf{t}) \propto (2\pi)^{\frac{1}{2}p} |C^{-1}|^{\frac{1}{2}} \exp\left(-\tfrac{1}{2}\mathbf{t}'\mathbf{B}\mathbf{C}^{-1}\mathbf{B}'\mathbf{t}\right). \tag{15.10}$$

Since, by (15.6),

$$\mathbf{B}\mathbf{C}^{-1}\mathbf{B}' = \mathbf{B}(\mathbf{B}'\mathbf{A}\mathbf{B})^{-1}\mathbf{B}' = \mathbf{B}\mathbf{B}^{-1}\mathbf{A}^{-1}(\mathbf{B}')^{-1}\mathbf{B}' = \mathbf{A}^{-1}, \tag{15.11}$$

[(*)] If $\mathbf{A}$ has rank $r < p$, the distribution degenerates into r dimensions, i.e. one or more variates are redundant.

we have, since $\mathbf{B}$ is orthogonal, $\mathbf{BB}' = \mathbf{I}$ and

$$|A^{-1}| = |BC^{-1}B'| = |B|\,|C^{-1}|\,|B'| = |C^{-1}|. \tag{15.12}$$

We may substitute (15.11) and (15.12) into (15.10) to give

$$\phi(\mathbf{t}) \propto (2\pi)^{\frac{1}{2}p}|A^{-1}|^{\frac{1}{2}} \exp\left(-\tfrac{1}{2}\mathbf{t}'\mathbf{A}^{-1}\mathbf{t}\right). \tag{15.13}$$

Since $\phi(\mathbf{0}) = 1$, in (15.13) we obtain immediately for the proportionality constant of the multivariate normal (multinormal) distribution

$$(2\pi)^{-\frac{1}{2}p}|A^{-1}|^{-\frac{1}{2}} = (2\pi)^{-\frac{1}{2}p}|A|^{\frac{1}{2}}.$$

Thus (15.2) becomes

$$f(\mathbf{x}) = (2\pi)^{-\frac{1}{2}p}|A|^{\frac{1}{2}} \exp\{-\tfrac{1}{2}(\mathbf{x}-\boldsymbol{\mu})'\mathbf{A}(\mathbf{x}-\boldsymbol{\mu})\} \tag{15.14}$$

with c.f.

$$\phi(\mathbf{t}) = \exp\left(-\tfrac{1}{2}\mathbf{t}'\mathbf{A}^{-1}\mathbf{t}\right)\exp(i\mathbf{t}'\boldsymbol{\mu}), \tag{15.15}$$

the second factor on the right of (15.15) arising if the location vector $\boldsymbol{\mu}$ is non-null, as may be seen from the effect on (15.5) of putting $(\mathbf{x}-\boldsymbol{\mu})$ for $\mathbf{x}$ in the second term of the exponent of (15.4) and in (15.3).

15.3 From the c.f. (15.15), the moments of the distribution, or, more conveniently, its cumulants, are easily obtainable. The c.g.f. is

$$\psi(\mathbf{t}) = -\tfrac{1}{2}\mathbf{t}'\mathbf{A}^{-1}\mathbf{t} + i\mathbf{t}'\boldsymbol{\mu}. \tag{15.16}$$

It is clear from the form of (15.16), which has no terms of degree above 2, that all cumulants of order greater than 2 are equal to zero, thus generalizing the univariate property of the normal distribution. The mean of the jth variate is

$$\left[\frac{\partial\psi(\mathbf{t})}{\partial(it_j)}\right]_{\mathbf{t}=\mathbf{0}} = \mu_j,$$

so that the location vector $\boldsymbol{\mu}$ is the vector of means of the variables. The variance of the jth variate is

$$\left[\frac{\partial^2\psi(\mathbf{t})}{\partial(it_j)^2}\right]_{\mathbf{t}=\mathbf{0}} = a^{jj}, \tag{15.17}$$

the element in the jth row and jth column of $\mathbf{A}^{-1}$. The covariance of the jth and kth variates is

$$\left[\frac{\partial^2\psi(\mathbf{t})}{\partial(it_j)\,\partial(it_k)}\right]_{\mathbf{t}=\mathbf{0}} = a^{jk}. \tag{15.18}$$

(15.17) and (15.18) show that $\mathbf{A}^{-1}$ is the matrix of variances and covariances of the variates, or the *covariance matrix*, which we denote by $\boldsymbol{\Sigma}$. Thus, the final expression for the p.d.f. p-variate normal distribution, in terms of its first and second order moments,

$$f(\mathbf{x}) = (2\pi)^{-\frac{1}{2}p}|\mathbf{\Sigma}^{-1}|^{\frac{1}{2}}\exp\{-\tfrac{1}{2}(\mathbf{x}-\boldsymbol{\mu})'\mathbf{\Sigma}^{-1}(\mathbf{x}-\boldsymbol{\mu})\} \tag{15.19}$$

with c.f.

$$\phi(\mathbf{t}) = \exp(-\tfrac{1}{2}\mathbf{t}'\mathbf{\Sigma}\mathbf{t})\exp(i\mathbf{t}'\boldsymbol{\mu}). \tag{15.20}$$

If and only if $\mathbf{\Sigma}$ is a diagonal matrix, the variates are independent.

The covariance matrix $\mathbf{\Sigma}$ of any (not necessarily multinormal) variates is non-negative definite since the quadratic form in $u_1, \ldots, u_p$,

$$\mathbf{u}'\mathbf{\Sigma}\mathbf{u} = E\left\{\sum_{j=1}^{p}(x_j-\mu_j)u_j\right\}^2 \geqslant 0.$$

In the non-singular case, $\mathbf{\Sigma}$ is strictly positive definite.

Example 15.1 (The bivariate normal (binormal) distribution)

We have already encountered the bivariate specializations of (15.19) and (15.20) in Example 3.19 and elsewhere. In this case the covariance matrix is

$$\mathbf{\Sigma} = \begin{pmatrix} \sigma_1^2 & \rho\sigma_1\sigma_2 \\ \rho\sigma_1\sigma_2 & \sigma_2^2 \end{pmatrix},$$

where ρ, the correlation coefficient, is defined by $\mu_{11}/\sigma_1\sigma_2$. Hence

$$\mathbf{\Sigma}^{-1} = \begin{pmatrix} \dfrac{1}{\sigma_1^2(1-\rho^2)} & \dfrac{-\rho}{\sigma_1\sigma_2(1-\rho^2)} \\ \dfrac{-\rho}{\sigma_1\sigma_2(1-\rho^2)} & \dfrac{1}{\sigma_2^2(1-\rho^2)} \end{pmatrix},$$

and the density function with zero means is, from (15.19),

$$f(\mathbf{x}) = \frac{1}{2\pi\sigma_1\sigma_2(1-\rho^2)^{\frac{1}{2}}}\exp\left\{-\frac{1}{2(1-\rho^2)}\left(\frac{x_1^2}{\sigma_1^2}-\frac{2\rho x_1x_2}{\sigma_1\sigma_2}+\frac{x_2^2}{\sigma_2^2}\right)\right\},$$

with c.f., from (15.20),

$$\phi(t_1, t_2) = \exp\{-\tfrac{1}{2}(\sigma_1^2t_1^2 + 2\rho\sigma_1\sigma_2t_1t_2 + \sigma_2^2t_2^2)\}.$$

independence is here equivalent to the condition $\rho = 0$.

Linear functions of normal variates

15.4 We have seen (Example 11.2) that a linear function of independent univariate normal variates is itself normally distributed. We may now obtain a much more general result, to the effect that a set of variates, each of which is a linear function of a set of multinormal variates (i.e. variates that have a multivariate normal distribution), itself has a multinormal distribution. (15.20) is the c.f. of a p-variate non-singular normal distribution. It may be written (putting $\boldsymbol{\mu} = \mathbf{0}$)

$$E\{\exp(i\mathbf{t}'\mathbf{x})\} = \exp(-\tfrac{1}{2}\mathbf{t}'\mathbf{\Sigma}\mathbf{t}). \tag{15.21}$$

Suppose that we write

$$\mathbf{t}' = \mathbf{s}'\mathbf{A}. \tag{15.22}$$

where $\mathbf{s}'$ is a vector of q dummy variables, just as $\mathbf{t}'$ is a vector of p such variables, and $\mathbf{A}$ is any $(q \times p)$ matrix. (15.22) substituted into (15.21) gives

$$E\{\exp(i\mathbf{s}'\mathbf{A}\mathbf{x})\} = \exp(-\tfrac{1}{2}\mathbf{s}'\mathbf{A}\mathbf{\Sigma}\mathbf{A}'\mathbf{s}). \tag{15.23}$$

(15.23) shows that the set of q linear functions of the p x-variates defined by $\mathbf{y} = \mathbf{A}\mathbf{x}$ has a q-variate multinormal c.f. with covariance matrix $\mathbf{A}\mathbf{\Sigma}\mathbf{A}'$. If $\mathbf{A}$ is a non-singular matrix, the multivariate form of the Inversion Theorem (4.17) gives us a multinormal density function (15.19) for $\mathbf{y}$, with $\mathbf{A}\mathbf{\Sigma}\mathbf{A}'$ replacing $\mathbf{\Sigma}$.

The converse is also true: if (15.23) holds for every $\mathbf{A}$, (15.22) leads us back to (15.21) and the multinormal distribution of $\mathbf{x}$. In particular, putting $q = 1$ shows that the multinormal distribution may be characterized by the property that every linear function of its variates has a univariate normal distribution. Marginal normality of the variates is not sufficient for this—see Exercise 15.20.

If $\mathbf{A}$ is singular, $(\mathbf{A}\mathbf{\Sigma}\mathbf{A}')^{-1}$ is not defined, but (15.23) is nevertheless the c.f. of a degenerate multinormal distribution. In particular, this is so if $q > p$. Looking at the matter from this standpoint, it is evident that any degenerate multinormal distribution may be regarded, through its c.f., as having arisen by a singular transformation from a non-degenerate one. Such singularity causes no difficulty in practice, since a distribution can be made non-singular by discarding redundant variates.

An important corollary of the result of this section, obtained by making $\mathbf{A}$ an identity matrix augmented by zeros, with $q \leqslant p - 1$, is that the (marginal) distribution of any subset of the original variates is multinormal, with variances and covariances unchanged, although the inverse of their covariance matrix, and hence the coefficients in the exponent of their (marginal) density function, will of course be changed. Marginal normality, however, does not imply the higher-order normality of the joint distribution, not even if $\mathbf{\Sigma}$ is diagonal—see Exercise 15.20.

Exercise 15.1 discusses the conditional distributions. For a detailed discussion of the multinormal distribution, see Tong (1990).

The multinormal integral

15.5 Apart from the location and scale parameters, which may be removed by standardization, the density function (15.19) contains $\frac{1}{2}p(p-1)$ parameters, the covariances between the p variates. It is thus clear that tabulation of its distribution function is a formidable undertaking for $p > 2$. Bivariate tables are given in *Tables of the Bivariate Normal Distribution and Related Functions* (National Bureau of Standards, Applied Mathematics Series, 50, Washington, 1959) for $\pm\rho = 0(0.05)0.95(0.01)1$ and variates in the range 0(0.1)4, to 6 or 7 d.p. Zelen and Severo (1960) give charts for reading the binormal integral value with an error of 1 percent or less. C.f. Exercise 15.28 for a simplifying relationship.

D. B. Owen (1956, 1957) gives tables for computing the binormal integral over regions bounded by straight lines and certain other regions. Divgi (1979) uses an orthogonal polynomial expansion of $x\{1-F(x)\}$, where F is the univariate normal d.f., to approximate the binormal d.f. within 5×10^{-7} using 11 terms.

G. P. Steck (1958b) gives tables for computing the trinormal integral. S. S. Gupta (1963a, b) surveys the literature for $p \geqslant 2$ and gives a bibliography. For any p, Dutt (1973) gives rapidly convergent formulae by using integral transforms in tetrachoric series like (15.33) below.

Schervish (1984) gives an algorithm which uses numerical quadrature to compute multinormal probabilities over regions $\{a_i \leqslant x_i \leqslant b_i,\ i=1,\ldots,p\}$ for $p \leqslant 7$.

A direct approach to the evaluation of upper tail probabilities is to use Hermite polynomials (cf. 6.14) centred on the univariate densities. We find that

$$\begin{aligned}P(\mathbf{x}>\mathbf{h}) &= P(x_1>h_1,\ldots,x_p>h_p)\\ &= \sum\left(\prod_{i<j}\frac{\rho_{ij}^{n_{ij}}}{n_{ij}!}\right)\prod_{i=1}^{p} H_{n_i-1}(h_i)\alpha(h_i),\end{aligned} \tag{15.24}$$

where $\alpha(h)$ is the density of the standardized univariate normal, the summation is taken over all $0\leqslant n_{ij}<\infty$, $i<j$, and $n_i=\sum_{j\neq i} n_{ij}$. The method of derivation follows **15.8** below. At one time it was believed that the expansion converged for all $\{\rho_{ij}\}$, but Harris and Soms (1980) showed that the series is divergent whenever any $\rho_{ij}>(p-1)^{-1}$ for p even, or $\rho_{ij}>(p-2)^{-1}$ for p odd. Thus (15.24) always converges when $p\leqslant 3$ but is of limited utility for higher p.

An alternative expansion, due to Moran (1983), considers the integral in the form

$$\begin{aligned}P(\mathbf{x}>\mathbf{h}) &= P(x_i>h_i,\quad i=1,\ldots,p)\\ &= K\int_{h_1}^{\infty}\cdots\int_{h_p}^{\infty}\exp(-\tfrac{1}{2}\mathbf{x}'\mathbf{A}\mathbf{x})\Pi\,\mathrm{d}x,\end{aligned}$$

where $K=(2\pi)^{-p/2}|A|^{\frac{1}{2}}$ and the x-values are scaled so that the diagonal elements of $\mathbf{A}$ are unity. The cross-product terms in the exponent are now expanded in series form to give

$$P(\mathbf{x}>\mathbf{h}) = K\sum(-1)^n\prod_{i<j}\frac{a_{ij}^{n_{ij}}}{n_{ij}!}I_{n_1}(h_1)\ldots I_{n_p}(h_p), \tag{15.25}$$

where the summation is over the same range as before and

$$I_m(h)=\int_h^{\infty}x^m e^{-\frac{1}{2}x^2}\,\mathrm{d}x,$$

which can be expressed recursively so that only $I_0(h)$ and $I_1(h)=\exp(-\frac{1}{2}h^2)$ require explicit evaluation.

Moran shows that (15.25) often converges in regions of the parameter space where (15.24) is divergent, although such convergence may be slow.

15.6 If we write

$$P(x > h, y > k) = L(h, k, \rho) = \int_k^\infty \int_h^\infty f(x, y)\, dx\, dy$$

and let

$$u \equiv u(x, k) = (k - \rho x)/(1 - \rho^2)^{1/2}, \tag{15.26}$$

$$L(h, k, \rho) = G(h) \int_h^\infty \left\{ \int_u^\infty \frac{f_z(z)}{G(h)}\, dz \right\} f_x(x)\, dx \tag{15.27}$$

where $G(a) = P(x > h)$ and f_z, f_x denote standard normal p.d.fs. Equation (15.27) may be rewritten as

$$L = L(h, k, \rho) = G(h) E_x[F\{-u(x, k) \mid x > h\}] \tag{15.28}$$

Since $E(x|x > h) = f(h)/G(h) = \mu(h)$, say, the expectation in (15.28) may be replaced by a Taylor series expansion. Restricting attention to the leading term, Cox and Wermuth (1991) and Olson and Weissfeld (1991) suggest the approximation

$$L \doteq G(h)\, G(u^*) \tag{15.29}$$

where u^* is given by (15.26) with x replaced by $\mu(h)$.

Approximation (15.29) applies for $\rho > 0$; for $\rho < 0$ we may use the relationship

$$L(h, k, \rho) = G(a) - L(h, -k, -\rho). \tag{15.30}$$

15.7 The approximation (15.29) is more accurate if the variables are ordered so that $h \geqslant k$. The performance tends to deteriorate as $\rho \to 1$ but this difficulty can be ameliorated by using the relationship (cf. Henery, 1981)

$$L(h, k, \rho) = G(h) G[u(h, k)] + L[-u(h, k), k, -(1 - \rho^2)^{1/2}] \tag{15.31}$$

since (15.31) ensures that we may restrict attention to $\rho^2 < 1/2$.

Both Cox and Wermuth (1991) and Olson and Weissfeld (1991) show how the accuracy of the approximation may be improved by using further terms of the Taylor series expansion. They also extend the approximation to the trivariate case.

Wang (1987) has developed an iterative procedure to evaluate binormal probabilities using a discretized version. Drezner (1990) provides approximations for 3- and 4-variate distributions expressed in terms of lower order L-functions such as that defined in (15.27). Other useful relationships among multinormal integrals are given by D. B. Owen (1980).

15.8 A number of inequalities exists for the multinormal. In particular, it may be shown that

$$P(\mathbf{x} > \mathbf{h}) \geqslant \prod_{j=1}^{p} P(x_j > h_j) \tag{15.32}$$

provided $\mathbf{\Sigma}$ has all elements $\sigma_{jk} \geqslant 0$; (15.32) was demonstrated under these conditions by Pitt (1982), earlier authors having proved that (15.32) holds under the more restrictive conditions that all off-diagonal elements of $\boldsymbol{\Sigma}^{-1}$ were $\leqslant 0$. By setting all elements of $\mathbf{h}$ equal to h, it follows that

$$P(M_1 > h) \geqslant a^p \tag{15.33}$$

where $M_1 = \min(x_1, \ldots, x_p)$ and $a = P(x > h)$. Similarly, if $M_2 = \max(x_1, \ldots, x_p)$, it may be shown that

$$P(M_2 \leqslant h) \geqslant (1-a)^p. \tag{15.34}$$

Bølviken (1982) shows that the assumption of normality is critical for (15.32) since the inequality does not hold for the multivariate t distribution.

Steck (1979) provides inequalities for the multivariate Mills' ratio; see Exercise 15.25.

15.9 T. W. Anderson (1955) showed that if the set $\mathbf{C}$ is centrally symmetric (so that $\mathbf{C} \equiv -\mathbf{C}$) and is convex, then

$$P(\mathbf{x} \in \mathbf{C} \mid \boldsymbol{\Sigma}_1) \geqslant P(\mathbf{x} \in \mathbf{C} \mid \boldsymbol{\Sigma}_2) \tag{15.35}$$

for two multinormals with zero means and covariance matrices $\boldsymbol{\Sigma}_1$, $\boldsymbol{\Sigma}_2$ such that $(\boldsymbol{\Sigma}_2 - \boldsymbol{\Sigma}_1)$ is positive semi-definite.

(15.35) is termed a *concentration* inequality and is particularly useful in comparing vectors of estimators (cf Chapter 17, Volume 2A) where it is often the case that the sampling distributions are asymptotically normal. Eaton and Perlman (1991) provide a number of extensions to (15.35).

Orthant probabilities

15.10 A special problem that frequently arises is the evaluation of the multinormal integral over the range for which all p standardized variates are positive. Although this is a considerable simplification of the general problem, it still presents analytical difficulties for $p > 3$, and a number of special devices have been used to overcome these in particular cases.

We deal first with the two- and three-variate cases, the only ones for which analytical solutions of the problem are known.

In the bivariate case, we write the density function as the Fourier transform (4.31) of its c.f. (15.20) and evaluate its integral over the positive quadrant as

$$P_2^0 = \int_0^\infty \int_0^\infty \left\{ \frac{1}{(2\pi)^2} \int_{-\infty}^\infty \int_{-\infty}^\infty \exp\left(-i\mathbf{t}'\mathbf{x} - \tfrac{1}{2}\mathbf{t}'\boldsymbol{\Sigma}\mathbf{t}\right) \mathrm{d}t_1\, \mathrm{d}t_2 \right\} \mathrm{d}x_1\, \mathrm{d}x_2, \tag{15.36}$$

where the vectors and matrix are now of order two. Reversing the order of integration in (15.36), we integrate out x_1 and x_2 and obtain

$$4\pi^2 P_2^0 = \int_{-\infty}^\infty \int_{-\infty}^\infty \exp\left(-\tfrac{1}{2}\mathbf{t}'\boldsymbol{\Sigma}\mathbf{t}\right) \frac{\mathrm{d}t_1}{it_1} \frac{\mathrm{d}t_2}{it_2}.$$

The coefficient of $(-t_1 t_2)$ in the exponent of the integrand is the covariance of the two variates, which is (see Example 15.1) $\rho\sigma_1\sigma_2$. Differentiating with respect to ρ, we obtain

$$4\pi^2 \frac{\partial P_2^0}{\sigma_1\sigma_2\,\partial\rho} = \int_{-\infty}^\infty \int_{-\infty}^\infty \exp\left(-\tfrac{1}{2}\mathbf{t}'\boldsymbol{\Sigma}\mathbf{t}\right) \mathrm{d}t_1\, \mathrm{d}t_2. \tag{15.37}$$

The integral on the right of (15.37) is that of a binormal density in t_1 and t_2 apart from the constant, and therefore equals the reciprocal of that constant, so that (15.37) becomes

$$\frac{\partial P_2^0}{\sigma_1\sigma_2\partial\rho} = \frac{1}{4\pi^2}\cdot\frac{2\pi}{|\Sigma|^{\frac{1}{2}}} = \frac{1}{2\pi|\Sigma|^{\frac{1}{2}}} = \{2\pi\sigma_1\sigma_2(1-\rho^2)^{\frac{1}{2}}\}^{-1},$$

from Example 15.1. On integration, this gives $P_2^0 = \arcsin\rho/(2\pi) +$ a constant, and the constant is seen to be $\frac{1}{4}$ by putting $\rho = 0$. Thus

$$P_2^0 = \tfrac{1}{4} + \arcsin\rho/(2\pi). \tag{15.38}$$

(15.38) is usually called Sheppard's (1898) theorem on median dichotomy, although it was given earlier by Stieltjes. By symmetry, the same result holds for the negative quadrant, while the two "mixed" quadrants have content $(\frac{1}{2} - P_2^0)$ each. As is to be expected, P_2^0 does not depend on the variances, which cannot affect the proportion of the total probability in a quadrant. P_2^0 is tabulated by National Bureau of Standards (1959)—cf. reference in **15.5**—for $\rho = 0(0.01)1$.

For a simpler derivation, see Stigler (1989).

15.11 The corresponding three-variate result is easily derived from (15.38).

We have seen (Exercise 8.1) that the probability of occurrence of at least one of a set of p events is

$$\sum P_1 - \sum P_2 + \sum P_3 - \ldots + (-1)^{p-1}P_p, \tag{15.39}$$

the summations being of the probabilities of occurrence of the events singly, in pairs, and so on to the last term, which contains the probability that all p events occur jointly.

If these p events are the positiveness of p multinormal variates, the probability that at least one is positive is the complement of the probability that none is positive, and, by symmetry, also the complement of the probability that none is negative, which is what we seek. Using (15.39) and writing the zero superscript as before, we therefore have

$$1 - P_p^0 = \sum P_1^0 - \sum P_2^0 + \sum P_3^0 \ldots + (-1)^{p-1}P_p^0. \tag{15.40}$$

If p is even, the last term on the right has a negative sign, and (15.40) is of no use; but if p is odd, it becomes

$$P_p^0 = \frac{1}{2}\left\{1 - \sum P_1^0 + \sum P_2^0 \ldots + (-1)^{p-1}\sum P_{p-1}^0\right\}. \tag{15.41}$$

(15.41) enables us to obtain P_p^0 for odd p from lower-order probabilities.

With $p = 3$, (15.41) gives

$$P_3^0 = \frac{1}{2}\left(1 - \sum P_1^0 + \sum P_2^0\right) = \frac{1}{2}\left\{1 - 3\cdot\tfrac{1}{2} + \sum\left(\tfrac{1}{4} + \arcsin\rho/(2\pi)\right)\right\}$$

$$= \frac{1}{8} + \frac{1}{4\pi}(\arcsin\rho_{12} + \arcsin\rho_{13} + \arcsin\rho_{23}), \tag{15.42}$$

a simple generalization of Sheppard's formula.

15.12 No such simple result is available for $p > 3$. Abrahamson (1964) gives tables for calculating P_4^0—some special covariance structures are treated by Cheng (1969). Childs (1967) reduces P_4^0 to the evaluation of three single integrals and also obtains results for P_6^0; these results are extended by Sun (1988).

For general p, Kendall (1941) gave a power series expansion in the covariance parameters. The Fourier inversion of the c.f. (cf. (15.36)) can be expanded as a power series in the correlations ρ_{ij}. Thus

$$P_p^0 = \int_0^\infty \cdots \int_0^\infty \left\{ \frac{1}{(2\pi)^p} \int_{-\infty}^\infty \cdots \int_{-\infty}^\infty \exp\left(-i\mathbf{t}'\mathbf{x} - \tfrac{1}{2}\mathbf{t}'\boldsymbol{\Sigma}\mathbf{t}\right) \mathrm{d}t_1 \ldots \mathrm{d}t_p \right\} \mathrm{d}x_1 \ldots \mathrm{d}x_p$$

$$= \int_0^\infty \cdots \int_0^\infty \left\{ \frac{1}{(2\pi)^p} \int_{-\infty}^\infty \cdots \int_{-\infty}^\infty \exp\left(-i\mathbf{t}'\mathbf{x} - \tfrac{1}{2}\mathbf{t}'\mathbf{t}\right) \cdot \sum \left[(-1)^{n_{..}} \frac{\rho_{12}^{n_{12}} \rho_{13}^{n_{13}} \cdots}{n_{12}!\, n_{13}! \ldots} \right] \right.$$

$$\left. \times\, t_1^{n_{1.}} t_2^{n_{2.}} \ldots t_p^{n_{p.}}\, \mathrm{d}t_1 \ldots \mathrm{d}t_p \right\} \mathrm{d}x_1 \ldots \mathrm{d}x_p, \tag{15.43}$$

where the summation is over all possible sets of the ρ_{ij} taken over all non-negative values of the n_{ij}; $n_{i.} = \sum_j (n_{ij} + n_{ji})$ and $n_{..} = \sum_i n_{i.}$.

Interchanging summation and integration, we may rewrite (15.43) as

$$P_p^0 = \sum \left[(-1)^{n_{..}} \frac{\rho_{12}^{n_{12}} \rho_{13}^{n_{13}} \cdots}{n_{12}!\, n_{13}! \ldots} \right] \prod_{j=1}^{p} G_{n_{j.}}, \tag{15.44}$$

where

$$G_{n_{j.}} = \frac{1}{2\pi} \int_0^\infty \int_{-\infty}^\infty t_j^{n_{j.}} \exp\left(-it_j x_j - \tfrac{1}{2} t_j^2\right) \mathrm{d}t_j\, \mathrm{d}x_j. \tag{15.45}$$

If $n_{j.}$ is a positive integer, we integrate for x_j as at (15.36) to obtain

$$G_{n_{j.}} = \frac{1}{2\pi i} \int_{-\infty}^\infty t_j^{n_{j.}-1} e^{-\frac{1}{2}t_j^2}\, \mathrm{d}t_j,$$

and by the formulae for the moments of the univariate normal distribution (5.97), we have, since $G_{n_{j.}} = \frac{1}{2}$ if $n_{j.} = 0$ by (15.45),

$$G_{n_{j.}} = \begin{cases} \frac{1}{2} & \text{if } n_{j.} = 0 \\ 0 & \text{if } n_{j.} = 2m \quad (m = 1, 2, \ldots) \\ \dfrac{(2m)!}{\sqrt{(2\pi)}\, i\, 2^m m!} & \text{if } n_{j.} = 2m+1 \quad (m = 0, 1, 2, \ldots). \end{cases} \tag{15.46}$$

Thus partitions vanish from (15.44) which contain any t_i an even number of times.

(15.44) and (15.46) formally solve the problem. But although the series (15.44) for P_p^0 always converges, it does so only slowly if the ρ_{ij} are not small. Moreover, the terms in (15.44) are not all positive. In the four-variate case (cf. Moran (1948)), (15.44) becomes

$$P_4^0 = \frac{1}{16} + \frac{1}{8\pi}\sum_{i,j}\rho_{ij} + \frac{1}{4\pi^2}\sum_{ijkl}\rho_{ij}\rho_{kl} + \ldots$$

but the coefficient of terms such as $\rho_{12}\rho_{13}\rho_{14}$ is negative, being $-1/(4\pi^2)$.

For higher order p, Moran (1984) provides a transformation to polar co-ordinates that enables the Monte Carlo evaluation of the orthant probabilities to be performed more efficiently.

15.13 In the special case when all the correlations are equal, a remarkable simplification is possible.

Let $x_0, x_1, \ldots, x_p$ be a set of $p+1$ independent standardized normal variates. Consider the derived set of variables

$$y_i = x_i - bx_0, \qquad i = 1, \ldots, p, \qquad b > 0. \tag{15.47}$$

These will be multinormally distributed with zero means, variances $(1+b^2)$ and covariances b^2. The correlation between any pair of ys is thus

$$\rho = b^2/(1+b^2).$$

This ranges from zero when b is zero to 1 when $b \to \infty$. The y_i can thus be taken to represent the general multinormal distribution with equal (positive) correlations. The value of P_p^0 is immediately deducible from the construction of the distribution; it is simply

$$P_p^0 = \int_{-\infty}^{\infty} (2\pi)^{-\frac{1}{2}} e^{-\frac{1}{2}x^2} \left\{ \int_{bx}^{\infty} (2\pi)^{-\frac{1}{2}} e^{-\frac{1}{2}t^2} dt \right\}^p dx, \tag{15.48}$$

the probability that p independent normal variates exceed bx, integrated over a standard normal distribution. This is an integral of the Hojo type (cf. **14.5**) and can be simply calculated with little error by replacing the integration by a summation. Evidently, the argument leading to the result is basically unaffected if some y_i are defined by $y_i = x_i + bx_0$ so that some of the y_i have correlations $-\rho$ with the others. In this case, of course, (15.48) must be suitably modified.

Ruben (1954) has tabulated P_p^0 in the equal-correlations case for $\rho = 1/j$, $p = 1, 2, \ldots, 51-j$, where $j = 2, 3, \ldots, 12$.

The case of equal negative correlations is covered by the generalization of (15.47) given in Exercise 15.11, which deals with the multinormal order statistics for any ρ. Gupta *et al.* (1973) tabulate the upper percentiles of $y_{(p)}$ for 17 positive values of ρ and $p = 1(1)10(2)50$. Odeh (1982b) gives tables of two-sided percentage points for the same values of ρ and $p = 2(1)40(2)50$.

G. P. Steck (1962) reviews the equal-correlations case and gives new results, especially for $p = 4$. See also Bacon (1963), S. S. Gupta (1963a), Dutt (1973), Bechhofer and Tamhane (1974) and Six (1981).

Exercise 15.12 generalizes (15.48) to the case $\rho_{ij} = a_i a_j$.

Example 15.2 (The evaluation of orthant probabilities)

To illustrate the approximation of the outer integral in (15.48) by a sum, we consider the case $p=2$ when $b=1$ in (15.47), i.e. $\rho=\frac{1}{2}$. Replacing the integral by the sum of the values at $x=\pm 3, \pm 2, \pm 1, 0$, we find from Appendix Tables 1 and 2 the following results:

x	$(1)=\int_x^\infty (2\pi)^{-\frac{1}{2}}e^{-\frac{1}{2}t^2}\,dt$	$(2)=\{(1)\}^2$	(3) = Sums *of* (2)	$(4)=(2\pi)^{-\frac{1}{2}}e^{-\frac{1}{2}x^2}$	(3) × (4)
−3 +3	0.998 65 0.001 35	0.997 30 0.000 00	0.997 30	0.004 43	0.004 42
−2 +2	0.977 25 0.022 75	0.955 02 0.000 52	0.955 54	0.053 99	0.051 59
−1 +1	0.841 34 0.158 66	0.707 85 0.025 17	0.733 02	0.241 97	0.177 37
0	0.500 00	0.250 00	0.250 00	0.398 94	0.099 73
					0.333 11

The value obtained is 0.3331 to 4 d.p. The exact value, given by (15.38), is $\frac{1}{3}$. Greater accuracy could have been obtained by using a sum of more than seven values. For example, if we add in a further pair of terms for $x=\pm 4$, they contribute 0.000 13 to the sum, giving a total of 0.333 24. To 5 d.p., no further gain in accuracy results from adding terms for $x=\pm 5$, but accuracy as great as is desired could be obtained by spacing the values at narrower intervals.

Moran (1956) gives a bound for the relative error in the sum compared to the integral. It is, in our notation

$$2\exp\left\{-\frac{\pi^2}{h^2(1+pb^2)}\right\},$$

where h is the interval of summation. In our example, this is

$$2\exp\left(-\frac{\pi^2}{3h^2}\right).$$

With $h=1$, this gives a bound considerably greater than the error obtained above with nine values. With smaller h, the bound decreases sharply and also seems to be closer to the actual error.

Quadratic forms in normal variates

15.14 Suppose that $\mathbf{x}$ is distributed with the multinormal density

$$f(\mathbf{x}) \propto \exp(-\tfrac{1}{2}\mathbf{x}'\mathbf{A}\mathbf{x})$$

where **A** is not necessarily non-singular. Consider the distribution of the quadratic form $\mathbf{x}'\mathbf{A}\mathbf{x}$ itself. Its characteristic function is, writing $\theta = it$,

$$\phi(t) \propto \int_{-\infty}^{\infty} \cdots \int_{-\infty}^{\infty} \exp\{-\tfrac{1}{2}\mathbf{x}'\mathbf{A}\mathbf{x}(1-2\theta)\} \prod_j dx_j. \tag{15.49}$$

The transformation (15.3) orthogonalizes the quadratic form as in **15.2** and (15.49) becomes

$$\phi(t) \propto \int_{-\infty}^{\infty} \cdots \int_{-\infty}^{\infty} \exp\{-\tfrac{1}{2}\mathbf{y}'\mathbf{C}\mathbf{y}(1-2\theta)\} \prod_j dy_j. \tag{15.50}$$

The number of non-zero diagonal elements in **C** is the number of non-zero latent roots of **A**, say r, so (15.50) becomes an r-fold integral, to which we apply the transformation

$$z_j = y_j c_{jj}^{\frac{1}{2}}(1-2\theta)^{\frac{1}{2}}$$

to obtain

$$\phi(t) \propto (1-2\theta)^{-\frac{1}{2}r} \int_{-\infty}^{\infty} \cdots \int_{-\infty}^{\infty} \exp(-\tfrac{1}{2}\mathbf{z}'\mathbf{z}) \prod_j dz_j. \tag{15.51}$$

The integral in (15.51) is independent of θ. Thus

$$\phi(t) = (1-2\theta)^{-\frac{1}{2}r}, \tag{15.52}$$

the elimination of constants being verified by the fact that $\phi(0) = 1$. By the Inversion Theorem and Example 3.6, (15.52) shows that the quadratic form in the exponent of a multinormal distribution has a distribution known as the χ^2 distribution with r degrees of freedom, defined by (11.8) with $n = r$ (see also Chapter 16), where r is the rank (of the quadratic form) of the distribution. This generalizes the result of (11.8), which deals effectively with the case where **A** is diagonal and non-singular.

Example 15.3 (The χ^2 goodness-of-fit distribution)

Given a sample of n observations from a multinomial distribution (7.7) with probabilities p_i $(i = 1, \ldots, k)$, it follows from the multivariate form of the Central Limit theorem that the observed numbers x_i in the k groups will tend, with increasing n, to have a multinormal distribution, with means np_i and covariance matrix given (as at (7.18)) by

$$\boldsymbol{\Sigma} = n\begin{pmatrix} p_1(1-p_1) & -p_1p_2 & -p_1p_3 & \cdots & -p_1p_k \\ -p_2p_1 & p_2(1-p_2) & -p_2p_3 & \cdots & -p_2p_k \\ \vdots & \vdots & \vdots & \vdots & \vdots \\ -p_kp_1 & -p_kp_2 & \cdots & \cdots & p_k(1-p_k) \end{pmatrix}$$

Because of the linear restriction $\sum_{j=1}^{k} x_j = n$, the multinormal distribution is singular, of rank $(k-1)$, so $|\boldsymbol{\Sigma}| = 0$ and we cannot invert $\boldsymbol{\Sigma}$. However, we may omit one of the x_i (say x_k), as it is redundant, the remaining $(k-1)$ x_i being multinormally distributed with means np_i and covariance matrix $\boldsymbol{\Sigma}^*$ which is simply $\boldsymbol{\Sigma}$ with its last row and column eliminated. It is easily verified by a direct matrix multiplication that

$$(\boldsymbol{\Sigma}^*)^{-1} = \frac{1}{n}\begin{pmatrix} \frac{1}{p_1}+\frac{1}{p_k} & \frac{1}{p_k} & \frac{1}{p_k} & \cdots & \frac{1}{p_k} \\ \frac{1}{p_k} & \frac{1}{p_2}+\frac{1}{p_k} & \frac{1}{p_k} & \cdots & \frac{1}{p_k} \\ \vdots & \vdots & \vdots & \vdots\vdots\vdots & \vdots \\ \frac{1}{p_k} & \frac{1}{p_k} & \frac{1}{p_k} & \cdots & \frac{1}{p_{k-1}}+\frac{1}{p_k} \end{pmatrix}.$$

The exponent of the multinormal distribution thus has the quadratic form

$$\begin{aligned}(\mathbf{x}-\boldsymbol{\mu})'(\boldsymbol{\Sigma}^*)^{-1}(\mathbf{x}-\boldsymbol{\mu}) &= \sum_{i=1}^{k-1}\frac{(x_i-np_i)^2}{np_i}+\frac{1}{np_k}\sum_{i=1}^{k-1}\sum_{j=1}^{k-1}(x_i-np_i)(x_j-np_j) \\ &= \sum_{i=1}^{k-1}\frac{(x_i-np_i)^2}{np_i}+\frac{1}{np_k}\left\{\sum_{i=1}^{k-1}(x_i-np_i)\right\}^2 \\ &= \sum_{i=1}^{k}\frac{(x_i-np_i)^2}{np_i}.\end{aligned}$$

By the result of **15.14**, this variate is distributed like χ^2 with $(k-1)$ degrees of freedom.

This result, due to Karl Pearson (1900), is the basis of the χ^2 goodness-of-fit test which we shall discuss in Chapter 30, Volume 2A. Here, we observe that for the data of Table 1.4 discussed in this context in **9.13** we find that the quadratic form equals $(26^2+107^2+\ldots+147^2)/1000 = 58.5$, a very extreme value for a χ^2 distribution with 9 degrees of freedom—we shall see in **16.5** that this is more than 10 s.d. above the mean.

15.15 We now suppose that $\mathbf{x}$ is a vector of p independent standardized normal variates, and consider the distribution of the general quadratic form

$$Q = \mathbf{x}'\mathbf{A}\mathbf{x},$$

where the real matrix $\mathbf{A}$ may be assumed symmetric without loss of generality. By the orthogonal transformation (15.3), $\mathbf{x}$ is transformed to $\mathbf{y}$, a vector of independent standardized normal variates by Example 11.2, and using (15.6)

$$Q = \mathbf{y}'\mathbf{C}\mathbf{y} = \sum_{i=1}^{p} a_i y_i^2, \tag{15.53}$$

where the a_i are latent roots of $\mathbf{A}$—only the r ($\leqslant p$) non-zero a_i contribute to (15.53), where r is the rank of $\mathbf{A}$, so that the distribution of Q depends only upon them. The c.g.f. of $a_i y_i^2$, a constant multiple of a χ^2 distribution with 1 d.fr., is (cf. **15.14** and **16.3** below)

$$\psi_i(t) = -\tfrac{1}{2}\log(1-2\theta a_i), \tag{15.54}$$

where $\theta = it$. By the additive property of c.g.f.s, that of Q is therefore

$$\psi(t) = -\tfrac{1}{2}\sum_{i=1}^{p}\log(1-2\theta a_i). \tag{15.55}$$

Expanding the logarithm in series near $t = 0$, it follows that the cumulants of Q are

$$\kappa_s = 2^{s-1}(s-1)! \sum_{i=1}^{p} a_i^s. \tag{15.56}$$

Now $\sum_{i=1}^{p} a_i^s$ is the trace (i.e. the sum of elements in the leading diagonal) of the sth power of **A**, and we denote this by $\text{tr}\,\mathbf{A}^s$. Thus (15.56) is

$$\kappa_s = 2^{s-1}(s-1)!\,\text{tr}\,\mathbf{A}^s \tag{15.57}$$

and the c.g.f. (15.55) may therefore be written in the form

$$\psi(t) = \sum_{s=1}^{\infty} 2^{s-1}\,\text{tr}\,(\theta\mathbf{A})^s/s. \tag{15.58}$$

The c.f. is, using (15.55),

$$\phi(t) = E\{\exp(\mathbf{x}'\theta\mathbf{A}\mathbf{x})\} = \prod_{i=1}^{p}(1-2\theta a_i)^{-\frac{1}{2}}$$

$$= |\mathbf{I} - 2\theta\mathbf{A}|^{-\frac{1}{2}}, \tag{15.59}$$

since $1 - 2\theta a_i$ is a latent root of $\mathbf{I} - 2\theta\mathbf{A}$ and the determinant of a matrix is the product of its latent roots.

If (15.59) is to be of the form (15.52) of the c.f. of a χ^2 distribution with ν degrees of freedom, we must therefore have, identically in θ,

$$\prod_{i=1}^{p}(1-2\theta a_i) = (1-2\theta)^{\nu}.$$

Thus ν of the a_i must be equal to 1, $\nu \leqslant p$, and the other $p - \nu$ equal to 0, whence ν must equal r, the rank of **A**, and (15.56) becomes $\kappa_s = 2^s(s-1)!r$.

From (15.6), $\mathbf{A} = \mathbf{BCB}'$, so $\mathbf{A}^2 = \mathbf{BCB}'\mathbf{BCB}' = \mathbf{BC}^2\mathbf{B}' = \mathbf{BCB}'$ since **C** is a diagonal matrix containing only elements 0 and 1. Thus $\mathbf{A} = \mathbf{A}^2$. Conversely, $\mathbf{A} = \mathbf{A}^2$ implies that r of the $a_i = 1$ and the others 0. A matrix $\mathbf{A} = \mathbf{A}^2$ is called *idempotent*. We have established that Q is distributed as χ^2 with r d.fr. if and only if **A** is idempotent of rank r.

Series representations of the distribution of Q for any positive definite **A** are unified by Kotz *et al.* (1967a, b), who use the more general assumptions of **15.25** below. See also Gideon and Gurland (1976).

Tables of the distribution of Q were given by Solomon (1960) for $r = 2$ and 3 and by Johnson and Kotz (1967a, b, 1968) for $r = 4$ and 5. B. K. Shah (in Shah and Odeh (1986)) gives tables for $r = 1(1)10$ that permit interpolation accurate to within 1 percent. A. W. Davis (1977) gives a method of computation using differential equations and J. J. Walker (1979) one that is useful when the a_i differ widely. Sankaran (1959) and Jensen and Solomon (1972) used the method of **16.7** below, as generalized by Haldane (1938), to find the fractional power h of Q that is most closely normally distributed and obtained $h = 1 - 2\,\text{tr}\,\mathbf{A}\,\text{tr}\,\mathbf{A}^3/\{3((\text{tr}\,\mathbf{A}^2)^2\}$, reducing to $h = \frac{1}{3}$ when **A** is idempotent and distributed as χ^2, agreeing with **16.7**.

Solomon and Stephens (1977, 1978) give new approximations using the moments of Q, and tabulate some percentage points for $r = 6$, 8, 10. A good approximation is obtained by

equating the first three moments of Q with those of a multiple of a χ^2 variate; indeed, the simple approximation treating $(r/\sum a_i)Q$ as a χ_r^2 variate (so that Q has the correct mean (15.57)) is remarkably effective, variation in the a_i having surprisingly little effect on the value of the d.f. at any fixed point.

If the a_i are equal in pairs, (15.59) becomes $\phi(t)=\prod_{i=1}^{\frac{1}{2}p}(1-2\theta a_i)^{-1}$, which may be expanded in partial fractions as $\phi(t)=\sum_{i=1}^{\frac{1}{2}p}c_i(1-2\theta a_i)^{-1}$, whence the Inversion Theorem for c.f.s gives $\text{Prob}\{Q>d\}=\sum_{i=1}^{\frac{1}{2}p}c_i\,\text{Prob}\{z_i>d/a_i\}$, where the z_i are independent χ^2 variates with 2 d.fr., so that $\frac{1}{2}z_i$ are independent exponential variates. A more general result of this kind is given in Exercise 15.29.

15.16 If we now turn to the joint distribution of two real quadratic forms

$$Q_1=\mathbf{x}'\mathbf{A}\mathbf{x},\qquad Q_2=\mathbf{x}'\mathbf{B}\mathbf{x},$$

we see at once that its c.f. $\phi(t_1,t_2)=E\{\exp[\mathbf{x}'(\theta_1\mathbf{A}+\theta_2\mathbf{B})\mathbf{x}]\}$ is simply (15.59) with $\theta\mathbf{A}$ replaced by $(\theta_1\mathbf{A}+\theta_2\mathbf{B})$. Thus we have

$$\phi(t_1,t_2)=|\mathbf{I}-2\theta_1\mathbf{A}-2\theta_2\mathbf{B}|^{-\frac{1}{2}} \tag{15.60}$$

and for the c.g.f. similarly, from (15.58),

$$\psi(t_1,t_2)=\sum_{s=1}^{\infty}2^{s-1}\,\text{tr}\,(\theta_1\mathbf{A}+\theta_2\mathbf{B})^s/s. \tag{15.61}$$

In particular, (15.61) gives with $s=2$, 4 respectively the product cumulants

$$\left.\begin{aligned}\kappa_{11}&=\text{tr}\,(\mathbf{AB}+\mathbf{BA})=2\,\text{tr}\,\mathbf{AB},\\ \kappa_{22}&=8\,\text{tr}\,\{\mathbf{A}^2\mathbf{B}^2+(\mathbf{AB})^2+\mathbf{AB}^2\mathbf{A}+\mathbf{BA}^2\mathbf{B}+(\mathbf{BA})^2+\mathbf{B}^2\mathbf{A}^2\}\\ &=16\,\text{tr}\,\{2\mathbf{A}^2\mathbf{B}^2+(\mathbf{AB})^2\}.\end{aligned}\right\} \tag{15.62}$$

From **4.16–17**, Q_1 and Q_2 will be independently distributed if and only if their joint c.f. factorizes into their individual c.f.s. From (15.59–60), this condition is

$$|\mathbf{I}-2\theta_1\mathbf{A}-2\theta_2\mathbf{B}|=|\mathbf{I}-2\theta_1\mathbf{A}|\,|\mathbf{I}-2\theta_2\mathbf{B}|, \tag{15.63}$$

a result due to Cochran (1934).

15.17 The right-hand side of (15.63) is equal to

$$|\mathbf{I}-2\theta_1\mathbf{A}-2\theta_2\mathbf{B}+4\theta_1\theta_2\mathbf{AB}| \tag{15.64}$$

and this is equal to the left-hand side of (15.63) if

$$\mathbf{AB}=\mathbf{0}. \tag{15.65}$$

The converse is also true, as we shall see below. This is A. T. Craig's (1943) theorem: Q_1 and Q_2 are independent if and only if (15.65) holds.

Before we prove the converse, we observe that (15.65) implies

$$\text{tr}\,(\mathbf{A}^2\mathbf{B}^2) = \text{tr}\,(\mathbf{A}\cdot\mathbf{AB}\cdot\mathbf{B}) = 0, \tag{15.66}$$

and it is easy to see that (15.66) also implies (15.65), for we always have

$$\text{tr}\,(\mathbf{A}^2\mathbf{B}^2) = \text{tr}\,(\mathbf{BA}\cdot\mathbf{AB}) = \text{tr}\,\{(\mathbf{AB})'\mathbf{AB}\} \geqslant 0,$$

since tr $\{(\mathbf{AB})'\mathbf{AB}\}$ is the sum of the squares of the elements of **AB**. If this sum is zero, all the elements of **AB** must also be zero. Hence (15.66) is equivalent to (15.65), and is itself a necessary and sufficient condition for the independence of Q_1 and Q_2, due to Lancaster (1954).

Now the converse of Craig's theorem follows easily, for if Q_1 and Q_2 are independent, their product cumulants are zero by Example 12.7, and in particular (15.62) gives

$$\kappa_{22} = 16\,\text{tr}\,\{2\mathbf{A}^2\mathbf{B}^2 + (\mathbf{AB})^2\} = 0.$$

Thus we may write

$$0 = 2\,\text{tr}\,\{2\mathbf{A}^2\mathbf{B}^2 + (\mathbf{AB})^2\} = \text{tr}\,(\mathbf{AB}+\mathbf{BA})^2 + 2\,\text{tr}\,\{(\mathbf{AB})'\mathbf{AB}\}.$$

Since **A** and **B** are symmetric, so is (**AB** + **BA**), and its square has non-negative real latent roots and trace; and we have seen above that the trace of $(\mathbf{AB})'\mathbf{AB} \geqslant 0$. Thus both traces on the right must be zero and (15.66) must hold, a proof essentially given by Ogasawara and Takahashi (1951).

15.18 Further, since **A** and **B** are symmetric, we may write $\mathbf{A} = \mathbf{TT}'$, $\mathbf{B} = \mathbf{UU}'$, and $\mathbf{AB} = \mathbf{TT}'\mathbf{UU}'$. Hence

$$\begin{aligned}\text{tr}\,(\mathbf{AB}) &= \text{tr}\,(\mathbf{TT}'\mathbf{U}\cdot\mathbf{U}') = \text{t}\,(\mathbf{U}'\mathbf{T}\cdot\mathbf{T}'\mathbf{U})\\ &= \text{tr}\,\{(\mathbf{T}'\mathbf{U})'(\mathbf{T}'\mathbf{U})\}.\end{aligned}$$

The leading diagonal of $(\mathbf{T}'\mathbf{U}')'(\mathbf{T}'\mathbf{U})$ contains only the squares of all the elements of $(\mathbf{T}'\mathbf{U})$, but these are not necessarily all real. But if **A** and **B** are *non-negative* forms, the elements are all real, and hence

$$\text{tr}\,(\mathbf{AB}) = \text{tr}\,\{(\mathbf{T}'\mathbf{U})'(\mathbf{T}'\mathbf{U})\} = 0 \tag{15.67}$$

implies $\mathbf{T}'\mathbf{U} = \mathbf{0}$ and hence $\mathbf{AB} = \mathbf{T}\cdot\mathbf{T}'\mathbf{U}\cdot\mathbf{U}' = \mathbf{0}$.

Thus for non-negative quadratic forms, (15.67) implies (15.65) and independence, a result due to Matérn (1949). From (15.62), we therefore see that for non-negative quadratic forms, $\kappa_{11} = 0$ is equivalent to independence.

In particular, **A** and **B** are non-negative if Q_1 and Q_2 are χ^2 variates. In this case, the equivalence of (15.67) and (15.65) may be seen directly since **A**, **B** are idempotent and $\mathbf{AB} = \mathbf{A}^2\mathbf{B}^2$, so that (15.67) implies (15.66) and hence (15.65) and independence.

Example 15.4 (The independence of sample mean and variance in normal samples)

We have seen (Examples 11.3, 11.7) that in samples of size n from a standardized univariate normal distribution, the sample mean $\bar{x}$ is itself normally distributed with mean 0 and variance $1/n$, and the sample sum of squares $\sum_{i=1}^{n}(x_i-\bar{x})^2$ is distributed like χ^2 with $(n-1)$ degrees of freedom. It follows that $n\bar{x}^2$ is also distributed like χ^2 with 1 degree of freedom. The quadratic forms $n\bar{x}^2$ and $\sum_{i=1}^{n}(x_i-\bar{x})^2$ have matrices

$$\mathbf{A}=\frac{1}{n}\mathbf{U},\qquad \mathbf{B}=\mathbf{I}-\frac{1}{n}\mathbf{U},$$

where $\mathbf{U}$ is the matrix with unity for each of its elements. Thus

$$\text{tr}\,(\mathbf{AB})=\text{tr}\,(\mathbf{A}-\mathbf{A}^2)=0$$

by the idempotency of χ^2 variates. Hence, from (15.67), $n\bar{x}^2$ and $\sum_{i=1}^{n}(x_i-\bar{x})^2$ are independently distributed, as we have already seen in Example 11.3, where we obtained the result in a slightly different form.

15.19 As a consequence of Craig's condition (15.65), we may obtain a necessary and sufficient condition for the independence of a quadratic form $Q=\mathbf{x}'\mathbf{A}\mathbf{x}$ and a linear form $L=\mathbf{b}'\mathbf{x}$. For Q and L are independent if and only if Q and L^2 are independent, i.e. if $\mathbf{x}'\mathbf{A}\mathbf{x}$ and $(\mathbf{b}'\mathbf{x})'\mathbf{b}'\mathbf{x}$ are. From (15.65), this will be so if and only if

$$\mathbf{Abb}'=\mathbf{0}. \tag{15.68}$$

From (15.68) it follows that

$$\mathbf{Abb}'\mathbf{b}=\mathbf{0}, \tag{15.69}$$

and as $\mathbf{b}'\mathbf{b}$ is a non-zero scalar, we may drop it from (15.69) to give

$$\mathbf{Ab}=\mathbf{0}. \tag{15.70}$$

This establishes the necessity of (15.70) for (15.68). Its sufficiency follows immediately.

Example 15.5 (The independence of sample mean and variance using Craig's result)

We may apply (15.70) to obtain the result of Example 15.4. Here the quadratic form is, using $\mathbf{1}$ to denote a vector of ones,

$$\sum_{i=1}^{n}(x_i-\bar{x})^2=\mathbf{x}'\left(\mathbf{I}-\frac{1}{n}\mathbf{U}\right)\mathbf{x},$$

and the linear form is $\bar{x}=(1/n)\mathbf{1}'\mathbf{x}$. The product of these matrices is

$$\left(\mathbf{I}-\frac{1}{n}\mathbf{U}\right)\cdot\frac{1}{n}\mathbf{1}=\frac{1}{n}\mathbf{1}-\frac{1}{n^2}\cdot n\mathbf{1}=\mathbf{0},$$

so (15.70) is satisfied and $\sum_{i=1}^{n}(x_i-\bar{x})^2$ and $\bar{x}$ are independent.

The decomposition of quadratic forms in independent normal variates

15.20 We now prove a remarkable result due to Cochran (1934), whose fundamental theorem has been amplified by James (1952) and Lancaster (1954). With $\mathbf{x}$ a vector of p independent standardized normal variates as before, suppose that the sum of squares $Q = \mathbf{x}'\mathbf{x}$ is decomposed into k quadratic forms $Q_i = \mathbf{x}'\mathbf{A}_i\mathbf{x}$ with ranks r_i, i.e. that

$$\sum_{i=1}^{k} \mathbf{x}'\mathbf{A}_i\mathbf{x} = \mathbf{x}'\mathbf{I}\mathbf{x}. \tag{15.71}$$

Then any one of the following three conditions implies the other two:

(a) The ranks r_i of the Q_i add to that of Q.
(b) Each of the Q_i is distributed like χ^2.
(c) Each Q_i is independent of every other.

15.21 First, we deduce (b) from (a) and conversely.

Select an arbitrary Q_i, say Q_1. If we make an orthogonal transformation (15.3) that diagonalizes $\mathbf{A}_1$, we obtain

$$\mathbf{y}'\mathbf{B}'\mathbf{A}_1\mathbf{B}\mathbf{y} + \mathbf{y}'\mathbf{B}'(\mathbf{I} - \mathbf{A}_1)\mathbf{B}\mathbf{y} = \mathbf{y}'\mathbf{I}\mathbf{y}, \tag{15.72}$$

the identity matrix, and the ranks on the left, remaining invariant under orthogonal transformation. Since the first and last quadratic forms in (15.72) are diagonal, so must the second be. Moreover, since $(p - r_1)$ of the leading diagonal elements of $\mathbf{B}'\mathbf{A}_1\mathbf{B}$ are zero, the corresponding elements of $\mathbf{B}'(\mathbf{I} - \mathbf{A}_1)\mathbf{B} = \mathbf{D}$ are unity since they must add to those of $\mathbf{I}$. If the rank of $\mathbf{D}$ is q, (15.72) shows that $r_1 + q \geqslant p$, and if condition (a) holds, we similarly have $q \leqslant \sum_{i=2}^{k} r_i = p - r_1$. Thus $q = (p - r_1)$, the other elements of the leading diagonal of $\mathbf{D}$ are zero, and the corresponding elements of $\mathbf{B}'\mathbf{A}_1\mathbf{B}$ are unity. Hence, from **15.11**, Q_1 is a χ^2 variable, $\mathbf{A}_1$ being idempotent. The same result holds for the other $\mathbf{A}_i$. Thus we have established (b) from (a).

Further, from (15.71)

$$\mathbf{I} = \sum_i \mathbf{A}_i. \tag{15.73}$$

If (b) holds, $\mathbf{A}_i$ has r_i latent roots of unity and $(p - r_i)$ of zero, and from (15.73) it follows, on taking traces of both sides, that

$$p = \sum_{i=1}^{k} r_i,$$

which establishes (a) from (b).

15.22 Now we deduce (c) from (b) and conversely.

Since the $\mathbf{A}_i$ are idempotent if (b) holds, (15.73) squared gives

$$\mathbf{I} = \sum_i \mathbf{A}_i + \sum_{i \neq j} \mathbf{A}_i\mathbf{A}_j$$

or

$$\sum_{i \neq j} \mathbf{A}_i \mathbf{A}_j = \mathbf{0}. \tag{15.74}$$

Taking traces on both sides of (15.74),

$$\sum_{i \neq j} \operatorname{tr}(\mathbf{A}_i \mathbf{A}_j) = 0. \tag{15.75}$$

But since the $\mathbf{A}_i$ are idempotent,

$$\operatorname{tr}(\mathbf{A}_i \mathbf{A}_j) = \operatorname{tr}(\mathbf{A}_i^2 \mathbf{A}_j^2) \geqslant 0, \qquad i \neq j, \tag{15.76}$$

as we saw below (15.66). Thus, as there, (15.75–6) imply

$$\operatorname{tr}(\mathbf{A}_i^2 \mathbf{A}_j^2) = 0, \qquad (\text{all } i \neq j), \tag{15.77}$$

which implies $\mathbf{A}_i \mathbf{A}_j = \mathbf{0}$ (all $i \neq j$), and the independence of Q_i from Q_j. Thus (c) is established from (b).

Conversely, taking powers of (15.73) gives, if $\mathbf{A}_i \mathbf{A}_j = \mathbf{0}$ for all $i \neq j$,

$$\sum_i \mathbf{A}_i^s = \mathbf{I} \quad (\text{all } s), \tag{15.78}$$

and taking traces of both sides of (15.78) gives

$$\operatorname{tr} \sum \mathbf{A}_i^s = p \quad (\text{all } s). \tag{15.79}$$

(15.79) can only hold if every non-zero latent root of each $\mathbf{A}_i$ is unity, i.e. if each Q_i is distributed like χ^2. This establishes (b) from (c).

15.23 We have thus seen that (a) implies (b) and conversely, and also that (b) implies (c) and conversely. This establishes the sufficiency of any one of the conditions for the other two.

15.24 Generalizing (15.71), suppose now that Q on its right-hand side may be any quadratic form $\mathbf{x}'\mathbf{A}\mathbf{x}$ where $\mathbf{A}$ is idempotent of rank $r \leqslant p$. If we define $P = \mathbf{x}'(\mathbf{I} - \mathbf{A})\mathbf{x}$ we have $Q + P = \mathbf{x}'\mathbf{x}$ and write this as

$$\sum_{i=1}^{k} \mathbf{x}'\mathbf{A}_i\mathbf{x} + \mathbf{x}'(\mathbf{I} - \mathbf{A})\mathbf{x} = \mathbf{x}'\mathbf{I}\mathbf{x},$$

which is precisely of the original form of (15.71), so that **15.20–24** apply here. Now $(\mathbf{I} - \mathbf{A})$ is idempotent since $\mathbf{A}$ is, so by **15.20** P and Q are independent and P has rank $(p - r)$. Also, if each $\mathbf{A}_i \mathbf{A}_j = \mathbf{0}$, we see as at (15.79) that $\operatorname{tr} \sum_{i=1}^{k} \mathbf{A}_i^s = \operatorname{tr} \mathbf{A}^s = r$, all s, so each $\mathbf{A}_i$ is idempotent and $\mathbf{A}_i(\mathbf{I} - \mathbf{A}) = \mathbf{A}_i - \mathbf{A}_i^2 = \mathbf{0}$ also. Thus any one of the conditions (a) to (c) of **15.20** that holds for the k Q_i also holds for the Q_i and P together and hence implies the other two conditions for the Q_i and P, and thus for the Q_i alone. Cochran's theorem is therefore unchanged.

Exercise 15.16 shows that for $k = 2$, idempotency of $\mathbf{A}_1$ and *non-negativity* of $\mathbf{A}_2$ suffice for Cochran's theorem to hold.

It is not true generally that if two variates are distributed like χ^2 and their sum is also distributed like χ^2 with degrees of freedom equal to the sum of their individual degrees of freedom, then the variates are independent: Example 8.14 above is a counter-example. The restriction to quadratic forms in standardized normal variates is crucial here.

Example 15.6 (Joint distribution of the sample mean and variance)

Reverting to the distribution of mean and variance in normal samples discussed in Example 15.4, we see that the decomposition of the sum of squares

$$\sum_{i=1}^{n} (x_i - \bar{x})^2 + n\bar{x}^2 = \sum_{i=1}^{n} x_i^2$$

obeys the conditions for the application of Cochran's theorem. Condition (b) of **15.20** was established in Examples 11.3 and 11.7, and immediately implies the independence of the two sums on the left. It is easily verified that their ranks are $(n-1)$ and 1 respectively (agreeing with the degrees of freedom in their χ^2 distributions), and this satisfaction of condition (a) is enough to imply both their χ^2 distributions and their independence.

15.25 From **15.15** onwards, we have supposed $\mathbf{x}$ to be a vector of independent standardized normal variates, but no essential difference is made to our results if we take $\mathbf{x}$ to be multinormally and non-singularly distributed with mean $\mathbf{0}$ and covariance matrix $\mathbf{\Sigma}$. For since $\mathbf{\Sigma}$ is positive definite, we may write

$$\mathbf{\Sigma} = \mathbf{TT}' \tag{15.80}$$

where $\mathbf{T}$ has real elements, and the transformation $\mathbf{x} = \mathbf{Ty}$ transforms the exponent of the distribution of $\mathbf{x}$ from $\mathbf{x}'\mathbf{\Sigma}^{-1}\mathbf{x}$ to

$$\mathbf{y}'\mathbf{T}'\mathbf{\Sigma}^{-1}\mathbf{Ty} = \mathbf{y}'\mathbf{T}'(\mathbf{T}')^{-1}\mathbf{T}^{-1}\mathbf{Ty} = \mathbf{y}'\mathbf{y},$$

using (15.80), and we may treat the vector of independent variates $\mathbf{y}$ as before. Thus the quadratic forms $\mathbf{x}'\mathbf{Ax}$, $\mathbf{x}'\mathbf{Bx}$ become $\mathbf{y}'\mathbf{T}'\mathbf{ATy}$ and $\mathbf{y}'\mathbf{T}'\mathbf{BTy}$ respectively, and Craig's condition (15.65) becomes, using (15.80),

$$\mathbf{0} = \mathbf{T}'\mathbf{AT}\cdot\mathbf{T}'\mathbf{BT} = \mathbf{T}'\cdot\mathbf{A\Sigma B}\cdot\mathbf{T}$$

or

$$\mathbf{A\Sigma B} = \mathbf{0}, \tag{15.81}$$

a generalization due to Aitken (1950), while (15.70) generalizes as in Exercise 15.13. Similarly, Lancaster's condition (15.66) becomes

$$0 = \text{tr}\,\{(\mathbf{T}'\mathbf{AT})^2(\mathbf{T}'\mathbf{BT})^2\} = \text{tr}\,\{(\mathbf{A\Sigma})^2(\mathbf{B\Sigma})^2\}, \tag{15.82}$$

and Matérn's condition (15.67) for non-negative forms becomes

$$0 = \text{tr}\,\{(\mathbf{T}'\mathbf{AT})(\mathbf{T}'\mathbf{BT})\} = \text{tr}\,(\mathbf{A\Sigma B\Sigma}). \tag{15.83}$$

Similarly, Cochran's theorem holds with obvious modifications (Exercise 15.17).

Exercises 15.14 and 15.18 generalize even further to the singular case.

Subrahmaniam (1966) investigates the effect of non-normality on the distributions of quadratic forms.

There are theorems, analogous to those concerning quadratic forms in normal variables, for linear forms in order statistics from an exponential distribution—see Tanis (1964). Khatri (1977) extends Cochran's theorem to the multivariate case.

Ratios of quadratic forms

15.26 As we shall see in Volume 2A, many test statistics involve ratios of quadratic forms and it is therefore of interest to examine the sampling distributions of such ratios. Here we assume that $\mathbf{x} \sim MVN(\mathbf{0}, \sigma^2 \mathbf{I})$, leaving the *non-central* distributions, where the mean vector may be non-null, to Chapter 23.

If we consider the ratio

$$R = \mathbf{x}'\mathbf{A}\mathbf{x}/\mathbf{x}'\mathbf{B}\mathbf{x}, \tag{15.84}$$

the simplest case arises when $\mathbf{A}$ and $\mathbf{B}$ are idempotent and $\mathbf{AB} = \mathbf{0}$ so that Cochran's theorem holds. R is then distributed as the ratio of two independent χ^2 variates, a topic we pursue in **16.15–21**.

15.27 Suppose now that $\mathbf{A}$ is not idempotent, but $\mathbf{B} = \mathbf{I}$. Pitman (1937b) showed that $Q = \mathbf{x}'\mathbf{x}$ and $h(\mathbf{x})$, where h is any scale-free function of $\mathbf{x}$, are independent, by the following argument.

Let $\theta_k = it_k$, $k = 1, 2$ so that the joint c.f. of Q and h is

$$\begin{aligned} \phi(\theta_1, \theta_2) &= E[\exp(\theta_1 h + \theta_2 Q)] \\ &\propto \int_{(n)} \exp(\theta_1 h) \exp[-\tfrac{1}{2}(1 - 2\theta_2)Q] \prod dx_j, \end{aligned} \tag{15.85}$$

where the n-fold integral is taken over $-\infty < x_j < \infty$ for $j = 1, \ldots, n$. Setting $\mathbf{y} = (1 - 2\theta_2)^{1/2}\mathbf{x}$ we have, after the transformation,

$$\phi(\theta_1, \theta_2) = (1 - 2\theta_2)^{-n/2} E[\exp(\theta_1 h)] \tag{15.86}$$

since h is scale-free; thus h and Q are independent. It then follows directly from (15.86) that

$$E[(hQ)^j] = E(h^j)E(Q^j), \qquad j = 1, 2, \ldots; \tag{15.87}$$

when h corresponds to R in (15.85) with $\mathbf{B} = \mathbf{I}$, (15.87) reduces to

$$E(R^j) = E[(\mathbf{x}'\mathbf{A}\mathbf{x})^j]/E[(\mathbf{x}'\mathbf{x})^j]. \tag{15.88}$$

Thus, the moments of R can be derived using (15.57); in particular,

$$\left.\begin{aligned} E(R) &= \operatorname{tr}(\mathbf{A})/n, \\ \text{and} \qquad E(R^2) &= \{\operatorname{tr}(\mathbf{A}^2) + [\operatorname{tr}(\mathbf{A})]^2\}/n(n+2). \end{aligned}\right\} \tag{15.89}$$

Result (15.88) and a converse are considered further in Exercise 15.22.

15.28 Let $\mathbf{T}$ be an orthogonal matrix such that

$$\mathbf{T} = \begin{pmatrix} \mathbf{T}_1 \\ \mathbf{T}_2 \end{pmatrix} \begin{matrix} (n-p)\times n \\ p \times n \end{matrix}$$

and $\mathbf{B} = \mathbf{T}_1'\mathbf{T}_1$ is idempotent of rank $(n-p)$. Thus $\mathbf{y}_1 = \mathbf{T}_1\mathbf{x}$ is a p-dimensional variate that is $MVN\,(\mathbf{0}, \sigma^2\mathbf{I}_{n-p})$ and independent of $\mathbf{y}_2 = \mathbf{T}_2\mathbf{x}$, of order p. If (15.85) can be rewritten as

$$R = \mathbf{x}'\mathbf{B}'\mathbf{A}\mathbf{B}\mathbf{x}/\mathbf{x}'\mathbf{B}\mathbf{x} \tag{15.90}$$

it follows that (15.87) and (15.89) continue to hold with n replaced by $(n-p)$ and $\mathbf{A}$ replaced by $\mathbf{T}_1\mathbf{A}\mathbf{T}_1'$. After some reduction we have

$$E(R) = \text{tr}(\mathbf{AB})/(n-p),$$

and (15.91)

$$E(R^2) = \{\text{tr}\,[(\mathbf{AB})^2] + [\text{tr}(\mathbf{AB})]^2\}/(n-p)(n-p+2).$$

Results (15.91) are particularly useful when dealing with deviations about the means or, more generally, regression residuals; cf Chapter 28, Volume 2A.

15.29 When B in (15.85) is not idempotent, the results are more complex. Morin-Wahhab (1985) gives the exact moments for such ratios as extended hypergeometric series expansions; see also M. D. Smith (1989). Magnus (1990) gives detailed results on the existence and expectation of ratios of general quadratic forms.

Characterizations of the normal distribution

15.30 Many, although not all, of the properties of the normal distribution that we have discussed are characterizing properties, shared by no other distribution. Thus, we saw in **15.4** that the normality of all linear functions of the variates is a characterizing property; on the other hand, Exercises 11.23 and 11.29 showed that the results of Example 11.21 and (11.8) are not unique to the normal distribution. We now discuss some of the other important characterizations of normality.

15.31 We have seen in Example 12.7 that the independence of sample mean and variance characterizes the univariate normal distribution among distributions with all cumulants finite. This result may be generalized to the multivariate case: the independence of the sample mean vector and the elements of the sample covariance matrix characterizes the multinormal distribution among all multivariate distributions with finite covariance matrices. It will be noticed that the existence of higher-order cumulants is not necessary to the result, which is due to Lukacs (1942). We first give his proof for the univariate case, and sketch the generalization.

We suppose a population to have finite mean μ and variance σ^2, and c.f. $\phi(t)$. The joint c.f. of the sample mean $\bar{x}$ and variance s^2 in samples of size n is

$$\phi_{12}(t_1, t_2) = \int \cdots \int \exp\,(it_1\bar{x} + it_2 s^2)\,dF_n. \tag{15.92}$$

A necessary and sufficient condition for the independence of $\bar{x}$ and s^2 is that (15.92) factorizes, i.e. that

$$\phi_{12}(t_1, t_2) = \phi_1(t_1)\phi_2(t_2). \tag{15.93}$$

Differentiating (15.93) with respect to t_2 we obtain

$$\left[\frac{\partial \phi_{12}}{\partial t_2}\right]_{t_2=0} = \phi_1(t_1)\left[\frac{\partial \phi_2}{\partial t_2}\right]_{t_2=0}. \tag{15.94}$$

Now in (15.94), we may use the relation connecting the c.f. of the sample mean with that of the original distribution

$$\phi_1(t_1) = \left\{\phi\left(\frac{t_1}{n}\right)\right\}^n \tag{15.95}$$

Further

$$\left[\frac{\partial \phi_2}{\partial t_2}\right]_{t_2=0} = iE(s^2) = i.\frac{(n-1)}{n}\sigma^2. \tag{15.96}$$

Finally, from (15.92),

$$\frac{\partial \phi_{12}}{\partial t_2} = i\int \dots \int s^2 \exp(it_1\bar{x} + it_2 s^2)\,\mathrm{d}F_n, \tag{15.97}$$

and since

$$s^2 = \frac{1}{n}\sum_i x_i^2 - \bar{x}^2 = \frac{1}{n^2}\left\{(n-1)\sum_i x_i^2 - \sum_{i\neq j}\sum x_i x_j\right\}, \tag{15.98}$$

(15.97) and (15.98) give

$$\begin{aligned}
\left[\frac{\partial \phi_{12}}{\partial t_2}\right]_{t_2=0} &= \frac{i}{n^2}\int \dots \int \left\{(n-1)\sum_i x_i^2 - \sum_{i\neq j}\sum x_i x_j\right\}\exp\left(it_1\sum_i x_i/n\right)\mathrm{d}F_n \\
&= \frac{i(n-1)}{n}\left\{\int x^2 \exp(it_1 x/n)\,\mathrm{d}F_1 \cdot \left[\phi\left(\frac{t_1}{n}\right)\right]^{n-1}\right. \\
&\quad \left. - \left[\int x \exp(it_1 x/n)\,\mathrm{d}F_1\right]^2 \left[\phi\left(\frac{t_1}{n}\right)\right]^{n-2}\right\} \\
&= \frac{i(n-1)}{n}\left[\phi\left(\frac{t_1}{n}\right)\right]^{n-2} \cdot \left\{-\phi\left(\frac{t_1}{n}\right)\cdot \frac{\partial^2 \phi\left(\frac{t_1}{n}\right)}{\partial\left(\frac{t_1}{n}\right)^2} + \left[\frac{\partial \phi\left(\frac{t_1}{n}\right)}{\partial\left(\frac{t_1}{n}\right)}\right]^2\right\}.
\end{aligned} \tag{15.99}$$

Writing t for t_1/n, we put (15.99), (15.95) and (15.96) into (15.94) and obtain

$$-\phi(t)\phi''(t) + \{\phi'(t)\}^2 = \{\phi(t)\}^2\sigma^2. \tag{15.100}$$

(15.100) may be rewritten

$$\frac{\mathrm{d}}{\mathrm{d}t}\left\{\frac{d}{dt}\log\phi(t)\right\} = -\sigma^2, \tag{15.101}$$

so that integration of (15.101) gives

$$\phi'(t)/\phi(t) = -\sigma^2 t + c. \tag{15.102}$$

From the initial conditions $\phi(0) = 1$, $\phi'(0) = i\mu$, we find in (15.102)

$$c = i\mu,$$

and (15.102) becomes

$$\frac{d}{dt}\log\phi(t) = i\mu - \sigma^2 t. \tag{15.103}$$

Integrating (15.103), we have

$$\log\phi(t) = i\mu t - \tfrac{1}{2}\sigma^2 t^2,$$

the constant of integration being zero since $\log\phi(0) = 0$. Thus finally

$$\phi(t) = \exp\left(i\mu t - \tfrac{1}{2}\sigma^2 t^2\right), \tag{15.104}$$

which is the c.f. unique to the normal distribution. (15.104) thus establishes the independence of sample mean and variance as a characterization of the normal distribution alone among distributions with finite variance.

Kawata and Sakamoto (1949) relax even the restriction of requiring a finite variance.

15.32 The generalization of the result of **15.31** to the multivariate case is straightforward. If the distribution of the sample mean vector is independent of $s_{l,m}$, the sample covariance of x_l and x_m, we obtain, by the argument leading to (15.100),

$$-\phi''_{l,m}/\phi + \phi'_l\phi'_m/\phi^2 = \sigma_{lm}, \tag{15.105}$$

where ϕ is the c.f. of the population distribution and its suffixes denote the t-variable with respect to which differentiation has been carried out. If (15.105) holds for l, $m = 1, 2, \ldots, p$, we have a system of partial differential equations which may be written in matrix form, corresponding to (15.101),

$$\left(\frac{\partial}{\partial t_l}\frac{\partial}{\partial t_m}\log\phi\right) = -\mathbf{\Sigma}, \tag{15.106}$$

so that on applying the initial conditions, (15.106) yields after two integrations

$$\phi(\mathbf{t}) = \exp\left(i\mathbf{t}'\boldsymbol{\mu} - \tfrac{1}{2}\mathbf{t}'\mathbf{\Sigma}\mathbf{t}\right),$$

the c.f. of the multinormal distribution.

Parthasarathy (1976) shows that if the sample mean vector is independent of any non-negative homogeneous polynomial in the deviations of the variates from their sample means, the population must be multinormal.

15.33 In our discussion of the geometrical method of deriving sampling distributions in Chapter 11, we saw that the probability density of a sample of independent observations from a standardized univariate normal distribution is constant on the sphere $\sum x^2 = \text{constant}$, since the density is a function of $\sum x^2$ only. The question arises whether this property of radial symmetry is possessed by any other distribution. The following argument, due to Bartlett (1934), shows that it is not.

If the density may be written

$$L = \prod_{i=1}^{n} f(x_i) \propto g\left(\sum x^2\right), \tag{15.107}$$

we must have

$$\frac{\partial L}{\partial x_i} = \frac{\partial \log L}{\partial x_i} = 0$$

for

$$\sum_i x_i^2 = \text{constant}.$$

Using Lagrange's method with a multiplier λ, we therefore have

$$\frac{\partial \log f(x_i)}{\partial x_i} + \lambda x_i = 0 \quad (\text{all } i). \tag{15.108}$$

Integration yields

$$\log f(x_i) = -\tfrac{1}{2}\lambda x_i^2 + k_i$$

or

$$f(x_i) \propto \exp\left(-\tfrac{1}{2}\lambda x_i^2\right). \tag{15.109}$$

(15.107) thus characterizes the normal distribution.

15.34 The argument generalizes at once to the multivariate case. Instead of radial symmetry with respect to a single vector of length $(\Sigma x^2)^{\frac{1}{2}}$, we now consider a group of p vectors. If the distribution has a density that is a function only of the squared lengths of these vectors (the sums of squares) and the angles between them (the correlation coefficients—cf. **16.24**), it must be multinormal. For if we define $\mathbf{x}_i$ as the $(p \times 1)$ vector containing the ith set of observations on the p variates,[*] and $\mathbf{X}'$ as the $(p \times n)$ matrix formed by these vectors, the sums of squares and products are the elements of $\mathbf{X}'\mathbf{X}$, and we have

$$L = \prod_i f(\mathbf{x}_i) \propto g(\mathbf{X}'\mathbf{X}). \tag{15.110}$$

For $\mathbf{X}'\mathbf{X}$ constant, we now introduce a $(p \times p)$ matrix $\mathbf{\Lambda}$ of Lagrange multipliers, and differentiate logarithmically to obtain

$$\frac{\partial \log f(\mathbf{x}_i)}{\partial \mathbf{x}_i} + \mathbf{\Lambda}\mathbf{x}_i = \mathbf{0} \quad (\text{all } i),$$

whence

$$f(\mathbf{x}_i) \propto \exp\left(-\tfrac{1}{2}\mathbf{x}_i'\mathbf{\Lambda}\mathbf{x}_i\right),$$

so that (15.110) characterizes the multinormal distribution.

---

[*] These new vectors are, of course, distinct from the $(n \times 1)$ vectors of observations on single variates just discussed.

15.35 Next we prove a characterization of the normal distribution that illuminates the position of central importance that it holds in statistical theory. We have seen (Examples 11.2 and 11.3) that an orthogonal transformation from a set of independent standardized normal variates yields a set of linear functions of the original variates which are also independent standardized normal variates. We now prove that if $\mathbf{x}$ is a vector of (not necessarily identical) independent standardized variates with finite cumulants, and the non-trivial transformation

$$\mathbf{x} = \mathbf{C}\mathbf{y} \tag{15.111}$$

gives a vector $\mathbf{y}$ of independent standardized variates, then each of the x_i is normal (and hence so is each y_i) and the transformation is orthogonal. The result is due to Lancaster (1954).

The orthogonality of $\mathbf{C}$ follows at once, for since the variates are standardized

$$\sum_j c_{ij}^2 = 1, \tag{15.112}$$

while since they are independent

$$\sum_j c_{ij} c_{kj} = 0 \quad (i \neq k). \tag{15.113}$$

(15.112) and (15.113) are the conditions for orthogonality. Thus $\mathbf{C}' = \mathbf{C}^{-1}$ and

$$\mathbf{y} = \mathbf{C}'\mathbf{x}. \tag{15.114}$$

Since the y_i are independent, (15.111) gives for the sth cumulant of x_i, κ_{si},

$$\kappa_{si} = \sum_k c_{ik}^s \lambda_{sk} \quad \text{(all } s\text{)}, \tag{15.115}$$

where λ_{sk} is the sth cumulant of y_k. Conversely, from (15.114),

$$\lambda_{si} = \sum_j (c'_{ij})^s \kappa_{sj} \quad \text{(all } s\text{)}. \tag{15.116}$$

Combination of (15.115) and (15.116) gives

$$\kappa_{si} = \sum \kappa_{sj} \sum_k c_{ik}^s c_{jk}^s \quad \text{(all } s\text{)}. \tag{15.117}$$

For $s = 1, 2$, (15.117) reduces to $\kappa_{1i} = 0$, $\kappa_{2i} = 1$, as it must for standardized variates. Now

$$\sum_j \left| \sum_k c_{ik}^s c_{jk}^s \right| \leqslant \sum_j \sum_k |c_{ik}^s| \, |c_{jk}^s|, \tag{15.118}$$

and using (15.102), (15.118) becomes, for $s \geqslant 3$,

$$\sum_j \left| \sum_k c_{ik}^s c_{jk}^s \right| < 1. \tag{15.119}$$

Taking absolute values in (15.117), we have

$$|\kappa_{si}| \leqslant \sum_j |\kappa_{sj}| \left| \sum_k c_{ik}^s c_{jk}^s \right|. \tag{15.120}$$

(15.120) has on the right-hand side a weighted sum of the κ_{sj}, the weights adding to less than 1 by (15.119). If for fixed s we take that value of i for which κ_{si} is largest, (15.120) cannot hold unless that $\kappa_{si}=0$, i.e. all $\kappa_{si}=0$. This held for all s. Thus

$$\kappa_{si}=0 \quad (\text{all } s \geqslant 3) \tag{15.121}$$

and each x_i is normal. The normality of the y_i follows at once.

15.36 Although Exercises 11.9 and 11.26 (and 16.22 below) give instances of products of independent variates that are exactly normally distributed, only normal independent variates can have a normally distributed sum. For if

$$\phi_1(t)\phi_2(t) = \exp(\mu it - \tfrac{1}{2}\sigma^2 t^2), \tag{15.122}$$

neither ϕ_1 nor ϕ_2 can have any zeros, and since (15.122) holds identically in t, it follows that we must have

$$\phi_j(t) = \exp\{a_j it + \tfrac{1}{2} b_j (it)^2\},$$

so that the components are normal. This characterization is usually known as Cramér's theorem—cf. Cramér (1937) for a detailed proof.

15.37 Lukacs (1956) reviews characterization methods and results generally. See also Patil and Seshadri (1964). Lancaster (1960) reviews the characterizations of normality by independence properties, and shows in particular that the latter imply the existence of all moments, which therefore need not be assumed. A general exposition of characterizations is given in the book by Kagan *et al.* (1973). Mathai and Pederzoli (1977) discuss characterizations of the normal distribution, while Galambos and Kotz (1978) emphasize characterizations of the exponential. The volumes edited by Patil *et al.* (1975, 1985) and Taillie *et al.* (1981) describe characterizations for a variety of different distributions.

Exercise 15.22 characterizes Gamma distributions.

EXERCISES

15.1 From a multinormal distribution, a conditional distribution is found by fixing the values of certain of the variates. Show that this distribution is itself multinormal, with mean vector a function of the eliminated variates and with the same elements in the inverse covariance matrix as the variates possessed in the original distribution.

15.2 A trivariate distribution, not necessarily normal, has correlations $\rho_{12}, \rho_{13}, \rho_{23}$. Show that $1+2\rho_{12}\rho_{13}\rho_{23} \geqslant \rho_{12}^2+\rho_{13}^2+\rho_{23}^2$.

15.3 In a p-variate distribution all the correlations ρ are equal. Show that this is possible if, and only if, $\rho \geqslant -1/(p-1)$.

15.4 Show that the density function of the multinormal distribution satisfies the system of equations

$$\frac{\partial f}{\partial \rho_{ij}} = \frac{\partial^2 f}{\partial x_i \, \partial x_j}$$

where ρ_{ij} is the correlation coefficient of x_i and x_j.

15.5 By considering the binormal distribution with independent variates, show that

$$\int_{-x}^{x} (2\pi)^{-\frac{1}{2}} \exp\left(-\tfrac{1}{2}t^2\right) dt \leqslant \{1-\exp(-2x^2/\pi)\}^{\frac{1}{2}}.$$

(The error in taking the equality to hold is always less than $\frac{3}{4}$ percent.)
(Pólya, 1945; J. D. Williams, 1946)

15.6 In a standardized binormal distribution with covariance ρ, show that

$$E\{|x_1|\,|x_2|\} = 4\int_0^\infty\int_0^\infty x_1 x_2 \, dF - \rho = 4I-\rho, \quad \text{say.}$$

Show as in **15.10** that

$$4\pi^2 I = \int_{-\infty}^{\infty}\int_{-\infty}^{\infty} \exp\left(-\tfrac{1}{2}\mathbf{t}'\Sigma\mathbf{t}\right)\frac{dt_1\,dt_2}{(it_1)^2(it_2)^2}$$

and hence that $\partial^2 I/\partial\rho^2 = \partial P_2^0/\partial\rho$, so that

$$E\{|x_1|\,|x_2|\} = \frac{2}{\pi}\{\rho \arcsin\rho + (1-\rho^2)^{\frac{1}{2}}\}.$$

Use this result to establish the exact variance (10.40) of the mean deviation in normal samples, and to derive the exact variance of the mean difference at (10.46) by evaluating $\mathcal{J}$ in (10.44).

15.7 Shots are fired at a vertical circular target of unit radius. The distribution of horizontal and vertical deviations from the centre of the target is binormal, with zero means, equal variances v and correlation ρ. Show that the probability of hitting the target is

$$\frac{(1-\rho^2)^{\frac{1}{2}}}{2\pi}\int_0^{2\pi}\left[1-\exp\left\{-\frac{1+\rho\cos 2\theta}{2v(1-\rho^2)}\right\}\right]\frac{d\theta}{1+\rho\cos 2\theta},$$

reducing to $1-\exp\{-1/(2v)\}$ when $\rho=0$.

15.8 For standardized binormal variates x and y, show that $x^2+2axy+y^2$ cannot be independent of $x^2+2bxy+y^2$ unless $|a|=|b|=1$ and $a+b=0$.

15.9 From **15.13**, show that for a standardized p-variate multinormal distribution with correlations all equal to $\frac{1}{2}$, the probability that all the variates are positive is $1/(p+1)$.

(Cf. Foster and Stuart, 1954)

15.10 In the previous exercise, if, instead of the correlations being $\frac{1}{2}$, the covariance matrix Σ has elements

$$\sigma_{ij} = \begin{cases} 1, & i=j, \\ -\frac{1}{2}, & |i-j|=1, \\ 0 & \text{otherwise,} \end{cases}$$

show that the probability becomes $1/(p+1)!$.

15.11 In **15.13**, let x_0 have correlation coefficient λ with each of the other p x_i, which remain mutually independent, and replace (15.47) by $y_i = ax_0 + bx_i$. Show that if the variances of the y_i are equated to unity, they are multinormally distributed with equal correlations $\rho = 1 - b^2$. By putting $\lambda = 0$ for $\rho \geqslant 0$ and $\lambda = -a/b$ for $\rho < 0$, show that $y_i = |\rho|^{\frac{1}{2}}x_0 + (1-\rho)^{\frac{1}{2}}x_i$ and that if $y_{(i)}$, $x_{(i)}$ denote the order statistics of the p ys and of the p xs, the c.f. of $y_{(i)}$ is related to that of $x_{(i)}$ by

$$\phi_{y(i)}(t) = \exp\left(-\tfrac{1}{2}\rho t^2\right)\phi_{x(i)}\left(t(1-\rho)^{\frac{1}{2}}\right),$$

so that in particular we have

$$E\left(y_{(i)}\right) = (1-\rho)^{\frac{1}{2}}E\left(x_{(i)}\right),$$
$$\text{var}\left(y_{(i)}\right) = \rho + (1-\rho)\,\text{var}\left(x_{(i)}\right),$$

connecting the order statistics of p equally correlated multinormal variates with univariate normal order statistics in samples of size p, discussed in **14.21**.

(Similar results are available for the product-moments of pairs of order statistics—cf. Owen and Steck (1962). For $p = 2$, Clark (1961) gave the first four moments of the larger of bivariate normal variates with arbitary means and variances.)

15.12 By considering the variates $y_i = x_i - b_i x_0$, where the x_i are independent standardized normal variates with density function f, generalize (15.48) by showing that for a multinormal distribution with correlations $\rho_{ij} = a_i a_j (i \neq j)$ the probability that all the variables are positive is

$$\int_{-\infty}^{\infty}\left\{\prod_i \int_{b_i x}^{\infty} f(t)\,dt\right\} f(x)\,dx.$$

(Stuart, 1958)

15.13 If $\mathbf{x}$ has a multinormal distribution with mean $\mathbf{0}$ and non-singular covariance matrix Σ, show that a necessary and sufficient condition for the independence of a quadratic form $\mathbf{x'Ax}$ and a linear form $\mathbf{b'x}$ is that $\mathbf{A\Sigma b} = \mathbf{0}$.

(Aitken, 1950)

15.14 If $\mathbf{x'Ax}$ and $\mathbf{x'Bx}$ are quadratic forms, and $\mathbf{a'x}$, $\mathbf{b'x}$ are linear forms, in multinormal variables $\mathbf{x}$ whose covariance matrix Σ is not necessarily non-singular, show by using c.f.s that the independence conditions in (15.81) and Exercise 15.13 are replaced by $\mathbf{\Sigma A \Sigma B \Sigma} = \mathbf{0}$ and by $\mathbf{\Sigma A \Sigma b} = \mathbf{0}$ respectively, and that the linear forms are independent if and only if $\mathbf{a'\Sigma b} = \mathbf{0}$.

(Good, 1963, 1966)

15.15 If in **15.16** Q_1 and Q_2 are distributed like χ^2 with n and $m(<n)$ degrees of freedom, show that a necessary and sufficient condition that $Q_1 = Q_2 + Q_3$, where Q_3 is also a χ^2 variate, is that the correlation coefficient between Q_1 and Q_2 is $(m/n)^{1/2}$.

(Lancaster, 1954)

15.16 Show that if $\mathbf{x'Ax} = \mathbf{x'A_1x} + \mathbf{x'A_2x}$, where $\mathbf{A}$ and $\mathbf{A}_1$ are idempotent and $\mathbf{A}_2$ is non-negative, then $\mathbf{A}_2$ is idempotent, so that we obtain a decomposition as in **15.20**.

(Hogg and Craig, 1958)

15.17 If $\mathbf{x}$ is multinormal with mean $\mathbf{0}$ and non-singular covariance matrix $\mathbf{\Sigma}$, show as in **15.15** that the c.f. of $\mathbf{x'Ax}$ is (15.52) if and only if $\mathbf{A\Sigma}$ is idempotent of rank r. If $\sum_{i=1}^{k} \mathbf{x'A}_i\mathbf{x} = \mathbf{x'Ax}$, where $\mathbf{A\Sigma}$ is idempotent, show that any of the conditions (a), (b), (c) of **15.20** implies the other two, where (b) here requires the idempotency of each $\mathbf{A}_i\mathbf{\Sigma}$ and (c) requires that $\mathbf{A}_i\mathbf{\Sigma}\mathbf{A}_j = \mathbf{0}$, all $i \neq j$.

15.18 If $\mathbf{\Sigma}$ may be singular in Exercise 15.17, show that necessary and sufficient conditions that the c.f. of $\mathbf{x'Ax}$ is (15.52) are:

(a) $\mathbf{\Sigma A \Sigma A \Sigma} = \mathbf{\Sigma A \Sigma}$ and rank $(\mathbf{\Sigma A \Sigma}) = \text{tr}(\mathbf{A\Sigma}) = r$;
or (b) $\mathbf{A\Sigma A\Sigma} = \mathbf{A\Sigma}$, if and only if rank $(\mathbf{A\Sigma}) = \text{tr}(\mathbf{A\Sigma}) = r$;
or (c) $\mathbf{A\Sigma A\Sigma} = \mathbf{A\Sigma}$, if and only if rank $(\mathbf{A\Sigma}) = \text{rank}(\mathbf{\Sigma A\Sigma}) = r$.

(Cf. Styan, 1970; see also Khatri, 1978.)

15.19 $x_1, x_2, \ldots, x_n$ are distributed in multivariate normal form, the covariance of x_i and x_j being σ_{ij}. Show that their covariance, conditional upon $\mathbf{l'x} = \text{constant}$, is

$$\sigma_{ij} - (\mathbf{l}'\boldsymbol{\sigma}_i)(\mathbf{l}'\boldsymbol{\sigma}_j)/\mathbf{l}'\mathbf{\Sigma}\mathbf{l}$$

where $\boldsymbol{\sigma}_i$ denotes the ith column of $\mathbf{\Sigma}$ and $\mathbf{l}$ is any vector of constants.

15.20 x and y have a standardized binormal distribution. Whenever the signs of x and y differ, the sign of x is changed to give a new variate $x' = x\,\text{sgn}\,(xy)$. Show that x' and y are marginally normally distributed as before, but that their joint distribution is not binormal since $\text{sgn}\,(x') = \text{sgn}\,(y)$. Show further that linear functions of x' and y are not in general normally distributed.

If x and y are independent ($\rho = 0$), show that by "folding over" four half-quadrants, instead of two whole quadrants as above, we may construct x'' and y'' which are marginally normal and have zero correlation, but are not binormal. Generalize to show that, for a set of n variates, it is possible for any subset of $p < n$ variates to be multinormally distributed without the n variates being so. (Exercise 16.22 gives another instance.)

15.21 Show that the joint distribution of k independent Poisson variables, conditional upon the value of their sum, is a multinomial distribution. Hence establish the asymptotic χ^2 distribution of Example 15.3.

15.22 If $x_1, \ldots, x_k$ are independent variables having (not necessarily identical) Gamma-distributions (6.18), show by using m.g.f.s that any scale-free function of them, $h(x_1, \ldots, x_k)$, is distributed independently of $S = \sum_{i=1}^{k} x_i$. In particular this holds for $h(x_1, \ldots, x_k) = \sum_{i=1}^{k} a_i x_i \Big/ \sum_{i=1}^{k} x_i = R$. Hence show that for a vector $\mathbf{x}$ of independent standardized normal variates, $\mathbf{x'Ax}/\mathbf{x'x}$ is independent of $\mathbf{x'x}$, as in **15.27**.

(Pitman (1937b). Laha (1954) gives a converse: if the x_i are independently *and identically* distributed with finite variance, the independence of R and S implies that each x_i has a Gamma distribution.)

15.23 As in **15.10**, write the d.f. of the binormal distribution as

$$F_\rho(u, v) = \int_{-\infty}^{u}\int_{-\infty}^{v}\left\{(2\pi)^{-2}\int_{-\infty}^{\infty}\int_{-\infty}^{\infty}\exp\left(-i\mathbf{t}'\mathbf{x} - \tfrac{1}{2}\mathbf{t}'\mathbf{\Sigma}\mathbf{t}\right)\mathrm{d}t_1\,\mathrm{d}t_2\right\}\mathrm{d}x_1\,\mathrm{d}x_2$$

and show that for any fixed (u, v), $\dfrac{\partial}{\partial\rho}F_\rho(u, v) > 0$. Hence show that if a binormal distribution is doubly dichotomized at any point (u, v), with frequencies $\begin{array}{c|c} a & b \\ \hline c & d \end{array}$ in the four resulting quadrants, the function

$$R_{u,v.\rho} \equiv bc - ad$$

is a monotone increasing function of ρ, always having the same sign as ρ.

(The result is originally due to G. U. Yule, who called any bivariate distribution *isotropic* if R does not change sign as a function of u and v.)

15.24 Using Exercise 7.25, show that the bivariate distributions defined by (7.83) are isotropic. Show also that the distributions in Exercise 1.22 are isotropic.

15.25 Let $R(\mathbf{h}, \mathbf{A}) = [f(\mathbf{h})]^{-1}P(\mathbf{x} \geqslant \mathbf{h} \mid \boldsymbol{\mu} = \mathbf{0}, \mathbf{\Sigma} = \mathbf{A}^{-1})$ denote the multivariate Mills' ratio for the p-variate normal (cf. **5.38** when $p = 1$). Let $\mathbf{d} = \mathbf{A}\mathbf{h}$, where it is assumed that $d_i \geqslant 0$, $i = 1, \ldots, p$. Show that

$$\frac{1 - \sum_i a_{ii}/d_i^2 - \sum\sum_{i<j} a_{ij}/d_i d_j}{\pi d_i} \leqslant R(\mathbf{h}, \mathbf{A}) \leqslant \frac{1}{\pi d_i}.$$

(I. R. Savage, 1962; for improved bounds, see Steck (1979).)

15.26 If x, y are standardized binormal variates with correlation ρ, show from Exercise 15.6 that $|x|$, $|y|$ have correlation

$$R = \frac{2}{\pi - 2}\{\rho \arcsin \rho + (1 - \rho^2)^{\frac{1}{2}} - 1\}$$

and that $|\rho| > R > 0$ unless $|\rho| = R = 1$ or 0. Show that $|\rho| - R$ is maximized at $|\arcsin\rho| = \frac{1}{2}\pi - 1$, when $|\rho| = 0.54$, $R = 0.26$, and that $R/|\rho|$ is monotone increasing in $|\rho|$ and $\to 0$ as $|\rho| \to 0$.

15.27 In Exercise 1.22, let Δ_j be the mean difference (2.24) and σ_j the s.d. of the marginal distributions $f_j(x_j)$, $j = 1, 2$. Using Exercise 2.9, show that the correlation coefficient in the bivariate distribution $f(x_1, x_2)$ is exactly $\rho = \dfrac{\theta}{4}\dfrac{\Delta_1\Delta_2}{\sigma_1\sigma_2}$ and hence that $|\rho| \leqslant \frac{1}{3}$.

15.28 Assume (x, y) has a standardized binormal distribution with correlation ρ and let $L(h, k, \rho) = \text{Prob}(x \geqslant h, y \geqslant k)$. Let

$$M = L\left(h, 0, \frac{(\rho h - k)\,\text{sgn}(h)}{(h^2 - 2\rho hk + k^2)^{\frac{1}{2}}}\right) + L\left(0, k, \frac{(\rho k - h)\,\text{sgn}(k)}{(h^2 - 2\rho hk + k^2)^{\frac{1}{2}}}\right).$$

Show that

$$L(h, k, \rho) = M, \quad \text{if } hk > 0 \text{ or if } hk = 0 \text{ with } h + k > 0,$$
$$= M - \tfrac{1}{2}, \quad \text{otherwise.}$$

(cf Zelen and Severo, 1960)

15.29 If we denote by z_{g_j} independent χ^2 variables with $2g_j$ d.fr., where the g_j are positive integers, show by resolving the c.f. of $\sum_{j=1}^{r} c_j z_{g_j}$ into partial fractions that

$$\text{Prob}\left\{\sum_{j=1}^{r} c_j z_{g_j} > x\right\} = \sum_{j=1}^{r}\sum_{s=1}^{g_j} \alpha_{js}\,\text{Prob}\{z_s > x/c_j\}.$$

Writing $1 - 2itc_k = y$ in the c.f., show that the constants on the right are given by

$$\alpha_{k, g_k - h} = f_k^{(h)}(0)/h!$$

where

$$f_k(y) = \prod_{\substack{j=1 \\ j \neq k}}^{r} \{1 + (y-1)c_j/c_k\}^{-g_k}.$$

(G.E.P. Box, 1954)

15.30 Let $f(\mathbf{x})$ be the p.d.f. of the p-dimensional variate $\mathbf{x}$. Define the entropy of the distribution by $E = -\int f(x) \ln f(\mathbf{x}) \Pi dx_j$. If we maximize E subject to the conditions that $E(\mathbf{x}) = \boldsymbol{\mu}$ and $\text{var}(\mathbf{x}) = \boldsymbol{\Sigma}$, show that the distribution is multinormal.

(C.R. Rao, 1973, chapter 8)

CHAPTER 16

DISTRIBUTIONS ASSOCIATED WITH THE NORMAL

16.1 In the course of our investigations in earlier chapters, we have encountered several distributions, related to the normal distribution, that play important parts in the theory of Statistics precisely because they are the forms taken by the sampling distributions of various statistics in samples from normal populations. The special position that the normal distribution holds, mainly because of the Central Limit theorem in one or other of its forms, is reflected in the positions of special importance occupied by these related distributions. Whenever some statistic has a sampling distribution that is asymptotically normal, other statistics will have the corresponding related distributions. We have already encountered an instance of this in discussing the χ^2 goodness-of-fit test in Example 15.3.

In the first part of the chapter, the three important distributions related to the univariate normal, namely the χ^2 distribution, Student's t distribution, and Fisher's variance ratio (F) distribution, will be examined in turn, and their salient properties described. We then consider sampling distributions arising from the bivariate normal distribution, particularly that for the correlation coefficient.

In the second half of the chapter the theory is extended to the multivariate case. In particular, we will consider the Wishart distribution for the sample covariance matrix, the distribution of the generalized variance (the determinant of the sample covariance matrix) and Hotelling's T^2, the multivariate analogue of Student's t distribution. The chapter concludes with consideration of some asymptotic results for the latent roots and vectors of the covariance matrix.

The χ^2 distribution

16.2 We have already seen (cf. **11.2**, Example 11.6) that the sum z of the squares of n independent standardized normal variates is distributed with density function

$$dF = \frac{1}{2^{\frac{1}{2}n}\Gamma(\frac{1}{2}n)} \exp(-\tfrac{1}{2}z) z^{\frac{1}{2}n-1} dz \qquad 0 \leqslant z < \infty, \tag{16.1}$$

and that if we impose p independent linear constraints upon these variates, i.e. consider the distribution of z conditional upon $p(<n)$ linear relations among the normal variates being satisfied, the effect of the constraints is simply to replace n in (16.1) by $(n-p)$. Further, we have also seen (Example 11.7) that if the sum of squares is taken about the sample mean instead of the population mean, the effect on the distribution of z is to impose one linear constraint, replacing n in (16.1) by $(n-1)$.

The distribution (16.1) is called the χ^2 distribution with n degrees of freedom (d.fr.) and sometimes, concisely, the χ^2_n distribution. The term "degrees of freedom" arises naturally in normal variation, being the effective dimensionality of the variation, as Examples 11.6–7 showed. If rewritten, with $z=\chi^2$ and $n=\nu$, as

$$dF = \frac{1}{\Gamma(\frac{1}{2}\nu)} \exp(-\tfrac{1}{2}\chi^2)(\tfrac{1}{2}\chi^2)^{\frac{1}{2}\nu-1}\, d(\tfrac{1}{2}\chi^2), \qquad \nu>0;\quad 0\leqslant\chi^2<\infty, \tag{16.2}$$

it is at once evident that $\frac{1}{2}\chi^2$ has the Gamma distribution with parameter $\frac{1}{2}\nu$. (16.2) is a Pearson Type III distribution. Although we have so far considered only cases with ν a positive integer, we have taken the opportunity here to widen the definition of the distribution to all $\nu>0$, since the integral of (16.2) still converges, so that the distribution is always properly defined.

By differentiation of (16.2), we find that for $\nu>2$, the χ^2 distribution rises from zero at the origin to a unique mode at $\nu-2$ and then falls away to zero asymptotically, with positive skewness; despite this skewness, a further differentiation shows that for $\nu>4$ the two points of inflexion are equidistant from the mode, c.f. 6.3. For $\nu=2$, the distribution is J-shaped, with a peak at the origin, while for $0<\nu<2$ the distribution is J-shaped and infinite at the origin.

Since $\mu_1'=\nu$, we see that for $\nu>2$ (2.13) gives the median as approximately $\nu-\frac{2}{3}$, and Appendix Table 3, in the column $P=0.50$, shows that this approximation is in error by only 0.033 at $\nu=3$ and by 0.01 or less for $\nu\geqslant 9$. Exercise 6.20 applies here when the distribution is standardized.

Properties of the χ^2 distribution

16.3 The characteristic function of χ^2 is obtained at once from (16.1) (see Example 3.6) as

$$\phi(t) = (1-2it)^{-\frac{1}{2}\nu} \tag{16.3}$$

whence, for the cumulants, we have

$$\kappa_r = \nu 2^{r-1}(r-1)! \tag{16.4}$$

so that $\mu_1'=\nu$ and for moments about the mean

$$\left.\begin{aligned} \mu_2 &= 2\nu \\ \mu_3 &= 8\nu \\ \mu_4 &= 12\nu(\nu+4) \\ \mu_5 &= 32\nu(5\nu+12) \\ \mu_6 &= 40\nu(3\nu^2+52\nu+96). \end{aligned}\right\} \tag{16.5}$$

Since κ_r is linear in ν, μ_r, which can contain only $[\frac{1}{2}r]$ powers of μ_2, must be of degree $[\frac{1}{2}r]$ in ν, i.e. $\frac{1}{2}r$ if r is even and $\frac{1}{2}(r-1)$ if r is odd.

Exercise 16.3 gives relations among the central moments.

As ν tends to infinity the χ^2 distribution tends to normality, for on standardizing we have

$$\phi(t) = e^{-\nu it/\sqrt{(2\nu)}}\left(1-\frac{2it}{\sqrt{(2\nu)}}\right)^{-\frac{1}{2}\nu}$$

and

$$\log\phi(t)=\frac{-\nu it}{\sqrt{(2\nu)}}-\frac{\nu}{2}\left\{\frac{-2it}{\sqrt{(2\nu)}}-\frac{1}{2}\left(\frac{2it}{\sqrt{(2\nu)}}\right)^2+o(\nu^{-1})\right\}\to-\tfrac{1}{2}t^2.$$

The tendency is, however, rather slow, and there are better approximation, as we shall see in **16.5-9**.

The d.f. of (16.2) is an incomplete Gamma function. We have

$$F(\zeta)=\int_0^{\zeta}\frac{1}{2^{\frac{1}{2}\nu}\Gamma(\frac{1}{2}\nu)}\exp\left(-\tfrac{1}{2}z\right)z^{\frac{1}{2}\nu-1}\,dz=\Gamma_{\frac{1}{2}\zeta}(\tfrac{1}{2}\nu)/\Gamma(\tfrac{1}{2}\nu),$$

or, in the notation of Pearson's tables,

$$=I\left(\frac{\frac{1}{2}\zeta}{\sqrt{(\frac{1}{2}\nu)}},\ \tfrac{1}{2}\nu-1\right).$$

Series expansions of the d.f. are given in Exercise 16.7, which also gives the complementary relation with the Poisson d.f. implicit in (5.23).

Tables of the χ^2 distribution

16.4 (a) E. S. Pearson and Hartley (1950) give the d.f. of χ^2 to 5 d.p. for $\nu=1(1)20(2)70$ and $\chi^2=0.001(0.001)0.01(0.01)0.1(0.1)2.0(0.2)10(0.5)20(1)40(2)$ 120. This table is fully reproduced in the *Biometrika Tables.*

(b) Thompson (1941) gives a table of the percentage points of the distribution, i.e. the values of χ^2 corresponding to the values $(1-P)$ of the d.f. for $P=0.995$, 0.990, 0.975, 0.950, 0.900, 0.750, 0.500, 0.250, 0.100, 0.050, 0.025, 0.010, 0.005 and $\nu=1(1)30(10)100$. This table, too, is reproduced in the *Biometrika Tables*, with the addition of entries for $P=0.001$.

(c) Fisher and Yates (1963) give percentage points for $P=0.99$, 0.98, 0.95, 0.90 (0.10) 0.10, 0.05, 0.02, 0.01, 0.001 and $\nu=1(1)30$. A selection from this table is given as Appendix Table 3.

(d) A table by Yule, reproduced in Appendix Table 4, gives P for $\nu=1$, $\chi^2=0$ (0.01)1(0.1)10.

(e) Harter (1964a) gives a 9 d.p. table of the d.f. and 23 percentage points to 6 significant figures for $\nu=1(1)150(2)330$. For $\nu=1(1)100$, the latter appear in Harter (1964b). The *Biometrika Tables*, Vol. II, extend the percentage points to the range $\nu=0.1(0.1)3.0(0.2)10.0(1)100$.

(f) Khamis and Rudert (1965) give a 10 d.p. table of the d.f. for $\nu=$ 0.1(0.1)20(0.2)40(0.5)140 at the χ^2 values 0.0001(0.0001)0.001(0.001)0.01(0.01)1(0.05) 6(0.1)16(0.5)66(1)166(2)250.

(g) Mardia and Zemroch (1978) give tables of percentage points for $\nu=$ 0.1(0.1)3(0.2)7(0.5)11(1)30(5)60(10)120 for 19 values of P in each tail, from 0.0001 to 0.40.

Lackritz (1983) gives another method of obtaining the percentage points.

Square-root and cube-root transformations of χ^2

16.5 Two approximations to the distribution of χ^2 are in common use, each of which transforms to a fractional power of χ^2 that is normally distributed to a closer approximation than is χ^2 itself. They are:

(a) Fisher's result that $\sqrt{(2\chi^2)}$ is approximately normally distributed with mean $\sqrt{(2\nu-1)}$ and unit variance;

(b) Wilson and Hilferty's (1931) result that $(\chi^2/\nu)^{\frac{1}{3}}$ is approximately normally distributed with mean $1-2/(9\nu)$ and variance $2/(9\nu)$.

The second of these is the more accurate approximation, as we now show.

16.6 The relative speed of approach to normality of χ^2 and $\sqrt{(2\chi^2)}$ may be compared through their skewness and kurtosis.

For χ^2 we have, from (16.5), the coefficients

$$\gamma_1 = \sqrt{\beta_1} = (8/\nu)^{\frac{1}{2}}, \quad \gamma_2 = \beta_2 - 3 = 12/\nu. \tag{16.6}$$

For Fisher's square-root transformation, we need the moments of χ, for which we have

$$\begin{aligned} \mu_r' &= \frac{1}{2^{\frac{1}{2}(\nu-2)}\Gamma(\frac{1}{2}\nu)} \int_0^\infty e^{-\frac{1}{2}\chi^2}\chi^{\nu+r-1}\,d\chi \\ &= \frac{2^{\frac{1}{2}r}\Gamma\{\frac{1}{2}(\nu+r)\}}{\Gamma(\frac{1}{2}\nu)}. \end{aligned} \tag{16.7}$$

It should be noticed that (16.7) remains valid for negative r if $\nu+r>0$, and in particular that the reciprocal of a χ_ν^2 variate has expectation $\mu_{-2}' = (\nu-2)^{-1}$ for $\nu>2$—cf. Example 3.17.

Thus

$$\mu_1' = \sqrt{2}\frac{\Gamma\{\frac{1}{2}(\nu+1)\}}{\Gamma(\frac{1}{2}\nu)}.$$

Using Stirling's expansion (3.63), we find after substitution and reduction

$$\mu_1' = \nu^{\frac{1}{2}}\left(1 - \frac{1}{4\nu} + \frac{1}{32\nu^2} + \frac{5}{128\nu^3} + \dots\right),$$

so that

$$\mu_1'^2 = \nu\left(1 - \frac{1}{2\nu} + \frac{1}{8\nu^2} + \frac{1}{16\nu^3} + \dots\right),$$

and

$$\mu_2' = \nu, \qquad \mu_3' = (\nu+1)\mu_1', \qquad \mu_4' = (\nu+2)\nu.$$

Thus, the moments about the mean are

$$\mu_2=\frac{1}{2}-\frac{1}{8\nu}+o(\nu^{-1}),\qquad \mu_3=\frac{1}{4\sqrt{\nu}}+o(\nu^{-\frac{1}{2}}),\qquad \mu_4=\frac{3}{4}-\frac{3}{8\nu}+o(\nu^{-1})$$

so that the moments of $\sqrt{(2\chi^2)}$ are

$$\left.\begin{aligned}\mu_1' &= (2\nu)^{\frac{1}{2}}\left(1-\frac{1}{4\nu}+\dots\right)=(2\nu-1)^{\frac{1}{2}}+o(\nu^{\frac{1}{2}}),\\ \mu_2 &= 1-\frac{1}{4\nu}+o(\nu^{-1}),\\ \gamma_1 &= \frac{1}{\sqrt{(2\nu)}}+o(\nu^{-\frac{1}{2}}),\\ \gamma_2 &= \frac{3}{4\nu^2}+o(\nu^{-2}).\end{aligned}\right\} \tag{16.8}$$

A comparison of (16.8) with (16.6) indicates that $\sqrt{(2\chi^2)}$ tends to normality with considerably greater rapidity than χ^2, with mean $(2\nu-1)^{\frac{1}{2}}$ and variance unity to the first order of approximation.

The *Biometrika Tables* give μ_1', $\mu_2^{\frac{1}{2}}$, β_1 and β_2 of $\chi/\nu^{\frac{1}{2}}$ to at least 4 d.p. for $\nu=1(1)20(5)50(10)100$. Hodges and Lehmann (1967) study the moments in detail.

16.7 For the Wilson–Hilferty cube-root transformation, consider the distribution of χ^2 about its mean value ν. Let us find the distribution of $(\chi^2/\nu)^h=y$, say, h as yet being undetermined. Write $\xi=\chi^2-\nu$. Then

$$y=\left(1+\frac{\xi}{\nu}\right)^h=\sum_{j=0}^{\infty}\binom{h}{j}\left(\frac{\xi}{\nu}\right)^j \tag{16.9}$$

Taking expectations and using the results of (16.5), we find, after some reduction,

$$\begin{aligned}\mu_1'(y) &= 1+\sum_{j=2}^{\infty}\binom{h}{j}\mu_j(\chi^2)/\nu^j\\ &= 1+\frac{h(h-1)}{\nu}+\frac{h(h-1)(h-2)(3h-1)}{6\nu^2}\\ &\quad+\frac{h^2(h-1)^2(h-2)(h-3)}{6\nu^3}+O(\nu^{-4}).\end{aligned} \tag{16.10}$$

If in (16.9) we put rh for h, we obtain $\mu_r'(y)$. We the find in particular, using (3.9),

$$\mu_3(y)=\frac{4(3h-1)}{\nu^2}+O(\nu^{-3}).$$

Thus if we take $h=\frac{1}{3}$, μ_3 will be zero to order ν^{-2}, and the distribution will presumably be brought closer to symmetry and normality. We then find, with $h=\frac{1}{3}$,

$$\left.\begin{aligned}
\mu_1'(y) &= 1-\frac{2}{9\nu}+\frac{80}{3^7\nu^3}+O(\nu^{-4}),\\
\mu_2'(y) &= 1-\frac{2}{9\nu}+\frac{4}{3^4\nu^2}+\frac{56}{3^7\nu^3}+O(\nu^{-4}),\\
\mu_3'(y) &= 1,\\
\mu_4'(y) &= 1+\frac{4}{9\nu}-\frac{4}{3^3\nu^2}+\frac{80}{3^7\nu^3}+O(\nu^{-4}),
\end{aligned}\right\} \tag{16.11}$$

or, about the mean,

$$\left.\begin{aligned}
\mu_2(y) &= \frac{2}{9\nu}-\frac{104}{3^7\nu^3}+O(\nu^{-4})\\
\mu_3(y) &= \frac{32}{3^6\nu^3}+O(\nu^{-4}),\\
\mu_4(y) &= \frac{4}{3^3\nu^2}-\frac{16}{3^6\nu^3}+O(\nu^{-4}).
\end{aligned}\right\} \tag{16.12}$$

We now find for the skewness and kurtosis coefficients

$$\left.\begin{aligned}
\gamma_1 &= \frac{2^{\frac{7}{2}}}{3^3\nu^{\frac{1}{2}}}+o(\nu^{-\frac{1}{2}})\\
\gamma_2 &= -\frac{4}{9\nu}+o(\nu^{-1}).
\end{aligned}\right\} \tag{16.13}$$

Comparison with (16.6) and (16.8) shows that $\left(\frac{\chi^2}{\nu}\right)^{\frac{1}{3}}$ tends to symmetry, as measured by γ_1, more rapidly than either χ^2 or $\sqrt{(2\chi^2)}$. To order ν^{-2} the variance is, from (16.12), equal to $\frac{2}{9\nu}$ and the mean, from (16.11), to $1-\frac{2}{9\nu}$. The result may also be expressed by saying that

$$\left\{\left(\frac{\chi^2}{\nu}\right)^{\frac{1}{3}}+\frac{2}{9\nu}-1\right\}\left(\frac{9\nu}{2}\right)^{\frac{1}{2}} \tag{16.14}$$

is approximately a standardized normal variate.

Since the mean and median of a normal variate coincide, (16.14) implies that the median of χ^2 is

$$\nu\left(1-\frac{2}{9\nu}\right)^3 = \nu-\frac{2}{3}+\frac{4}{27\nu}-\frac{8}{729\nu^2},$$

a closer approximation than the first two terms used at the end of **16.2**.

Exercise 28.25 (in Volume 2A) will give Haldane's (1938) generalization of the method of this section to a class of variates. In general, $h = 1-\kappa_1\kappa_3/(3\kappa_2^2)$, reducing to $\frac{1}{3}$ in the χ^2 case on substitution from (16.4).

16.8 The following table, quoted from Garwood (1936), shows some comparisons of the square-root and cube-root approximations with the exact values. The lower and upper 5% and 1% points of the approximate distributions of $\frac{1}{2}\chi^2$ are compared with the exact values.

Table 16.1 Comparison of approximations to the χ^2 d.f.
(Garwood (1936))

	ν	Exact (1)	Square-root (2)	(1)−(2) (3)	Cube-root (4)	(1)−(4) (5)
$P=0.99$	40	11.082	10.764	0.318	11.070	0.012
	60	18.742	18.414	0.328	18.732	0.010
	80	26.770	26.436	0.334	26.761	0.009
	100	35.032	34.694	0.338	35.025	0.007
$P=0.95$	40	13.255	13.116	0.139	13.254	0.001
	60	21.594	21.455	0.139	21.594	0.000
	80	30.196	30.056	0.140	30.196	0.000
	100	38.965	38.825	0.140	38.965	0.000
$P=0.05$	42	29.062	28.919	0.143	29.060	0.002
	62	40.691	40.548	0.143	40.689	0.002
	82	52.069	51.926	0.143	52.068	0.001
	102	63.287	63.144	0.143	63.286	0.001
$P=0.01$	42	33.103	32.700	0.403	33.113	−0.010
	62	45.401	45.003	0.398	45.409	−0.008
	82	57.347	56.953	0.394	57.355	−0.008
	102	69.067	68.676	0.391	69.074	−0.007

The cube-root approximation is evidently very good and the square-root approximation is fair.

Mathur (1961) has compared the three normal approximations that we have discussed in respect of the maximum error in the d.f. committed over the range of the variate—cf. Exercise 16.24.

Hawkins and Wixley (1986) show that the fourth-root transformation $\chi^{\frac{1}{2}}$ is close to normality with mean $(\nu-\frac{1}{2})^{\frac{1}{4}}$ and variance $(8\nu^{\frac{1}{2}})^{-1}$ for any ν, and a closer approximation than the cube-root when $\nu=1$ or 2.

Slutsky (1950) tabulates the χ^2 probabilities using the approximate normality of $\{\sqrt{(2\chi^2)}-\sqrt{(2\nu)}\}$ for $\sqrt{(2/\nu)}=0$ to 0.25, i.e. from $\nu=\infty$ down to $\nu=32$, to 5 d.p.

16.9 Another approximation may be obtained by the Cornish–Fisher method of **6.25–6**, and our Example 6.4 was in effect based on the χ^2 distribution. The Gamma variate there with parameter λ corresponds to χ^2 with 2λ degrees of freedom.

A variety of other approximations have been suggested; many of these are evaluated in the studies by Ling (1978) and El Lozy (1982). The best normal approximation appears to be that given by Peizer and Pratt (1968) and Pratt (1968), although the generalized Cornish-Fisher expansion (cf. 6.26) is even more accurate.

Student's t distribution

16.10 In Example 11.8, we have seen that, in samples from a normal population, the ratio of the difference between the sample and population means to its estimated standard error,

$$t = (\bar{x}-\mu)\sqrt{\{n(n-1)\}}/\sqrt{\sum(x-\bar{x})^2} = (\bar{x}-\mu)\sqrt{(n/s^2)},$$

where $s^2 = \sum(x-\bar{x})^2/(n-1)$, has a sampling distribution given by

$$f(t) = 1\bigg/\left\{\nu^{\frac{1}{2}}B(\tfrac{1}{2},\tfrac{1}{2}\nu)\left(1+\frac{t^2}{\nu}\right)^{\frac{1}{2}(\nu+1)}\right\}, \qquad \nu>0;\ -\infty<t<\infty. \tag{16.15}$$

As before we write ν for the degrees of freedom (d.fr.), here equal to $n-1$, of the statistic s^2 entering into the denominator of t. ν may thus, by an extension of the term, be called the number of degrees of freedom of t.

(16.15) is a Pearson Type VII distribution. It is symmetrical about the origin and unimodal, and extends to infinity in both directions. It is known as Student's t distribution (the name "Student" having been the pseudonym of its discoverer, W. S. Gosset) and is sometimes called a t_ν distribution. The case $\nu=1$ is the Cauchy distribution $\{\pi(1+t^2)\}^{-1}$—see Example 16.1 below.

It is clear, from the fact that the estimated standard error of a sample mean converges to the true standard error in large samples, that the distribution (16.15) must tend with large ν to the standardized normal distribution, for the mean of a normal sample is exactly normally distributed (Example 11.12). Compare Example 4.8.

Properties of Student's t distribution

16.11 The c.f. of the distribution has already been examined in Example 3.13, where for our present notation we replace m by $\frac{1}{2}(\nu+1)$. The moments μ_r of the distribution exist only for $r<\nu$, and are then equal to zero by symmetry for odd-order moments, while for even moments (Example 3.3)

$$\mu_{2r} = \nu^r\frac{\Gamma(r+\frac{1}{2})\Gamma(\frac{1}{2}\nu-r)}{\Gamma(\frac{1}{2})\Gamma(\frac{1}{2}\nu)}, \qquad 2r<\nu. \tag{16.16}$$

Thus the variance is $\nu/(\nu-2)$ for $\nu>2$ and the skewness and kurtosis coefficients are

$$\gamma_1=0, \qquad \gamma_2 = 6/(\nu-4), \qquad \nu>4.$$

The d.f. of t is

$$F(\tau) = \int_{-\infty}^{\tau} f(t)\,dt.$$

Thus, using the symmetry of (16.15),

$$2F-1=\frac{2}{B\left(\frac{\nu}{2},\frac{1}{2}\right)}\int_0^{\tau}\frac{1}{\left(1+\frac{t^2}{\nu}\right)^{\frac{1}{2}(\nu+1)}}\frac{\mathrm{d}t}{\sqrt{\nu}},$$

and by putting $\xi=\left(1+\frac{t^2}{\nu}\right)^{-1}$ this is seen to be

$$2F-1=\frac{1}{B\left(\frac{\nu}{2},\frac{1}{2}\right)}\int_{\xi(\tau)}^{1}\xi^{\frac{1}{2}\nu-1}(1-\xi)^{-\frac{1}{2}}\mathrm{d}\xi=1-I_{\xi}\left(\frac{\nu}{2},\frac{1}{2}\right),$$

where I is the Incomplete Beta Function of **5.7**, so that

$$F=1-\tfrac{1}{2}I_{\xi}\left(\frac{\nu}{2},\frac{1}{2}\right). \tag{16.17}$$

The values of ξ for which I_{ξ} has the values 0.50, 0.25, 0.10, 0.05, 0.025, 0.01, 0.005 and $\nu=1(1)30$, 40, 60, 120, ∞, have been tabled to five significant figures by C. M. Thompson *et al.* (1941) and used to derive the values of t for the corresponding percentage points for $F=0.75$, 0.875, 0.95, 0.975, 0.9875, 0.995 and 0.9975. These tables are reproduced in the *Biometrika Tables*. Lackritz (1984) gives another method of obtaining the percentage points.

Example 16.1 (The d.f. of the Cauchy distribution)

In the Cauchy case $\nu=1$, (16.17) may be evaluated in closed form, for the Incomplete Beta Function then is

$$\frac{1}{B(\frac{1}{2},\frac{1}{2})}\int_0^{(1+\tau^2)^{-1}}u^{-\frac{1}{2}}(1-u)^{-\frac{1}{2}}\mathrm{d}u.$$

The substitution $v=(1-u)^{\frac{1}{2}}$ reduces this to

$$\begin{aligned}\frac{2}{B(\frac{1}{2},\frac{1}{2})}\int_{\{\tau^2/(1+\tau^2)\}^{\frac{1}{2}}}^{1}(1-v^2)^{-\frac{1}{2}}\mathrm{d}v&=\frac{2}{\pi}\left[\arcsin v\right]_{\{\tau^2/(1+\tau^2)\}^{\frac{1}{2}}}^{1}\\&=1-\frac{2}{\pi}\arcsin\left(\frac{\tau^2}{1+\tau^2}\right)^{\frac{1}{2}}\\&=1-\frac{2}{\pi}\arctan\tau.\end{aligned}$$

Thus (16.17) becomes simply

$$F(t)=\tfrac{1}{2}+\frac{1}{\pi}\arctan t.$$

16.12 The d.f. of the t distribution may be evaluated using a series expansion, as we now show.

We have

$$-\log\left(1+\frac{t^2}{\nu}\right)=\frac{t^2}{\nu}+\frac{t^4}{2\nu^2}-\ldots+\frac{(-t^2)^j}{j\nu^j}+\ldots$$

and hence

$$-\tfrac{1}{2}(\nu+1)\log\left(1+\frac{t^2}{\nu}\right)=-\tfrac{1}{2}t^2+\ldots-\frac{j(-t^2)^{j+1}+(j+1)(-t^2)^j}{2j(j+1)\nu^j}+\ldots \tag{16.18}$$

Further, from the expansion (3.63) we find

$$\log\left\{\frac{\Gamma\{\frac{1}{2}(\nu+1)\}}{\Gamma(\frac{1}{2}\nu)}\sqrt{\frac{2}{\nu}}\right\}=-\frac{1}{4\nu}+\frac{1}{24\nu^3}-\frac{1}{20\nu^5}+\ldots \tag{16.19}$$

Now as ν tends to infinity, t tends to the standardized normal form, as we have already seen. Writing

$$y=\frac{1}{\sqrt{(2\pi)}}\exp\left(-\tfrac{1}{2}t^2\right), \tag{16.20}$$

we use (16.18), (16.19) and (16.20) to find, for the logarithm of $f(t)$ in (16.15), in descending powers of ν,

$$\log y+\frac{1}{4\nu}(t^4-2t^2-1)-\frac{1}{12\nu^2}(2t^6-3t^4)+\frac{1}{24\nu^3}(3t^8-4t^6+1)$$
$$-\frac{1}{40\nu^4}(4t^{10}-5t^8)+\frac{1}{60\nu^5}(5t^{12}-6t^{10}-3)-\ldots \tag{16.21}$$

Taking the exponential of (16.21) and integrating from t to ∞, we find

$$1-F=\int_t^\infty y\,dt+y\left\{\frac{1}{4\nu}t(t^2+1)+\frac{1}{96\nu^2}(3t^6-7t^4-5t^2-3)t\right.$$
$$+\frac{1}{384\nu^3}(t^{10}-11t^8+14t^6+6t^4-3t^2-15)t$$
$$\left.+\frac{1}{92\,160\nu^4}(15t^{14}-375t^{12}+2225t^{10}-2141t^8-939t^6-213t^4+915t^2+945)t+\ldots\right\}. \tag{16.22}$$

(16.22) is the expression, due to Fisher, that was used by Student himself in calculating the d.f. of t. For values of $\nu>20$ the first four terms of (16.22) give F with a maximum error of 0.000 005.

Exercise 16.10 gives a finite series expansion for the d.f. for even ν.

The Cornish-Fisher and generalized Cornish-Fisher expansions (cf. 6.26) provide excellent approximations to the t distribution, particularly as ν increases. For numerical comparisons of a variety of approximations, see Ling (1978), Bailey (1980), Gaver and Kafadar (1984) and El Lozy (1982). George and Sivaram (1987) provide modified Cornish-Fisher expansions for both the d.f. and the percentiles; these series appear to yield even greater accuracy.

Shenton and Carpenter (1965) gave a continued fraction yielding an asymptotic expansion for the Mills' ratio (cf. **5.38**) of the t distribution. Soms (1976) gives an elementary derivation and (1980) gives bounds for the ratio.

Tables of Student's t distribution

16.13

(a) The *Biometrika Tables* give the d.f. to 5 d.p. for $t=0(0.1)4(0.2)8$ and $\nu=1(1)20$ and for $t=0(0.05)2(0.1)4$, 5 for $\nu=20$, 24, 30, 40, 60, 120 and ∞, with some extreme percentage points for $\nu=1(1)10$.
(b) Fisher and Yates (1963) give percentage points of the distribution, i.e. the values of t corresponding to the values of the d.f. F, satisfying $2(1-F)=0.9(0.1)0.1$, 0.05, 0.02, 0.01, 0.001 for $\nu=1(1)30$, 40, 60, 120 and ∞. This table is given as Appendix Table 5.
(c) Thompson *et al.*'s (1941) tables of percentage points were discussed in **16.11** above.
(d) Baldwin (1946) gives percentage points for $2(1-F)=0.05$, 0.01 with $\nu=1(1)30(2)100$.
(e) Federighi (1959) gives percentage points for 20 values of $(1-F)$ between 0.25 and 0.000 000 1 and $\nu=1(1)30(5)60(10)100$, 200, 500, 1000, 2000, 10 000 and ∞.
(f) Smirnov (1961) gives the d.f. and p.d.f. to 6 d.p. at $t=0(0.1)3.00(0.02)4.50(0.05)6.50$ for $\nu=1(1)12$, and at $t=0(0.01)2.50(0.02)3.50(0.05)6.50$ for $\nu=13(1)25$; and also the d.f. at $t=0(0.01)2.50(0.02)3.50(0.05)5.00$ for $\nu=25(1)35$. There are auxiliary tables for larger values of t and ν. Tables of percentage points of the d.f. are given for $2(1-F)=0.4$, 0.25, 0.10, 0.05, 0.01, 0.005, 0.0025, 0.001, 0.0005 and $\nu=1(1)30(10)100$, 120, 150(50)500.
(g) Lempers and Louter (1971) give the percentage points of t for $F=k/16$, $k=1(1)15$ and $\nu=1(1)30$, 40, 60, 120, ∞.
(h) Mardia and Zemroch (1978) give percentage points of t for $\nu=0.1(0.1)3(0.2)7(0.5)11(1)40$, 60, 120, ∞.
(i) Kafadar and Tukey (1988) give easily-interpolated single-tail tables of the percentage points for $\alpha>10^{-4}$, $\nu=1(1)10$ and for $\alpha>10^{-6}$, $\nu=10(1)20$, 24, 30, 40, 60, 120 and ∞.

16.14 Before leaving the t distribution, we should mention that in many applications it is the distribution of t^2, rather than that of t, which is of interest. There is, of course, no trouble in passing from t to t^2 in (16.15). It turns out, however, that the distribution of t^2 is merely a special case of a more general distribution, Fisher's F or z, and it is to this distribution, the last of the three basic sampling distributions related to the normal, that we now turn.

Fisher's F and z distributions

16.15 In discussing the distribution of a ratio, we saw in Example 11.20 that in independent samples from two normal populations with equal variances, the sampling distribution of the sample variance ratio (using (n_p-1) as divisors)

$$F = \frac{\sum_{i=1}^{n_1} (x_{1i}-\bar{x}_1)^2/(n_1-1)}{\sum_{j=1}^{n_2} (x_{2j}-\bar{x}_2)/(n_2-1)}, \tag{16.23}$$

was given at (11.81) by

$$dG = \frac{\nu_1^{\frac{1}{2}\nu_1}\nu_2^{\frac{1}{2}\nu_2}F^{\frac{1}{2}(\nu_1-2)}\,dF}{B(\frac{1}{2}\nu_1,\frac{1}{2}\nu_2)(\nu_1F+\nu_2)^{\frac{1}{2}(\nu_1+\nu_2)}}, \qquad \nu_1,\nu_2>0;\quad 0\leqslant F<\infty, \tag{16.24}$$

where we have written $\nu_1 = n_1-1$, $\nu_2 = n_2-1$, for the degrees of freedom as before. If we put $x = (\nu_1/\nu_2)F$ in (16.24), we reduce it to a Beta distribution of the second kind as at (6.16), with parameters $p=\frac{1}{2}\nu_1$, $q=\frac{1}{2}\nu_2$. We also refer to (16.24) as an F_{ν_1,ν_2} distribution.

On comparing (16.24) with (16.15), it is evident that if we put $\nu_1 = 1$, $\nu_2 = \nu$, (16.24) becomes identical with the distribution of t^2 obtainable immediately from (16.15). Further, it will be seen that in Example 11.20 the only facts made use of in deriving (16.24) were that $\sum_i (x_{1i}-\bar{x}_1)^2\Big/\sum_j (x_{2j}-\bar{x}_2)^2$ be a ratio of two independently distributed χ^2 variates with ν_1 and ν_2 degrees of freedom for numerator and denominator respectively. It follows that (16.24) holds for the distribution of any such ratio. For example, we saw below (11.80) that if the normal populations we are sampling have different variances σ_1^2 and σ_2^2, it will be $(\sigma_2^2/\sigma_1^2)F$, where F is as at (16.23), that has the distribution (16.24). For, in each sample, it is the quantity

$$\sum (x_{pi}-\bar{x}_p)^2/\sigma_p^2, \qquad p=1,2,$$

which is distributed like χ^2, since σ_p is the scale factor.

When $\nu_1 = \nu_2$, another relation between (16.24) and (16.15) is given in Exercise 16.2.

16.16 The discussion of Example 11.20 introduced the simple transformation of (16.23)

$$z = \tfrac{1}{2}\log F, \tag{16.25}$$

whose distribution, given at (11.82) or immediately obtained from (16.24), is

$$dH = \frac{2\nu_1^{\frac{1}{2}\nu_1}\nu_2^{\frac{1}{2}\nu_2}}{B(\frac{1}{2}\nu_1,\frac{1}{2}\nu_2)}\cdot\frac{\exp(\nu_1 z)\,dz}{\{\nu_1\exp(2z)+\nu_2\}^{\frac{1}{2}(\nu_1+\nu_2)}}, \qquad \nu_1,\nu_2>0;\ -\infty<z<\infty. \tag{16.26}$$

It was in this form that the distribution was first obtained by R. A. Fisher. Modern practice is almost entirely in favour of using the simpler statistic F, although z was tabulated earlier, and the original mathematical investigations of the distribution were carried out in terms of z.

Properties of the F and z distributions

16.17 The general shape of the F distribution (16.24) is indicated by the fact that, while it always ranges from zero to infinity, it is J-shaped if $\nu_1 \leqslant 2$, while if $\nu_1 > 2$ it is a unimodal skew distribution with modal value at

$$\tilde{F} = \frac{\nu_1 - 2}{\nu_1} \frac{\nu_2}{\nu_2 + 2}. \tag{16.27}$$

It is easily verified directly from (16.24) that the mean and variance of F are

$$\left.\begin{aligned} \mu_1' &= \nu_2/(\nu_2 - 2), \qquad \nu_2 > 2, \\ \mu_2 &= 2\nu_2^2(\nu_1 + \nu_2 - 2)/\{\nu_1(\nu_2 - 2)^2(\nu_2 - 4)\}, \qquad \nu_2 > 4, \end{aligned}\right\} \tag{16.28}$$

the conditions on ν_2 being required for the existence of these moments.

Comparison of (16.27) and (16.28) shows that for $\nu_1, \nu_2 > 2$ the F distribution always has its modal value below $F = 1$ and its mean value above $F = 1$, confirming that the distribution is always positively skew, as is otherwise obvious. That $E(F) > 1$ irrespective of the form of the distribution is a consequence of Exercise 9.15.

Higher moments of (16.24) are given in Exercise 16.1. The c.f. is

$$\phi(t) = [\Gamma\{\tfrac{1}{2}(\nu_1 + \nu_2)\}\Gamma(\tfrac{1}{2}\nu_2)]\psi(\tfrac{1}{2}\nu_1, 1 - \tfrac{1}{2}\nu_2; -it\nu_2/\nu_1),$$

where ψ is the confluent hypergeometric function of the second kind—see Phillips (1982).

The d.f. of F (or of its monotonic function z) may be obtained from tables of the Incomplete Beta Function by making the transformation

$$\xi = (1 + \nu_1 F/\nu_2)^{-1}$$

in (16.24). We may then, as in **16.11** above, use Incomplete Beta Function tables, the d.f. of F being given by $1 - I_\xi\left(\frac{\nu_2}{2}, \frac{\nu_1}{2}\right)$, of which (16.17) is a special case—the factor $\frac{1}{2}$ in (16.17) disappears if we work in terms of t^2 instead of t. Thompson *et al.*'s (1941) tables discussed in **16.11** were used to obtain the percentage points at 0.50, 0.75, 0.90, 0.95, 0.975, 0.99 and 0.995 for $\nu_1, \nu_2 = 1(1)30, 40, 60, 120, \infty$.

Lackritz (1984) gives another method of obtaining the percentage points.

16.18 The c.f. of z is most easily obtained from the F distribution as $E(e^{\theta\frac{1}{2}\log F}) = E(F^{\frac{1}{2}\theta})$ where $\theta = it$. This is formally the $(\frac{1}{2}\theta)$th moment of the F distribution, and Exercise 16.1 shows this to be

$$\phi(t) = \left(\frac{\nu_2}{\nu_1}\right)^{\frac{1}{2}\theta} \frac{\Gamma\{\frac{1}{2}(\nu_1 + \theta)\}\Gamma\{\frac{1}{2}(\nu_2 - \theta)\}}{\Gamma(\frac{1}{2}\nu_1)\Gamma(\frac{1}{2}\nu_2)}. \tag{16.29}$$

All the moments of the z distribution are finite, as may be seen from (16.26). We shall only obtain their asymptotic forms here. Using the Stirling expansion (3.63), we find from (16.29), for the c.g.f. of z,

$$\log \phi(t) = \frac{\theta}{2}\left(\frac{1}{\nu_2} - \frac{1}{\nu_1}\right) + \frac{\theta^2}{4}\left(\frac{1}{\nu_1} + \frac{1}{\nu_2}\right) + o\left(\frac{1}{\nu}\right).... \tag{16.30}$$

For large ν_1 and ν_2, z is thus distributed approximately normally with moments

$$\mu_1' = \frac{1}{2}\left(\frac{1}{\nu_2} - \frac{1}{\nu_1}\right), \qquad \mu_2 = \frac{1}{2}\left(\frac{1}{\nu_1} + \frac{1}{\nu_2}\right). \tag{16.31}$$

The asymptotic cumulants of z are given in more detail at (16.34) below.

The z distribution may be symmetric or skew with sign depending on the values of ν_1 and ν_2, as Example 16.2 shows.

Example 16.2 (The distribution of Fisher's z)

If we put $\nu_1 = \nu_2 = 2$, (16.26) becomes

$$dH = 2e^{2z}\,dz/(e^{2z}+1)^2,$$

and thus $2z$ has the logistic distribution of Exercise 4.21. Here, and more generally whenever $\nu_1 = \nu_2$, (16.26) is symmetric about the origin, being proportional to $(\cosh z)^{-\nu}$.

When $\nu_1 \neq \nu_2$, the symmetry is lost, but (16.26) always has a unique mode at the origin, since its derivative consists of positive terms multiplied by the factor $(e^{2z}-1)$.

Since the mode is zero, (2.13) implies that the median is approximately $\frac{2}{3}$ of the mean, so that from (16.31), the median and the skewness have the same sign as $\nu_1 - \nu_2$.

Tables of F and z

16.19 Because there are two degrees of freedom entries in any table of F, a full tabulation of the distribution would require a table of triple entry, and it is therefore more convenient to use inverse tables of the quantiles of the distribution.

(a) Fisher and Yates (1963) give the percentage points of F (and z) corresponding to the values $(1-P)$ of the d.f. for $P = 0.20$, 0.10, 0.05, 0.01 and 0.001; $\nu_1 = 1(1)6$; 8, 12, 24, ∞ and $\nu_2 = 1(1)30$; 40, 60, 120, ∞. The upper 5% and 1% points tables are reproduced as Appendix Tables 6–9 see pp 607–10.

(b) The *Biometrika Tables* give the percentage points of F to 2 d.p. for $P = 0.25$, 0.10, 0.05, 0.025, 0.01, 0.005 and 0.001; $\nu_1 = 1(1)10$; 12, 15, 20, 24, 30, 40, 60, 120, ∞ and $\nu_2 = 1(1)30$; 40, 60, 120, ∞; Vol. II of the *Biometrika Tables* give 5 significant figure tables, including also $P = 0.5$ and 0.0025, with an auxiliary table for fractional d.fr.

(c) Thompson *et al.*'s (1941) tables of percentage points were discussed in **16.17** and **16.11** above.

(d) Harter (1964a) gives 23 percentage points to 7 significant figures for *even* $\nu_1, \nu_2 \leqslant 80$.

(e) Vogler (1964) gives 17 percentage points to 6 significant figures for $\nu_1 = 1(1)10$, 12, 15, 20, 24, 30, 40, 60, 120 and $\nu_2 = 1(0.1)2$, 2.2, 2.5(0.5)5, 6(1)10, 12, 15, 20, 24, 30, 40, 60, 120, ∞.

(f) Mardia and Zemroch (1978) give 20 percentage points for $\nu_1 = 0.1(0.1)1(0.2)2(0.5)5(1)16$, 18, 20, 24, 30, 40, 60, 120, ∞, and $\nu_2 = 0.1(0.1)3(0.2)7(0.5)11(1)40$, 60, 80, 120, ∞.

16.20 As is obvious from the definition of F as a ratio, we may interchange its numerator and denominator and obtain a distribution of exactly the same form, but with

ν_1 and ν_2 interchanged. This may be verified by making the transformation $F = 1/y$ in (16.24). In terms of the distribution function of F, say $G_{\nu_1,\nu_2}(F)$, this implies

$$G_{\nu_1,\nu_2}(F) = 1 - G_{\nu_2,\nu_1}\left(\frac{1}{F}\right). \tag{16.32}$$

Since there is a one-one correspondence between F and z, and $\log 1/F = -\log F$, this is equivalent to

$$H_{\nu_1,\nu_2}(z) = 1 - H_{\nu_2,\nu_1}(-z). \tag{16.33}$$

The relations (16.32) and (16.33) make it unnecessary to tabulate the distributions at both extremes. In fact, if we are interested in the d.f. for fractional values of F (i.e. negative values of z) we have only to invert the ratio (i.e. change the sign of z) and use these relations to obtain the value of the d.f. that we are seeking.

16.21 Various approximations have been given for the case when ν_1 and ν_2 are not large enough to justify the use of the normal approximation of **16.18**.

(a) (Cornish and Fisher, 1937) The method is that of **6.25** and depends on the expansion of the distribution in a Gram–Charlier series. From the successive derivatives of $\log\Gamma(1+x)$ we can find those of $\log\phi(t)$, and hence ascertain the cumulants of z. Writing $r_1 = 1/\nu_1$ and $r_2 = 1/\nu_2$, we find the leading terms in the cumulants are, to order $2 + [\frac{1}{2}j]$ in r for the jth cumulant:

$$\left.\begin{aligned} \kappa_1 &= -\tfrac{1}{2}(r_1 - r_2) - \tfrac{1}{6}(r_1^2 - r_2^2), & \kappa_4 &= r_1^3 + r_2^3 + 3(r_1^4 + r_2^4), \\ \kappa_2 &= \tfrac{1}{2}(r_1 + r_2) + \tfrac{1}{2}(r_1^2 + r_2^2) + \tfrac{1}{3}(r_1^3 + r_2^3), & \kappa_5 &= -3(r_1^4 - r_2^4), \\ \kappa_3 &= -\tfrac{1}{2}(r_1^2 - r_2^2) - (r_1^3 - r_2^3), & \kappa_6 &= 12(r_1^5 + r_2^5). \end{aligned}\right\} \tag{16.34}$$

It will be seen that the value of κ_3 confirms the conclusion of Example 16.2, that skewness takes the sign of $(\nu_1 - \nu_2)$.

Hence, putting $\sigma = r_1 + r_2$ and $\delta = r_1 - r_2$, we find for the ls of **6.25** ($m = 0$, variance $= \frac{1}{2}\sigma$):

$$l_1 = -\left(\frac{2}{\sigma}\right)^{\frac{1}{2}}(\tfrac{1}{2}\delta + \tfrac{1}{6}\delta\sigma), \qquad l_2 = \tfrac{1}{2}\left(\sigma + \frac{\delta^2}{\sigma}\right) + \tfrac{1}{6}(\sigma^2 + 3\delta^2),$$

and so on. After some reduction we find, for the value of z corresponding to a probability α (which in turn corresponds to a standardized normal variate value ξ)—

$$\begin{aligned} z = \xi\left(\frac{\sigma}{2}\right)^{\frac{1}{2}} &- \tfrac{1}{6}\delta(\xi^2 + 2) + \left(\frac{\sigma}{2}\right)^{\frac{1}{2}}\left\{\frac{\sigma}{24}(\xi^3 + 3\xi) + \frac{1}{72}\frac{\delta^2}{\sigma}(\xi^3 + 11\xi)\right\} \\ &- \frac{\delta\sigma}{120}(\xi^4 + 9\xi^2 + 8) + \frac{\delta^3}{3240\sigma}(3\xi^4 + 7\xi^2 - 16) + \left(\frac{\sigma}{2}\right)^{\frac{1}{2}}\left\{\frac{\sigma^2}{1920}(\xi^5 + 20\xi^3 + 15\xi)\right. \\ &\left. + \frac{\delta^4}{2880}(\xi^5 + 44\xi^3 + 183\xi) + \frac{\delta^4}{155\,520\sigma^2}(9\xi^5 - 284\xi^3 - 1513\xi)\right\}. \end{aligned} \tag{16.35}$$

(b) (Fisher, extended by Cochran (1940).) Writing n indifferently for ν_1 and ν_2, we have, from (16.35), to order $n^{-3/2}$:

$$z = \xi\left(\frac{\sigma}{2}\right)^{\frac{1}{2}} - \tfrac{1}{6}\delta(\xi^2+2) + \left(\frac{\sigma}{2}\right)^{\frac{1}{2}}\left\{\frac{\sigma}{24}(\xi^3+3\xi) + \frac{1}{72}\frac{\delta^2}{\sigma}(\xi^3+11\xi)\right\}.$$

Put $h = 2/\sigma$. Then

$$z = \frac{\xi}{\sqrt{h}} - \tfrac{1}{6}\delta(\xi+2) + \frac{1}{\sqrt{h}}\left\{\frac{\xi^3+3\xi}{12h} + \frac{\xi^3+11\xi}{144}h\delta^2\right\}. \tag{16.36}$$

Now

$$\frac{\xi}{\sqrt{(h-\lambda)}} = \frac{\xi}{\sqrt{h}} + \frac{\lambda\xi}{2h\sqrt{h}} + o(h^{-\frac{3}{2}}).$$

Hence, if we put $\lambda = \dfrac{\xi^2+3}{6}$ and

$$z = \frac{\xi}{\sqrt{(h-\lambda)}} - \tfrac{1}{6}\delta(\xi^2+2), \tag{16.37}$$

the difference of (16.37) from (16.36) is

$$\frac{(\xi^3+11\xi)\delta^2\sqrt{h}}{144},$$

and this is small because of the large denominator and the factor $\delta^2 = \left(\dfrac{1}{\nu_1} - \dfrac{1}{\nu_2}\right)^2$ which is small if ν_1 and ν_2 are not too different. Thus we may take z as approximately given by (16.37). The values of λ for various values of the d.f. $(1-P)$ are:

$100P\%$	40%	30%	20%	10%	5%	1%	0.1%
λ	0.51	0.55	0.62	0.77	0.95	1.40	2.09

For the commoner values of P the form taken by (16.37) is

$$\text{20 percent: } \frac{0.8416}{\sqrt{(h-\lambda)}} - 0.4514\delta; \tag{16.38}$$

$$\text{5 percent: } \frac{1.6449}{\sqrt{(h-\lambda)}} - 0.7843\delta; \tag{16.39}$$

$$\text{1 percent: } \frac{2.3263}{\sqrt{(h-\lambda)}} - 1.235\delta; \tag{16.40}$$

$$\text{0.1 percent: } \frac{3.0902}{\sqrt{(h-\lambda)}} - 1.925\delta. \tag{16.41}$$

The accuracy of the approximation for $\nu_1 = 24$, $\nu_2 = 60$ may be judged from the following comparison:

100 P %	Value of z from (16.37)	Exact value
20	0.1337	0.1338
1	0.3748	0.3746
0.1	0.4966	0.4955

(c) (Paulson, 1942) Let $p = s_1^2/s_2^2$ be the ratio of two independent quantities distribution as χ^2/ν with ν_1 and ν_2 d.fr. The cube-root approximation to χ^2 of **16.7** indicates that $s_j^{2/3}$ is distributed normally about mean $\mu_j = 1 - \dfrac{2}{9\nu_j}$ with variance $\sigma_j^2 = \dfrac{2}{9\nu_j}$. Exercise 11.11 then shows that

$$u = \frac{\mu_1 - \mu_2 p^{1/3}}{(\sigma_1^2 + \sigma_2^2 p^{2/3})^{1/2}} = \frac{\left(1 - \dfrac{2}{9\nu_1}\right) - \left(1 - \dfrac{2}{9\nu_2}\right)\left(\dfrac{s_1}{s_2}\right)^{2/3}}{\left\{\dfrac{2}{9\nu_1} + \dfrac{2}{9\nu_2}\left(\dfrac{s_1}{s_2}\right)^{4/3}\right\}^{1/2}} \tag{16.42}$$

is approximately standardized normal. The approximation seems remarkably good. For instance, for $\nu_1 = 6$, $\nu_2 = 12$, the table below shows the value of s_1^2/s_2^2 obtained by equating (16.42) to the value of the appropriate upper percentile of, the normal distribution, together with the exact percentiles of s_1^2/s_2^2 obtained from the F distribution in Appendix Tables 7 and 9 and in Fisher and Yates' *Tables* (1963).

100 P %	Approximate percentile of s_1^2/s_2^2, from (16.42)	Exact percentile of the F distribution
20	1.72	1.72
5	3.00	3.00
1	4.85	4.82
0.1	8.58	8.38

Comparative studies by Ling (1978) and El Lozy (1982) suggest that the normal approximation of Peizer and Pratt (1968) and Pratt (1968) and the generalized Cornish-Fisher expansion of Hill and Davis (1968) provide the most accurate approximations.

Wise (1960) shows that $(-\log \zeta)^{1/3}$ is approximately normal, ζ being the Beta variable obtained from F by the transformation in **16.17**.

Relations among the distributions

16.22 We have now completed our study of the three fundamental sampling distributions arising from the univariate normal. Before proceeding to the development of the sampling theory of the bivariate normal distribution, we shall summarize the various relations existing between these three derived distributions and the normal distribution. In brief, these are as follows:

(1) The sum of squares of n standardized normal variates about the population (or sample) mean has a χ^2 distribution with $\nu = n$ (or $(n-1)$) d.fr.

(2) The χ^2 distribution itself is approximately of the normal form for increasing d.fr.

(3) The ratio of the square of a normal variate with zero mean to an independent estimator of its variance σ^2 (distributed like $\chi^2\sigma^2/\nu$) has Student's t^2 distribution, and its square root a Student's t distribution.

(4) Student's t distribution tends to the standardized normal form as its degrees of freedom increase.

(5) The ratio of two independent variates, each distributed like χ^2/ν, has a Fisher's F distribution, and half its natural logarithm has the z distribution. The t^2 distribution is the special case of the F distribution where the numerator variate has only one d.fr.—see (3) and (1) above.

(6) If denominator d.fr., ν_2, alone tend to infinity, it is easily seen that $\nu_1 F$ tends to a χ^2 distribution with ν_1 d.fr., for then we effectively return from (5) to (1) above.

(7) As d.fr. increase for both numerator and denominator, the F distribution tends to normality, from (6) and (2).

Use of these relationships between the distributions often makes it possible to use one set of tables for many purposes.

The bivariate normal (binormal) distribution

16.23 We have seen (Example 15.1) that the binormal distribution has p.d.f.

$$f(x,y) = \frac{1}{2\pi\sigma_1\sigma_2(1-\rho^2)^{\frac{1}{2}}}\exp\left\{-\frac{1}{2(1-\rho^2)}\left(\frac{x^2}{\sigma_1^2}-\frac{2\rho xy}{\sigma_1\sigma_2}+\frac{y^2}{\sigma_2^2}\right)\right\}, \quad |\rho|<1, \quad (16.43)$$

where σ_1^2, σ_2^2 are the variances of the variates, μ_{11} their covariance and $\rho = \mu_{11}/(\sigma_1\sigma_2)$ is called the *correlation coefficient* between x and y in the distribution. In (16.43), we have measured each variate from its mean, a simplification which affects none of the results that we shall discuss.

By rewriting the exponent in (16.43) alternatively as

$$-\frac{1}{2(1-\rho^2)}\left\{\left(\frac{x}{\sigma_1}-\rho\frac{y}{\sigma_2}\right)^2+\frac{y^2}{\sigma_2^2}(1-\rho^2)\right\} \quad (16.44)$$

and

$$-\frac{1}{2(1-\rho^2)}\left\{\left(\frac{y}{\sigma_2}-\rho\frac{x}{\sigma_1}\right)^2+\frac{x^2}{\sigma_1^2}(1-\rho^2)\right\}, \quad (16.45)$$

we see from (16.44) that the conditional distribution of x, given any fixed value of y, is univariate normal with mean and variance

$$\left.\begin{aligned} E(x|y) &= \rho y\sigma_1/\sigma_2, \\ \text{var}\,(x|y) &= \sigma_1^2(1-\rho^2). \end{aligned}\right\} \quad (16.46)$$

From (16.45), analogous results follow for y, given x, which are (16.46) with x, y and subscripts 1, 2 interchanged.

It follows from (16.46) that the conditional mean of either variate, given the value of the other, is a linear function of that value, and that its conditional variance is constant, independent of that value. We may express this by saying that the *regression* of x on y (and of y on x) is linear and homoscedastic.(*)

In 26.44, Volume 2A it is shown that if both the regressions of x on y and of y on x are linear and homoscedastic, the distribution can only be binormal.

In Chapters 26–28, Volume 2A we shall discuss the theory of correlation and regression in detail. We now formally derive the sampling distributions of the sample correlation and regression coefficients in samples from a bivariate normal population.

The distribution of the sample correlation coefficient

16.24 The joint density of n sample values $(x_1, y_1) \dots (x_n, y_n)$ from a bivariate normal population with means μ_1 and μ_2 is

$$dF = \frac{1}{(2\pi)^n \sigma_1^n \sigma_2^n (1-\rho^2)^{\frac{1}{2}n}} \exp\left[-\frac{1}{2(1-\rho^2)}\left\{\sum\left(\frac{x-\mu_1}{\sigma_1}\right)^2 - 2\rho\sum\frac{(x-\mu_1)(y-\mu_2)}{\sigma_1\sigma_2} + \sum\left(\frac{y-\mu_2}{\sigma_2}\right)^2\right\}\right] dx_1\, dy_1 \dots dx_n\, dy_n. \quad (16.47)$$

The exponent in (16.47) may be expressed solely in terms of the five parameters of the distribution $(\mu_1, \mu_2, \sigma_1^2, \sigma_2^2, \rho)$ and the corresponding sample statistics

$$\left.\begin{aligned} \bar{x} &= \frac{1}{n}\sum x, \bar{y} = \frac{1}{n}\sum y, \\ s_1^2 &= \frac{1}{n}\sum (x-\bar{x})^2,\ s_2^2 = \frac{1}{n}\sum (y-\bar{y})^2, \\ r &= \frac{1}{n}\sum (x-\bar{x})(y-\bar{y})/(s_1 s_2) \end{aligned}\right\} \quad (16.48)$$

the last of these being the sample correlation coefficient. For example, leaving aside the factor $-\dfrac{1}{2(1-\rho^2)}$, the first term in the exponent of (16.47) is

$$\sum\left(\frac{x-\mu_1}{\sigma_1}\right)^2 = \sum\left\{\frac{(x-\bar{x})+(\bar{x}-\mu_1)}{\sigma_1}\right\}^2 = \frac{n}{\sigma_1^2}\{s_1^2 + (\bar{x}-\mu)^2\}, \quad (16.49)$$

the cross-product term vanishing since $\sum (x - \bar{x}) = 0$. The other two terms in the exponent of (16.47) can similarly be resolved into forms like (16.49), so we obtain for

(*) The terms "homoscedastic" for "of constant variance", and its opposite "heteroscedastic", are due to K. Pearson.

the term in braces in the exponent

$$n\left\{\left(\frac{\bar{x}-\mu_1}{\sigma_1}\right)^2-2\rho\left(\frac{\bar{x}-\mu_1}{\sigma_1}\right)\left(\frac{\bar{y}-\mu_2}{\sigma_2}\right)+\left(\frac{\bar{y}-\mu_2}{\sigma_2}\right)^2\right\}+n\left(\frac{s_1^2}{\sigma_1^2}-\frac{2\rho r s_1 s_2}{\sigma_1\sigma_2}+\frac{s_2^2}{\sigma_2^2}\right). \tag{16.50}$$

We proceed to find the joint sampling distribution of the five statistics defined at (16.48). To do this, we must transform (16.47) suitably. With the aid of (16.50) we can express the whole of the non-differential part of (16.47) in terms of the five statistics and the five parameters. We now consider the element $dv = dx_1\, dy_1 \ldots dx_n\, dy_n$.

Generalizing the geometrical approach of Chapter 11, we consider a sample space of n dimensions for x, and another such space for y, superimposed upon the first. The sample point varies in each of the two spaces, but not independently so. In fact, if P represents the point $(x_1, \ldots, x_n)$ in the x-space and Q the point $(y_1, \ldots, y_n)$ in the y-space, and if O_1, O_2 are the points $(\bar{x}, \ldots, \bar{x})$, $(\bar{y}, \ldots, \bar{y})$, then for any given r we have

$$r=\frac{\sum(x-\bar{x})(y-\bar{y})}{ns_1s_2}=\frac{\sum(x-\bar{x})(y-\bar{y})}{\{\sum(x-\bar{x})^2\sum(y-\bar{y})^2\}^{\frac{1}{2}}}$$

and thus r is the cosine of the angle, say θ, between O_1P and O_2Q, so that if P and r are fixed, Q varies on the cone in the y-space obtained by rotating O_2Q so that the angle made with O_1P is constant.

The element in the x-space is proportional to $s_1^{n-2}\, ds_1\, d\bar{x}$, as was seen in Example 11.7. For given $\bar{y}$ and s_2 the point Q likewise varies on the $(n-2)$-dimensional surface of the hypersphere of radius $s_2\sqrt{n}$, centre $\bar{y}$ and $(n-1)$ dimensions, but another dimension is lost by keeping θ constant, so that Q lies on a surface sphere of $(n-3)$ dimensions. This $(n-3)$-dimensional surface has radius $s_2\sqrt{n}\sin\theta = s_2\sqrt{n}(1-r^2)^{\frac{1}{2}}$ and "width" $s_2\sqrt{n}\, d\theta = \dfrac{s_2\sqrt{n}\, dr}{(1-r^2)^{\frac{1}{2}}}$ and thus its content is proportional to

$$\{s_2\sqrt{n}(1-r^2)^{\frac{1}{2}}\}^{n-3}\frac{s_2\sqrt{n}\, dr}{(1-r^2)^{\frac{1}{2}}},$$

that is, to $s_2^{n-2}(1-r^2)^{\frac{1}{2}(n-4)}$.

Thus the volume element may be written

$$\begin{aligned} dv &\propto s_1^{n-2}\, ds_1\, d\bar{x}\cdot s_2^{n-2}(1-r^2)^{\frac{1}{2}(n-4)}\, ds_2\, d\bar{y}\, dr \\ &\propto s_1^{n-2}s_2^{n-2}\, ds_1\, ds_2 (1-r^2)^{\frac{1}{2}(n-4)}\, dr\, d\bar{x}\, d\bar{y}. \end{aligned} \tag{16.51}$$

From (16.47), (16.50) and (16.51), the joint density of the five variables is proportional to

$$\exp\left(-\frac{n}{2(1-\rho^2)}\left[\left\{\frac{(\bar{x}-\mu_1)^2}{\sigma_1^2}-2\rho\frac{(\bar{x}-\mu_1)(\bar{y}-\mu_2)}{\sigma_1\sigma_2}+\frac{(\bar{y}-\mu_2)^2}{\sigma_2^2}\right\}+\left\{\frac{s_1^2}{\sigma_1^2}-2\rho r\frac{s_1s_2}{\sigma_1\sigma_2}+\frac{s_2^2}{\sigma_2^2}\right\}\right]\right)dv. \tag{16.52}$$

This fundamental result is due to R. A. Fisher (1915).

16.25 One important property of (16.52) may be noted. The distribution may be factorized into two parts, one containing only $\bar{x}$ and $\bar{y}$ and the other only s_1, s_2 and r, namely

$$dF \propto \exp\left[-\frac{n}{2(1-\rho^2)}\left\{\frac{(\bar{x}-\mu_1)^2}{\sigma_1^2} - 2\rho\frac{(\bar{x}-\mu_1)(\bar{y}-\mu_2)}{\sigma_1\sigma_2} + \frac{(\bar{y}-\mu_1)^2}{\sigma_2^2}\right\}\right] d\bar{x}\, d\bar{y} \tag{16.53}$$

and

$$dF \propto \exp\left[-\frac{n}{2(1-\rho^2)}\left\{\frac{s_1^2}{\sigma_1^2} - \frac{2\rho r s_1 s_2}{\sigma_1\sigma_2} + \frac{s_2^2}{\sigma_2^2}\right\}\right] s_1^{n-2} s_2^{n-2}(1-r^2)^{\frac{1}{2}(n-4)}\, ds_1\, ds_2\, dr. \tag{16.54}$$

Thus we see that in normal samples the (bivariate) distribution of means is entirely independent of the (trivariate) distribution of the variances and correlation coefficient. This is a characteristic property of multivariate normality—cf. **15.24**.

Before leaving (16.53), we may also note that the means are themselves distributed exactly in the binormal form, with $E(\bar{x}) = \mu_1$, $E(\bar{y}) = \mu_2$, $\text{var}\,\bar{x} = \frac{\sigma_1^2}{n}$, $\text{var}\,\bar{y} = \frac{\sigma_2^2}{n}$ (all of which results are already familiar), and

$$\text{cov}(\bar{x}, \bar{y}) = \frac{\sigma_1\sigma_2\rho}{n}, \tag{16.55}$$

so that the correlation between $\bar{x}$ and $\bar{y}$ is ρ, the correlation in the population. We indicated a more general result in Exercise 13.2.

16.26 We may now use (16.54) to obtain the distribution of the sample correlation coefficient by integrating with respect to s_1 and s_2 from 0 to ∞. We first integrate (16.54) over its whole domain to evaluate its constant of integration.

Make the variate transformation

$$\left.\begin{aligned} a &= \frac{s_1^2}{\sigma_1^2}\cdot\frac{n}{2(1-\rho^2)} \\ b &= \frac{r s_1 s_2}{\sigma_1\sigma_2}\cdot\frac{n}{2(1-\rho^2)} \\ c &= \frac{s_2^2}{\sigma_2^2}\cdot\frac{n}{2(1-\rho^2)}. \end{aligned}\right\} \tag{16.56}$$

We have for the reciprocal of the Jacobian

$$\frac{\partial(a, b, c)}{\partial(s_1, r, s_2)} = \begin{vmatrix} \frac{2s_1}{\sigma_1^2}\cdot\frac{n}{2(1-\rho^2)} & 0 & 0 \\ \frac{r s_2}{\sigma_1\sigma_2}\cdot\frac{n}{2(1-\rho^2)} & \frac{s_1 s_2}{\sigma_1\sigma_2}\cdot\frac{n}{2(1-\rho^2)} & \frac{r s_1}{\sigma_1\sigma_2}\cdot\frac{n}{2(1-\rho^2)} \\ 0 & 0 & \frac{2s_2}{\sigma_2^2}\cdot\frac{n}{2(1-\rho^2)} \end{vmatrix}$$

$$= \frac{s_1^2 s_2^2 n^3}{2\sigma_1^3\sigma_2^3(1-\rho^2)^3} = \frac{2acn}{\sigma_1\sigma_2(1-\rho^2)}$$

and also the relation

$$r^2 = b^2/ac.$$

The integral of (16.54) then becomes

$$\int \exp[-a+2\rho b-c]\cdot\left\{\frac{2a\sigma_1^2(1-\rho^2)}{n}\right\}^{\frac{1}{2}(n-2)}$$

$$\times\left\{\frac{2c\sigma_2^2(1-\rho^2)}{n}\right\}^{\frac{1}{2}(n-2)}\left(1-\frac{b^2}{ac}\right)^{\frac{1}{2}(n-4)}\frac{\sigma_1\sigma_2(1-\rho^2)}{2acn}\,da\,db\,dc$$

$$=\frac{2^{n-3}\sigma_1^{n-1}\sigma_2^{n-1}(1-\rho^2)^{n-1}}{n^{n-1}}\int \exp[-a+2\rho b-c](ac-b^2)^{\frac{1}{2}(n-4)}\,da\,db\,dc \quad (16.57)$$

where the limits of a and c are 0 to ∞ and those of b are $\pm\sqrt{ac}$ since by the Cauchy inequality $r^2 \leqslant 1$. This integral may be evaluated in terms of the Γ function. Putting $\xi = a-\frac{b^2}{c}$ we find

$$\int \exp(-\xi)\exp\left(+2\rho b-c-\frac{b^2}{c}\right)\xi^{\frac{1}{2}(n-4)}c^{\frac{1}{2}(n-4)}\,d\xi\,db\,dc,$$

$$0\leqslant\xi<\infty.\quad -\infty<b<\infty,\ 0\leqslant c<\infty,$$

$$=\Gamma\{\tfrac{1}{2}(n-2)\}\int \exp\left[-\frac{1}{c}\{(b-\rho c)^2+(1-\rho^2)c^2\}\right]c^{\frac{1}{2}(n-4)}\,db\,dc$$

$$=\Gamma\{\tfrac{1}{2}(n-2)\}\sqrt{\pi}\int \exp\{-(1-\rho^2)c\}c^{\frac{1}{2}(n-3)}\,dc$$

$$=\sqrt{\pi}\frac{\Gamma\{\frac{1}{2}(n-2)\}\Gamma\{\frac{1}{2}(n-1)\}}{(1-\rho^2)^{\frac{1}{2}(n-1)}}$$

$$=\frac{\pi\Gamma(n-2)}{2^{n-3}(1-\rho^2)^{\frac{1}{2}(n-1)}}, \quad (16.58)$$

using (3.66). The reciprocal of the product of the constants in (16.57) and (16.58) is the constant that we require for (16.54). Thus the joint distribution of s_1, s_2 and r is

$$dF=\frac{n^{n-1}}{\pi\sigma_1^{n-1}\sigma_2^{n-1}(1-\rho^2)^{\frac{1}{2}(n-1)}\Gamma(n-2)}\exp\left[-\frac{n}{2(1-\rho^2)}\left\{\frac{s_1^2}{\sigma_1^2}-\frac{2\rho r s_1 s_2}{\sigma_1\sigma_2}+\frac{s_2^2}{\sigma_2^2}\right\}\right]$$

$$\times s_1^{n-2}s_2^{n-2}(1-r^2)^{\frac{1}{2}(n-4)}\,ds_1\,ds_2\,dr. \quad (16.59)$$

16.27 Now put

$$\zeta=\frac{s_1 s_2}{\sigma_1\sigma_2},\qquad \beta=\log\frac{\sigma_2 s_1}{\sigma_1 s_2},\qquad r=r.$$

We find, for the reciprocal of the Jacobian,

$$\frac{\partial(\zeta, \beta, r)}{\partial(s_1, s_2, r)} = \begin{vmatrix} \dfrac{s_2}{\sigma_1\sigma_2} & \dfrac{s_1}{\sigma_1\sigma_2} & 0 \\ \dfrac{1}{s_1} & -\dfrac{1}{s_2} & 0 \\ 0 & 0 & 1 \end{vmatrix} = -\frac{2}{\sigma_1\sigma_2}.$$

The exponent in (16.59) becomes

$$\exp\left[-\frac{n}{2(1-\rho^2)}\{\zeta e^{\beta} - 2\rho r\zeta + \zeta e^{-\beta}\}\right]$$

and after a little reduction the distribution becomes

$$dF = \frac{n^{n-1}}{2\pi(1-\rho^2)^{\frac{1}{2}(n-1)}\Gamma(n-2)}$$
$$\times \exp\left[-\frac{n}{(1-\rho^2)}\zeta(\cosh\beta - \rho r)\right]\zeta^{n-2}\, d\zeta\, d\beta(1-r^2)^{\frac{1}{2}(n-4)}\, dr.$$

On integration with respect to ζ over its range $(0, \infty)$, we have

$$dF = \frac{(1-\rho^2)^{\frac{1}{2}(n-1)}}{2\pi\Gamma(n-2)} \frac{(1-r^2)^{\frac{1}{2}(n-4)}\Gamma(n-1)}{(\cosh\beta - \rho r)^{n-1}}\, d\beta\, dr. \tag{16.60}$$

(16.60) is an even function of β, so we integrate out β from 0 to ∞, dropping the factor 2 from the denominator to compensate.

Putting $-\rho r = \cos\theta$ we have, since

$$\int_0^\infty \frac{d\beta}{\cosh\beta + \cos\theta} = \frac{\theta}{\sin\theta},$$

$$dF = \frac{(1-\rho^2)^{\frac{1}{2}(n-1)}}{\pi\Gamma(n-2)}(1-r^2)^{\frac{1}{2}(n-4)} \frac{d^{n-2}}{d(-\cos\theta)^{n-2}}\left(\frac{\theta}{\sin\theta}\right) dr$$
$$= \frac{(1-\rho^2)^{\frac{1}{2}(n-1)}}{\pi\Gamma(n-2)}(1-r^2)^{\frac{1}{2}(n-4)} \frac{d^{n-2}}{d(\rho r)^{n-2}}\left\{\frac{\arccos(-\rho r)}{\sqrt{(1-\rho^2 r^2)}}\right\} dr, \quad -1 \leqslant r \leqslant 1. \tag{16.61}$$

These results are due to Fisher (1915).

16.28 In the particular case $\rho = 0$, (16.61) reduces, as can also be seen directly from the factorization of (16.59), into three independent parts, to

$$dF = \frac{1}{B\{\frac{1}{2}, \frac{1}{2}(n-2)\}}(1-r^2)^{\frac{1}{2}(n-4)}\, dr, \tag{16.62}$$

a form surmised by Student (1908b). Its distribution function may be obtained from tables of the Incomplete Beta Function, or more directly by putting

$$t = \{(n-2)r^2/(1-r^2)\}^{\frac{1}{2}}, \tag{16.63}$$

which reduces (16.62) to a Student's t distribution (16.15) with $\nu = n-2$ d.fr.

16.29 When $n=2$ the distribution of (16.61) becomes nugatory because of the factor $\Gamma(n-2)$. This is understandable because, for samples of two, r must be either $+1$ or -1. In such a case, then, we have a discrete distribution which we may regard as an extreme case of a U-shaped distribution.

When $n=3$ we find for the density (16.61), with $\cos\theta = -\rho r$ as before,

$$f \propto \left\{\frac{1}{\sin^2\theta} - \frac{\theta\cos\theta}{\sin^3\theta}\right\}\frac{1}{(1-r^2)^{\frac{1}{2}}},$$

again a U-shaped distribution. For $n=4$,

$$f \propto \frac{1}{\sin^3\theta}(\theta - 3\cot\theta + 3\theta\cot^2\theta).$$

If $\rho=0$ this reduces to the uniform distribution $f=\frac{1}{2}$; if $\rho\neq 0$, the distribution is J-shaped.

For $n>4$ the density function is unimodal and increasingly skew as $|\rho|$ increases, as follows from the facts that the mode moves with ρ and that r, being the cosine of an angle, satisfies $r^2 \leqslant 1$. For any ρ, as $n\to\infty$, the distribution of r tends to normality, though slowly. Some interesting photographs of models of these curves are given in the *Co-operative Study* (1917).

16.30 We may express the density function (16.61) as a hypergeometric function, as Hotelling (1953) pointed out. Consider the integral

$$I_{n-1} = \int_0^\infty \frac{d\beta}{(\cosh\beta - \rho r)^{n-1}}. \tag{16.64}$$

The substitution $\cosh\beta = (1-\rho r z)/(1-z)$ transforms (16.64) to

$$I_{n-1} = \frac{1}{\sqrt{2}(1-\rho r)^{n-\frac{3}{2}}}\int_0^1 z^{-\frac{1}{2}}(1-z)^{n-2}\{1-\tfrac{1}{2}(1+\rho r)z\}^{-\frac{1}{2}}dz,$$

and if we expand the factor in braces under the integral sign into a uniformly convergent series, and integrate the series term-by-term, we obtain

$$I_{n-1} = \frac{B(\frac{1}{2}, n-1)}{\sqrt{2}(1-\rho r)^{n-\frac{3}{2}}}F(\tfrac{1}{2}, \tfrac{1}{2}, n-\tfrac{1}{2}, \tfrac{1}{2}(1+\rho r)) \tag{16.65}$$

in the usual hypergeometric notation, so that (16.60), (16.64) and (16.65) yield the alternative expression for (16.61)

$$dF = \frac{(n-2)\,dr}{(n-1)\sqrt{2}B(\frac{1}{2}, n-\frac{1}{2})} \times (1-\rho^2)^{\frac{1}{2}(n-1)}(1-r^2)^{\frac{1}{2}(n-4)}(1-\rho r)^{3/2-n}F(\tfrac{1}{2}, \tfrac{1}{2}, n-\tfrac{1}{2}, \tfrac{1}{2}(1+\rho r)). \quad (16.66)$$

(16.66) converges rapidly even for small n. For large n, the first term is often enough. The error caused by stopping at any stage is considerably less than $2/(1-\rho r)$ times the last term used.

A recurrence relation and a differential equation for the density of r are given in Exercises 16.14–15.

16.31 Term-by-term integration of the uniformly convergent series (16.66) gives us an expression for the d.f. of r, since the powers of $(1+\rho r)$ can be expressed as powers of $\{2-(1-\rho r)\}$ and expanded binomially as a power series in $(1-\rho r)$. For $r > \rho$, Hotelling (1953) obtained the result

$$1 - F(r) = \frac{(n-2)}{(n-1)\sqrt{2}B(\frac{1}{2}, n-\frac{1}{2})}\left\{M_0 + \frac{2M_0 - M_1}{4(2n-1)} + \frac{9(4M_0 - 4M_1 + M_2)}{32(2n-1)(2n+1)} + \ldots\right\} \quad (16.67)$$

where

$$M_k = \int_r^1 (1-\rho^2)^{\frac{1}{2}(n-1)}(1-x^2)^{\frac{1}{2}(n-4)}(1-\rho x)^{3/2+k-n}\,dx, \qquad k = 0, 1, 2, \ldots,$$

and the asymptotic formula, for $r > \rho$,

$$1 - F(r) = \frac{(n-2)}{(n-1)^2\sqrt{2}B(\frac{1}{2}, n-\frac{1}{2})} \cdot \frac{(1-\rho^2)^{\frac{1}{2}(n-1)}(1-r^2)^{\frac{1}{2}(n-2)}}{(r-\rho)(1-\rho r)^{n-\frac{3}{2}}}\{1 + O(n^{-1})\}. \quad (16.68)$$

The probability that r is positive is particularly simple—Exercise 16.6 gives the exact result in terms of a Student's t variate

$$P(r \geqslant 0) = P\left\{t_{n-1} \geqslant -\left[\frac{(n-1)\rho^2}{1-\rho^2}\right]^{\frac{1}{2}}\right\}.$$

The density and d.f. of the correlation coefficient have been tabulated by F. N. David (1938) for values of $n = 3(1)25$, 50, 100, 200 and 400; for $\rho = 0.0(0.1)0.9$; and for $r = -1.00(0.05)1.00$, with finer intervals in places—Boomsma (1975) corrects errors in the d.f. table for $n = 100$, $\rho = 0.4$. Lee (1972) gives 5% and 1% points of $|r|$ for $|\rho| = 0(0.1)0.9$ and $(n-1)^{\frac{1}{2}} = 60/v$ with $v = 1(1)6(2)20$. Odeh (1982a) gives further tables.

16.32 The moments of the distribution are also expressible in terms of hypergeometric functions. Returning to (16.59), let us write

$$\alpha_1^2 = \frac{\sigma_1^2(1-\rho^2)}{n}, \qquad \alpha_2^2 = \frac{\sigma_2^2(1-\rho^2)}{n}.$$

After a little rearrangement (16.59) becomes

$$dF = \frac{(1-\rho^2)^{\frac{1}{2}(n-1)}}{\pi\Gamma(n-2)} \exp\left\{-\frac{1}{2}\frac{s_1^2}{\alpha_1^2} - \frac{1}{2}\frac{s_2^2}{\alpha_2^2} + \rho r \frac{s_1 s_2}{\alpha_1 \alpha_2}\right\}$$
$$\times \left(\frac{s_1}{\alpha_1}\right)^{n-2} \left(\frac{s_2}{\alpha_2}\right)^{n-2} (1-r^2)^{\frac{1}{2}(n-4)} \left(\frac{ds_1}{\alpha_1}\right)\left(\frac{ds_2}{\alpha_2}\right) dr. \tag{16.69}$$

Putting $u_1 = \frac{s_1}{\alpha_1}$, $u_2 = \frac{s_2}{\alpha_2}$ and expanding the term in $\exp\left(\rho r \frac{s_1 s_2}{\alpha_1 \alpha_2}\right)$ in (16.69), we have

$$dF = \frac{(1-\rho^2)^{\frac{1}{2}(n-1)}}{\pi\Gamma(n-2)} \exp\left(-\tfrac{1}{2}u_1^2 - \tfrac{1}{2}u_2^2\right) u_1^{n-2} u_2^{n-2} (1-r^2)^{\frac{1}{2}(n-4)}$$
$$\times \sum_{j=0}^{\infty} \frac{(\rho r u_1 u_2)^j}{j!} du_1\, du_2\, dr. \tag{16.70}$$

Integrating u_2 from 0 to ∞ in (16.70) we find for the distribution of u_1 and r

$$dF = \frac{(1-\rho^2)^{\frac{1}{2}(n-1)}}{\pi\Gamma(n-2)} \exp\left(-\tfrac{1}{2}u_1^2\right) u_1^{n-2} (1-r^2)^{\frac{1}{2}(n-4)}$$
$$\times \sum \frac{(\rho r u_1)^j}{j!} \Gamma\left(\frac{n-1+j}{2}\right) \cdot 2^{\frac{1}{2}(n+j-3)}\, du_1\, dr. \tag{16.71}$$

Multiplying the right-hand side of (16.71) by r and integrating from -1 to $+1$ we find,

$$\frac{(1-\rho^2)^{\frac{1}{2}(n-1)}}{\pi\Gamma(n-2)} \exp\left(-\tfrac{1}{2}u_1^2\right) u_1^{n-2} \times \sum_{j=0}^{\infty} \frac{(\rho u_1)^{2j+1}}{(2j+1)!} \Gamma\left(\frac{n+2j}{2}\right) B\left(\frac{n-2}{2}, \frac{2j+3}{2}\right) \cdot 2^{\frac{1}{2}(n+2j-2)}\, du_1$$

and finally, integrating with respect to u_1, we obtain the expectation of r,

$$\mu_1'(r) = \frac{(1-\rho^2)^{\frac{1}{2}(n-1)}}{\pi\Gamma(n-2)} \sum_{j=0}^{\infty} \frac{\rho^{2j+1}}{(2j+1)!} \Gamma\left(\frac{n+2j}{2}\right) B\left(\frac{n-2}{2}, \frac{2j+3}{2}\right) \Gamma\left(\frac{n+2j}{2}\right) 2^{n+2j-2}. \tag{16.72}$$

Substituting for the Beta function in terms of Gamma functions and using (3.66), we find

$$\mu_1'(r) = \frac{\rho(1-\rho^2)^{\frac{1}{2}(n-1)}\Gamma^2(\frac{1}{2}n)}{\Gamma\{\frac{1}{2}(n-1)\}\Gamma\{\frac{1}{2}(n+1)\}} \left\{1 + \frac{\frac{1}{2}n \cdot \frac{1}{2}n}{\frac{1}{2}(n+1)} \frac{\rho^2}{1!} + \frac{\frac{1}{2}n(\frac{1}{2}n+1) \cdot \frac{1}{2}n(\frac{1}{2}n+1)}{\frac{1}{2}(n+1) \cdot \frac{1}{2}(n+3)} \frac{\rho^4}{2!} + \dots\right\}$$
$$= \frac{\rho(1-\rho^2)^{\frac{1}{2}(n-1)}\Gamma^2(\frac{1}{2}n)}{\Gamma\{\frac{1}{2}(n-1)\}\Gamma\{\frac{1}{2}(n+1)\}} F\left(\tfrac{1}{2}n, \tfrac{1}{2}n, \tfrac{1}{2}(n+1), \rho^2\right)$$

and since $F(\alpha, \beta, \gamma, x) = (1-x)^{\gamma-\alpha-\beta} F(\gamma-\alpha, \gamma-\beta, \gamma, x)$

$$\mu_1'(r) = \frac{\rho\Gamma^2(\frac{1}{2}n)}{\Gamma\{\frac{1}{2}(n-1)\}\Gamma\{\frac{1}{2}(n+1)\}} F\left(\tfrac{1}{2}, \tfrac{1}{2}, \tfrac{1}{2}(n+1), \rho^2\right)$$
$$= \rho\left\{1 - \frac{(1-\rho^2)}{2n} + O\left(\frac{1}{n^2}\right)\right\}. \tag{16.73}$$

Similarly (cf. Hotelling (1953)) we find

$$\mu_2(r) = \frac{(1-\rho^2)^2}{n-1}\left(1+\frac{11\rho^2}{2n}\right) + O\left(\frac{1}{n^3}\right), \tag{16.74}$$

and

$$\gamma_1 = \frac{-6n}{n^{\frac{1}{2}}} + o(n^{-\frac{1}{2}}), \qquad \gamma_2 = \frac{6(12\rho^2-1)}{n} + o(n^{-1}).$$

Thus the skewness increases with $|\rho|$ and declines only as $n^{-\frac{1}{2}}$.

B. K. Ghosh (1966) gives the first four moments of r as power series in $(n+6)^{-1}$ to 4 or more terms. Exercise 16.17 shows that when $\rho = 0$, (16.73-4) agree with the exact results for *any* continuous independent variates.

Fisher's transformation of r

16.33 Fisher (1921b) found (cf. Exercise 16.18) a remarkable transformation of r which tends to normality very much faster than r, with a variance almost independent of ρ. Putting

$$r = \tanh z, \qquad z = \tfrac{1}{2}\log\frac{1+r}{1-r}, \qquad \rho = \tanh\zeta, \qquad \zeta = \tfrac{1}{2}\log\frac{1+\rho}{1-\rho}, \tag{16.75}$$

(which is easily done using a special table of the transformation in the *Biometrika Tables*) we may expand the density function of r in powers of $z-\zeta$, $=x$ say, and inverse powers of n. Fisher gives the expansion

$$\begin{aligned} f = \frac{n-2}{\sqrt{\{2\pi(n-1)\}}} e^{-\frac{1}{2}(n-1)x^2}\Bigg\{1 + \tfrac{1}{2}\rho x + \left(\frac{2+\rho^2}{8(n-1)} + \frac{4-\rho^2}{8}x^2 + \frac{n-1}{12}x^4\right) \\ + \rho x\left(\frac{4-\rho^2}{16(n-1)} + \frac{4+3\rho^2}{48}x^2 + \frac{n-1}{24}x^4\right) + \left(\frac{4+12\rho^2+9\rho^4}{128(n-1)^2} + \frac{8-2\rho^2+3\rho^4}{64(n-1)}x^2\right. \\ \left. + \frac{8+4\rho^2-5\rho^4}{128}x^4 + \frac{28-15\rho^2}{1440}x^6(n-1)\right) + \frac{(n-1)^2}{288}x^8 + \dots\Bigg\}. \end{aligned} \tag{16.76}$$

Taking moments about $x = 0$ we find, on transferring to the mean,(*)

$$\left.\begin{aligned} \mu_1' &= \frac{\rho}{2(n-1)}\left\{1 + \frac{5+\rho^2}{4(n-1)} + \frac{11+2\rho^2+3\rho^4}{8(n-1)^2} + \dots\right\} \\ \mu_2 &= \frac{1}{n-1}\left\{1 + \frac{4-\rho^2}{2(n-1)} + \frac{22-6\rho^2-3\rho^4}{6(n-1)^2} + \dots\right\} \\ \mu_3 &= \frac{\rho^3}{(n-1)^3} + \dots \\ \mu_4 &= \frac{1}{(n-1)^2}\left\{3 + \frac{14-3\rho^2}{n-1} + \frac{184-48\rho^2-21\rho^4}{4(n-1)^2} + \dots\right\} \end{aligned}\right\} \tag{16.77}$$

(*) Equations (16.76-77), as given in Fisher's original paper, contained some errors which were corrected by A. H. Pollard (1945, unpublished) and by Gayen (1951). The main result, equation (16.80), remains unaffected.

whence

$$\left.\begin{aligned}\gamma_1 &= \frac{\rho^3}{(n-1)^{\frac{3}{2}}}+\dots\\ \gamma_2 &= \frac{2}{n-1}+\frac{4+2\rho^2-3\rho^4}{(n-1)^2}+\dots\end{aligned}\right\} \tag{16.78}$$

Thus the variance of $z-\zeta$ is almost independent of ρ, while γ_1 declines as $n^{-3/2}$ and we may take $z-\zeta$ to be approximately normally distributed with mean and variance as given by (16.77). Using leading terms only, we may take

$$\mu_1'(z-\zeta) \doteq \frac{\rho}{2(n-1)}, \qquad \text{var}\,(z-\zeta) \doteq \frac{1}{n-1}+\frac{4-\rho^2}{2(n-1)^2}; \tag{16.79}$$

for small ρ the variance is

$$\frac{1}{n-1}+\frac{2}{(n-1)^2} = \frac{1}{n-3} \text{ approximately.} \tag{16.80}$$

When n is moderately large we may take a still rougher approximation by assuming $z-\zeta$ to be normally distributed about zero mean with variance $\dfrac{1}{n-3}$. Some comparisons of the various approximations are given in the introduction to F. N. David's (1938) tables, and it appears that for $n > 50$ the forms (16.79) and (16.80) are adequate. The approximation given by (16.77) appears to be reasonably close for values of n as low as 11. See also Gurland (1968) and Gurland and Milton (1970).

When $\rho = 0$, the exact distribution of z is given in Exercise 16.21.

A much closer approximation is obtained from

$$P\{m^{\frac{1}{2}}(z-\zeta) < y\} = G(y) - \frac{1}{2}\left(\frac{\rho}{m^{\frac{1}{2}}}+\frac{y^3}{6m}\right)g(y), \tag{16.81}$$

where $m = n - \frac{5}{2} + \frac{1}{4}\rho^2$ and $G(y)$, $g(y)$ are respectively the standardized normal d.f. and density. Konishi (1978a, b) showed that (16.81) is accurate within 0.001 for $\rho \leqslant 0.9$ and $n \geqslant 11$, and is very much more accurate than even the best z approximation based on (16.77).

Ruben (1966) used the identity given in Exercise 16.6 and the device of **12.19** to obtain an approximation to the distribution of r that is much more accurate than z, and about as accurate as z^* in Exercise 16.19. Gurland (1968) gives a finite sum for the d.f. of $r^2/(1-r^2)$ when n is even, infinite sums for odd or even n, and an approximation to the density. See also Gurland and Milton (1970).

Samiuddin (1970) shows that if (16.63) is generalized to

$$t = (r-\rho)\{(n-2)/[(1-r^2)(1-\rho^2)]\}^{\frac{1}{2}}, \tag{16.82}$$

t remains approximately distributed in Student's form with $(n-2)$ d.fr. for $\rho \neq 0$ and $n \geqslant 8$. See also Kraemer (1973).

16.34 Just as r is defined as the ratio of sample covariance to the product of sample standard deviations, and ρ as the corresponding population parameter, we now define (see (16.48))

$$b_1 = \frac{1}{n}\sum (x-\bar{x})(y-\bar{y})/s_2^2, \qquad b_2 = \frac{1}{n}\sum (x-\bar{x})(y-\bar{y})/s_1^2, \tag{16.83}$$

the population analogues being $\beta_1 = \mu_{11}/\sigma_2^2$ and $\beta_2 = \mu_{11}/\sigma_1^2$ respectively. b_1 and b_2 are called the sample *regression coefficients*, of x on y and of y on x respectively. β_1 and β_2 are the corresponding population regression coefficients.

Referring to (16.46), we see that

$$E(x|y) = \beta_1 y, \qquad E(y|x) = \beta_2 x.$$

We proceed to consider the sampling distribution of b_2. That for b_1 is immediately obtainable from this by permutation of subscripts.

Distribution of regression coefficients in binormal samples

16.35 Turning again to equation (16.59) we have, substituting

$$b_2 = \sum (x-\bar{x})(y-\bar{y})/ns_1^2 = \frac{rs_2}{s_1},$$

the joint density function of s_1, s_2 and b_2,

$$dF \propto \exp\left[-\frac{n}{2(1-\rho^2)}\left\{\frac{s_1^2}{\sigma_1^2} - \frac{2\rho s_1^2 b_2}{\sigma_1\sigma_2} + \frac{s_2^2}{\sigma_2^2}\right\}\right] s_1^{n-1} s_2^{n-3}\left(1 - \frac{s_1^2 b_2^2}{s_1^2}\right)^{\frac{1}{2}(n-4)} ds_1\, ds_2\, db_2. \tag{16.84}$$

Integration of (16.84) with respect to s_2 gives, for the distribution of s_1 and b_2,

$$dF \propto \exp\left[-\frac{ns_1^2}{2(1-\rho^2)}\left\{\frac{1}{\sigma_1^2} - \frac{2\rho b_2}{\sigma_1\sigma_2} + \frac{b_2^2}{\sigma_2^2}\right\}\right] s_1^{n-1}\, ds_1\, db_2. \tag{16.85}$$

A further integration of (16.85) with respect to s_1 gives for the distribution of b_2

$$dF \propto \frac{db_2}{\left(1 - \frac{2\rho\sigma_1}{\sigma_2} b_2 + \frac{\sigma_1^2}{\sigma_2^2} b_2^2\right)^{\frac{1}{2}n}}$$

or, on evaluation of the constant,

$$dF = \frac{\Gamma(\frac{1}{2}n)\sigma_2^{n-1}(1-\rho^2)^{\frac{1}{2}(n-1)}}{\sqrt{\pi}\,\Gamma\left(\frac{n-1}{2}\right)\sigma_1^{n-1}} \frac{db_2}{\left\{\frac{\sigma_2^2}{\sigma_1^2}(1-\rho^2) + \left(b_2 - \frac{\rho\sigma_2}{\sigma_1}\right)^2\right\}^{\frac{1}{2}n}}. \tag{16.86}$$

It follows that if we write

$$g = (b_2 - \beta_2)\sigma_1/\{\sigma_2(1-\rho^2)^{\frac{1}{2}}\}, \tag{16.87}$$

where $\beta_2 = \dfrac{\rho\sigma_2}{\sigma_1}$, the population regression coefficient, then $(n-1)^{\frac{1}{2}}g$ has Student's t distribution (16.15) with $\nu = n-1$. Thus b_2 is distributed symmetrically about β_2. It tends to normality fairly rapidly, and the use of the standard error for regressions is therefore valid for lower values of n than in the case of the correlation coefficient. For small samples, however, (16.87) is useless since it depends on the unknown parameters σ_1, σ_2 and ρ.

16.36 To avoid this difficulty, we substitute s_1, s_2 and r for σ_1, σ_2 and ρ in (16.87), obtaining $h = (b_2 - \beta_2)s_1/\{s_2(1-r^2)^{\frac{1}{2}}\}$. To find its distribution, we return to that of the quantities a, b, c of equation (16.56), namely, from (16.57),

$$dF \propto \exp[-a + 2\rho b - c](ac - b^2)^{\frac{1}{2}(n-4)}\, da\, db\, dc. \tag{16.88}$$

Now

$$h = \frac{b - \rho a}{\sqrt{(ac - b^2)}},$$

and on transforming from c to h in (16.88) we have, after a little reduction,

$$dF \propto \frac{\exp\left[-a\left(1+\dfrac{\rho^2}{h^2}\right)\right] da\, dh}{ah^{n-1}} \cdot \exp\left[\frac{h^2+1}{h^2}\left(2\rho b - \frac{b^2}{a}\right)\right](b - \rho a)^{n-2}\, db. \tag{16.89}$$

Integration of the second factor on the right of (16.89) will be found to give a result proportional to

$$\exp\left\{\frac{a\rho^2(h^2+1)}{h^2}\right\} \cdot \left(\frac{ah^2}{h^2+1}\right)^{\frac{1}{2}(n-1)} \tag{16.90}$$

and hence for the distribution of a and h we find, from (16.89) and (16.90),

$$dF \propto a^{\frac{1}{2}(n-3)} \exp\{a(\rho^2 - 1)\}\, da \cdot \frac{dh}{(1+h^2)^{\frac{1}{2}(n-1)}}. \tag{16.91}$$

Thus a and h are independent, as may also be seen from the argument in **15.27**. The density for h becomes

$$dF \propto \frac{dh}{(1+h^2)^{\frac{1}{2}(n-1)}}. \tag{16.92}$$

This distribution does not contain any of the population parameters. It follows that

$$t = h\sqrt{(n-2)} = \frac{(b_2 - \beta_2)s_1\sqrt{(n-2)}}{s_2(1-r^2)^{\frac{1}{2}}} \tag{16.93}$$

is distributed in Student's form (16.15) with $\nu = n-2$ degrees of freedom. Exercise 16.6 proves our results differently.

Only one d.fr. has been lost in passing from g of (16.87) to h here, since s_2 is the only statistic not already used in b_2. When $\beta_2 = 0$, (16.93) coincides with (16.63), since then $\rho = 0$ also.

A conditional approach

16.37 We may use the results of Chapter 15 to arrive at this result by a different route. If we examine the distribution of b_2 in (16.83) conditionally upon the values of the xs, we see that b_2 is a linear function of the $(y - \bar{y})$ and hence of the ys. It follows directly that

$$E(b_2|\mathbf{x}) = \frac{1}{ns_1^2}\sum (x - \bar{x})E(y - \bar{y}|\mathbf{x}) = \frac{1}{ns_1^2}\sum (x - \bar{x})\{E(y|\mathbf{x}) - E(\bar{y}|\mathbf{x})\}$$

and using **16.34** this is

$$= \sum (x - \bar{x}) \cdot \beta_2(x - \bar{x})/ns_1^2$$

$$= \beta_2. \tag{16.94}$$

In like fashion, we may show that

$$V(b_2|\mathbf{x}) = V(y|\mathbf{x})/ns_1^2. \tag{16.95}$$

Since b_2 is a linear function of the ys, it follows that b_2 is conditionally normally distributed with mean and variance given by (16.94) and (16.95). It follows from an application of Cochran's theorem (**15.16–21**) that

$$\hat{V} = \frac{1}{n-2}\sum [y - \bar{y} - b_2(x - \bar{x})]^2 = \frac{n}{n-2}(s_2^2 - b_2^2 s_1^2)$$

$$= \frac{n}{n-2}s_2^2(1 - r^2) \tag{16.96}$$

is χ^2-distributed with $(n-2)$ degrees of freedom. Further, $E(\hat{V}) = V(y|x)$, so that $\hat{V}/ns_1^2$ is an unbiased estimator of $V(b_2|\mathbf{x})$. From **15.15**, b_2 and $\hat{V}$ are independent. It then follows from Example 11.8 that t in (16.93) has a Student's t distribution with $(n-2)$ degrees of freedom.

In this derivation, we have made no distributional assumptions about $\mathbf{x}$. We shall pursue this approach systematically in Chapter 19, Volume 2A, in the manner of Exercise 16.5.

Robustness

16.38 Our discussion throughout this chapter has focused, deliberately, upon exact results for sampling distributions when the underlying population is normal. In practice, the effect of non-normality upon these distributions must be assessed. For the most part, we leave such a discussion to our study of robust methods for the linear model (**28.71** ff, Volume 2A); however, a few comments are in order here.

16.39 Bondesson (1983) has shown that, when the underlying distribution has all moments finite, the statistic $t = (\bar{x} - \mu)(n/s^2)^{\frac{1}{2}}$ has a Student's t distribution with

$(n-1)$ degrees of freedom if and only if the underlying distribution is normal. Nevertheless, it is found that the distribution is often close to Student's t distribution for non-normal populations.

More accurate approximations for non-normal populations are given by Bowman *et al.* (1977). They use moments expansions to suggest approximations based on Pearson distributions (6.1–11).

16.40 By contrast, the χ^2 and F results are sensitive to departures from normality. For a non-normal distribution with kurtosis coefficient $\gamma_2 = \beta_2 - 3$, we saw in Example 12.1 that, if $s_a^2 = \sum(x-\bar{x})^2/(n-1)$,

$$V(s_a^2) = \kappa_2^2\left(\frac{2}{n-1} + \frac{\gamma_2}{n}\right) \tag{16.97}$$

which does not approach the value for the normal distribution (with $\gamma_2 = 0$) as n increases. G. E. P. Box (1953) proposed use of the adjusted variate $u = cs_a^2$, where

$$c = (1+\tfrac{1}{2}\gamma_2)^{-1}; \tag{16.98}$$

u is approximately χ^2 with $\nu = c(n-1)$ degrees of freedom; the value of c is selected to make the first two moments of u agree exactly with those of a χ^2 distribution. Solomon and Stephens (1983) propose a more accurate approximation based upon equating the first three moments of $u = cs_a^k$ to those of a χ^2 distribution.

The same problems apply, *a fortiori*, to the F distribution. The approach of the previous paragraph suggests that we use F as in (16.23) but with degrees of freedom given by $\nu_i = c_i(n_i - 1)$, $i = 1, 2$, where c_i is given in (16.98).

16.41 The discussion in **16.37** suggests that the distiibution of r, given $\rho = 0$, will be robust to non-normality, since then $\rho = \beta_2 = 0$, and this proves to be so. In particular, when $\rho = 0$ we may consider the permutations distribution for r by treating the $n!$ possible pairings of the values $(y_1, \ldots, y_n)$ with $(x_1, \ldots, x_n)$ as equally likely. The resulting distribution is very close to (16.62); see **26.22–3**, Volume 2A. In a series of studies, Kocherlakota and others (see Kocherlakota and Singh (1982) and the references given therein) show that Fisher's z transformation and Samiuddin's t in (16.82) are both fairly robust to non-normality.

Multinormal distribution theory

16.42 In (15.1) and (15.9), we wrote the p-dimensional multivariate normal distribution in two forms:

$$dF \propto \exp\left\{-\tfrac{1}{2}\sum_{j=1}^{p}\sum_{k=1}^{p}\alpha_{jk}\left(\frac{x_j-\mu_j}{\sigma_j}\right)\left(\frac{x_k-\mu_k}{\sigma_k}\right)\right\}\prod_{j=1}^{p}\frac{dx_j}{\sigma_j},$$

$$-\infty < x_j < \infty, \quad j = 1, 2, \ldots, p \tag{16.99}$$

and

$$dF = \frac{|\alpha|^{\frac{1}{2}}}{(2\pi)^{\frac{1}{2}p}}\exp\{-\tfrac{1}{2}(\mathbf{x}-\boldsymbol{\mu})'\boldsymbol{\alpha}(\mathbf{x}-\boldsymbol{\mu})\}\prod_j dx_j, \tag{16.100}$$

where μ_j, σ_j^2 are the mean and variance of the jth variable, $\mathbf{x}$ is a $(p \times 1)$ column vector, and α is the inverse of the covariance matrix, $\boldsymbol{\Sigma}$.

Further, we recall from (15.20) that the characteristic function of (16.100) is given by

$$\phi(\mathbf{t}) = \exp\left(-\tfrac{1}{2}\mathbf{t}'\boldsymbol{\Sigma}\mathbf{t}\right)\exp(i\mathbf{t}'\boldsymbol{\mu}). \qquad (16.101)$$

The Wishart distribution

16.43 We first investigate the joint distribution of means and covariances for the multivariate normal distribution. Suppose we have a random sample of size n. Writing x_{jl} for the lth observation of the jth variate, the matrix of observations is

$$\mathbf{x} = (x_{jl}) = \begin{pmatrix} x_{11} & x_{12} & \dots & x_{1n} \\ x_{21} & x_{22} & \dots & x_{2n} \\ \cdot & \cdot & \cdot & \cdot \\ x_{p1} & x_{p2} & \dots & x_{pn} \end{pmatrix} = (\mathbf{x}_1, \dots, \mathbf{x}_n). \qquad (16.102)$$

The sample means will be denoted by the vector $\bar{\mathbf{x}} = (\bar{x}_1, \dots, \bar{x}_p)'$ and the sample covariance matrix by $\mathbf{S} = \{s_{jk}\}$, where

$$s_{jk} = r_{jk}s_j s_k = \sum_l (x_{jl} - \bar{x}_j)(x_{kl} - \bar{x}_k)/n; \qquad (16.103)$$

s_{jk} denotes the sample covariance for x_j and x_k, and r_{jk} is the corresponding sample correlation. Note that, throughout, we use n in the denominator rather than $(n-1)$.

The joint density for the sample is given by

$$\mathrm{d}F = \frac{|\alpha|^{\frac{1}{2}n}}{(2\pi)^{\frac{1}{2}np}} \exp\left\{-\tfrac{1}{2}\sum_{j=1}^{n} (\mathbf{x}_j - \boldsymbol{\mu})'\alpha(\mathbf{x}_j - \boldsymbol{\mu})\right\} \prod_{j=1}^{n}\prod_{k=1}^{p} \mathrm{d}x_{jk}. \qquad (16.104)$$

We already know from Example 11.7 and **16.25** that for $p = 1, 2$, the distribution can be split into two independent distributions, one for the means and the other for the covariances. We prove, first of all, that the same is true of any value of p. We have the familiar algebraic identity

$$\sum_j (\mathbf{x}_j - \boldsymbol{\mu})'\alpha(\mathbf{x}_j - \boldsymbol{\mu}) = \sum_j (\mathbf{x}_j - \bar{\mathbf{x}})'\alpha(\mathbf{x}_j - \bar{\mathbf{x}}) + n(\bar{\mathbf{x}} - \boldsymbol{\mu})'\alpha(\bar{\mathbf{x}} - \boldsymbol{\mu}). \qquad (16.105)$$

Let $\mathrm{tr}(\mathbf{B}) = \sum_j b_{jj}$ denote the brace operator for matrix $\mathbf{B}$. Since $\mathbf{a}'\mathbf{B}\mathbf{a} = \mathrm{tr}(\mathbf{B}\mathbf{a}\mathbf{a}')$ for any matrix $\mathbf{B}$ and vector $\mathbf{a}$, the first term on the right in (16.105) becomes

$$\mathrm{tr}\left[\alpha\sum_j (\mathbf{x}_j - \bar{\mathbf{x}})(\mathbf{x}_j - \bar{\mathbf{x}})'\right] = n\,\mathrm{tr}(\alpha\mathbf{S}). \qquad (16.106)$$

Thus the exponent in (16.104) factorizes into two components. We now have to consider the differential elements. It will be convenient to make on each variable an orthogonal Helmert transformation of the type used in Example 11.3:

$$\left.\begin{aligned} y_1 &= \frac{1}{\sqrt{2}}(x_1 - x_2) \\ y_2 &= \frac{1}{\sqrt{6}}(x_1 + x_2 - 2x_3) \\ &\cdots\cdots \\ y_{n-1} &= \frac{1}{\sqrt{\{n(n-1)\}}}\{x_1 + x_2 + \ldots + x_{n-1} - (n-1)x_n\} \\ y_n &= \frac{1}{\sqrt{n}}(x_1 + x_2 + \ldots + x_n) = \sqrt{n}\bar{x}, \end{aligned}\right\} \quad (16.107)$$

where the subscripts refer to the observations. The inverse relationship is

$$\left.\begin{aligned} x_1 &= \frac{1}{\sqrt{2}}y_1 + \frac{1}{\sqrt{6}}y_2 + \ldots + \frac{1}{\sqrt{\{n(n-1)\}}}y_{n-1} + \frac{1}{\sqrt{n}}y_n \\ x_2 &= -\frac{1}{\sqrt{2}}y_1 + \frac{1}{\sqrt{6}}y_2 + \ldots + \frac{1}{\sqrt{\{n(n-1)\}}}y_{n-1} + \frac{1}{\sqrt{n}}y_n \\ x_n &= \qquad -\frac{n-1}{\sqrt{\{n(n-1)\}}}y_{n-1} + \frac{1}{\sqrt{n}}y_n \end{aligned}\right\} \quad (16.108)$$

$|J| = 1$ and the differential element is (16.104) becomes simply

$$\prod_{l=1}^{n}\prod_{j=1}^{p} dy_{jl}. \quad (16.109)$$

Returning to (16.105) we note that $y_n = \sqrt{n}\bar{x}$ and we see that the second factor is a function only of y_{jn}, $(j = 1, 2, \ldots, p)$. From (16.107), the first factor depends only on $y_1, \ldots, y_{n-1}$.

Hence we may factorize the original density into two terms whose exponents are given by (16.105), multiplied by $-\frac{1}{2}$. With an appropriate adjustment to the constant factor we then obtain for the means

$$dF = \frac{|\alpha|}{(2\pi)^{\frac{1}{2}p}} \exp\{-\tfrac{1}{2}n(\mathbf{\bar{x}} - \boldsymbol{\mu})'\boldsymbol{\alpha}(\mathbf{\bar{x}} - \boldsymbol{\mu})\} \prod_{j=1}^{p} d(\sqrt{n}\bar{x}_j). \quad (16.110)$$

The joint distribution of means is, in fact, the same as that of the original variables, apart from the factor in n, a result more directly established from (16.101) by writing the c.f. for $\mathbf{\bar{x}}$ as

$$\phi_{\bar{\mathbf{x}}}(\mathbf{t}) = [\phi(\mathbf{t}/n)]^n = \exp\{i\mathbf{t}'\boldsymbol{\mu} - \tfrac{1}{2}n\mathbf{t}'\boldsymbol{\Sigma}\mathbf{t}\}.$$

16.44 The distribution of the sample variances and covariances is thus confined to the $(n-1)$-spaces orthogonal to the sample means. Since the orthogonal transformation is simply a rotation of axes, it leaves distances and angles invariant. Since variances and covariances are functions of these alone (cf. Example 11.7 and **16.24**), they too are invariant. From (16.106) the remaining non-differential component of (16.103) may be written as

$$f = \frac{|\alpha|^{\frac{1}{2}(n-1)}}{(2\pi)^{\frac{1}{2}(n-1)p}} \exp[-\tfrac{1}{2}n \operatorname{tr}(\alpha \mathbf{S})\}. \tag{16.111}$$

Our principal problem is to evaluate the differential elements in terms of the $\{s_{jk}, k \leqslant j\}$.

Let $\mathbf{u}_j$ denote the $(n-1)$ vector with elements

$$u_{ji} = y_{ji} - \bar{y}_j, \qquad i = 1, \ldots, n-1 \tag{16.112}$$

so that the covariance matrix has elements

$$s_{jk} = \mathbf{u}_j' \mathbf{u}_k / n. \tag{16.113}$$

Hence,

$$n\mathbf{S} = \mathbf{U}'\mathbf{U} \tag{16.114}$$

where $\mathbf{U} = [\mathbf{u}_1, \ldots, \mathbf{u}_p]$. Note that here and throughout we use n as the sample size, not the degrees of freedom of the covariances.

Generalizing the argument of **16.24**, we take p flat spaces of $n-1$ dimensions, one for each u, and let the sample points be represented by $P_1, P_2, \ldots, P_p$. We consider in turn the variation in P_1, then that of P_2 for fixed P_1, then that of P_3 for fixed P_1 and P_2, and so on. We then multiply all these together to find the variation of $P_1, P_2, \ldots, P_p$. We require that

$$n - 1 \geqslant p,$$

so that the dimensionality is high enough to accommodate the p-space thus generated.

Consider then P_m given $P_1, P_2, \ldots, P_{m-1}$. Let O be the origin and imagine the spaces superposed on one another. For fixed OP_m and fixed angles P_mOP_1, P_mOP_2, $\ldots, P_mOP_{m-1}$, the point P_m varies on a hypersphere of $n-m$ dimensions. Let the length of the perpendicular line from P_m on to the hyperplane determined by O, P_1, $\ldots, P_{m-1}$ be t_m. Then the "content" (volume) of variation permissible to P_m is that of the "surface" of a hypersphere in $n-m$ dimensions with radius t_m, which is (cf M. G. Kendall, 1961, p. 42)

$$\frac{2\pi^{\frac{1}{2}(n-m)} t_m^{n-m-1}}{\Gamma\{\frac{1}{2}(n-m)\}}. \tag{16.115}$$

We need to multiply this by the element of variation perpendicular to the hypersphere. Consider the transformation in the m-space determined by $O, P_1, \ldots, P_m$, based on (16.113),

$$\zeta_{mj} = ns_{mj} = \mathbf{u}_m' \mathbf{u}_j \quad j = 1, 2, \ldots, m. \tag{16.116}$$

The Jacobian of the transformation is given by

$$J = \frac{\partial(\xi_{m1}, \ldots, \xi_{mm})}{\partial(u_{m1}, \ldots, u_{mm})} = \begin{vmatrix} u_{11} & u_{12} & \ldots & u_{1m} \\ u_{21} & u_{22} & \ldots & u_{2m} \\ \ldots & \ldots & \ldots & \ldots \\ 2u_{m1} & 2u_{m2} & \ldots & 2u_{mm} \end{vmatrix} \tag{16.117}$$

and this is equal to $2v_m$ where v_m is the volume of the parallelotope (the m-dimensional parallelogram) determined by $O, P_1, \ldots, P_m$. Thus the different element is

$$\frac{1}{2v_m} \prod_{j=1}^{m} d\xi_{mj}. \tag{16.118}$$

On multiplication by (16.114) the total variation of P_m is then given by

$$\frac{\pi^{\frac{1}{2}(n-m)} t_m^{n-m-1}}{\Gamma\{\frac{1}{2}(n-m)\} v_m} \prod_{j=1}^{m} d\xi_{mj}. \tag{16.119}$$

But we have

$$t_m = v_m / v_{m-1}$$

so that the element of variation (16.119) becomes

$$\frac{\pi^{\frac{1}{2}(n-m)}}{\Gamma\{\frac{1}{2}(n-m)\}} \frac{v_m^{n-m-2}}{v_{m-1}^{n-m-1}} \prod d\xi_{mj}. \tag{16.120}$$

We now multiply together expressions of type (16.120) for $m = 1, 2, \ldots, p$. The terms in v cancel except for those in v_0 (which is unity) and v_p, and we obtain

$$\frac{\pi^{\frac{1}{2}p(2n-p-1)}}{\prod_{j=1}^{p} \Gamma\{\frac{1}{2}(n-j)\}} v_p^{n-p-2} \prod_{j=1}^{p} \prod_{k=1}^{j} d\xi_{jk}. \tag{16.121}$$

Let ξ_p denote the $p \times p$ matrix of elements $\{\xi_{jk}\}$ then, from (16.115) and (16.116) we have

$$v_p^2 = |\xi_p| = |U'U| = n^p |S|. \tag{16.122}$$

Putting (16.121–2) and (16.111) together, we then have for the distribution of covariances

$$dF = \frac{(\frac{1}{2}n)^{\frac{1}{2}p(n-1)} |\alpha|^{\frac{1}{2}(n-1)} |S|^{\frac{1}{2}(n-p-2)}}{\pi^{\frac{1}{4}p(p-1)} \prod_{j=1}^{p} \Gamma\{\frac{1}{2}(n-j)\}} \exp(-\tfrac{1}{2} n \operatorname{tr}(\alpha S)) \prod_{k \leq j}^{p} ds_{jk}. \tag{16.123}$$

This is Wishart's (1928a, b) distribution, the generalization of the χ^2 distribution to p-dimensional variation. (11.41) in Example 11.7 is the case $p = 1$; (16.54) is a simple transformation of the case $p = 2$.

Readers who find the geometrical line of proof hard to follow may prefer to consult the review of alternative proofs in Wishart's articles (1948a, b).

16.45 Let us note some minor but not unimportant details concerning the distribution:

(a) In the differential element there are $\frac{1}{2}p(p+1)$ terms, not p^2. Thus, for $p=2$ the element is $ds_{11}ds_{12}ds_{22}$.

(b) The domain of variation for each s_{jj} is 0 to ∞, but we cannot easily specify those for the other ss, which are conditioned by the fact that the matrix (s_{jk}) must be positive definite. This makes it extremely difficult to integrate out certain variables s in order to obtain the marginal distributions of others.

(c) We have defined the sample variances and covariances by dividing the appropriate product sum by n. We may, if we prefer, divide by $n-1$, in which case appropriate adjustments have to be made in (16.123). The reader should watch this point in consulting the literature because usage varies.

(d) The covariance matrix **S** may be expressed as **TT′**, where **T** is a lower triangular matrix. This expression is known as the *Bartlett decomposition*; see Exercises 16.39–40 for details. The decomposition is particularly useful for simulation purposes.

16.46 We can now derive the characteristic function of the Wishart distribution. Writing a single integral sign to denote integration over the domain of **S**, we have from (16.123)

$$\int |S|^{\frac{1}{2}(n-p-2)}\exp\left(-\tfrac{1}{2}n\,\mathrm{tr}\,(\alpha\mathbf{S})\right)\prod ds_{jk} = k|\alpha|^{-\frac{1}{2}(n-1)} \tag{16.124}$$

where k is some constant. If we replace

$$\alpha_{jj} \text{ by } \alpha_{jj}-2\theta_{jj}/n, \quad \alpha_{jk} \text{ by } \alpha_{jk}-\theta_{jk}/n, \quad j\neq k,$$

the exponent under the integral sign gives us the c.f. of **S** with θ_{jk} for the usual imaginary dummy variable it_{jk}. Making the substitutions on the right in (16.124), and adjusting the constant to make ϕ unity of when $\boldsymbol{\theta}=0$, we have

$$\begin{aligned}\phi(\boldsymbol{\theta}) &= |\alpha|^{\frac{1}{2}(n-1)}/|\alpha-n^{-1}\theta|^{\frac{1}{2}(n-1)}\\ &= |I-n^{-1}\theta\Sigma|^{-\frac{1}{2}(n-1)}\end{aligned} \tag{16.125}$$

where $\boldsymbol{\theta}$ has elements θ_{jk} if $j\neq k$ and $2\theta_{jj}$ on the main diagonal.

This substitutional device, avoiding the problem of integrating over the domain of variation, is a useful one which we shall employ again.

Example 16.3 (The bivariate Wishart distribution)

In the bivariate case $(p=2)$ we have, in the usual notation (with unit variances),

$$\boldsymbol{\Sigma}=\begin{pmatrix}1 & \rho\\ \rho & 1\end{pmatrix}, \quad |\boldsymbol{\Sigma}|=1-\rho^2,$$

$$\boldsymbol{\alpha} = \begin{vmatrix} \dfrac{1}{1-\rho^2} & -\dfrac{\rho}{1-\rho^2} \\ -\dfrac{\rho}{1-\rho^2} & \dfrac{1}{1-\rho^2} \end{vmatrix}, \quad |\boldsymbol{\alpha}| = (1-\rho^2)^{-1}.$$

Thus (16.123) reduces to

$$dF = \frac{(\frac{1}{2}n)^{n-1}}{(1-\rho^2)^{\frac{1}{2}(n-1)}} \frac{s_1^{n-4}s_2^{n-4}(1-r^2)^{\frac{1}{2}(n-4)}}{\pi^{\frac{1}{2}}\Gamma\{\frac{1}{2}(n-1)\}\Gamma\{\frac{1}{2}(n-2)\}}$$
$$\times \exp\left\{\frac{-n}{2(1-\rho^2)}(s_1^2 - 2\rho r s_1 s_2 + s_2^2)\right\} d(s_1^2)d(s_2^2)d(rs_1s_2).$$

On using the duplication formula

$$\Gamma\{\tfrac{1}{2}(n-1)\}\Gamma\{\tfrac{1}{2}(n-2)\} = \frac{\pi^{\frac{1}{2}}\Gamma(n-2)}{2^{n-3}} \tag{16.126}$$

we find

$$dF = \frac{n^{n-1}}{(1-\rho^2)^{\frac{1}{2}(n-1)}} \frac{s_1^{n-4}s_2^{n-4}(1-r^2)^{\frac{1}{2}(n-4)}}{4\pi\Gamma(n-2)} \tag{16.127}$$
$$\times \exp\left\{\frac{-n}{2(1-\rho^2)}(s_1^2 - 2\rho r s_1 s_2 + s_2^2)\right\} d(s_1^2)d(s_2^2)d(rs_1s_2).$$

which reduces to the form found in **16.26**.

Example 16.4 (Moments of the covariance)

Let us now consider the moments of the distribution of the covariance when $p=2$. From (16.125) we have, putting $\theta_{11} = \theta_{22} = 0$ and $\theta_{12} = \theta$,

$$\phi(\theta) = \begin{vmatrix} 1-\rho\theta/n & -\theta/n \\ -\theta/n & 1-\rho\theta/n \end{vmatrix}^{-\frac{1}{2}(n-1)} = \left\{1 - \frac{2\rho\theta}{n} - \frac{(1-\rho^2)\theta^2}{n^2}\right\}^{-\frac{1}{2}(n-1)}. \tag{16.128}$$

Taking logarithms and evaluating coefficients of $\theta^s/s!$ we find

$$\kappa_1 = \frac{n-1}{n}\rho, \quad \kappa_2 = \frac{n-1}{n^2}(1+\rho^2), \quad \kappa_3 = \frac{2(n-1)}{n^3}\rho(3+\rho^2), \quad \kappa_4 = \frac{6(n-1)}{n^4}(1+6\rho^2+\rho^4).$$

In standard measure the distribution tends to normality as n tends to infinity. But for finite n

$$\beta_1^2 = \frac{4}{n-1}\frac{\rho^2(3+\rho^2)^2}{(1+\rho^2)^3}, \tag{16.129}$$

$$\beta_2 = 3 + \frac{6}{n-1}\frac{1+6\rho^2+\rho^4}{(1+\rho^2)^2}. \tag{16.130}$$

Thus, even for $\rho = 0$, the distribution, though symmetrical, is not normal. We have derived these results differently in Example 13.3.

Wishart (1929) gave the formulae explicitly for $p \leqslant 8$ as far as those of the fourth order.

An asymptotic series expansion for the Wishart, based on the multinormal is given by Nel (1978).

The additive property of Wishart distributions

16.47 One property of the Wishart distribution, analogous to that of χ^2 in one dimension, is worth noting.

Suppose we have two samples, n_1 and n_2 in number, from the same multivariate distribution. If we pool them, of course, the covariances of the total sample will follow the Wishart distribution with $n = n_1 + n_2$. But we may also consider the joint distribution of the covariances from each sample. If we form a new covariance matrix by adding corresponding covariances, i.e.

$$\mathbf{S} = \mathbf{S}_1 + \mathbf{S}_2 \tag{16.131}$$

where the subscripts refer to the first and second sample, then **S** is distributed in Wishart form with $(n_1 - 1) + (n_2 - 1)$ instead of $(n - 1)$, i.e. $n_1 + n_2 - 1$ instead on n.

This is perhaps most easily seen from the characteristic function; the product of two c.f.s of form (16.125) remains of the same form.

For $p = 1$, this result is familiar: each sample sum of squares about the mean is a multiple of a χ^2 variable with $(n_j - 1)$ d.fr., so the pooled sum will correspond to $(n_1 + n_2 - 1)$ d.fr., but the added sums to $(n_1 - 1) + (n_2 - 1)$ d.fr.

16.48 It would add a pleasant completeness to the sampling theory of covariances if we could proceed from the Wishart distribution and deduce the distribution of particular functions of the variances and covariances by integrating over appropriate domains to eliminate unwanted variables. Unfortunately this is prohibitively complicated in general. As in Example 16.4 we can derive moments and product moments of sample covariances; finding the explicit distributions is then usually unnecessary. There are, however, a few cases in which we can proceed further.

Moments of the generalized variance

16.49 Consider the distribution of the covariance determinant $|S|$, which is usually known as the generalized variance, first studied by Wilks (1932). From (16.123), the integral being taken over the variances and covariances,

$$\int \frac{(\frac{1}{2}n)^{\frac{1}{2}p(n-1)}|\alpha|^{\frac{1}{2}(n-1)}|S|^{\frac{1}{2}(n-p-2)}}{\pi^{\frac{1}{4}p(p-1)}\prod_{j=1}^{p}\Gamma\{\frac{1}{2}(n-j)\}} \exp\{-\tfrac{1}{2}n \operatorname{tr}(\boldsymbol{\alpha}\mathbf{S})\} \prod_{j \leqslant k}^{p} \mathrm{d}s_{jk} = 1. \tag{16.132}$$

Put $n\boldsymbol{\alpha} = \boldsymbol{\beta}$. We then have

$$\int \frac{|\boldsymbol{\beta}|^{\frac{1}{2}(n-1)}|S|^{\frac{1}{2}(n-p-2)}}{\pi^{\frac{1}{4}p(p-1)}} \exp\{-\tfrac{1}{2}\operatorname{tr}(\boldsymbol{\beta}\mathbf{S})\} \prod \mathrm{d}s_{jk} = 2^{\frac{1}{2}p(n-1)} \prod_{j=1}^{p} \Gamma\{\tfrac{1}{2}(n-j)\}. \tag{16.133}$$

If we replace n by $n+2t$ and divide by $|\beta|^t$, we obtain

$$\int \frac{|\beta|^{\frac{1}{2}(n-1)}|S|^{\frac{1}{2}(n-p-2)+t}}{\pi^{\frac{1}{4}p(p-1)}} \exp\{-\tfrac{1}{2}\operatorname{tr}(\beta S)\} \prod ds_{jk} = \frac{2^{\frac{1}{2}p(n-1)+pt}}{|\beta|^t} \prod_{j=1}^{p} \Gamma\{\tfrac{1}{2}(n-j)+t\}. \tag{16.134}$$

If we now replace β by $n\alpha$ and divide by $2^{\frac{1}{2}p(n-1)} \prod \Gamma\{\frac{1}{2}(n-j)\}$ we have on the left the expectation of $|S|^t$. Thus

$$E(|S|^t) = \frac{2^{pt}}{n^{pt}|\alpha|^t} \frac{\prod_{j=1}^{p} \Gamma\{\frac{1}{2}(n-j)+t\}}{\prod_{j=1}^{p} \Gamma\{\frac{1}{2}(n-j)\}}$$

$$= \frac{2^{pt}}{n^{pt}} \prod_{j=1}^{p} \frac{\Gamma\{\frac{1}{2}(n-j)+t\}}{\Gamma\{\frac{1}{2}(n-j)\}} |\Sigma|^t. \tag{16.135}$$

From this we can determine as many moments or cumulants as we wish. Again the substitutional device, obviating the awkward integrations, is to be noted.

16.50 If we write $V = n|S|/|\Sigma|$ we have

$$E(V^t) = 2^{pt} \prod_{j=1}^{p} \frac{\Gamma\{\frac{1}{2}(n-j)+t\}}{\Gamma\{\frac{1}{2}(n-j)\}}. \tag{16.136}$$

Now a χ^2 variable with ν degrees of freedom has moments about zero given by

$$\mu_t' = 2^t \frac{\Gamma(\frac{1}{2}\nu + t)}{\Gamma(\frac{1}{2}\nu)}.$$

The right-hand side of (16.136) is the product of p such factors with $\nu = n-1$, $n-2, \ldots, n-p$. Since the moment of a product of independent variables is the product of their moments, we see that V can be represented as the product of p independent factors, distributed as χ^2 with $n-1$, $n-2, \ldots, n-p$ degrees of freedom.

Example 16.5 (Moments of the generalized variance)

When $p=1$ we have the familiar result that a sum of squares, standardized by division by the variance, is distributed as χ^2 with $n-1$ d.fr.

For $p=2$ we find from (16.136) for the moments of V

$$\mu_t' = \frac{\Gamma\{\frac{1}{2}(n-1)+t\}}{\Gamma\{\frac{1}{2}(n-1)\}} \frac{\Gamma\{\frac{1}{2}(n-2)+t\}}{\Gamma\{\frac{1}{2}(n-2)\}} 2^{2t}. \tag{16.137}$$

Using the duplication formula for the Gamma function, (16.126), we may reduce this to

$$\mu_t' = \frac{\Gamma(n-2+2t)}{\Gamma(n-2)}. \tag{16.138}$$

This is the $(2t)$th moment of a Gamma variable with parameter $(n-2)$—cf. Exercise 4.6. Thus the tth moment of $2V$ is equal to the $(2t)$th moment of a χ^2 variable with $2(n-2)$ d.fr. This is a special case of the result of Exercise 11.9.

Exercise 16.34 gives a large-sample result for general p, which reduces to these exact results for $p=1, 2$.

The correlation determinant

16.51 There is considerable interest in the joint distribution of the sample correlations. In the general case (non-vanishing population correlations) distributions are complicated, even for $p=2$—cf. **16.26–7**. However, we can make some progress when all the population correlations are zero. The Wishart distribution (16.123) then reduces to

$$dF = \frac{(\frac{1}{2}n)^{\frac{1}{2}p(n-1)}|S|^{\frac{1}{2}(n-p-2)}}{\pi^{\frac{1}{4}p(p-1)}\prod\Gamma\{\frac{1}{2}(n-j)\}}\exp\{-\tfrac{1}{2}nS\}\prod_{j\leq k} ds_{jk}. \tag{16.139}$$

Transforming to the new variables

$$r_{jk} = s_{jk}/s_j s_k \tag{16.140}$$

we find that the Jacobian is given by

$$J = 2^p \prod_{j=1}^{p} s_j^p. \tag{16.141}$$

This is independent of the r_{jk}, as is the exponent in (16.139), and consequently terms in s can be factored out, leaving the distribution of correlations

$$dF = \frac{|R|^{\frac{1}{2}(n-p-2)}[\Gamma\{\frac{1}{2}(n-1)\}]^p}{\pi^{\frac{1}{4}p(p-1)}\prod_{j=1}^{p}\Gamma\{\frac{1}{2}(n-j)\}}\prod_{j<k} dr_{jk}, \tag{16.142}$$

where $|R|$ is the sample correlation determinant and the constant has been adjusted so that the density function integrates to unity. We are again hindered from making progress by the boundaries of the domain of variation. In the manner of **16.49** however, we can find the moments of $|R|$ itself. They are

$$E|R|^t = \frac{[\Gamma\{\frac{1}{2}(n-1)\}]^p}{[\Gamma\{\frac{1}{2}(n-1)+t\}]^p}\,\frac{\prod_{j=1}^{p}\Gamma\{\frac{1}{2}(n-j)+t\}}{\prod_{j=1}^{p}\Gamma\{\frac{1}{2}(n-j)\}}. \tag{16.143}$$

Example 16.6 (Moments of the logarithm of the correlation determinant)

Writing $L=\log|R|$, $\nu=\frac{1}{2}(n-1)$, $m=\frac{1}{2}(j-1)$ we have from (16.143)

$$E(e^{Lt}) = \prod_{j=1}^{p}\frac{\Gamma(\nu)}{\Gamma(\nu+t)}\,\frac{\Gamma(\nu-m+t)}{\Gamma(\nu-m)}. \tag{16.144}$$

This was shown to be true for integral t only, but by analytic continuation may be shown to be generally true for all t. We may therefore interpret t as an imaginary number, and

(16.144) then gives us the characteristic function of L. For the corresponding cumulant generating function we then have

$$\psi = \sum_{j=1}^{p} \{\log \Gamma(\nu) + \log \Gamma(\nu - m + t) - \log \Gamma(\nu + t) - \log \Gamma(\nu - m)\}. \quad (16.145)$$

Using Stirling's approximation (cf 3.24)

$$\log \Gamma(x) = (x - \tfrac{1}{2}) \log x - x + \tfrac{1}{2} \log (2\pi) + \frac{1}{12x} - \frac{1}{360x^3} + \dots$$

substitution into (16.145) gives us for the term in braces

$$-\frac{mt}{\nu} + \frac{1}{\nu^2}\{-\tfrac{1}{2}m(m+1)t + \tfrac{1}{2}mt^2\} + O(\nu^{-3}). \quad (16.146)$$

Now put $m = \frac{1}{2}(j-1)$ and sum from $j = 1$ to p. We find

$$\psi = -\tfrac{1}{4}p(p-1)\frac{t}{\nu} - O\left(\frac{t}{\nu^2}\right) + \tfrac{1}{4}p(p-1)\frac{t^2}{2\nu^2} + O(\nu^{-3}). \quad (16.147)$$

Thus

$$E(L) = -\tfrac{1}{4}p(p-1)/\nu + O(\nu^{-2}), \quad (16.148)$$

$$\text{var } L = \tfrac{1}{4}p(p-1)/\nu^2 + O(\nu^{-3}). \quad (16.149)$$

Furthermore we have

$$\left(\frac{\mathrm{d}}{\mathrm{d}x}\right)^k \log \Gamma(x) = \frac{(-1)^{k-2}(k-2)!}{x^{k-1}} + o(x^{-k+1}), \quad k > 1. \quad (16.150)$$

Hence, from (16.145), to the greatest power in ν,

$$\left.\frac{\mathrm{d}^k\psi}{\mathrm{d}t^k}\right|_{t=0} = \sum_{j=1}^{p} (-1)^{k-2}(k-2)!\left\{\frac{1}{(\nu - m + t)^{k-1}} - \frac{1}{(\nu + t)^{k-1}}\right\}_{t=0}$$

$$= \sum_{j=1}^{p} \frac{(-1)^{k-2}(k-1)!}{\nu^k} m, \quad \text{where } m = \tfrac{1}{2}(j-1)$$

so that the kth cumulant is

$$\kappa_k = (-2\nu)^{-k} 2^{k-1}(k-1)! \tfrac{1}{2}p(p-1) + o(\nu^{-k}). \quad (16.151)$$

Comparison with equation (16.4) shows that, with a suitable choice of origin, $-2\nu \log |R|$ is asymptotically distributed as χ^2 with $\frac{1}{2}p(p-1)$ degrees of freedom. To order ν^{-1} the origin, from (16.148), is seen to be zero.

Thus $-(n-1) \log |R|$ is asymptotically distributed as χ^2 with $\frac{1}{2}p(p-1)$d.fr. Bartlett (1951) gave the more refined result, that

$$-\{n - \tfrac{1}{6}(2p+1)\} \log |R|$$

is χ^2 with $\frac{1}{2}p(p-1)$ d.fr. The extra term derives from an allowance for the terms of order n^{-1} in the mean.

Hotelling's T^2

16.52 We now derive a generalization of Student's t, due to Hotelling (1931). Define $\bar{\mathbf{x}}$ and $\mathbf{S}$ as in **16.42** and let

$$T^2 = (n-1)\bar{\mathbf{x}}'\mathbf{S}^{-1}\bar{\mathbf{x}}. \tag{16.152}$$

When $p=1$, T reduces to Student's t since

$$T^2 = (n-1)\bar{x}^2/s^2 = t^2.$$

Now let

$$\mathbf{D} = n\mathbf{S}, \quad \mathbf{z} = n^{\frac{1}{2}}\bar{\mathbf{x}} \quad \text{and} \quad \mathbf{M} = \mathbf{D} + \mathbf{z}\mathbf{z}'. \tag{16.153}$$

Further, let

$$\mathbf{M}_1 = \begin{pmatrix} 1 & \mathbf{z}' \\ \mathbf{0} & \mathbf{D}+\mathbf{z}\mathbf{z}' \end{pmatrix}; \tag{16.154}$$

it follows that $|M_1| = |M|$, given the structure of the first column. If we pre-multiply $\mathbf{M}_1$ by

$$\begin{pmatrix} 1 & \mathbf{0} \\ -\mathbf{z} & \mathbf{1} \end{pmatrix}$$

and post-multiply by

$$\begin{pmatrix} 1 & \mathbf{0} \\ \mathbf{D}^{-1}\mathbf{z} & \mathbf{I} \end{pmatrix},$$

we obtain

$$\begin{aligned} \mathbf{M}_2 &= \begin{pmatrix} 1 & \mathbf{0} \\ -\mathbf{z} & \mathbf{I} \end{pmatrix} \begin{pmatrix} 1 & \mathbf{z}' \\ \mathbf{0} & \mathbf{D}+\mathbf{z}\mathbf{z}' \end{pmatrix} \begin{pmatrix} 1 & \mathbf{0} \\ \mathbf{D}^{-1}\mathbf{z} & \mathbf{I} \end{pmatrix} \\ &= \begin{pmatrix} 1 & \mathbf{z}' \\ -\mathbf{z} & \mathbf{D} \end{pmatrix} \begin{pmatrix} 1 & \mathbf{0} \\ \mathbf{D}^{-1}\mathbf{z} & \mathbf{I} \end{pmatrix} = \begin{pmatrix} 1+\mathbf{z}'\mathbf{D}^{-1}\mathbf{z} & \mathbf{z}' \\ \mathbf{0} & \mathbf{D} \end{pmatrix} \end{aligned} \tag{16.155}$$

so that

$$|M_2| = |M_1| = |M| = (1+\mathbf{z}'\mathbf{D}^{-1}\mathbf{z}) \cdot |D| \tag{16.156}$$

or

$$\frac{|D|}{|M|} = \frac{1}{1+T^2/(n-1)}, \tag{16.157}$$

which we shall denote by h.

16.53 In order to derive the distribution of T^2, we first demonstrate the multivariate extension of Pitman's result, introduced in **15.27**.

Suppose $\mathbf{x} \doteq MVN\ (\mathbf{0},\mathbf{\Sigma})$ and we have a random sample of n observations. Let $\mathbf{S}$ be the sample covariance matrix and let $h = h(\mathbf{x})$ be any scale-free function of the $\mathbf{x}$s. We now show that h and $\mathbf{S}$ are independent. The proof follows similar lines to that given in **15.27**.

The joint characteristic function of h and **S** is, with $\delta = id$, $\boldsymbol{\theta} = i\mathbf{t}$,

$$\phi(\delta, \theta) = E[\exp\{h\delta + \tfrac{1}{2}\operatorname{tr}(\boldsymbol{\theta}\mathbf{S})\}]$$

$$\propto \int \exp(h\delta)|\alpha|^{n/2}\exp[-\tfrac{n}{2}\operatorname{tr}\{(\alpha - n^{-1}\boldsymbol{\theta})\mathbf{S}\}] \prod_j \prod_k dx_{jk} \qquad (16.158)$$

where θ is defined as in **16.46**. Following the same line of argument we have

$$\phi(\delta, \theta) = \frac{|\alpha|}{|\alpha - n^{-1}\theta|} E\{\exp(h\delta)\} \qquad (16.159)$$

showing that h and **S** are independent. As for the univariate case, if h can be expressed in the form $h = Q(\mathbf{x})/g(\mathbf{S})$, it follows that

$$E(h^u) = E(Q^u)/E(g^u), \qquad u = 1, 2, \ldots \qquad (16.160)$$

For a detailed discussion of multivariate quadratic forms, see Khatri (1977) and de Gunst (1987).

16.54 Returning to Hotelling's T^2, we recall from (16.157) that $h = |D|/|M|$, where **M** follows the Wishart distribution with index n. Further **D** is Wishart with index $(n-1)$ so using (16.135) in conjunction with (16.160) we have

$$\begin{aligned} E(h^u) &= \prod_{j=1}^{p} \frac{\Gamma\{\frac{1}{2}(n+1-j)\}\Gamma\{\frac{1}{2}(n-j)+u\}}{\Gamma\{\frac{1}{2}(n+1-j)+u\}\Gamma\{\frac{1}{2}(n-j)\}} \\ &= \frac{\Gamma(\frac{1}{2}n)\Gamma\{\frac{1}{2}(n-p)+u\}}{\Gamma(\frac{1}{2}n+u)\Gamma\{\frac{1}{2}(n-p)\}} \\ &= \frac{B\{\frac{1}{2}(n-p)+u, \frac{1}{2}p\}}{B\{\frac{1}{2}(n-p), \frac{1}{2}p\}}. \end{aligned} \qquad (16.161)$$

This is the uth moment of the Beta distribution

$$dF = \frac{1}{B\{\frac{1}{2}(n-p), \frac{1}{2}p\}} x^{\frac{1}{2}(n-p-2)}(1-x)^{\frac{1}{2}(p-2)}\, dx, \qquad (16.162)$$

which is uniquely determined by its moments. This then is the distribution of the ratio h, and on substitution for T from (16.157) we arrive at

$$dF = \frac{1}{B\{\frac{1}{2}(n-p), \frac{1}{2}p\}} \frac{\{T^2/(n-1)\}^{\frac{1}{2}(p-2)}}{\{1 + T^2/(n-1)\}^{\frac{1}{2}n}}\, d\left(\frac{T^2}{n-1}\right). \qquad (16.163)$$

which is a Beta distribution of the second kind, cf **6.7**; (16.163) defines *Hotelling's T^2 distribution* with $n-1$ d.fr. Equivalently, we may say that the variable

$$F = \frac{(n-p)T^2}{p(n-1)}$$

has an $F(p, n-p)$ distribution.

As for the Wishart distribution in **16.44**, we require $n-1 \geqslant p$ for the validity of the above results.

It will be seen that the essential feature of the T^2 statistic is that it is of form $\mathbf{z}'\hat{\mathbf{V}}^{-1}\mathbf{z}$, where $\mathbf{z}$ is a multinormal vector with means zero and covariance matrix $\mathbf{V}$, while $\hat{\mathbf{V}}$ is the ML estimator of $\mathbf{V}$, adjusted to be unbiased. Above, we had $\mathbf{z}=\bar{\mathbf{x}}$, $\hat{\mathbf{V}}=\mathbf{S}/(n-1)$. Similarly, we may define a test statistic for the two-sample case (generalizing Student's two-sample t^2 test) with $\mathbf{z}=(\bar{\mathbf{x}}_1-\bar{\mathbf{x}}_2)$, $\hat{\mathbf{V}}=\mathbf{S}_p\left(\frac{1}{n_1}+\frac{1}{n_2}\right)$, where $\mathbf{S}_p$ is the unbiased estimator of the common covariance matrix of the two populations, calculated from the pooled samples. The condition corresponding to $n-1 \geqslant p$ becomes $n_1+n_2-2 \geqslant p$. For the contrary case, see Dempster (1958).

The Mahalanobis distance, introduced by Mahalanobis (1930) is

$$D^2=(\bar{\mathbf{x}}_1-\bar{\mathbf{x}}_2)'\mathbf{S}_p^{-1}(\bar{\mathbf{x}}_1-\bar{\mathbf{x}}_2)$$

which may be rewritten as $D^2=(n_1+n_2)T^2/n_1n_2$, where T^2 is the two-sample statistic.

16.55 So far we have generalized to p dimensions the normal distribution theory of means, covariances, and the Student ratio. It is natural to enquire whether there is a similar generalization of the Fisher test for the ratio of two independent variances. We defer a discussion of this topic to Volume 2A, where it arises naturally in connection with tests of hypotheses. We may remark here, however, that exact distributions in closed form corresponding to Fisher's z, for example, are difficult to derive, but moments can usually be obtained in the manner of **16.49**. A further point of interest is that, in generalizations of variance analysis, it is not the ratio of two independent covariance determinants which arises, but the ratio $|S_1|/|S_1+S_2|$ where $\mathbf{S}_1$ and $\mathbf{S}_2$ are independent; the moments of these ratios may be obtainable using (16.160).

Large-sample results

16.56 Even for normal variation, where we are not concerned with cumulants of order higher than the second, there is an embarrassing profusion of parameters, p means, p variances, and $\frac{1}{2}p(p-1)$ covariances (or correlations), $\frac{1}{2}p(p+3)$ in all. For p as low as 5 there are 20, and for $p=10$ there are 65. The distributional problems are correspondingly intractable unless we assume away many of these parameters, e.g. by considering the case of independent variables, for which all correlations vanish.

In such circumstances large-sample results are not to be despised, though they are often neglected. To the first order in n we have

$$E(\mathbf{S})=\mathbf{\Sigma}+O(n^{-1}) \tag{16.164}$$

To the same order

$$E(s_{jk}s_{lm})=\frac{1}{n^2}E\left\{\sum_{\alpha=1}^{n}x_{j\alpha}x_{k\alpha}\sum_{\beta=1}^{n}x_{l\beta}x_{m\beta}\right\}. \tag{16.165}$$

If $\alpha \neq \beta$ two sums on the right are independent. There are $n(n-1)$ such cases and the expectation is $\sigma_{jk}\sigma_{lm}$. If $\alpha = \beta$ there are n terms such as $E(x_{j\alpha}x_{k\alpha}x_{l\alpha}x_{m\alpha})$. In general this involves fourth-order cumulants, but for normal variation we see from the c.f. of the xs,

$$\phi = \exp\left(-\tfrac{1}{2}\sum_j \sum_k \sigma_{jk}t_j t_k\right),$$

that

$$E(x_{j\alpha}x_{k\alpha}x_{l\alpha}x_{m\alpha}) = \sigma_{jk}\sigma_{lm} + \sigma_{jm}\sigma_{kl} + \sigma_{jl}\sigma_{km}. \tag{16.166}$$

Substitution in (16.165) then gives us

$$E(s_{jk}s_{lm}) = \sigma_{jk}\sigma_{lm} + \frac{1}{n}(\sigma_{jm}\sigma_{kl} + \sigma_{jl}\sigma_{km})$$

and hence

$$\text{cov}(s_{jk}, s_{lm}) = \frac{1}{n}(\sigma_{jm}\sigma_{kl} + \sigma_{jl}\sigma_{km}). \tag{16.167}$$

In particular, if $j = k = l = m = 1$ we have the known result

$$\text{var}\, s_{11} = \frac{2\sigma_{11}^2}{n} = \frac{2\sigma_1^4}{n}.$$

Example 16.7 (Large sample moments of correlations)

Let $r_{1j} = s_{1j}/(s_{11}s_{jj})^{\frac{1}{2}}$; if we take logarithms, the differential elements are

$$\frac{dr_{1j}}{r_{1j}} = \frac{ds_{1j}}{s_{1j}} - \frac{1}{2}\frac{ds_{11}}{s_{11}} - \frac{1}{2}\frac{ds_{jj}}{s_{jj}}. \tag{16.168}$$

Multiplying terms like (16.168) for $j = 2, 3$ and taking expectations, we have

$$\begin{aligned}\frac{\text{cov}(r_{12}, r_{13})}{\rho_{12}\rho_{13}} &= \frac{\text{cov}(s_{12}, s_{13})}{\sigma_{12}\sigma_{13}} - \frac{1}{2}\frac{\text{cov}(s_{11}, s_{12})}{\sigma_{11}\sigma_{12}} - \frac{1}{2}\frac{\text{cov}(s_{11}, s_{13})}{\sigma_{11}\sigma_{13}} \\ &\quad - \frac{1}{2}\frac{\text{cov}(s_{22}, s_{13})}{\sigma_{22}\sigma_{13}} - \frac{1}{2}\frac{\text{cov}(s_{33}, s_{12})}{\sigma_{33}\sigma_{12}} + \frac{1}{4}\frac{\text{var}\, s_{11}}{\sigma_{11}^2} \\ &\quad + \frac{1}{4}\frac{\text{cov}(s_{11}, s_{33})}{\sigma_{11}\sigma_{33}} + \frac{1}{4}\frac{\text{cov}(s_{11}, s_{22})}{\sigma_{11}\sigma_{22}} + \frac{1}{4}\frac{\text{cov}(s_{22}, s_{33})}{\sigma_{22}\sigma_{33}} + O(n^{-2}).\end{aligned} \tag{16.169}$$

Relations of type (16.167) reduce this to

$$n\,\text{cov}(r_{12}, r_{13}) = \rho_{23}(1 - \rho_{12}^2 - \rho_{13}^2) + \tfrac{1}{2}\rho_{12}\rho_{13}(\rho_{12}^2 + \rho_{13}^2 + \rho_{23}^2 - 1) + O(n^{-1}). \tag{16.170}$$

The large sample correlation of r_{12} and r_{13} is easily determined from the consideration that n var $r = (1 - \rho^2)^2$.

The latent roots of a covariance matrix

16.57 In multivariate analysis we are interested in the latent roots of a stochastic matrix. If **C** is a covariance matrix we shall wish to study the behaviour of the roots of the pth order polynomial in λ given by

$$|\mathbf{C}-\lambda\mathbf{I}|=0, \tag{16.171}$$

Equation (16.171) is a particular case of the more general form

$$|\mathbf{C}-\lambda\mathbf{B}|=0 \tag{16.172}$$

where **B**, **C** are multiples of independent covariance matrices based on m, n observations respectively, estimating the same population covariance matrix. We may write equivalently

$$|u(\mathbf{C}+\mathbf{B})-\mathbf{B}|=0 \tag{16.173}$$

with

$$u=\frac{1}{1+\lambda},\quad \lambda=\frac{1-u}{u}.$$

The us have the advantage of lying between 0 and 1. The complexity of the distributional problem arises from the fact that the roots in λ or u are not algebraic functions of the covariances. It is easier to derive sampling distributions of symmetric functions of the roots than of the individual roots themselves.

16.58 We assume that the underlying distribution is normal with unit variances and zero covariances. The joint distribution of **C** and **B** then has the density function, as at (16.139),

$$f=\frac{(\frac{1}{2}n)^{\frac{1}{2}p(n-1)}(\frac{1}{2}m)^{\frac{1}{2}p(m-1)}}{\pi^{\frac{1}{2}p(p-1)}}\times\frac{|C|^{\frac{1}{2}(n-p-2)}|B|^{\frac{1}{2}(m-p-2)}}{\prod_{j=1}^{p}[\Gamma\{\frac{1}{2}(n-j)\}\Gamma\{\frac{1}{2}(m-j)\}]}\exp\{-\tfrac{1}{2}\operatorname{tr}(n\mathbf{C}+m\mathbf{B})\}. \tag{16.174}$$

To remove the effect of the differing multipliers m and n in the exponent, we redefine our symbols so that $\mathbf{c}=n\mathbf{C}$, $\mathbf{b}=m\mathbf{B}$. The Jacobian of this transformation is $(nm)^{\frac{1}{2}p(p+1)}$ so that the constants $n^{\frac{1}{2}p(n-1)}$, $m^{\frac{1}{2}p(m-1)}$ in (16.174) are exactly absorbed. The density becomes

$$f=k\frac{|c|^{\frac{1}{2}(n-p-2)}|b|^{\frac{1}{2}(m-p-2)}}{\prod[\Gamma\{\frac{1}{2}(n-j)\}\Gamma\{\frac{1}{2}(m-j)\}]}\exp\{-\tfrac{1}{2}\operatorname{tr}(\mathbf{c}+\mathbf{b})\}, \tag{16.175}$$

where $k^{-1}=2^{\frac{1}{2}p(m+n-2)}\pi^{\frac{1}{2}p(p-1)}$.

To each root u_j of (16.173) there corresponds a latent vector $\mathbf{l}_j$ such that

$$\{\mathbf{b}-u_j(\mathbf{b}+\mathbf{c})\}\mathbf{l}_j=\mathbf{0} \tag{16.176}$$

and we may choose the $\mathbf{l}_j$ so that

$$\mathbf{l}_j'(\mathbf{b}+\mathbf{c})\,\mathbf{l}_j = 1. \tag{16.177}$$

If $\mathbf{L}$ is the $p \times p$ matrix of latent vectors we have from (16.176)

$$\mathbf{bL} = (\mathbf{b}+\mathbf{c})\mathbf{Lu} \tag{16.178}$$

where $\mathbf{u}$ is a diagonal matrix whose elements are $u_1, u_2, \ldots, u_p$. We now suppose that the us are arranged in descending order of magnitude, $u_1 \geqslant u_2 \geqslant \ldots \geqslant u_p$.

We also have from (16.176)

$$\mathbf{l}_k'\mathbf{b}\mathbf{l}_j = u_j \mathbf{l}_k'(\mathbf{b}+\mathbf{c})\,\mathbf{l}_j \tag{16.179}$$

and by transposition ($\mathbf{b}$ and $\mathbf{c}$ being symmetric)

$$\mathbf{l}_k'\mathbf{b} = \mathbf{l}_k' u_k(\mathbf{b}+\mathbf{c})$$

so that

$$\mathbf{l}_k'\mathbf{b}\mathbf{l}_j = u_k \mathbf{l}_k'(\mathbf{b}+\mathbf{c})\,\mathbf{l}_j. \tag{16.180}$$

It follows from (16.179) and (16.180) that, if $u_j \neq u_k$,

$$\mathbf{l}_k'(\mathbf{b}+\mathbf{c})\,\mathbf{l}_j = 0. \tag{16.181}$$

Multiplying (16.178) on the left by $\mathbf{L}'$ and using (16.177) we have

$$\mathbf{L}'\mathbf{bL} = \mathbf{u}. \tag{16.182}$$

From (16.177) and (16.181),

$$\mathbf{L}'(\mathbf{b}+\mathbf{c})\mathbf{L} = \mathbf{I} \tag{16.183}$$

and hence

$$\mathbf{b}+\mathbf{c} = (\mathbf{L}')^{-1}\mathbf{L}^{-1}. \tag{16.184}$$

Likewise from (16.182)

$$\mathbf{b} = (\mathbf{L}')^{-1}\mathbf{u}\mathbf{L}^{-1} \tag{16.185}$$

and hence

$$\mathbf{c} = (\mathbf{L})^{-1}(\mathbf{I}-\mathbf{u})\mathbf{L}^{-1}. \tag{16.186}$$

16.59 Looking back to (16.176), we see that with the transformation (16.185–6) the density function is given by

$$f \propto \frac{\left\{\prod_{j=1}^{p}(1-u_j)\right\}^{\frac{1}{2}(n-p-2)}\left\{\prod_{j=1}^{p}u_j\right\}^{\frac{1}{2}(m-p-2)}}{\prod[\Gamma\{\frac{1}{2}(n-j)\}\Gamma\{\frac{1}{2}(m-j)\}]} \times \text{a function of } \mathbf{L}. \tag{16.187}$$

We now consider the Jacobian of the transformation. It will appear that the distribution of us is independent of that of the ls and the contribution from the latter may be factored out.

We observe from (16.185) and (16.186) that there are $\frac{1}{2}p(p+1)$ variables in each of $\mathbf{b}$ and $\mathbf{c}$, $p(p+1)$ in all; and p^2 variables in $\mathbf{L}$ and p us, again making $p(p+1)$.

The Jacobian of the transformation to **b**, **c** is the same as that to **b**, **b** + **c**. Writing **g** for $\mathbf{L}^{-1}$ in (16.184–5), we then have to consider the Jacobian of the transformation

$$\mathbf{b} = \mathbf{g}'\mathbf{u}\mathbf{g}$$
$$\mathbf{b} + \mathbf{c} = \mathbf{g}'\mathbf{g}.$$

We prove that $u_1 - u_2$ is a factor of the Jacobian. Although the argument is general, it will be clearer if we write it explicitly for $p = 2$.

The matrix **b** then becomes

$$\begin{pmatrix} g_{11}^2 u_1 + g_{21}^2 u_2 & g_{11}g_{12}u_1 + g_{21}g_{22}u_2 \\ g_{12}g_{11}u_1 + g_{22}g_{21}u_2 & g_{12}^2 u_1 + g_{22}^2 u_2 \end{pmatrix}. \tag{16.188}$$

$\mathbf{b} + \mathbf{c}$ is the same with the us put equal to unity.

For the Jacobian we have

$$\frac{\partial\{b_{11}, b_{12}, b_{22}, (b+c)_{11}, (b+c)_{12}, (b+c)_{22}\}}{\partial(u_1, u_2, g_{11}, g_{12}, g_{21}, g_{22})}$$

$$= \begin{vmatrix} g_{11}^2 & g_{21}^2 & 2g_{11}u_1 & 0 & 2g_{21}u_2 & 0 \\ g_{11}g_{12} & g_{21}g_{22} & g_{12}u_1 & g_{11}u_1 & g_{22}u_2 & g_{21}u_2 \\ g_{12}^2 & g_{22}^2 & 0 & 2g_{12}u_1 & 0 & 2g_{22}u_2 \\ 0 & 0 & 2g_{11} & 0 & 2g_{21} & 0 \\ 0 & 0 & g_{12} & g_{11} & g_{22} & g_{21} \\ 0 & 0 & 0 & 2g_{12} & 0 & 2g_{22} \end{vmatrix}$$

If $u_1 = u_2$ we can subtract multiples of the bottom three rows from the top three to obliterate all except the first two terms in these rows, and the determinant vanishes.

It follows that every $u_j - u_k$, $j < k$, is a factor of the Jacobian. The product of these factors is of degree $\frac{1}{2}p(p-1)$ in the us. There can be no other terms involving u because the Jacobian can be of no higher degree.

Thus, for the us alone we have from (16.187)

$$dF = k \frac{\left\{\prod (1-u_j)\right\}^{\frac{1}{2}(n-p-2)} \left(\prod u_j\right)^{\frac{1}{2}(m-p-2)}}{\prod [\Gamma\{\frac{1}{2}(n-j)\}\Gamma\{\frac{1}{2}(m-j)\}]} \prod_{j<k} (u_j - u_k) \prod du_j, \tag{16.189}$$

where k is some constant. To evaluate it by explicit integration would be very difficult. The following indirect route suffices:

k arises from terms in (16.175) and the Jacobian involving p and $m+n$, but not m separately. Write it then as $k(p, m+n)$. Note that $|b| = \prod u_j$. If we increase m by $2t$ in (16.189) and integrate we have the tth moment of $|b|$ except for a factor $k(p, m+n+2t)$. As in **16.49** we can find this expectation and it follows that

$$\frac{k(p, m+n+2t)}{k(p, m+n)} = \prod_{j=1}^{p} \frac{\Gamma\{\frac{1}{2}(m+n-1-j)+t\}}{\Gamma\{\frac{1}{2}(m+n-1-j)\}}, \tag{16.190}$$

so that

$$k(p, m+n) = K(p) \prod_{j=1}^{p} \Gamma\{\tfrac{1}{2}(m+n-1-j)\}, \tag{16.191}$$

where $K(p)$ is a function of p only. To evaluate it, make the substitution in (16.189) of $u_j = 2v_j/n$ and let n tend to infinity. Our distribution becomes

$$\begin{aligned} dF = \left(\prod v_j\right)^{\frac{1}{2}(m-p-2)} \exp\left(-\sum v_j\right) \prod (v_j - v_k) \prod dv_j \\ \times \prod \frac{\Gamma\{\frac{1}{2}(m+n-1-j)\}K(p)}{\Gamma\{\frac{1}{2}(m-j)\}\Gamma\{\frac{1}{2}(n-j)\}}. \end{aligned} \tag{16.192}$$

This may be evaluated by step-by-step substitutions of the type

$$\begin{aligned} v_1 &= w_1 \\ v_j &= w_j + v_1, \quad j > 1, \end{aligned}$$

and choosing m at each stage so that the coefficient in $\prod v_j$ vanishes, as we may since the result is independent of m. We find

$$\frac{K(p)}{2^{\frac{1}{2}p(p-1)}} \prod \frac{\Gamma(p+1-j)}{\Gamma\{\frac{1}{2}(p-j)\}} = 1.$$

Use of the duplication formula (16.126) for the Gamma function gives us

$$K(p) = \frac{\pi^{\frac{1}{2}p}}{\prod \Gamma\{\frac{1}{2}(p+1-j)\}}. \tag{16.193}$$

Using (16.191) and (16.193) in (16.189), we find finally the distribution

$$dF = \pi^{\frac{1}{2}p} \prod_{j=1}^{p} \frac{\Gamma\{\frac{1}{2}(m+n-1-j)\} u_j^{\frac{1}{2}(m-p-2)} (1-u_j)^{\frac{1}{2}(n-p-2)}}{\Gamma\{\frac{1}{2}(m-j)\}\Gamma\{\frac{1}{2}(n-j)\}\Gamma\{\frac{1}{2}(p+1-j)\}} \prod_{j<k} (u_j - u_k) \prod du_j, \tag{16.194}$$

a remarkable form discovered in 1939 by Fisher, P. L. Hsu, S. N. Roy, Girshick, and Mood, independently—cf. Mood (1951).

The distribution of the λs is, of course, given by a simple substitution

$$u = 1/(1+\lambda).$$

16.60 In the case of (16.171), when the matrix **B** reduces to the identity matrix, a slightly different result is obtained. We will quote the result for the distribution of the roots λ in this case. The reader may care to run through the foregoing proof and modify it where necessary to obtain this result:

$$dF = \frac{\pi^{\frac{1}{2}p}}{2^{\frac{1}{2}p(n-1)}} \exp\left\{-\tfrac{1}{2}\sum_{j=1}^{p} \lambda_j\right\} \prod_{j=1}^{p} \frac{\lambda_j^{\frac{1}{2}(n-p-2)}}{\Gamma\{\frac{1}{2}(n-j)\}\Gamma\{\frac{1}{2}(p+1-j)\}} \prod_{j<k} (\lambda_j - \lambda_k) \prod d\lambda_j, \tag{16.195}$$

where now the λ_j are in descending order.

Example 16.8 (Distribution of the latent roots, $p=2$)

These distributions are very intractable except in simple cases. Let us consider the case when $p=2$. From (16.195) we have

$$dF=\frac{\pi^{\frac{1}{2}}}{2^{n-1}}\frac{(\lambda_1\lambda_2)^{\frac{1}{2}(n-4)}\exp\{-\frac{1}{2}(\lambda_1+\lambda_2)\}}{\Gamma\{\frac{1}{2}(n-1)\}\Gamma\{\frac{1}{2}(n-2)\}}(\lambda_1-\lambda_2)d\lambda_1\,d\lambda_2. \tag{16.196}$$

The duplication formula (16.126) for Gamma functions reduces the density function to

$$dF=\frac{(\lambda_1\lambda_2)^{\frac{1}{2}(n-4)}\exp\{-\frac{1}{2}(\lambda_1+\lambda_2)\}(\lambda_1-\lambda_2)d\lambda_1 d\lambda_2}{4\Gamma(n-2)}.$$

If we try to integrate for λ_2 over the range 0 to λ_1, we obtain for λ_1 an Incomplete Gamma function. On the other hand, for the functions

$$x=\lambda_1\lambda_2,\quad y=\lambda_1+\lambda_2$$

we find $1/J=|\lambda_1-\lambda_2|$ and the distribution becomes

$$dF=\frac{x^{\frac{1}{2}(n-4)}e^{-\frac{1}{2}y}\,dx\,dy}{4\Gamma(n-2)},\quad 0\leqslant x\leqslant\tfrac{1}{4}y^2, \tag{16.197}$$

and on integration for x,

$$dF=\frac{(\frac{1}{2}y)^{n-2}e^{-\frac{1}{2}y}\,d(\frac{1}{2}y)}{\Gamma(n-1)},\quad 0\leqslant y<\infty. \tag{16.198}$$

In fact, in this case the determinant reduces to

$$\begin{vmatrix} ns_1^2-\lambda & ns_1s_2r \\ ns_1s_2r & ns_2^2-\lambda\end{vmatrix}=\lambda^2-n(s_1^2+s_2^2)\lambda+n^2s_1^2s_2^2(1-r^2)=0.$$

The sum of the roots is thus equal to the sum of two independent variables ns_1^2, ns_2^2, each of which is a χ^2_{n-1} variable, and has a χ^2 distribution with $2n-2$ degrees of freedom, agreeing with (16.198). Also, it follows from (16.197) that $x^{\frac{1}{2}}$ has a Gamma distribution with index $n-2$.

Large sample results for latent roots

16.61 We now consider the large sample standard errors and covariances for the latent roots when all the population latent roots are distinct. These results were first obtained by Girshick (1939).

As in **16.58**, $\mathbf{L}=(\mathbf{l}_1,\ldots,\mathbf{l}_p)$ is the $(p\times p)$ matrix of latent vectors such that

$$\mathbf{L}'\mathbf{L}=\mathbf{I}\quad\text{and}\quad\mathbf{SL}=\mathbf{La},$$

where $\mathbf{a}$ is the diagonal matrix of latent roots, $\mathbf{a}=\text{diag}\,(a_1,\ldots,a_p)$. If the $\mathbf{x}$ are independent and normally distributed with covariance matrix $\Lambda=\text{diag}(\lambda_1,\ldots,\lambda_p)$, it follows that

$$E(a_j)=\lambda_j+O(n^{-1}) \tag{16.199}$$

and, using (16.167)

$$\begin{aligned}\text{cov}\,(a_j, a_k) &= \text{cov}\,(s_{jj}, s_{kk}) \\ &= 2\lambda_j^2\delta_{jk}/n + O\,(n^{-2}) \qquad (16.200)\end{aligned}$$

where δ_{jk} is the Kronecker delta. Thus a_j and a_k are asymptotically uncorrelated and

$$\text{var}\,(a_j) = 2\lambda_j^2/n + O(n^{-2}),\ \text{var}\,(\log a_j) = 2/n + O(n^{-2}).$$

Let $\otimes$ denote the operation of element-by-element multiplication so that $\mathbf{l}\otimes\mathbf{l}$ has elements $\{l_i^2\}$ and rescale $\mathbf{L}$ so that $\mathbf{L}'\mathbf{L} = \mathbf{\Lambda}$. If

$$\mathbf{l}_{j(2)} = \mathbf{l}_j \otimes \mathbf{l}_j, \quad \mathbf{R}_{(2)} = \mathbf{R}\otimes\mathbf{R}$$

we have that

$$\begin{aligned}\text{cov}\,(a_j, a_k) = 2n^{-1}[\lambda_j\lambda_k\delta_{jk} + \mathbf{l}'_{j(2)}\mathbf{R}_{(2)}\mathbf{l}_{k(2)} \\ - (\lambda_j + \lambda_k)\mathbf{l}'_{j(2)}\mathbf{l}_{k(2)}] + O(n^{-2}). \qquad (16.201)\end{aligned}$$

16.62 The same method may be used to derive variances and covariances for the coefficients of the latent vectors. For what they are worth, we quote the results; but it must be remembered that in practice we should rarely wish to test an individual direction cosine.

$$\text{cov}\,(l_{jk}, l_{mt}) = \frac{-\lambda_k\lambda_t l_{jt} l_{mk}}{n(\lambda_k - \lambda_t)^2}, \quad k \neq t, \qquad (16.202)$$

$$\text{cov}\,(l_{jk}, l_{mk}) = \frac{1}{n}\sum_{s\neq k}\frac{\lambda_k\lambda_s l_{js} l_{ms}}{(\lambda_k - \lambda_s)^2}, \qquad (16.203)$$

so that var l_{jk} is given by (16.203) with $m = j$.

These results were obtained by T. W. Anderson (1963) who demonstrated the asymptotic normality of the distribution of latent roots and vectors, including the case where some of the population roots are equal. James (1964) has found the exact distribution of the latent roots in terms of zonal polynomials, although the resulting series expansion is generally slow to converge. Assuming distinct population roots, G. A. Anderson (1965) gave an approximation up to $O(n^{-2})$, although the n^{-2} term was conjectural for $p \geq 4$. Khatri and Srivastava (1978) confirmed these results and have obtained more general results which allow some roots to be equal.

Sugiura (1976), Fujikoshi (1977) and Krishnaiah and Lee (1979) all provide asymptotic expansions for the distributions of the latent roots, allowing some of the population roots to be equal. See Muirhead (1978) for a detailed review.

Unfortunately, at this point, it becomes necessary to introduce a jarring note into the discussions. Reworking Girshick's original results for non-normal populations with finite fourth cumulants $\kappa_4(j)$ and $\kappa_{22}(j, k)$ for the jth and kth variables, Waternaux (1976) finds that, to $O(n^{-1})$,

$$\text{var}\,(u_j) = (\kappa_4(j) + 2\lambda_j^2)/n$$

and

$$\text{cov}\,(u_j, u_k) = \kappa_{22}(j, k)/n.$$

This lack of robustness to non-normality in second-order moments was noted in **16.40** and poses a major problem for other areas of multivariate analysis also. See also Fujikoshi (1980).

Further results

16.63 In conclusion we may note some points which we do not have space to develop in detail.

(1) The distribution of latent roots (16.194) reduces for $p = 1$ to a Beta distribution. Foster and Rees (1957) and Foster (1957, 1958) therefore called it a "generalized Beta distribution"; they tabulated percentiles of the largest root for $p = 2$, 3 and 4. Pillai (1966, 1967a) has improved the method and tabulated (1964, 1965, 1967b) up to $p = 20$. Sugiyama (1967a, b) Konishi (1978b), and Konishi and Sugiyama (1981) give expressions for the exact distribution of the largest root in (16.195) and the largest and smallest roots in (16.194). Muirhead and Chikuse (1975) give asymptotic expansions for the joint and the marginal distributions of the latent roots.

The *Biometrika Tables* , Vol. II, reproduce both the Foster–Rees and the Pillai tables in modified form.

(2) Krishnaiah (1980) has drawn together a considerable number of relevant tables. These include the percentage points, typically $\alpha = 0.01, 0.025, 0.05, 0.10$, for various values of n up to 50 of

(i) the largest latent root, $p = 2(1)15$;
(ii) the smallest latent root, $p = 2(1)15$;
(iii) individual latent roots, $p = 2(1)7$;
(iv) the ratio of individual latent roots to the trace, $p = 2(1)6$;
(v) ratio of largest and smallest roots, $p = 2$ (n up to 119) with lesser ranges for $p = 3, 4, 5$.

(3) If $\mathbf{S}_1$, $\mathbf{S}_2$ follow a Wishart distribution based on p variables with sample sizes n_1 and n_2, there exists a lower triangular matrix $\mathbf{V}$ such that $\mathbf{S}_1 + \mathbf{S}_2 = \mathbf{V}\mathbf{V}'$ (cf. Exercise 16.39). If $\mathbf{L}$ is defined by $\mathbf{S}_1 = \mathbf{V}\mathbf{L}\mathbf{V}'$, then $\mathbf{V}$ and $\mathbf{L}$ are independently distributed and $\mathbf{L}$ has density function

$$f \propto |\mathbf{L}|^{\frac{1}{2}(n_1-p-2)}|\mathbf{I} - \mathbf{L}|^{\frac{1}{2}(n_2-p-2)}.$$

This result is originally due to Hsu (1939). See also Kshirsagar (1961). The distribution of $\mathbf{L}$ is sometimes called "the multivariate Beta distribution."

(4) J. M. Chambers (1967) gives methods of asymptotic approximations for multivariate distributions, including Edgeworth expansions (cf. **6.18**) and perturbation approximations which generalize the methods of **10.6**.

EXERCISES

16.1 Show that the rth moment of the variance-ratio distribution at (16.24) is

$$\mu'_r = \left(\frac{\nu_2}{\nu_1}\right)^r \frac{\Gamma(\frac{1}{2}\nu_1 + r)\Gamma(\frac{1}{2}\nu_2 - r)}{\Gamma(\frac{1}{2}\nu_1)\Gamma(\frac{1}{2}\nu_2)}$$

if $2r < \nu_2$, and does not exist if $2r \geqslant \nu_2$. Hence show that

$$\beta_1 = \mu_3^2/\mu_2^3 = \frac{8(\nu_2 - 4)(2\nu_1 + \nu_2 - 2)^2}{\nu_1(\nu_2 - 6)^2(\nu_1 + \nu_2 - 2)}$$

and that

$$\beta_2 = \mu_4/\mu_2^2 = \frac{3}{\nu_2 - 8}\{\nu_2 - 4 + \tfrac{1}{2}(\nu_2 - 6)\beta_1\}.$$

16.2 In the variance-ratio distribution (16.24), show that when $\nu_1 = \nu_2 = \nu$, the variate $\frac{1}{2}\nu^{\frac{1}{2}}(F^{\frac{1}{2}} - F^{-\frac{1}{2}})$ has Student's t distribution (16.15).

(Cf. Cacoullos, 1965)

16.3 Show that the c.f. of a χ^2_ν variate about its mean satisfies the equation

$$2it\nu\phi(t) = (1 - 2it)\frac{\partial\phi(t)}{\partial(it)}$$

and hence by equating coefficients that the moments about the mean satisfy the relation

$$\mu_{j+1} = 2j(\mu_j + \nu\mu_{j-1}), \quad j \geqslant 1.$$

Use this to verify (16.5). Show further that

$$\frac{\partial\phi(t)}{\partial\nu} = -\phi(t)\{it + \tfrac{1}{2}\log(1 - 2it)\}$$

and hence that

$$\frac{\partial\mu_r}{\partial\nu} = \sum_{s=2}^{r}\binom{r}{s}2^{s-1}(s-1)!\,\mu_{r-s}.$$

16.4 n variates $x_1, \ldots, x_n$ are independently distributed in the uniform distribution $f(x) = 1$, $0 < x \leqslant 1$. Show that $-2\log(x_1 x_2 \ldots x_n)$ is distributed as χ^2 with $2n$ d.fr.

16.5 $X_i (i = 1, \ldots, n)$ are a given set of constants with zero mean and ε_i $(i = 1, \ldots, n)$ a set of independent values of a normal random variable with zero mean and variance σ^2. Show that if

$$y_i = \beta_0 + \beta_1 X_i + \varepsilon_i,$$

and the constants β_0, β_1 are estimated as b_0, b_1 where $b_0 = \bar{y}$, $b_1 = \sum(y_i - \bar{y})X_i / \sum X_i^2$, then if

$$y'_i = y_i - (b_0 + b_1 X_i),$$

we have

$$\sum(y_i - \bar{y})^2 = \sum(y'_i)^2 + \frac{\{\sum X_i(y_i - \bar{y})\}^2}{\sum X_i^2}.$$

Hence show that $\sum(y'_i)^2$ is distributed as $\sigma^2\chi^2$ with $(n-2)$ d.fr. independently of $b_1^2\sum X_i^2$ which is distributed as $\sigma^2\chi^2$ with one d.fr.

If x and y are distributed normally and independently, deduce that $(n-2)r^2/(1-r^2)$ is distributed as Student's t^2 with $(n-2)$ d.fr.

16.6 In the notation of **16.23-36**, define

$$v^2 = ns_1^2/\sigma_1^2, \qquad u^2 = \frac{\sigma_1^2(b_2-\beta_2)^2}{\sigma_2^2(1-\rho^2)}v^2, \qquad w^2 = \frac{ns_2^2(1-r^2)}{\sigma_2^2(1-\rho^2)},$$

$$\tilde{r} = r/(1-r^2)^{\frac{1}{2}}, \qquad \tilde{\rho} = \rho/(1-\rho^2)^{\frac{1}{2}}.$$

Show that $\tilde{r} \equiv (u+\tilde{\rho}v)/w$ for any bivariate population. For the bivariate normal case, show that u^2, v^2 and w^2 are independent χ^2 variables, with respectively 1, $(n-1)$ and $(n-2)$ d.fr. Derive the results of **16.35-6**.

Show that $r \geqslant$ if and only if $\dfrac{u}{v} \geqslant -\tilde{\rho}$ and hence that

$$P(r \geqslant 0) = P\{t_{n-1} \geqslant -(n-1)^{\frac{1}{2}}\tilde{\rho}\}$$

where t_{n-1} is a Student's t variate with $(n-1)$ d.fr.

(Cf. Ruben, 1966.)

16.7 Denoting the χ^2 density (16.1) by $f_\nu(z)$ and its d.f. by $F_\nu(z)$, and writing $G(x)$ for the d.f. of a standardized normal distribution, show that for positive integers s,

$$F_{\nu+2s}(z) = F_\nu(z) - 2\sum_{r=1}^{s} f_{\nu+2r}(z),$$

and hence

$$F_{2s-1}(z) = 2G(z^{\frac{1}{2}}) - 1 - 2\sum_{r=2}^{s} f_{2r-1}(z),$$

$$F_{2s}(z) = 1 - 2\sum_{r=1}^{s} f_{2r}(z),$$

so that $F_{2s}(z)$ is the complement to unity of the sum of the first s terms of a Poisson distribution with parameter $\frac{1}{2}z$.

Letting $s \to \infty$, show that for any value $z \geqslant 0$

$$\sum_{r=2}^{\infty} f_{2r-1}(z) = G(z^{\frac{1}{2}}) - \tfrac{1}{2},$$

$$\sum_{r=1}^{\infty} f_{2r}(z) = \tfrac{1}{2}, \quad \text{and hence that}$$

$$\sum_{\nu=2}^{\infty} f_\nu(z) = G(z^{\frac{1}{2}}).$$

16.8 If x, y are standardized binormal variates with correlation parameter ρ, show that the c.f. of their product is

$$\phi(t) = \{(1-\rho it)^2 + t^2\}^{-\frac{1}{2}}$$

and hence that $E(xy) = \rho$, $\text{var}(xy) = 1 + \rho^2$—cf. Example 11.22 when $\rho = 0$.

Verify var (xy) from Exercise 10.23. Use Exercise 4.3 to show that when $\rho = 0$ we may obtain $g(z)$ in Exercise 11.21. Show that the case $\rho = 1$ reduces to Exercise 4.18.

16.9 Show directly from (16.15) that Student's t tends to normality as ν tends to infinity.

16.10 By the substitution $q = (1 + t^2/\nu)^{-1}$ show that the d.f. $F(t)$ of Student's t is, for even ν, given by

$$2F(t) - 1 = (1-q)^{\frac{1}{2}}\left\{1 + \sum_{s=1}^{\frac{1}{2}(\nu-2)} \frac{q^s}{2^{2s-1}\cdot s \cdot B(s,s)}\right\},$$

the expression in braces on the right-hand side being the cumulative sum of terms of the binomial expansion $(1-q)^{-\frac{1}{2}}$.

(Fisher, 1935)

16.11 x, y, z are multinormally distributed with all population correlations zero, so that in a sample of n observations, each of the three sample correlations is distributed in the form (16.62). Show that each of these correlations is distributed independently of each of the others, but that the three correlations are not completely independent of each other.

16.12 Verify (16.29), the c.f. of the z-distribution, and by expanding $\log \phi(t)$ verify the cumulants at (16.34). Deduce the asymptotic normality as at (16.31).

16.13 Using the methods of Chapter 10, show that in large samples from a binormal population the variance of a sample regression coefficient is approximately

$$\text{var}\left(\frac{m_{11}}{m_{20}}\right) = \frac{\sigma_2^2}{\sigma_1^2}\frac{(1-\rho^2)}{n}.$$

16.14 By differentiation of (16.64) with respect to (ρr) show that

$$\{1 - (\rho r)^2\}I_2 = 1 + \rho r I_1$$

and hence that

$$n\{1-(\rho r)^2\}I_{n+1} - (2n-1)\rho r I_n - (n-1)I_{n-1} = 0.$$

Hence show that the density function, f_n, of r based on a sample of size n satisfies the recurrence formula

$$(n-1)\{1-(\rho r)^2\}f_{n+2} = (2n-1)\rho r(1-\rho^2)^{\frac{1}{2}}(1-r^2)^{\frac{1}{2}}f_{n+1} + (n-1)^2(1-\rho^2)(1-r^2)f_n/(n-2).$$

(*Co-operative Study*, 1917; Hotelling, 1953)

16.15 Verify that the density function of r satisfies the relation

$$r\frac{\partial f_n}{\partial r} + \frac{(n-4)r^2 f_n}{1-r^2} = \rho\frac{\partial f_n}{\partial \rho} + \frac{(n-1)\rho^2 f_n}{1-\rho^2}.$$

Hence show that for $n \geqslant 5$,

$$(n-4)E\{r^2/(1-r^2)\} = 1 + (n-1)\rho^2/(1-\rho^2)$$

and verify this for $\rho = 0$ from (16.63).

(Cf. Hotelling, 1953)

16.16 From (16.60), show that the distribution of $v = \frac{s_1}{\sigma_1}\Big/\frac{s_2}{\sigma_2}$ and r is given by

$$dF \propto \frac{v^{n-2}(1-r^2)^{\frac{1}{2}(n-4)}}{(1-2\rho rv+v^2)^{n-1}}\,dr\,dv.$$

Integrating for r from -1 to $+1$ by putting

$$r = \frac{u(\lambda+\mu)-(1-u)(\lambda-\mu)}{u(\lambda+\mu)+(1-u)(\lambda-\mu)}, \qquad \lambda = 1+v^2, \qquad \mu = 2\rho v,$$

show that the distribution of v is

$$dF = \frac{2(1-\rho^2)^{\frac{1}{2}(n-1)}}{B\left(\frac{n-1}{2},\frac{n-1}{2}\right)}\frac{v^{n-2}}{(1+v^2)^{n-1}}\left\{1-\frac{4\rho^2v^2}{(1+v^2)^2}\right\}^{-\frac{1}{2}n}dv,$$

and that when $\rho = 0$, the distribution of $F = v^2$ is (16.24) with $\nu_1 = \nu_2 = n-1$.

(S. S. Bose, 1935; Finney, 1938)

16.17 If x and y are independent continuous variates, show that in samples of size n from their joint distribution the sample correlation coefficient r has zero mean and variance $(n-1)^{-1}$.

(Pitman, 1937a)

16.18 If t is a statistic which in large samples tends to a population parameter θ, and the variance of t is some function of θ, say $f(\theta)/n + o(n^{-1})$, show using (10.14) that the variate

$$z = \int_0^t \{f(\theta)\}^{-\frac{1}{2}}\,d\theta$$

has variance $1/n + o(n^{-1})$.

Hence, from the large-sample result obtained from (16.74), $\text{var}\, r = (1-\rho^2)^2/(n-1)$, derive the transformation $z = \tanh^{-1} r$, with variance $(n-1)^{-1} + o(n^{-1})$.

16.19 Noting that, from (16.77), the variance of z is

$$\frac{1}{n-1}\left(1+\frac{4-\rho^2}{2n}\right)+o(n^{-2}),$$

show that z^*, defined by

$$z^* \propto \int_0^r \frac{d\rho}{(1-\rho^2)\left(1+\frac{4-\rho^2}{2n}\right)^{\frac{1}{2}}} = z - \frac{3z+r}{4n} + o(n^{-1}),$$

has variance

$$\text{var}\, z^* = \frac{1}{n-1} + o(n^{-2}).$$

(Hotelling, 1953; Kendall, 1953)

16.20 In (16.54), transform from (s_1, s_2, r) to the variables

$$u = \frac{2s_1s_2r}{s_1^2+s_2^2}, \qquad v = s_1^2+s_2^2, \qquad w = s_2^2,$$

and then from (u, v, w) to (u, v, t), where $t = (2w-v)/\{v(1-u^2)^{\frac{1}{2}}\}$. Integrate with respect to v and transform the resulting distribution of u and t to that of u and z, where $t = (1-z)/(1+z)$. Finally, integrate with respect to z to obtain the distribution of u:

$$p(u) = \frac{(n-2)2^{n-3}}{\pi} B(\tfrac{1}{2}n, \tfrac{1}{2}n - 1)\left(\frac{1-\rho^2}{\sigma_1^2\sigma_2^2}\right)^{\frac{1}{2}(n-1)} \frac{a}{(a^2-b^2)^{\frac{1}{2}n}} (1-u^2)^{\frac{1}{2}(n-3)},$$

where $a = \frac{1}{2}\left(\frac{1}{\sigma_1^2} + \frac{1}{\sigma_2^2}\right) - \frac{\rho u}{\sigma_1\sigma_2}$ and $b = \frac{1}{2}\left(\frac{1}{\sigma_2^2} - \frac{1}{\sigma_1^2}\right)(1-u^2)^{\frac{1}{2}}$.

Show that $|u| \leqslant |r|$ and that if $\sigma_1 = \sigma_2$ and $\rho = 0$, $p(u)$ is exactly (16.62) with $(n+1)$ written for n. (DeLury, 1938; Mehta and Gurland, 1969)

16.21 Using Exercise 16.2 on (16.63) show that when $\rho = 0$ the exact distribution of Fisher's z defined at (16.75) is the variance-ratio z distribution (16.26) with $\nu_1 = \nu_2 = n-2$. From its c.f. (16.29), use the relation for positive integers p

$$\frac{\Gamma''(p)}{\Gamma(p)} - \left\{\frac{\Gamma'(p)}{\Gamma(p)}\right\}^2 = \frac{\pi^2}{6} - \sum_{r=1}^{p-1} \frac{1}{r^2}$$

to show that for even $n \geqslant 6$ the exact variance of z is

$$\mu_2 = \pi^2/12 - \frac{1}{2} \sum_{r=1}^{\frac{1}{2}n-2} \frac{1}{r^2},$$

which for $n = 6$ equals 0.32 against the value 0.33 obtained from (16.80).

16.22 If u is a Gamma variable with parameter p, and v is independently a Beta variable distributed as at (6.14) with parameters $(q, p-q)$, show that $w = uv$ and $y = u(1-v)$ are independent Gamma variables with parameters q and $p-q$. (If q or $p - q = \frac{1}{2}$, $(2w)^{\frac{1}{2}}$ or $(2y)^{\frac{1}{2}}$ is normal.) Show that the result of Exercise 11.26 is a special case with $p = 1$, $q = \frac{1}{2}$.

If λt is exponential as in (5.82) and θ/λ is independently a Beta variable (6.14) with parameters $(r, 1-r)$ show that θt is a Gamma variable with parameter r and that, as in Exercise 4.24, this distribution of the ratio of λt to λ/θ is the unconditional distribution of the mixture of exponentials.

Use this result in (16.59) to show that in binormal samples when $\rho = 0$, the distribution of $z_i = rs_i/\sigma_i$ is exactly normal with mean zero and variance $\frac{1}{n}$ $(i = 1, 2)$, but that z_1 and z_2 are not binormally distributed.

(Stuart (1962); the product of other independent variables can have a Gamma or a normal distribution—cf. Kotlarski (1965) and Exercise 11.9. See also Aitchison (1963).)

16.23 In Exercise 16.22, show that conversely if the distributions of u and w are as stated, that of v follows if it is independent of u.

Let $x_1, x_2, \ldots, x_n$ be a random sample from a normal distribution with variance σ^2, $\bar{x} = \sum_{i=1}^{n} x_i/n$, $d_i^2 = (x_i - \bar{x})^2$ and $s^2 = \sum_{i=1}^{n} d_i^2/(n-1)$. Show that $w_i = d_i^2 n/\{(n-1)2\sigma^2\}$ and $u = (n-1)s^2/(2\sigma^2)$ are Gamma variables with parameters $\frac{1}{2}$ and $\frac{1}{2}(n-1)$ respectively, and that $v_i = w_i/u$ is distributed independently of u. Hence show that for any $i = 1, 2, \ldots, n$,

$$v_i = \frac{n}{(n-1)^2} \frac{(x_i - \bar{x})^2}{s^2}$$

is distributed in the Beta form (6.14) with parameters $(\frac{1}{2}, \frac{1}{2}(n-2))$, so that $t = \{(n-2)v_i/(1-v_i)\}^{\frac{1}{2}}$ is distributed in Student's form with $\nu = n-2$.

16.24 Using the Gram–Charlier method of Exercise 6.11, show that for the cube-root normal approximation to the χ^2 distribution, with standardized cumulants (16.13), the maximum absolute error in the d.f. resulting from the approximation satisfies

$$|e(x)| < 0.001, \qquad n \geqslant 15,$$
$$|e(x)| < 0.007, \qquad n \geqslant 3.$$

Similarly, use (16.8) to show that for the square-root normal approximation, we must have $n \geqslant 23$ for $|e(x)| < 0.01$, while from (16.6), the approximation of χ^2 itself requires $n \geqslant 354$ for $|e(x)| < 0.001$.

(Mathur, 1961)

16.25 Given σ^2, x is normally distributed with mean zero and variance σ^2, where ν/σ^2 is distributed in the χ^2 form with ν d.fr. Show that the distribution of x is Student's t with ν d.fr. (Exercise 5.25 ends with the special case $\nu = 1$, as may be seen from Exercise 11.25.)

16.26 If t_m and t_n are independent Student's t variables, distributed as at (16.15) with m, n d.fr., respectively, show, using **11.8**, that their difference $d_{m,n} = t_m - t_n$ has the same distribution f as their sum, and that f is symmetrical about its unique mode zero and also symmetric in (m, n).

Writing $t_j = y_j/(u_j/j)^{\frac{1}{2}}$, where the y_j are standardized normal and independently the u_j are χ^2 variables with j d.fr., show that $x = u_m/(u_m + u_n)$ has the Beta distribution (6.14) with parameters $(\frac{1}{2}m, \frac{1}{2}n)$ and that the conditional distribution of $d_{m,n}$ given x is a $\left\{\dfrac{m(1-x)+nx}{2(m+n)x(1-x)}\right\}^{\frac{1}{2}} t_{m+n}$ variable, so that the distribution of $d_{m,n}$ may be represented as the product of two independent random variables

$$t_{m+n} \times \left\{\frac{(1+F)(m+nF)}{2(m+n)F}\right\}^{\frac{1}{2}},$$

where F has the variance-ratio distribution (16.24) with n and m d.fr.

(Ruben, 1960; B. K. Ghosh (1975) gives tables when $m = n$ and a good approximation for the general case. Walker and Saw (1978) consider general linear combinations of t-variates.)

16.27 Generalizing Exercise 16.25, show that if ν/σ^2 is distributed as there in the joint distribution of the mean $\bar{x}$ and variance s^2 in samples from $N(0, \sigma^2)$ given in Examples 11.3 and 11.7, the resulting conditional distribution of $\bar{x}$ given the value of s^2 is such that

$$\left\{\frac{n(n-1+\nu)}{ns^2+\nu}\right\}^{\frac{1}{2}} \bar{x}$$

has a Student's t distribution with $(n-1+\nu)$ d.fr. (Exercise 16.25 is the case $n = 1$.)

16.28 Show from Exercise 4.21 that the logistic distribution ($2z$ with $\nu_1 = \nu_2 = 2$ in Example 16.2) has kurtosis as for Student's t with $\nu = 9$.

(Mudholkar and George (1978) verify that the d.f.s never differ by as much as 0.005.)

16.29 Show that in samples of size n from a χ^2_ν distribution, the large-sample variance of the skewness coefficient $b_1^{1/2} = m_3/m_2^{3/2}$ given in Exercise 10.26 equals $\dfrac{6}{n}\left(1 + \dfrac{20}{\nu^2}\right)$, reducing to the normal value $6/n$ as $\nu \to \infty$. Using the recurrence relation for central moments in Exercise 16.3, show similarly that the corresponding result in Exercise 10.27 for b_2 tends to the normal value $24/n$ as $\nu \to \infty$.

16.30 In a four-variate normal distribution show that the correlation between the covariances s_{12} and s_{34} is

$$\frac{\rho_{13}\rho_{24}+\rho_{14}\rho_{23}}{\{(1+\rho_{12}^2)(1+\rho_{34}^2)\}^{\frac{1}{2}}}.$$

(Wishart, 1929)

16.31 A multivariate normal population has means $\boldsymbol{\mu}$, all variances equal to σ^2 and all correlations equal to ρ. Defining

$$(n-1)ps^2=\sum_{i,j}(x_{ij}-\bar{x}_i)^2,$$

$$(n-1)(p-1)ps^2r=\sum_{i\neq k}\sum_{j}(x_{ij}-\bar{x}_i)(x_{kj}-\bar{x}_k),$$

show that the joint density of

$$u=s^2\{1+(p-1)r\}$$
$$v=s^2(1-r)$$

is given by

$$f\propto u^{\frac{1}{2}(n-3)}v^{\frac{1}{2}(p-1)(n-1)-1}\exp\left\{\tfrac{1}{2}(n-1)\left(\frac{u}{\alpha}+\frac{(p-1)v}{\theta}\right)\right\}$$

where

$$\alpha=\sigma^2\{1+(p-1)\rho\},$$
$$\theta=\sigma^2(1-\rho).$$

Hence show that $u\theta/(v\alpha)$ is distributed in the F form with $n-1$, $(p-1)(n-1)$ d.fr.

(Geisser, 1964)

16.32 $\mathbf{S}$ is distributed in the Wishart form in samples from a population with covariance matrix $\boldsymbol{\Sigma}$. Show that $\mathbf{h}'\mathbf{S}\mathbf{h}$ is distributed with corresponding parameter $\mathbf{h}'\boldsymbol{\Sigma}\mathbf{h}$, $\mathbf{h}$ being an arbitrary non-singular $p\times p$ matrix.

16.33 If each column of $\mathbf{x}$ at (16.102) is, independently of the others, normal with zero mean and covariance matrix $\boldsymbol{\Sigma}$, and $\mathbf{M}$ is an orthogonal matrix, show that each column of $\mathbf{y}=\mathbf{M}\mathbf{x}$ is also $N(\mathbf{0},\boldsymbol{\Sigma})$. In particular, if $\mathbf{M}$ is the Helmert matrix of **16.43** show that the sample covariance matrix $\mathbf{S}$ may be represented as

$$\sum_{j=1}^{n-1}\mathbf{y}_j\mathbf{y}_j'$$

where the $\mathbf{y}_j$ are independent $N(\mathbf{0},\boldsymbol{\Sigma})$ vectors. Deduce the additive property of Wishart matrices of **16.47**.

16.34 In **16.50** show that the density function of $\{|\mathbf{S}|/|\boldsymbol{\Sigma}|\}^{1/p}$, say y, is given approximately by

$$\frac{\alpha^{\frac{1}{2}p(n-p)}y^{\frac{1}{2}p(n-p)-1}e^{-\alpha y}}{\Gamma\{\frac{1}{2}p(n-p)\}},$$

where

$$\alpha = np\left\{1 - \frac{(p-1)(p-2)}{2n}\right\}^{1/p}.$$

(P. G. Hoel, 1937)

16.35 Show that for a sample covariance matrix **S**, $n^{\frac{1}{2}}\{|S|/|\Sigma| - 1\}$ is distributed about zero with variance $2p$ for large samples.

(T. W. Anderson, 1984)

16.36 Show algebraically that Hotelling's T^2 is invariant under linear transformations of the p variates.

16.37 For a pair of normal variates with correlation ρ, show that, defining v by

$$v = \frac{ns_{12}}{\sigma_1\sigma_2(1-\rho^2)},$$

we have for the density function of v

$$f(v) = \frac{(1-\rho^2)^{\frac{1}{2}(n-1)}e^{\rho v}}{\pi^{\frac{1}{2}}2^{\frac{1}{2}n-1}\Gamma\{\frac{1}{2}(n-1)\}}\{v^{\frac{1}{2}n-1}K_{\frac{1}{2}n-1}(v)\},$$

for $v > 0$ and a similar expression with $-v$ for v inside the curly brackets if $v < 0$. Here K is the Bessel function of second kind with imaginary argument.

(Wishart and Bartlett, 1933)

16.38 Verify equation (16.143).

16.39 (Bartlett decomposition.) Let x_{jk}, $j = 1, 2, \ldots, p$; $k = 1, 2, \ldots, n$, be independent $N(0, 1)$ variables. Take

$$\begin{aligned}
\mathbf{y}_1 &= \mathbf{x}_1 \\
\mathbf{y}_2 &= \mathbf{x}_2 - b'_{21}\mathbf{y}_1 \\
&\cdots\cdots\cdots\cdots\cdots\cdots \\
\mathbf{y}_p &= \mathbf{x}_p - b'_{p1}\mathbf{y}_1 - \ldots - b'_{p.p-1}\mathbf{y}_{p-1}
\end{aligned}$$

and take the ys to be orthogonal so that $\mathbf{y}'_j\mathbf{y}_k = 0$, $j \neq k$. Then $b'_{jk} = \mathbf{y}'_k\mathbf{x}_j/\mathbf{y}'_k\mathbf{y}_k$. Take $b_{jk} = (\mathbf{y}'_k\mathbf{y}_k)^{\frac{1}{2}}b'_{jk}$. Show that

$$\mathbf{x}'\mathbf{x} = \mathbf{BB}'$$

where **B** is the triangular matrix

$$\begin{bmatrix}
(\mathbf{y}'_1\mathbf{y}_1)^{\frac{1}{2}} & 0 & 0 & \ldots & 0 \\
b_{21} & (\mathbf{y}'_2\mathbf{y}_2)^{\frac{1}{2}} & 0 & \ldots & 0 \\
\cdots & \cdots & \cdots & \cdots & \cdots \\
b_{p1} & b_{p2} & \ldots & \ldots & (\mathbf{y}'_p\mathbf{y}_p)^{\frac{1}{2}}
\end{bmatrix}.$$

Show that the b_{jk}, $k = 1, \ldots, (p-1)$ are independent $N(0, 1)$ variables and hence that each $\mathbf{y}'_k\mathbf{y}_k$ is a χ^2 variable with $n - k + 1$ degrees of freedom, independent of the bs and the other $\mathbf{y}'_j\mathbf{y}_j$.

(Bartlett, 1933. Cf. Kshirsagar, 1959)

16.40 In the foregoing exercise, by taking determinants in the equation $\mathbf{x}'\mathbf{x} = \mathbf{BB}'$, show that $\mathbf{y}_k'\mathbf{y}_k$ is the ratio of two product-sum determinants. Hence show that the kth diagonal element of the inverse of $\mathbf{x}'\mathbf{x}$ is distributed as the reciprocal of a χ^2 variable with $n-k+1$ degrees of freedom.

(Wijsman, 1957; see also Kshirsagar, 1959)

16.41 Use the previous exercise to prove the result of **16.50**, that $V = n|S|/|\Sigma|$ is the product of p independent χ^2 factors with degrees of freedom $n-1, n-2, \ldots, n-p$.

(Wijsman, 1957)

APPENDIX TABLES

Appendix Table 1 Density function of the normal distribution, $f=\frac{1}{\sqrt{(2\pi)}}e^{-\frac{1}{2}x^2}$ with first and second differences

x	f	$\Delta^1(-)$	Δ^2	x	y	$\Delta^1(-)$	Δ^2
0.0	0.39894	199	−392	2.5	0.01753	395	+79
0.1	0.39695	591	−374	2.6	0.01358	316	+66
0.2	0.39104	965	−347	2.7	0.01042	250	+53
0.3	0.38139	1312	−308	2.8	0.00792	197	+45
0.4	0.36827	1620	−265	2.9	0.00595	152	+36
0.5	0.35207	1885	−212	3.0	0.00443	116	+27
0.6	0.33322	2097	−159	3.1	0.00327	89	+23
0.7	0.31225	2256	−104	3.2	0.00238	66	+17
0.8	0.28969	2360	− 52	3.3	0.00172	49	+13
0.9	0.26609	2412	0	3.4	0.00123	36	+10
1.0	0.24197	2412	+ 46	3.5	0.00087	26	+ 7
1.1	0.21785	2366	+ 84	3.6	0.00061	19	+ 6
1.2	0.19419	2282	+118	3.7	0.00042	13	+ 4
1.3	0.17137	2164	+143	3.8	0.00029	9	+ 2
1.4	0.14973	2021	+161	3.9	0.00020	7	+ 3
1.5	0.12952	1860	+173	4.0	0.00013	4	—
1.6	0.11092	1687	+177	4.1	0.00009	3	—
1.7	0.09405	1510	+177	4.2	0.00006	2	—
1.8	0.07895	1333	+170	4.3	0.00004	2	—
1.9	0.06562	1163	+162	4.4	0.00002	—	—
2.0	0.05399	1001	+150	4.5	0.00002	—	—
2.1	0.04398	851	+137	4.6	0.00001	—	—
2.2	0.03547	714	+120	4.7	0.00001	—	—
2.3	0.02833	594	+108	4.8	0.00000	—	—
2.4	0.02239	486	+ 91				

Appendix Table 2 Distribution function of the normal distribution

The table shows the area under the curve $f = (2\pi)^{-\frac{1}{2}}e^{-\frac{1}{2}x^2}$ lying to the left of specified deviates x; e.g. the area corresponding to a deviate 1.86 (=1.5 + 0.36) is 0.9686.

Deviate	0.0+	0.5+	1.0+	1.5+	2.0+	2.5+	3.0+	3.5+
0.00	5000	6915	8413	9332	9772	$9^2$379	$9^2$865	$9^3$77
0.01	5040	6950	8438	9345	9778	$9^2$396	$9^2$869	$9^3$78
0.02	5080	6985	8461	9357	9783	$9^2$413	$9^2$874	$9^3$78
0.03	5120	7019	8485	9370	9788	$9^2$430	$9^2$878	$9^3$79
0.04	5160	7054	8508	9382	9793	$9^2$446	$9^2$882	$9^3$80
0.05	5199	7088	8531	9394	9798	$9^2$461	$9^2$886	$9^3$81
0.06	5239	7123	8554	9406	9803	$9^2$477	$9^2$889	$9^3$81
0.07	5279	7157	8577	9418	9808	$9^2$492	$9^2$893	$9^3$82
0.08	5319	7190	8599	9429	9812	$9^2$506	$9^2$897	$9^3$83
0.09	5359	7224	8621	9441	9817	$9^2$520	$9^2$900	$9^3$83
0.10	5398	7257	8643	9452	9821	$9^2$534	$9^3$03	$9^3$84
0.11	5438	7291	8665	9463	9826	$9^2$547	$9^3$06	$9^3$85
0.12	5478	7324	8686	9474	9830	$9^2$560	$9^3$10	$9^3$85
0.13	5517	7357	8708	9484	9834	$9^2$573	$9^3$13	$9^3$86
0.14	5557	7389	8729	9495	9838	$9^2$585	$9^3$16	$9^3$86
0.15	5596	7422	8749	9505	9842	$9^2$598	$9^3$18	$9^3$87
0.16	5636	7454	8770	9515	9846	$9^2$609	$9^3$21	$9^3$87
0.17	5675	7486	8790	9525	9850	$9^2$621	$9^3$24	$9^3$88
0.18	5714	7517	8810	9535	9854	$9^2$632	$9^3$26	$9^3$88
0.19	5753	7549	8830	9545	9857	$9^2$643	$9^3$29	$9^3$89
0.20	5793	7580	8849	9554	9561	$9^2$653	$9^3$31	$9^3$89
0.21	5832	7611	8869	9564	9864	$9^2$664	$9^3$34	$9^3$90
0.22	5871	7642	8888	9573	9868	$9^2$674	$9^3$36	$9^3$90
0.23	5910	7673	8907	9582	9871	$9^2$683	$9^3$38	$9^4$04
0.24	5948	7704	8925	9591	9875	$9^2$693	$9^3$40	$9^4$08
0.25	5987	7738	8944	9599	9878	$9^2$702	$9^3$42	$9^4$12
0.26	6026	7764	8962	9608	9881	$9^2$711	$9^3$44	$9^4$15
0.27	6064	7794	8980	9616	9884	$9^2$720	$9^3$46	$9^4$18
0.28	6103	7823	8997	9625	9887	$9^2$728	$9^3$48	$9^4$22
0.29	6141	7852	9015	9633	9890	$9^2$736	$9^3$50	$9^4$25
0.30	6179	7881	9032	9641	9893	$9^2$744	$9^3$52	$9^4$28
0.31	6217	7910	9049	9649	9896	$9^2$752	$9^3$53	$9^4$31
0.32	6255	7939	9066	9656	9898	$9^2$760	$9^3$55	$9^4$33
0.33	6293	7967	9082	9664	9901	$9^2$767	$9^3$57	$9^4$36
0.34	6331	7995	9099	9671	9904	$9^2$774	$9^3$58	$9^4$39
0.35	6368	8023	9115	9678	9906	$9^2$781	$9^3$60	$9^4$41
0.36	6406	8051	9131	9686	9909	$9^2$788	$9^3$61	$9^4$43
0.37	6443	8078	9147	9693	9911	$9^2$795	$9^3$62	$9^4$46
0.38	6480	8106	9162	9699	9913	$9^2$801	$9^3$64	$9^4$48
0.39	6517	8133	9177	9706	9916	$9^2$807	$9^3$65	$9^4$50
0.40	6554	8159	9192	9713	9918	$9^2$813	$9^3$66	$9^4$52
0.41	6591	8186	9207	9719	9920	$9^2$819	$9^3$68	$9^4$54
0.42	6628	8212	9222	9726	9922	$9^2$825	$9^3$69	$9^4$56
0.43	6664	8238	9236	9732	9925	$9^2$831	$9^3$70	$9^4$58
0.44	6700	8264	9251	9738	9927	$9^2$836	$9^3$71	$9^4$59
0.45	6736	8289	9265	9744	9929	$9^2$841	$9^3$72	$9^4$61
0.46	6772	8315	9279	9750	9931	$9^2$846	$9^3$73	$9^4$63
0.47	6808	8340	9292	9756	9932	$9^2$851	$9^3$74	$9^4$64
0.48	6844	8365	9306	9761	9934	$9^2$856	$9^3$75	$9^4$66
0.49	6879	8389	9319	9767	9936	$9^2$861	$9^3$76	$9^4$67

Note—Decimal points in the body of the table are omitted. Repeated 9s are indicated by powers, e.g. $9^3$71 stands for 0.999 71.

Appendix Table 3 Quantiles of the d.f. of χ^2

(Reproduced from Table III of Fisher (1970), by kind permission of the author and publishers)

$P = 1 - F$	0.99	0.98	0.95	0.90	0.80	0.70	0.50	0.30	0.20	0.10	0.05	0.02	0.01
$\nu = 1$	0.0^3157	0.0^3628	0.0^2393	0.0158	0.0642	0.148	0.455	1.074	1.642	2.706	3.841	5.412	6.635
2	0.0201	0.0404	0.103	0.211	0.446	0.713	1.386	2.408	3.219	4.605	5.991	7.824	9.210
3	0.115	0.185	0.352	0.584	1.005	1.424	2.366	3.665	4.642	6.251	7.815	9.837	11.345
4	0.297	0.429	0.711	1.064	1.649	2.195	3.357	4.878	5.989	7.779	9.488	11.668	13.277
5	0.554	0.752	1.145	1.160	2.343	3.000	4.351	6.064	7.289	9.236	11.070	13.388	15.086
6	0.872	1.134	1.635	2.204	3.070	3.828	5.348	7.231	8.558	10.645	12.592	15.033	16.812
7	1.239	1.564	2.167	2.833	3.822	4.671	6.346	8.383	9.803	12.017	14.067	16.622	18.475
8	1.646	2.032	2.733	3.490	4.594	5.527	7.344	9.524	11.030	13.362	15.507	18.168	20.090
9	2.088	2.532	3.325	4.168	5.380	6.393	8.343	10.656	12.242	14.684	16.919	19.679	21.666
10	2.358	3.059	3.940	4.865	6.179	7.267	9.342	11.781	13.422	15.987	18.307	21.161	23.209
11	3.053	3.609	4.575	5.578	6.989	8.148	10.341	12.899	14.631	17.275	19.675	22.618	24.725
12	3.571	4.178	5.226	6.304	7.807	9.034	11.340	14.011	15.821	18.549	21.026	24.054	26.217
13	4.107	4.765	5.892	7.042	8.634	9.926	12.340	15.119	16.985	19.812	22.362	25.472	27.688
14	4.660	5.368	6.571	7.790	9.467	10.821	13.339	16.222	18.151	21.064	23.685	26.873	29.141
15	5.229	5.985	7.261	8.547	10.307	11.721	14.339	17.322	19.311	22.307	24.996	28.259	30.578
16	5.812	6.614	7.962	9.312	11.152	12.624	15.338	18.418	20.465	23.542	26.296	29.633	32.000
17	6.408	7.255	8.672	10.085	12.002	13.531	16.338	19.511	21.615	24.769	27.587	30.995	33.499
18	7.015	7.906	9.390	10.865	12.857	14.440	17.338	20.601	22.760	25.989	28.869	32.346	34.805
19	7.633	8.567	10.117	11.651	13.716	15.352	18.338	21.689	23.900	27.204	30.144	33.687	36.191
20	8.260	9.237	10.851	12.443	14.578	16.266	19.337	22.775	25.038	28.412	31.410	35.020	37.566
21	8.897	9.915	11.591	13.240	15.445	17.182	20.337	23.858	26.171	29.615	32.671	36.343	38.932
22	9.542	10.600	12.338	14.041	16.314	18.101	21.337	24.939	27.301	30.813	33.924	37.659	40.289
23	10.196	11.293	13.091	14.848	17.187	19.021	22.337	26.018	28.429	32.007	35.172	38.968	41.638
24	10.856	11.992	13.848	15.659	18.062	19.943	23.337	27.096	29.553	33.196	36.415	40.270	42.980
25	11.524	12.697	14.611	16.473	18.940	20.867	24.337	28.172	30.675	34.382	37.652	41.566	44.314
26	12.198	13.409	15.379	17.292	19.820	21.792	25.336	29.246	31.795	35.563	38.885	42.856	45.642
27	12.879	14.125	16.151	18.114	20.703	22.719	26.336	30.319	32.912	36.741	40.113	44.140	46.963
28	13.565	14.847	16.928	18.939	21.588	23.647	27.336	31.391	34.027	37.916	41.337	45.419	48.278
29	14.256	15.574	17.708	19.768	22.475	24.577	28.336	32.461	35.139	39.087	42.557	46.693	49.588
30	14.953	16.306	18.493	20.599	23.364	25.508	29.336	33.530	36.250	40.256	43.773	47.962	50.892

Note—For values of ν greater than 30 the quantity $\sqrt{(2\chi^2)}$ may be taken to be distributed normally about mean $\sqrt{(2\nu - 1)}$ with unit variance.

Appendix Table 4a Distribution function of χ^2 for one degree of freedom for values of χ^2 from 0 to 1 by steps of 0.01

χ^2	$P = 1 - F$	Δ	χ^2	$P = 1 - F$	Δ
0	1.00000	7966	0.50	0.47950	436
0.01	0.92034	3280	0.51	0.47514	430
0.02	0.88754	2505	0.52	0.47084	423
0.03	0.86249	2101	0.53	0.46661	418
0.04	0.84148	1842	0.54	0.46243	411
0.05	0.82306	1656	0.55	0.45832	406
0.06	0.80650	1516	0.56	0.45426	400
0.07	0.79134	1404	0.57	0.45026	395
0.08	0.77730	1312	0.58	0.44631	389
0.09	0.76418	1235	0.59	0.44242	384
0.10	0.75183	1169	0.60	0.43858	379
0.11	0.74014	1111	0.61	0.43479	374
0.12	0.72903	1060	0.62	0.43105	369
0.13	0.71843	1015	0.63	0.42736	365
0.14	0.70828	974	0.64	0.42371	360
0.15	0.69854	938	0.65	0.42011	355
0.16	0.68916	905	0.66	0.41656	351
0.17	0.68011	874	0.67	0.41305	346
0.18	0.67137	845	0.68	0.40959	343
0.19	0.66292	820	0.69	0.40616	338
0.20	0.65472	795	0.70	0.40278	334
0.21	0.64677	773	0.71	0.39944	330
0.22	0.63904	752	0.72	0.39614	326
0.23	0.63152	731	0.73	0.39288	322
0.24	0.62421	713	0.74	0.38966	318
0.25	0.61708	696	0.75	0.38648	315
0.26	0.61012	679	0.76	0.38333	311
0.27	0.60333	663	0.77	0.38022	308
0.28	0.59670	648	0.78	0.37714	304
0.29	0.59022	634	0.79	0.37410	301
0.30	0.58388	620	0.80	0.37109	297
0.31	0.57768	607	0.81	0.36812	294
0.32	0.57161	595	0.82	0.36518	291
0.33	0.56566	583	0.83	0.36227	287
0.34	0.55983	572	0.84	0.35940	285
0.35	0.55411	560	0.85	0.35655	281
0.36	0.54851	551	0.86	0.35374	278
0.37	0.54300	540	0.87	0.35096	276
0.38	0.53760	530	0.88	0.34820	272
0.39	0.53230	521	0.89	0.34548	270
0.40	0.52709	512	0.90	0.34278	267
0.41	0.52197	503	0.91	0.34011	264
0.42	0.51694	495	0.92	0.33747	261
0.43	0.51199	487	0.93	0.33486	258
0.44	0.50712	479	0.94	0.33228	256
0.45	0.50233	471	0.95	0.32972	253
0.46	0.49762	463	0.96	0.32719	251
0.47	0.49299	457	0.97	0.32468	248
0.48	0.48842	449	0.98	0.32220	246
0.49	0.48393	443	0.99	0.31974	243
0.50	0.47950	436	1.00	0.31731	241

Appendix Table 4b Distribution function of χ^2 for one degree of freedom for values of χ^2 from 1 to 10 by steps of 0.1

χ^2	$P = 1 - F$	Δ	χ^2	$P = 1 - F$	Δ
1.0	0.31731	2304	5.5	0.01902	106
1.1	0.29427	2095	5.6	0.01796	99
1.2	0.27332	1911	5.7	0.01697	94
1.3	0.25421	1749	5.8	0.01603	89
1.4	0.23672	1605	5.9	0.01514	83
1.5	0.22067	1477	6.0	0.01431	79
1.6	0.20590	1361	6.1	0.01352	74
1.7	0.19229	1258	6.2	0.01278	71
1.8	0.17971	1163	6.3	0.01207	66
1.9	0.16808	1078	6.4	0.01141	62
2.0	0.15730	1000	6.5	0.01079	59
2.1	0.14730	929	6.6	0.01020	56
2.2	0.13801	864	6.7	0.00964	52
2.3	0.12937	803	6.8	0.00912	50
2.4	0.12134	749	6.9	0.00862	47
2.5	0.11385	699	7.0	0.00815	44
2.6	0.10686	651	7.1	0.00771	42
2.7	0.10035	609	7.2	0.00729	39
2.8	0.09426	568	7.3	0.00690	38
2.9	0.08858	532	7.4	0.00652	35
3.0	0.08326	497	7.5	0.00617	33
3.1	0.07829	465	7.6	0.00584	32
3.2	0.07364	436	7.7	0.00552	30
3.3	0.06928	408	7.8	0.00522	28
3.4	0.06520	383	7.9	0.00494	26
3.5	0.06137	359	8.0	0.00468	25
3.6	0.05778	337	8.1	0.00443	24
3.7	0.05441	316	8.2	0.00419	23
3.8	0.05125	296	8.3	0.00396	21
3.9	0.04829	279	8.4	0.00375	20
4.0	0.04550	262	8.5	0.00355	19
4.1	0.04288	246	8.6	0.00336	18
4.2	0.04042	231	8.7	0.00318	17
4.3	0.03811	217	8.8	0.00301	16
4.4	0.03594	205	8.9	0.00285	15
4.5	0.03389	192	9.0	0.00270	14
4.6	0.03197	181	9.1	0.00256	14
4.7	0.03016	170	9.2	0.00242	13
4.8	0.02846	160	9.3	0.00229	12
4.9	0.02686	151	9.4	0.00217	12
5.0	0.02535	142	9.5	0.00205	10
5.1	0.02393	134	9.6	0.00195	11
5.2	0.02259	126	9.7	0.00184	10
5.3	0.02133	119	9.8	0.00174	9
5.4	0.02014	112	9.9	0.00165	8
5.5	0.01902	106	10.0	0.00157	8

Appendix Table 5 Quantiles of the d.f. of t

(Reproduced from Fisher and Yates (1963), by kind permission of the authors and publishers)

$P = 2(1 - F)$	0.9	0.8	0.7	0.6	0.5	0.4	0.3	0.2	0.1	0.05	0.02	0.01	0.001
$\nu = 1$	0.158	0.325	0.510	0.727	1.000	1.376	1.963	3.078	6.314	12.706	31.821	63.657	636.619
2	0.142	0.289	0.445	0.617	0.816	1.061	1.386	1.886	2.920	4.303	6.965	9.925	31.598
3	0.137	0.277	0.424	0.584	0.765	0.978	1.250	1.638	2.353	3.182	4.541	5.841	12.924
4	0.134	0.271	0.414	0.569	0.741	0.941	1.190	1.533	2.132	2.776	3.747	4.604	8.610
5	0.132	0.267	0.408	0.559	0.727	0.920	1.156	1.476	2.015	2.571	3.365	4.032	6.869
6	0.131	0.265	0.404	0.553	0.718	0.906	1.134	1.440	1.943	2.447	3.143	3.707	5.959
7	0.130	0.263	0.402	0.549	0.711	0.896	1.119	1.415	1.895	2.365	2.998	3.499	5.408
8	0.130	0.262	0.399	0.546	0.706	0.889	1.108	1.397	1.860	2.306	2.896	3.355	5.041
9	0.129	0.261	0.398	0.543	0.703	0.883	1.100	1.383	1.833	2.262	2.821	3.250	4.781
10	0.129	0.260	0.397	0.542	0.700	0.879	1.093	1.372	1.812	2.228	2.764	3.169	4.587
11	0.129	0.260	0.396	0.540	0.697	0.876	1.088	1.363	1.796	2.201	2.718	3.106	4.437
12	0.128	0.259	0.395	0.539	0.695	0.873	1.083	1.356	1.782	2.179	2.681	3.055	4.318
13	0.128	0.259	0.394	0.538	0.694	0.870	1.079	1.350	1.771	2.160	2.650	3.012	4.221
14	0.128	0.258	0.393	0.537	0.692	0.868	1.076	1.345	1.761	2.145	2.624	2.977	4.140
15	0.128	0.258	0.393	0.536	0.691	0.866	1.074	1.341	1.753	2.131	2.602	2.947	4.073
16	0.128	0.258	0.392	0.535	0.690	0.865	1.071	1.337	1.746	2.120	2.583	2.921	4.015
17	0.128	0.257	0.392	0.534	0.689	0.863	1.069	1.333	1.740	2.110	2.567	2.898	3.965
18	0.127	0.257	0.392	0.534	0.688	0.862	1.067	1.330	1.734	2.101	2.552	2.878	3.922
19	0.127	0.257	0.391	0.533	0.688	0.861	1.066	1.328	1.729	2.093	2.539	2.861	3.883
20	0.127	0.257	0.391	0.533	0.687	0.860	1.064	1.325	1.725	2.086	2.528	2.845	3.850
21	0.127	0.257	0.391	0.532	0.686	0.859	1.063	1.323	1.721	2.080	2.518	2.831	3.819
22	0.127	0.256	0.390	0.532	0.686	0.858	1.061	1.321	1.717	2.074	2.508	2.819	3.792
23	0.127	0.256	0.390	0.532	0.685	0.858	1.060	1.319	1.714	2.069	2.500	2.807	3.767
24	0.127	0.256	0.390	0.531	0.685	0.857	1.059	1.318	1.711	2.064	2.492	2.797	3.745
25	0.127	0.256	0.390	0.531	0.684	0.856	1.058	1.316	1.708	2.060	2.485	2.787	3.725
26	0.127	0.256	0.390	0.531	0.684	0.856	1.058	1.315	1.706	2.056	2.479	2.779	3.707
27	0.127	0.256	0.389	0.531	0.684	0.855	1.057	1.314	1.703	2.052	2.473	2.771	3.690
28	0.127	0.256	0.389	0.530	0.683	0.855	1.056	1.313	1.701	2.048	2.467	2.763	3.674
29	0.127	0.256	0.389	0.530	0.683	0.854	1.055	1.311	1.699	2.045	2.462	2.756	3.659
30	0.127	0.256	0.389	0.530	0.683	0.854	1.055	1.310	1.697	2.042	2.457	2.750	3.646
40	0.126	0.255	0.388	0.529	0.681	0.851	1.050	1.303	1.684	2.021	2.423	2.704	3.551
60	0.126	0.254	0.387	0.527	0.679	0.848	1.046	1.296	1 671	2.000	2.390	2.660	3.460
120	0.126	0.254	0.386	0.526	0.677	0.845	1.041	1.289	1.658	1.980	2.358	2.617	3.373
∞	0.126	0.253	0.385	0.524	0.674	0.842	1.036	1.282	1.645	1.960	2.326	2.576	3.291

Appendix Table 6 5 percent points of the distribution of z (values at which the d.f. = 0.95) (Reprinted from Table VI of Fisher (1970), by kind permission of the author and publishers)

Values of ν_2	Values of ν_1: 1	2	3	4	5	6	8	12	24	∞
1	2.5421	2.6479	2.6870	2.7071	2.7194	2.7276	2.7380	2.7484	2.7588	2.7693
2	1.4592	1.4722	1.4765	1.4787	1.4800	1.4808	1.4819	1.4830	1.4840	1.4851
3	1.1577	1.1284	1.1137	1.1051	1.0994	1.0953	1.0899	1.0842	1.0781	1.0716
4	1.0212	0.9690	0.9429	0.9272	0.9168	0.9093	0.8993	0.8885	0.8767	0.8639
5	0.9441	0.8777	0.8441	0.8236	0.8097	0.7997	0.7862	0.7714	0.7750	0.7368
6	0.8948	0.8188	0.7798	0.7558	0.7394	0.7274	0.7112	0.6931	0.6729	0.6499
7	0.8606	0.7777	0.7347	0.7080	0.6896	0.6761	0.6576	0.6369	0.6134	0.5862
8	0.8355	0.7475	0.7014	0.6725	0.6525	0.6378	0.6175	0.5945	0.5682	0.5371
9	0.8163	0.7242	0.6757	0.6450	0.6238	0.6080	0.5862	0.5613	0.5324	0.4979
10	0.8012	0.7058	0.6553	0.6232	0.6009	0.5843	0.5611	0.5346	0.5035	0.4657
11	0.7889	0.6909	0.6387	0.6055	0.5822	0.5648	0.5406	0.5126	0.4795	0.4387
12	0.7788	0.6786	0.6250	0.5907	0.5666	0.5487	0.5234	0.4941	0.4592	0.4156
13	0.7703	0.6682	0.6134	0.5783	0.5535	0.5350	0.5089	0.4785	0.4419	0.3957
14	0.7630	0.6594	0.6036	0.5677	0.5423	0.5233	0.4964	0.4649	0.4269	0.3782
15	0.7568	0.6518	0.5950	0.5585	0.5326	0.5131	0.4855	0.4532	0.4138	0.3628
16	0.7514	0.6451	0.5876	0.5505	0.5241	0.5042	0.4760	0.4428	0.4022	0.3490
17	0.7466	0.6393	0.5811	0.5434	0.5166	0.4964	0.4676	0.4337	0.3919	0.3366
18	0.7424	0.6341	0.5753	0.5371	0.5099	0.4894	0.4602	0.4255	0.3827	0.3253
19	0.7386	0.6295	0.5701	0.5315	0.5040	0.4832	0.4535	0.4182	0.3743	0.3151
20	0.7352	0.6254	0.5654	0.5265	0.4986	0.4776	0.4474	0.4116	0.3668	0.3057
21	0.7322	0.6216	0.5612	0.5219	0.4938	0.4725	0.4420	0.4055	0.3599	0.2971
22	0.7294	0.6182	0.5574	0.5178	0.4894	0.4679	0.4370	0.4001	0.3536	0.2892
23	0.7269	0.6151	0.5540	0.5140	0.4854	0.4636	0.4325	0.3950	0.3478	0.2818
24	0.7246	0.6123	0.5508	0.5106	0.4817	0.4598	0.4283	0.3904	0.3425	0.2749
25	0.7225	0.6097	0.5478	0.5074	0.4783	0.4562	0.4244	0.3862	0.3376	0.2685
26	0.7205	0.6073	0.5451	0.5045	0.4752	0.4529	0.4209	0.3823	0.3330	0.2625
27	0.7187	0.6051	0.5427	0.5017	0.4723	0.4499	0.4176	0.3786	0.3287	0.2569
28	0.7171	0.6030	0.5403	0.4992	0.4696	0.4471	0.4146	0.3752	0.3248	0.2516
29	0.7155	0.6011	0.5382	0.4969	0.4671	0.4444	0.4117	0.3720	0.3211	0.2466
30	0.7141	0.5994	0.5362	0.4947	0.4648	0.4420	0.4090	0.3691	0.3176	0.2419
60	0.6933	0.5738	0.5073	0.4632	0.4311	0.4064	0.3702	0.3255	0.2654	0.1644
∞	0.6729	0.5486	0.4787	0.4319	0.3974	0.3706	0.3309	0.2804	0.2085	0

Appendix Table 7 5 percent points of the variance ratio F (values at which the d.f. = 0.95)
(Reproduced from Fisher and Yates (1963), by kind permission of the authors and publishers)

ν_2 \ ν_1	1	2	3	4	5	6	8	12	24	∞
1	161.40	199.50	215.70	224.60	230.20	234.00	238.90	243.90	249.00	254.30
2	18.51	19.00	19.16	19.25	19.30	19.33	19.37	19.41	19.45	19.50
3	10.13	9.55	9.28	9.12	9.01	8.94	8.84	8.74	8.64	8.53
4	7.71	6.94	6.59	6.39	6.26	6.16	6.04	5.91	5.77	5.63
5	6.61	5.79	5.41	5.19	5.05	4.95	4.82	4.68	4.53	4.36
6	5.99	5.14	4.76	4.53	4.39	4.28	4.15	4.00	3.84	3.67
7	5.59	4.74	4.35	4.12	3.97	3.87	3.73	3.57	3.41	3.23
8	5.32	4.46	4.07	3.84	3.69	3.58	3.44	3.28	3.12	2.93
9	5.12	4.26	3.86	3.63	3.48	3.37	3.23	3.07	2.90	2.71
10	4.96	4.10	3.71	3.48	3.33	3.22	3.07	2.91	2.74	2.54
11	4.84	3.98	3.59	3.36	3.20	3.09	2.95	2.79	2.61	2.40
12	4.75	3.88	3.49	3.26	3.11	3.00	2.85	2.69	2.50	2.30
13	4.67	3.80	3.41	3.18	3.02	2.92	2.77	2.60	2.42	2.21
14	4.60	3.74	3.34	3.11	2.96	2.85	2.70	2.53	2.35	2.13
15	4.54	3.68	3.29	3.06	2.90	2.79	2.64	2.48	2.29	2.07
16	4.49	3.63	3.24	3.01	2.85	2.74	2.59	2.42	2.24	2.01
17	4.45	3.59	3.20	2.96	2.81	2.70	2.55	2.38	2.19	1.96
18	4.41	3.55	3.16	2.93	2.77	2.66	2.51	2.34	2.15	1.92
19	4.38	3.52	3.13	2.90	2.74	2.63	2.48	2.31	2.11	1.88
20	4.35	3.49	3.10	2.87	2.71	2.60	2.45	2.28	2.08	1.84
21	4.32	3.47	3.07	2.84	2.68	2.57	2.42	2.25	2.05	1.81
22	4.30	3.44	3.05	2.82	2.66	2.55	2.40	2.23	2.03	1.78
23	4.28	3.42	3.03	2.80	2.64	2.53	2.38	2.20	2.00	1.76
24	4.26	3.40	3.01	2.78	2.62	2.51	2.36	2.18	1.98	1.73
25	4.24	3.38	2.99	2.76	2.60	2.49	2.34	2.16	1.96	1.71
26	4.22	3.37	2.98	2.74	2.59	2.47	2.32	2.15	1.95	1.69
27	4.21	3.35	2.96	2.73	2.57	2.46	2.30	2.13	1.93	1.67
28	4.20	3.34	2.95	2.71	2.56	2.44	2.29	2.12	1.91	1.65
29	4.18	3.33	2.93	2.70	2.54	2.43	2.28	2.10	1.90	1.64
30	4.17	3.32	2.92	2.69	2.53	2.42	2.27	2.09	1.89	1.62
40	4.08	3.23	2.84	2.61	2.45	2.34	2.18	2.00	1.79	1.51
60	4.00	3.15	2.76	2.52	2.37	2.25	2.10	1.92	1.70	1.39
120	3.92	3.07	2.68	2.45	2.29	2.17	2.02	1.83	1.61	1.25
∞	3.84	2.99	2.60	2.37	2.21	2.09	1.94	1.75	1.52	1.00

Lower 5 percent points are found by interchange of ν_1 and ν_2, i.e. ν_1 must always correspond to the greater mean square.

Appendix Table 8 1 percent points of the distribution of z (values at which the d.f. = 0.99)
(Reprinted from Table VI of Fisher (1970), by kind permission of the author and publishers)

Values of ν_2	Values of ν_1: 1	2	3	4	5	6	8	12	24	∞
1	4.1535	4.2585	4.2974	4.3175	4.3297	4.3379	4.3482	4.3585	4.3689	4.3794
2	2.2950	2.2976	2.2984	2.2988	2.2991	2.2992	2.2994	2.2997	2.2999	2.3001
3	1.7649	1.7140	1.6915	1.6786	1.6703	1.6645	1.6569	1.6489	1.6404	1.6314
4	1.5270	1.4452	1.4075	1.3856	1.3711	1.3609	1.3473	1.3327	1.3170	1.3000
5	1.3943	1.2929	1.2449	1.2164	1.1974	1.1838	1.1656	1.1457	1.1239	1.0997
6	1.3103	1.1955	1.1401	1.1068	1.0843	1.0680	1.0460	1.0218	0.9948	0.9643
7	1.2526	1.1281	1.0672	1.0300	1.0048	0.9864	0.9614	0.9335	0.9020	0.8658
8	1.2106	1.0787	1.0135	0.9734	0.9459	0.9259	0.8983	0.8673	0.8319	0.7904
9	1.1786	1.0411	0.9724	0.9299	0.9006	0.8791	0.8494	0.8157	0.7769	0.7305
10	1.1535	1.0114	0.9399	0.8954	0.8646	0.8419	0.8104	0.7744	0.7324	0.6816
11	1.1333	0.9874	0.9136	0.8674	0.8354	0.8116	0.7785	0.7405	0.6958	0.6408
12	1.1166	0.9677	0.8919	0.8443	0.8111	0.7864	0.7520	0.7122	0.6649	0.6061
13	1.1027	0.9511	0.8737	0.8248	0.7907	0.7652	0.7295	0.6882	0.6386	0.5761
14	1.0909	0.9370	0.8581	0.8082	0.7732	0.7471	0.7103	0.6675	0.6159	0.5500
15	1.0807	0.9249	0.8448	0.7939	0.7582	0.7314	0.6937	0.6496	0.5961	0.5269
16	1.0719	0.9144	0.8331	0.7814	0.7450	0.7177	0.6791	0.6339	0.5786	0.5064
17	1.0641	0.9051	0.8229	0.7705	0.7335	0.7057	0.6663	0.6199	0.5630	0.4879
18	1.0572	0.8970	0.8138	0.7607	0.7232	0.6950	0.6549	0.6075	0.5491	0.4712
19	1.0511	0.8897	0.8057	0.7521	0.7140	0.6854	0.6447	0.5964	0.5366	0.4560
20	1.0457	0.8831	0.7985	0.7443	0.7058	0.6768	0.6355	0.5864	0.5253	0.4421
21	1.0408	0.8772	0.7920	0.7372	0.6984	0.6690	0.6272	0.5773	0.5150	0.4294
22	1.0363	0.8719	0.7860	0.7309	0.6916	0.6620	0.6196	0.5691	0.5056	0.4176
23	1.0322	0.8670	0.7806	0.7251	0.6855	0.6555	0.6127	0.5615	0.4969	0.4068
24	1.0285	0.8626	0.7757	0.7197	0.6799	0.6496	0.6064	0.5545	0.4890	0.3967
25	1.0251	0.8585	0.7712	0.7148	0.6747	0.6442	0.6006	0.5481	0.4816	0.3872
26	1.0220	0.8548	0.7670	0.7103	0.6699	0.6392	0.5952	0.5422	0.4748	0.3784
27	1.0191	0.8513	0.7631	0.7062	0.6655	0.6346	0.5902	0.5367	0.4685	0.3701
28	1.0164	0.8481	0.7595	0.7023	0.6614	0.6303	0.5856	0.5316	0.4626	0.3624
29	1.0139	0.8451	0.7562	0.6987	0.6576	0.6263	0.5813	0.5269	0.4570	0.3550
30	1.0116	0.8423	0.7531	0.6954	0.6540	0.6226	0.5773	0.5224	0.4519	0.3481
60	0.9784	0.8025	0.7086	0.6472	0.6028	0.5687	0.5189	0.4574	0.3746	0.2352
∞	0.9462	0.7636	0.6651	0.5999	0.5522	0.5152	0.4604	0.3908	0.2913	0

Appendix Table 9 1 percent points of the variance ratio F (values at which the d.f. = 0.99) (Reproduced from Fisher and Yates (1963), by kind permission of the authors and publishers)

ν_2 \ ν_1	1	2	3	4	5	6	8	12	24	∞
1	4052	4999	5403	5625	5764	5859	5981	6106	6234	6366
2	98.49	99.00	99.17	99.25	99.30	99.33	99.36	99.42	99.46	99.50
3	34.12	30.81	29.46	28.71	28.24	27.91	27.49	27.05	26.60	26.12
4	21.20	18.00	16.69	15.98	15.52	15.21	14.80	14.37	13.93	13.46
5	16.26	13.27	12.06	11.39	10.97	10.67	10.27	9.89	9.47	9.02
6	13.74	10.92	9.78	9.15	8.75	8.47	8.10	7.72	7.31	6.88
7	12.25	9.55	8.45	7.85	7.46	7.19	6.84	6.47	6.07	5.65
8	11.26	8.65	7.59	7.01	6.63	6.37	6.03	5.67	5.28	4.86
9	10.56	8.02	6.99	6.42	6.06	5.80	5.47	5.11	4.73	4.31
10	10.04	7.56	6.55	5.99	5.64	5.39	5.06	4.71	4.33	3.91
11	9.65	7.20	6.22	5.67	5.32	5.07	4.74	4.40	4.02	3.60
12	9.33	6.93	5.95	5.41	5.06	4.82	4.50	4.16	3.78	3.36
13	9.07	6.70	5.74	5.20	4.86	4.62	4.30	3.96	3.59	3.16
14	8.86	6.51	5.56	5.03	4.69	4.46	4.14	3.80	3.43	3.00
15	8.68	6.36	5.42	4.89	4.56	4.32	4.00	3.67	3.29	2.87
16	8.53	6.23	5.29	4.77	4.44	4.20	3.89	3.55	3.18	2.75
17	8.40	6.11	5.18	4.67	4.34	4.10	3.79	3.45	3.08	2.65
18	8.28	6.01	5.09	4.58	4.25	4.01	3.71	3.37	3.00	2.57
19	8.18	5.93	5.01	4.50	4.17	3.94	3.63	3.30	2.92	2.49
20	8.10	5.85	4.94	4.43	4.10	3.87	3.56	3.23	2.86	2.42
21	8.02	5.78	4.87	4.37	4.04	3.81	3.51	3.17	2.80	2.36
22	7.94	5.72	4.82	4.31	3.99	3.76	3.45	3.12	2.75	2.31
23	7.88	5.66	4.76	4.26	3.94	3.71	3.41	3.07	2.70	2.26
24	7.82	5.61	4.72	4.22	3.90	3.67	3.36	3.03	2.66	2.21
25	7.77	5.57	4.68	4.18	3.86	3.63	3.32	2.99	2.62	2.17
26	7.72	5.53	4.64	4.14	3.82	3.59	3.29	2.96	2.58	2.13
27	7.68	5.49	4.60	4.11	3.78	3.56	3.26	2.93	2.55	2.10
28	7.64	5.45	4.57	4.07	3.75	3.53	3.23	2.90	2.52	2.06
29	7.60	5.42	4.54	4.04	3.73	3.50	3.20	2.87	2.49	2.03
30	7.56	5.39	4.51	4.02	3.70	3.47	3.17	2.84	2.47	2.01
40	7.31	5.18	4.31	3.83	3.51	3.29	2.99	2.66	2.29	1.80
60	7.08	4.98	4.13	3.65	3.34	3.12	2.82	2.50	2.12	1.60
120	6.85	4.79	3.95	3.48	3.17	2.96	2.66	2.34	1.95	1.38
∞	6.64	4.60	3.78	3.32	3.02	2.80	2.51	2.18	1.79	1.00

Lower 1 percent points are found by interchange of ν_1 and ν_2, i.e. ν_1 must always correspond to the greater mean square.

Appendix Table 10 Augmented symmetric functions in terms of power-sums and vice versa
(Reproduced from F. N. David and Kendall (1949) by kind permission of Prof. David and the editors of *Biometrika*)

weight 1 (1) = [1]

weight 2

	[2]	[1^2]
(2)	1	−1
$(1)^2$	1	1

weight 3

	[3]	[21]	[1^3]
(3)	1	−1	2
(2)(1)	1	1	−3
$(1)^3$	1	3	1

weight 4

	[4]	[31]	[2^2]	[21^2]	[1^4]
(4)	1	−1	−1	2	−6
(3)(1)	1	1	·	−2	8
$(2)^2$	1	·	1	−1	3
$(2)(1)^2$	1	2	1	1	−6
$(1)^4$	1	4	3	6	1

weight 5

	[5]	[41]	[32]	[31^2]	[2^21]	[21^3]	[1^5]
(5)	1	−1	−1	2	2	−6	24
(4)(1)	1	1	·	−2	−1	6	−30
(3)(2)	1	·	1	−1	−2	5	−20
$(3)(1)^2$	1	2	1	1	·	−3	20
$(2)^2(1)$	1	1	2	·	1	−3	15
$(2)(1)^3$	1	3	4	3	3	1	−10
$(1)^5$	1	5	10	10	15	10	1

weight 6

	[6]	[51]	[42]	[41^2]	[3^2]	[321]	[31^3]	[2^3]	[2^21^2]	[21^4]	[1^6]
(6)	1	−1	−1	2	−1	2	−6	2	−6	24	−120
(5)(1)	1	1	·	−2	·	−1	6	·	4	−24	144
(4)(2)	1	·	1	−1	·	−1	3	−3	5	−18	90
$(4)(1)^2$	1	2	1	1	·	·	−3	·	−1	12	−90
$(3)^2$	1	·	·	·	1	−1	2	·	2	−8	40
(3)(2)(1)	1	1	1	·	1	1	−3	·	−4	20	−120
$(3)(1)^3$	1	3	3	3	1	3	1	·	·	−4	40
$(2)^3$	1	·	3	·	·	·	·	1	−1	3	−15
$(2)^2(1)^2$	1	2	3	1	2	4	·	1	1	−6	45
$(2)(1)^4$	1	4	7	6	4	16	4	3	6	1	−15
$(1)^6$	1	6	15	15	10	60	20	15	45	15	1

To express the [ ] functions in terms of ( ), read downwards as far as and including the main diagonal, e.g. $[41]^2 = 2(6) - 2(5)(1) - (4)(2) + (4)(1)^2$. To express the ( ) functions in terms of [ ], read across as far as and including the main diagonal, e.g. $(4)(1)^2 = [6] + 2[51] + [42] + [41^2]$.

Appendix Table 11 Multiple *k*-statistics in terms of augmented symmetric functions and vice versa

(Reproduced from Wishart (1952) by kind permission of the late author and the editors of *Biometrika*)

1st order $l_1 = [1]/n$

2nd order

	l_{11}	l_2
$[1^2]/n^{[2]}$	1	−1
$[2]/n$	1	1

3rd order

	l_{111}	l_{21}	l_3
$[1^3]/n^{[3]}$	1	−1	2
$[21]/n^{[2]}$	1	1	−3
$[3]/n$	1	3	1

4th order

	l_{1111}	l_{211}	l_{22}	l_{31}	l_4
$[1^4]/n^{[4]}$	1	−1	1	2	−6
$[21^2]/n^{[3]}$	1	1	−2	−3	12
$[2^2]/n^{[2]}$	1	2	1	·	−3
$[31]/n^{[2]}$	1	3	·	1	−4
$[4]/n$	1	6	3	4	1

5th order

	l_{11111}	l_{2111}	l_{221}	l_{311}	l_{32}	l_{41}	l_5
$[1^5]/n^{[5]}$	1	−1	1	2	−2	−6	24
$[21^3]/n^{[4]}$	1	1	−2	−3	5	12	−60
$[2^21]/n^{[3]}$	1	2	1	·	−3	−3	30
$[31^2]/n^{[3]}$	1	3	·	1	−1	−4	20
$[32]/n^{[2]}$	1	4	3	1	1	·	−10
$[41]/n^{[2]}$	1	6	3	4	·	1	−5
$[5]/n$	1	10	15	10	10	5	1

6th order

	l_{111111}	l_{21111}	l_{2211}	l_{3111}	l_{222}	l_{321}	l_{411}	l_{33}	l_{42}	l_{51}	l_6
$[1^6]/n^{[6]}$	1	−1	1	2	−1	−2	−6	4	6	24	−120
$[21^4]/n^{[5]}$	1	1	−2	−3	3	5	12	−12	−18	−60	360
$[2^21^2]/n^{[4]}$	1	2	1	·	−3	−3	−3	9	15	30	−270
$[31^3]/n^{[4]}$	1	3	·	1	·	−1	−4	4	4	20	−120
$[2^3]/n^{[3]}$	1	3	3	·	1	·	·	·	−3	·	30
$[321]/n^{[3]}$	1	4	3	1	·	1	·	−6	−4	−10	120
$[41^2]/n^{[3]}$	1	6	3	4	·	·	1	·	−1	−5	30
$[3^2]/n^{[2]}$	1	6	9	2	·	6	·	1	·	·	−10
$[42]/n^{[2]}$	1	7	9	4	3	4	1	·	1	·	−15
$[51]/n^{[2]}$	1	10	15	10	·	10	5	·	·	1	−6
$[6]/n$	1	15	45	20	15	60	15	10	15	6	1

To express *l*s in terms of symmetrics read downwards as far as and including the main diagonal, e.g. $l_{222} = -[1^6]/n^{[6]} + 3[21^4]/n^{[5]} - 3[2^21^2]/n^{[4]} + [2^3]/n^{[3]}$.

To express symmetrics in terms of *l*s read across as far as and including the diagonal, e.g. $[2^21^2]/n^{[4]} = l_{111111} + 2l_{21111} + l_{2211}$.

REFERENCES

Note References to W. G. Cochran, R. A. Fisher, M. G. Kendall, J. Neyman, K. Pearson, and Student that are marked with an asterisk are reproduced in the following collections:
W. G. Cochran, *Contributions to Statistics*, Wiley, New York, 1982.
R. A. Fisher, *Contributions to Mathematical Statistics*, Wiley, New York, 1950.
Statistics: Theory and Practice, Selected Papers by Sir Maurice Kendall, Griffin, High Wycombe, 1984.
A Selection of Early Statistical Papers of J. Neyman, Cambridge Univ. Press, London, 1967.
Karl Pearson's Early Statistical Papers, Cambridge Univ. Press, London, 1948.
"Student's" Collected Papers, Biometrika Office, University College London, 1942.
Statistical Papers of George Udny Yule, Griffin, London, 1971.

ABDEL-ATY, S. H. (1954). Tables of generalised *k*-statistics. *Biometrika*, **41**, 253.

ABRAHAMSON, I. G. (1964). Orthant probabilities for the quadrivariate normal distribution. *Ann. Math. Statist.*, **35**, 1685.

AITCHISON, J. (1963). Inverse distributions and independent gamma-distributed products of random variables. *Biometrika*, **50**, 505.

AITCHISON, J. and BROWN, J. A. C. (1957). *The Lognormal Distribution with Special Reference to its Uses in Economics*. Cambridge Univ. Press.

AITCHISON, J. and HO, C. H. (1989). The multivariate Poisson-lognormal distribution. *Biometrika*, **76**, 643.

AITKEN, A. C. (1950). Statistical independence of quadratic forms in normal variates. *Biometrika*, **37**, 93.

ALI, M. M. (1974). Stochastic ordering and kurtosis measure. *J. Amer. Statist. Ass.*, **69**, 543.

ALI, M. M. and MIKHAIL, N. N. (1979). Bivariate moments and product moments of moment statistics for samples from finite populations. *J. Statist. Res.*, **13**, 31.

ALTHAM, P. M. E. (1978). Two generalizations of the binomial distribution. *Appl. Statist.*, **27**, 162.

AMOS, D. E. (1963). Additional percentage points for the incomplete beta distribution. *Biometrika*, **50**, 449.

ANDERSON, G. A. (1965). An asymptotic expansion for the distribution of the latent roots of the estimated covariance matrix. *Ann. Math. Statist.*, **36**, 1153.

ANDERSON, G. L. and DE FIGUEIREDO, R. J. P. (1980). An adaptive orthogonal-series estimator for probability density functions. *Ann. Statist*, **8**, 347.

ANDERSON, T. W. (1955). The integral of a symmetric unimodal function over a symmetric convex set and some probability inequalities. *Proc. Amer. Math. Soc.*, **6**, 170.

ANDERSON, T. W. (1963). Asymptotic theory for principal component analysis. *Ann. Math. Statist.*, **34**, 122.

ANDERSON, T. W. (1984). *An Introduction to Multivariate Statistical Analysis* (Second Edition). Wiley, New York.

ANDREWS, D. F., BICKEL, P. J., HAMPEL, F. R., HUBER, P. J., ROGERS, W. H. and TUKEY, J. W. (1972). *Robust Estimates of Location: Survey and Advances*. Princeton Univ. Press, Princeton.

ANSCOMBE, F. J. (1950). Sampling theory of the negative-binomial and logarithmic series distributions. *Biometrika*, **37**, 358.

ANSCOMBE, F. J., and GLYNN, W. J. (1983). Distribution of the kurtosis statistic b_2 for normal samples. *Biometrika*, **70**, 227.

ARNOLD, B. C. and BALAKRISHNAN, N. (1989). *Relations, Bounds and Approximations for Order Statistics*. Springer, Berlin and New York.

ARNOLD, B. C. and GROENEVELD, R. A. (1981). Maximal deviation between sample and population means in finite populations. *J. Amer. Statist. Ass.*, **76**, 443.

ARNOLD, B. C. and STRAUSS, D. J. (1988). Bivariate distributions with exponential conditionals. *J. Amer. Statist. Ass.*, **83**, 522.

ARNOLD, S. and NORTON, R. M. (1985). A theorem on moments. *Amer. Statistician*, **39**, 106.

ATKINSON, A. C. (1979a). A family of switching algorithms for the computer generation of beta random variables. *Biometrika*, **66**, 141.

ATKINSON, A. C. (1979b). The computer generation of Poisson random variables. *Appl. Statist.*, **28**, 29.

ATKINSON, A. C. (1979c). Recent developments in the computer generation of Poisson random variables. *Appl. Statist.*, **28**, 260.

ATKINSON, A. C. (1980). Tests of pseudo-random numbers. *Appl. Statist.*, **29**, 164.

ATKINSON, A. C. and PEARCE, M. C. (1976). The computer generation of beta, gamma and normal random variables. *J. R. Statist. Soc.*, **A**, **139**, 431.

ATKINSON, A. C. and WHITTAKER, J. (1979). The generation of beta random variables with one parameter greater and one parameter less than one. *Appl. Statist.*, **28**, 90.

BABU, G. J. and SINGH, K. (1983). Inference on means using the bootstrap. *Ann. Statist.*, **11**, 999.

BACON, R. H. (1963). Approximations to multivariate normal orthant probabilities. *Ann. Math. Statist.*, **34**, 191.

BAI, Z. D. and RAO, C. R. (1991). Edgeworth expansions for functions of means. *Ann. Statist.*, **19**, 1295.

BAILEY, B. J. R., (1980). Accurate normalizing transformations of a Student's t variate. *Appl. Statist.*, **29**, 304.

BAIN, L. J. and ENGELHARDT, M. (1975). A two-moment chi-square approximation for the statistic log $(\bar{x}/\tilde{x})$ *J. Amer. Statist. Ass.*, **70**, 948.

BALAKRISHNAN, N. (1986) Order statistics from discrete distributions. *Commun. Statist.*, **A15**, 657.

BALAKRISHNAN, N. (1987). A note on moments of order statistics from exchangeable variates. *Commun. Statist.*, **A16**, 855.

BALAKRISHNAN, N., MALIK, H. J., and AHMED, S. E. (1988). Recurrence relations and identities for moments of order statistics, II. *Commun. Statist.*, **A17**, 2657.

BALANDA, K. P. and MACGILLIVRAY, H. (1988). Kurtosis: A critical review. *Amer. Statist.*, **42**, 111.

BALANDA, K. P. and MACGILLIVRAY, H. (1990). Kurtosis and spread. *Canad. J. Statist.*, **18**, 17.

BALDWIN, E. M. (1946). Percentage points of the t distribution. *Biometrika*, **33**, 362.

BALITSKAYA, E. O. and ZOLOTUHINA, L. A. (1988). On the representation of a density by an Edgeworth series. *Biometrika*, **75**, 185.

BARLOW, R. E. and PROSCHAN, F. (1975). *Statistical Analysis of Reliability and Life Testing*. New York, Holt, Rinehart and Winston.

BARGMANN, R. E. and PARRISH, R. S. (1981). A method for the evaluation of cumulative probabilities of bivariate distributions using the Pearson family. *In* Taillie *et al.* (1981), **5**, 241.

BARNDORFF-NIELSEN, O. and COX, D. R. (1979). Edgeworth and saddle-point approximations, with statistical applications. *J. R. Statist. Soc.*, **B**, **41**, 279.

BARNDORFF-NIELSEN, O. and PEDERSEN, B. V. (1979). The bivariate Hermite polynomials up to order six. *Scand. J. Statist.*, **6**, 127.

BARNETT, V. (1976). The ordering of multivariate data. *J. R. Statist, Soc.*, **A**, **139**, 318.

BARNETT, V. (1980). Some bivariate uniform distributions. *Commun. Statist.*, **A9**, 453.

BARNETT, V. (1982). *Comparative Statistical Inference*, 2nd edn. Wiley, Chichester.

BARNETT, V. (1985). The bivariate exponential distribution: a review and some new results. *Statist. Neerl.*, **39**, 343.

BARTLETT, M. S. (1933). On the theory of statistical regression. *Proc. Roy. Soc. Edin.*, **53**, 260.

BARTLETT, M. S. (1934). The vector representation of a sample. *Proc. Camb. Phil. Soc.*, **30**, 327.

BARTLETT, M. S. (1951). The effect of standardization on a χ^2 approximation in factor analysis. *Biometrika*, **38**, 337.

BARTLETT, M. S. (1975). *Probability, Statistics and Time*. Methuen, London.

BARTON, D. E. (1957). The modality of Neyman's contagious distribution of Type A. *Trab. Estadist.*, **8**, 13.

BARTON, D. E. and DENNIS, K. E. (1952). The conditions under which Gram-Charlier and Edgeworth curves are positive definite and unimodal. *Biometrika*, **39**, 425.

BASU, A. P. (1971). Bivariate failure rate. *J. Amer Statist. Ass.*, **66**, 103.

BAYES, T. (1764). An essay towards solving a problem in the doctrine of chances. *Phil. Trans.*, **53**, 370. (Reproduced in *Biometrika*, **45**, 293 (1958), edited and introduced by G. A. Barnard.)

BECHHOFER, R. E. and TAMHANE, A. C. (1974). An iterated integral representation for a multivariate normal integral having block covariance structure. *Biometrika*, **61**, 615.

BECKER, P. J. and ROUZ, J. J. J. (1981). A bivariate extension of the gamma distribution. *S. African J. Statist.*, **15**, 1.

BEESACK, P. R. (1984). Inequalities for absolute moments of a distribution: From Laplace to Von Mises. *J. Math. Anal.*, **A**, **98**, 435.

BEHBOODIAN, J. (1970). On the modes of a mixture of two normal distributions. *Technometrics*, **12**, 131.

BELZ, M. H. (1947). Note on the Liapounoff inequality for absolute moments. *Ann. Math. Statist.*, **18**, 604.

BERGSTRÖM, H. (1952). On some expansions of stable distribution functions. *Arkiv Math.*, **2**, 375.

BERNDT, G. D. (1957). The regions of unimodality and positivity in the abbreviated Edgeworth and Gram-Charlier series. *J. Amer. Statist. Ass.*, **52**, 253.

BERRY, A. C. (1941). The accuracy of the Gaussian approximation to the sum of independent variables. *Trans. Amer. Math. Soc.*, **49**, 122.

BESAG, J. E. (1974). Spatial interaction and the statistical analysis of lattice systems. *J. Roy. Statist. Soc.*, **B**, **36**, 192.

BHATTACHARYA, R. N. and GHOSH, J. K. (1978). On the validity of the formal Edgeworth expansion. *Ann. Statist.*, **6**, 434.

BHATTACHARYA, R. N. and GHOSH, J. K. (1988). On moment conditions for valid formal Edgeworth expansions. *J. Multiv. Anal.*, **27**, 68.

BHATTACHARYYA, B. B. (1987). One sided Chebyshev inequality when the first four moments are known. *Commun. Statist.*, **A**, **16**, 2789.

BICKEL, P. J. (1967). Some contributions to the theory of order statistics. *Proc. 5th Berkeley Symp. Math. Statist. and Prob.*, **1**, 575.

BOHRNSTEDT, G. W. and GOLDBERGER, A. S. (1969). On the exact covariance of products of random variables. *J. Amer. Statist. Ass.*, **64**, 1439.

BØLVIKEN, E. (1982). Probability inequalities for the multivariate normal with non-negative partial correlations. *Scand. J. Statist.*, **9**, 49.

BONDESSON, L. (1979). A general result on infinite divisibility. *Ann. Prob.*, **7**, 965.

BONDESSON, L. (1983). When is the t-statistic t-distributed? *Sankhyā*, **A**, **45**, 338.

BOOKSTEIN, F. L. (1986). Size and shape spaces for landmark data in two dimensions. *Statistical Science*, **1**, 181.

BOOMSMA, A. (1975). An error in F. N. David's tables of the correlation coefficient. *Biometrika*, **62**, 711.

BORGES, R. (1966). A characterization of the normal distribution. *Zeit. Wahrsch. Vere. Geb.*, **5**, 244.

BOSE, S. S. (1935). On the distribution of the ratio of variances of two samples drawn from a given normal bivariate correlated population. *Sankhyā*, **2**, 65.

BOSWELL, M. T. and DE ANGELIS, R. J. (1981). A rejection technique for the generation of random variables with the beta distribution. *In* Taillie *et al.* (1981), **4**, 305.

BOWMAN, K. O. and SHENTON, L. R. (1979a). Approximate percentage points for Pearson distributions. *Biometrika*, **66**, 147.

BOWMAN, K. O. and SHENTON, L. R. (1979b). Further approximate Pearson percentage points and Cornish-Fisher. *Commun. Statist.*, **B**, **8**, 231.

BOWMAN, K. O. and SHENTON, L. R. (1980). Evaluation of the parameters of S_U by rational functions. *Commun. Statist.*, **B**, **9**, 127.

BOWMAN, K. O. and SHENTON, L. R. (1988). Solutions to Johnson's S_B and S_U. *Commun. Statist.*, **B17**, 343.

BOWMAN, K. O. and SHENTON, L. R. (1989). S_B and S_U distributions fitted by percentiles: A general criterion. *Commun. Statist.*, **B18**, 1.

BOWMAN, K. O., BEAUCHAMP, J. J. and SHENTON, L. R. (1977). The distribution of the *t*-statistic under non-normality. *Int. Statist. Rev.*, **45**, 233.

BOWMAN, K. O., SERBIN, C. A. and SHENTON, L. R. (1981). Explicit approximate solutions for S_B. *Commun. Statist.*, **B**, **10**, 1.

BOWMAN, K. O., SHENTON, L. R. and KASTENBAUM, M. A. (1991). *Discrete Pearson Distributions*. Oak Ridge National Labs. Tech. Report ORNL/TM-11899.

BOX, G. E. P. (1953). Non-normality and tests on variances. *Biometrika*, **40**, 318.

BOX, G. E. P. (1954). Some theorems on quadratic forms applied to the study of analysis of variance problems, I, II. *Ann. Math. Statist.*, **25**, 290, 484.

BOX, G. E. P. and COX, D. R. (1964). An analysis of transformations. *J. R. Statist. Soc.*, **B**, **26**, 211.

BOX, G. E. P. and MULLER, M. E. (1958). A note on the generation of random normal deviates. *Ann. Math. Statist.*, **29**, 610.

BROCKWELL, P. J. and BROWN, B. M. (1978). Expansions for the positive stable laws. *Zeit. Wahr.*, **45**, 213.

BROWN, G. W. and TUKEY, J. W. (1946). Some distributions of sample means. *Ann. Math. Statist.*, **17**, 1.

BROWN, M. and SOLOMON, H. (1979). On combining pseudorandom number generators. *Ann. Statist.*, **7**, 691.

BRYSON, M. C. and SIDDIQUI, M. M. (1969). Some criteria for ageing. *J. Amer. Statist. Ass.*, **64**, 1472.

BUCKLE, N., KRAFT, C. and VAN EEDEN, C. (1969). An approximation to the Wilcoxon-Mann-Whitney distribution. *J. Amer. Statist. Ass.*, **64**, 591.

BUKAČ, J. (1972). Fitting S_B curves using symmetrical percentile points. *Biometrika*, **59**, 688.

BURR, I. W. (1942). Cumulative frequency functions. *Ann. Math. Statist.*, **13**, 215.

BURR, I. W. (1955). Calculation of exact sampling distribution of ranges from a discrete population. *Ann. Math. Statist.*, **26**, 530.

BURR, I. W. and CISLAK, P. J. (1968). On a general system of distributions. I. Its curve-shape characteristics. II. The sample median. *J. Amer. Statist. Ass.*, **63**, 627.

BURRELL, Q. L. and CANE, V. R. (1982). The analysis of library data. *J. R. Statist. Soc.*, **A**, **145**, 439.

CACOULLOS, T. (1965). A relation between *t* and *F*-distributions. *J. Amer. Statist. Ass.*, **60**, 528.

CAMPBELL, J. W. and TSOKOS, C. P. (1973). The asymptotic distribution of maxima in bivariate samples. *J. Amer. Statist. Ass.*, **68**, 734.

CARLEMAN, T. (1925). *Les fonctions quasi-analytiques*. Gauthier-Villars, Paris.

CHAKRABARTI, M. C. (1948). On the ratio of mean deviation to standard deviation. *Calcutta Statist. Ass. Bull.*, **1**, 187.

CHAMBERS, J. M. (1967). On methods of asymptotic approximation for multivariate distributions. *Biometrika*, **54**, 367.

CHAN, L. K. (1967). On a characterization of distributions by expected values of extreme order statistics. *Amer. Math. Monthly*, **74**, 950.

CHARLIER, C. V. L. (1931). *Applications [de la théorie des probabilités] à l'astronomie*. (Part of the *Traité* edited by Borel.) Gauthier-Villars, Paris.

CHENG, M. C. (1969). The orthant probabilities of four Gaussian variates. *Ann. Math. Statist.*, **40**, 152.

CHERNOFF, H., GASTWIRTH, J. L. and JOHNS, M. V., Jr. (1967). Asymptotic distribution of linear combinations of functions of order statistics with applications to estimation. *Ann. Math. Statist.*, **38**, 52.

CHILDS, D. R. (1967). Reduction of the multivariate normal integral to characteristic form. *Biometrika*, **54**, 293.

CHU, J. T. (1955). The "inefficiency" of the sample median for many familiar symmetric distributions. *Biometrika*, **42**, 520.

CHURCHILL, E. (1946). Information given by odd moments. *Ann. Math. Statist.*, **17**, 244.

CLARK, C. E. (1961). The greatest of a finite set of random variables. *Operations Research*, **9**, 145.

COBB, L., KOPPSTEIN, P. and CHEN, N. H. (1983). Estimation and moment recursion relations for multimodal distributions of the exponential family. *J. Amer. Statist. Ass.*, **78**, 124.

COCHRAN, W. G. (1934).* The distribution of quadratic forms in a normal system, with applications to the analysis of covariance. *Proc. Camb. Phil. Soc.*, **30**, 178.

COCHRAN, W. G. (1940).* Note on an approximative formula for significance levels of z. *Ann. Math. Statist.*, **11**, 93.

COHEN, A. C. and WHITTEN, B. J. (1980). Estimation in the three-parameter lognormal distribution. *J. Amer. Statist. Ass.*, **75**, 399.

CONNOR, R. J. (1969). The sampling distribution of the range from discrete uniform finite populations and a range test for homogeneity. *J. Amer. Statist. Ass.*, **64**, 1443.

COOK, M. B. (1951). Bivariate k-statistics and cumulants of their joint sampling distribution. *Biometrika*, **38**, 179.

COOK, R. D. and JOHNSON, M. E. (1981). A family of distributions for modelling non-elliptically symmetric multivariate data. *J. R. Statist. Soc.*, **B**, **43**, 210.

COOK, R. D. and JOHNSON, M. E. (1986). Generalized Burr-Pareto-Logistic distributions with applications to a uranium exploration data set. *Technometrics*, **28**, 123.

Co-operative Study (1917). On the distribution of the correlation coefficient in small samples. *Biometrika*, **11**, 328.

CORNISH, E. A., and FISHER, R. A. (1937).* Moments and cumulants in the specification of distributions. *Rev. Int. Statist. Inst.*, **5**, 307.

COX, D. R. (1948). Asymptotic distribution of the range. *Biometrika*, **35**, 310.

COX, D. R. and MILLER, H. D. (1965). *The Theory of Stochastic Processes*. Methuen, London.

COX, D. R. and WERMUTH, N. (1991). A simple approximation for bivariate and trivariate normal integrals. *Int. Statist. Rev.*, **59**, 263.

CRAIG, A. T. (1943). Note on the independence of certain quadratic forms. *Ann. Math. Statist.*, **14**, 195.

CRAIG, C. C. (1936a). A new exposition and chart for the Pearson system of frequency curves. *Ann. Math. Statist.*, **7**, 16.

CRAIG, C. C. (1936b). Sheppard's corrections for a discrete variable. *Ann. Math. Statist.*, **7**, 55.

CRAMÉR, H. (1925). On some classes of series used in mathematical statistics. *Trans. 6th Congr. Scand. Math.*, 399.

CRAMÉR, H. (1928). On the composition of elementary errors. *Skand. Aktuartidskr.*, **11**, 13 and 141.

CRAMÉR, H. (1937). *Random Variables and Probability Distributions*. Cambridge Univ. Press, London.

CRAMÉR, H., and WOLD, H. (1936). Some theorems on distribution functions. *J. Lond. Math. Soc.*, **11**, 290.

CRESSIE, N., DAVIS, A. S., FOLKS, J. L. and POLICELLO, G. E. II (1981). The moment-generating function and negative integer moments. *Amer. Statist.*, **35**, 148.

CROW, E. L. and SHIMIZU, K. eds., (1988). *Lognormal Distributions: Theory and Applications*. Dekker, New York.

D'AGOSTINO, R. and PEARSON, E. S. (1973). Tests for departure from normality. Empirical results for the distributions of b_2 and $\sqrt{b_1}$. *Biometrika*, **60**, 613.

DALE, A. I. (1991). *A History of Inverse Probability from Thomas Bayes to Karl Pearson*. Springer, Berlin and New York.

DALL'AGLIO, G., KOTZ, S. and SALINETTI, G. eds., (1991). *Advances in Probability Distributions with Given Marginals, Beyond the Copulas*. Kluwer, Dordrecht and Boston.

DANIELS, H. E. (1954). Saddlepoint approximations in statistics. *Ann. Math. Statist.*, **25**, 631.

DANIELS, H. E. (1980). Exact saddlepoint approximations. *Biometrika*, **67**, 59.

DANIELS, H. E. (1983). Saddlepoint approximations for estimating equations. *Biometrika*, **70**, 89.

DANIELS, H. E. (1987). Tail probability approximations. *Int. Statist. Rev.*, **55**, 37.

DAS, S. C. (1956). The numerical evaluation of a class of integral. II. *Proc. Camb. Phil. Soc.*, **52**, 442.

DAVID, F. N. (1938). *Tables of the Correlation Coefficient*. Cambridge Univ. Press, London.

DAVID, F. N. (1949a). Note on the application of Fisher's k-statistics. *Biometrika*, **36**, 383.

DAVID, F. N. (1949b). Moments of the z and F distributions. *Biometrika*, **36**, 394

DAVID, F. N. (1962). *Games, Gods and Gambling: A History of Probability and Statistical Ideas*. Griffin, London.

DAVID, F. N., and JOHNSON, N. L. (1951). The effect of non-normality on the power function of the F-test. *Biometrika*, **38**, 43.

DAVID, F. N., and JOHNSON, N. L. (1954). Statistical treatment of censored data. Part I. Fundamental formulae. *Biometrika*, **41**, 225.

DAVID, F. N., and KENDALL, M. G. (1949, 1951, 1953, 1955). Tables of symmetric ftinctions. *Biometrika*, **36**, 431; **38**, 435; **40**, 427; **42**, 223.

DAVID, F. N., BARTON, D. E., GANESHALINGAM, S., HARTER, H. L., KIM, P. J., MERRINGTON, M. and WALLEY, D. (1968). *Normal Centroids, Medians and Scores for Ordinal Data*. Cambridge University Press, London.

DAVID, H. A. (1968). Gini's mean difference rediscovered. *Biometrika*, **55**, 573.

DAVID, H. A. (1981). *Order Statistics*, 2nd edn. Wiley, New York.

DAVID, H. A. and GROENEVELD, R. A. (1982). Measures of local variation in a distribution: Expected length of spacings and variances of order statistics. *Biometrika*, **69**, 227.

DAVID, H. A. and JOSHI, P. C. (1968). Recurrence relations between moments of order statistics for exchangeable variates. *Ann. Math. Statist.*, **39**, 272.

DAVID, H. A. and MISHRIKY, R. S. (1968). Order statistics for discrete populations and for grouped samples. *J. Amer. Statist. Ass.*, **63**, 1390.

DAVID, H. A., O'CONNELL, M. J. and YANG, S. S. (1977). Distribution and expected value of the rank of a concomitant of an order statistic. *Ann. Statist.*, **5**, 216.

DAVIS, A. W. (1971). Percentile approximations for a class of likelihood criteria. *Biometrika*, **58**, 349.

DAVIS, A. W. (1976). Statistical distributions in univariate and multivariate Edgeworth populations. *Biometrika*, **63**, 661.

DAVIS, A. W. (1977). A differential equation approach to linear combinations of independent chi-squares. *J. Amer. Statist. Ass.*, **72**, 212.

DAVIS, C. S. and STEPHENS, M. A. (1978). Approximating the covariance matrix of normal order statistics. *Appl. Statist.*, **27**, 206.

DAVIS, C. S. and STEPHENS, M. A. (1983). Approximate percentage points using Pearson curves. *Appl. Statist.*, **32**, 322.

DAWID, A. P., STONE, M. and ZIDEK, J. V. (1973). Marginalization paradoxes in Bayesian and structural inference. *J. R. Statist. Soc.*, **B**, **35**, 189.

DE FINETTI, B. (1974, 1975). *Theory of Probability*. Volumes 1 and 2. Wiley, Chichester and New York.

DE GUNST, M. C. M. (1987). On the distribution of general quadratic functions in normal vectors. *Statist. Neerl.*, **41**, 245.

DE HAAN, L. (1970). *On Regular Variation and Its Application to the Weak Convergence of Sample Extremes*. Math. Centre Tracts, No. 32. Math. Centrum, Amsterdam, Netherlands.

DELURY, D. B. (1938). Note on correlations. *Ann. Math. Statist.*, **9**, 149.

DEMING, W. E. and GLASSER, G. J. (1959). On the problem of matching lists by samples. *J. Amer. Statist. Ass.*, **54**, 403.

DEMPSTER, A. P. (1958). A high dimensional two sample significance test. *Ann. Math. Statist.*, **29**, 995.

D'ESTE, G. M. (1981). A Morgenstern-type bivariate gamma distribution. *Biometrika*, **68**, 339.

DIVGI, D. R. (1979). Calculation of univariate and bivariate normal probability functions. *Ann. Statist.*, **7**, 903.

DOUGLAS, J. B. (1980). *Analysis with Standard Contagious Distributions*. International Co-operative Publishing House, Burtonsville, Maryland.

DRAPER, J. (1952). Properties of distributions resulting from certain simple transformations of the normal distribution. *Biometrika*, **39**, 290.

DRAPER, N. R. and TIERNEY, D. E. (1972). Regions of positive and unimodal series expansion of the Edgeworth and Gram-Charlier approximations. *Biometrika*, **59**, 463.

DREZNER, Z. (1990). Approximations to the multivariate normal integral. *Commun. Statist.*, **B19**, 527.

DRONKERS, J. J. (1958). Approximate formulae for the statistical distributions of extreme values. *Biometrika*, **45**, 447.

DUNNING, K. A. and HANSON, J. N. (1977). Generalized Pearson distributions and non-linear programming. *J. Statist. Comput. Simul.*, **6**, 115.

DURBIN, J. and WATSON, G. S. (1951). Testing for serial correlation in least squares regression. II. *Biometrika*, **38**, 159.

DUTT, J. E. (1973). A representation of multivariate normal probability integrals by integral transforms. *Biometrika*, **60**, 637.

DWYER, P. S. (1964). Properties of polykays of deviates. *Ann. Math. Statist.*, **35**, 1167.

DWYER, P. S. and TRACY, D. S. (1964). A combinatorial method for products of two polykays with some general formulae. *Ann. Math. Statist.*, **35**, 1174.

DWYER, P. S. and TRACY, D. S. (1980). Expectation and estimation of product moments in sampling from a finite population. *J. Amer. Statist. Ass.*, **75**, 431.

DYSON, F. J. (1943). A note on kurtosis. *J. Roy. Statist. Soc.*, **A**, **106**, 30.

EASTON, G. S. and RONCHETTI, E. (1986). General saddlepoint approximations with applications to L statistics. *J. Amer. Statist. Ass.*, **81**, 420.

EATON, M. L. (1982). A review of selected topics in multivariate probability inequalities. *Ann. Statist.*, **10**, 11.

EATON, M. L. and PERLMAN, M. D. (1991). Concentration inequalities for multivariate distributions: I. Multivariate normal distributions. *Statist. Prob. Letters*, **12**, 487.

EDDY, W. F. (1980). Optimal kernel estimators of the mode. *Ann. Statist.*, **6**, 870.

EDGEWORTH, F. Y. (1904). The Law of Error. *Trans. Camb. Phil. Soc.*, **20**, 36 and 113 (with an Appendix not printed in the *T.C.P.S.* but issued with reprints).

EDWARDS, C. B. and GURLAND, J. (1961). A class of distributions applicable to accidents. *J. Amer. Statist. Ass.*, **56**, 503.

EFRON, B. (1979). Bootstrap methods: Another look at the jackknife. *Ann. Statist.*, **7**, 1.

EFRON, B. (1981). Nonparametric estimates of standard error: the jackknife, the bootstrap and other methods. *Biometrika*, **68**, 589.

EFRON, B. (1982). *The Jackknife, the Bootstrap and Other Resampling Plans*. Soc. Ind. Appl. Math., Philadelphia.

EFRON, B. and TIBSHIRANI, R. (1986). Bootstrap methods for standard errors, confidence intervals and other measures of statistical accuracy. *Statist. Sci.*, **1**, 54.

ELDERTON, W. P. (1938). Correzioni dei momenti quando la curva è simmetrica. *Giorn. Ist. Ital Attuari*, **9**, 145.

ELDERTON, W. P. and JOHNSON, N. L. (1969). *Systems of Frequency Curves*. Cambridge University Press, London.

ELFVING, G. (1947). Asymptotical distribution of range in samples from a normal population. *Biometrika*, **34**, 111.

EL LOZY, M. (1982). Efficient computation of the distribution functions of Student's t, chi-squared and F to moderate accuracy. *J. Statist. Comput. Simul.*, **14**, 179.

ESSEEN, C.-G. (1945). Fourier analysis of distribution functions. A mathematical study of the Laplace-Gaussian law. *Acta Mathematica*, **77**, 1.

EVERITT, B. S. and HAND, D. J. (1981). *Finite Mixture Distributions*. Chapman and Hall, London.

FAHMY, S. and PROSCHAN, F. (1981). Bounds on differences of order statistics. *Amer. Statist.*, **35**, 46.

FAMA, E. F. and ROLL, R. (1968). Some properties of symmetric stable distributions. *J. Amer. Statist. Ass.*, **63**, 817.

FANG, K.-T., KOTZ, S. and NG, K.-W. (1989). *Symmetric Multivariate and Related Distributions*. Chapman and Hall, London.

FAREBROTHER, R. W. (1990). The cumulants of the logarithm of a gamma variable. *J. Statist. Comp. Simul.*, **36**, 243.

FARLIE, D. J. G. (1960). The performance of some correlation coefficients for a general bivariate distribution. *Biometrika*, **47**, 307.

FEDERIGHI, E. T. (1959). Extended tables of the percentage points of Student's t-distribution. *J. Amer. Statist. Ass.*, **54**, 683.

FELLER, W. (1968, 1971). *An Introduction to the Theory of Probability and its Applications*. Vol. 1 (3rd edn) and Vol. 2 (2nd edn). Wiley, New York.

FEUERVERGER, A. and MCDUNNOUGH, P. (1981). On the efficiency of empirical characteristic function procedures. *J. R. Statist. Soc.*, **B**, **43**, 26.

FINNEY, D. J. (1938). The distribution of the ratio of estimates of the two variances in a sample from a normal bivariate population. *Biometrika*, **30**, 190.

FINUCAN, H. M. (1964). A note on kurtosis. *J. R. Statist. Soc.*, **B**, **26**, 111.

FISHER, R. A. (1915). Frequency-distribution of the values of the correlation coefficient in samples from an indefinitely large population. *Biometrika*, **10**, 507.

FISHER, R. A. (1920).* A mathematical examination of the methods of determining the accuracy of an observation by the mean error and by the mean square error. *Month. Not. R. Astr. Soc.*, **80**, 758.

FISHER, R. A. (1921a).* On the mathematical foundations of theoretical statistics. *Phil. Trans. Roy. Soc.*, **A**, **222**, 309.

FISHER, R. A. (1921b). On the probable error of a coefficient of correlation deduced from a small sample. *Metron*, **1**, No. 4, 1.

FISHER, R. A. (1929).* Moments and product moments of sampling distributions. *Proc. Lond. Math. Soc.*, (2), **30**, 199.

FISHER, R. A. (1930).* The moments of the distribution for normal samples of measures of departure from normality. *Proc. Roy. Soc.*, **A**, **130**, 16.

FISHER, R. A. (1935).* The mathematical distributions used in the common tests of significance. *Econometrica*, **3**, 353.

FISHER, R. A. (1939).* The sampling distribution of some statistics obtained from non-linear equations. *Ann. Eugen.*, **9**, 238.

FISHER, R. A. (1970) *Statistical Methods for Research Workers*. Oliver and Boyd, Edinburgh. (Fourteenth Edition, first published in 1925).

FISHER, R. A. and CORNISH, E. A. (1960). The percentile points of distributions having known cumulants. *Technometrics*, **2**, 209.

FISHER, R. A. and TIPPETT, L. H. C. (1928).* Limiting forms of the frequency-distribution of the largest or smallest member of a sample. *Proc. Camb. Phil. Soc.*, **24**, 180.

FISHER, R. A. and WISHART, J. (1931). The derivation of the pattern formulae of two-way partitions from those of simpler patterns. *Proc. Lond. Math. Soc.*, **33**, 195.

FISHER, R. A. and YATES, F. (1963). *Statistical Tables for Biological, Agricultural and Medical Research*. 6th edition. Oliver and Boyd, Edinburgh.

FISHER, R. A., CORBET, A. S. and WILLIAMS, C. B. (1943).* The relation between the number of species and the number of individuals. *J. Animal Ecology*, **12**, 42.

FISHMAN, G. S. (1979). Sampling from the binomial distribution on a computer. *J. Amer. Statist. Ass.*, **74**, 418.

FISHMAN, G. S. and MOORE, L. R. (1982). A statistical evaluation of the multiplicative congruential random number generators with modulus $2^{31}-1$. *J. Amer. Statist. Ass.*, **77**, 129.

FOLKS, J. L. and CHHIKARA, R. S. (1978). The Inverse Gaussian distribution and its statistical application—a review. *J. R. Statist. Soc.*, **B**, **40**, 263.

FORSYTHE, G. E. (1972). Von Neumann's comparison method for random sampling from the normal and other distributions. *Math. Comput.*, **26**, 817.

FOSTER, F. G. (1957, 1958). Upper percentage points of the generalized beta distribution. *Biometrika*, **44**, 441 and **45**, 492.

FOSTER, F. G. and REES, D. H. (1957). Upper percentage points of the generalized beta distribution. *Biometrika*, **44**, 237.

FOSTER, F. G. and STUART, A. (1954). Distribution-free tests in time-series based on the breaking of records. *J. R. Statist. Soc.*, **B**, **16**, 1.

FRAME, J. S. (1945). Mean deviation of the binomial distribution. *Amer. Math. Monthly*, **52**, 377.

FRÉCHET, M. (1927). Sur la loi de probabilité de l'écart maximum. *Annales de la Soc. Polonaise. de Math.*, **6**, 92.

FRÉCHET, M. (1937). *Recherches théoriques modernes.* (Part of the *Traité* edited by Borel.) Gauthier-Villars, Paris.

FRÉCHET, M. (1951). Sur les tableaux de corrélation dont les marges sont données. *Ann. Univ. Lyon*, **A**, **(3)**, **14**, 53.

FRENCH, S. (1980). Updating of belief in light of someone else's opinion. *J. R. Statist. Soc.*, **A**, **143**, 43.

FREUND, R. J. (1961). A bivariate extension of the exponential distribution. *J. Amer. Statist. Ass.*, **56**, 971.

FRISCH, R. (1925). Recurrence formulae for moments of the point binomial. *Biometrika*, **17**, 165.

FRISCH, R. (1926). Sur les semi-invariants et moments employés dans l'étude des distributions statistiques. Oslo, *Skrifter af det Norske Videnskaps Academie, II, Hist.-Filos. Klasse*, No. 3.

FU, J. C. (1984). The moments do not determine a distribution. *Amer. Statistician*, **38**, 294.

FUJIKOSHI, Y. (1977). An asymptotic expansion for the distributions of the latent roots of the Wishart matrix with multiple population roots. *Ann. Inst. Statist. Math.*, **29**, 379.

FUJIKOSHI, Y. (1980). Asymptotic expression for the distributions of the sample roots under non-normality. *Biometrika*, **67**, 45.

GALAMBOS, J. (1975). Order statistics of samples from multivariate distributions. *J. Amer. Statist. Ass.*, **70**, 674.

GALAMBOS, J. (1984). Order statistics. In *Handbook of Statistics, IV: Nonparametric Statistics* (P. R. Krishnaiah and P. K. Sen, eds), Elsevier, Amsterdam.

GALAMBOS, J. (1987). *The Asymptotic Theory of Extreme Order Statistics.* (Second Edition) Wiley, New York.

GALAMBOS, J. and KOTZ, S. (1978). *Characterizations of Probability Distributions.* Lecture Notes in Mathematics, No 675. Springer-Verlag, New York.

GARWOOD, F. (1936). Fiducial limits for the Poisson distribution. *Biometrika*, **28**, 437.

GASTWIRTH, J. L. (1974). Large sample theory of some measures of income inequality. *Econometrica*, **42**, 191.

GAVER, D. P. and KAFADAR, K. (1984). A retrievable recipe for inverse *t*. *Amer. Statistician*, **38**, 308.

GAYEN, A. K. (1951). The frequency distribution of the product-moment correlation coefficient in random samples of any size drawn from non-normal universes. *Biometrika*, **38**, 219.

GEARY, R. C. (1930). The frequency distribution of the quotient of two normal variables. *J. R. Statist. Soc.*, **93**, 442.

GEARY, R. C. (1936a). The distribution of "Student's" ratio for non-normal samples. *Supp. J. R. Statist. Soc.*, **3**, 178.

GEARY, R. C. (1936b). Moments of the ratio of the mean deviation to the standard deviation for normal samples. *Biometrika*, **28**, 295.

GEARY, R. C. (1944). Extension of a theorem by Harald Cramér on the frequency distribution of the quotient of two variables. *J. R. Statist. Soc.*, **107**, 56.

GEBHARDT, F. (1969). Some numerical comparisons of several approximations to the binomial distribution. *J. Amer. Statist. Ass.*, **64**, 1638.

GEISSER, S. (1964). Estimation in the uniform covariance case. *J. R. Statist. Soc.*, **B, 26**, 477.

GENEST, C. and MACKAY, R. J. (1986a). Copules Archimédiennes et familles de lois bidimensionelles dont les marges sont données. *Canad. J. Statist.*, **14**, 145.

GENEST, C. and MACKAY, R. J. (1986b). The joy of copulas: bivariate distributions with uniform marginals. *Amer. Statist.*, **40**, 280.

GEORGE, E. O. and SIVARAM, M. (1987). A modification of the Fisher–Cornish approximation for the student *t* percentiles. *Commun. Statist.*, **B16**, 1123.

GHOSH, B. K. (1966). Asymptotic expansions for the moments of the distribution of correlation coefficient. *Biometrika*, **53**, 258.

GHOSH, B. K. (1975). On the distribution of the difference of two *t*-variables. *J. Amer. Statist. Ass.*, **70**, 463.

GHOSH, B. K. (1980). Two normal approximations to the binomial distribution. *Commun. Statist.*, **A9**, 427.

GIDEON, R. A. and GURLAND, J. (1976). Series expansions for quadratic forms in normal variables. *J. Amer. Statist. Ass.*, **71**, 227.

GINI, C. (1912). Variabilità e Mutabilità, contributo allo studio delle distribuzioni e relazioni statistiche. *Studi Economico-Giuridici dell' Univ. di Cagliari*, **3**, part 2, 1–158.

GIRSHICK, M. A. (1939). On the sampling theory of the roots of determinantal equations. *Ann. Math. Statist.*, **10**, 203.

GLASER, R. E. (1976). The ratio of the geometric mean to the arithmetic mean for a random sample from a Gamma distribution. *J. Amer. Statist. Ass.*, **71**, 480.

GNEDENKO, B. V. (1943). Sur la distribution limite du terme maximum d'une série aléatoire. *Ann. Math.*, (2), **44**, 423.
GODWIN, H. J. (1945). On the distribution of the estimate of mean deviation obtained from samples from a normal population. *Biometrika*, **33**, 254.
GODWIN, H. J. (1948). A further note on the mean deviation. *Biometrika*, **35**, 304.
GODWIN, H. J. (1955). Generalisations of Tchebycheff's inequality. *J. Amer. Statist. Ass.*, **50**, 923.
GODWIN, H. J. (1964). *Inequalities on Distribution Functions*. Griffin, London.
GOOD, I. J. (1963, 1966). On the independence of quadratic expressions. *J. R. Statist. Soc.*, **B**, **25**, 377; corrigenda, **28**, 584.
GOOD, I. J. and GASKINS, R. A. (1971). Nonparametric roughness penalties for probability densities. *Biometrika*, **58**, 255.
GOODMAN, L. A. (1952). On the analysis of samples from *k* lists. *Ann. Math. Statist.*, **23**, 632.
GOODMAN, L. A. (1960). On the exact variance of products. *J. Amer. Statist. Ass.*, **55**, 708.
GOODMAN, L. A. (1962). The variance of the product of *K* random variables. *J. Amer. Statist. Ass.*, **57**, 54.
GOVINDARAJULU, Z. (1963). On moments of order statistics and quasi-ranges from normal populations. *Ann. Math. Statist.*, **34**, 633.
GOWER, J. C. (1974). The mediancentre. *Appl. Statist.*, **23**, 466.
GRAY, H. L. and ODELL, P. L. (1966). On sums and products of rectangular variates. *Biometrika*, **53**, 615.
GRAY, H. L., COBERLY, W. A. and LEWIS, T. O. (1975). On Edgeworth expansions with unknown cumulants. *Ann. Statist.*, **3**, 741.
GREENWOOD, M. and YULE, G. U. (1920).* An inquiry into the nature of frequency-distributions of multiple happenings, etc. *J. R. Statist. Soc.*, **83**, 255.
GRIFFITHS, R. C. and MILNE, R. K. (1987). A class of infinitely divisible multivariate negative binomial distributions. *J. Multiv. Anal.*, **22**, 13.
GRIFFITHS, R. C., MILNE, R. K. and WOOD, R. (1979). Aspects of correlation in bivariate Poisson distributions and processes. *Austral. J. Statist.*, **21**, 238.
GROENEVELD, R. A. and MEEDEN, G. (1984). Measuring skewness and kurtosis. *Statistician*, **33**, 391.
GULDBERG, S. (1935). Recurrence formulae for the semi-invariants of some discontinuous frequency distributions of *n* variables. *Skand. Aktuartidskr.*, **18**, 270.
GUMBEL, E. J. (1934). Les valeurs extrêmes des distributions statistiques. *Ann. Inst. H. Poincaré*, **5**, 115.
GUMBEL, E. J. (1947). The distribution of the range. *Ann. Math. Statist.*, **18**, 384.
GUMBEL, E. J. (1949). Probability tables for the range. *Biometrika*, **36**, 142.
GUMBEL, E. J. (1954). *Statistical Theory of Extreme Values and some Practical Applications.* National Bureau of Standards, A.M.S. 33.
GUMBEL, E. J. (1958). *Statistics of Extremes.* Columbia Univ. Press, New York.
GUMBEL, E. J. (1960). Bivariate exponential distributions. *J. Amer. Statist. Ass.*, **55**, 698.
GUPTA, S. S. (1961). Percentage points and modes of order statistics from the normal distribution. *Ann. Math. Statist.*, **32**, 888.
GUPTA, S. S. (1963a). Probability integrals of multivariate normal and multivariate *t*. *Ann. Math. Statist.*, **34**, 792.
GUPTA, S. S. (1963b). Bibliography on the multivariate normal integrals and related topics. *Ann. Math. Statist.*, **34**, 829.
GUPTA, S. S. and HUANG, W.-T. (1981). On mixtures of distributions: A survey and some new results on ranking and selection. *Sankhyā*, **B**, **43**, 245.
GUPTA, S. S. and PANCHAPAKESAN, S. (1974). On moments of order statistics from independent binomial populations. *Ann. Inst. Statist. Math.*, **8**, 95.

GUPTA, S. S. and SHAH, B. K. (1965). Exact moments and percentage points of the order statistics and the distribution of the range from the logistic distribution. *Ann. Math. Statist.*, **36**, 907.

GUPTA, S. S., NAGEL, K. and PANCHAPAKESAN, S. (1973). On the order statistics from equally correlated normal random variables. *Biometrika*, **60**, 403.

GURLAND, J. (1968). A relatively simple form of the distribution of the multiple correlation coefficient. *J. R. Statist. Soc.*, **B**, **30**, 276.

GURLAND, J. and MILTON, R. (1970). Further consideration of the distribution of the multiple correlation coefficient. *J. R. Statist. Soc.*, **B**, **32**, 381.

HALD, A. (1989). *A History of Probability and Statistics and their Applications before 1750.* Wiley, New York and London.

HALDANE, J. B. S. (1938). The approximate normalization of a class of frequency distributions. *Biometrika*, **29**, 392.

HALDANE, J. B. S. (1939). Cumulants and moments of the binomial distribution. *Biometrika*, **31**, 392.

HALDANE, J. B. S. (1942). Mode and median of a nearly normal distribution with given cumulants. *Biometrika*, **32**, 294.

HALDANE, J. B. S. (1948). Note on the median of a multivariate distribution. *Biometrika*, **35**, 414.

HALDANE, J. B. S. and JAYAKAR, S. D. (1963). The distribution of extremal and nearly extremal values in samples from a normal distribution. *Biometrika*, **50**, 89.

HALL, P. (1927). The distribution of means for samples of size N drawn from a population in which the variate takes values between 0 and 1, all such values being equally probable. *Biometrika*, **19**, 240.

HALL, P. (1978). Some expansions of moments of order statistics. *Stoch. Proc.*, **7**, 265.

HALL, P. (1982). Limit theorems for estimators based on inverses of spacings of order statistics. *Ann. Prob.*, **10**, 992.

HALL, P. (1983). Inverting an Edgeworth expansion. *Ann. Statist.*, **11**, 569.

HALL, P. (1988). Theoretical comparison of bootstrap confidence intervals. *Ann. Statist.*, **16**, 927.

HALL, P. (1990). On the relative performance of bootstrap and Edgeworth expansions of a distribution function. *J. Multiv. Anal.*, **35**, 108.

HALL, P. (1992). *The Bootstrap and Edgeworth Expansion*. Springer-Verlag, New York and Berlin.

HAMBURGER, H. (1920, 1921). Über eine Erweiterung des Stieltjesschen Moment-problems. *Math. Ann.*, **81**, 235; **82**, 120 and 168.

HAMDAN, M. A. and JENSEN, D. R. (1976). A bivariate binomial distribution and some applications. *Austral. J. Statist.*, **18**, 163.

HAMPEL, F. R. (1974). The influence curve and its role in robust estimation. *J. Amer. Statist. Ass.*, **69**, 383.

HANNAN, J. F. (1985). Sufficient and necessary conditions for finite negative moments. *Amer. Statist.*, **39**, 326.

HARRIS, B. and SOMS, A. P. (1980). The use of tetrachoric series for evaluating multivariate normal probabilities. *J. Mult. Anal.*, **10**, 252.

HARTER, H. L. (1960). Tables of range and studentized range. *Ann. Math. Statist.*, **31**, 1122.

HARTER, H. L. (1961). Expected values of normal order statistics. *Biometrika*, **48**, 151, 476.

HARTER, H. L. (1964a). *New Tables of the Incomplete Gamma-Function Ratio and of Percentage Points of the Chi-square and Beta Distributions.* Aerospace Research Laboratories, Office of Aerospace Research, U.S.A., Report 64–123, Wright-Patterson A.F.B., Ohio.

HARTER, H. L. (1964b). A new table of percentage points of the chi-square distribution. *Biometrika*, **51**, 231.

HARTER, H. L. (1969). *Order Statistics and their Use in Testing and Estimation.* Volume 1, *Tests Based on Range and Studentized Range of Samples from a Normal Population.* Volume 2,

Estimates Based on Order Statistics of Samples from Various Populations. Aerospace Research Laboratories, U.S. Air Force; U.S. Government Printing Office, Washington.

HARTER, H. L. (1978, 1983, 1991a,b). *The Chronological Annotated Bibliography of Order Statistics.* Vol. 1: Pre-1950, Vol. 2: 1950-59, Vol. 3: 1960-61, Vol. 4: 1962-63. Amer. Scientific Press, Ohio.

HARTLEY, H. O. (1945). Note on the calculation of the distribution of the estimate of mean deviation in normal samples. Tables of the probability integral of the mean deviation in normal samples. *Biometrika,* **33,** 257.

HARTLEY, H. O. and DAVID, H. A. (1954). Universal bounds for mean range and extreme observation. *Ann. Math. Statist.,* **25,** 85.

HASTINGS, C. Jr. (1955). *Approximations for Digital Computers.* Princeton U.P. and Oxford U.P.

HASTINGS, C., MOSTELLER, F., TUKEY, J. W. and WINSOR, C. P. (1947). Low moments for small samples: a comparative study of order statistics. *Ann. Math. Statist.* **18,** 413.

HATKE, M. A. (1949). A certain cumulative probability function. *Ann. Math. Statist.,* **20,** 461.

HAWKINS, D. M. and WIXLEY, R. A. J. (1986). A note on the transformation of chi-squared variables to normality. *Amer. Statist.,* **40,** 296.

HELMERT, F. R. (1876). Die Genauigkeit der Formel von Peters zur Berechnung des wahrscheinlichen Beobachtungsfehlers direkter Beobachtungen gleicher Genauigkeit. *Astronomische Nachrichten,* **88,** No. 2096, 113.

HENERY, R. J. (1981). An approximation to certain multivariate normal probabilities. *J. R. Statist. Soc.,* **B, 43,** 81.

HEYDE, C. C. (1963). On a property of the lognormal distribution. *J. R. Statist. Soc.,* **B, 25,** 392.

HILL, B. M. (1963). The three parameter lognormal distribution and Bayesian analysis of a point source epidemic. *J. Amer. Statist. Ass.,* **58,** 72.

HILL, G. W. (1976). New approximations to the von Mises distribution. *Biometrika,* **63,** 673.

HILL, G. W. and DAVIS, A. W. (1968). Generalised asymptotic expansions of Cornish-Fisher type. *Ann. Math. Statist.,* **39,** 1264.

HILL, I. D., HILL, R. and HOLDER, R. L. (1976). Algorithm AS99. Fitting Johnson curves by moments. *Appl. Statist.,* **25,** 180.

HINKLEY, D. V. (1969). On the ratio of two correlated normal random variables. *Biometrika,* **56,** 635; correction, **57,** 683.

HODGES, J. L., Jr and LEHMANN, E. L. (1967). Moments of chi and power of *t*. *Proc. 5th Berkeley Symp. Math. Statist. and Prob.,* **1,** 187.

HODGES, J. S. (1987). Assessing the accuracy of normal approximations. *J. Amer. Statist. Ass.,* **82,** 149.

HOEL, P. G. (1937). A significance test for component analysis. *Ann. Math. Statist.,* **8,** 149.

HOGG, R. V. and CRAIG, A. T. (1958). On the decomposition of certain variables. *Ann. Math. Statist.,* **29,** 608.

HOJO, T. (1931, 1933). Distribution of the median, quartiles and interquartile distance in samples from a normal population. *Biometrika,* **23,** 315; and: A further note on the relation between the median and the quartiles in small samples from a normal population. *Biometrika,* **25,** 79.

HOLGATE, P. (1966). Bivariate generalizations of Neyman's Type A distribution. *Biometrika,* **53,** 241.

HOLGATE, P. (1970). The modality of some compound Poisson distributions. *Biometrika,* **57,** 666.

HOLGATE, P. (1989). The lognormal characteristic function. *Commun. Statist.,* **A, 18,** 4539.

HOLST, L. (1979). Asymptotic normality of sum-functions of spacings. *Ann. Prob.,* **7,** 1066.

HOLT, D. R. and CROW, E. L. (1973). Tables and graphs of the stable probability density functions. *J. Research. Nat. Bureau Standards,* **77B,** 143.

HOTELLING, H. (1931). The generalization of Student's ratio. *Ann. Math. Statist.,* **2,** 360.

Hotelling, H. (1953). New light on the correlation coefficient and its transforms. *J. R. Statist. Soc.* **B, 15,** 193.

Hotelling, H. and Solomons, L. M. (1932). The limits of a measure of skewness. *Ann. Math. Statist.*, **3,** 141.

Hsu, C. T. and Lawley, D. N. (1939). The derivation of the fifth and sixth moments of b_2 in samples from a normal population. *Biometrika*, **31,** 238.

Hsu, P. L. (1939). On the distribution of roots of certain determinantal equations. *Ann. Eugen.*, **9,** 250.

Huang, J. S. and Kotz, S. (1984). Correlation structure in iterated Farlie-Gumbel-Morgenstern distributions. *Biometrika*, **71,** 633.

Huber, P. J. (1981). *Robust Statistics.* Wiley, New York.

Ibragimov, I. A. and Linnik, Y. V. (1971). *Independent and Stationary Sequences of Random Variables.* Wolters–Noordhoff, Groningen.

Inoueh, H., Kumahora, H. and Yoshizawa, Y. (1983). Random numbers generated by a physical device. *Appl. Statist.*, **32,** 115.

Irwin, J. O. (1927). On the frequency-distribution of the means of samples from a population having any law of frequency with finite moments, etc. *Biometrika*, **19,** 225; correction, **21,** 431.

Irwin, J. O. (1937). The frequency-distribution of the difference between two independent variates following the same Poisson distribution. *J. R. Statist. Soc.*, **100,** 415.

Irwin, J. O. and Kendall, M. G. (1944).* Sampling moments of moments for a finite population. *Ann. Eugen.*, **12,** 138.

Jackson, D. (1921). Note on the median of a set of numbers. *Bull. Amer. Math. Soc.*, **27,** 160.

Jacobson, H. I. (1969). The maximum variance of restricted unimodal distributions. *Ann. Math. Statist.*, **40,** 1746.

Jaeckel, L. A. (1972). Estimating regression coefficients by minimizing the dispersion of the residuals. *Ann. Math. Statist.*, **43,** 1449.

James, A. T. (1964). Distributions of matrix variates and latent roots derived from normal samples. *Ann. Math. Statist.*, **35,** 475.

James, G. S. (1952). Notes on a theorem of Cochran. *Proc. Camb. Phil. Soc.*, **48,** 443.

James, G. S. (1955). Cumulants of a transformed variate. *Biometrika*, **42,** 529.

James, G. S. (1958). On moments and cumulants of systems of statistics. *Sankhyā*, **20,** 1.

James, G. S. and Mayne, A. J. (1962). Cumulants of functions of random variables. *Sankhyā*, **A, 24,** 47.

James, I. R. (1972). Products of independent Beta variables. . . . *J. Amer. Statist. Ass.*, **67,** 910.

James, I. R. (1975). Multivariate distributions which have beta conditional distributions. *J. Amer. Statist. Ass.*, **70,** 681.

Jayachandran, T. and Barr, D. R. (1970). On the distribution of the difference of two scaled chi-square random variables. *Amer. Statistician*, **24,** (5), 29.

Jeffreys, H. (1961). *Theory of Probability*, 3rd edn, Oxford U.P.

Jenkinson, A. F. (1955). The frequency-distribution of the annual maximum (or minimum) values of meteorological elements. *Quart. J. R. Met. Soc.*, **81,** 158.

Jensen, D. R. and Solomon, H. (1972). A Gaussian approximation to the distribution of a definite quadratic form. *J. Amer. Statist. Ass.*, **67,** 898.

John, J. A. and Draper, N. R. (1980). An alternative family of transformations. *Appl. Statist.*, **29,** 190.

Johnson, M. E. (1987). *Multivariate Statistical Simulations.* Wiley, New York.

Johnson, M. E. and Lowe, V. W. (1979). Bounds on the sample skewness and kurtosis. *Technometrics*, **21,** 377.

JOHNSON, N. L. (1949a). Systems of frequency curves generated by methods of translation. *Biometrika,* **36**, 149.

JOHNSON, N. L. (1949b). Bivariate distributions based on simple translation systems. *Biometrika*, **36**, 297.

JOHNSON, N. L. (1954). Systems of frequency curves derived from the first law of Laplace. *Trabajos Estadistica*, **5**, 283.

JOHNSON, N. L. (1957). A note on the mean deviation of the binomial distribution. *Biometrika*, **44**, 532; corrigenda, **45**, 587.

JOHNSON, N. L. (1965). Tables to facilitate fitting S_U frequency curves. *Biometrika*, **52**, 547; extensions and corrections, **61**, 203.

JOHNSON, N. L. and KITCHEN, J. O. (1971a, b). Tables to facilitate fitting S_B curves: some notes on *Biometrika*, **58**, 223; II, Both terminals known, *Biometrika*, **58**, 657.

JOHNSON, N. L. and KOTZ, S. (1967a, b, 1968). Tables of distributions of quadratic forms in central normal variables, I, II. *Institute of Statistics Mimeo Series Nos. 534, 547, University of N. Carolina*. Tables of distributions of positive definite quadratic forms in central normal variables. *Sankhyā*, **B**, **30**, 303.

JOHNSON, N. L. and KOTZ, S. (1969, 1970a, b, 1972). *Distributions in Statistics. I: Discrete Distributions; II, III: Continuous Univariate Distributions, 1, 2; IV: Continuous Multivariate Distributions*. Houghton Mifflin, New York.

JOHNSON, N. L. and KOTZ, S. (1975a). A vector multivariate hazard rate. *J. Multiv. Anal.*, **5**, 53.

JOHNSON, N. L. and KOTZ, S. (1975b, 1977a). On some generalized Farlie-Gumbel-Morgenstern distributions, I, II. *Commun. Statist.*, **A4**, 415 and **A6**, 485.

JOHNSON, N. L. and KOTZ, S. (1977b). *Urn Models and their Applications, an Approach to Modern Discrete Probability Theory*. Wiley, New York.

JOHNSON, N. L. and KOTZ, S. (1981). Dependent revelations: time to failure under dependence. *Amer. J. Math. Management Sci.*, **1**, 155.

JOHNSON, N. L. and KOTZ, S. (1982). Developments in discrete distributions, 1969–80. *Int. Statist. Rev.*, **50**, 70.

JOHNSON, N. L., KOTZ, S. and KEMP, A. W. (1993). *Univariate Discrete Distributions*. (2nd edition) Wiley, New York.

JOHNSON, N. L. and ROGERS, C. A. (1951). The moment problem for unimodal distributions. *Ann. Math. Statist.*, **22**, 433.

JOHNSON, N. L., NIXON, E., AMOS, D. E. and PEARSON, E. S. (1963). Table of percentage points of Pearson curves, for given $\sqrt{\beta_1}$ and β_2, expressed in standard measure. *Biometrika*, **50**, 459.

JOHNSON, R. A., LADALLA, J. and LIU, S. T. (1979). Differential relations, in the original parameters, which determine the first two moments of the multiparameter exponential family. *Ann. Statist.*, **7**, 232.

JOSHI, P. C. (1969). Bounds and approximations for the moments of order statistics. *J. Amer. Statist. Ass.*, **64**, 1617.

JOSHI, P. C. (1973). Two identities involving order statistics. *Biometrika*, **60**, 428.

KAFADAR, K. and TUKEY, J. W. (1988). A bidec *t* table. *J. Amer. Statist. Ass.*, **83**, 532.

KAGAN, A. M., LINNIK, Yu. V. and RAO, C. R. (1973). *Characterization Problems in Mathematical Statistics*. Wiley, New York.

KAIGH, W. D. and CHENG, C. (1991). Subsampling quantile estimator standard errors with applications. *Commun. Statist.*, **A20**, 977.

KAMAT, A. R. (1959). Contributions to the theory of Gini's mean difference. *Volume in onore di Corrado Gini* (Istituto di Statistica, Roma), **1**, 285.

KAMAT, A. R. (1965). A property of the mean deviation for a class of continuous distributions. *Biometrika*, **52**, 288.

KAPLAN, E. L. (1952). Tensor notation and the sampling cumulants of *k*-statistics. *Biometrika*, **39**, 319.

KAPLANSKY, I. (1945). A common error concerning kurtosis. *J. Amer. Statist. Ass.*, **40**, 259.

KATSNELSON, J. and KOTZ, S. (1957). On the upper limits of some measures of variability. *Archiv. f. Meteor., Geophys. u. Bioklimat.*, **(B)**, **8**, 103.

KAWATA, T. and SAKAMOTO, H. (1949). On the characterisation of the normal population by the independence of the sample mean and the sample variance. *J. Math. Soc. Japan*, **1**, 111.

KEMP, A. W. (1981a). Efficient generation of logarithmically distributed pseudo-random variables. *Appl. Statist.*, **30**, 249.

KEMP, A. W. (1981b). Frugal methods for generating bivariate discrete random variables. *In* Taillie *et al.* (1981), **4**, 321.

KEMP, A. W. and KEMP, C. D. (1991). Weldon's dice data revisited. *Amer. Statist.*, **45**, 216.

KEMP, C. D. and KEMP, A. W. (1956). Generalized hypergeometric distributions. *J. R. Statist. Soc.*, **B**, **18**, 202.

KEMP, C. D. and KEMP, A. W. (1965). Some properties of the Hermite distribution. *Biometrika*, **52**, 381.

KEMP, C. D. and KEMP, A. W. (1987). Rapid generation of frequency tables. *Appl. Statist.*, **36**, 277.

KEMP, C. D. and LOUKAS, S. (1978). The computer generation of bivariate random variables. *J. Roy. Statist. Soc.*, **A**, **141**, 513.

KEMP, C. D. and PAPAGEORGIOU, H. (1982). Bivariate Hermite distributions. *Sankhyā*, **A44**, 269.

KEMPERMAN, J. H. B. (1969). On the optimum rate of transmitting information. *Prob. & Info. Theory Series, Springer Lecture Notes in Math*, **89**, 126.

KENDALL, D. G. (1984). Shape manifolds, procrustean metrics and complex projective spaces. *Bull. Lond. Math. Soc.*, **16**, 81.

KENDALL, D. G. (1991). The Mardia-Dryden shape distribution for triangles: a stochastic calculus approach. *J. Appl. Prob.*, **28**, 225.

KENDALL, D. G. and RAO, K. S. (1950). On the generalized second limit theorem in the theory of probabilities. *Biometrika*, **37**, 224.

KENDALL, M. G. (1938). The conditions under which Sheppard's corrections are valid. *J. R. Statist. Soc.*, **101**, 592.

KENDALL, M. G. (1940a, b, c).* Some properties of *k*-statistics. *Ann. Eugen.*, **10**, 106. Proof of Fisher's rules for ascertaining the sampling semi-invariants of *k*-statistics. *ibid.*, **10**, 215. The derivation of multivariate sampling formulae from univariate formulae by symbolic operation. *Ibid.*, **10**, 392.

KENDALL, M. G. (1941). Relations connected with the tetrachoric series and its generalisation. *Biometrika*, **32**, 196.

KENDALL, M. G. (1942). * On seminvariant statistics. *Ann. Eugen.*, **11**, 300.

KENDALL, M. G. (1949a). Reconciliation of theories of probability. *Biometrika*, **36**, 101.

KENDALL, M. G. (1949b). Rank and product-moment correlation. *Biometrika*, **36**, 177.

KENDALL, M. G. (1952).* Moment-statistics in samples from a finite population. *Biometrika*, **39**, 14.

KENDALL, M. G. (1953). Discussion of Hotelling (1953).

KENDALL, M. G. (1961). *A Course in the Geometry of n Dimensions*. Griffin, London.

KENDALL, M. G. and GIBBONS, J. D. (1990). *Rank Correlation Methods*. 5th edition. Edward Arnold, London.

KENDALL, M. G. and ORD, J. K. (1990). *Time Series*. (3rd edition). Edward Arnold, London.

KENDALL, M. G. and PLACKETT, R. L. (eds) (1977). *Studies in the History of Statistics and Probability*, Vol. 2. Griffin, London.

KENDALL, M. G. and SMITH, B. B. (1938–9). Randomness and random sampling numbers. *J. R. Statist. Soc.*, **101**, 147, and *Supp. J. R. Statist. Soc.*, **6**, 51.

Kennedy, J. E. and Quine, M. P. (1989). The total variation distance between the binomial and Poisson distributions. *Ann. Probab.*, **17**, 396.

Kennedy, W. J. and Gentle, J. E. (1980). *Statistical Computing*. Dekker, New York.

Kerridge, D. F. and Cook, G. W. (1976). Yet another series for the normal integral. *Biometrika*, **63**, 401.

Keynes, J. M. (1921). *A Treatise on Probability*. Macmillan, London.

Khamis, S. H. and Rudert, W. (1965). *Tables of the Incomplete Gamma Function Ratio: Chi-square Integral, Poisson Distribution*. Justus von Liebig, Darmstadt, Germany.

Khatri, C. G. (1959). On certain properties of power series distributions. *Biometrika*, **46**, 486.

Khatri, C. G. (1977). Quadratic forms and extension of Cochran's theorem to normal vector variables. In *Multivariate Analysis, IV* (P. R. Krishnaiah, ed.), North-Holland, Amsterdam, p. 79.

Khatri, C. G. (1978). A remark on the necessary and sufficient conditions for a quadratic form to be distributed as chi-squared. *Biometrika*, **65**, 239.

Khatri, C. G. and Srivastava, M. S. (1978). Asymptotic expansions for distributions of characteristic roots of covariances. *S. African J. Statist.*, **12**, 161.

Kingman, J. F. C. and Taylor, S. J. (1966). *Introduction to Measure Theory and Probability*. Cambridge University Press, London.

Knuth, D. E. (1969). *The Art of Computer Programming. Vol 2: Seminumerical Algorithms*. Addison-Wesley, Reading, Mass.

Kocherlakota, S. (1989). A note on the bivariate binomial distribution. *Statist. Prob. Letters*, **8**, 21.

Kocherlakota, S. and Kocherlakota, K. (1992). *Bivariate Discrete Distributions*. Dekker, New York.

Kocherlakota, S. and Singh, M. (1982). On the behaviour of some transforms of the sample correlation coefficient in samples from the bivariate t and the bivariate χ^2 distribution. *Commun. Statist.*, **A**, **11**, 2045.

Kolassa, J. E. and McCullagh, P. (1990). Edgeworth series for lattice distributions. *Ann. Statist.*, **18**, 981.

Kolmogorov, A. N. (1933). Grundbegriffe der Wahrscheinlichkeitsrechnung, *Ergebnisse der Math.*, **2**, no. 3.

Konishi, S. (1978a). An approximation to the distribution of the sample correlation coefficient. *Biometrika*, **65**, 654.

Konishi, S. (1978b). Asymptotic expansions for the distributions of statistics based on a correlation matrix. *Canad. J. Statist.*, **6**, 49.

Konishi, S. and Sugiyama, T. (1981). Improved approximations to distributions of the largest and smallest latent roots of a Wishart matrix. *Ann. Inst. Statist. Math.*, **33**, 27.

Kotlarski, I. (1962). On groups of n independent random variables whose product follows the beta distribution. *Colloq. Math.*, **9**, 325.

Kotlarski, I. (1964). On bivariate random variables where the quotient of their coordinates follows some known distribution. *Ann. Math. Statist.*, **35**, 1673.

Kotlarski, I. (1965). On pairs of independent random variables whose product follows the Gamma distribution. *Biometrika*, **52**, 289.

Kotz, S. and Shanbhag, D. N. (1980). Some new approaches to probability distributions. *Adv. Appl. Prob.*, **12**, 903.

Kotz, S., Johnson, N. L. and Boyd, D. W. (1967a). Series representations of distributions of quadratic forms in normal variables. I. Central Case. *Ann. Math. Statist.* **38**, 823.

Kotz, S., Johnson, N. L. and Boyd, D. W. (1967b). Series representations of distributions of quadratic forms in normal variables. II. Non-central case. *Ann. Math. Statist.*, **38**, 838.

Koutrouvelis, I. A. (1982). Estimation of location and scale in Cauchy distributions using the empirical characteristic function. *Biometrika*, **69**, 205.

Kraemer, H. C. (1973). Improved approximation to the non-null distribution of the correlation coefficient. *J. Amer. Statist. Ass.*, **68**, 1004.

Kratky, J., Reinfelds, J., Hutcheson, K. and Shenton, L. R. (1972). *Tables of Crude Moments Expressed in Terms of Cumulants.* Computer Center Report No. 7201, University of Georgia, Athens, Georgia.

Krieger, A. M. L. (1979). Bounding moments, the Gini index and the Lorenz curve from grouped data for unimodal density functions. *J. Amer. Statist. Ass.*, **74**, 375.

Krishnaiah, P. R. (1980). Computations of some multivariate distributions. In *Handbook of Statistics, I: Analysis of Variance* (P. R. Krishnaiah, ed.) North-Holland, Amsterdam, p. 745.

Krishnaiah, P. R. and Lee, J. C. (1979). On the asymptotic joint distributions of certain functions of the eigenvalues of four random matrices. *J. Multiv. Anal.*, **9**, 248.

Kruger, L., Daston, L. J. and Heidelberger, M. (1987, eds.). *The Probabilistic Revolution*, 2 Volumes. M. I. T. Press, Cambridge, Mass. and London.

Kshirsagar, A. M. (1959). Bartlett decomposition and Wishart distribution. *Ann. Math. Statist.*, **30**, 239.

Kshirsagar, A. M. (1961). The non-central multivariate beta distribution. *Ann. Math. Statist.*, **32**, 104.

Krishna Iyer, P. V. and Sinha, P. S. (1967). *Tables of Bivariate Random Normal Deviates.* Defence Science Laboratory, Delhi, India.

Kullback, S. (1934). An application of characteristic functions to the distribution problem of statistics. *Ann. Math. Statist.*, **5**, 264.

Kyburg, H. E. and Smokler, H. E. (eds) (1964). *Studies in Subjective Probability.* Wiley, New York.

Lackritz, J. (1983). Exact *p*-values for chi-squared tests. *Proc. Sec. Statist. Educ., Amer. Statist. Ass.*, 130.

Lackritz, J. (1984). Exact *p*-values for *F* and *t* tests. *Amer. Statistician*, **38**, 312.

Laha, R. G. (1954). On a characterisation of the gamma distribution. *Ann. Math. Statist.*, **25**, 784.

Laha, R. G. (1959). On the laws of Cauchy and Gauss. *Ann. Math. Statist.*, **30**, 1165.

Lancaster, H. O. (1954). Traces and cumulants of quadratic forms in normal variables. *J. R. Statist. Soc.*, **B**, **16**, 247.

Lancaster, H. O. (1960). The characterisation of the normal distribution. *J. Austral. Math. Soc.*, **1**, 368.

Laue, G. (1980). Remarks on the relation between fractional moments and fractional derivatives of characteristic functions. *J. Appl. Prob.*, **17**, 456.

Lawrence, M. J. (1975). Inequalities on *s*-ordered distributions. *Ann. Statist.*, **3**, 413.

LeCam, L. (1986). The central limit theorem around 1935. *Statist. Science*, **1**, 78.

Lee, Y. S. (1972). Tables of upper percentage points of the multiple correlation coefficient. *Biometrika*, **59**, 175.

Leipnik, R. (1981). The lognormal distribution and strong non-uniqueness of the moment problem. *Theory Prob. Applic.*, **26**, 850.

Lempers, F. B. and Louter, A. S. (1971). An extension of the table of the Student distribution. *J. Amer. Statist. Ass.*, **66**, 503.

Lentner, M. M. and Buehler, R. J. (1963). Some inferences about gamma parameters with an application to a reliability problem. *J. Amer. Statist. Ass.*, **58**, 670.

Leonard, T. (1978). Density estimation, stochastic processes and prior information. *J. R. Statist. Soc.*, **B**, **40**, 113.

Lévy, P. (1925). *Calcul des Probabilités.* Gauthier-Villars, Paris.

Lewis, T. G. and Payne, W. H. (1973). Generalized feedback shift register pseudorandom number generators. *J. Assoc. Comp. Mach.*, **20**, 456.

LIAPUNOV, A. M. (1900). Sur une proposition de la théorie des probabilités. *Bull. Acad. Imp. Sci. St Petersbg.*, (5), **13**, (4), 359.

LIAPUNOV, A. M. (1901). Nouvelle forme du théorème sur la limite de probabilité. *Mem. Acad. Imp. Sci. St. Pétersbg.*, (8), **12**, (5), 1.

LIEBERMAN, G. J. and OWEN, D. B. (1961). *Tables of the Hypergeometric Probability Distribution*. Stanford U.P.

LINDEBERG, J. W. (1922). Eine neue Herleitung des Exponentialgestezes in der Wahrscheinlichkeitsrechnung. *Math. Zeitschrift*, **15**, 211.

LINDLEY, D. V., TVERSKY, A. and BROWN, R. V. (1979). On the reconciliation of probability assessments. *J. R. Statist. Soc.*, **A**, **142**, 146.

LINDSAY, B. G. (1989). Moment matrices: Applications in mixtures. *Ann. Statist.*, **17**, 722.

LINDSAY, B. G. and BASAK, P. (1993). Multivariate normal mixtures: A fast consistent method of moments. *J. Amer. Statist. Ass.*, **88**, 468.

LING, R. F. (1978). A study of the accuracy of some approximations for t, χ^2 and F tail probabilities. *J. Amer. Statist. Ass.*, **73**, 274.

LING, R. F. and PRATT, J. W. (1984). The accuracy of Peizer approximations to the hypergeometric distribution, with comparisons to some other approximations. *J. Amer. Statist. Ass.*, **79**, 49.

LOMNICKI, Z. A. (1952). The standard error of Gini's mean difference. *Ann. Math. Statist.*, **23**, 635.

LOMNICKI, Z. A. (1967). On the distribution of products of random variables. *J. R. Statist. Soc.*, **B**, **29**, 513.

LUGANNANI, R. and RICE, S. O. (1980). Saddlepoint approximations for the distribution of the sum of independent random variables. *Adv. Appl. Prob.*, **12**, 475.

LUKACS, E. (1942). A characterization of the normal distribution. *Ann. Math. Statist.*, **13**, 91.

LUKACS, E. (1952). An essential property of the Fourier transforms of distribution functions. *Proc. Amer. Math. Soc.*, **3**, 508.

LUKACS, E. (1956). Characterization of populations by properties of suitable statistics. *Proc. 3rd Berkeley Symp. Math. Statist. and Prob.*, **2**, 195.

LUKACS, E. (1970). *Characteristic Functions*, 2nd edn. Griffin, London.

LUKACS, E. (1983). *Developments in Characteristic Function Theory*. Griffin, London.

LUKACS, E. and LAHA, R. G. (1964). *Applications of Characteristic Functions*. Griffin, London.

LUKACS, E. and SZÁSZ, O. (1952). On analytic characteristic functions. *Pacific J. Math.*, **2**, 615.

MCCORD, J. R. (1964). On asymptotic moments of extreme statistics. *Ann. Math. Statist.*, **35**, 1738.

MCCULLAGH, P. (1984). Tensor notation and cumulants of polynomials. *Biometrika*, **71**, 461.

MCCULLAGH, P. and PREGIBON, D. (1987). *K*-statistics and dispersion effects in regression. *Ann. Statist.*, **15**, 202.

MACGILLIVRAY, H. L. (1981). The mean, median, mode inequality and skewness for a class of densities. *Austral. J. Statist.*, **23**, 247.

MACGILLIVRAY, H. (1986). Skewness and asymmetry: measures and orderings. *Ann. Statist.*, **14**, 994.

MACGILLVRAY, H. and BALANDA, K. P. (1988). Mixtures, myths and kurtosis. *Commun. Statist.*, **B**, **17**, 789.

MCKAY, A. T. (1935). The distribution of the difference between the extreme observation and the sample mean in samples of n from a normal universe. *Biometrika*, **27**, 466.

MACLAREN, H. D. and MARSAGLIA, G. (1965). Uniform random number generators. *J. Ass. Comput. Mach.*, **12**, 83.

MAGE, D. T. (1980). An explicit solution for S_B parameters using four percentile points. *Technometrics*, **22**, 247.

MAGNUS, J. R. (1990). On certain moments relating to ratios of quadratic forms in normal variables: Further results. *Sankhyā*, **B52**, 1.

MAISTROV, L. E. (1974). *Probability Theory: a Historical Sketch*. Academic Press, New York.

MAJINDAR, K. N. (1962). Improved bounds on a measure of skewness. *Ann. Math. Statist.*, **33**, 1192.

MALIK, H. J., BALAKRISHNAN, N. and AHMED, S. E. (1988). Recurrence relations and identities for moments of order statistics, I. *Commun. Statist.*, **A17**, 2623.

MARCINKIEWICZ, J. (1938). Sur une propriété de la loi de Gauss. *Math. Zeitschr.*, **44**, 612.

MARDIA, K. V. (1965). Tippett's formulas and other results on sample range and extremes. *Ann. Inst. Statist. Math.*, **17**, 85.

MARDIA, K. V. (1967). Some contributions to contingency-type bivariate distributions. *Biometrika*, **54**, 235; corrections, **55**, 597.

MARDIA, K. V. (1970a). *Families of Bivariate Distributions*. Griffin, London.

MARDIA, K. V. (1970b). Some problems of fitting for contingency-type bivariate distributions. *J. R. Statist. Soc.*, **B**, **32**, 254.

MARDIA, K. V. (1970c). Measures of multivariate skewness and kurtosis with applications. *Biometrika*, **57**, 519.

MARDIA, K. V. (1972). *Statistics of Directional Data*. Academic Press, London.

MARDIA, K. V. (1974). Applications of some measures of multivariate skewness and kurtosis in testing normality and robustness studies. *Sankhyā*, **B36**, 115.

MARDIA, K. V. (1981). Recent directional distributions with applications. *In* Taillie *et al.* (1981), **6**, 1.

MARDIA, K. V. and DRYDEN, I. L. (1989). Shape distributions for landmark data. *Adv. Appl. Prob.*, **21**, 742.

MARDIA, K. V. and ZEMROCH, P. J. (1978). *Tables of the F and Related Distributions with Algorithms*. Academic Press, New York.

MARSAGLIA, G. (1965). Ratios of normal variables and ratios of sums of uniform variables. *J. Amer. Statist. Ass.*, **60**, 193.

MARSAGLIA, G. (1972). The structure of linear congruential sequences. *In* S. K. Zaremba (ed.), *Applications of Number Theory to Numerical Analysis*, 249. Academic Press, New York.

MARSAGLIA, G. (1984). The exact-approximation method for generating random variables in a computer. *J. Amer. Statist. Ass.*, **79**, 218.

MARSAGLIA, G. and BRAY, T. A. (1964). A convenient method for generating normal variables. *Rev. Soc. Ind. Appl. Math.*, **6**, 260.

MARSAGLIA, G. and ZAMAN, A. (1991). A new class of random number generators. *Ann. Appl. Prob.*, **1**, 462.

MARSHALL, A. W. and OLKIN, I. (1967). A multivariate exponential distribution. *J. Amer. Statist. Ass.*, **62**, 30.

MARSHALL, A. W. and OLKIN, I. (1983). Domains of attraction of multivariate extreme value distributions. *Ann. Prob.*, **11**, 168.

MARSHALL, A. W. and OLKIN, I. (1985). A family of bivariate distributions generated by the bivariate Bernoulli distribution. *J. Amer. Statist. Ass.*, **80**, 332.

MARSHALL, A. W. and OLKIN, I. (1988). Families of multivariate distributions. *J. Amer. Statist. Ass.*, **83**, 834.

MARSHALL, A. W. and SHAKED, M. (1986). Multivariate new better than used distributions: a survey. *Scand. J. Statist.*, **13**, 277.

MARTIN, E. S. (1934). On the correction for the moment coefficients of frequency-distributions when the start of the frequency is one of the characteristics to be determined. *Biometrika*, **26**, 12.

MATÉRN, B. (1949). Independence of non-negative quadratic forms in normally correlated variables. *Ann. Math. Statist.*, **20**, 119.

MATHAI, A. M. and PEDERZOLI, G. (1977). *Characterizations of the Normal Probability Law.* Wiley, New York.

MATHUR, R. K. (1961). A note on Wilson-Hilferty transformation of χ^2. *Bull. Calcutta Statist. Ass.*, **10**, 102.

MEHTA, J. S. and GURLAND, J. (1969). Some properties and an application of a statistic arising in testing correlation. *Ann. Math. Statist.*, **40**, 1736.

MIKHAIL, N. N. (1968). Multivariate tables of symmetric function. *Egypt. Statist. J.*, **12**, 17.

MIKHAIL, N. N. and ALI, M. M. (1980). Moments and products of moments of the k-statistics, h-statistics and L-statistics for infinite and finite populations. *J. Statist. Res.*, **14**, 39.

MILLER, J. C. P. (1954). Tables of binomial coefficients. *Roy. Soc. Math. Tables,* vol. 3.

MILLER, R. G. (1974). The jackknife—a review. *Biometrika*, **61**, 1.

MILLS, J. P. (1926). Table of the ratio: Area to bounding ordinate for any portion of the normal curve. *Biometrika*, **18**, 395.

MOLINA, E. C. (1942). *Poisson's Exponential Binomial Limit.* Van Nostrand, New York.

MOOD, A. M. (1951). On the distribution of the characteristic roots of normal second-moment matrices. *Ann. Math. Statist.*, **22**, 266.

MORAN, P. A. P. (1948). Rank correlation and product moment correlation. *Biometrika*, **35**, 203.

MORAN, P. A. P. (1956). The numerical evaluation of a class of integrals. *Proc. Camb. Phil. Soc.*, **52**, 230.

MORAN, P. A. P. (1967). Testing for correlation between non-negative variates. *Biometrika*, **54**, 385.

MORAN, P. A. P. (1980). Calculation of the normal distribution function. *Biometrika*, **67**, 675.

MORAN, P. A. P. (1983). A new expansion for the multivariate normal. *Austral. J. Statist.*, **25**, 339.

MORAN, P. A. P. (1984). The Monte Carlo evaluation of orthant probabilities for multivariate normal distributions. *Austr. J. Statist.*, **26**, 39.

MORGENSTERN, D. (1956). Einfache Beispiele zweidimensionaler Verteilungen. *Mitteil. f. Math. Statistik.*, **8**, 234.

MORIGUTI, S. (1951). Extremal properties of extreme value distributions. *Ann. Math. Statist.*, **22**, 523.

MORIGUTI, S. (1952). A lower bound for a probability moment of any absolutely continuous distribution with finite variance. *Ann. Math. Statist.*, **23**, 286.

MORIN-WAHHAB, D. (1985). Moments of a ratio of two quadratic forms. *Commun. Statist.*, **A14**, 499.

MORRIS, C. N. (1982, 1983). Natural exponential families with quadratic variance functions. *Ann. Statist.*, **10**, 65 and **11**, 515.

MOSIMANN, J. E. (1962). On the compound multinomial distribution, the multivariate β-distribution and correlations among proportions. *Biometrika*, **49**, 65.

MOSIMANN, J. E. (1963). On the compound negative multinomial distribution and correlations among inversely sampled pollen counts. *Biometrika*, **50**, 47.

MUDHOLKAR, G. S. and GEORGE, E. O. (1978). A remark on the shape of the logistic distribution. *Biometrika*, **65**, 667.

MUIRHEAD, R. J. (1978). Latent roots and matrix variates: a review of some asymptotic results. *Ann. Statist.*, **6**, 5.

MUIRHEAD, R. J. and CHIKUSE, Y. (1975). Asymptotic expansions for the joint and marginal distributions of the latent roots of the covariance matrix. *Ann. Statist.*, **3**, 1011.

MULHOLLAND, H. P. (1977). On the null distribution of $\sqrt{b_1}$ for samples of size at most 25, with tables. *Biometrika*, **64**, 401.

MÜLLER, P. H. and VAHL, H. (1976). Pearson's system of frequency curves whose left boundary and first three moments are known. *Biometrika*, **63**, 191.

NAGAMBAL, P. N. and TRACY, D. S. (1970). Products of two polykays when one has weight 5. *Ann. Math. Statist.*, **41**, 1114.

NAIR, U. S. (1936). The standard error of Gini's mean difference. *Biometrika*, **28**, 428.

NANCE, R. E. and OVERSTREET, C. (1978). Some experimental observations on the behaviour of composite random number generators. *Operations Res.*, **26**, 915.

NAYLOR, J. C. and SMITH, A. F. M. (1988). Econometric illustrations of novel numerical integration strategies for Bayesian inference. *J. Econometrics*, **38**, 103.

NEL, D. G. (1978). On the symmetric multivariate normal distribution and the asymptotic expansion of a Wishart matrix. *S. African J. Statist.*, **12**, 145.

NEWMAN, T. G. and ODELL, P. L. (1971). *The Generation of Random Variates*. Griffin, London.

NEYMAN, J. (1939).* On a new class of "contagious" distributions, applicable in entomology and bacteriology. *Ann. Math. Statist.*, **10**, 35.

ODEH, R. E. (1982a). Critical values of the sample product moment correlation coefficient in the bivariate normal distribution. *Commun. Statist.*, **B**, **11**, 1.

ODEH, R. E. (1982b). Tables of percentage points of the distribution of the maximum absolute value of equally correlated normal random variables. *Commun. Statist.*, **B11**, 65.

OGASAWARA, T. and TAKAHASHI, M. (1951). Independence of quadratic quantities in a normal system. *J. Sci. Hiroshima University*, **A**, **15**, 1.

O'HAGAN, A. (1994). *Kendall's Advanced Theory of Statistics. Bayesian Inference Volume*. Arnold, London.

OJA, H. (1981). On location, scale, skewness and kurtosis of univariate distributions, *Scand. J. Statist.*, **8**, 154.

OJA, H. (1983). Descriptive statistics for multivariate distributions. *Statist. Prob. Letters*, **1**, 327.

OLSON, J. M. and WEISSFELD, L. A. (1991). Approximation of certain multivariate integrals. *Statist. Prob. Letters.*, **11**, 309.

ORD, J. K. (1967a). Graphical methods for a class of discrete distributions. *J. R. Statist. Soc.*, **A**, **130**, 232.

ORD, J. K. (1967b). On a system of discrete distributions. *Biometrika*, **54**, 649.

ORD, J. K. (1968a). Approximations to distribution functions which are hypergeometric series, *Biometrika*, **55**, 243.

ORD, J. K. (1968b). The discrete Student's t distribution. *Ann. Math. Statist.*, **39**, 1513.

ORD, J. K. (1972). *Families of Frequency Distributions*. Griffin, London.

ORD, J. K. (1980). Tests of significance using non-normal data. *Geog. Anal.*, **12**, 387.

ORD, J. K. and WHITMORE, G. A. (1986). The Poisson-inverse Gaussian distribution as a model for species abundance. *Commun. Statist.*, **A15**, 853.

OWEN, D. B. (1956). Tables for computing bivariate normal probabilities. *Ann. Math. Statist.*, **27**, 1075.

OWEN, D. B. (1957). The bivariate normal probability integral. Research Report, Sandia Corp., Washington.

OWEN, D. B. (1980). A table of normal integrals. *Commun. Statist.*, **B9**, 389.

OWEN, D. B. and STECK, G. P. (1962). Moments of order statistics from the equicorrelated multivariate normal distribution. *Ann. Math. Statist.*, **33**, 1286.

PAIRMAN, E. and PEARSON, K. (1919). On the corrections for moment coefficients of limited-range frequency-distributions when there are finite or infinite ordinates and any slopes at the terminals of the range. *Biometrika*, **12**, 231.

PARRISH, R. S. (1981). Evaluation of bivariate cumulative probabilities using moments of the fourth order. *J. Statist. Comput. Simul.*, **13**, 181.

PARRISH, R. S. (1983). On an integrated approach to member selection and parameter estimation for Pearson distributions. *Comput. Statist. Data Anal.*, **1**, 239.

PARTHASARATHY, K. R. (1976). Characterization of the normal law through local independence of certain statistics. *Sankhyā*, **A, 38**, 174.

PARZEN, E. (1962). On estimation of a probability density function and mode. *Ann. Math. Statist.*, **33**, 1065.

PATIL, G. P. (1962a). Certain properties of the generalized power series distributions. *Ann. Inst. Statist. Math.*, **14**, 179.

PATIL, G. P. (1962b). Maximum likelihood estimation for generalized power series distributions and its application to a truncated binomial distribution. *Biometrika*, **49**, 227.

PATIL, G. P. (1965). On the multivariate generalized power series distribution and its application to the multinomial and negative multinomial. In *Classical and Contagious Discrete Distributions* (G. P. Patil, ed.). Pergamon, New York.

PATIL, G. P. and BILDIKAR, S. (1967). Multivariate logarithmic series distribution as a probability model in population and community ecology and some of its statistical properties. *J. Amer. Statist. Ass.*, **62**, 655.

PATIL, G. P. and ORD, J. K. (1976). On size-biased sampling and related form-invariant weighted distributions. *Sankhyā*, **B, 38**, 48.

PATIL, G. P. and SESHADRI, V. (1964). Characterization theorems for some univariate probability distributions. *J. R. Statist. Soc.*, **B, 26**, 286.

PATIL, G. P., KAMAT, A. R. and WANI, J. K. (1964). *Certain Studies on the Structure and Statistics of the Logarithmic Series Distribution and Related Tables.* U.S. Dept of Commerce, Washington.

PATIL, G. P., KOTZ, S. and ORD, J. K. (eds) (1975). *Statistical Distributions in Scientific Work*, Vols 1–3. Reidel, Dordrecht.

PATIL, G. P., BOSWELL, M. T. and RATNAPARKHI, M. V. (1985a, b, c). *Dictionary and Classified Bibliography of Statistical Distributions in Scientific Work.* Vol. I (with S. W. JOSHI): *Discrete Models*; Vol. 2: *Univariate Continuous Models*; Vol. 3 (with J. J. J. ROUX: Multivariate Models. International Cooperative Publishing House, Burtonsville, Maryland.

PAULSON, E. (1942). An approximate normalisation of the analysis of variance distribution. *Ann. Math. Statist.*, **13**, 233.

PEARSON, E. S. (1926). A further note on the distribution of range in samples taken from a normal population. *Biometrika*, **18**, 173.

PEARSON, E. S. (1930). A further development of tests for normality. *Biometrika*, **22**, 239.

PEARSON, E. S. (1932). The percentage limits for the distribution of range in samples from a normal population. *Biometrika*, **24**, 404.

PEARSON, E. S. (1963). Some problems arising in approximating to probability distributions, using moments. *Biometrika*, **50**, 95.

PEARSON, E. S. and HARTLEY, H. O. (1942). The probability integral of the range in samples of n observations from a normal population. *Biometrika*, **32**, 301.

PEARSON, E. S. and HARTLEY, H. O. (1950). Tables of the χ^2 integral and of the cumulative Poisson distribution. *Biometrika*, **37**, 313.

PEARSON, E. S. and KENDALL, M. G. (eds) (1970). *Studies in the History of Statistics and Probability.* Griffin, London.

PEARSON, E. S. and TUKEY, J. W. (1965). Approximate means and standard deviations based on distances between percentage points of frequency curves. *Biometrika*, **52**, 533.

PEARSON, E. S., JOHNSON, N. L. and BURR, I. W. (1979). Comparison of the percentage points of distributions with the same first four moments, chosen from eight different systems of frequency curves. *Commun. Statist.*, **B, 8**, 191.

PEARSON, K. (1894).* Contributions to the mathematical study of evolution. *Phil. Trans. Roy. Soc.*, **A, 185**, 71.

PEARSON, K. (1895).* Skew variation in homogeneous material. *Phil. Trans. Roy. Soc.*, **A, 186**, 393.

PEARSON, K. (1900).* On a criterion that a given system of deviations from the probable in the case of a correlated system of variables is such that it can be reasonably supposed to have arisen in random sampling. *Phil. Mag.*, (5), **50**, 157.

PEARSON, K. (1919). On generalised Tchebycheff theorems in the mathematical theory of statistics. *Biometrika*, **12**, 284.

PEARSON, K. (1923). On non-skew frequency surfaces. *Biometrika*, **15**, 231.

PEARSON, K. (1924). On the moments of the hypergeometrical series. *Biometrika*, **16**, 157.

PEARSON, K. (1931). Appendix to a paper by Professor Tokishige Hojo. On the standard error of the median to a third approximation, etc. *Biometrika*, **23**, 361.

PEARSON, K., STOUFFER, S. A. and DAVID, F. N. (1932). Further applications in statistics of the $T_m(x)$ Bessel function. *Biometrika*, **24**, 293.

PEIZER, D. B. and PRATT, J. W. (1968). A normal approximation for binomial, *F*, beta, and other common, related tail probabilities, I. *J. Amer. Statist. Ass.*, **63**, 1416.

PEREZ-ABREU, V. (1991). Poisson approximation to power series distributions. *Amer. Statist.*, **45**, 42.

PETERSON, A. W. and KRONMAL, R. A. (1983). Analytic comparison of three general purpose methods for the computer generation of discrete random variables. *Appl. Statist.*, **32**, 276.

PETROV, V. V. (1974). *Sums of Independent Random Variables.* Springer-Verlag, Berlin.

PETTIGREW, H. M., GART, J. J. and THOMAS, D. G. (1986). The bias and higher cumulants of the logarithm of a binomial variate. *Biometrika*, **73**, 425.

PHILLIPS, P. C. B. (1982). The true characteristic function of the *F* distribution. *Biometrika*, **69**, 261.

PICKANDS, J. (1986). The continuous and differentiable domains of attraction of the extreme value distributions. *Ann. Prob.*, **14**, 996.

PILLAI, K. C. S. (1964). On the distribution of the largest of seven roots of a matrix in multivariate analysis. *Biometrika*, **51**, 270.

PILLAI, K. C. S. (1965). On the distribution of the largest characteristic root of a matrix in multivariate analysis. *Biometrika*, **52**, 405.

PILLAI, K. C. S. (1966). On the non-central multivariate beta distribution and the moments of traces of some matrices. *Proc. Symp. Multiv. Analysis.* Academic Press, New York.

PILLAI, K. C. S. (1967a). On the distribution of the largest root of a matrix in multivariate analysis. *Ann. Math. Statist.*, **38**, 616.

PILLAI, K. C. S. (1967b). Upper percentage points of the largest roots of a matrix in multivariate analysis. *Biometrika*, **54**, 189.

PITMAN, E. J. G. (1937a). Significance tests which may be applied to samples from any population. II. The correlation coefficient test. *Supp. J. R. Statist. Soc.*, **4**, 225.

PITMAN, E. J. G. (1937b). The "closest" estimates of statistical parameters. *Proc. Camb. Phil. Soc.*, **33**, 212.

PITMAN, E. J. G. (1956). On the derivatives of a characteristic function at the origin. *Ann. Math. Statist.*, **27**, 1156.

PITT, L. D. (1982). Positively correlated normal variables are associated. *Ann. Prob.*, **10**, 496.

PLACHKY, D. and THOMSEN, W. (1981). On a theorem of Pólya. *In* Taillie *et al.* (1981), **4**, 287.

PLACKETT, R. L. (1947). Limits of the ratio of mean range to standard deviation. *Biometrika*, **34**, 120.

PLACKETT, R. L. (1965). A class of bivariate distributions. *J. Amer. Statist. Ass.*, **60**, 516.

PLACKETT, R. L. (1968). Random permutations. *J. R. Statist. Soc.*, **B**, **30**, 517.

PÓLYA, G. (1945). Remarks on computing the probability integral in one and two dimensions. *Proc. 1st Berkeley Symp. Math. Statist. and Prob.*, 63.

POOR, H. V. (1991). The maximum difference between the binomial and Poisson distributions. *Statist. Prob. Letters.*, **11**, 103.

POPPER, K. R. (1968). *The Logic of Scientific Discovery.* Hutchinson, London.

POSNER, E. C., RODEMICH, E. R., ASHLOCK, J. C. and LUBIN, S. (1969). Application of an estimator of high efficiency in bivariate extreme value theory. *J. Amer. Statist. Ass.*, **64**, 1403.

PRATT, J. W. (1968). A normal approximation for binomial, *F*, beta and other common, related tail probabilities, II. *J. Amer. Statist. Ass.*, **63**, 1457.

PRESCOTT, P. (1974). Variances and covariances of order statistics from the gamma distribution. *Biometrika*, **61**, 607.

PRESS, S. J. (1972). Estimation in univariate and multivariate stable distributions. *J. Amer. Statist. Ass.*, **67**, 842.

PRETORIUS, S. J. (1930). Skew bivariate frequency surfaces, examined in the light of numerical illustrations. *Biometrika*, **22**, 109.

PURI, P. S. and RUBIN, H. (1970). A characterization based on the absolute difference of two i.i.d. random variables. *Ann. Math. Statist.*, **41**, 2113.

PYKE, R. (1965). Spacings. *J. R. Statist. Soc.*, **B**, **27**, 395.

QUENOUILLE, M. H. (1949). A relation between the logarithmic, Poisson, and negative binomial series. *Biometrics*, **5**, 162.

QUENOUILLE, M. H. (1956). Notes on bias in estimation. *Biometrika*, **43**, 353.

QUENOUILLE, M. H. (1959). Tables of random observations from standard distributions, *Biometrika*, **46**, 178.

RABINOWITZ, P. (1969). New Chebyshev polynomial approximations to Mills' ratio. *J. Amer. Statist. Ass.*, **64**, 647.

RAFF, M. S. (1956). On approximating the point binomial. *J. Amer. Statist. Ass.*, **51**, 293.

RAMASUBBAN, T. A. (1958). The mean difference and the mean deviation of some discontinuous distributions. *Biometrika*, **45**, 549.

RAMBERG, J. S., TADIKAMALLA, P. R., DUDEWICZ, E. J. and MYKYTKA, C. F. (1979). A probability distribution and its uses in fitting data. *Technometrics*, **21**, 201.

RAMSEY, F. P. (1931). *The Foundations of Mathematics and Other Essays*. Kegan Paul, Trench, Trubner, London.

RAO, C. R. (1973). *Linear Statistical Inference and Its Applications. 2nd edition*. Wiley, New York.

RAO, C. R. and RUBIN, H. (1964). On a characterization of the Poisson distribution. *Sankhyā*, **A**, **26**, 295.

RAY, W. D. and PITMAN, A. E. N. T. (1963). Chebyshev polynomial and other new approximations to Mills' Ratio. *Ann. Math. Statist.*, **34**, 892.

REID, N. (1988). Saddlepoint methods and statistical inference. *Statist. Science*, **3**, 213.

RÉNYI, A. (1953). On the theory of order statistics. *Acta Math. Acad. Sci. Hung.*, **4**, 191.

RIPLEY, B. D. (1983). Computer generation of random variables: a tutorial. *Int. Statist. Rev.*, **51**, 301.

ROBERTSON, W. H. (1960). *Tables of the Binomial Distribution Function for Small Values of P*. Sandia Corp., Albuquerque, New Mexico, U.S.A.

ROBINSON, J. (1978). An asymptotic expansion for samples from a finite population. *Ann. Statist.*, **6**, 1005.

RODRIGUEZ, R. N. (1977). A guide to the Burr Type XII distributions. *Biometrika*, **64**, 129.

ROHATGI, V. K. and SZEKELY, G. J. (1989). Sharp inequalities between skewness and kurtosis. *Statist. Prob. Letters*, **8**, 297.

ROMANOVSKY, V. (1923). Note on the moments of a binomial $(p+q)^n$ about its mean. *Biometrika*, **15**, 410.

ROMANOVSKY, V. (1925). On the moments of the hypergeometrical series. *Biometrika*, **17**, 57.

ROMANOWSKA, M. (1977). A note on the upper bound for the distance in total variation between the binomial and the Poisson distribution. *Statist. Neerl.*, **31**, 127.

ROMIG, H. G. (1953). *50–100 Binomial Tables.* Wiley, New York.
ROSENBLATT, M. (1956). Remarks on some nonparameric estimators of a density function. *Ann. Math. Statist.*, **27**, 832.
ROY, S. N. (1939). *p*-statistics, or some generalizations in analysis of variance appropriate to multivariate problems. *Sankhyā*, **4**, 381.
ROYSTON, J. P. (1982). Expected normal order statistics (exact and approximate). *Appl. Statist.*, **31**, 161.
RUBEN, H. (1954). On the moments of order statistics in samples from normal populations. *Biometrika*, **41**, 200.
RUBEN, H. (1960). On the distribution of the weighted difference of two independent Student variables. *J. R. Statist. Soc.*, **B**, **22**, 188.
RUBEN, H. (1962). A new asymptotic expansion for the normal probability integral and Mills' ratio. *J. R. Statist. Soc.*, **B**, **24**, 177.
RUBEN, H. (1963). A convergent asymptotic expansion for Mills' ratio and the normal probability integral in terms of rational functions. *Math. Ann.*, **151**, 355.
RUBEN, H. (1964). Irrational fraction approximations to Mills' ratio. *Biometrika*, **51**, 339.
RUBEN, H. (1966). Some new results on the distribution of the sample correlation coefficient. *J. R. Statist. Soc.*, **B**, **28**, 513.

SAHAI, H. (1980). A supplement to Sowey's bibliography on random number generation and related topics. *Biom. J.*, **22**, 41.
SAMIUDDIN, M. (1970). On a test for an assigned value of correlation in a bivariate normal distribution. *Biometrika*, **57**, 461.
SAMPFORD, M. (1953). Some inequalities on Mill's ratio and related functions. *Ann. Math. Statist.*, **24**, 130.
SANDIFORD, P. J. (1960). A new binomial approximation for use in sampling from finite populations. *J. Amer. Statist. Ass.*, **55**, 718.
SANDON, F. (1924). Note on the simplification of the calculation of abruptness coefficients to correct crude moments. *Biometrika*, **16**, 193.
SANKARAN, M. (1959). On the non-central chi-square distribution. *Biometrika*, **46**, 235.
SARHAN, A. E. and GREENBERG, B. G. (1956, 1958). Estimation of location and scale parameters by order statistics from singly and doubly censored samples. I. *Ann. Math. Statist.*, **27**, 427. II. Same *Ann.*, **29**, 79.
SARHAN, A. E. and GREENBERG, B. G., eds (1962). *Contributions to Order Statistics.* Wiley, New York.
SATHE, Y. S. and DIXIT, U. J. (1990). On a recurrence relation for order statistics. *Statist. Prob. Letters*, **9**, 1.
SAVAGE, I. R. (1962). Mills' ratio for multivariate normal integrals. *J. Res. Nat. Bur. Standards*, **B**, **66**, 93.
SAVAGE, L. J. (1954). *The Foundations of Statistics.* Wiley, New York.
SAVAGE, L. J. (1961). The foundations of statistics reconsidered. *Proc. 4th Berkeley Symp. Math. Statist. and Prob.*, **1**, 575.
SAVAGE, L. J. (1962). *The Foundations of Statistical Inference: a Discussion . . . at a Meeting of the Joint Statistics Seminar, Birkbeck and Imperial Colleges, in the University of London.* Methuen, London.
SAVAGE, L. J. (1971). Elicitation of personal probabilities and expectations. *J. Amer. Statist. Ass.*, **66**, 783.
SAW, J. G., YANG, M. C. K. and MO, T. C. (1984). Chebyshev inequality with estimated mean and variance. *Amer. Statist.*, **38**, 130.
SCHERVISH, M. J. (1984). Multivariate normal probabilities with error bound. *Appl. Statist.*, **33**, 81 (corr., **34**, 103.)
SCHMEISER, B. W. and SHALABY, M. A. (1980). Acceptance/rejection methods for beta variate generation. *J. Amer. Statist. Ass.*, **75**, 673.

SCOTT, D. W. (1979). On optimal and data-based histograms. *Biometrika*, **66**, 605.

SCOTT, D. W. (1985). Frequency polygons: theory and application. *J. Amer. Statist. Ass.*, **80**, 348.

SCOTT, D. W., TAPIA, R. A. and THOMPSON, J. R. (1980). Nonparametric density estimation by discrete maximum penalized-likelihood criteria. *Ann. Statist.*, **8**, 820.

SEN, P. K. (1961). On some properties of the asymptotic variance of the sample quantiles and mid-ranges. *J. R. Statist. Soc.*, **B**, **23**, 453.

SHAH, B. K. and ODEH, R. E. (1986). *Selected Tables in Mathematical Statistics*, Vol. X. Amer. Math. Soc., Providence, R.I., U.S.A.

SHANNON, C. E. and WEAVER, W. W. (1949). *The Mathematical Theory of Communication*. University of Illinois Press, Champaign, Illinois.

SHANTARAM, R. and HARKNESS, W. L. (1972). On a certain class of limit distributions. *Ann. Math. Statist.*, **43**, 2067.

SHEA, B. L. and SCALLON, A. J. (1988). Approximating the covariance matrix of normal order statistics. *Appl. Statist.*, **37**, 151.

SHENTON, L. R. (1954). Inequalities for the normal integral, including a new continued fraction. *Biometrika*, **41**, 177.

SHENTON, L. R. and CARPENTER, J. A. (1965). The Mills ratio and the probability integral for a Pearson Type IV distribution. *Biometrika*, **52**, 119.

SHENTON, L. R., BOWMAN, K. O. and LAM, H. K. (1979). Comments on a paper by R. C. Geary on standardized mean deviation. *Biometrika*, **66**, 400.

SHEPPARD, W. F. (1898). On the application of the theory of error to cases of normal distributions and normal correlations. *Phil. Trans*, **A**, **192**, 101, and *Proc. Roy. Soc.*, **62**, 170.

SHEPPARD, W. F. (1939). *The Probability Integral*. British Ass. Math. Tables, vol. 7, Cambridge Univ. Press, London.

SHIMIZU, R. (1972). On the decomposition of stable characteristic functions. *Ann. Inst. Statist. Math.*, **24**, 347.

SHOHAT, J. A. and TAMARKIN, J. D. (1943). *The Problem of Moments*. Amer. Math. Soc., New York.

SHORACK, G. R. (1969, 1972). Asymptotic normality of linear combinations of functions of order statistics; *and* Functions of order statistics. *Ann. Math. Statist.*, **40**, 2041, and **43**, 412.

SIBUYA, M. and SHIMIZU, R. (1981). The generalized hypergeometric family of distributions. *Ann. Inst. Statist. Math.*, **33**, 177.

SICHEL, H. S. (1949). The method of frequency moments and its application to Type VII distributions. *Biometrika*, **36**, 404.

SICHEL, H. S. (1980). A class of multiparameter power series distributions. *S. African J. Statist.*, **14**, 1.

SILLITTO, G. P. (1951). Interrelations between certain linear systematic statistics of samples from any continuous population. *Biometrika*, **38**, 377.

SILLITTO, G. P. (1969). Derivation of approximants to the inverse distribution function of a continuous univariate population from the order statistics of a sample. *Biometrika*, **56**, 641.

SILVERMAN, B. (1978). Choosing the window width when estimating a density. *Biometrika*, **65**, 1.

SILVERMAN, B. W. (1986). *Density Estimation for Statistics and Data Analysis*. Chapman & Hall, London.

SINGH, C. (1967). On the extreme values and range of samples from non-normal populations. *Biometrika*, **54**, 541.

SINGH, C. (1970). On the distribution of range of samples from nonnormal populations. *Biometrika*, **57**, 451.

SINGH, C. (1972). Order statistics from nonnormal populations. *Biometrika*, **59**, 229.

SINGH, C. (1976). Moments of the range of samples from nonnormal populations. *J. Amer. Statist. Ass.*, **71**, 988.

SIX, F. B. (1981). Representations of multivariate normal distributions with special correlation structures. *Commun. Statist.*, **A**, **10**, 1285.

SKOUGAARD, I. M. (1986). On multivariate Edgeworth expansions. *Inter. Statist. Rev.*, **54,** 169.

SLIFKER, J. F. and SHAPIRO, S. S. (1980). The Johnson system: Selection and parameter estimation. *Technometrics*, **22,** 239.

SLUTSKY, E. E. (1950). *Tables for the Calculation of the Incomplete Gamma Function and the* χ^2 *Probability Function*. Moscow, Isdatelstvo Akad. Nauk S.S.S.R.

SMIRNOV, N. V. (1949). Limit distribution for the terms of a variational series. *Amer. Math. Soc. Transl.*, 67.

SMIRNOV, N. V. (1961). *Tables for the Distribution and Density Functions of t-distribution*. Pergamon, Oxford.

SMIRNOV, N. V. (1965). *Tables of the Normal Probability Integral, the Normal Density and its Normalized Derivatives*. Transl. from Russian by D. E. Brown. Pergamon, Oxford.

SMITH, C. A. B. (1965). Personal probability and statistical analysis. *J. R. Statist. Soc.*, **A, 128,** 469.

SMITH, M. D. (1989). On the expectation of a ratio of quadratic forms in normal variables. *J. Multiv. Anal.*, **31,** 244.

SOBEL, M., UPPULURI, V. R. R. and FRANKOWSKI, K. (1976). *Selected Tables in Mathematical Statistics,* Vol. IV, ed. D. B. Owen and R. E. Odeh. Amer. Math. Soc., Providence, R. I., U.S.A.

SOLOMON, H. (1960). Distribution of quadratic forms—tables and applications. *Appl. Math. and Statist. Labs., Tech. Report 45, Stanford University*.

SOLOMON, H. and STEPHENS, M. A. (1977). Distribution of a sum of weighted chi-square variables. *J. Amer. Statist. Ass.*, **72,** 881.

SOLOMON, H. and STEPHENS, M. A. (1978). Approximations to density functions using Pearson curves. *J. Amer. Statist. Ass.*, **73,** 153.

SOLOMON, H. and STEPHENS, M. A. (1983). An approximation to the distribution of the sample variance. *Canad. J. Statist.*, **11,** 149.

SOMS, A. P. (1976). An asymptotic expansion for the tail area of the *t*-distribution. *J. Amer. Statist. Ass.*, **71,** 728.

SOMS, A. P. (1980). Rational bounds for the *t*-tail area. *J. Amer. Statist. Ass.*, **75,** 438.

SOWEY, E. R. (1972). A chronological and classified bibliography on random number generation and testing. *Int. Statist. Rev.*, **40,** 355.

SOWEY, E. R. (1978). A second classified bibliography on random number generation and testing. *Int. Statist. Rev.*, **46,** 89.

SOWEY, E. R. (1986). A third classified bibliography on random number generation and testing. *J. R. Statist. Soc.*, **A, 149,** 83.

SPEED, T. P. (1983). Cumulants and partition lattices, I. *Aust. J. Statist.*, **25,** 378.

SPEED, T. P. (1986a, b, c). Cumulants and partition lattices. II, III, IV. *J. Austr. Math. Soc.*, **A40,** 34, 161 and **A41,** 79.

SPEED, T. P. and SILCOCK, H. L. (1988). Cumulants and partition lattices, V. *J. Austr. Math. Soc.*, **A44,** 171.

STECK, G. P. (1958a). A uniqueness property not enjoyed by the normal distribution. *Ann. Math. Statist.*, **29,** 604.

STECK, G. P. (1958b). A table for computing trivariate normal probabilities. *Ann. Math. Statist.*, **29,** 780.

STECK, G. P. (1962). Orthant probabilities for the equicorrelated multivariate normal distribution. *Biometrika*, **49,** 433.

STECK, G. P. (1968). A note on contingency-type bivariate distributions. *Biometrika*, **55,** 262.

STECK, G. P. (1979). Lower bounds for the multivariate Mills ratio. *Ann. Prob.*, **7,** 547.

STEIN, G. Z. and JURITZ, J. M. (1987). Bivariate compound Poisson distributions. *Commun. Statist.*, **A16,** 3591.

STEPHENS, M. A. (1966). Statistics connected with the uniform distribution: percentage points and applications to testing for randomness of directions. *Biometrika*, **53,** 235.

STEYN, H. S. (1955). On discrete multivariate probability distributions of hypergeometric type. *Proc. Kon. Akad. Wetensch.*, **A58**, 588.

STEYN, H. S. (1957). On regression properties of discrete systems of probability functions. *Proc. Kon. Akad. Wetensch.*, **A60**, 119.

STEYN, H. S. (1960). On regression properties of multivariate probability functions of Pearson's type. *Proc. Kon. Ned. Akad. Wetensch., Amsterdam*, **A63**, 302.

STIELTJES, T. J. (1918). Recherches sur les fractions continues. *Oeuvres*, Groningen.

STIGLER, S. M. (1986). *The History of Statistics: The Measurement of Uncertainty Before 1900.* Havard University, Cambridge, Mass. and London.

STIGLER, S. M. (1989). Normal orthant probabilities. *Amer. Statist.*, **43**, 291.

STUART, A. (1958). Equally correlated variates and the multinormal integral. *J. R. Statist. Soc.*, **B, 20**, 373.

STUART, A. (1962). Gamma-distributed products of independent random variables. *Biometrika*, **49**, 564.

STUDENT (1908a).* On the probable error of a mean. *Biometrika*, **6**, 1.

STUDENT (1908b).* On the probable error of a correlation coefficient. *Biometrika*, **6**, 302.

STYAN, G. P. H. (1970). Notes on the distribution of quadratic forms in singular normal variables. *Biometrika*, **57**, 567.

SUBRAHMANIAM, K. (1966). Some contributions to the theory of non-normality—I (Univariate case). *Sankyhā*, **A, 28**, 389

SUBRAHMANIAM, K. (1969). Order statistics from a class of non-normal distributions. *Biometrika*, **56**, 415; correction, **57**, 226.

SUGIURA, N. (1976). Asymptotic expansions of the distributions of the latent roots and latent vectors of the Wishart and multivariate F matrices. *J. Multiv. Anal.*, **6**, 500.

SUGIYAMA, T. (1967a). On the distribution of the largest latent root of the covariance matrix. *Ann. Math. Statist.*, **38**, 1148.

SUGIYAMA, T. (1967b). Distribution of the largest latent root and the smallest latent root of the generalized B statistic and F statistic in multivariate analysis. *Ann. Math. Statist.*, **38**, 1152.

SUKHATME, P. V. (1937). Tests of significance for samples of the χ^2-population with two degrees of freedom. *Ann. Eugen.*, **8**, 52.

SUN, H.-J. (1988). A general reduction method for n-variate normal orthant probability. *Commun. Statist.*, **A17**, 3913.

TADIKAMALLA, P. R. (1980). A look at the Burr and related distributions. *Int. Statist. Rev.*, **48**, 337.

TADIKAMALLA, P. R. and JOHNSON, N. L. (1982a). Systems of frequency curves generated by transformations of logistic variables. *Biometrika*, **69**, 461.

TADIKAMALLA, P. R. and JOHNSON, N. L. (1982b). Tables to facilitate fitting L_U distributions. *Commun. Statist.*, **B11**, 249.

TADIKAMALLA, P. R. and JOHNSON, N. L. (1990). Tables to facilitate fitting Tadikamalla and Johnson's L_B distributions. *Commun. Statist.*, **B19**, 1201.

TAILLIE, C., PATIL, G. P. and BALDESSARI, B. (eds) (1981). *Statistical Distributions in Scientific Work*, Vols. 4–6. Reidel, Dordrecht.

TANIS, E. A. (1964). Linear forms in the order statistics from an exponential distribution. *Ann. Math. Statist.*, **35**, 270.

TAWN, J. A. (1988). Bivariate extreme value theory: Models and estimation. *Biometrika*, **75**, 397.

TAWN, J. A. (1990). Modelling multivariate extreme value distributions. *Biometrika*, **77**, 245.

TEICHER, H. (1954). On the multivariate Poisson distribution. *Skand. Aktuarietidskrift.*, **37**, 1.

TEICHROEW, D. (1956). Tables of expected values of order statistics and products of order statistics. *Ann. Math. Statist.*, **27**, 410.

TEUGELS, J. L. (1990). Some representations of the multivariate Bernoulli and binomial distributions. *J. Multiv. Anal.*, **32**, 256.

THIELE, T. N. (1903). *Theory of Observations*. Reprint (1931) in *Ann. Math. Statist.*, **2**, 165, of the English version published in 1903; the original (Danish) appeared in 1889 and 1897.

THOMPSON, C. M. (1941). Tables of percentage points of the χ^2-distribution. *Biometrika*, **32**, 187.

THOMPSON, C. M., PEARSON, E. S., COMRIE, L. J. and HARTLEY, H. O. (1941). Tables of percentage points of the incomplete beta-function. *Biometrika*, **32**, 151.

TIAGO DE OLIVEIRA, J. (1962–3). Structure theory of bivariate extremes: extensions. *Estudos de Matematica, Estatistica e Econometria*, **7**, 165.

TIAGO DE OLIVEIRA, J. (1980). Bivariate extremes: foundations and statistics. In *Multivariate Analysis, V* (P. R. Krishnaiah, ed). North-Holland, Amsterdam, p. 349.

TIETJEN, G. L., KAHANER, D. K. and BECKMAN, R. J. (1977). Variances and covariances of the normal order statistics for sample sizes 2 to 50. In *Selected Tables in Mathematical Statistics*, Vol. V, ed. D. B. Owen and R. E. Odeh, Amer. Math. Soc., Providence, R. I., U.S.A.

TIPPETT, L. H. C. (1925). On the extreme individuals and the range of samples taken from a normal population. *Biometrika*, **17**, 364.

TODHUNTER, I. (1949). *A History of the Mathematical Theory of Probability*. (Orig. pub. 1865). Chelsea, New York.

TONG, Y. L. (1981). *Probability Inequalities in Multivariate Distributions*. Academic Press, New York.

TONG, Y. L. (1984 ed.). *Inequalities in Statistics and Probability*. Inst. Math. Statistics, Hayward, Calif.

TONG, Y. L. (1990). *The Multivariate Normal Distribution*. Springer-Verlag, Berlin and New York.

TRACY, D. S. (1968). Some rules for a combinatorial method for multiple products of generalized k-statistics. *Ann. Math. Statist.*, **39**, 983.

TRACY, D. S. (1969). Some multiple products of polykays. *Ann. Math. Statist.*, **40**, 1297.

TRACY, D. S. (1972). *Symmetric Functions in Statistics*. Proceedings of a Symposium in Honour of Professor Paul S. Dwyer. University of Windsor, Ontario.

TRACY, D. S. and GUPTA, B. C. (1973). Multiple products of polykays using ordered partitions. *Ann. Statist.*, **1**, 913.

TRACY, D. S. and GUPTA, B. C. (1974). Generalized h-statistics and other symmetric functions. *Ann. Statist.*, **2**, 837.

TUKEY, J. W. (1950). Some sampling simplified. *J. Amer. Statist. Ass.*, **45**, 501.

TUKEY, J. W. (1958). Bias and confidence in not-quite large samples. (Abstract.) *Ann. Math. Statist.*, **29**, 614.

TUKEY, J. W. (1977). *Exploratory Data Analysis*. Addison–Wesley, Reading, Mass.

TVERSKY, A. (1974). Assessing uncertainty. *J. R. Statist. Soc.*, **B**, **36**, 148.

TWEEDIE, M. C. K. (1957). Statistical properties of Inverse Gaussian distributions, 1. *Ann. Math. Statist.*, **28**, 362.

VAN UVEN, M. J. (1947–8). Extensions of Pearson's probability distributions to two variables, I, II, III, IV. *Proc. Kon. Ned. Akad. Wetensch., Amsterdam*, **A**, **50**, 1063, 1252 and **51**, 41, 191.

VAN ZWET, W. R. (1964). *Convex Transformations of Random Variables*. Tract No. 7, Math. Centrum, Amsterdam.

VAUGHAN, R. J. and VENABLES, W. N. (1972). Permanent expressions for order statistic densities. *J. R. Statist. Soc.*, **B**, **34**, 308.

VOGLER, L. E. (1964). *Percentage Points of the Beta Distribution*. U.S. Govt. Printing Office, Washington.

VON MISES, R. (1936). La distribution de la plus grande de n valeurs. *Revue de l'Union Interbalkanique*, **1**, 1.

VON MISES, R. (1957). *Probability, Statistics and Truth*, 2nd edn. Allen and Unwin, London.

WAHBA, G. (1975). Interpolating splines for density estimation, 1. Equi-spaced knots. *Ann. Statist.*, **3**, 30.

WALKER, A. M. (1968). A note on the asymptotic distribution of sample quantiles. *J. R. Statist. Soc.*, **B**, **30**, 570.

WALKER, G. A. and SAW, J. G. (1978). The distribution of linear combinations of t-variables. *J. Amer. Statist. Ass.*, **73**, 876.

WALKER, J. J. (1979). An asymptotic expansion for unbalanced quadratic forms in normal variables. *J. Amer. Statist. Ass.*, **74**, 389.

WALLACE, D. L. (1958). Asymptotic approximations to distributions. *Ann. Math. Statist.*, **29**, 635.

WALLACE, N. D. (1974). Computer generation of gamma random variables with non-integral shape parameters. *Commun. Ass. Comput. Mach.*, **17**, 691.

WALTER, S. D. and STITT, L. W. (1988). Extended tables for moments of gamma distribution order statistics. *Commun. Statist.*, **B17**, 471.

WANG, Y. J. (1987). The probability integrals of bivariate normal distributions: A contingency table approach. *Biometrika*, **74**, 185.

WATERNAUX, C. M. (1976). Asymptotic distribution of the sample roots for a nonnormal population. *Biometrika*, **63**, 639.

WEIBULL, W. (1951). A statistical distribution function of wide applicability. *J. Appl. Mech.*, **18**, 293.

WEINTRAUB, S. (1963). *Tables of the Cumulative Binomial Probability Distribution for Small Values of p*. Free Press of Glencoe, New York.

WHEELER, R. E. (1980). Quantile estimators of Johnson curve parameters. *Biometrika*, **67**, 725.

WHITE, J. S. (1970). Tables of normal percentile points. *J. Amer. Statist. Ass.*, **65**, 635.

WHITMORE, G. A. and LEE, M.-L. T. (1991). A multivariate survival distribution generated by an inverse Gaussian mixture of exponentials. *Technometrics*, **33**, 39.

WICHMANN, B. A. and HILL, I. D. (1982). An efficient and portable pseudorandom number generator. *Appl. Statist.*, **31**, 188.

WICKSELL, S. D. (1917). On logarithmic correlation with an application to the distribution of ages at first marriage. *Medd. Lunds Astr. Obs.*, No. 84.

WIDDER, D. V. (1941). *The Laplace Transform*. Princeton U. P.

WIJSMAN, R. A. (1957). Random orthogonal transformations and their use in some classical distribution problems in multivariate analysis. *Ann. Math. Statist.*, **28**, 415.

WILKS, S. S. (1932). Certain generalizations in the analysis of variance. *Biometrika*, **24**, 471.

WILLIAMS, J. D. (1946). An approximation to the probability integral. *Ann. Math. Statist.*, **17**, 363.

WILLIAMS, J. S. (1966). An example of the misapplication of conditional densities. *Sankhyā*, **A**, **28**, 297.

WILLIAMS, P. (1935). Note on the sampling distribution of $\sqrt{\beta_1}$ when the population is normal. *Biometrika*, **27**, 269.

WILLIAMSON, E. and BRETHERTON, M. H. (1963). *Tables of the Negative Binomial Probability Distribution*. Wiley, London.

WILLIAMSON, E. and BRETHERTON, M. H. (1964). Tables of the logarithmic series distribution. *Ann. Math. Statist.*, **35**, 284.

WILSON, E. B. and HILFERTY, M. M. (1931). The distribution of chi-square. *Proc. Nat. Acad. Sci., U.S.A.*, **17**, 684.

WISE, M. E. (1954). A quickly convergent expansion for cumulative hypergeometric probabilities, direct and inverse. *Biometrika*, **41**, 317.

WISE, M. E. (1960). On normalizing the incomplete beta-function for fitting to dose-response curves. *Biometrika*, **47**, 173.

WISHART, J. (1928a, b). The generalized product moment distribution in samples from a multinormal population. *Biometrika*, **20A**, 32 and 424.

WISHART, J. (1929). The correlation between product moments of any order in samples from a normal population. *Proc. Roy. Soc. Edin.*, **49**, 78.

WISHART, J. (1948a, b). Proofs of the distribution law of the second order moment statistics. *Biometrika*, **35**, 55 and 422.

WISHART, J. (1949). Cumulants of multivariate multinomial distributions. *Biometrika*, **36**, 47.

WISHART, J. (1952). Moment coefficients of the k-statistics in samples from a finite population. *Biometrika*, **39**, 1.

WISHART, J. and BARTLETT, M. S. (1933). The generalized product moment distribution in a normal system. *Proc. Camb. Phil. Soc.*, **29**, 260.

WITHERS, C. S. (1984). Asymptotic expansions for distributions and quantiles with power series cumulants. *J. R. Statist. Soc.*, **B**, **46**, 389.

WOLD, H. (1934a). Sulla correzione di Sheppard. *Giorn. Ist. Ital. Attuari*, **5**, 304.

WOLD, H. (1934b). Sheppard's correction formulae in several variables. *Skand. Aktuartidskr.*, **17**, 248.

WOLKOWICZ, H. and STYAN, G. P. H. (1979). Extensions of Samuelson's inequality. *Amer. Statist.*, **33**, 143.

WOODROOFE, M. (1970). On choosing a delta sequence. *Ann. Math. Statist.*, **41**, 1665.

WOODWARD, W. A. (1976). Approximation of Pearson Type IV tail probabilities. *J. Amer. Statist. Ass.*, **71**, 513.

WOOFF, D. A. (1985). Bounds on reciprocal moments with applications and developments in Stein estimation and post-stratification. *J. R. Statist. Soc.*, **B**, **47**, 362.

WU, C. F. J. (1986). Jackknife, bootstrap and other resampling methods in regression analysis. *Ann. Statist.*, **14**, 1261.

YATES, F. (1935). Some examples of biassed sampling. *Ann. Eugen.*, **6**, 202.

YOUNG, D. H. (1970). The order statistics of the negative binomial distribution. *Biometrika*, **57**, 181.

YUAN, P. T. (1933). On the logarithmic frequency distribution. *Ann. Math. Statist.*, **4**, 30.

YULE, G. U. (1927). On reading a scale. *J. R. Statist. Soc.*, **90**, 570.

ZELEN, M. and SEVERO, N. C. (1960). Graphs for bivariate normal probabilities. *Ann. Math. Statist.*, **31**, 619.

ZIA UD-DIN, M. (1954). Expression of the k-statistics, k_9 and k_{10}, in terms of power sums and sample moments. *Ann. Math. Statist.*, **25**, 800.

ZIA UD-DIN, M. (1959). The expression of k-statistic k_{11} in terms of power sums and sample moments. *Ann. Math. Statist.*, **30**, 825.

INDEX OF EXAMPLES IN TEXT

Chapter	1	2	3	4	5	6	7	8	9	10	11	12	13	14	15	16
Examples																
.1	**.34**	**.3**	**.3**	**.6**	**.20**	**.11**	**.15**	**.10**	**.5**	**.4**	**.3**	**.5**	**.2**	**.9**	**.3**	**.11**
.2	**.37**	**.3**	**.4**	**.6**	**.21**	**.20**	**.15**	**.10**	**.5**	**.4**	**.3**	**.13**	**.3**	**.9**	**.13**	**.18**
.3	**.38**	**.4**	**.4**	**.6**	**.25**	**.23**	**.16**	**.10**	**.5**	**.6**	**.3**	**.14**	**.17**	**.17**	**.14**	**.46**
.4		**.5**	**.4**	**.6**	**.26**	**.26**	**.18**	**.17**	**.5**	**.6**	**.4**	**.14**		**.28**	**.18**	**.46**
.5		**.8**	**.5**	**.6**	**.27**	**.31**	**.19**	**.17**	**.13**	**.7**	**.5**	**.15**		**.28**	**.19**	**.50**
.6		**.14**	**.5**	**.15**	**.44**	**.34**	**.24**	**.19**	**.15**	**.9**	**.5**	**.15**			**.24**	**.51**
.7		**.19**	**.10**	**.23**	**.44**	**.35**	**.25**	**.19**	**.15**	**.10**	**.5**	**.15**				**.56**
.8		**.19**	**.11**	**.28**	**.47**		**.30**	**.19**	**.20**	**.11**	**.5**	**.21**				
.9		**.19**	**.16**	**.30**	**.47**		**.33**	**.24**	**.24**	**.22**	**.5**	**.21**				
.10		**.26**	**.16**	**.33**	**.47**		**.34**	**.27**	**.24**	**.22**	**.6**	**.22**				
.11			**.16**	**.33**	**.47**		**.36**	**.30**	**.29**		**.6**	**.22**				
.12			**.16**	**.36**	**.47**			**.32**	**.30**		**.6**	**.23**				
.13			**.16**	**.37**	**.47**			**.42**	**.33**		**.6**					
.14			**.17**	**.42**	**.48**			**.51**	**.37**		**.6**					
.15			**.19**	**.42**	**.48**				**.37**		**.7**					
.16			**.19**								**.7**					
.17			**.26**								**.8**					
.18			**.26**								**.8**					
.19			**.27**								**.9**					
.20			**.32**								**.10**					
.21			**.32**								**.11**					
.22			**.32**								**.11**					
.23			**.33**								**.16**					
.24			**.40**													
.25			**.41**													
.26			**.43**													
.27			**.44**													

Each entry in the table gives the section number that contains the Example numbered in the left margin, in the chapter at the head of the column. Thus the entry **.48**, with coordinates (.15,5), means that Example 5.15 appears in section **5.48**.

INDEX

(References are to chapter sections, displayed at the tops of pages. Examples in the text are indexed by the chapter section in or immediately after which they appear. Exercises appear at the ends of chapters.)

Printed in the USA/Agawam, MA
November 20, 2024

876860.016